Mathematics with Allied Health Applications

Richard N. Aufmann
Palomar College

Joanne S. Lockwood
Nashua Community College

with additional contributions by

Catherine W. Johnson
Alamance Community College

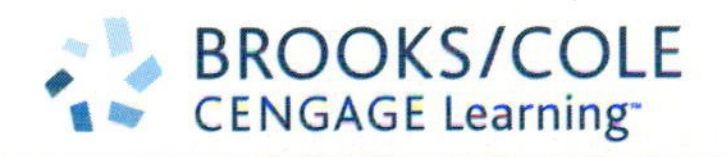

Australia • Brazil • Japan • Korea • Mexico • Singapore • Spain • United Kingdom • United States

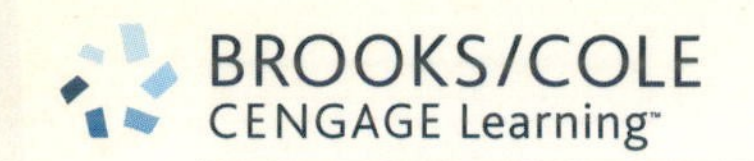

Mathematics with Allied Health Applications
Richard N. Aufmann and Joanne S. Lockwood

Acquisitions Editor: Marc Bove
Developmental Editor: Stefanie Beeck
Assistant Editor: Shaun Williams
Editorial Assistant: Zachary Crockett
Media Editors: Heleny Wong and Bryon Spencer
Marketing Manager: Gordon Lee
Marketing Coordinator: Shannon Maier
Marketing Communications Manager: Mary Anne Payumo
Content Project Manager: Cheryll Linthicum
Art Director: Vernon Boes
Print Buyer: Becky Cross
Rights Acquisitions Specialist: Roberta Broyer
Production Service: Graphic World Inc.
Text Designer: The Davis Group
Photo Researcher: Chris Althof
Copy Editor: Jean Bermingham
Illustrator: Graphic World Inc.
Cover Designer: Irene Morris
Cover Image: Panoramic Images/Panoramic Images Collection/Getty Images
Compositor: Graphic World Inc.

Library of Congress Control Number: 2011939827

ISBN-13: 978-1-111-98674-2

ISBN-10: 1-111-98674-6

Brooks/Cole
20 Davis Drive
Belmont, CA 94002-3098
USA

Cengage Learning is a leading provider of customized learning solutions with office locations around the globe, including Singapore, the United Kingdom, Australia, Mexico, Brazil, and Japan. Locate your local office at **www.cengage.com/global.**

Cengage Learning products are represented in Canada by Nelson Education, Ltd.

To learn more about Brooks/Cole, visit **www.cengage.com/brookscole**

Purchase any of our products at your local college store or at our preferred online store **www.CengageBrain.com**

Printed in the United States of America
1 2 3 4 5 6 7 15 14 13 12 11

Contents

CHAPTER 2 Fractions 63

CHAPTER 3 Decimals 125

CHAPTER 7 Statistics and Probability 293

CHAPTER 8 U.S. Customary Units of Measurement 339

CHAPTER 9 The Metric System of Measurement 371

CHAPTER 10 Rational Numbers 403

Preface

The goal of this textbook is to introduce students to arithmetic in the context of the mathematics used in the allied health field. Many examples and problems relate to topics that various health care providers encounter on a daily basis. Those students enrolled in allied health programs to become nurses; medical laboratory technicians; medical office, nursing, and dental assistants; and nutritionists will find the arithmetic in this textbook useful as they study and master topics in their various health care curricula.

Our authors have built their reputations on a successful, objective-based approach to learning mathematics called the **Aufmann Interactive Method (AIM)**. Featuring *How Tos* and paired *Example/You Try Its,* AIM engages students by asking them to practice the mathematics associated with concepts *as* they are presented. Providing students with worked examples and then affording them the opportunity to immediately work similar problems helps them build their confidence and eventually master the concepts. Our new Allied Health series furthers the approach by introducing examples and problem sets more oriented for the student undertaking a career in allied health.

To this point, simplicity plays a key factor in the organization of this edition. All lessons, exercise sets, tests, and supplements are organized around a carefully constructed hierarchy of objectives. This "objective-based" approach serves the needs of not only students, in terms of helping them to clearly organize their thoughts around the content, but instructors as well, as they work to design syllabi, lesson plans, and other administrative documents.

Features of This Edition

- Enhanced WebAssign accompanies the text.
- Dana Mosely's Text-Specific Videos are available for both instructor and student support.
- Exercise sets with applications relate to allied heath careers.
- **Think About It** exercises promote conceptual understanding.
- **In the News** application exercises are based on information found in popular media sources.
- **Concept Reviews, Chapter Review Exercises,** and **Chapter Tests** are available at the end of each chapter.
- **Prep Tests** are included at the beginning of each chapter.

Examples and exercises include real-life scenarios that allied health students will encounter in their allied health careers. The Examples and exercises demonstrate how relevant mastering the skills of mathematics are in the working world. Other topics that students will encounter in their personal lives, such as finances and consumerism, are also featured in the Examples and exercises.

Nutrition and Weight Loss

Medication Dosages

99. **Weight Loss** A physician advises an overweight patient to lose $\frac{1}{5}$ of her current weight, or 49.8 pounds, over the next 12 months. How many pounds should the patient try to lose each month in order to reach this goal?

Point of Interest

Not all tablets of medication can be safely divided in half. If a tablet is scored with a line down the middle, it is probably safe to cut into two pieces. However, always check with your doctor or pharmacist.

100. **Medication** A doctor prescribes a dose of 1.25 milligrams of a medication for a patient. The medication is in liquid form and each milliliter of liquid contains 0.25 milligram of medication. How many milliliters of the liquid should a nurse give the patient?

101. **Medication** A certain pain medication is in tablet form and contains 0.15 gram of medication per tablet. If the doctor has ordered a patient to have 0.3 gram of pain medication, how many tablets should a nurse give the patient?

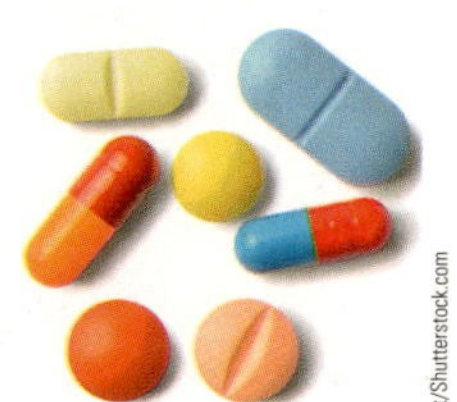

ajt/Shutterstock.com

102. **Medication** A doctor orders 0.5 gram of a medication for a patient. The pharmacy has scored tablets on hand and each one contains 0.2 gram of medication. How many tablets will be required for one dose of medication?

103. **Investments** A pharmaceutical company has issued 3,541,221,500 shares of stock. The company paid $6,090,990,120 in dividends. Find the dividend for each share of stock. Round to the nearest cent.

104. **Insurance** Earl is 52 years old and is buying $70,000 of life insurance for an annual premium of $703.80. If he pays the annual premium in 12 equal installments, how much is each monthly payment?

105. **Fuel Efficiency** A car with an odometer reading of 17,814.2 is filled with 9.4 gallons of gas. At an odometer reading of 18,130.4, the tank is empty and the car is filled with 12.4 gallons of gas. How many miles does the car travel on 1 gallon of gasoline?

106. **Salaries** A survey of four regional hospitals revealed different annual salaries for operating room nurses with less than 5 years of experience. Find the average salary for an operating room nurse with less than 5 years of experience in this region if the four annual salaries are $52,505, $55,250, $52,355, and $53,265.

In the News

"Green" Banking Has Far-Reaching Effects

Banking and paying bills online not only saves trees; it cuts down on the amount of fuel used by vehicles that transport paper checks. According to Javelin Strategy and Research, if every household in the United States paid its bills online, solid waste would be reduced by 1.6 billion tons a year and greenhouse-gas emissions would be cut by 2.1 million tons a year.

Source: Time, April 9, 2008

107. **Hemoglobin Levels** A patient with a low hemoglobin count is anemic. A normal hemoglobin range for a woman after middle age is 11.7 to 13.8 grams of hemoglobin per 100 milliliters of blood. Sonja's hemoglobin numbers for the past 6 months are 10.9, 11.5, 12.1, 11.8, 13.1, and 11.0. Find her average hemoglobin level for the 6-month period. Round to the nearest tenth.

108. **Going Green** See the news clipping at the right.

a. Find the reduction in solid waste per month if every U.S. household viewed and paid its bills online.

b. Find the reduction in greenhouse gas emissions per month if every household viewed and paid its bills online.

Write your answers in standard form, rounded to the nearest whole number.

Lab Test Results

Health Care Salaries and Business Operations

Take AIM and Succeed!

Mathematics with Allied Health Applications is organized around a carefully constructed hierarchy of **OBJECTIVES**. This "objective-based" approach provides an integrated learning environment that allows students and professors to find resources such as assessment (both within the text and online), videos, tutorials, and additional exercises.

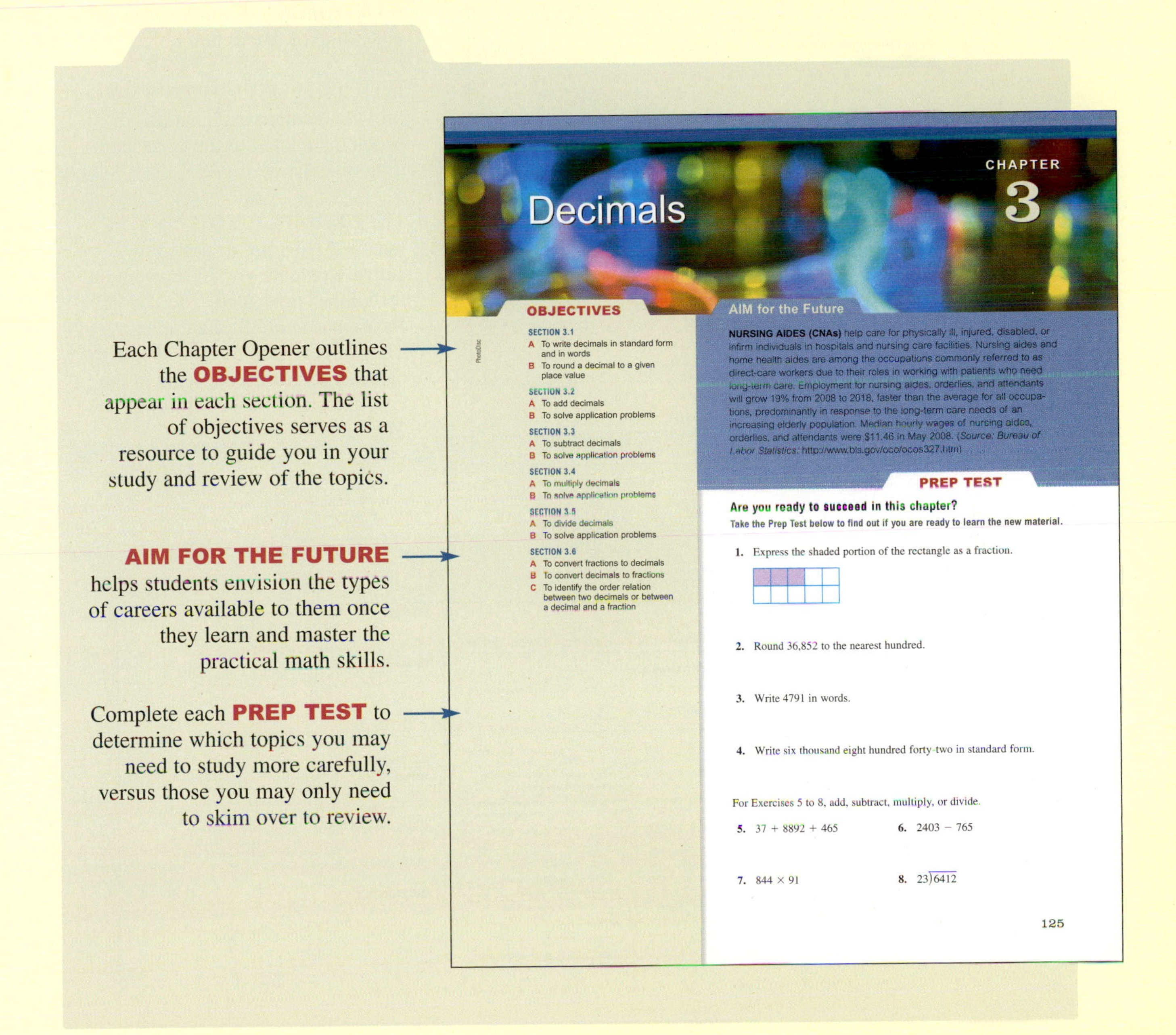

Decimals

CHAPTER 3

OBJECTIVES

SECTION 3.1
A To write decimals in standard form and in words
B To round a decimal to a given place value

SECTION 3.2
A To add decimals
B To solve application problems

SECTION 3.3
A To subtract decimals
B To solve application problems

SECTION 3.4
A To multiply decimals
B To solve application problems

SECTION 3.5
A To divide decimals
B To solve application problems

SECTION 3.6
A To convert fractions to decimals
B To convert decimals to fractions
C To identify the order relation between two decimals or between a decimal and a fraction

AIM for the Future

NURSING AIDES (CNAs) help care for physically ill, injured, disabled, or infirm individuals in hospitals and nursing care facilities. Nursing aides and home health aides are among the occupations commonly referred to as direct-care workers due to their roles in working with patients who need long-term care. Employment for nursing aides, orderlies, and attendants will grow 19% from 2008 to 2018, faster than the average for all occupations, predominantly in response to the long-term care needs of an increasing elderly population. Median hourly wages of nursing aides, orderlies, and attendants were $11.46 in May 2008. (*Source: Bureau of Labor Statistics*; http://www.bls.gov/oco/ocos327.htm)

PREP TEST

Are you ready to succeed in this chapter?
Take the Prep Test below to find out if you are ready to learn the new material.

1. Express the shaded portion of the rectangle as a fraction.
2. Round 36,852 to the nearest hundred.
3. Write 4791 in words.
4. Write six thousand eight hundred forty-two in standard form.

For Exercises 5 to 8, add, subtract, multiply, or divide.

5. $37 + 8892 + 465$
6. $2403 - 765$
7. 844×91
8. $23\overline{)6412}$

125

Each Chapter Opener outlines the **OBJECTIVES** that appear in each section. The list of objectives serves as a resource to guide you in your study and review of the topics.

AIM FOR THE FUTURE helps students envision the types of careers available to them once they learn and master the practical math skills.

Complete each **PREP TEST** to determine which topics you may need to study more carefully, versus those you may only need to skim over to review.

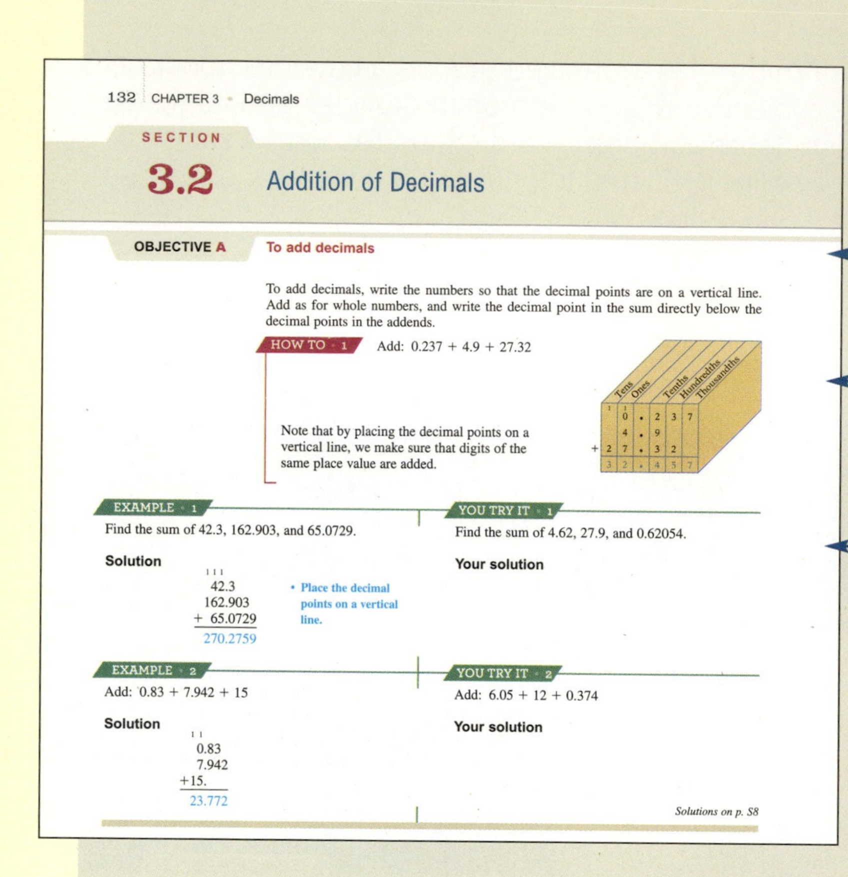

132 CHAPTER 3 • Decimals

SECTION

3.2 Addition of Decimals

OBJECTIVE A **To add decimals**

To add decimals, write the numbers so that the decimal points are on a vertical line. Add as for whole numbers, and write the decimal point in the sum directly below the decimal points in the addends.

HOW TO • 1 Add: 0.237 + 4.9 + 27.32

Note that by placing the decimal points on a vertical line, we make sure that digits of the same place value are added.

EXAMPLE • 1

Find the sum of 42.3, 162.903, and 65.0729.

Solution

42.3
162.903
+ 65.0729
270.2759

• Place the decimal points on a vertical line.

YOU TRY IT • 1

Find the sum of 4.62, 27.9, and 0.62054.

Your solution

EXAMPLE • 2

Add: 0.83 + 7.942 + 15

Solution

0.83
7.942
+15.
23.772

YOU TRY IT • 2

Add: 6.05 + 12 + 0.374

Your solution

Solutions on p. S8

In each section, **OBJECTIVE STATEMENTS** introduce each new topic of discussion.

In each section, the **HOW TOS** provide detailed explanations of problems related to the corresponding objectives.

The **EXAMPLE/YOU TRY IT** matched pairs are designed to actively involve you in learning the techniques presented. The You Try Its are based on the Examples. They appear side by side so you can easily refer to the steps in the Examples as you work through the You Try Its.

Complete, **WORKED-OUT SOLUTIONS** to the You Try It problems are found in an appendix at the back of the text. Compare your solutions to the solutions in the appendix to obtain immediate feedback and reinforcement of the concept(s) you are studying.

SOLUTIONS TO CHAPTER 3 "YOU TRY IT"

SECTION 3.1

You Try It 1 The digit 4 is in the thousandths place.

You Try It 2 $\frac{501}{1000} = 0.501$ (five hundred one thousandths)

You Try It 3 $0.67 = \frac{67}{100}$ (sixty-seven hundredths)

You Try It 4 Fifty-five and six thousand eighty-three ten-thousandths

You Try It 5 806.00491 • 1 is in the hundred-thousandths place.

You Try It 6 3.675849 — Given place value; 4 < 5

3.675849 rounded to the nearest ten-thousandth is 3.6758.

You Try It 7 48.907 — Given place value; 0 < 5

48.907 rounded to the nearest tenth is 48.9.

Strategy To determine the number, add the numbers of hearing-impaired Americans ages 45 to 54, 55 to 64, 65 to 74, and 75 and over.

Solution

4.48
4.31
5.41
+3.80
18.00

18 million Americans ages 45 and older are hearing-impaired.

You Try It 4

Strategy To find the total income, add the pay for working nights (81.20, 60.90, 40.60, 81.20) to her regular weekly pay (480.15).

Solution 480.15 + 81.20 + 60.90 + 40.60 + 81.20 = 744.05

The nurse's aide's total income for the week was $744.05.

SECTION 3.3

You Try It 1

72.039
− 8.47
63.569

Check: 8.47 + 63.569 = 72.039

Mathematics with Allied Health Applications contains **A WIDE VARIETY OF EXERCISES** that promote skill building, skill maintenance, concept development, critical thinking, and problem solving.

THINK ABOUT IT exercises promote conceptual understanding. Completing these exercises will deepen your understanding of the concepts being addressed.

3.4 EXERCISES

OBJECTIVE A To multiply decimals

For Exercises 1 to 73, multiply.

1. 0.9×0.4 **2.** 0.7×0.9 **3.** 0.5×0.5 **4.** 0.7×0.7 **5.** 7.7×0.9

For Exercises 25 to 28, without actually doing any division, state whether the decimal equivalent of the given fraction is greater than 1 or less than 1.

25. $\frac{54}{57}$ **26.** $\frac{176}{129}$ **27.** $\frac{88}{80}$ **28.** $\frac{2007}{2008}$

APPLYING THE CONCEPTS exercises may involve further exploration of topics, or they may involve analysis. They may also integrate concepts introduced earlier in the text. **Optional** scientific calculator exercises are included, denoted by .

Applying the Concepts

109. **Nutrition** You are on a 1500-calorie per day diet. You consume 375 calories at breakfast, 450 calories at lunch, and 630 calories at supper. Using the list of healthy snacks on page 19, which snack(s) can you eat before bedtime and not exceed your daily goal of 1500 calories?

110. Answer true or false.
a. The phrases "the difference between 9 and 5" and "5 less than 9" mean the same thing.
b. $9 - (5 - 3) = (9 - 5) - 3$
c. Subtraction is an associative operation. *Hint:* See part (b) of this exercise.

Working through the application exercises that contain **REAL DATA** will help prepare you to answer questions and/or solve problems based on your own experiences, using facts or information you gather.

Health Insurance The table at the right shows the number of Americans without health insurance by age in 2009. Use this table for Exercises 110 to 112.

Americans Without Health Insurance by Age in 2009

Age	Number Uninsured (in thousands)
Under 18 years	75.04
18 to 24 years	29.31
25 to 34 years	41.09
35 to 44 years	40.45
45 to 64 years	79.78
65 years and older	38.61

Source: U.S. Census Bureau, Current Population Survey, 2009 and 2010 Annual Social and Economic Supplements

110. Find the difference between the number of Americans without health insurance who are aged under 18 and those aged 65 and older.

111. How many times greater is the number of Americans under 18 who are uninsured than the number of uninsured Americans who are 18 to 24 years old? Round to the nearest hundredth.

112. What was the total number of Americans without health insurance in 2009?

113. **Population Growth** The U.S. population of people ages 85 and over is expected to grow from 4.2 million in 2000 to 8.9 million in 2030. How many times greater is the population of this segment expected to be in 2030 than in 2000? Round to the nearest tenth.

Completing the **WRITING** exercises will help you to improve your communication skills while increasing your understanding of mathematical concepts.

114. Explain how the decimal point is moved when a number is divided by 10, 100, 1000, 10,000, etc.

115. **Grade Point Average** Explain how to calculate a Grade Point Average (GPA) for a semester. Then calculate the GPA for a student taking four 3-hour courses and earning 2 Cs, 1 B, and 1 A.

116. Explain how the decimal point is placed in the quotient when a number is divided by a decimal.

Mathematics with Allied Health Applications addresses students' broad range of study styles by offering
A WIDE VARIETY OF TOOLS FOR REVIEW

At the end of each chapter you will find a **SUMMARY** with **KEY WORDS** and **ESSENTIAL RULES AND PROCEDURES**. Each entry includes an example of the summarized concept, an objective reference, and a page reference to show where each concept was introduced.

CHAPTER 3

SUMMARY

KEY WORDS	EXAMPLES
A number written in *decimal notation* has three parts: a *whole-number part,* a *decimal point,* and a *decimal part.* The decimal part of a number represents a number less than 1. A number written in decimal notation is often simply called a *decimal.* [3.1A, p. 126]	For the decimal 31.25, 31 is the whole-number part and 25 is the decimal part.

ESSENTIAL RULES AND PROCEDURES	EXAMPLES
To write a decimal in words, write the decimal part as if it were a whole number. Then name the place value of the last digit. The decimal point is read as "and." [3.1A, p. 126]	The decimal 12.875 is written in words as twelve and eight hundred seventy-five thousandths.
To write a decimal in standard form when it is written in words, write the whole-number part, replace the word *and* with a decimal point, and write the decimal part so that the last digit is in the given place-value position. [3.1A, p. 127]	The decimal forty-nine and sixty-three thousandths is written in standard form as 49.063.

CHAPTER 3

CONCEPT REVIEW

Test your knowledge of the concepts presented in this chapter. Answer each question. Then check your answers against the ones provided in the Answer Section.

1. How do you round a decimal to the nearest tenth?

2. How do you write the decimal 0.37 as a fraction?

3. How do you write the fraction $\frac{173}{10{,}000}$ as a decimal?

CONCEPT REVIEWS actively engage you as you study and review the contents of a chapter. The **ANSWERS** to the questions are found in an appendix at the back of the text. After each answer, look for an objective reference that indicates where the concept was introduced.

By completing the chapter **REVIEW EXERCISES**, you can practice working problems that appear in an order that is different from the order they were presented in the chapter. The **ANSWERS** to these exercises include references to the section objectives upon which they are based. This will help you to quickly identify where to go to review the concepts if needed.

CHAPTER 3

REVIEW EXERCISES

1. Find the quotient of 3.6515 and 0.067.

2. Find the sum of 369.41, 88.3, 9.774, and 366.474.

3. Place the correct symbol, < or >, between the two numbers.
   0.055   0.1

4. Write 22.0092 in words.

5. Round 0.05678235 to the nearest hundred-thousandth.

6. Convert $2\frac{1}{3}$ to a decimal. Round to the nearest hundredth.

7. Convert 0.375 to a fraction.

8. Add: $3.42 + 0.794 + 32.5$

Each chapter **TEST** is designed to simulate a possible test of the concepts covered in the chapter. The **ANSWERS** include references to section objectives. References to How Tos, worked Examples, and You Try Its, which provide solutions to similar problems, are also included.

CHAPTER 3

TEST

1. Place the correct symbol, < or >, between the two numbers.
   0.66 0.666
2. Subtract: $\begin{array}{r} 13.027 \\ -\ 8.94 \\ \hline \end{array}$
3. Write 45.0302 in words.
4. Convert $\frac{9}{13}$ to a decimal. Round to the nearest thousandth.
5. Convert 0.825 to a fraction.
6. Round 0.07395 to the nearest ten-thousandth.

CUMULATIVE REVIEW EXERCISES

1. Divide: $89\overline{)20{,}932}$
2. Simplify: $2^3 \cdot 4^2$
3. Simplify: $2^2 - (7 - 3) \div 2 + 1$
4. Find the LCM of 9, 12, and 24.
5. Write $\frac{22}{5}$ as a mixed number.
6. Write $4\frac{5}{8}$ as an improper fraction.
7. Write an equivalent fraction with the given denominator.
   $\frac{5}{12} = \frac{\ \ }{60}$
8. Add: $\frac{3}{8} + \frac{5}{12} + \frac{9}{16}$

CUMULATIVE REVIEW EXERCISES, which appear at the end of each chapter (beginning with Chapter 2), help you maintain skills you previously learned. The **ANSWERS** include references to the section objectives upon which the exercises are based.

FINAL EXAM

1. Subtract: 100,914 − 97,655
2. Find 34,821 divided by 657.
3. Find 90,001 decreased by 29,796.
4. Simplify: $3^2 \cdot (5 - 3)^2 \div 3 + 4$
5. Find the LCM of 9, 12, and 16.
6. Add: $\frac{3}{8} + \frac{5}{6} + \frac{1}{5}$
7. Subtract: $7\frac{5}{12} - 3\frac{13}{16}$
8. Find the product of $3\frac{5}{8}$ and $1\frac{5}{7}$.
9. Divide: $1\frac{2}{3} \div 3\frac{3}{4}$
10. Simplify: $\left(\frac{2}{3}\right)^3 \cdot \left(\frac{3}{4}\right)^2$
11. Simplify: $\left(\frac{2}{3}\right)^2 \div \left(\frac{3}{4} + \frac{1}{3}\right) - \frac{1}{3}$
12. Add: $\begin{array}{r} 4.972 \\ 28.6 \\ 1.88 \end{array}$

A **FINAL EXAM** appears after the last chapter in the text. It is designed to simulate a possible examination of all the concepts covered in the text. The **ANSWERS** to the exam questions are provided in the answer appendix at the back of the text and include references to the section objectives upon which the questions are based.

Other Key Features

MARGINS Within the margins, students can find the following.

Take Note boxes alert students to concepts that require special attention.

Point of Interest boxes, which may be historical in nature or be of general interest, relate to topics under discussion.

Integrated Technology boxes, which are offered as optional instruction in the proper use of the scientific calculator, appear for selected topics under discussion.

Tips for Success boxes outline good study habits.

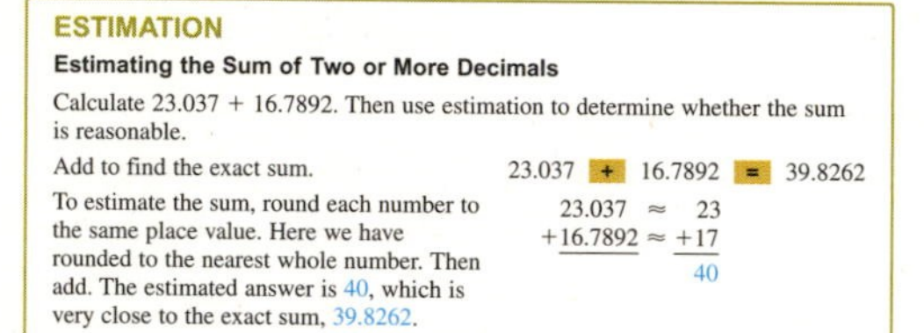

ESTIMATION

Estimating the Sum of Two or More Decimals

Calculate 23.037 + 16.7892. Then use estimation to determine whether the sum is reasonable.

Add to find the exact sum. 23.037 + 16.7892 = 39.8262

To estimate the sum, round each number to the same place value. Here we have rounded to the nearest whole number. Then add. The estimated answer is 40, which is very close to the exact sum, 39.8262.

$$\begin{array}{r} 23.037 \approx 23 \\ +16.7892 \approx +17 \\ \hline 40 \end{array}$$

ESTIMATION Throughout the textbook, Estimation boxes appear where appropriate. Tied to relevant content, the Estimation boxes demonstrate how estimation may be used to check answers for reasonableness.

PROBLEM-SOLVING STRATEGIES The text features a carefully developed approach to problem solving that encourages students to develop a Strategy for a problem and then to create a Solution based on the Strategy.

EXAMPLE 7

A pediatrician prescribes 1 teaspoon of amoxicillin every 4 hours. The patient is required to take this 1 teaspoon (0.167 ounce/teaspoon) of medication four times a day for 10 days. Find the total number of ounces of amoxicillin ingested during the 10 days. Round your answer to the nearest tenth.

Strategy

- To find the ounces ingested per day, multiply 1 teaspoon in ounces (0.167 ounce) by the times per day (4).
- To find the total ounces ingested for the prescription, multiply the ounces ingested per day (answer from step 1) by the total number of days (10).
- Round the final answer to one decimal place.

Solution

$$\begin{array}{r} 0.167 \\ \times \quad 4 \\ \hline 0.668 \end{array}$$ ounces per day

$$\begin{array}{r} 0.668 \\ \times \quad 10 \\ \hline 6.668 \end{array}$$ total ounces per 10-day prescription

Round 6.668 to 6.7.

The patient has ingested 6.7 ounces of amoxicillin for the 10 days.

YOU TRY IT 7

The cost to operate an x-ray processor in July 2010 was $0.1201 per kilowatt hour. The x-ray processor ran for 56 hours during the first week of July. Find the total cost to operate the x-ray processor during that week. Round to the nearest cent.

Your strategy

Your solution

FOCUS ON PROBLEM SOLVING At the end of each chapter, the Focus on Problem Solving fosters further discovery of new problem-solving strategies, such as applying solutions to other problems, working backward, inductive reasoning, and trial and error.

FOCUS ON PROBLEM SOLVING

Relevant Information

Problems in mathematics or real life involve a question or a need and information or circumstances related to that question or need. Solving problems in the sciences usually involves a question, an observation, and measurements of some kind.

One of the challenges of problem solving in the sciences is to separate the information that is relevant to the problem from other information. Following is an example from the physical sciences in which some relevant information was omitted.

Hooke's Law states that the distance that a weight will stretch a spring is directly proportional to the weight on the spring. That is, $d = kF$, where d is the distance the spring is stretched and F is the force. In an experiment to verify this law, some physics students were continually getting inconsistent results. Finally, the instructor discovered that the heat produced when the lights were turned on was affecting the experiment. In this case, relevant information was omitted—namely, that the temperature of the spring can affect the distance it will stretch.

A physician drove 8 miles to the train station. After a 35-minute ride of 18 miles, the

PROJECTS AND GROUP ACTIVITIES appear at the end of each chapter. Your instructor may assign these to you individually, or you may be asked to work through the activity in groups.

PROJECTS AND GROUP ACTIVITIES

Fractions as Terminating or Repeating Decimals

The fraction $\frac{3}{4}$ is equivalent to 0.75. The decimal 0.75 is a **terminating decimal** because there is a remainder of zero when 3 is divided by 4. The fraction $\frac{1}{3}$ is equivalent to 0.333 . . . . The three dots mean the pattern continues on and on. 0.333 . . . is a **repeating decimal.** To determine whether a fraction can be written as a terminating decimal, first write the fraction in simplest form. Then look at the denominator of the fraction. If it contains prime factors of only 2s and/or 5s, then it can be expressed as a terminating decimal. If it contains prime factors other than 2s or 5s, it represents a repeating decimal.

Take Note If the denominator of a fraction in simplest form is 20, then it can be written as a terminating decimal because $20 = 2 \cdot 2 \cdot 5$ (only prime factors of 2 and 5). If the denominator of a fraction in simplest form is 6, it

1. Assume that each of the following numbers is the denominator of a fraction written in simplest form. Does the fraction represent a terminating or repeating decimal?
   a. 4 **b.** 5 **c.** 7 **d.** 9 **e.** 10 **f.** 12 **g.** 15
   h. 16 **i.** 18 **j.** 21 **k.** 24 **l.** 25 **m.** 28 **n.** 40

Acknowledgments

The authors would like to thank the people who have reviewed this manuscript and provided many valuable suggestions:

Joyce Brasfield Adams, *Hinds Community College*
Natasha Baer, *Anoka-Ramsey Community College*
Stephanie J. S. Baiyasi, *Delta College*
Alfred Basta
Lorri Christiansen, *Centura College*
Sheilla Coplin, *National College of Martinsville*
Steven Coppola, *The Permanente Medical Group*
Bobby F. Crawford, *Heald College-Hayward*
Leuda Forrester, *Monroe College*
Lynnette Garetz, *Heald College-Hayward*
Richard Gerstin, *Brown Mackie College-Atlanta*
Laura Ristrom Goodman, *Pima Medical Institute*
Randall Green, *Sanford-Brown College-Hillside*
Christina Havlin, *ECPI College of Technology*
Melissa Hibbard, *Miami Jacobs Career College*
Andrew Jaffe, *Goodwin College*
Patricia Hoskins Jones, *Sanford-Brown College-Milwaukee*
Lane Miller, *Medical Careers Institute*
Ali Mustafa, *Hesser College*
Lori Padgett, *Skyline College*
Shelle Ridings, *Sanford-Brown College-Collinsville*
Sami Segale, *Heald College-Hayward*
Jason Shea, *Goodwin College*
Terrina Thomas, *Sentara Healthcare*
Yuriy Tolstykh, *Heald College-Hayward*
Kasey Waychoff, *Centura College*
Michele Wootton, *Carrington College California*
Mindy Wray, *ECPI College of Technology*
Frederick Zingeser, *Heald College-Hayward*

Special thanks go to Pat Foard for preparing the solutions and to Shelle Ridings for her work in ensuring the accuracy of the text. We would also like to thank the many people at Cengage Learning who worked to guide the manuscript from development through production.

Instructor Resources

Print Supplements

Complete Solutions Manual
(ISBN: 978-1-133-11231-0)

The Complete Solutions Manual contains fully worked-out solutions to all of the exercises in the text.

Instructor's Resource Binder with Appendix
(ISBN: 978-1-133-11250-1)

Each section of the main text is discussed in uniquely designed Teaching Guides, containing tips, examples, activities, worksheets, overheads, assessments, and solutions to all worksheets and activities.

Electronic Supplements

Enhanced WebAssign (ISBN: 978-0-538-73810-1)

Exclusively from Cengage Learning, Enhanced WebAssign® combines the exceptional mathematics content that you know and love with the most powerful online homework solution, WebAssign. Enhanced WebAssign engages students with immediate feedback, rich tutorial content, and interactive eBooks, which help students to develop a deeper conceptual understanding of their subject matter. Online assignments can be built by selecting from thousands of text-specific problems or supplemented with problems from any Cengage Learning textbook.

Enhanced WebAssign: Start Smart Guide for Students (ISBN: 978-0-495-38479-3)

The Enhanced WebAssign Student Start Smart Guide helps students get up and running quickly with Enhanced WebAssign so they can study smarter and improve their performance in class.

Text-Specific Videos:
Author: Dana Mosely

These text-specific instructional videos provide students with visual reinforcement of concepts and explanations given in easy-to-understand terms with detailed examples and sample problems. A flexible format offers versatility for quickly accessing topics or catering lectures to self-paced, online, or hybrid courses. Closed captioning is provided for the hearing impaired. Available in YouBook and CourseMate.

PowerLecture with Diploma®
(ISBN: 978-1-133-11233-4)

This CD-ROM provides the instructor with dynamic media tools for teaching. Create, deliver, and customize tests (both print and online) in minutes with Diploma's Computerized Testing, featuring algorithmic equations. Easily build solution sets for homework or exams using Solution Builder's online solutions manual.

Solution Builder

This online instructor database offers complete worked-out solutions to all exercises in the text, which allows you to create customized, secure solutions printouts (in PDF format), matched exactly to the problems you assign in class. For more information, visit www.cengage.com/solutionbuilder.

Syllabus Creator
(Included on the PowerLecture)

Easily write, edit, and update your syllabus with the Aufmann/Lockwood Syllabus Creator. This software program allows you to create your new syllabus in six easy steps: select the required course objectives, and add your contact information, course information, student expectations, the grading policy, dates and location, and course outline. And now you have your syllabus!

Printed Access Card for CourseMate with eBook for Mathematics for the Allied Health Professional, 1st Edition (ISBN: 978-1-133-51127-4)

Instant Access Card for CourseMate with eBook for Mathematics for the Allied Health Professional, 1st Edition (ISBN: 978-1-133-51126-7)

Complement your text and course content with study and practice materials. Cengage Learning's Developmental Mathematics CourseMate brings course concepts to life with interactive learning, study, and exam preparation tools that support the printed textbook. Watch student comprehension soar as your class works with the printed textbook and the textbook-specific website. Developmental Mathematics CourseMate goes beyond the book to deliver what you need!

Student Resources

Print Supplements

Student Solutions Manual
(ISBN: 978-1-133-11230-3)

Go beyond the answers—see what it takes to get there and improve your grade! This manual provides worked-out, step-by-step solutions to the odd-numbered problems in the text. This gives you the information you need to truly understand how these problems are solved.

Student Workbook (ISBN: 978-1-133-11232-7)

Get a head start. The Student Workbook contains assessments, activities, and worksheets for classroom discussions, in-class activities, and group work.

AIM for Success Student Practice Sheets
(ISBN: 978-1-133-11234-1)

Practice with additional problems to help you learn the material.

Electronic Supplements

Enhanced WebAssign (ISBN: 978-0-538-73810-1)

Enhanced WebAssign (assigned by the instructor) provides instant feedback on homework assignments. This online homework system is easy to use and includes helpful links to textbook sections, video examples, and problem-specific tutorials.

Enhanced WebAssign: Start Smart Guide for Students (ISBN: 978-0-495-38479-3)

If your instructor has chosen to package Enhanced WebAssign with your text, this manual will help you get up and running quickly with the Enhanced WebAssign system so you can study smarter and improve your performance in class.

Text-Specific Videos:
Author: Dana Mosely

These text-specific instructional videos provide you with visual reinforcement of concepts and explanations given in easy-to-understand terms with detailed examples and sample problems. A flexible format offers versatility for quickly accessing topics or catering lectures to self-paced, online, or hybrid courses. Closed captioning is provided for the hearing impaired. Available in YouBook and CourseMate.

Printed Access Card for CourseMate with eBook for Mathematics for the Allied Health Professional, 1st Edition (ISBN: 978-1-133-51127-4)

Instant Access Card for CourseMate with eBook for Mathematics for the Allied Health Professional, 1st Edition (ISBN: 978-1-133-51126-7)

The more you study, the better the results. You can make the most of your study time by accessing everything you need to succeed in one place: read the textbook, take notes, review flashcards, watch videos, and take practice quizzes—online with CourseMate.

AIM for Success: Getting Started

Welcome to *Mathematics with Allied Health Applications!* Students come to this course with varied backgrounds and different experiences in learning math. We are committed to your success in learning mathematics and have developed many tools and resources to support you along the way. Want to excel in this course? Read on to learn the skills you'll need and how best to use this book to get the results you want.

Motivate Yourself

In this book you'll find many real-life problems relating to various health occupations. We hope that these topics will help you understand how you will use mathematics in your real life. However, to learn all of the necessary skills and how you can apply them to your life outside this course, you need to stay motivated.

Take Note

Motivation alone won't lead to success. For example, suppose a person who cannot swim is rowed out to the middle of a lake and thrown overboard. That person has a lot of motivation to swim but will most likely drown without some help. You'll need motivation and learning in order to succeed.

THINK ABOUT WHY YOU WANT TO SUCCEED IN THIS COURSE. LIST THE REASONS HERE (NOT IN YOUR HEAD . . . ON THE PAPER!):

____

____

We also know that this course may be a requirement for you to graduate or complete your major. That's OK. If you have a goal for the future, such as becoming a nurse or a teacher, you will need to succeed in mathematics first. Picture yourself where you want to be, and use this image to stay on track.

Photodisc

Make the Commitment

Stay committed to success! With practice, you will improve your math skills. Skeptical? Think about when you first learned to ride a bike or drive a car. You probably felt self-conscious and worried that you might fail. But with time and practice, it became second nature to you.

You will also need to put in the time and practice to do well in mathematics. Think of us as your "driving" instructors. We'll lead you along the path to success, but we need you to stay focused and energized along the way.

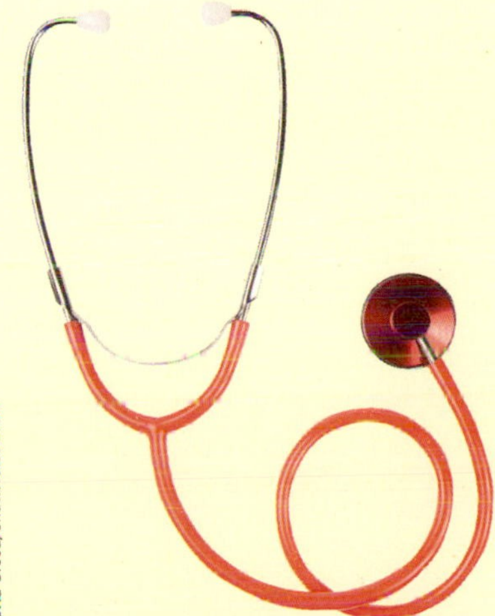

David Orcea/Shutterstock.com

LIST A SITUATION IN WHICH YOU ACCOMPLISHED YOUR GOAL BY SPENDING TIME PRACTICING AND PERFECTING YOUR SKILLS (SUCH AS LEARNING TO PLAY THE PIANO OR PLAYING BASKETBALL):

____

____

____

____

If you spend time learning and practicing the skills in this book, you will also succeed in math.

Photodisc

Think You Can't Do Math? Think Again!

You can do math! When you first learned the skills you just listed, you may have not done them well. With practice, you got better. With practice, you will be better at math. Stay focused, motivated, and committed to success.

It is difficult for us to emphasize how important it is to overcome the I Can't Do Math Syndrome. If you listen to interviews of very successful athletes after a particularly bad performance, you will note that they focus on the positive aspect of what they did, not the negative. Sports psychologists encourage athletes to always be positive—to have a "can-do" attitude. Develop this attitude toward math and you will succeed.

Skills for Success

GET THE BIG PICTURE If this were an English class, we wouldn't encourage you to look ahead in the book. But this is mathematics—go right ahead! Take a few minutes to read the table of contents. Then, look through the entire book. Move quickly: scan titles, look at pictures, notice diagrams.

Getting this big picture view will help you see where this course is going. To reach your goal, it's important to get an idea of the steps you will need to take along the way.

As you look through the book, find topics that interest you. What's your preference? Nutrition? Medication dosages? Office management? Find the Index of Applications at the back of the book and pull out three subjects that interest you. Then, flip to the pages in the book where the topics are featured and read the exercises or problems where they appear.

Stephen VanHorn/Shutterstock.com

WRITE THE TOPIC HERE:	WRITE THE CORRESPONDING EXERCISE/PROBLEM HERE:
____________	____________
____________	____________
____________	____________

You'll find it's easier to work at learning the material if you are interested in how it can be used in your everyday life.

Use the following activities to think about more ways you might use mathematics in your daily life. Flip open your book to the following exercises to answer the questions.

- (see p. 83, #82) I just started a new job and will be paid hourly, but my hours change every week. I need to use mathematics to . . .

____

____

____

____

- (see p. 228, #22) I'd like to eat a healthier diet, but I need to know what to eat. I need mathematics to . . .

____

____

- (see p. 345, #32) I'd like to know if my height qualifies me for the Marine Corps. I need mathematics to . . .

____

____

You know that the activities you just completed are from daily life, but do you notice anything else they have in common? That's right—they are **word problems.** Try not to be intimidated by word problems. You just need a strategy. It's true that word problems can be challenging because we need to use multiple steps to solve them:

- Read the problem.
- Determine the quantity we must find.
- Think of a method to find it.
- Solve the problem.
- Check the answer.

In short, we must come up with a **strategy** and then use that strategy to find the **solution.**

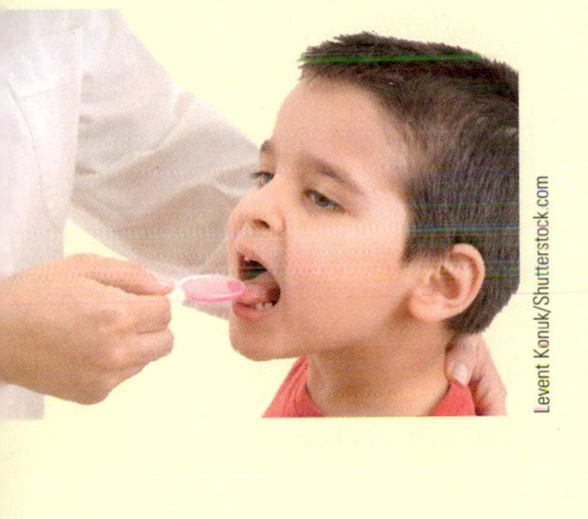
Levent Konuk/Shutterstock.com

We'll teach you about strategies for tackling word problems that will make you feel more confident in branching out to these problems from daily life. After all, even though no one will ever come up to you on the street and ask you to solve a multiplication problem, you will need to use math every day to balance your checkbook, evaluate credit card offers, etc.

Take a look at the following example. You'll see that solving a word problem includes finding a *strategy* and using that strategy to find a *solution.* If you find yourself struggling with a word problem, try writing down the information you know about the problem. Be as specific as you can. Write out a phrase or a sentence that states what you are trying to find. Ask yourself whether there is a formula that expresses the known and unknown quantities. Then, try again!

EXAMPLE 7

A pediatrician prescribes 1 teaspoon of amoxicillin every 4 hours. The patient is required to take this 1 teaspoon (0.167 ounce/teaspoon) of medication four times a day for 10 days. Find the total number of ounces of amoxicillin ingested during the 10 days. Round your answer to the nearest tenth.

Strategy

- To find the ounces ingested per day, multiply 1 teaspoon in ounces (0.167 ounce) by the times per day (4).
- To find the total ounces ingested for the prescription, multiply the ounces ingested per day (answer from step 1) by the total number of days (10).
- Round the final answer to one decimal place.

Solution

$$\begin{array}{r} 0.167 \\ \times \quad 4 \\ \hline 0.668 \end{array} \text{ ounces per day} \qquad \begin{array}{r} 0.668 \\ \times \quad 10 \\ \hline 6.668 \end{array} \text{ total ounces per 10-day prescription}$$

Round 6.668 to 6.7.

The patient has ingested 6.7 ounces of amoxicillin for the 10 days.

YOU TRY IT 7

The cost to operate an x-ray processor in July 2010 was $0.1201 per kilowatt hour. The x-ray processor ran for 56 hours during the first week of July. Find the total cost to operate the x-ray processor during that week. Round to the nearest cent.

Your strategy

Your solution

Page 143

Take Note

Take a look at your syllabus to see if your instructor has an **attendance policy** that is part of your overall grade in the course.

The attendance policy will tell you:

- How many classes you can miss without a penalty
- What to do if you miss an exam or quiz
- If you can get the lecture notes from the professor if you miss a class

GET THE BASICS On the first day of class, your instructor will hand out a **syllabus** listing the requirements of your course. Think of this syllabus as your personal road map to success. It shows you the destinations (topics you need to learn) and the dates you need to arrive at those destinations (by when you need to learn the topics). Learning mathematics is a journey. But, to get the most out of this course, you'll need to know what the important stops are and what skills you'll need to learn for your arrival at those stops.

You've quickly scanned the table of contents, but now we want you to take a closer look. Flip open to the table of contents and look at it next to your syllabus. Identify when your major exams are and what material you'll need to learn by those dates. For example, if you know you have an exam in the second month of the semester, how many chapters of this text will you need to learn by then? What homework do you have to do during this time? Managing this important information will help keep you on track for success.

Take Note

When planning your schedule, give some thought to how much time you realistically have available each week. For example, if you work 40 hours a week, take 15 units, spend the recommended study time given at the right, and sleep 8 hours a day, you will use over 80% of the available hours in a week. That leaves less than 20% of the hours in a week for family, friends, eating, recreation, and other activities.

Visit http://college.cengage.com/masterstudent/shared/content/time_chart/chart.html and use the Interactive Time Chart to see how you're spending your time—you may be surprised.

MANAGE YOUR TIME We know how busy you are outside of school. Do you have a full-time or a part-time job? Do you have children? Visit your family often? Play basketball or write for the school newspaper? It can be stressful to balance all of the important activities and responsibilities in your life. Making a **time management plan** will help you create a schedule that gives you enough time for everything you need to do.

Photodisc

Let's get started! Create a weekly schedule.

First, list all of your responsibilities that take up certain set hours during the week. Be sure to include:

- each class you are taking
- time you spend at work
- any other commitments (child care, tutoring, volunteering, etc.)

Then, list all of your responsibilities that are more flexible. Remember to make time for:

- **STUDYING** You'll need to study to succeed, but luckily you get to choose what times work best for you. Keep in mind:
  - Most instructors ask students to spend twice as much time studying as they do in class (3 hours of class = 6 hours of study).
  - Try studying in chunks. We've found it works better to study an hour each day, rather than studying for 6 hours on one day.
  - Studying can be even more helpful if you're able to do it right after your class meets, when the material is fresh in your mind.
- **MEALS** Eating well gives you energy and stamina for attending classes and studying.
- **ENTERTAINMENT** It's impossible to stay focused on your responsibilities 100% of the time. Giving yourself a break for entertainment will reduce your stress and help keep you on track.
- **EXERCISE** Exercise contributes to overall health. You'll find you're at your most productive when you have both a healthy mind and a healthy body.

Here is a sample of what part of your schedule might look like:

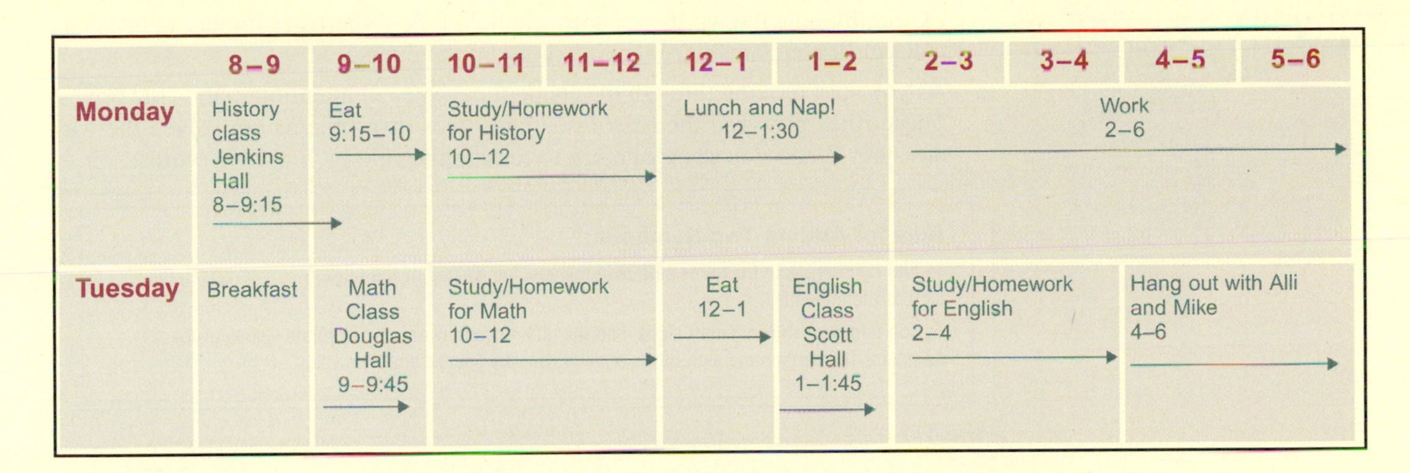

	8–9	9–10	10–11	11–12	12–1	1–2	2–3	3–4	4–5	5–6
Monday	History class Jenkins Hall 8–9:15	Eat 9:15–10	Study/Homework for History 10–12		Lunch and Nap! 12–1:30		Work 2–6			
Tuesday	Breakfast	Math Class Douglas Hall 9–9:45	Study/Homework for Math 10–12		Eat 12–1	English Class Scott Hall 1–1:45	Study/Homework for English 2–4		Hang out with Alli and Mike 4–6	

Features for Success in This Text

ORGANIZATION Let's look again at the table of contents. There are 11 chapters in this book. You'll see that every chapter is divided into **sections,** and each section contains a number of **learning objectives.** Each learning objective is labeled with a letter from A to D. Knowing how this book is organized will help you locate important topics and concepts as you're studying.

PREPARATION Ready to start a new chapter? Take a few minutes to be sure you're ready, using some of the tools in this book.

- **CUMULATIVE REVIEW EXERCISES:** You'll find these exercises after every chapter, starting with Chapter 2. The questions in the Cumulative Review Exercises are taken from the previous chapters. For example, the Cumulative Review for Chapter 3 will test all of the skills you have learned in Chapters 1, 2, and 3. Use this to refresh yourself before moving on to the next chapter or to test what you know before a big exam.

Here's an example of how to use the Cumulative Review:

- Turn to page 171 and look at the questions for the Chapter 3 Cumulative Review, which are taken from the current chapter and the previous chapters.
- We have the answers to all of the Cumulative Review Exercises in the back of the book. Flip to page A10 to see the answers for this chapter.
- Got the answer wrong? We can tell you where to go in the book for help! For example, scroll down page A10 to find the answer for the first exercise, which is 235 r17. You'll see that after this answer, there is an **objective reference** [1.5C]. This means that the question was taken from Chapter 1, Section 5, Objective C. Go here to restudy the objective.

- **PREP TESTS:** These tests are found at the beginning of every chapter and will help you see if you've mastered all of the skills needed for the new chapter.

Here's an example of how to use the Prep Test:

- Turn to page 173 and look at the Prep Test for Chapter 4.
- All of the answers to the Prep Tests are in the back of the book. You'll find them in the first set of answers in each answer section for a chapter. Turn to page A10 to see the answers for this Prep Test.
- Restudy the objectives if you need some extra help.

Photodisc

- Before you start a new section, take a few minutes to read the **Objective Statement** for that section. Then, browse through the objective material. Especially note the words or phrases in bold type—these are important concepts that you'll need as you're moving along in the course.
- As you start moving through the chapter, pay special attention to the **rule boxes.** These rules give you the reasons certain types of problems are solved the way they are. When you see a rule, try to rewrite the rule in your own words.

Rule for Adding Two Numbers

To add numbers with the same sign, add the absolute values of the numbers. Then attach the sign of the addends.

To add numbers with different signs, find the difference between the absolute values of the numbers. Then attach the sign of the addend with the greater absolute value.

Page 411

Knowing what to pay attention to as you move through a chapter will help you study and prepare.

INTERACTION We want you to be actively involved in learning mathematics and have given you many ways to get hands-on with this book.

- **HOW TO EXAMPLES** Take a look at page 86 shown here. See the HOW TO example? This contains an explanation by each step of the solution to a sample problem.

HOW TO • 6 The outside diameter of a laboratory beaker is $5\frac{1}{8}$ inches and the thickness of the glass is $\frac{1}{8}$ inch. Find the inside diameter of the beaker.

$$\frac{1}{8}+\frac{1}{8}=\frac{2}{8}=\frac{1}{4}$$

• Add $\frac{1}{8}$ and $\frac{1}{8}$ to find the total thickness of the two walls of the beaker.

$$\begin{array}{r} 5\frac{1}{8} = 5\frac{1}{8} = 4\frac{9}{8} \\ -\frac{1}{4} = \frac{2}{8} = \frac{2}{8} \\ \hline 4\frac{7}{8} \end{array}$$

• Subtract the total thickness of the two walls from the outside diameter to find the inside diameter.

Page 86

Grab a paper and pencil and work along as you're reading through each example. When you're done, get a clean sheet of paper. Write down the problem and try to complete the solution without looking at your notes or at the book. When you're done, check your answer. If you got it right, you're ready to move on.

- **EXAMPLE/YOU TRY IT PAIRS** You'll need hands-on practice to succeed in mathematics. When we show you an example, work it out beside our solution. Use the Example/You Try It pairs to get the practice you need.

  Take a look at page 178, Example 2 and You Try It 2 shown here:

EXAMPLE • 2

Write "2 grams of medication per 5 milliliters of solution" as a unit rate.

Solution

$$\frac{2 \text{ grams}}{5 \text{ milliliters}} = 5\overline{)2.0}\ \ (0.4)$$

YOU TRY IT • 2

Write "150 milligrams of medication in 3 tablets" as a unit rate.

Your solution

Page 178

You'll see that each Example is fully worked-out. Study this Example carefully by working through each step. Then, try your hand at it by completing the You Try It. If you get stuck, the solutions to the You Try Its are provided in the back of the book. There is a page number following the You Try It, which shows you where you can find the completely worked-out solution. Use the solution to get a hint for the step on which you are stuck. Then, try again!

When you've finished the solution, check your work against the solution in the back of the book. Turn to page S11 to see the solution for You Try It 2.

Remember that sometimes there can be more than one way to solve a problem. But, your answer should always match the answers we've given in the back of the book. If you have any questions about whether your method will always work, check with your instructor.

Kletr/Shutterstock.com

REVIEW We have provided many opportunities for you to practice and review the skills you have learned in each chapter.

- **SECTION EXERCISES** After you're done studying a section, flip to the end of the section and complete the exercises. If you immediately practice what you've learned, you'll find it easier to master the core skills. Want to know if you answered the questions correctly? The answers to the odd-numbered exercises are given in the back of the book.

- **CHAPTER SUMMARY** Once you've completed a chapter, look at the Chapter Summary. This is divided into two sections: *Key Words* and *Essential Rules and Procedures.* Flip to page 395 to see the Chapter Summary for Chapter 9. This summary shows all of the important topics covered in the chapter. See the reference following each topic? This shows you the objective reference and the page in the text where you can find more information on the concept.

- **CONCEPT REVIEW** Following the Chapter Summary for each chapter is the Concept Review. Flip to page 396 to see the Concept Review for Chapter 9. When you read each question, jot down a reminder note on the right about whatever you feel will be most helpful to remember if you need to apply that concept during an exam. You can also use the space on the right to mark what concepts your instructor expects you to know for the next test. If you are unsure of the answer to a concept review question, flip to the answers appendix at the back of the book.

- **CHAPTER REVIEW EXERCISES** You'll find the Chapter Review Exercises after the Concept Review. Flip to page 333 to see the Chapter Review Exercises for Chapter 7. When you do the review exercises, you're giving yourself an important opportunity to test your understanding of the chapter. The answer to each review exercise is given at the back of the book, along with the objective the question relates to. When you're done with the Chapter Review Exercises, check your answers. If you had trouble with any of the questions, you can restudy the objectives and retry some of the exercises in those objectives for extra help.

Photodisc

- **CHAPTER TESTS** The Chapter Tests can be found after the Chapter Review Exercises and can be used to prepare for your exams. The answer to each test question is given at the back of the book, along with a reference to a How To, Example, or You Try It that the question relates to. Think of these tests as "practice runs" for your in-class tests. Take the test in a quiet place and try to work through it in the same amount of time you will be allowed for your exam.

Here are some strategies for success when you're taking your exams:

- Scan the entire test to get a feel for the questions (get the big picture).
- Read the directions carefully.
- Work the problems that are easiest for you first.
- Stay calm, and remember that you will have lots of opportunities for success in this class!

EXCEL Visit **www.cengage.com/math/aufmann** to learn about additional study tools!

- *Enhanced* ***WebAssign***® online practice exercises and homework problems match the textbook exercises.
- **Videos** These text-specific instructional videos provide you with visual reinforcement of concepts and explanations given in easy-to-understand terms with detailed examples and sample problems. Available in YouBook and CourseMate.

Get Involved Have a question? Ask! Your professor and your classmates are there to help. Here are some tips to help you jump into the action:

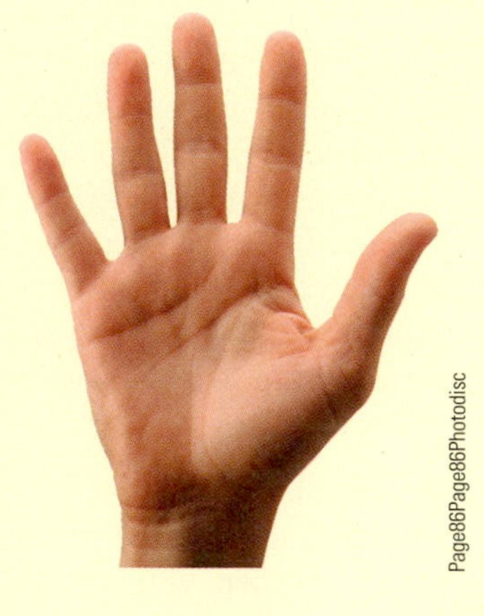

Page86Page86Photodisc

- Raise your hand in class.
- If your instructor prefers, e-mail or call your instructor with your question. If your professor has a website where you can post your question, also look there for answers to previous questions from other students. Take advantage of these ways to get your questions answered.
- Visit a **math center.** Ask your instructor for more information about the math center services available on your campus.
- Your instructor will have **office hours** where he or she will be available to help you. Take note of where and when your instructor holds office hours. Use this time for one-on-one help, if you need it.
- Form a **study group** with students from your class. This is a great way to prepare for tests, catch up on topics you may have missed, or get extra help on problems you're struggling with. Here are a few suggestions to make the most of your study group:
  - **Test each other by asking questions.** Have each person bring a few sample questions when you get together.

Alexander Raths/Shutterstock.com

- **Practice teaching each other.** We've found that you can learn a lot about what you know when you have to explain it to someone else.
- **Compare class notes.** Couldn't understand the last five minutes of class? Missed class because you were sick? Chances are someone in your group has the notes for the topics you missed.

PhotoDisc

- **Brainstorm test questions.**
- **Make a plan for your meeting.** Agree on what topics you'll talk about and how long you'll be meeting. When you make a plan, you'll be sure that you make the most of your meeting.

Ready, Set, Succeed! It takes hard work and commitment to succeed, but we know you can do it! Doing well in mathematics is just one step you'll take along the path to success.

I succeeded in Mathematics with Allied Health Applications!

We are confident that if you follow our suggestions, you will succeed. Good luck!

Rubberball

CHAPTER 1

Whole Numbers

OBJECTIVES

SECTION 1.1
- **A** To identify the order relation between two numbers
- **B** To write whole numbers in words and in standard form
- **C** To write whole numbers in expanded form
- **D** To round a whole number to a given place value

SECTION 1.2
- **A** To add whole numbers
- **B** To solve application problems

SECTION 1.3
- **A** To subtract whole numbers without borrowing
- **B** To subtract whole numbers with borrowing
- **C** To solve application problems

SECTION 1.4
- **A** To multiply a number by a single digit
- **B** To multiply larger whole numbers
- **C** To solve application problems

SECTION 1.5
- **A** To divide by a single digit with no remainder in the quotient
- **B** To divide by a single digit with a remainder in the quotient
- **C** To divide by larger whole numbers
- **D** To solve application problems

SECTION 1.6
- **A** To simplify expressions that contain exponents
- **B** To use the Order of Operations Agreement to simplify expressions

SECTION 1.7
- **A** To factor numbers
- **B** To find the prime factorization of a number

AIM for the Future

HOME HEALTH AIDES help people who are disabled, chronically ill, or cognitively impaired. They also work with older adults who may need assistance to live in their own homes or in residential facilities. Most aides work with elderly or with physically or mentally disabled clients who need more care than family or friends can provide. Employment of home health aides is projected to grow by 50% between 2008 and 2018, which is much faster than the average for all occupations. Median hourly wages of wage-and-salary home health aides were $9.22 in May 2008, with the highest 10% earning more than $12.33 an hour. (*Source: Bureau of Labor Statistics: http://www.bls.gov/oco/ocos326.htm*)

PREP TEST

Are you ready to succeed in this chapter?

Take the Prep Test below to find out if you are ready to learn the new material.

1. Name the number of ♦s shown below.

♦♦♦♦♦♦♦♦

2. Write the numbers from 1 to 10.

1 ____ ____ ____ ____ ____ ____ ____ ____ 10

3. Match the number with its word form.

a. 4	**A.** five
b. 2	**B.** one
c. 5	**C.** zero
d. 1	**D.** four
e. 3	**E.** two
f. 0	**F.** three

SECTION

1.1 Introduction to Whole Numbers

OBJECTIVE A To identify the order relation between two numbers

The **whole numbers** are 0, 1, 2, 3, 4, 5, 6, 7, 8, 9, 10, 11, 12, 13, 14, ...

The three dots mean that the list continues on and on and that there is no largest whole number.

Just as distances are associated with the markings on the edge of a ruler, the whole numbers can be associated with points on a line. This line is called the **number line.** The arrow on the number line below indicates that there is no largest whole number.

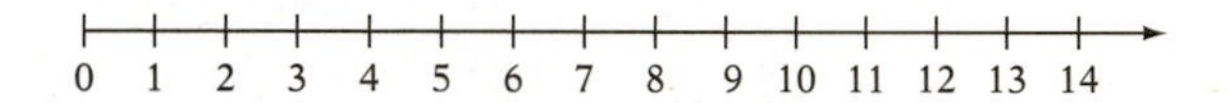

The **graph of a whole number** is shown by placing a heavy dot directly above that number on the number line. Here is the graph of 7 on the number line:

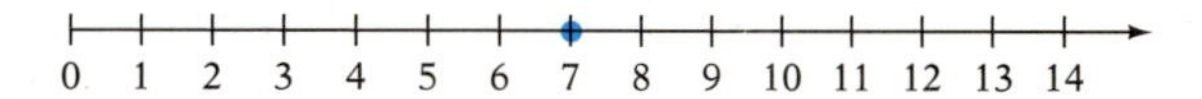

The number line can be used to show the order of whole numbers. A number that appears to the left of a given number **is less than** (<) the given number. A number that appears to the right of a given number **is greater than** (>) the given number.

Four is less than seven.
$4 < 7$

Twelve is greater than seven.
$12 > 7$

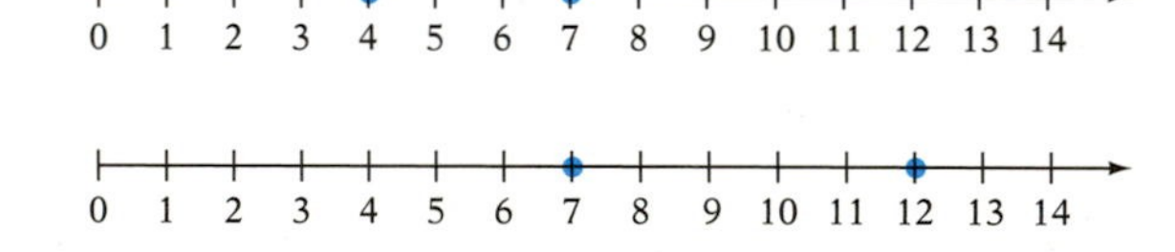

EXAMPLE 1

Graph 11 on the number line.

Solution

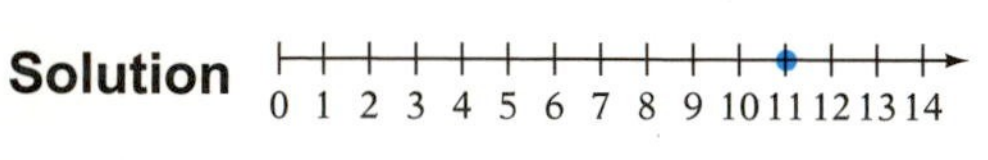

YOU TRY IT 1

Graph 6 on the number line.

Your solution

0 1 2 3 4 5 6 7 8 9 10 11 12 13 14

EXAMPLE 2

Place the correct symbol, < or >, between the two numbers.

a. 39 24

b. 0 51

Solution

a. $39 > 24$

b. $0 < 51$

YOU TRY IT 2

Place the correct symbol, < or >, between the two numbers.

a. 45 29

b. 27 0

Your solution

a.

b.

Solutions on p. S1

OBJECTIVE B To write whole numbers in words and in standard form

Point of Interest

The Babylonians had a place-value system based on 60. Its influence is still with us in angle measurement and time: 60 seconds in 1 minute, 60 minutes in 1 hour. It appears that the earliest record of a base-10 place-value system for natural numbers dates from the 8th century.

When a whole number is written using the digits 0, 1, 2, 3, 4, 5, 6, 7, 8, and 9, it is said to be in **standard form.** The position of each digit in the number determines the digit's **place value.** The diagram below shows a **place-value chart** naming the first 12 place values. The number 37,462 is in standard form and has been entered in the chart.

In the number 37,462, the position of the digit 3 determines that its place value is ten-thousands.

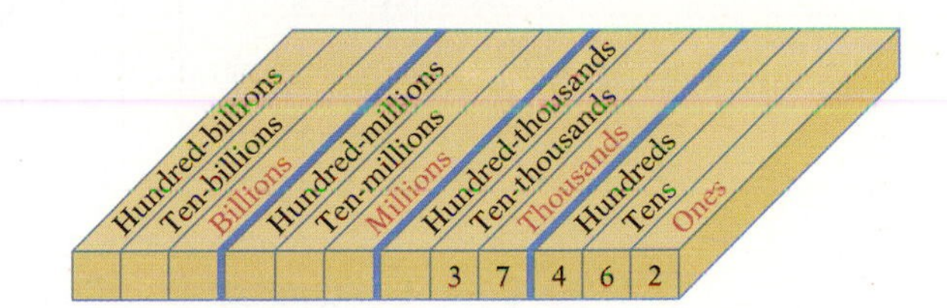

When a number is written in standard form, each group of digits separated from the other digits by a comma (or commas) is called a **period.** The number 3,786,451,294 has four periods. The period names are shown in red in the place-value chart above.

To write a number in words, start from the left. Name the number in each period. Then write the period name in place of the comma.

3,786,451,294 is read "three billion seven hundred eighty-six million four hundred fifty-one thousand two hundred ninety-four."

To write a whole number in standard form, write the number named in each period, and replace each period name with a comma.

Four million sixty-two thousand five hundred eighty-four is written 4,062,584. The zero is used as a place holder for the hundred-thousands place.

EXAMPLE 3

Write 25,478,083 in words.

Solution

Twenty-five million four hundred seventy-eight thousand eighty-three

YOU TRY IT 3

Write 36,462,075 in words.

Your solution

EXAMPLE 4

Write three hundred three thousand three in standard form.

Solution

303,003

YOU TRY IT 4

Write four hundred fifty-two thousand seven in standard form.

Your solution

Solutions on p. S1

OBJECTIVE C To write whole numbers in expanded form

The whole number 26,429 can be written in **expanded form** as

20,000 + 6000 + 400 + 20 + 9.

The place-value chart can be used to find the expanded form of a number.

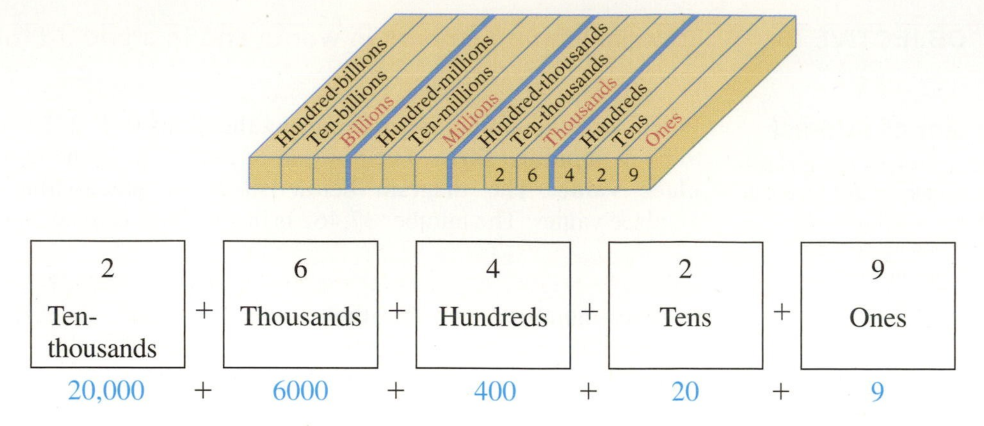

The number 420,806 is written in expanded form below. Note the effect of having zeros in the number.

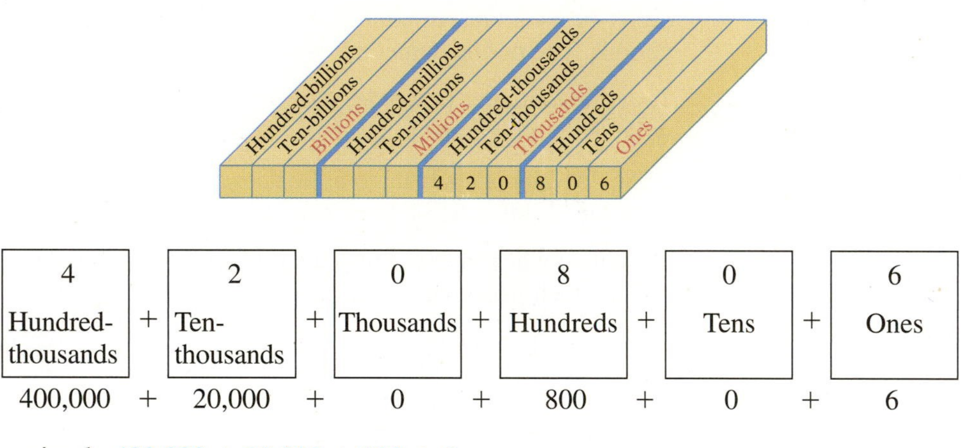

or simply 400,000 + 20,000 + 800 + 6.

EXAMPLE • 5

Write 23,859 in expanded form.

Solution

20,000 + 3000 + 800 + 50 + 9

YOU TRY IT • 5

Write 68,281 in expanded form.

Your solution

EXAMPLE • 6

Write 709,542 in expanded form.

Solution

700,000 + 9000 + 500 + 40 + 2

YOU TRY IT • 6

Write 109,207 in expanded form.

Your solution

Solutions on p. S1

OBJECTIVE D To round a whole number to a given place value

When the distance to the moon is given as 240,000 miles, the number represents an approximation to the true distance. Taking an approximate value for an exact number is called **rounding.** A rounded number is always rounded to a given place value.

37 is closer to 40 than it is to 30. 37 rounded to the nearest ten is 40.

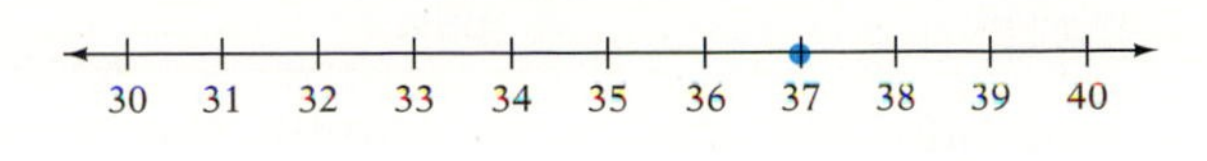

673 rounded to the nearest ten is 670. 673 rounded to the nearest hundred is 700.

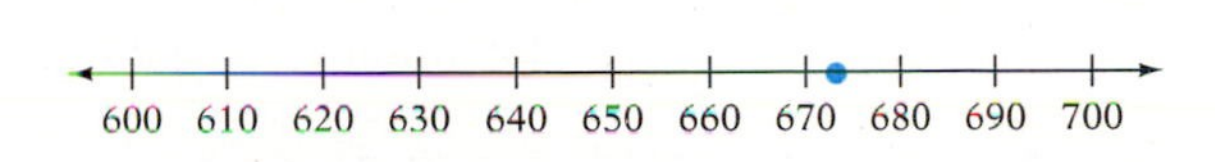

A whole number is rounded to a given place value without using the number line by looking at the first digit to the right of the given place value.

HOW TO • 1 Round 13,834 to the nearest hundred.

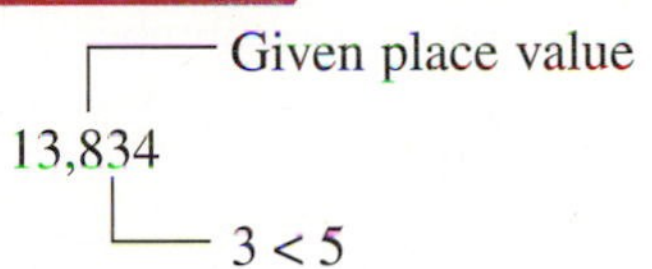

• If the digit to the right of the given place value is less than 5, that digit and all digits to the right are replaced by zeros.

13,834 rounded to the nearest hundred is 13,800.

HOW TO • 2 Round 386,217 to the nearest ten-thousand.

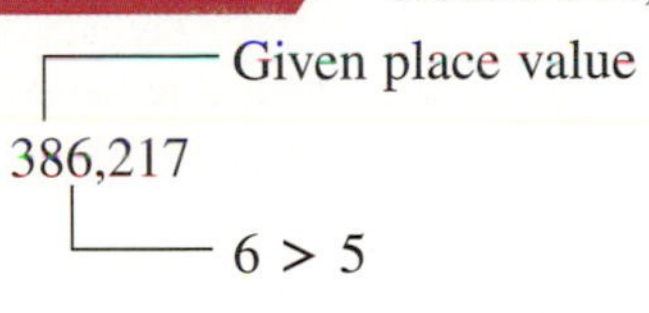

• If the digit to the right of the given place value is greater than or equal to 5, increase the digit in the given place value by 1, and replace all other digits to the right by zeros.

386,217 rounded to the nearest ten-thousand is 390,000.

EXAMPLE • 7

Round 525,453 to the nearest ten-thousand.

Solution

Given place value

525,453

$5 = 5$

525,453 rounded to the nearest ten-thousand is 530,000.

YOU TRY IT • 7

Round 368,492 to the nearest ten-thousand.

Your solution

EXAMPLE • 8

Round 1972 to the nearest hundred.

Solution

Given place value

1972

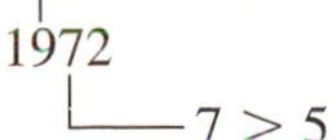

$7 > 5$

1972 rounded to the nearest hundred is 2000.

YOU TRY IT • 8

Round 3962 to the nearest hundred.

Your solution

Solutions on p. S1

1.1 EXERCISES

OBJECTIVE A To identify the order relation between two numbers

For Exercises 1 to 4, graph the number on the number line.

1. 3 0 1 2 3 4 5 6 7 8 9 10 11 12

2. 5 0 1 2 3 4 5 6 7 8 9 10 11 12

3. 9 0 1 2 3 4 5 6 7 8 9 10 11 12

4. 0 0 1 2 3 4 5 6 7 8 9 10 11 12

For Exercises 5 to 12, place the correct symbol, $<$ or $>$, between the two numbers.

5. 37 49 **6.** 58 21 **7.** 101 87 **8.** 245 158

9. 2701 2071 **10.** 0 45 **11.** 107 0 **12.** 815 928

13. Do the inequalities $21 < 30$ and $30 > 21$ express the same order relation?

OBJECTIVE B To write whole numbers in words and in standard form

For Exercises 14 to 17, name the place value of the digit 3.

14. 83,479 **15.** 3,491,507 **16.** 2,634,958 **17.** 76,319,204

For Exercises 18 to 25, write the number in words.

18. 2675 **19.** 3790 **20.** 42,928 **21.** 58,473

22. 356,943 **23.** 498,512 **24.** 3,697,483 **25.** 6,842,715

For Exercises 26 to 31, write the number in standard form.

26. Eighty-five **27.** Three hundred fifty-seven

28. Three thousand four hundred fifty-six **29.** Sixty-three thousand seven hundred eighty

30. Six hundred nine thousand nine hundred forty-eight

31. Seven million twenty-four thousand seven hundred nine

32. What is the place value of the first number on the left in a seven-digit whole number?

OBJECTIVE C To write whole numbers in expanded form

For Exercises 33 to 40, write the number in expanded form.

33. 5287

34. 6295

35. 58,943

36. 453,921

37. 200,583

38. 301,809

39. 403,705

40. 3,000,642

41. The expanded form of a number consists of four numbers added together. Must the number be a four-digit number?

OBJECTIVE D To round a whole number to a given place value

For Exercises 42 to 53, round the number to the given place value.

42. 926 Tens

43. 845 Tens

44. 1439 Hundreds

45. 3973 Hundreds

46. 43,607 Thousands

47. 52,715 Thousands

48. 389,702 Thousands

49. 629,513 Thousands

50. 647,989 Ten-thousands

51. 253,678 Ten-thousands

52. 36,702,599 Millions

53. 71,834,250 Millions

54. True or false? If a number rounded to the nearest ten is less than the original number, then the ones digit of the original number is greater than 5.

Applying the Concepts

55. If 3846 is rounded to the nearest ten and then that number is rounded to the nearest hundred, is the result the same as what you get when you round 3846 to the nearest hundred? If not, which of the two methods is correct for rounding to the nearest hundred?

SECTION

1.2 Addition of Whole Numbers

OBJECTIVE A To add whole numbers

Addition is the process of finding the total of two or more numbers.

Take Note
The numbers being added are called **addends.** The result is the **sum.**

By counting, we see that the total of \$3 and \$4 is \$7.

\$3	+	\$4	=	\$7
Addend		**Addend**		**Sum**

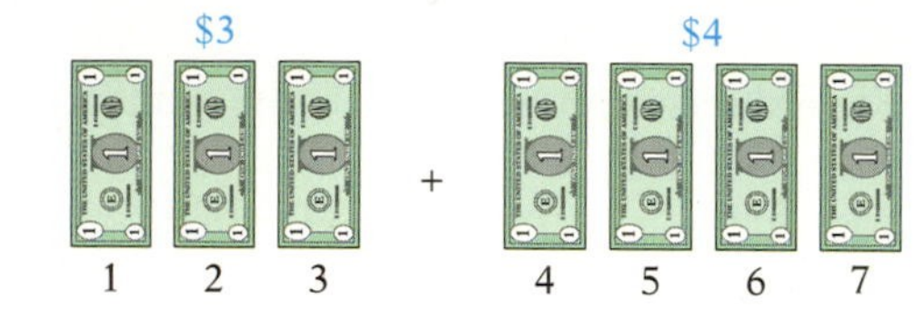

Addition can be illustrated on the number line by using arrows to represent the addends. The size, or magnitude, of a number can be represented on the number line by an arrow.

The number 3 can be represented anywhere on the number line by an arrow that is 3 units in length.

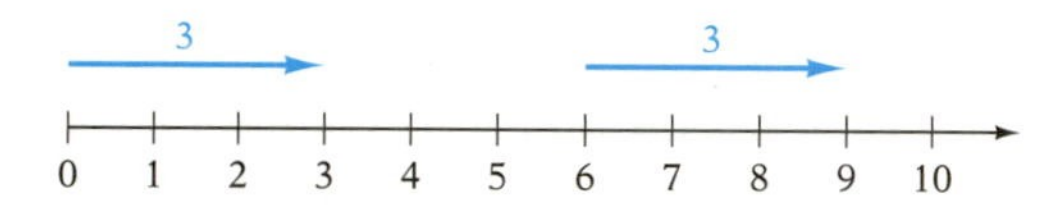

Point of Interest
The first use of the plus sign appeared in 1489 in *Mercantile Arithmetic.* It was used to indicate a surplus, not as the symbol for addition. That use did not appear until about 1515.

To add on the number line, place the arrows representing the addends head to tail, with the first arrow starting at zero. The sum is represented by an arrow starting at zero and stopping at the tip of the last arrow.

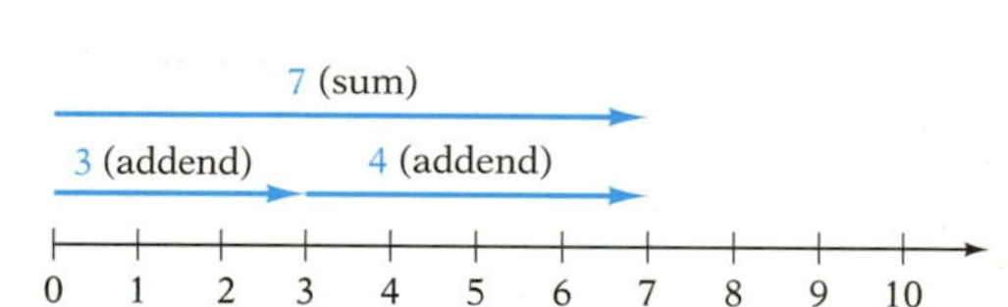

$$3 + 4 = 7$$

More than two numbers can be added on the number line.

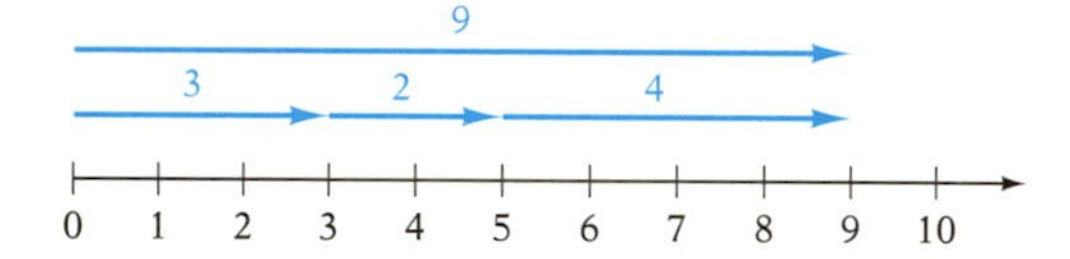

$$3 + 2 + 4 = 9$$

Some special properties of addition that are used frequently are given below.

Addition Property of Zero

Zero added to a number does not change the number.

$$4 + 0 = 4$$
$$0 + 7 = 7$$

Commutative Property of Addition

Two numbers can be added in either order; the sum will be the same.

$$4 + 8 = 8 + 4$$
$$12 = 12$$

Take Note
This is the same addition problem shown on the number line above.

Associative Property of Addition

Grouping the addition in any order gives the same result. The parentheses are grouping symbols and have the meaning "Do the operations inside the parentheses first."

$$\underbrace{(3 + 2)}_{5} + 4 = 3 + \underbrace{(2 + 4)}_{6}$$
$$5 + 4 = 3 + 6$$
$$9 = 9$$

The number line is not useful for adding large numbers. The basic addition facts for adding one digit to one digit should be memorized. Addition of larger numbers requires the repeated use of the basic addition facts.

To add large numbers, begin by arranging the numbers vertically, keeping the digits of the same place value in the same column.

HOW TO 1 Add: 321 + 6472

	THOUSANDS	HUNDREDS	TENS	ONES
		3	2	1
+	6	4	7	2
	6	7	9	3

• **Add the digits in each column.**

There are several words or phrases in English that indicate the operation of addition. Here are some examples:

added to	3 added to 5	$5 + 3$
more than	7 more than 5	$5 + 7$
the sum of	the sum of 3 and 9	$3 + 9$
increased by	4 increased by 6	$4 + 6$
the total of	the total of 8 and 3	$8 + 3$
plus	5 plus 10	$5 + 10$

HOW TO 2 What is the sum of 24 and 71?

$$\begin{array}{r} 24 \\ +\ 71 \\ \hline 95 \end{array}$$

• **The phrase *the sum of* means to add.**

The sum of 24 and 71 is 95.

Integrating Technology

Most scientific calculators use *algebraic logic:* the add (+), subtract (−), multiply (x), and divide (÷) keys perform the indicated operation on the number in the display and the next number keyed in. For instance, for the example at the right, enter 24 + 71 = . The display reads 95.

When the sum of the digits in a column exceeds 9, the addition will involve **carrying.**

HOW TO 3 Add: 487 + 369

	HUNDREDS	TENS	ONES
		1	
	4	8	7
+	3	6	9
			6

• **Add the ones column.**
7 + 9 = 16 (1 ten + 6 ones).
Write the 6 in the ones column and carry the 1 ten to the tens column.

$$\begin{array}{r} \scriptstyle 1\ \ 1\ \ \ \\ 4\ 8\ 7 \\ +\ 3\ 6\ 9 \\ \hline 5\ 6 \end{array}$$

• **Add the tens column.**
1 + 8 + 6 = 15 (1 hundred + 5 tens).
Write the 5 in the tens column and carry the 1 hundred to the hundreds column.

$$\begin{array}{r} \scriptstyle 1\ \ 1\ \ \ \\ 4\ 8\ 7 \\ +\ 3\ 6\ 9 \\ \hline 8\ 5\ 6 \end{array}$$

• **Add the hundreds column.**
1 + 4 + 3 = 8 (8 hundreds).
Write the 8 in the hundreds column.

EXAMPLE • 1

Find the total of 17, 103, and 8.

Solution

$$\begin{array}{r} \overset{1}{1}7 \\ 103 \\ +\ \ \ 8 \\ \hline 128 \end{array}$$

• $7 + 3 + 8 = 18$ Write the 8 in the ones column. Carry the 1 to the tens column.

YOU TRY IT • 1

What is 347 increased by 12,453?

Your solution

EXAMPLE • 2

Add: $89 + 36 + 98$

Solution

$$\begin{array}{r} \overset{2}{8}9 \\ 36 \\ +\ 98 \\ \hline 223 \end{array}$$

• $9 + 6 + 8 = 23$ Write the 3 in the ones column. Carry the 2 to the tens column.

YOU TRY IT • 2

Add: $95 + 88 + 67$

Your solution

EXAMPLE • 3

Add:
$$\begin{array}{r} 41{,}395 \\ 4{,}327 \\ 497{,}625 \\ +\ 32{,}991 \\ \hline \end{array}$$

Solution

$$\begin{array}{r} \scriptsize{1\,1\,2\ \ 2\,1} \\ 41{,}395 \\ 4{,}327 \\ 497{,}625 \\ +\ 32{,}991 \\ \hline 576{,}338 \end{array}$$

YOU TRY IT • 3

Add:
$$\begin{array}{r} 392 \\ 4{,}079 \\ 89{,}035 \\ +\ 4{,}992 \\ \hline \end{array}$$

Your solution

Solutions on p. S1

ESTIMATION

Estimation and Calculators

At some places in the text, you will be asked to use your calculator. Effective use of a calculator requires that you estimate the answer to the problem. This helps ensure that you have entered the numbers correctly and pressed the correct keys.

For example, if you use your calculator to find 22,347 + 5896 and the answer in the calculator's display is 131,757,912, you should realize that you have entered some part of the calculation incorrectly. In this case, you pressed **x** instead of **+**. By estimating the answer to a problem, you can help ensure the accuracy of your calculations. We have a special symbol for **approximately equal to (≈).**

For example, to estimate the answer to 22,347 + 5896, round each number to the same place value. In this case, we will round to the nearest thousand. Then add.

$$\begin{array}{rcr} 22{,}347 & \approx & 22{,}000 \\ +\ \ 5{,}896 & \approx & +\ \ \ 6{,}000 \\ \hline & & 28{,}000 \end{array}$$

The sum 22,347 + 5896 is approximately 28,000. Knowing this, you would know that 131,757,912 is much too large and is therefore incorrect.

To estimate the sum of two numbers, first round each whole number to the same place value and then add. Compare this answer with the calculator's answer.

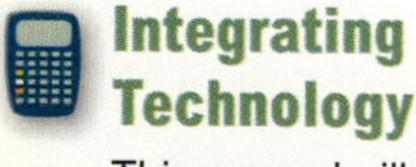

Integrating Technology

This example illustrates that estimation is important when one is using a calculator.

OBJECTIVE B To solve application problems

To solve an application problem, first read the problem carefully. The **strategy** involves identifying the quantity to be found and planning the steps that are necessary to find that quantity. The **solution of an application problem** involves performing each operation stated in the strategy and writing the answer.

HOW TO • 4

Nurse's aides (CNAs) must record the amount of fluid intake for some patients. All liquids or foods that melt at room temperature are recorded. This table gives the estimated amounts of liquid for each item in milliliters.

Item	*Estimated Volume*	*Item*	*Estimated Volume*
Popsicle, 3 ounces	90 milliliters	Ice cream, 4 ounces	120 milliliters
Milk, 8 ounces	240 milliliters	Bowl of soup/broth	150 milliliters
Coffee, 6 ounces	60 milliliters	Gelatin, 1 cup	120 milliliters

Find the total amount of fluid intake for a patient whose afternoon snack is a bowl of ice cream and glass of milk.

Strategy To find the total fluid intake, read the table to find the estimated volume for each item. Then add the numbers.

Solution

$$\begin{array}{r} 120 \\ +\ 240 \\ \hline 360 \end{array}$$

The patient's total fluid intake from the bowl of ice cream and the glass of milk is 360 milliliters.

EXAMPLE • 4

After surgery, a patient is restricted to a liquid diet for 24 hours. Use the table above to total the amount of fluid the patient receives at lunch if she consumes the following items: one bowl of soup, one popsicle, and one cup of coffee.

Strategy

To find the total fluid intake, read the table to find the estimated volume for all of the items listed. Then add the numbers.

Solution

$$\begin{array}{r} 150 \\ 90 \\ +\ \ 60 \\ \hline 300 \end{array}$$

The patient's total fluid intake at lunch is 300 milliliters.

YOU TRY IT • 4

For supper, your patient consumes a cup of broth, one bowl of ice cream, and a glass of milk. Use the table above to calculate the total fluid intake for this patient's supper.

Your strategy

Your solution

Solution on p. S1

1.2 EXERCISES

OBJECTIVE A To add whole numbers

For Exercises 1 to 32, add.

1. $\begin{array}{r} 17 \\ +\ 11 \\ \hline \end{array}$

2. $\begin{array}{r} 25 \\ +\ 63 \\ \hline \end{array}$

3. $\begin{array}{r} 83 \\ +\ 42 \\ \hline \end{array}$

4. $\begin{array}{r} 63 \\ +\ 94 \\ \hline \end{array}$

5. $\begin{array}{r} 77 \\ +\ 25 \\ \hline \end{array}$

6. $\begin{array}{r} 63 \\ +\ 49 \\ \hline \end{array}$

7. $\begin{array}{r} 56 \\ +\ 98 \\ \hline \end{array}$

8. $\begin{array}{r} 86 \\ +\ 68 \\ \hline \end{array}$

9. $\begin{array}{r} 658 \\ +\ 831 \\ \hline \end{array}$

10. $\begin{array}{r} 842 \\ +\ 936 \\ \hline \end{array}$

11. $\begin{array}{r} 735 \\ +\ 93 \\ \hline \end{array}$

12. $\begin{array}{r} 189 \\ +\ 50 \\ \hline \end{array}$

13. $\begin{array}{r} 859 \\ +\ 725 \\ \hline \end{array}$

14. $\begin{array}{r} 637 \\ +\ 829 \\ \hline \end{array}$

15. $\begin{array}{r} 470 \\ +\ 749 \\ \hline \end{array}$

16. $\begin{array}{r} 427 \\ +\ 690 \\ \hline \end{array}$

17. $\begin{array}{r} 36{,}925 \\ +\ 65{,}392 \\ \hline \end{array}$

18. $\begin{array}{r} 56{,}772 \\ +\ 51{,}239 \\ \hline \end{array}$

19. $\begin{array}{r} 50{,}873 \\ +\ 28{,}453 \\ \hline \end{array}$

20. $\begin{array}{r} 34{,}872 \\ +\ 46{,}079 \\ \hline \end{array}$

21. $\begin{array}{r} 878 \\ 737 \\ +\ 189 \\ \hline \end{array}$

22. $\begin{array}{r} 768 \\ 461 \\ +\ 669 \\ \hline \end{array}$

23. $\begin{array}{r} 319 \\ 348 \\ +\ 912 \\ \hline \end{array}$

24. $\begin{array}{r} 292 \\ 579 \\ +\ 315 \\ \hline \end{array}$

25. $\begin{array}{r} 9409 \\ 3253 \\ +\ 7078 \\ \hline \end{array}$

26. $\begin{array}{r} 8188 \\ 8020 \\ +\ 7104 \\ \hline \end{array}$

27. $\begin{array}{r} 2038 \\ 2243 \\ +\ 3139 \\ \hline \end{array}$

28. $\begin{array}{r} 4252 \\ 6882 \\ +\ 5235 \\ \hline \end{array}$

29. $\begin{array}{r} 67{,}428 \\ 32{,}171 \\ +\ 20{,}971 \\ \hline \end{array}$

30. $\begin{array}{r} 52{,}801 \\ 11{,}664 \\ +\ 89{,}638 \\ \hline \end{array}$

31. $\begin{array}{r} 76{,}290 \\ 43{,}761 \\ +\ 87{,}402 \\ \hline \end{array}$

32. $\begin{array}{r} 43{,}901 \\ 98{,}301 \\ +\ 67{,}943 \\ \hline \end{array}$

For Exercises 33 to 40, add.

33. 20,958 + 3218 + 42

34. 80,973 + 5168 + 29

35. 392 + 37 + 10,924 + 621

36. 694 + 62 + 70,129 + 217

37. 294 + 1029 + 7935 + 65

38. 692 + 2107 + 3196 + 92

39. 97 + 7234 + 69,532 + 276

40. 87 + 1698 + 27,317 + 727

41. What is 9874 plus 4509?

42. What is 7988 plus 5678?

43. What is 3487 increased by 5986?

44. What is 99,567 increased by 126,863?

45. What is 23,569 more than 9678?

46. What is 7894 more than 45,872?

47. What is 479 added to 4579?

48. What is 23,902 added to 23,885?

49. Find the total of 659, 55, and 1278.

50. Find the total of 4561, 56, and 2309.

51. Find the sum of 34, 329, 8, and 67,892.

52. Find the sum of 45, 1289, 7, and 32,876.

For Exercises 53 to 56, use a calculator to add. Then round the numbers to the nearest hundred, and use estimation to determine whether the sum is reasonable.

53. 1234 + 9780 + 6740

54. 919 + 3642 + 8796

55. 241 + 569 + 390 + 1672

56. 107 + 984 + 1035 + 2904

For Exercises 57 to 60, use a calculator to add. Then round the numbers to the nearest thousand, and use estimation to determine whether the sum is reasonable.

57.
$$\begin{array}{r} 32{,}461 \\ 9{,}844 \\ +\ 59{,}407 \\ \hline \end{array}$$

58.
$$\begin{array}{r} 29{,}036 \\ 22{,}904 \\ +\ 7{,}903 \\ \hline \end{array}$$

59.
$$\begin{array}{r} 25{,}432 \\ 62{,}941 \\ +\ 70{,}390 \\ \hline \end{array}$$

60.
$$\begin{array}{r} 66{,}541 \\ 29{,}365 \\ +\ 98{,}742 \\ \hline \end{array}$$

For Exercises 61 to 64, use a calculator to add. Then round the numbers to the nearest ten-thousand, and use estimation to determine whether the sum is reasonable.

61. 67,421
82,984
66,361
10,792
+ 34,037

62. 21,896
4,235
62,544
21,892
+ 1,334

63. 281,421
9,874
34,394
526,398
+ 94,631

64. 542,698
97,327
7,235
73,667
+ 173,201

65. Which property of addition (see page 8) allows you to use either arrangement shown at the right to find the sum of 691 and 452?

691
+ 452

452
+ 691

OBJECTIVE B To solve application problems

66. Fluid Intake At snack time, a patient consumed an item containing 60 milliliters of fluid and an item containing 90 milliliters of fluid. Use the table on page 11 to determine what two items the patient had.

67. Demographics In a recent year, according to the U.S. Department of Health and Human Services, there were 110,670 twin births in this country, 6,919 triplet births, 627 quadruplet deliveries, and 79 quintuplet and other higher-order multiple births. Find the total number of multiple births during the year.

Laura Dwight/PhotoEdit, Inc.

68. Demographics The Census Bureau estimates that the U.S. population will grow by 296 million people from 2000 to 2100. Given that the U.S. population in 2000 was 281 million, find the Census Bureau's estimate of the U.S. population in 2100.

The Film Industry The graph shows the box office income for four movies with a medical theme. Use this information for Exercises 69 to 71.

69. What is the total income produced by *Patch Adams* and *The Awakening*?

70. Find the total income for all four movies.

71. Find the total income from the two movies with the lowest box office incomes.

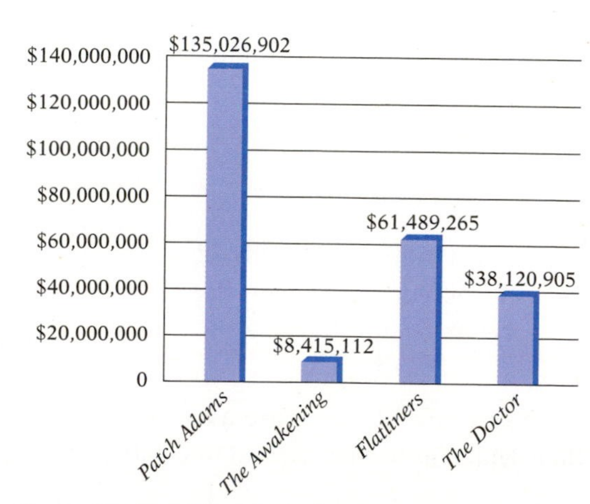

Source: http://boxofficemojo.com/movies

72. **Calories** The number of calories for each item in a typical fast food lunch is a quarter-pound hamburger, 420 calories; large French fries, 540 calories; a small vanilla milkshake, 570 calories. Find the total number of calories for this fast food lunch.

73. **Walking** One week, a hospital orderly wore a pedometer at work. The orderly walked 3 miles on Monday, 4 miles on both Tuesday and Wednesday, 5 miles on Thursday, and 2 miles on Friday.

a. How many miles did he walk at work from Monday through Wednesday?

b. How many miles did he walk this week at work?

74. **Computer Files** The Southern Hospital System consists of two hospitals. At Hospital A, there are 983,000 online medical files. Hospital B contains 1,017,000 online medical files. How many online medical files are present in the hospital system?

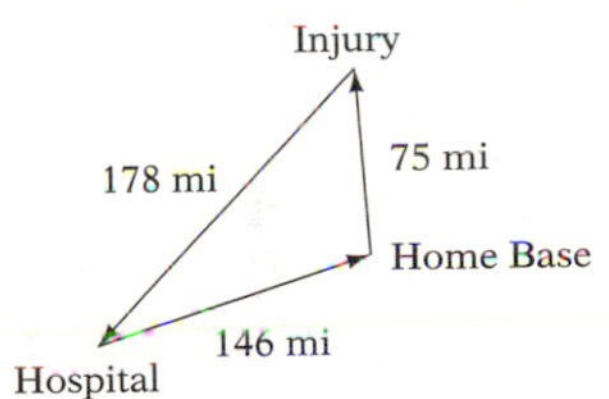

75. **Air Ambulance** A medical air rescue helicopter must travel a distance of 75 miles to the scene of an injury, and then must fly 178 miles to the nearest hospital and then another 146 miles to return to its home base. Find the total mileage flown by the medical helicopter.

Ivan Cholakov Gostock-dot-net/Shutterstock.com

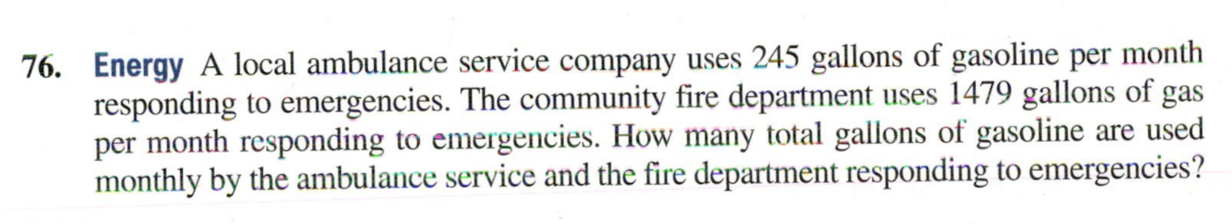

76. **Energy** A local ambulance service company uses 245 gallons of gasoline per month responding to emergencies. The community fire department uses 1479 gallons of gas per month responding to emergencies. How many total gallons of gasoline are used monthly by the ambulance service and the fire department responding to emergencies?

Applying the Concepts

77. If you roll two ordinary six-sided dice and add the two numbers that appear on top, how many different sums are possible?

78. If you add two *different* whole numbers, is the sum always greater than either one of the numbers? If not, give an example.

79. If you add two whole numbers, is the sum always greater than either one of the numbers? If not, give an example. (Compare this with the previous exercise.)

80. Make up a word problem for which the answer is the sum of 34 and 28.

81. Call a number "lucky" if it ends in a 7. How many lucky numbers are less than 100?

SECTION

1.3 Subtraction of Whole Numbers

OBJECTIVE A To subtract whole numbers without borrowing

Take Note

The **minuend** is the number from which another number is subtracted. The **subtrahend** is the number that is subtracted from another number. The result is the **difference.**

Subtraction is the process of finding the difference between two numbers.

By counting, we see that the difference between \$8 and \$5 is \$3.

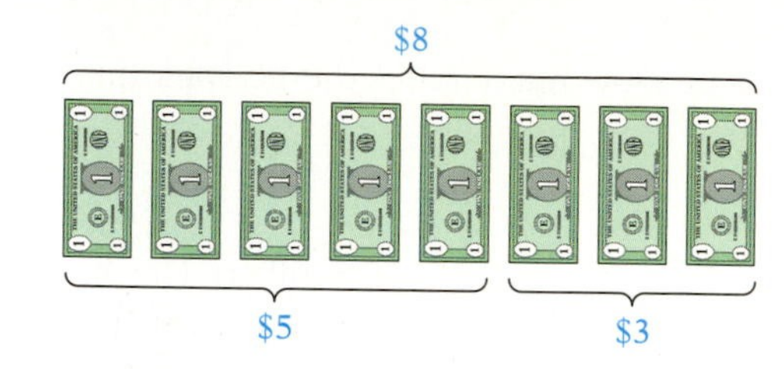

\$8	−	\$5	=	\$3
Minuend		**Subtrahend**		**Difference**

The difference 8 − 5 can be shown on the number line.

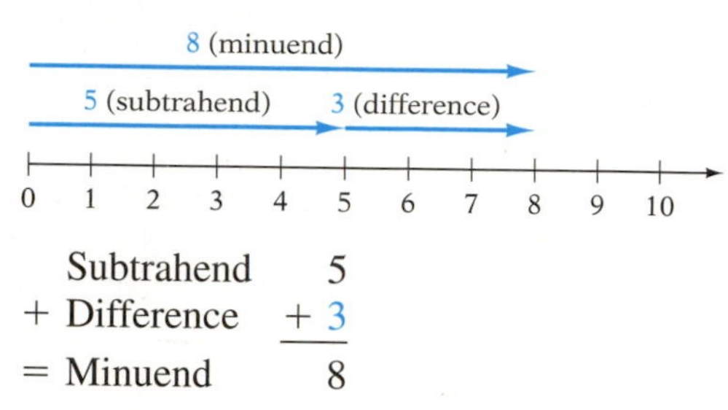

Note from the number line that addition and subtraction are related.

Subtrahend	5
+ Difference	+ 3
= Minuend	8

Point of Interest

The use of the minus sign dates from the same period as the plus sign, around 1515.

The fact that the sum of the subtrahend and the difference equals the minuend can be used to check subtraction.

To subtract large numbers, begin by arranging the numbers vertically, keeping the digits that have the same place value in the same column. Then subtract the digits in each column.

HOW TO • 1 Subtract 8955 − 2432 and check.

	THOUSANDS	HUNDREDS	TENS	ONES
	8	9	5	5
−	2	4	3	2
	6	5	2	3

Check:

Subtrahend	2432
+ Difference	+ 6523
= Minuend	8955

EXAMPLE • 1

Subtract 6594 − 3271 and check.

Solution

$$\begin{array}{r} 6594 \\ -\ 3271 \\ \hline 3323 \end{array} \qquad \textit{Check:} \quad \begin{array}{r} 3271 \\ +\ 3323 \\ \hline 6594 \end{array}$$

YOU TRY IT • 1

Subtract 8925 − 6413 and check.

Your solution

EXAMPLE • 2

Subtract 15,762 − 7541 and check.

Solution

$$\begin{array}{r} 15{,}762 \\ -\ 7{,}541 \\ \hline 8{,}221 \end{array} \qquad \textit{Check:} \quad \begin{array}{r} 7{,}541 \\ +\ 8{,}221 \\ \hline 15{,}762 \end{array}$$

YOU TRY IT • 2

Subtract 17,504 − 9302 and check.

Your solution

Solutions on p. S1

OBJECTIVE B To subtract whole numbers with borrowing

In all the subtraction problems in the previous objective, for each place value the lower digit was not larger than the upper digit. When the lower digit is larger than the upper digit, subtraction will involve **borrowing.**

HOW TO • 2 Subtract: 692 − 378

HUNDREDS	TENS	ONES
	8+1	
6	9	2
− 3	7	8

Because 8 > 2, borrowing is necessary. 9 tens = 8 tens + 1 ten.

HUNDREDS	TENS	ONES
	8+1	10
6	9	2
− 3	7	8

Borrow 1 ten from the tens column and write 10 in the ones column.

HUNDREDS	TENS	ONES
	8	12
6	9	2
− 3	7	8

Add the borrowed 10 to 2.

HUNDREDS	TENS	ONES
	8	12
6	9	2
− 3	7	8
3	1	4

Subtract the digits in each column.

The phrases below are used to indicate the operation of subtraction. An example is shown at the right of each phrase.

minus	8 minus 5	8 − 5
less	9 less 3	9 − 3
less than	2 less than 7	7 − 2
the difference between	the difference between 8 and 2	8 − 2
decreased by	5 decreased by 1	5 − 1

HOW TO • 3 Find the difference between 1234 and 485, and check.

"The difference between 1234 and 485" means 1234 − 485.

```
        2 14          1 12 14        0 11 12 14
  1  2  3  4          1  2  3  4     1  2  3  4
−    4  8  5      −      4  8  5  −     4  8  5
           9                4  9        7  4  9
```

Check:

```
   1 1
   485
 + 749
  1234
```

Subtraction with a zero in the minuend involves repeated borrowing.

HOW TO • 4 Subtract: 3904 − 1775

```
     8  10             9                 9
                    8  10 14          8  10 14
  3  9  0  4     3  9  0  4        3  9  0  4
− 1  7  7  5   − 1  7  7  5      − 1  7  7  5
                                   2  1  2  9
```

5 > 4
There is a 0 in the tens column. Borrow 1 hundred (= 10 tens) from the hundreds column and write 10 in the tens column.

Borrow 1 ten from the tens column and add 10 to the 4 in the ones column.

Subtract the digits in each column.

Tips for Success

The HOW TO feature indicates an example with explanatory remarks. Using paper and pencil, you should work through the example. See *AIM for Success* at the front of the book.

EXAMPLE • 3

Subtract 4392 − 678 and check.

Solution

$$\begin{array}{r} \overset{3}{\cancel{4}}\,\overset{13}{\cancel{3}}\,\overset{8}{\cancel{9}}\,\overset{12}{\cancel{2}} \\ -\quad 6\,7\,8 \\ \hline 3\,7\,1\,4 \end{array} \qquad \textit{Check:}\ \begin{array}{r} 678 \\ +\,3714 \\ \hline 4392 \end{array}$$

YOU TRY IT • 3

Subtract 3481 − 865 and check.

Your solution

EXAMPLE • 4

Find 23,954 less than 63,221 and check.

Solution

$$\begin{array}{r} \overset{5}{\cancel{6}}\,\overset{12}{\cancel{3}},\overset{11}{\cancel{2}}\,\overset{11}{\cancel{2}}\,\overset{11}{\cancel{1}} \\ -\,2\,3,9\,5\,4 \\ \hline 3\,9,2\,6\,7 \end{array} \qquad \textit{Check:}\ \begin{array}{r} 23{,}954 \\ +\,39{,}267 \\ \hline 63{,}221 \end{array}$$

YOU TRY IT • 4

Find 54,562 decreased by 14,485 and check.

Your solution

EXAMPLE • 5

Subtract 46,005 − 32,167 and check.

Solution

$$\begin{array}{r} 4\,\overset{5}{\cancel{6}},\overset{10}{\cancel{0}}\,0\,5 \\ -\,3\,2,1\,6\,7 \\ \hline \end{array}$$

- There are two zeros in the minuend. Borrow 1 thousand from the thousands column and write 10 in the hundreds column.

$$\begin{array}{r} 4\,\overset{5}{\cancel{6}},\overset{9}{\cancel{10}}\,\overset{10}{\cancel{0}}\,5 \\ -\,3\,2,1\,6\,7 \\ \hline \end{array}$$

- Borrow 1 hundred from the hundreds column and write 10 in the tens column.

$$\begin{array}{r} 4\,\overset{5}{\cancel{6}},\overset{9}{\cancel{10}}\,\overset{9}{\cancel{10}}\,\overset{15}{\cancel{5}} \\ -\,3\,2,1\,6\,7 \\ \hline 1\,3,8\,3\,8 \end{array}$$

- Borrow 1 ten from the tens column and add 10 to the 5 in the ones column.

$$\textit{Check:}\ \begin{array}{r} 32{,}167 \\ +\,13{,}838 \\ \hline 46{,}005 \end{array}$$

YOU TRY IT • 5

Subtract 64,003 − 54,936 and check.

Your solution

Solutions on pp. S1–S2

ESTIMATION

Estimating the Difference Between Two Whole Numbers

Calculate 323,502 − 28,912. Then use estimation to determine whether the difference is reasonable.

Subtract to find the exact difference. To estimate the difference, round each number to the same place value. Here we have rounded to the nearest ten-thousand. Then subtract. The estimated answer is 290,000, which is very close to the exact difference 294,590.

$$\begin{array}{r} 323{,}502 \\ -\ 28{,}912 \\ \hline 294{,}590 \end{array} \begin{array}{c} \approx \\ \approx \\ \\ \end{array} \begin{array}{r} 320{,}000 \\ -\ 30{,}000 \\ \hline 290{,}000 \end{array}$$

OBJECTIVE C To solve application problems

The table at the right provides calorie counts for healthy snacks. Use the information in this table for Example 6 and You Try It 6.

Snack	*Calories*	*Snack*	*Calories*
1 cup air-popped popcorn	31	Nonfat fruit yogurt (6 ounces)	160
1 medium apple	72	Low-fat granola bar	90
1 ounce dry roasted peanuts	166	1 ounce almonds	206

EXAMPLE 6

In order for Maria to lose 1.5 pounds per week, a dietician advises her to eat a maximum of 1426 calories per day. Today, Maria's breakfast was 350 calories, and her lunch was 480 calories. If she has a 6-ounce container of nonfat fruit yogurt for her snack, how many calories will she have left for her supper?

Strategy

Find the total number of calories that Maria will consume by adding the calories for breakfast, lunch, and snack. Then subtract this from her maximum allowance of 1426 calories.

Solution

350 breakfast
480 lunch
+160 yogurt
990 total

1426 daily allowance
− 990 total calories used
436 remaining

Maria has 436 calories remaining for her supper.

YOU TRY IT 6

If Maria increases her exercise routine, she will be allowed to eat 1550 calories per day and lose 1.5 pounds weekly. If she attempts to eat 400 calories for breakfast, 450 calories for lunch, and 3 cups of popcorn for her snack, how many calories will remain for her supper?

Your strategy

Your solution

EXAMPLE 7

You had a balance of $415 on your student debit card. You then used the card, deducting $197 for books, $48 for laboratory supplies, and $24 for a medical assisting student uniform. What is your new student debit card balance?

Strategy

To find your new debit card balance:

- Add to find the total of the three deductions (197 + 48 + 24).
- Subtract the total of the three deductions from the old balance (415).

Solution

197
48
+ 24
269 total deductions

415
− 269
146

Your new debit card balance is $146.

YOU TRY IT 7

Your total weekly salary is $638. Deductions of $127 for taxes, $18 for insurance, and $35 for savings are taken from your pay. Find your weekly take-home pay.

Your strategy

Your solution

Solutions on p. S2

1.3 EXERCISES

OBJECTIVE A To subtract whole numbers without borrowing

For Exercises 1 to 35, subtract.

1. $9 - 5$ **2.** $8 - 7$ **3.** $8 - 4$ **4.** $7 - 3$ **5.** $10 - 0$

6. $11 - 4$ **7.** $12 - 8$ **8.** $19 - 8$ **9.** $15 - 6$ **10.** $16 - 7$

11. $25 - 3$ **12.** $55 - 4$ **13.** $68 - 8$ **14.** $77 - 3$ **15.** $89 - 23$

16. $54 - 21$ **17.** $88 - 57$ **18.** $1202 - 701$ **19.** $1305 - 404$ **20.** $1763 - 801$

21. $1497 - 706$ **22.** $8974 - 3972$ **23.** $2836 - 1711$ **24.** $8976 - 7463$ **25.** $9273 - 6142$

26. $77 - 36$ **27.** $129 - 82$ **28.** $132 - 61$ **29.** $969 - 44$ **30.** $1347 - 103$

31. $4865 - 304$ **32.** $1525 - 702$ **33.** $9999 - 6794$ **34.** $7806 - 3405$ **35.** $8843 - 7621$

36. Suppose three whole numbers, called *minuend, subtrahend,* and *difference,* are related by the subtraction statement *minuend* − *subtrahend* = *difference.* State whether the given relationship *must be true, might be true,* or *cannot be true.*

a. minuend > difference **b.** subtrahend < difference

OBJECTIVE B To subtract whole numbers with borrowing

For Exercises 37 to 80, subtract.

37. $71 - 18$ **38.** $93 - 28$ **39.** $47 - 18$ **40.** $44 - 27$

41. $37 - 29$ **42.** $50 - 27$ **43.** $70 - 33$ **44.** $993 - 537$

45. $\begin{array}{r} 250 \\ -\ 192 \\ \hline \end{array}$

46. $\begin{array}{r} 840 \\ -\ 783 \\ \hline \end{array}$

47. $\begin{array}{r} 768 \\ -\ 194 \\ \hline \end{array}$

48. $\begin{array}{r} 770 \\ -\ 395 \\ \hline \end{array}$

49. 674 − 337

50. 3526 − 387

51. 1712 − 289

52. 4350 − 729

53. 1702 − 948

54. 1607 − 869

55. 5933 − 3754

56. 7293 − 3748

57. 9407 − 2918

58. 3706 − 2957

59. 8605 − 7716

60. 8052 − 2709

61. 80,305 − 9176

62. 70,702 − 4239

63. 10,004 − 9306

64. 80,009 − 63,419

65. 70,618 − 41,213

66. 80,053 − 27,649

67. 70,700 − 21,076

68. 80,800 − 42,023

69. $\begin{array}{r} 2600 \\ -\ 1972 \\ \hline \end{array}$

70. $\begin{array}{r} 8400 \\ -\ 3762 \\ \hline \end{array}$

71. $\begin{array}{r} 9003 \\ -\ 2471 \\ \hline \end{array}$

72. $\begin{array}{r} 6004 \\ -\ 2392 \\ \hline \end{array}$

73. $\begin{array}{r} 8202 \\ -\ 3916 \\ \hline \end{array}$

74. $\begin{array}{r} 7050 \\ -\ 4137 \\ \hline \end{array}$

75. $\begin{array}{r} 7015 \\ -\ 2973 \\ \hline \end{array}$

76. $\begin{array}{r} 4207 \\ -\ 1624 \\ \hline \end{array}$

77. $\begin{array}{r} 7005 \\ -\ 1796 \\ \hline \end{array}$

78. $\begin{array}{r} 8003 \\ -\ 2735 \\ \hline \end{array}$

79. $\begin{array}{r} 20{,}005 \\ -\ 9{,}627 \\ \hline \end{array}$

80. $\begin{array}{r} 80{,}004 \\ -\ 8{,}237 \\ \hline \end{array}$

81. Which of the following phrases represent the subtraction 673 − 571?
(i) 571 less 673 **(ii)** 571 less than 673 **(iii)** 673 decreased by 571

82. Find 10,051 less 9027.

83. Find 17,031 less 5792.

84. Find the difference between 1003 and 447.

85. What is 29,874 minus 21,392?

86. What is 29,797 less than 68,005?

87. What is 69,379 less than 70,004?

88. What is 25,432 decreased by 7994?

89. What is 86,701 decreased by 9976?

For Exercises 90 to 93, use the relationship between addition and subtraction to complete the statement.

90. ___ + 39 = 104 **91.** 67 + ___ = 90 **92.** ___ + 497 = 862 **93.** 253 + ___ = 4901

For Exercises 94 to 99, use a calculator to subtract. Then round the numbers to the nearest ten-thousand and use estimation to determine whether the difference is reasonable.

94. $\begin{array}{r} 80{,}032 \\ -\ 19{,}605 \\ \hline \end{array}$

95. $\begin{array}{r} 90{,}765 \\ -\ 60{,}928 \\ \hline \end{array}$

96. $\begin{array}{r} 32{,}574 \\ -\ 10{,}961 \\ \hline \end{array}$

97. $\begin{array}{r} 96{,}430 \\ -\ 59{,}762 \\ \hline \end{array}$

98. $\begin{array}{r} 567{,}423 \\ -\ 208{,}444 \\ \hline \end{array}$

99. $\begin{array}{r} 300{,}712 \\ -\ 198{,}714 \\ \hline \end{array}$

OBJECTIVE C To solve application problems

100. Banking You have $304 in your checking account. If you write a check for $139, how much is left in your checking account?

101. Sales The X-Ray Radiological Company sells their x-ray units in the United States and Canada. The table displays the number of units sold in five states and one Canadian province last year.

Location	New York	Ohio	Florida	Texas	California	Quebec
Number Sold	289	53	167	175	391	234

a. How many units were sold in New York, California, and Quebec?

b. What is the difference between the number of units sold in the five states and the Canadian province?

c. How many more units were sold in California than in Florida?

d. What is the total number of units sold in the states and province listed in the table?

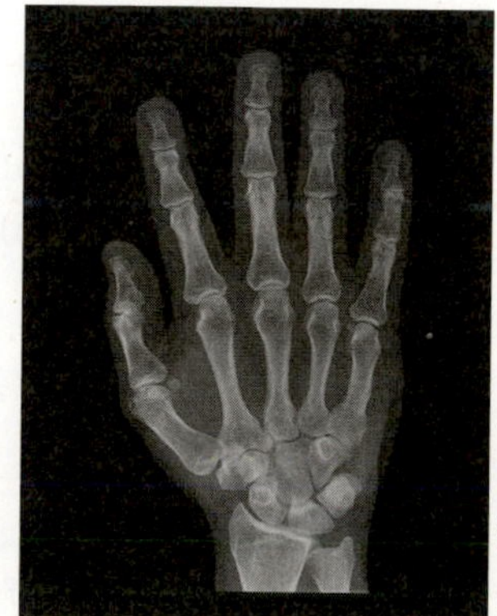

skyhawk/Shutterstock.com

102. Finances You had a credit card balance of $409 before you used the card to purchase books for $168, CDs for $36, and a pair of shoes for $97. You then made a payment of $350 to the credit card company. Find your new credit card balance.

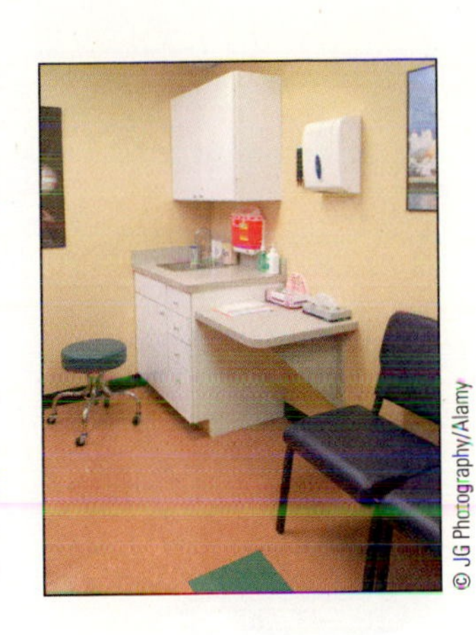

103. **Finances** As the office manager, you are required to order supplies and pay the office bills. The company already owes \$398 to Prescriptive Medical Supplies Company when you place another order with the company, ordering a new office desk chair for \$125, a file cabinet for \$457, and a box of examination table paper for \$87. You make a payment of \$650 to Prescriptive Medical Supplies Company. Find the amount your office still owes this company.

104. **Hydration** A certified nurse's aide (CNA) monitors both the intake (fluids consumed) and output (urine, vomitus, etc.) of patients to look for signs of dehydration. During her shift, the CNA records a patient's intake as two bowls of soup, a glass of milk, a cup of gelatin, and a popsicle. Use the table on page 11 to determine the volume of fluid for these foods. The patient's total urine output was 800 milliliters. By how much does the output exceed the intake for this patient?

105. **Education** In a recent year, 775,424 women and 573,079 men earned a bachelor's degree. How many more women than men earned a bachelor's degree in that year? (*Source:* The National Center for Education Statistics)

Demographics The graph at the right shows the expected U.S. population aged 100 and over for every 2 years from 2010 to 2020. Use this information for Exercises 106 to 108.

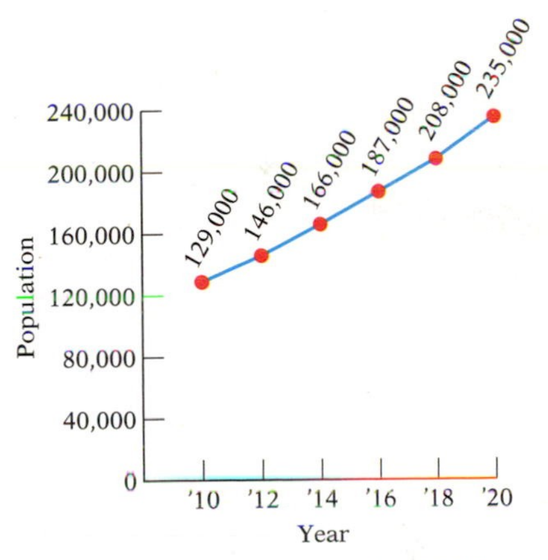

Expected U.S. Population Aged 100 and Over
Source: Census Bureau

106. What is the expected growth in the population aged 100 and over during the 10-year period?

107. **a.** Which 2-year period has the smallest expected increase in the number of people aged 100 and over?
b. Which 2-year period has the greatest expected increase?

108. What does the difference $208,000 - 166,000$ represent?

Applying the Concepts

109. **Nutrition** You are on a 1500-calorie per day diet. You consume 375 calories at breakfast, 450 calories at lunch, and 630 calories at supper. Using the list of healthy snacks on page 19, which snack(s) can you eat before bedtime and not exceed your daily goal of 1500 calories?

110. Answer true or false.
a. The phrases "the difference between 9 and 5" and "5 less than 9" mean the same thing.
b. $9 - (5 - 3) = (9 - 5) - 3$
c. Subtraction is an associative operation. *Hint:* See part (b) of this exercise.

111. Make up a word problem for which the difference between 15 and 8 is the answer.

SECTION

1.4 Multiplication of Whole Numbers

OBJECTIVE A To multiply a number by a single digit

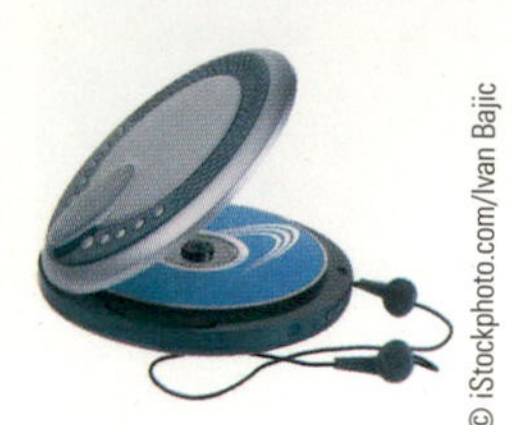

Six boxes of CD players are ordered. Each box contains eight CD players. How many CD players are ordered?

This problem can be worked by adding 6 eights.

$$8 + 8 + 8 + 8 + 8 + 8 = 48$$

This problem involves repeated addition of the same number and can be worked by a shorter process called **multiplication.** Multiplication is the repeated addition of the same number.

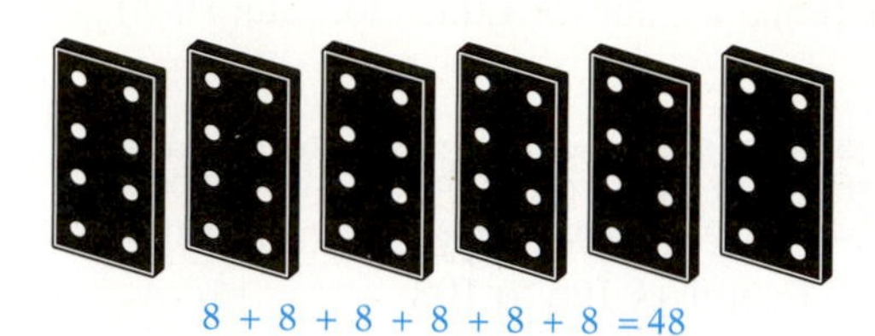

8 + 8 + 8 + 8 + 8 + 8 = 48

The numbers that are multiplied are called **factors.** The result is called the **product.**

or

6	×	8	=	48
Factor		**Factor**		**Product**

The product of 6 × 8 can be represented on the number line. The arrow representing the whole number 8 is repeated 6 times. The result is the arrow representing 48.

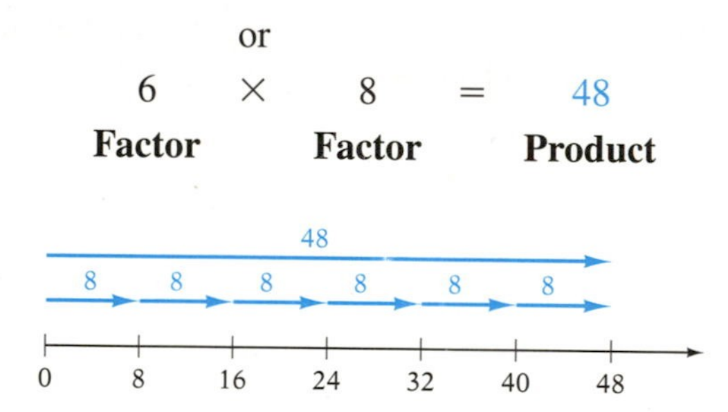

The times sign "×" is only one symbol that is used to indicate multiplication. Each of the expressions that follow represents multiplication.

$$7 \times 8 \qquad 7 \cdot 8 \qquad 7(8) \qquad (7)(8) \qquad (7)8$$

As with addition, there are some useful properties of multiplication.

Multiplication Property of Zero

The product of a number and zero is zero.

$0 \times 4 = 0$
$7 \times 0 = 0$

Multiplication Property of One

The product of a number and one is the number.

$1 \times 6 = 6$
$8 \times 1 = 8$

Commutative Property of Multiplication

Two numbers can be multiplied in either order. The product will be the same.

$4 \times 3 = 3 \times 4$
$12 = 12$

Associative Property of Multiplication

Grouping the numbers to be multiplied in any order gives the same result. Do the multiplication inside the parentheses first.

$$\underbrace{(4 \times 2)}_{8} \times 3 = 4 \times \underbrace{(2 \times 3)}_{6}$$
$$8 \times 3 = 4 \times 6$$
$$24 = 24$$

Tips for Success

Some students think that they can "coast" at the beginning of this course because the topic of Chapter 1 is whole numbers. However, this chapter lays the foundation for the entire course. Be sure you know and understand all the concepts presented. For example, study the properties of multiplication presented in this lesson.

The basic facts for multiplying one-digit numbers should be memorized. Multiplication of larger numbers requires the repeated use of the basic multiplication facts.

HOW TO • 1 Multiply: 37 × 4

$$\begin{array}{r} \overset{2}{3}\,7 \\ \times\quad 4 \\ \hline 8 \end{array}$$

- **4 × 7 = 28 (2 tens + 8 ones). Write the 8 in the ones column and carry the 2 to the tens column.**

$$\begin{array}{r} \overset{2}{3}\,7 \\ \times\quad 4 \\ \hline 14\,8 \end{array}$$

- **The 3 in 37 is 3 tens.**
  4 × 3 tens = 12 tens
  Add the carry digit. + 2 tens
  14 tens
- **Write the 14. The product is 148.**

The phrases below are used to indicate the operation of multiplication. An example is shown at the right of each phrase.

times	7 times 3	7 · 3
the product of	the product of 6 and 9	6 · 9
multiplied by	8 multiplied by 2	2 · 8

EXAMPLE • 1

Multiply: 735 × 9

Solution

$$\begin{array}{r} \overset{3\,4}{735} \\ \times\quad 9 \\ \hline 6615 \end{array}$$

- **9 × 5 = 45**
  Write the 5 in the ones column. Carry the 4 to the tens column.
  9 × 3 = 27, 27 + 4 = 31
  9 × 7 = 63, 63 + 3 = 66

YOU TRY IT • 1

Multiply: 648 × 7

Your solution

Solution on p. S2

OBJECTIVE B **To multiply larger whole numbers**

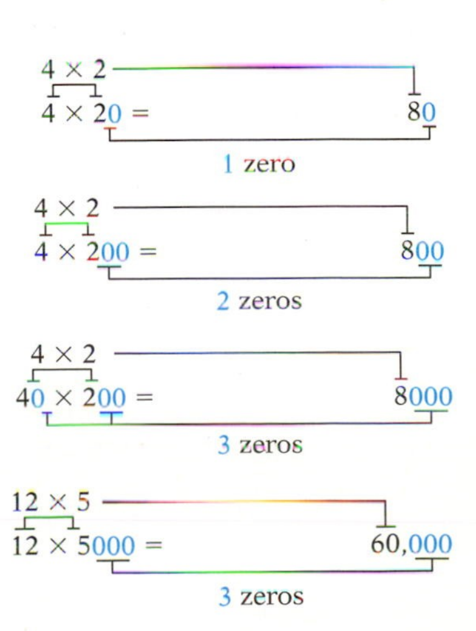

Note the pattern when the following numbers are multiplied.

Multiply the nonzero part of the factors.

Now attach the same number of zeros to the product as the total number of zeros in the factors.

HOW TO • 2 Find the product of 47 and 23.

Multiply by the ones digit.

$$\begin{array}{r} 47 \\ \times\ 23 \\ \hline 141 \end{array} \quad (= 47 \times 3)$$

Multiply by the tens digit.

$$\begin{array}{r} 47 \\ \times\ 23 \\ \hline 141 \\ 940 \end{array} \quad (= 47 \times 20)$$

Writing the 0 is optional.

Add.

$$\begin{array}{r} 47 \\ \times\ 23 \\ \hline 141 \\ 940 \\ \hline 1081 \end{array}$$

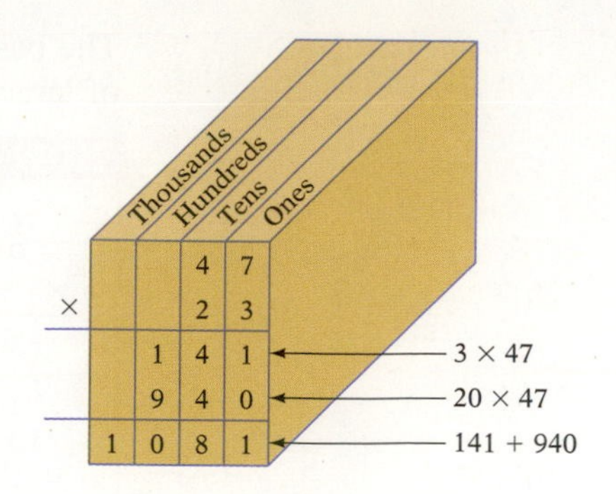

The place-value chart on the right above illustrates the placement of the products.

Note the placement of the products when we are multiplying by a factor that contains a zero.

HOW TO • 3 Multiply: 439 × 206

$$\begin{array}{r} 439 \\ \times\ 206 \\ \hline 2634 \\ 000 \\ 878 \\ \hline 90{,}434 \end{array} \quad 0 \times 439$$

When working the problem, we usually write only one zero. Writing this zero ensures the proper placement of the products.

$$\begin{array}{r} 439 \\ \times\ 206 \\ \hline 2634 \\ 8780 \\ \hline 90{,}434 \end{array}$$

EXAMPLE • 2

Find 829 multiplied by 603.

Solution

$$\begin{array}{r} 829 \\ \times\ 603 \\ \hline 2487 \\ 49740 \\ \hline 499{,}887 \end{array}$$

- 3 × 829 = 2487
- Write a zero in the tens column for 0 × 829.
- 6 × 829 = 4974

YOU TRY IT • 2

Multiply: 756 × 305

Your solution

Solution on p. S2

ESTIMATION

Estimating the Product of Two Whole Numbers

Calculate 3267 × 389. Then use estimation to determine whether the product is reasonable.

Multiply to find the exact product. 3267 **x** 389 **=** 1,270,863

To estimate the product, round each number so that it has only one nonzero digit. Then multiply. The estimated answer is 1,200,000, which is very close to the exact product 1,270,863.

$$\begin{array}{rr} 3267 \approx & 3000 \\ \times\ 389 \approx & \times\ \ 400 \\ \hline & 1{,}200{,}000 \end{array}$$

OBJECTIVE C To solve application problems

EXAMPLE • 3

A radiologist receives a salary of $1050 each week. How much does the radiologist earn in 4 weeks?

Strategy
To find the radiologist's earnings for 4 weeks, multiply the weekly salary (1050) by the number of weeks (4).

Solution

$$\begin{array}{r} 1050 \\ \times \quad 4 \\ \hline 4200 \end{array}$$

The radiologist earns $4200 in 4 weeks.

YOU TRY IT • 3

A surgical technician earns $24 per surgical hour. One week, the surgical technician worked 38 surgical hours. What was the surgical technician's income for the week?

Your strategy

Your solution

EXAMPLE • 4

One hospital pays its ER nurses $30 per hour for a regular 40-hour week. Overtime is paid at $45 per hour (time and a half). If one of the nurses works an extra 8-hour shift this week, for a total of 48 hours, find her total pay for the week.

Strategy
To find the nurse's pay for the week:
- Multiply the regular number of hours (40) times the regular hourly rate ($30).
- Multiply the overtime hours (8) by the overtime rate ($45).
- Add these two amounts to find the pay for this week.

Solution

$$\begin{array}{r} 30 \\ \times 40 \\ \hline 1200 \end{array} \text{ regular pay} \qquad \begin{array}{r} 45 \\ \times 8 \\ \hline 360 \end{array} \text{ overtime} \qquad \begin{array}{r} 1200 \\ +\ 360 \\ \hline 1560 \end{array}$$

The nurse is paid $1560 for this week after working 8 hours overtime.

YOU TRY IT • 4

A biomedical equipment technician earns $21 per hour for doing diagnostic evaluations on magnetic resonance imaging (MRI) machines. If he works 40 hours per week, how much does this technician earn in 3 weeks?

Your strategy

Your solution

Solutions on p. S2

1.4 EXERCISES

OBJECTIVE A To multiply a number by a single digit

For Exercises 1 to 4, write the expression as a product.

1. $2 + 2 + 2 + 2 + 2 + 2$ **2.** $4 + 4 + 4 + 4 + 4$ **3.** $7 + 7 + 7 + 7$ **4.** $18 + 18 + 18$

For Exercises 5 to 39, multiply.

5. $\begin{array}{r} 3 \\ \times 4 \\ \hline \end{array}$	**6.** $\begin{array}{r} 2 \\ \times 8 \\ \hline \end{array}$	**7.** $\begin{array}{r} 5 \\ \times 7 \\ \hline \end{array}$	**8.** $\begin{array}{r} 6 \\ \times 4 \\ \hline \end{array}$	**9.** $\begin{array}{r} 5 \\ \times 5 \\ \hline \end{array}$
10. $\begin{array}{r} 7 \\ \times 7 \\ \hline \end{array}$	**11.** $\begin{array}{r} 0 \\ \times 7 \\ \hline \end{array}$	**12.** $\begin{array}{r} 8 \\ \times 0 \\ \hline \end{array}$	**13.** $\begin{array}{r} 8 \\ \times 9 \\ \hline \end{array}$	**14.** $\begin{array}{r} 7 \\ \times 6 \\ \hline \end{array}$
15. $\begin{array}{r} 66 \\ \times\ 3 \\ \hline \end{array}$	**16.** $\begin{array}{r} 70 \\ \times\ 4 \\ \hline \end{array}$	**17.** $\begin{array}{r} 67 \\ \times\ 5 \\ \hline \end{array}$	**18.** $\begin{array}{r} 127 \\ \times\ 9 \\ \hline \end{array}$	**19.** $\begin{array}{r} 623 \\ \times\ 4 \\ \hline \end{array}$
20. $\begin{array}{r} 802 \\ \times\ 5 \\ \hline \end{array}$	**21.** $\begin{array}{r} 607 \\ \times\ 9 \\ \hline \end{array}$	**22.** $\begin{array}{r} 300 \\ \times\ 5 \\ \hline \end{array}$	**23.** $\begin{array}{r} 600 \\ \times\ 7 \\ \hline \end{array}$	**24.** $\begin{array}{r} 906 \\ \times\ 8 \\ \hline \end{array}$
25. $\begin{array}{r} 703 \\ \times\ 9 \\ \hline \end{array}$	**26.** $\begin{array}{r} 127 \\ \times\ 5 \\ \hline \end{array}$	**27.** $\begin{array}{r} 632 \\ \times\ 3 \\ \hline \end{array}$	**28.** $\begin{array}{r} 559 \\ \times\ 4 \\ \hline \end{array}$	**29.** $\begin{array}{r} 632 \\ \times\ 8 \\ \hline \end{array}$
30. $\begin{array}{r} 524 \\ \times\ 4 \\ \hline \end{array}$	**31.** $\begin{array}{r} 337 \\ \times\ 5 \\ \hline \end{array}$	**32.** $\begin{array}{r} 841 \\ \times\ 6 \\ \hline \end{array}$	**33.** $\begin{array}{r} 6709 \\ \times\ 7 \\ \hline \end{array}$	**34.** $\begin{array}{r} 3608 \\ \times\ 5 \\ \hline \end{array}$
35. $\begin{array}{r} 8568 \\ \times\ 7 \\ \hline \end{array}$	**36.** $\begin{array}{r} 5495 \\ \times\ 4 \\ \hline \end{array}$	**37.** $\begin{array}{r} 4780 \\ \times\ 4 \\ \hline \end{array}$	**38.** $\begin{array}{r} 3690 \\ \times\ 5 \\ \hline \end{array}$	**39.** $\begin{array}{r} 9895 \\ \times\ 2 \\ \hline \end{array}$

40. True or false? The product of two one-digit whole numbers must be a two-digit whole number.

41. Find the product of 5, 7, and 4.

42. Find the product of 6, 2, and 9.

43. What is 3208 multiplied by 7?

44. What is 5009 multiplied by 4?

45. What is 3105 times 6?

46. What is 8957 times 8?

OBJECTIVE B To multiply larger whole numbers

For Exercises 47 to 78, multiply.

47. $\begin{array}{r} 16 \\ \times\ 21 \\ \hline \end{array}$

48. $\begin{array}{r} 18 \\ \times\ 24 \\ \hline \end{array}$

49. $\begin{array}{r} 35 \\ \times\ 26 \\ \hline \end{array}$

50. $\begin{array}{r} 27 \\ \times\ 72 \\ \hline \end{array}$

51. $\begin{array}{r} 693 \\ \times\ \ 91 \\ \hline \end{array}$

52. $\begin{array}{r} 581 \\ \times\ \ 72 \\ \hline \end{array}$

53. $\begin{array}{r} 419 \\ \times\ \ 80 \\ \hline \end{array}$

54. $\begin{array}{r} 727 \\ \times\ \ 60 \\ \hline \end{array}$

55. $\begin{array}{r} 8279 \\ \times\ \ 46 \\ \hline \end{array}$

56. $\begin{array}{r} 9577 \\ \times\ \ 35 \\ \hline \end{array}$

57. $\begin{array}{r} 6938 \\ \times\ \ 78 \\ \hline \end{array}$

58. $\begin{array}{r} 8875 \\ \times\ \ 67 \\ \hline \end{array}$

59. $\begin{array}{r} 7035 \\ \times\ \ 57 \\ \hline \end{array}$

60. $\begin{array}{r} 6702 \\ \times\ \ 48 \\ \hline \end{array}$

61. $\begin{array}{r} 3009 \\ \times\ \ 35 \\ \hline \end{array}$

62. $\begin{array}{r} 6003 \\ \times\ \ 57 \\ \hline \end{array}$

63. $\begin{array}{r} 809 \\ \times\ 530 \\ \hline \end{array}$

64. $\begin{array}{r} 607 \\ \times\ 460 \\ \hline \end{array}$

65. $\begin{array}{r} 800 \\ \times\ 325 \\ \hline \end{array}$

66. $\begin{array}{r} 700 \\ \times\ 274 \\ \hline \end{array}$

67. $\begin{array}{r} 987 \\ \times\ 349 \\ \hline \end{array}$

68. $\begin{array}{r} 688 \\ \times\ 674 \\ \hline \end{array}$

69. $\begin{array}{r} 312 \\ \times\ 134 \\ \hline \end{array}$

70. $\begin{array}{r} 423 \\ \times\ 427 \\ \hline \end{array}$

71. $\begin{array}{r} 379 \\ \times\ 500 \\ \hline \end{array}$

72. $\begin{array}{r} 684 \\ \times\ 700 \\ \hline \end{array}$

73. $\begin{array}{r} 985 \\ \times\ 408 \\ \hline \end{array}$

74. $\begin{array}{r} 758 \\ \times\ 209 \\ \hline \end{array}$

75. $\begin{array}{r} 3407 \\ \times\ \ 309 \\ \hline \end{array}$

76. $\begin{array}{r} 5207 \\ \times\ \ 902 \\ \hline \end{array}$

77. $\begin{array}{r} 4258 \\ \times\ \ 986 \\ \hline \end{array}$

78. $\begin{array}{r} 6327 \\ \times\ \ 876 \\ \hline \end{array}$

79. Find a one-digit number and a two-digit number whose product is a number that ends in two zeros.

80. What is 5763 times 45?

81. What is 7349 times 27?

82. Find the product of 2, 19, and 34.

83. Find the product of 6, 73, and 43.

84. What is 376 multiplied by 402?

85. What is 842 multiplied by 309?

For Exercises 86 to 93, use a calculator to multiply. Then use estimation to determine whether the product is reasonable.

86. $\begin{array}{r} 8745 \\ \times \quad 63 \\ \hline \end{array}$

87. $\begin{array}{r} 4732 \\ \times \quad 93 \\ \hline \end{array}$

88. $\begin{array}{r} 2937 \\ \times \ 206 \\ \hline \end{array}$

89. $\begin{array}{r} 8941 \\ \times \ 726 \\ \hline \end{array}$

90. $\begin{array}{r} 3097 \\ \times \ 1025 \\ \hline \end{array}$

91. $\begin{array}{r} 6379 \\ \times \ 2936 \\ \hline \end{array}$

92. $\begin{array}{r} 32{,}508 \\ \times \quad 591 \\ \hline \end{array}$

93. $\begin{array}{r} 62{,}504 \\ \times \quad 923 \\ \hline \end{array}$

OBJECTIVE C To solve application problems

94. The cost of surgical 2.0 thread is $3.50 per foot. Each packet of thread contains 3 feet. You purchase four packets of surgical thread. Which of the following represents the price of the surgical thread?

(i) $2.0 \times 3.50 \times 3 \times 4$
(ii) $2.0 \times 3.50 \times 4$
(iii) $2.0 \times 3.50 \times 3$
(iv) $3.50 \times 3 \times 4$

95. Your patient is prescribed 25 milligrams of a medication to be given four times per day. What is the total daily dosage of medication for this patient?

96. LabTest Co. ordered 25 new lab jackets for its medical laboratory technicians at a cost of $45 each. What is the total cost of this order?

97. As a mobile x-ray technician, you use your car to travel to various nursing homes to take x-rays. Your car used a total of 20 gallons of fuel last week and gets 28 miles to the gallon. How many miles did you travel last week?

98. Hand sanitizer is packaged in cases of 24. If a clinic orders 12 cases, how many bottles of hand sanitizer will it receive?

99. If hand sanitizer is priced at $25 for a case of 24, find the cost of 12 cases.

100. **College Education** See the information at the right. **a.** Find the average cost of tuition and fees for the first two years of college at a public four-year college. **b.** Find the average cost of tuition and fees for the first two years of college at a public two-year college. **c.** Find the difference in cost of tuition and fees for the first two years of college at a four-year public and two-year public college.

In the News

Comparing Tuition Costs

- Public four-year colleges charge, on average, $7605 per year in tuition and fees.
- Public two-year colleges charge, on average, $2713 per year in tuition and fees.

Source: Trends in College Pricing 2010, http://trends.collegeboard.org

Medical Supplies The table below shows the cost of certain medical supplies at the supply company used by a walk-in clinic. Use this table for Exercises 101 to 103.

101. Calculate the cost of an order for two cases of latex-free exam gloves, 10 boxes of tongue depressors, and one case of alcohol prep pads.

102. Calculate the cost of an order for three cases of oval gauze pads, one case of 3″ × 3″ gauze pads, and five boxes of cloth adhesive tape.

103. If a case of cloth adhesive tape contains 24 boxes, how many rolls of tape would be in one case? How much would a case cost based on the cost per box?

Item	*Cost*
Latex-free exam gloves, 1000/case	$57
Oval gauze pads, 1000/case	$69
Gauze pads, 3″ x 3″, 2400/case	$68
Tongue depressors, 100/box	$7
Cloth adhesive tape, 1″ × 12 yd, 12/box	$18
Two-ply alcohol prep pads, 4000/case	$35

Applying the Concepts

104. Determine whether each of the following statements is always true, sometimes true, or never true.

a. A whole number times zero is zero.

b. A whole number times one is the whole number.

c. The product of two whole numbers is greater than either one of the whole numbers.

105. **Safety** According to the National Safety Council, in a recent year a death resulting from an accident occurred at the rate of 1 every 5 minutes. At this rate, how many accidental deaths occurred each hour? Each day? Throughout the year? Explain how you arrived at your answers.

106. **Demographics** According to the Population Reference Bureau, in the world today, 261 people are born every minute and 101 people die every minute. Using this statistic, what is the increase in the world's population every hour? Every day? Every week? Every year? Use a 365-day year. Explain how you arrived at your answers.

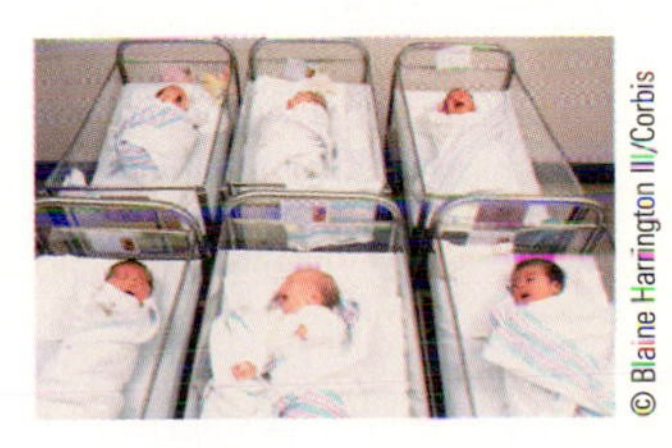

SECTION 1.5 Division of Whole Numbers

OBJECTIVE A To divide by a single digit with no remainder in the quotient

Division is used to separate objects into equal groups.

A store manager wants to display 24 new objects equally on 4 shelves. From the diagram, we see that the manager would place 6 objects on each shelf.

The manager's division problem can be written as follows:

Take Note
The **divisor** is the number that is divided into another number. The **dividend** is the number into which the divisor is divided. The result is the **quotient.**

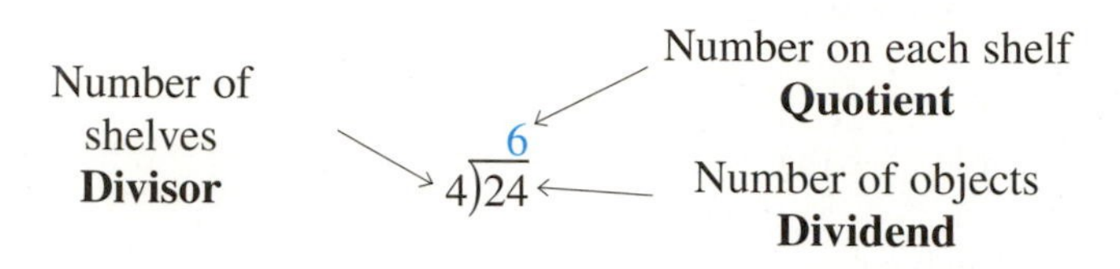

Note that the quotient multiplied by the divisor equals the dividend.

$4\overline{)24}$ with quotient 6 because 6 (Quotient) × 4 (Divisor) = 24 (Dividend)

$9\overline{)54}$ with quotient 6 because 6 × 9 = 54

$8\overline{)40}$ with quotient 5 because 5 × 8 = 40

Here are some important quotients and the properties of zero in division:

Properties of One in Division

Any whole number, except zero, divided by itself is 1. $8\overline{)8} = 1$ $14\overline{)14} = 1$ $10\overline{)10} = 1$

Any whole number divided by 1 is the whole number. $1\overline{)9} = 9$ $1\overline{)27} = 27$ $1\overline{)10} = 10$

Integrating Technology
Enter 8 ÷ 0 = on your calculator. An error message is displayed because division by zero is not allowed.

Properties of Zero in Division

Zero divided by any other whole number is zero. $7\overline{)0} = 0$ $13\overline{)0} = 0$ $10\overline{)0} = 0$

Division by zero is not allowed. $0\overline{)8} = \boxed{?}$ There is no number whose product with 0 is 8.

When the dividend is a larger whole number, the digits in the quotient are found in steps.

HOW TO • 1 Divide $4\overline{)3192}$ and check.

$$\begin{array}{r} 7 \\ 4\overline{)3192} \\ -28 \\ \hline 39 \end{array}$$

- Think $4\overline{)31}$.
- Subtract 7×4.
- Bring down the 9.

$$\begin{array}{r} 79 \\ 4\overline{)3192} \\ -28 \\ \hline 39 \\ -36 \\ \hline 32 \end{array}$$

- Think $4\overline{)39}$.
- Subtract 9×4.
- Bring down the 2.

$$\begin{array}{r} 798 \\ 4\overline{)3192} \\ -28 \\ \hline 39 \\ -36 \\ \hline 32 \\ -32 \\ \hline 0 \end{array}$$

- Think $4\overline{)32}$.
- Subtract 8×4.

Check: $\begin{array}{r} 798 \\ \times\quad 4 \\ \hline 3192 \end{array}$

The place-value chart can be used to show why this method works.

	HUNDREDS	TENS	ONES	
	7	9	8	

$$\begin{array}{r} 7\ 9\ 8 \\ 4\overline{)3\ 1\ 9\ 2} \\ -\ 2\ 8\ 0\ 0 \\ \hline 3\ 9\ 2 \\ -\ 3\ 6\ 0 \\ \hline 3\ 2 \\ -\ 3\ 2 \\ \hline 0 \end{array}$$

- $- 2800$: 7 hundreds $\times$ 4
- $- 360$: 9 tens $\times$ 4
- $- 32$: 8 ones $\times$ 4

There are other ways of expressing division.

54 divided by 9 equals 6.

54 $\div$ 9 equals 6.

$\frac{54}{9}$ equals 6.

EXAMPLE • 1

Divide $7\overline{)56}$ and check.

Solution

$$\begin{array}{r} 8 \\ 7\overline{)56} \end{array}$$

Check: $8 \times 7 = 56$

YOU TRY IT • 1

Divide $9\overline{)63}$ and check.

Your solution

EXAMPLE • 2

Divide $2808 \div 8$ and check.

Solution

$$\begin{array}{r} 351 \\ 8\overline{)\ 2808} \\ -24 \\ \hline 40 \\ -40 \\ \hline 08 \\ -\ 8 \\ \hline 0 \end{array}$$

Check: $351 \times 8 = 2808$

YOU TRY IT • 2

Divide $4077 \div 9$ and check.

Your solution

EXAMPLE • 3

Divide $7\overline{)2856}$ and check.

Solution

$$\begin{array}{r} 408 \\ 7\overline{)\ 2856} \\ -28 \\ \hline 05 \\ -0 \\ \hline 56 \\ -56 \\ \hline 0 \end{array}$$

- **Think $7\overline{)5}$. Place 0 in quotient.**
- **Subtract 0×7.**
- **Bring down the 6.**

Check: $408 \times 7 = 2856$

YOU TRY IT • 3

Divide $9\overline{)6345}$ and check.

Your solution

Solutions on pp. S2–S3

OBJECTIVE B To divide by a single digit with a remainder in the quotient

Sometimes it is not possible to separate objects into a whole number of equal groups.

A baker has 14 muffins to pack into 3 boxes. Each box holds 4 muffins. From the diagram, we see that after the baker places 4 muffins in each box, there are 2 left over. The 2 is called the **remainder.**

The baker's division problem could be written

$$\begin{array}{r} 4 \\ 3\overline{)\,14} \\ -12 \\ \hline 2 \end{array}$$

- Divisor (Number of boxes) → 3
- 4 ← Quotient (Number in each box)
- 14 ← Dividend (Total number of objects)
- 2 ← Remainder (Number left over)

The answer to a division problem with a remainder is frequently written

$$\begin{array}{r} 4 \text{ r2} \\ 3\overline{)14} \end{array}$$

Note that $\boxed{\begin{array}{c} 4 \qquad\quad 3 \\ \text{Quotient} \times \text{Divisor} \end{array}} + \boxed{\begin{array}{c} 2 \\ \text{Remainder} \end{array}} = \boxed{\begin{array}{c} 14 \\ \text{Dividend} \end{array}}$.

EXAMPLE • 4

Divide $4\overline{)2522}$ and check.

Solution

$$\begin{array}{r} 630 \text{ r2} \\ 4\overline{)\,2522} \\ -24 \\ \hline 12 \\ -12 \\ \hline 02 \\ -0 \\ \hline 2 \end{array}$$

- **Think $4\overline{)2}$. Place 0 in quotient.**
- **Subtract 0 × 4.**

Check: (630 × 4) + 2 =
2520 + 2 = 2522

YOU TRY IT • 4

Divide $6\overline{)5225}$ and check.

Your solution

EXAMPLE • 5

Divide $9\overline{)27{,}438}$ and check.

Solution

$$\begin{array}{r} 3{,}048 \text{ r6} \\ 9\overline{)\,27{,}438} \\ -27 \\ \hline 0\,4 \\ -0 \\ \hline 43 \\ -36 \\ \hline 78 \\ -72 \\ \hline 6 \end{array}$$

- **Think $9\overline{)4}$.**
- **Subtract 0 × 9.**

Check: (3048 × 9) + 6 =
27,432 + 6 = 27,438

YOU TRY IT • 5

Divide $7\overline{)21{,}409}$ and check.

Your solution

Solutions on p. S3

OBJECTIVE C To divide by larger whole numbers

When the divisor has more than one digit, estimate at each step by using the first digit of the divisor. If that product is too large, lower the guess by 1 and try again.

HOW TO • 2 Divide $34\overline{)1598}$ and check.

$$\begin{array}{r} 5 \\ 34\overline{)1598} \\ -170 \end{array}$$

- Think $3\overline{)15}$.
- Subtract 5 × 34.

170 is too large. Lower the guess by 1 and try again.

$$\begin{array}{r} 4 \\ 34\overline{)1598} \\ -136 \\ \hline 238 \end{array}$$

- Subtract 4 × 34.

$$\begin{array}{r} 47 \\ 34\overline{)1598} \\ -136 \\ \hline 238 \\ -238 \\ \hline 0 \end{array}$$

- Think $3\overline{)23}$.
- Subtract 7 × 34.

Check:

$$\begin{array}{r} 47 \\ \times 34 \\ \hline 188 \\ 141 \\ \hline 1598 \end{array}$$

Tips for Success

One of the key instructional features of this text is the Example/You Try It pairs. Each Example is completely worked. You are to solve the You Try It problems. When you are ready, check your solution against the one in the Solutions section. The solution for You Try It 6 below is on page S3 (see the reference at the bottom right of the You Try It). See *AIM for Success* at the front of the book.

The phrases below are used to indicate the operation of division. An example is shown at the right of each phrase.

the quotient of	the quotient of 9 and 3	9 ÷ 3
divided by	6 divided by 2	6 ÷ 2

EXAMPLE • 6

Find 7077 divided by 34 and check.

Solution

$$\begin{array}{r} 208 \text{ r5} \\ 34\overline{)7077} \\ -68 \\ \hline 27 \\ -0 \\ \hline 277 \\ -272 \\ \hline 5 \end{array}$$

- Think $34\overline{)27}$.
- Place 0 in quotient.
- Subtract 0 × 34.

Check: (208 × 34) + 5 =
7072 + 5 = 7077

YOU TRY IT • 6

Divide 4578 ÷ 42 and check.

Your solution

Solution on p. S3

EXAMPLE • 7

Find the quotient of 21,312 and 56 and check.

Solution

```
     380 r32
56)21,312        • Think 5)21.
  −16 8            4 × 56 is too large.
    4 51           Try 3.
   −4 48
       32
       −0
       32
```

Check: (380 × 56) + 32 =
21,280 + 32 = 21,312

YOU TRY IT • 7

Divide 18,359 ÷ 39 and check.

Your solution

EXAMPLE • 8

Divide 427)24,782 and check.

Solution

```
       58 r16
427)24,782
   −21 35
     3 432
    −3 416
        16
```

Check: (58 × 427) + 16 =
24,766 + 16 = 24,782

YOU TRY IT • 8

Divide 534)33,219 and check.

Your solution

EXAMPLE • 9

Divide 386)206,149 and check.

Solution

```
        534 r25
386)206,149
   −193 0
     13 14
    −11 58
      1 569
     −1 544
         25
```

Check: (534 × 386) + 25 =
206,124 + 25 = 206,149

YOU TRY IT • 9

Divide 515)216,848 and check.

Your solution

Solutions on p. S3

ESTIMATION

Estimating the Quotient of Two Whole Numbers

Calculate 36,936 ÷ 54. Then use estimation to determine whether the quotient is reasonable.

Divide to find the exact quotient. 36,936 ÷ 54 = 684

To estimate the quotient, round each number so that it contains one nonzero digit. Then divide. The estimated answer is 800, which is close to the exact quotient 684.

$36{,}936 \div 54 \approx$
$40{,}000 \div 50 = 800$

OBJECTIVE D To solve application problems

The **average** of several numbers is the sum of all the numbers divided by the number of those numbers.

$$\text{Average test score} = \frac{81 + 87 + 80 + 85 + 79 + 86}{6} = \frac{498}{6} = 83$$

HOW TO • 3

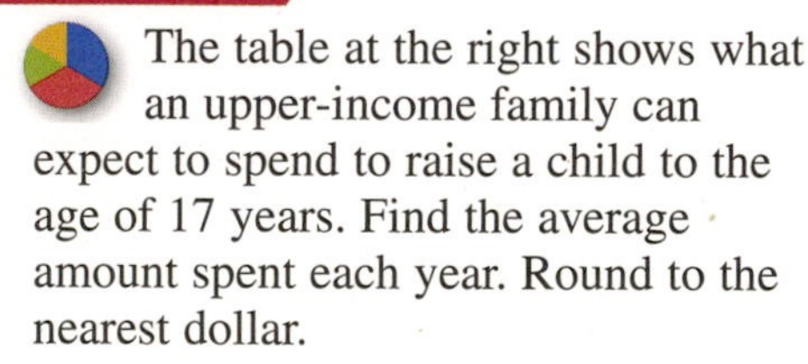

The table at the right shows what an upper-income family can expect to spend to raise a child to the age of 17 years. Find the average amount spent each year. Round to the nearest dollar.

Expenses to Raise a Child	
Housing	$89,580
Food	$35,670
Transportation	$32,760
Child care/education	$26,520
Clothing	$13,770
Health care	$13,380
Other	$30,090

Source: Department of Agriculture, *Expenditures on Children by Families*

Michelle D. Bridwell/PhotoEdit, Inc.

Strategy

To find the average amount spent each year:

- Add all the numbers in the table to find the total amount spent during the 17 years.
- Divide the sum by 17.

Solution

```
   89,580
   35,670
   32,760
   26,520
   13,770
   13,380
 + 30,090
 --------
  241,770   Sum of all the costs
```

```
       14,221
  17) 241,770
     -17
      --
       71
      -68
      ---
        37
       -34
       ---
         37
        -34
        ---
          30
         -17
         ---
          13
```

- **When rounding to the nearest whole number, compare twice the remainder to the divisor. If twice the remainder is less than the divisor, drop the remainder. If twice the remainder is greater than or equal to the divisor, add 1 to the units digit of the quotient.**
- **Twice the remainder is 2 × 13 = 26. Because 26 > 17, add 1 to the units digit of the quotient.**

The average amount spent each year to raise a child to the age of 17 is $14,222.

EXAMPLE • 10

A nurse gives a total of 1300 milligrams of a painkiller to a patient over a 24-hour time period by administering equal amounts every 6 hours. How many milligrams of medication are in each dose?

Strategy

To find the amount of medication in each dose, divide the total number of milligrams (1300) by the number of doses in a 24-hour period (4 because there is one every 6 hours).

Solution

```
    325
 4) 1300
   -12
     10
     -8
      20
     -20
       0
```

Each dose contains 325 milligrams of medication.

YOU TRY IT • 10

A shipment of 180 blood samples requires testing at a medical lab facility. The samples are divided among 12 employees. How many blood samples must each employee process?

Your strategy

Your solution

EXAMPLE • 11

A hospital purchases a computed tomography (CT) scanning machine for $486,600. A deposit of $97,800 is necessary to complete the purchase. The balance is to be paid in 72 equal monthly payments. How much is each monthly payment?

Strategy

To find the monthly payment:

- Find the balance by subtracting the deposit (97,800) from the cost of the CT machine (489,000).
- Divide the balance by the number of months required to pay off the CT machine (72).

Solution

```
  486,600
 - 97,800
  388,800
```

Remaining balance

```
          5400
 72) 388,800
    -360
      288
     -288
        0 0
       -  0
          0 0
         -  0
```

The monthly payment is $5400.

YOU TRY IT • 11

A patient's bill for his private room during a recent hospitalization at Memorial Hospital was $10,200. He was in the hospital for 12 days. What was the cost of the hospital room per day?

Your strategy

Your solution

Solutions on p. S3

1.5 EXERCISES

OBJECTIVE A To divide by a single digit with no remainder in the quotient

For Exercises 1 to 20, divide.

1. $4\overline{)8}$ **2.** $3\overline{)9}$ **3.** $6\overline{)36}$ **4.** $9\overline{)81}$

5. $7\overline{)49}$ **6.** $5\overline{)80}$ **7.** $6\overline{)96}$ **8.** $6\overline{)480}$

9. $4\overline{)840}$ **10.** $3\overline{)690}$ **11.** $7\overline{)308}$ **12.** $7\overline{)203}$

13. $9\overline{)6327}$ **14.** $4\overline{)2120}$ **15.** $8\overline{)7280}$ **16.** $9\overline{)8118}$

17. $3\overline{)64,680}$ **18.** $4\overline{)50,760}$ **19.** $6\overline{)21,480}$ **20.** $5\overline{)18,050}$

21. What is 7525 divided by 7?

22. What is 32,364 divided by 4?

23. If the dividend and the divisor in a division problem are the same number, what is the quotient?

For Exercises 24 to 27, use the relationship between multiplication and division to complete the multiplication problem.

24. ___ × 7 = 364 **25.** 8 × ___ = 376 **26.** 5 × ___ = 170 **27.** ___ × 4 = 92

OBJECTIVE B To divide by a single digit with a remainder in the quotient

For Exercises 28 to 50, divide.

28. $4\overline{)9}$ **29.** $2\overline{)7}$ **30.** $5\overline{)27}$ **31.** $9\overline{)88}$ **32.** $3\overline{)40}$

33. $6\overline{)97}$ **34.** $8\overline{)83}$ **35.** $5\overline{)54}$ **36.** $7\overline{)632}$ **37.** $4\overline{)363}$

38. $4\overline{)921}$

39. $7\overline{)845}$

40. $8\overline{)1635}$

41. $5\overline{)1548}$

42. $7\overline{)9432}$

43. $7\overline{)8124}$

44. $3\overline{)5162}$

45. $5\overline{)3542}$

46. $8\overline{)3274}$

47. $4\overline{)15,301}$

48. $7\overline{)43,500}$

49. $8\overline{)72,354}$

50. $5\overline{)43,542}$

51. What is 45,738 divided by 4? Round to the nearest ten.

52. What is 37,896 divided by 9? Round to the nearest hundred.

53. What is 3572 divided by 7? Round to the nearest ten.

54. What is 78,345 divided by 4? Round to the nearest hundred.

55. True or false? When a three-digit number is divided by a one-digit number, the quotient can be a one-digit number.

OBJECTIVE C To divide by larger whole numbers

For Exercises 56 to 83, divide.

56. $27\overline{)96}$

57. $44\overline{)82}$

58. $42\overline{)87}$

59. $67\overline{)93}$

60. $41\overline{)897}$

61. $32\overline{)693}$

62. $23\overline{)784}$

63. $25\overline{)772}$

64. $74\overline{)600}$

65. $92\overline{)500}$

66. $70\overline{)329}$

67. $50\overline{)467}$

68. $36\overline{)7225}$

69. $44\overline{)8821}$

70. $19\overline{)3859}$

71. $32\overline{)9697}$

72. $88\overline{)3127}$

73. $92\overline{)6177}$

74. $33\overline{)8943}$

75. $27\overline{)4765}$

76. $22\overline{)98,654}$

77. $77\overline{)83,629}$

78. $64\overline{)38,912}$

79. $78\overline{)31,434}$

80. $206\overline{)3097}$ **81.** $504\overline{)6504}$ **82.** $654\overline{)1217}$ **83.** $546\overline{)2344}$

84. Find the quotient of 5432 and 21.

85. Find the quotient of 8507 and 53.

86. What is 37,294 divided by 72?

87. What is 76,788 divided by 46?

88. Find 23,457 divided by 43. Round to the nearest hundred.

89. Find 341,781 divided by 43. Round to the nearest ten.

90. True or false? If the remainder of a division problem is 210, then the divisor was less than 210.

For Exercises 91 to 102, use a calculator to divide. Then use estimation to determine whether the quotient is reasonable.

91. $76\overline{)389{,}804}$ **92.** $53\overline{)117{,}925}$ **93.** $29\overline{)637{,}072}$ **94.** $67\overline{)738{,}072}$

95. $38\overline{)934{,}648}$ **96.** $34\overline{)906{,}304}$ **97.** $309\overline{)876{,}324}$ **98.** $642\overline{)323{,}568}$

99. $209\overline{)632{,}016}$ **100.** $614\overline{)332{,}174}$ **101.** $179\overline{)5{,}734{,}444}$ **102.** $374\overline{)7{,}712{,}254}$

OBJECTIVE D To solve application problems

Insurance Medical insurance claims for the previous year for one insurance company are listed in the table. Use the table for Exercises 103 to 106.

Surgery	*Claims Total*
Appendectomy	$320,000
Hernia	$256,000
Fracture repair	$684,000
Hip replacement	$879,000

103. What is the monthly average for hip replacement surgery?

104. What is the combined monthly average for hernia and appendectomy surgery?

105. Determine the average monthly cost for fracture repair.

106. Provide the monthly average expense for all surgical claims.

107. **Coins** The U.S. Mint estimates that about 114,000,000,000 of the 312,000,000,000 pennies it has minted over the last 30 years are in active circulation. That works out to how many pennies in circulation for each of the 300,000,000 people living in the United States?

108. **Patient Care** Three CNAs are working the floor at a nursing home. There are currently 15 rooms occupied on the floor, and each room contains two patients. If the patients are divided evenly among the three CNAs, how many patients are assigned to each assistant?

109. **Medical Billing** A local medical office processed and mailed 11,520 bills to patients last year. Find the average number of bills mailed each month.

110. **Arlington National Cemetery** There are approximately 10,200 funerals each year at Arlington National Cemetery. (*Source:* www.arlingtoncemetery.org) Calculate the average number of funerals each day at Arlington National Cemetery. Round to the nearest whole number.

Arlington National Cemetery

111. Which problems require division to solve?

(i) A medical practice spends $35,500 on treatment room supplies in a year. What is their monthly cost of treatment room supplies?

(ii) If five medical laboratory technicians need to complete 530 laboratory tests, what is the average number of laboratory tests each must perform?

(iii) A doctor's office requests 210 laboratory tests per week. How many tests are requested per month?

Applying the Concepts

112. **Wages** A medical assistant earns $500 per 40-hour workweek. If the assistant worked 4 hours overtime at $18 an hour, what is the total income earned for the week's pay?

113. Payroll Deductions Your paycheck shows deductions of $225 for savings, $98 for taxes, and $27 for insurance. Find the total of the three deductions.

Health Insurance Costs The bar graph shows the 2010 monthly health insurance costs for MedSupply's active employees. Use this graph for Exercises 114 to 116.

114. How much more is MedSupply's monthly cost for family coverage than for employee only coverage?

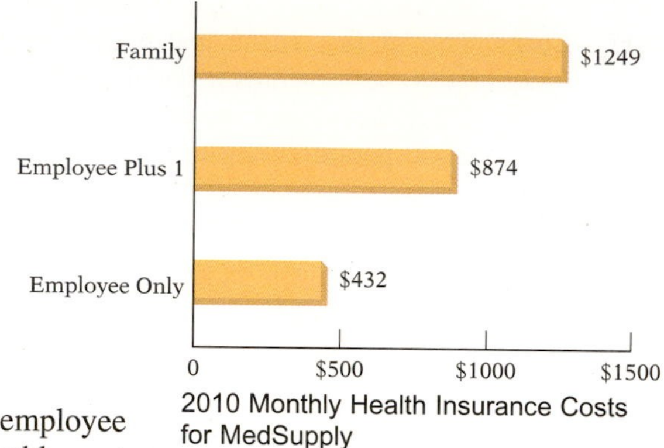

2010 Monthly Health Insurance Costs for MedSupply

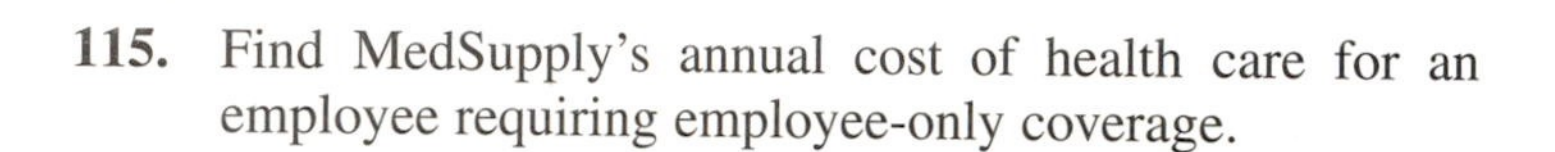

115. Find MedSupply's annual cost of health care for an employee requiring employee-only coverage.

116. If an employee wants to include a spouse on a policy, the employee must enroll in the family plan. What is the difference in monthly cost between the family plan coverage and the employee only coverage?

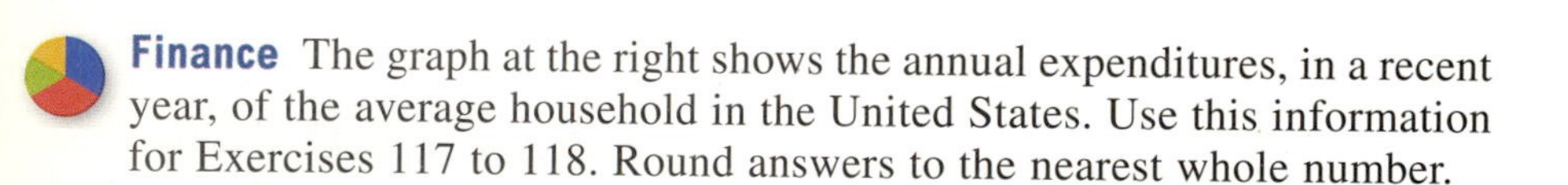

Finance The graph at the right shows the annual expenditures, in a recent year, of the average household in the United States. Use this information for Exercises 117 to 118. Round answers to the nearest whole number.

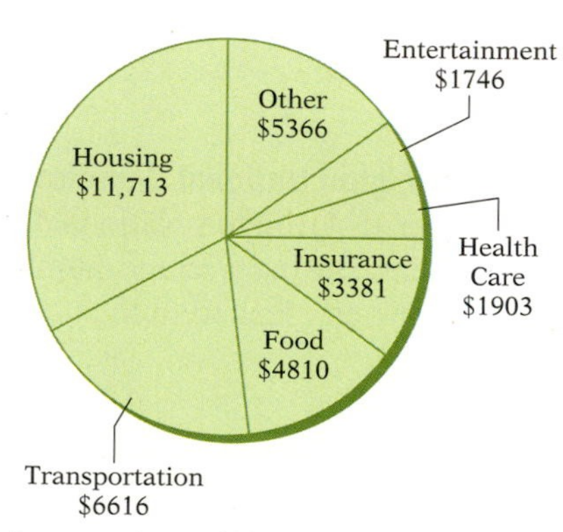

Average Annual Household Expenses

Source: Bureau of Labor Statistics Consumer Expenditure Survey

117. What is the total amount spent annually by the average household in the United States?

118. What is the average monthly expense for health care?

Dental Office Employees The graph shows annual salaries for employees in dental offices. Use this graph for Exercises 119 and 120.

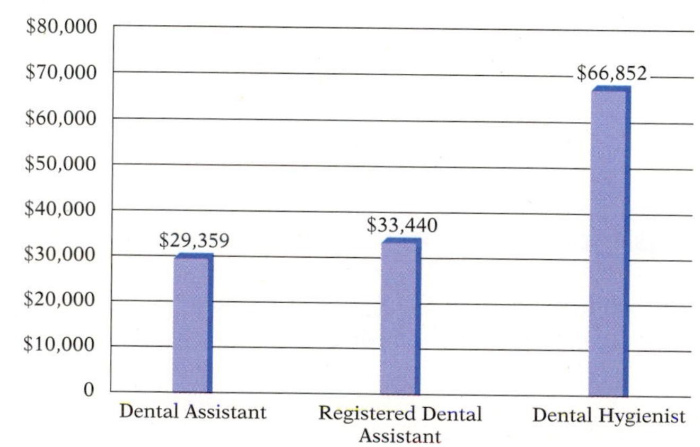

Average Annual Salaries of Dental Office Employees

Source: www.payscale.com

119. What is a dental hygienist's monthly pay?

120. What is the difference between a registered dental assistant's annual pay and a dental assistant's annual pay?

121. Finances You purchase a used car with a down payment of $2500 and monthly payments of $195 for 48 months. Find the total amount paid for the car.

SECTION 1.6 Exponential Notation and the Order of Operations Agreement

OBJECTIVE A To simplify expressions that contain exponents

Repeated multiplication of the same factor can be written in two ways:

$$3 \cdot 3 \cdot 3 \cdot 3 \cdot 3 \quad \text{or} \quad 3^5 \leftarrow \textbf{Exponent}$$

The **exponent** indicates how many times the factor occurs in the multiplication. The expression 3^5 is in **exponential notation.**

It is important to be able to read numbers written in exponential notation.

$6 = 6^1$	is read "six to the first **power**" or just "six." Usually the exponent 1 is not written.
$6 \cdot 6 = 6^2$	is read "six squared" or "six to the second power."
$6 \cdot 6 \cdot 6 = 6^3$	is read "six cubed" or "six to the third power."
$6 \cdot 6 \cdot 6 \cdot 6 = 6^4$	is read "six to the fourth power."
$6 \cdot 6 \cdot 6 \cdot 6 \cdot 6 = 6^5$	is read "six to the fifth power."

Each place value in the place-value chart can be expressed as a power of 10.

Ten =	10 =	10	$= 10^1$
Hundred =	100 =	$10 \cdot 10$	$= 10^2$
Thousand =	1000 =	$10 \cdot 10 \cdot 10$	$= 10^3$
Ten-thousand =	10,000 =	$10 \cdot 10 \cdot 10 \cdot 10$	$= 10^4$
Hundred-thousand =	100,000 =	$10 \cdot 10 \cdot 10 \cdot 10 \cdot 10$	$= 10^5$
Million =	1,000,000 =	$10 \cdot 10 \cdot 10 \cdot 10 \cdot 10 \cdot 10$	$= 10^6$

Integrating Technology

A calculator can be used to evaluate an exponential expression. The **yˣ** key (or, on some calculators, an **xʸ** or **^** key) is used to enter the exponent. For instance, for the example at the right, enter 4 **yˣ** 3 **=**. The display reads 64.

To simplify a numerical expression containing exponents, write each factor as many times as indicated by the exponent and carry out the indicated multiplication.

$$4^3 = 4 \cdot 4 \cdot 4 = 64$$

$$2^2 \cdot 3^4 = (2 \cdot 2) \cdot (3 \cdot 3 \cdot 3 \cdot 3) = 4 \cdot 81 = 324$$

EXAMPLE 1

Write $3 \cdot 3 \cdot 3 \cdot 5 \cdot 5$ in exponential notation.

Solution $3 \cdot 3 \cdot 3 \cdot 5 \cdot 5 = 3^3 \cdot 5^2$

YOU TRY IT 1

Write $2 \cdot 2 \cdot 2 \cdot 2 \cdot 3 \cdot 3 \cdot 3$ in exponential notation.

Your solution

EXAMPLE 2

Write as a power of 10: $10 \cdot 10 \cdot 10 \cdot 10$

Solution $10 \cdot 10 \cdot 10 \cdot 10 = 10^4$

YOU TRY IT 2

Write as a power of 10: $10 \cdot 10 \cdot 10 \cdot 10 \cdot 10 \cdot 10 \cdot 10$

Your solution

EXAMPLE 3

Simplify $3^2 \cdot 5^3$.

Solution

$$3^2 \cdot 5^3 = (3 \cdot 3) \cdot (5 \cdot 5 \cdot 5) = 9 \cdot 125 = 1125$$

YOU TRY IT 3

Simplify $2^3 \cdot 5^2$.

Your solution

Solutions on p. S4

OBJECTIVE B To use the Order of Operations Agreement to simplify expressions

More than one operation may occur in a numerical expression. The answer may be different, depending on the order in which the operations are performed. For example, consider $3 + 4 \times 5$.

Multiply first, then add.

$$3 + \underbrace{4 \times 5}$$
$$\underbrace{3 + 20}$$
$$23$$

Add first, then multiply.

$$\underbrace{3 + 4} \times 5$$
$$\underbrace{7 \times 5}$$
$$35$$

An Order of Operations Agreement is used so that only one answer is possible.

The Order of Operations Agreement

Step 1. Do all the operations inside parentheses.

Step 2. Simplify any number expressions containing exponents.

Step 3. Do multiplication and division as they occur from left to right.

Step 4. Do addition and subtraction as they occur from left to right.

Integrating Technology

Many scientific calculators have an **x²** key. This key is used to square the displayed number. For example, after the user presses 2 **x²** **=**, the display reads 4.

HOW TO • 1 $3 \times (2 + 1) - 2^2 + 4 \div 2$ by using the Order of Operations Agreement.

$$3 \times \underbrace{(2 + 1)} - 2^2 + 4 \div 2$$
$$3 \times 3 - \underbrace{2^2} + 4 \div 2$$
$$\underbrace{3 \times 3} - 4 + 4 \div 2$$
$$9 - 4 + \underbrace{4 \div 2}$$
$$\underbrace{9 - 4} + 2$$
$$\underbrace{5 + 2}$$
$$7$$

1. Perform operations in parentheses.
2. Simplify expressions with exponents.
3. Do multiplication and division as they occur from left to right.
4. Do addition and subtraction as they occur from left to right.

One or more of these steps may not be needed to simplify an expression. In that case, proceed to the next step in the Order of Operations Agreement.

HOW TO • 2 Simplify $5 + 8 \div 2$. There are no parentheses or exponents. Proceed to Step 3 of the agreement.

$$5 + \underbrace{8 \div 2}$$
$$\underbrace{5 + 4}$$
$$9$$

3. Do multiplication or division.
4. Do addition or subtraction.

EXAMPLE • 4

Simplify: $64 \div (8 - 4)^2 \cdot 9 - 5^2$

Solution

$64 \div (8 - 4)^2 \cdot 9 - 5^2$
$= 64 \div 4^2 \cdot 9 - 5^2$ • Parentheses
$= 64 \div 16 \cdot 9 - 25$ • Exponents
$= 4 \cdot 9 - 25$ • Division and multiplication
$= 36 - 25$ • Subtraction
$= 11$

YOU TRY IT • 4

Simplify: $5 \cdot (8 - 4)^2 \div 4 - 2$

Your solution

Solution on p. S4

1.6 EXERCISES

OBJECTIVE A To simplify expressions that contain exponents

For Exercises 1 to 12, write the number in exponential notation.

1. $2 \cdot 2 \cdot 2$

2. $7 \cdot 7 \cdot 7 \cdot 7 \cdot 7$

3. $6 \cdot 6 \cdot 6 \cdot 7 \cdot 7 \cdot 7 \cdot 7$

4. $6 \cdot 6 \cdot 9 \cdot 9 \cdot 9 \cdot 9$

5. $2 \cdot 2 \cdot 2 \cdot 3 \cdot 3 \cdot 3$

6. $3 \cdot 3 \cdot 10 \cdot 10$

7. $5 \cdot 7 \cdot 7 \cdot 7 \cdot 7 \cdot 7$

8. $4 \cdot 4 \cdot 4 \cdot 5 \cdot 5 \cdot 5$

9. $3 \cdot 3 \cdot 3 \cdot 6 \cdot 6 \cdot 6 \cdot 6$

10. $2 \cdot 2 \cdot 5 \cdot 5 \cdot 5 \cdot 8$

11. $3 \cdot 3 \cdot 3 \cdot 5 \cdot 9 \cdot 9 \cdot 9$

12. $2 \cdot 2 \cdot 2 \cdot 4 \cdot 7 \cdot 7 \cdot 7$

For Exercises 13 to 37, simplify.

13. 2^3

14. 2^6

15. $2^4 \cdot 5^2$

16. $2^6 \cdot 3^2$

17. $3^2 \cdot 10^2$

18. $2^3 \cdot 10^4$

19. $6^2 \cdot 3^3$

20. $4^3 \cdot 5^2$

21. $5 \cdot 2^3 \cdot 3$

22. $6 \cdot 3^2 \cdot 4$

23. $2^2 \cdot 3^2 \cdot 10$

24. $3^2 \cdot 5^2 \cdot 10$

25. $0^2 \cdot 4^3$

26. $6^2 \cdot 0^3$

27. $3^2 \cdot 10^4$

28. $5^3 \cdot 10^3$

29. $2^2 \cdot 3^3 \cdot 5$

30. $5^2 \cdot 7^3 \cdot 2$

31. $2 \cdot 3^4 \cdot 5^2$

32. $6 \cdot 2^6 \cdot 7^2$

33. $5^2 \cdot 3^2 \cdot 7^2$

34. $4^2 \cdot 9^2 \cdot 6^2$

35. $3^4 \cdot 2^6 \cdot 5$

36. $4^3 \cdot 6^3 \cdot 7$

37. $4^2 \cdot 3^3 \cdot 10^4$

38. Rewrite the expression using the numbers 3 and 5 exactly once. Then simplify the expression.

a. $3 + 3 + 3 + 3 + 3$

b. $3 \cdot 3 \cdot 3 \cdot 3 \cdot 3$

OBJECTIVE B To use the Order of Operations Agreement to simplify expressions

For Exercises 39 to 77, simplify by using the Order of Operations Agreement.

39. $4 - 2 + 3$

40. $6 - 3 + 2$

41. $6 \div 3 + 2$

42. $8 \div 4 + 8$

43. $6 \cdot 3 + 5$

44. $5 \cdot 9 + 2$

45. $3^2 - 4$

46. $5^2 - 17$

47. $4 \cdot (5 - 3) + 2$

48. $3 + (4 + 2) \div 3$

49. $5 + (8 + 4) \div 6$

50. $8 - 2^2 + 4$

51. $16 \cdot (3 + 2) \div 10$

52. $12 \cdot (1 + 5) \div 12$

53. $10 - 2^3 + 4$

54. $5 \cdot 3^2 + 8$

55. $16 + 4 \cdot 3^2$

56. $12 + 4 \cdot 2^3$

57. $16 + (8 - 3) \cdot 2$

58. $7 + (9 - 5) \cdot 3$

59. $2^2 + 3 \cdot (6 - 2)^2$

60. $3^3 + 5 \cdot (8 - 6)^3$

61. $2^2 \cdot 3^2 + 2 \cdot 3$

62. $4 \cdot 6 + 3^2 \cdot 4^2$

63. $16 - 2 \cdot 4$

64. $12 + 3 \cdot 5$

65. $3 \cdot (6 - 2) + 4$

66. $5 \cdot (8 - 4) - 6$

67. $8 - (8 - 2) \div 3$

68. $12 - (12 - 4) \div 4$

69. $8 + 2 - 3 \cdot 2 \div 3$

70. $10 + 1 - 5 \cdot 2 \div 5$

71. $3 \cdot (4 + 2) \div 6$

72. $(7 - 3)^2 \div 2 - 4 + 8$

73. $20 - 4 \div 2 \cdot (3 - 1)^3$

74. $12 \div 3 \cdot 2^2 + (7 - 3)^2$

75. $(4 - 2) \cdot 6 \div 3 + (5 - 2)^2$

76. $18 - 2 \cdot 3 + (4 - 1)^3$

77. $100 \div (2 + 3)^2 - 8 \div 2$

For Exercises 78 to 80, insert parentheses as needed in the expression $8 - 2 \cdot 3 + 1$ in order to make the statement true.

78. $8 - 2 \cdot 3 + 1 = 3$

79. $8 - 2 \cdot 3 + 1 = 0$

80. $8 - 2 \cdot 3 + 1 = 24$

Applying the Concepts

81. Explain the difference that the order of operations makes between **a.** $(14 - 2) \div 2 \cdot 3$ and **b.** $(14 - 2) \div (2 \cdot 3)$. Work the two problems. What is the difference between the larger answer and the smaller answer?

SECTION 1.7 Prime Numbers and Factoring

OBJECTIVE A To factor numbers

Whole-number **factors of a number** divide that number evenly (there is no remainder).

1, 2, 3, and 6 are whole-number factors of 6 because they divide 6 evenly. $1\overline{)6}$ = 6, $2\overline{)6}$ = 3, $3\overline{)6}$ = 2, $6\overline{)6}$ = 1

Note that both the divisor and the quotient are factors of the dividend.

To find the factors of a number, try dividing the number by 1, 2, 3, 4, 5, ... Those numbers that divide the number evenly are its factors. Continue this process until the factors start to repeat.

HOW TO • 1 Find all the factors of 42.

$42 \div 1 = 42$	1 and 42 are factors.
$42 \div 2 = 21$	2 and 21 are factors.
$42 \div 3 = 14$	3 and 14 are factors.
$42 \div 4$	Will not divide evenly
$42 \div 5$	Will not divide evenly
$42 \div 6 = 7$	6 and 7 are factors. } Factors are repeating; all the
$42 \div 7 = 6$	7 and 6 are factors. } factors of 42 have been found.

1, 2, 3, 6, 7, 14, 21, and 42 are factors of 42.

The following rules are helpful in finding the factors of a number.

2 is a factor of a number if the last digit of the number is 0, 2, 4, 6, or 8.

436 ends in 6; therefore, 2 is a factor of 436. $(436 \div 2 = 218)$

3 is a factor of a number if the sum of the digits of the number is divisible by 3.

The sum of the digits of 489 is $4 + 8 + 9 = 21$. 21 is divisible by 3. Therefore, 3 is a factor of 489. $(489 \div 3 = 163)$

5 is a factor of a number if the last digit of the number is 0 or 5.

520 ends in 0; therefore, 5 is a factor of 520. $(520 \div 5 = 104)$

EXAMPLE • 1

Find all the factors of 30.

Solution

$30 \div 1 = 30$	
$30 \div 2 = 15$	
$30 \div 3 = 10$	
$30 \div 4$	Will not divide evenly
$30 \div 5 = 6$	
$30 \div 6 = 5$	Factors repeating

1 2, 3, 5, 6, 10, 15, and 30 are factors of 30.

YOU TRY IT • 1

Find all the factors of 40.

Your solution

Solution on p. S4

OBJECTIVE B To find the prime factorization of a number

Point of Interest

Prime numbers are an important part of cryptology, the study of secret codes. To make it less likely that codes can be broken, cryptologists use prime numbers that have hundreds of digits.

A number is a **prime number** if its only whole-number factors are 1 and itself. 7 is prime because its only factors are 1 and 7. If a number is not prime, it is called a **composite number.** Because 6 has factors of 2 and 3, 6 is a composite number. The number 1 is not considered a prime number; therefore, it is not included in the following list of prime numbers less than 50.

2, 3, 5, 7, 11, 13, 17, 19, 23, 29, 31, 37, 41, 43, 47

The **prime factorization** of a number is the expression of the number as a product of its prime factors. We use a "T-diagram" to find the prime factors of 60. Begin with the smallest prime number as a trial divisor, and continue with prime numbers as trial divisors until the final quotient is 1.

	60	
2	30	$60 \div 2 = 30$
2	15	$30 \div 2 = 15$
3	5	$15 \div 3 = 5$
5	1	$5 \div 5 = 1$

The prime factorization of 60 is $2 \cdot 2 \cdot 3 \cdot 5$.

Finding the prime factorization of larger numbers can be more difficult. Try each prime number as a trial divisor. Stop when the square of the trial divisor is greater than the number being factored.

HOW TO • 2 Find the prime factorization of 106.

	106
2	53
53	1

• 53 cannot be divided evenly by 2, 3, 5, 7, or 11. Prime numbers greater than 11 need not be tested because 11^2 is greater than 53.

The prime factorization of 106 is $2 \cdot 53$.

EXAMPLE • 2

Find the prime factorization of 315.

Solution

	315	
3	105	• $315 \div 3 = 105$
3	35	• $105 \div 3 = 35$
5	7	• $35 \div 5 = 7$
7	1	• $7 \div 7 = 1$

$315 = 3 \cdot 3 \cdot 5 \cdot 7$

YOU TRY IT • 2

Find the prime factorization of 44.

Your solution

EXAMPLE • 3

Find the prime factorization of 201.

Solution

	201
3	67
67	1

• Try only 2, 3, 5, 7, and 11 because $11^2 > 67$.

$201 = 3 \cdot 67$

YOU TRY IT • 3

Find the prime factorization of 177.

Your solution

Solutions on p. S4

1.7 EXERCISES

OBJECTIVE A To factor numbers

For Exercises 1 to 40, find all the factors of the number.

1. 4

2. 6

3. 10

4. 20

5. 7

6. 12

7. 9

8. 8

9. 13

10. 17

11. 18

12. 24

13. 56

14. 36

15. 45

16. 28

17. 29

18. 33

19. 22

20. 26

21. 52

22. 49

23. 82

24. 37

25. 57

26. 69

27. 48

28. 64

29. 95

30. 46

31. 54

32. 50

33. 66

34. 77

35. 80

36. 100

37. 96

38. 85

39. 90

40. 101

41. True or false? A number can have an odd number of factors.

42. True or false? If a number has exactly four factors, then the product of those four factors must be the number.

OBJECTIVE B To find the prime factorization of a number

For Exercises 43 to 86, find the prime factorization.

43. 6

44. 14

45. 17

46. 83

47. 24 **48.** 12 **49.** 27 **50.** 9

51. 36 **52.** 40 **53.** 19 **54.** 37

55. 90 **56.** 65 **57.** 115 **58.** 80

59. 18 **60.** 26 **61.** 28 **62.** 49

63. 31 **64.** 42 **65.** 62 **66.** 81

67. 22 **68.** 39 **69.** 101 **70.** 89

71. 66 **72.** 86 **73.** 74 **74.** 95

75. 67 **76.** 78 **77.** 55 **78.** 46

79. 120 **80.** 144 **81.** 160 **82.** 175

83. 216 **84.** 400 **85.** 625 **86.** 225

87. True or false? The prime factorization of 102 is $2 \cdot 51$.

Applying the Concepts

88. In 1742, Christian Goldbach conjectured that every even number greater than 2 could be expressed as the sum of two prime numbers. Show that this conjecture is true for 8, 24, and 72. (*Note:* Mathematicians have not yet been able to determine whether Goldbach's conjecture is true or false.)

89. Explain why 2 is the only even prime number.

FOCUS ON PROBLEM SOLVING

Questions to Ask

© Brownie Harris/Corbis

You encounter problem-solving situations every day. Some problems are easy to solve, and you may mentally solve these problems without considering the steps you are taking in order to draw a conclusion. Others may be more challenging and may require more thought and consideration.

Suppose a friend suggests that you both take a trip over spring break. You'd like to go. What questions go through your mind? You might ask yourself some of the following questions:

How much will the trip cost? What will be the cost for travel, hotel rooms, meals, and so on?

Are some costs going to be shared by both me and my friend?

Can I afford it?

How much money do I have in the bank?

How much more money than I have now do I need?

How much time is there to earn that much money?

How much can I earn in that amount of time?

How much money must I keep in the bank in order to pay the next tuition bill (or some other expense)?

These questions require different mathematical skills. Determining the cost of the trip requires **estimation;** for example, you must use your knowledge of air fares or the cost of gasoline to arrive at an estimate of these costs. If some of the costs are going to be shared, you need to **divide** those costs by 2 in order to determine your share of the expense. The question regarding how much more money you need requires **subtraction:** the amount needed minus the amount currently in the bank. To determine how much money you can earn in the given amount of time requires **multiplication**—for example, the amount you earn per week times the number of weeks to be worked. To determine if the amount you can earn in the given amount of time is sufficient, you need to use your knowledge of **order relations** to compare the amount you can earn with the amount needed.

Facing the problem-solving situation described above may not seem difficult to you. The reason may be that you have faced similar situations before and, therefore, know how to work through this one. You may feel better prepared to deal with a circumstance such as this one because you know what questions to ask. An important aspect of learning to solve problems is learning what questions to ask. As you work through application problems in this text, try to become more conscious of the mental process you are going through. You might begin the process by asking yourself the following questions whenever you are solving an application problem.

1. Have I read the problem enough times to be able to understand the situation being described?

2. Will restating the problem in different words help me to understand the problem situation better?

3. What facts are given? (You might make a list of the information contained in the problem.)

4. What information is being asked for?

5. What relationship exists among the given facts? What relationship exists between the given facts and the solution?

6. What mathematical operations are needed in order to solve the problem?

Try to focus on the problem-solving situation, not on the computation or on getting the answer quickly. And remember, the more problems you solve, the better able you will be to solve other problems in the future, partly because you are learning what questions to ask.

PROJECTS AND GROUP ACTIVITIES

Order of Operations

Does your calculator use the Order of Operations Agreement? To find out, try this problem:

$$2 + 4 \cdot 7$$

If your answer is 30, then the calculator uses the Order of Operations Agreement. If your answer is 42, it does not use that agreement.

Even if your calculator does not use the Order of Operations Agreement, you can still correctly evaluate numerical expressions. The parentheses keys, (and), are used for this purpose.

Remember that $2 + 4 \cdot 7$ means $2 + (4 \cdot 7)$ because the multiplication must be completed before the addition. To evaluate this expression, enter the following:

Enter:	2	+	(	4	x	7	)	=
Display:	2	2	(	4	4	7	28	30

When using your calculator to evaluate numerical expressions, insert parentheses around multiplications and around divisions. This has the effect of forcing the calculator to do the operations in the order you want.

For Exercises 1 to 10, evaluate.

1. $3 \cdot 8 - 5$
2. $6 + 8 \div 2$
3. $3 \cdot (8 - 2)^2$
4. $24 - (4 - 2)^2 \div 4$
5. $3 + (6 \div 2 + 4)^2 - 2$
6. $16 \div 2 + 4 \cdot (8 - 12 \div 4)^2 - 50$
7. $3 \cdot (15 - 2 \cdot 3) - 36 \div 3$
8. $4 \cdot 2^2 - (12 + 24 \div 6) + 5$
9. $16 \div 4 \cdot 3 + (3 \cdot 4 - 5) + 2$
10. $15 \cdot 3 \div 9 + (2 \cdot 6 - 3) + 4$

Patterns in Mathematics

For the circle at the left, use a straight line to connect each dot on the circle with every other dot on the circle. How many different straight lines are there?

Follow the same procedure for each of the circles shown below. How many different straight lines are there in each?

Find a pattern to describe the number of dots on a circle and the corresponding number of different lines drawn. Use the pattern to determine the number of different lines that would be drawn in a circle with 7 dots and in a circle with 8 dots.

Now use the pattern to answer the following question. You are arranging a tennis tournament with 9 players. How many singles matches will be played among the 9 players if each player plays each of the other players only once?

Search the World Wide Web

Go to www.census.gov on the Internet.

Jonathan Nourak/PhotoEdit, Inc.

1. Find a projection for the total U.S. population 10 years from now and a projection for the total population 20 years from now. Record the two numbers.
2. Use the data from Exercise 1 to determine the expected growth in the population over the next 10 years.
3. Use the answer from Exercise 2 to find the average increase in the U.S. population per year over the next 10 years. Round to the nearest million.
4. Use data in the population table you found to write two word problems. Then state whether addition, subtraction, multiplication, or division is required to solve each of the problems.

CHAPTER 1

SUMMARY

KEY WORDS	EXAMPLES
The *whole numbers* are 0, 1, 2, 3, 4, 5, 6, 7, 8, 9, 10, ... [1.1A, p. 2]	
The *graph of a whole number* is shown by placing a heavy dot directly above that number on the number line. [1.1A, p. 2]	This is the graph of 4 on the number line. 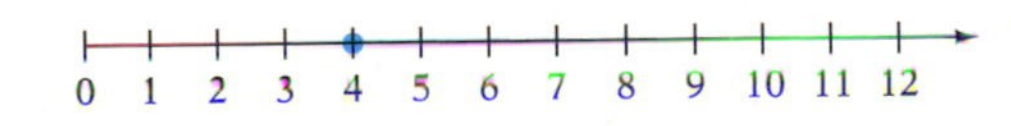
The symbol for *is less than* is $<$. The symbol for *is greater than* is $>$. These symbols are used to show the order relation between two numbers. [1.1A, p. 2]	$3 < 7$ $9 > 2$

When a whole number is written using the digits 0, 1, 2, 3, 4, 5, 6, 7, 8, and 9, it is said to be in *standard form.* The position of each digit in the number determines the digit's *place value.* The place values are used to write the expanded form of a number. [1.1B, p. 3]

The number 598,317 is in standard form. The digit 8 is in the thousands place. The number 598,317 is written in expanded form as 500,000 + 90,000 + 8000 + 300 + 10 + 7.

Addition is the process of finding the total of two or more numbers. The numbers being added are called *addends.* The result is the *sum.* [1.2A, p. 8]

$$\begin{array}{r} \scriptstyle 1\ \ 11 \\ 8{,}762 \\ +\ 1{,}359 \\ \hline 10{,}121 \end{array}$$

Subtraction is the process of finding the difference between two numbers. The *minuend* minus the *subtrahend* equals the *difference.* [1.3A, p. 16]

$$\begin{array}{r} \scriptstyle 4\ \ 11\ 11\ \ 6\ \ 13 \\ \not{5}\ \not{2},\not{1}\ \not{7}\ \not{3} \\ -\ 3\ 4{,}9\ 6\ 8 \\ \hline 1\ 7{,}2\ 0\ 5 \end{array}$$

Multiplication is the repeated addition of the same number. The numbers that are multiplied are called *factors.* The result is the *product.* [1.4A, p. 24]

$$\begin{array}{r} \scriptstyle 4\ 5\ \ \\ 358 \\ \times\quad 7 \\ \hline 2506 \end{array}$$

Division is used to separate objects into equal groups. The *dividend* divided by the *divisor* equals the *quotient.* [1.5A, p. 32]
For any division problem,
(*quotient* · *divisor*) + *remainder* = *dividend.* [1.5B, p. 35]

$$\begin{array}{r} 93\text{ r3} \\ 7\overline{)\ 654} \\ -63 \\ \hline 24 \\ -21 \\ \hline 3 \end{array}$$

Check: $(7 \cdot 93) + 3 = 651 + 3 = 654$

The expression 4^3 is in *exponential notation.* The *exponent,* 3, indicates how many times 4 occurs as a factor in the multiplication. [1.6A, p. 45]

$5^4 = 5 \cdot 5 \cdot 5 \cdot 5 = 625$

Whole-number *factors of a number* divide that number evenly (there is no remainder). [1.7A, p. 49]

$18 \div 1 = 18$
$18 \div 2 = 9$
$18 \div 3 = 6$
$18 \div 4$ 4 does not divide 18 evenly.
$18 \div 5$ 5 does not divide 18 evenly.
$18 \div 6 = 3$ The factors are repeating.
The factors of 18 are 1, 2, 3, 6, 9, and 18.

A number greater than 1 is a *prime number* if its only whole-number factors are 1 and itself. If a number is not prime, it is a *composite number.* [1.7B, p. 50]

The prime numbers less than 20 are 2, 3, 5, 7, 11, 13, 17, and 19.
The composite numbers less than 20 are 4, 6, 8, 9, 10, 12, 14, 15, 16, and 18.

The *prime factorization* of a number is the expression of the number as a product of its prime factors. [1.7B, p. 50]

$$\begin{array}{r|l} & 42 \\ \hline 2 & 21 \\ 3 & 7 \\ 7 & 1 \end{array}$$

The prime factorization of 42 is $2 \cdot 3 \cdot 7$.

ESSENTIAL RULES AND PROCEDURES	EXAMPLES
To round a number to a given place value: If the digit to the right of the given place value is less than 5, replace that digit and all digits to the right by zeros. If the digit to the right of the given place value is greater than or equal to 5, increase the digit in the given place value by 1, and replace all other digits to the right by zeros. [1.1D, p. 5]	36,178 rounded to the nearest thousand is 36,000. 4592 rounded to the nearest thousand is 5000.
Properties of Addition [1.2A, p. 8]	
Addition Property of Zero Zero added to a number does not change the number.	$7 + 0 = 7$
Commutative Property of Addition Two numbers can be added in either order; the sum will be the same.	$8 + 3 = 3 + 8$
Associative Property of Addition Numbers to be added can be grouped in any order; the sum will be the same.	$(2 + 4) + 6 = 2 + (4 + 6)$
To estimate the answer to an addition calculation: Round each number to the same place value. Perform the calculation using the rounded numbers. [1.2A, p. 10]	39,471 → 40,000 12,586 → + 10,000 50,000 50,000 is an estimate of the sum of 39,471 and 12,586.
Properties of Multiplication [1.4A, p. 24]	
Multiplication Property of Zero The product of a number and zero is zero.	$3 \cdot 0 = 0$
Multiplication Property of One The product of a number and one is the number.	$6 \cdot 1 = 6$
Commutative Property of Multiplication Two numbers can be multiplied in either order; the product will be the same.	$2 \cdot 8 = 8 \cdot 2$
Associative Property of Multiplication Grouping numbers to be multiplied in any order gives the same result.	$(2 \cdot 4) \cdot 6 = 2 \cdot (4 \cdot 6)$
Division Properties of Zero and One [1.5A, p. 32] Any whole number, except zero, divided by itself is 1. Any whole number divided by 1 is the whole number. Zero divided by any other whole number is zero. Division by zero is not allowed.	 $3 \div 3 = 1$ $3 \div 1 = 3$ $0 \div 3 = 0$ $3 \div 0$ is not allowed.
Order of Operations Agreement [1.6B, p. 46]	
Step 1 Do all the operations inside parentheses.	$5^2 - 3(2 + 4) = 5^2 - 3(6)$
Step 2 Simplify any number expressions containing exponents.	$= 25 - 3(6)$
Step 3 Do multiplications and divisions as they occur from left to right.	$= 25 - 18$
Step 4 Do addition and subtraction as they occur from left to right.	$= 7$

CHAPTER 1

CONCEPT REVIEW

Test your knowledge of the concepts presented in this chapter. Answer each question. Then check your answers against the ones provided in the Answer Section.

1. What is the difference between the symbols $<$ and $>$?

2. How do you round a four-digit whole number to the nearest hundred?

3. What is the difference between the Commutative Property of Addition and the Associative Property of Addition?

4. How do you estimate the sum of two numbers?

5. When is it necessary to borrow when performing subtraction?

6. What is the difference between the Multiplication Property of Zero and the Multiplication Property of One?

7. How do you multiply a whole number by 100?

8. How do you estimate the product of two numbers?

9. What is the difference between $0 \div 9$ and $9 \div 0$?

10. How do you check the answer to a division problem that has a remainder?

11. What are the steps in the Order of Operations Agreement?

12. How do you know if a number is a factor of another number?

13. What is a quick way to determine if 3 is a factor of a number?

CHAPTER 1

REVIEW EXERCISES

1. Simplify: $3 \cdot 2^3 \cdot 5^2$

2. Write 10,327 in expanded form.

3. Find all the factors of 18.

4. Find the sum of 5894, 6301, and 298.

5. Subtract: $\begin{array}{r} 4926 \\ -\ 3177 \\ \hline \end{array}$

6. Divide: $7\overline{)14,945}$

7. Place the correct symbol, $<$ or $>$, between the two numbers: 101 87

8. Write $5 \cdot 5 \cdot 7 \cdot 7 \cdot 7 \cdot 7 \cdot 7$ in exponential notation.

9. What is 2019 multiplied by 307?

10. What is 10,134 decreased by 4725?

11. Add: $\begin{array}{r} 298 \\ 461 \\ +\ 322 \\ \hline \end{array}$

12. Simplify: $2^3 - 3 \cdot 2$

13. Round 45,672 to the nearest hundred.

14. Write 276,057 in words.

15. Find the quotient of 109,763 and 84.

16. Write two million eleven thousand forty-four in standard form.

17. What is 3906 divided by 8?

18. Simplify: $3^2 + 2^2 \cdot (5 - 3)$

19. Simplify: $8 \cdot (6 - 2)^2 \div 4$

20. Find the prime factorization of 72.

21. What is 3895 minus 1762?

22. Multiply: $\begin{array}{r} 843 \\ \times\ 27 \\ \hline \end{array}$

23. **Blood Count** White blood cell (WBC) counts can indicate illness or infections. Before surgery, a patient's WBC count was 6478; the day after surgery, it had risen to 12,789. What was the difference in the count before and after surgery?

24. **Wages** A medical transcriptionist earns $660 for working 40 hours during the week. One week, the transcriptionist works an additional 20 hours at $24 per hour. What is the transcriptionist's total pay for the week?

25. **Consumerism** A massage therapist uses 6 ounces of massage oil on each client. During one month, the massage therapist used 426 ounces of massage oil. How many clients did the massage therapist see in that month?

26. **Consumerism** A medical clinic purchases an x-ray machine at a cost of $22,250. After a deposit of $2000, the balance is due in 50 equal monthly payments. Calculate the monthly payment.

27. **Banking** A medical laboratory has $5396 in its checking account. The lab then receives payments of $3894 and $2985 and deposits them in the checking account. Find the total amount deposited in the checking account and the new account balance.

28. **Compensation** You have a car payment of $246 per month. What is the total of the car payments over a 12-month period?

Employment The table shows the projected employment data for physical therapist assistants and occupational therapist assistants from the year 2008 to 2018. Use the table for Exercises 29 to 32.

Projection Data from the National Employment Matrix

Occupation Title	*Employment, 2008*	*Projected Employment, 2018*
Occupational therapist assistant (OTA)	26,600	34,600
Physical therapist assistant (PTA)	63,800	85,000

Source: Occupational Outlook Handbook, 2010–11 Edition, www.bls.gov

29. Find the difference between the 2008 figures for physical therapist assistant employment and occupational therapist assistant employment.

30. Find the difference between the physical therapist assistant employment figures for 2008 and 2018.

31. Determine the projected average yearly increase in the number of employed physical therapist assistants from 2008 to 2018.

32. Determine the projected average yearly increase in the number of employed occupational therapist assistants from 2008 to 2018.

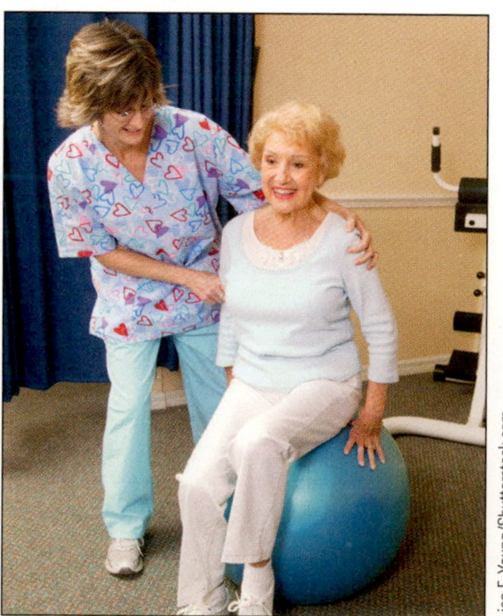
Lisa F. Young/Shutterstock.com

CHAPTER 1

TEST

1. Simplify: $3^3 \cdot 4^2$

2. Write 207,068 in words.

3. Subtract: $\begin{array}{r} 17{,}495 \\ -\ 8{,}162 \\ \hline \end{array}$

4. Find all the factors of 20.

5. Multiply: $\begin{array}{r} 9736 \\ \times\ 704 \\ \hline \end{array}$

6. Simplify: $4^2 \cdot (4 - 2) \div 8 + 5$

7. Write 906,378 in expanded form.

8. Round 74,965 to the nearest hundred.

9. Divide: $97\overline{)108{,}764}$

10. Write $3 \cdot 3 \cdot 3 \cdot 7 \cdot 7$ in exponential form.

11. Find the sum of 8756, 9094, and 37,065.

12. Find the prime factorization of 84.

13. Simplify: $16 \div 4 \cdot 2 - (7 - 5)^2$

14. Find the product of 8 and 90,763.

15. Write one million two hundred four thousand six in standard form.

16. Divide: $7\overline{)60{,}972}$

17. Place the correct symbol, $<$ or $>$, between the two numbers: 21 19

18. Find the quotient of 5624 and 8.

19. Add:

$$\begin{array}{r} 25{,}492 \\ +71{,}306 \\ \hline \end{array}$$

20. Find the difference between 29,736 and 9814.

Education The table at the right shows the projected enrollment in public and private elementary and secondary schools in the fall of 2013 and the fall of 2016. Use this information for Exercises 21 and 22.

Year	*Pre-Kindergarten through Grade 8*	*Grades 9 through 12*
2013	41,873,000	16,000,000
2016	43,097,000	16,684,000

Source: The National Center for Education Statistics

21. Find the difference between the total enrollment in 2016 and that in 2013.

22. Find the average enrollment in each of grades 9 through 12 in 2016.

23. **Dental Supplies** A local dentist office gives toothbrushes to its patients after each cleaning. A certain adult toothbrush is packaged 144 to a box and 10 boxes to a case. If the office manager orders three cases, how many toothbrushes are ordered?

qvist/Shutterstock.com

24. **Health Insurance** Your health insurance premiums are $237 each month. How much will your insurance cost over a 12-month period?

25. **Intake and Output** CNAs measure fluid intake and urine output of patients to help monitor their hydration. All fluids are measured in milliliters.

a. A patient consumes 120 milliliters, 210 milliliters, 90 milliliters, and 150 milliliters of fluid at various times during the day. Find the total intake of fluids.

b. The nurse's aide measures 330 milliliters, 225 milliliters, and 60 milliliters of urine output for the same patient. Find the total output of fluids.

c. Which is greater: the total input or total output for the patient? By how many milliliters?

CHAPTER 2

Fractions

OBJECTIVES

SECTION 2.1
- A To find the least common multiple (LCM)
- B To find the greatest common factor (GCF)

SECTION 2.2
- A To write a fraction that represents part of a whole
- B To write an improper fraction as a mixed number or a whole number, and a mixed number as an improper fraction

SECTION 2.3
- A To find equivalent fractions by raising to higher terms
- B To write a fraction in simplest form

SECTION 2.4
- A To add fractions with the same denominator
- B To add fractions with different denominators
- C To add whole numbers, mixed numbers, and fractions
- D To solve application problems

SECTION 2.5
- A To subtract fractions with the same denominator
- B To subtract fractions with different denominators
- C To subtract whole numbers, mixed numbers, and fractions
- D To solve application problems

SECTION 2.6
- A To multiply fractions
- B To multiply whole numbers, mixed numbers, and fractions
- C To solve application problems

SECTION 2.7
- A To divide fractions
- B To divide whole numbers, mixed numbers, and fractions
- C To solve application problems

SECTION 2.8
- A To identify the order relation between two fractions
- B To simplify expressions containing exponents
- C To use the Order of Operations Agreement to simplify expressions

AIM for the Future

PERSONAL AND HOME CARE AIDES—also called caregivers, companions, or personal attendants—work for various public and private agencies that provide home care services. In these agencies, caregivers are likely supervised by a licensed nurse, social worker, or other non-medical manager. Personal and home care aides work independently, with only periodic visits by their supervisors. Employment of personal and home care aides is expected to grow by 46% from 2008 to 2018, which is much faster than the average for all occupations. Median hourly wages of home health aides were $9.84 in May 2008, with the highest 10% earning more than $13.93 an hour. (*Source: Bureau of Labor Statistics:* http://www.bls.gov/oco/ocos326.htm)

PREP TEST

Are you ready to succeed in this chapter?

Take the Prep Test below to find out if you are ready to learn the new material.

For Exercises 1 to 6, add, subtract, multiply, or divide.

1. 4×5

2. $2 \cdot 2 \cdot 2 \cdot 3 \cdot 5$

3. 9×1

4. $6 + 4$

5. $10 - 3$

6. $63 \div 30$

7. Which of the following numbers divide evenly into 12?
1 2 3 4 5 6 7 8 9 10 11 12

8. Simplify: $8 \times 7 + 3$

9. Complete: $8 = ? + 1$

10. Place the correct symbol, $<$ or $>$, between the two numbers.
44 48

SECTION

2.1 The Least Common Multiple and Greatest Common Factor

OBJECTIVE A To find the least common multiple (LCM)

Tips for Success

Before you begin a new chapter, you should take some time to review previously learned skills. One way to do this is to complete the Prep Test. See page 63. This test focuses on the particular skills that will be required for the new chapter.

The **multiples of a number** are the products of that number and the numbers 1, 2, 3, 4, 5, ...

$3 \times 1 = 3$
$3 \times 2 = 6$
$3 \times 3 = 9$
$3 \times 4 = 12$ The multiples of 3 are 3, 6, 9, 12, 15, ...
$3 \times 5 = 15$
⋮

A number that is a multiple of two or more numbers is a **common multiple** of those numbers.

The multiples of 4 are 4, 8, 12, 16, 20, 24, 28, 32, 36, ...
The multiples of 6 are 6, 12, 18, 24, 30, 36, 42, ...
Some common multiples of 4 and 6 are 12, 24, and 36.

The **least common multiple (LCM)** is the smallest common multiple of two or more numbers.

The least common multiple of 4 and 6 is 12.

Listing the multiples of each number is one way to find the LCM. Another way to find the LCM uses the prime factorization of each number.

To find the LCM of 450 and 600, find the prime factorization of each number and write the factorization of each number in a table. Circle the greatest product in each column. The LCM is the product of the circled numbers.

	2	3	5
450 =	2	(3 · 3)	(5 · 5)
600 =	(2 · 2 · 2)	3	5 · 5

• In the column headed by 5, the products are equal. Circle just one product.

The LCM is the product of the circled numbers.
The LCM $= 2 \cdot 2 \cdot 2 \cdot 3 \cdot 3 \cdot 5 \cdot 5 = 1800$.

EXAMPLE • 1

Find the LCM of 24, 36, and 50.

Solution

	2	3	5
24 =	(2 · 2 · 2)	3	
36 =	2 · 2	(3 · 3)	
50 =	2		(5 · 5)

The LCM $= 2 \cdot 2 \cdot 2 \cdot 3 \cdot 3 \cdot 5 \cdot 5 = 1800$.

YOU TRY IT • 1

Find the LCM of 12, 27, and 50.

Your solution

Solution on p. S4

OBJECTIVE B To find the greatest common factor (GCF)

Recall that a number that divides another number evenly is a factor of that number. The number 64 can be evenly divided by 1, 2, 4, 8, 16, 32, and 64, so the numbers 1, 2, 4, 8, 16, 32, and 64 are factors of 64.

A number that is a factor of two or more numbers is a **common factor** of those numbers.

The factors of 30 are 1, 2, 3, 5, 6, 10, 15, and 30.
The factors of 105 are 1, 3, 5, 7, 15, 21, 35, and 105.
The common factors of 30 and 105 are 1, 3, 5, and 15.

The **greatest common factor (GCF)** is the largest *common factor* of two or more numbers.

The greatest common factor of 30 and 105 is 15.

Listing the factors of each number is one way of finding the GCF. Another way to find the GCF is to use the prime factorization of each number.

To find the GCF of 126 and 180, find the prime factorization of each number and write the factorization of each number in a table. Circle the least product in each column that does not have a blank. The GCF is the product of the circled numbers.

	2	3	5	7
126 =	(2)	(3 · 3)		7
180 =	2 · 2	3 · 3	5	

• In the column headed by 3, the products are equal. Circle just one product. Columns 5 and 7 have a blank, so 5 and 7 are not common factors of 126 and 180. Do not circle any number in these columns.

The GCF is the product of the circled numbers.
The GCF = 2 · 3 · 3 = 18.

EXAMPLE 2

Find the GCF of 90, 168, and 420.

Solution

	2	3	5	7
90 =	(2)	3 · 3	5	
168 =	2 · 2 · 2	(3)		7
420 =	2 · 2	3	5	7

The GCF = 2 · 3 = 6.

YOU TRY IT 2

Find the GCF of 36, 60, and 72.

Your solution

EXAMPLE 3

Find the GCF of 7, 12, and 20.

Solution

	2	3	5	7
7 =				7
12 =	2 · 2	3		
20 =	2 · 2		5	

Because no numbers are circled, the GCF = 1.

YOU TRY IT 3

Find the GCF of 11, 24, and 30.

Your solution

Solutions on p. S4

2.1 EXERCISES

OBJECTIVE A To find the least common multiple (LCM)

For Exercises 1 to 34, find the LCM.

1. 5, 8
2. 3, 6
3. 3, 8
4. 2, 5
5. 5, 6

6. 5, 7
7. 4, 6
8. 6, 8
9. 8, 12
10. 12, 16

11. 5, 12
12. 3, 16
13. 8, 14
14. 6, 18
15. 3, 9

16. 4, 10
17. 8, 32
18. 7, 21
19. 9, 36
20. 14, 42

21. 44, 60
22. 120, 160
23. 102, 184
24. 123, 234
25. 4, 8, 12

26. 5, 10, 15
27. 3, 5, 10
28. 2, 5, 8
29. 3, 8, 12
30. 5, 12, 18

31. 9, 36, 64
32. 18, 54, 63
33. 16, 30, 84
34. 9, 12, 15

35. True or false? If two numbers have no common factors, then the LCM of the two numbers is their product.

36. True or false? If one number is a multiple of a second number, then the LCM of the two numbers is the second number.

OBJECTIVE B To find the greatest common factor (GCF)

For Exercises 37 to 70, find the GCF.

37. 3, 5
38. 5, 7
39. 6, 9
40. 18, 24
41. 15, 25

42. 14, 49
43. 25, 100
44. 16, 80
45. 32, 51
46. 21, 44

47. 12, 80

48. 8, 36

49. 16, 140

50. 12, 76

51. 24, 30

52. 48, 144

53. 44, 96

54. 18, 32

55. 3, 5, 11

56. 6, 8, 10

57. 7, 14, 49

58. 6, 15, 36

59. 10, 15, 20

60. 12, 18, 20

61. 24, 40, 72

62. 3, 17, 51

63. 17, 31, 81

64. 14, 42, 84

65. 25, 125, 625

66. 12, 68, 92

67. 28, 35, 70

68. 1, 49, 153

69. 32, 56, 72

70. 24, 36, 48

71. True or false? If two numbers have a GCF of 1, then the LCM of the two numbers is their product.

72. True or false? If the LCM of two numbers is one of the two numbers, then the GCF of the numbers is the other of the two numbers.

Applying the Concepts

73. **Work Schedules** Joe Salvo, a nurse, works 3 days and then has a day off. Joe's friend works 5 days and then has a day off. How many days after Joe and his friend have a day off together will they have another day off together?

74. Find the LCM of each of the following pairs of numbers: 2 and 3, 5 and 7, and 11 and 19. Can you draw a conclusion about the LCM of two prime numbers? Suggest a way of finding the LCM of three distinct prime numbers.

75. Find the GCF of each of the following pairs of numbers: 3 and 5, 7 and 11, and 29 and 43. Can you draw a conclusion about the GCF of two prime numbers? What is the GCF of three distinct prime numbers?

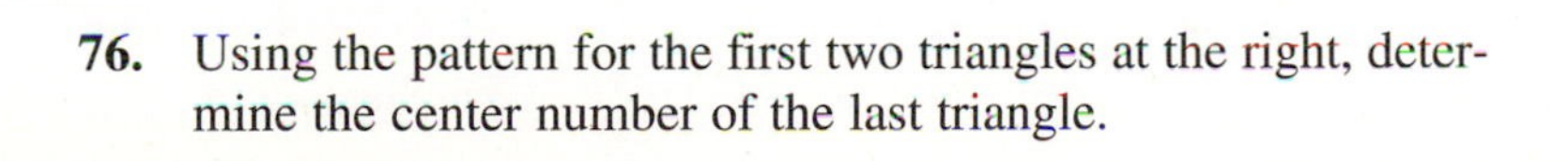

76. Using the pattern for the first two triangles at the right, determine the center number of the last triangle.

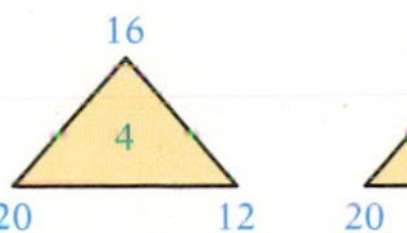

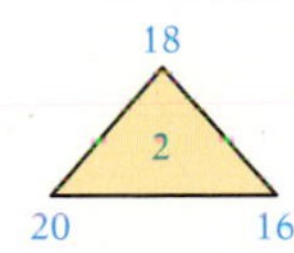

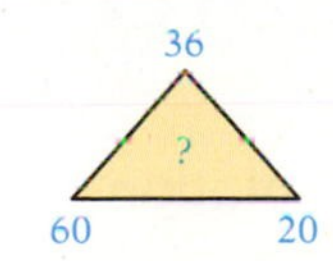

SECTION

2.2 Introduction to Fractions

OBJECTIVE A To write a fraction that represents part of a whole

Take Note

The **fraction bar** separates the numerator from the denominator. The **numerator** is the part of the fraction that appears above the fraction bar. The **denominator** is the part of the fraction that appears below the fraction bar.

Point of Interest

The fraction bar was first used in 1050 by al-Hassar. It is also called a vinculum.

A **fraction** can represent the number of equal parts of a whole.

The shaded portion of the circle is represented by the fraction $\frac{4}{7}$. Four of the seven equal parts of the circle (that is, four-sevenths of it) are shaded.

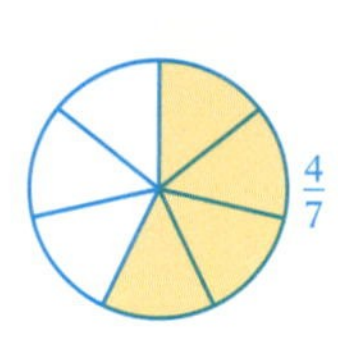

Each part of a fraction has a name.

Fraction bar $\rightarrow \frac{4}{7}$ $\leftarrow$ **Numerator** / $\leftarrow$ **Denominator**

A **proper fraction** is a fraction less than 1. The numerator of a proper fraction is smaller than the denominator. The shaded portion of the circle can be represented by the proper fraction $\frac{3}{4}$.

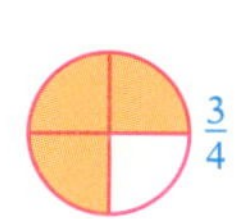

A **mixed number** is a number greater than 1 with a whole-number part and a fractional part. The shaded portion of the circles can be represented by the mixed number $2\frac{1}{4}$.

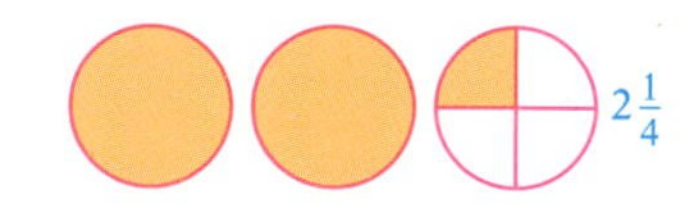

An **improper fraction** is a fraction greater than or equal to 1. The numerator of an improper fraction is greater than or equal to the denominator. The shaded portion of the circles can be represented by the improper fraction $\frac{9}{4}$. The shaded portion of the square can be represented by $\frac{4}{4}$.

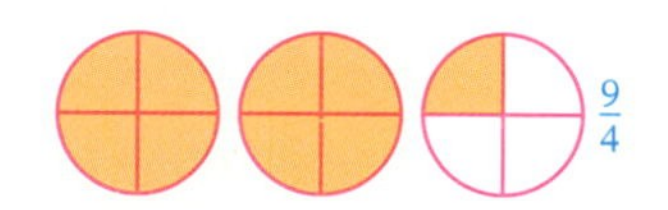

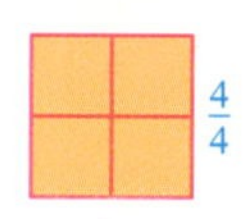

EXAMPLE 1

Express the shaded portion of the circles as a mixed number.

   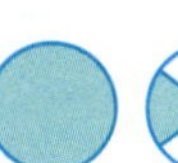

Solution $3\frac{2}{5}$

YOU TRY IT 1

Express the shaded portion of the circles as a mixed number.

 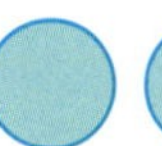  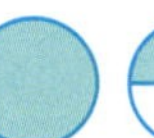 

Your solution

EXAMPLE 2

Express the shaded portion of the circles as an improper fraction.

   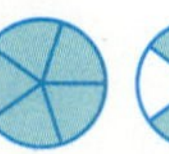 

Solution $\frac{17}{5}$

YOU TRY IT 2

Express the shaded portion of the circles as an improper fraction.

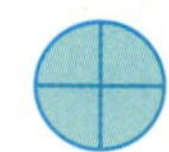 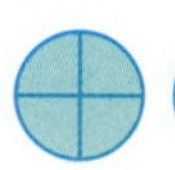 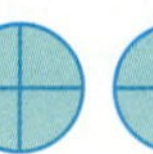  

Your solution

Solutions on p. S4

OBJECTIVE B **To write an improper fraction as a mixed number or a whole number, and a mixed number as an improper fraction**

Note from the diagram that the mixed number $2\frac{3}{5}$ and the improper fraction $\frac{13}{5}$ both represent the shaded portion of the circles.

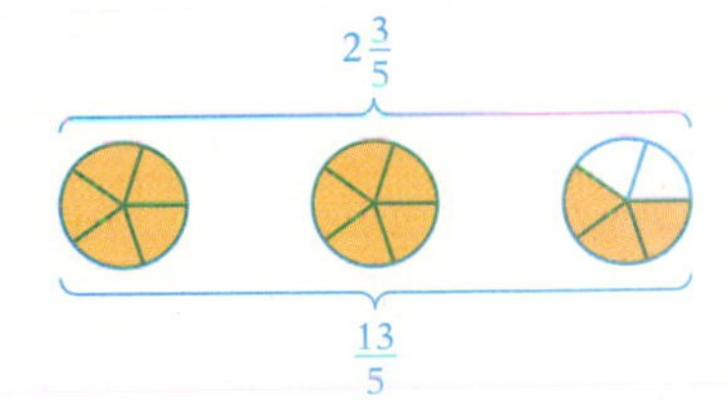

$$2\frac{3}{5} = \frac{13}{5}$$

An improper fraction can be written as a mixed number or a whole number.

Point of Interest

Archimedes (c. 287–212 B.C.) is the person who calculated that $\pi \approx 3\frac{1}{7}$. He actually showed that $3\frac{10}{71} < \pi < 3\frac{1}{7}$. The approximation $3\frac{10}{71}$ is more accurate but more difficult to use.

HOW TO • 1 Write $\frac{13}{5}$ as a mixed number.

Divide the numerator by the denominator.

$$\begin{array}{r} 2 \\ 5\overline{)\,13} \\ -10 \\ \hline 3 \end{array}$$

To write the fractional part of the mixed number, write the remainder over the divisor.

$$\begin{array}{r} 2\frac{3}{5} \\ 5\overline{)\,13} \\ -10 \\ \hline 3 \end{array}$$

Write the answer.

$$\frac{13}{5} = 2\frac{3}{5}$$

To write a mixed number as an improper fraction, multiply the denominator of the fractional part by the whole-number part. The sum of this product and the numerator of the fractional part is the numerator of the improper fraction. The denominator remains the same.

HOW TO • 2 Write $7\frac{3}{8}$ as an improper fraction.

$$7\frac{3}{8} = \frac{(8 \times 7) + 3}{8} = \frac{56 + 3}{8} = \frac{59}{8} \qquad 7\frac{3}{8} = \frac{59}{8}$$

EXAMPLE • 3

Write $\frac{21}{4}$ as a mixed number.

Solution

$$\begin{array}{r} 5 \\ 4\overline{)\,21} \\ -20 \\ \hline 1 \end{array} \qquad \frac{21}{4} = 5\frac{1}{4}$$

YOU TRY IT • 3

Write $\frac{22}{5}$ as a mixed number.

Your solution

EXAMPLE • 4

Write $\frac{18}{6}$ as a whole number.

Solution $\frac{18}{6} = 18 \div 6 = 3$

YOU TRY IT • 4

Write $\frac{28}{7}$ as a whole number.

Your solution

EXAMPLE • 5

Write $21\frac{3}{4}$ as an improper fraction.

Solution $21\frac{3}{4} = \frac{84 + 3}{4} = \frac{87}{4}$

YOU TRY IT • 5

Write $14\frac{5}{8}$ as an improper fraction.

Your solution

Solutions on p. S4

2.2 EXERCISES

OBJECTIVE A To write a fraction that represents part of a whole

For Exercises 1 to 4, identify the fraction as a proper fraction, an improper fraction, or a mixed number.

1. $\frac{12}{7}$ **2.** $5\frac{2}{11}$ **3.** $\frac{29}{40}$ **4.** $\frac{19}{13}$

For Exercises 5 to 8, express the shaded portion of the circle as a fraction.

5. 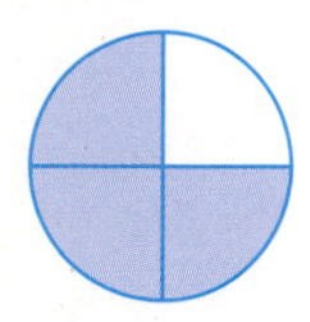

6. 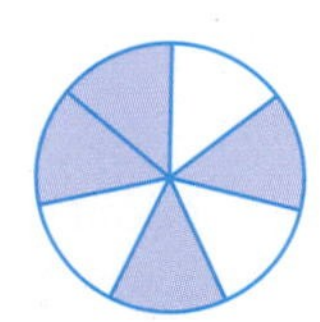

7. 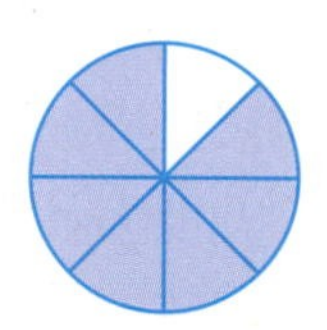

8. 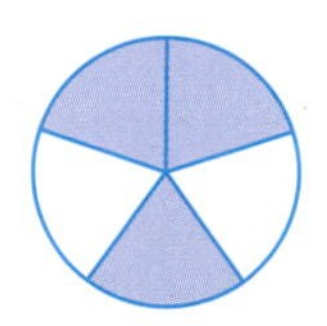

For Exercises 9 to 14, express the shaded portion of the circles as a mixed number.

9. 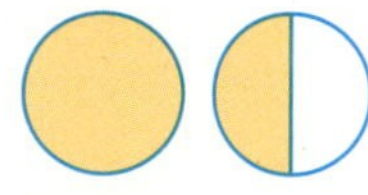

10. 

11. 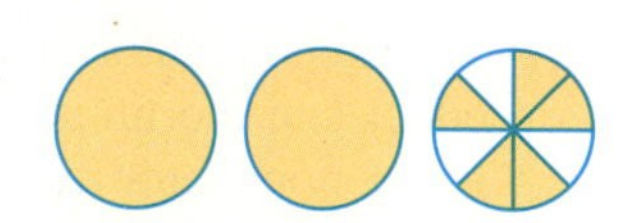

12. 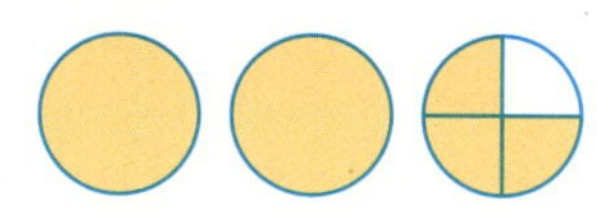

13. 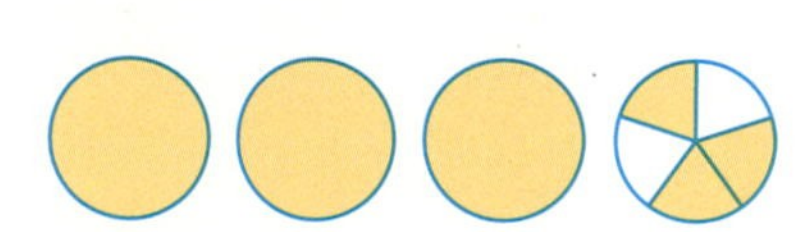

14. 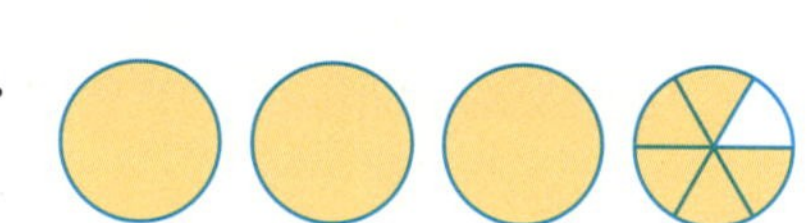

For Exercises 15 to 20, express the shaded portion of the circles as an improper fraction.

15. 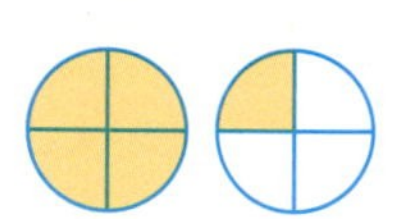

16. 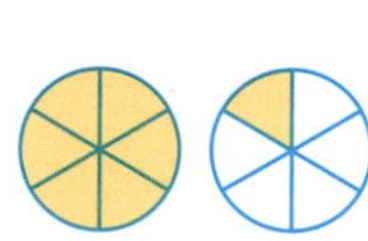

17. 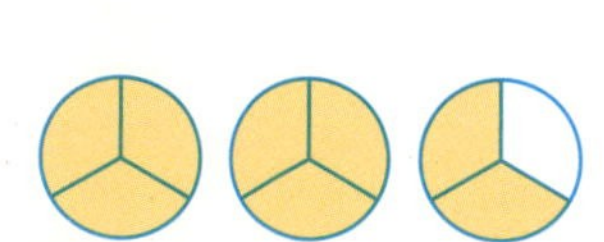

18. 

19. 

20. 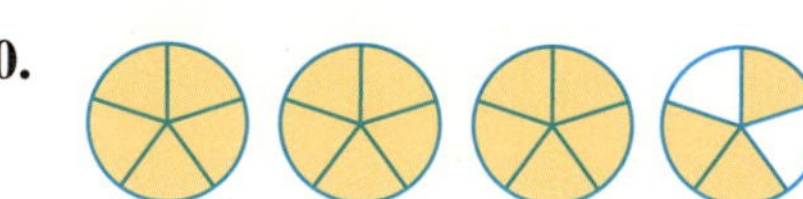

21. Shade $1\frac{2}{5}$ of 

22. Shade $1\frac{3}{4}$ of 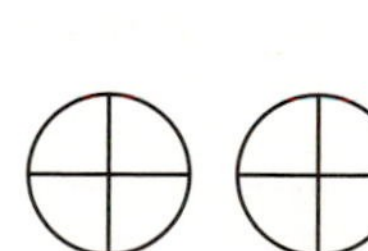

23. Shade $\frac{6}{5}$ of

24. Shade $\frac{7}{3}$ of 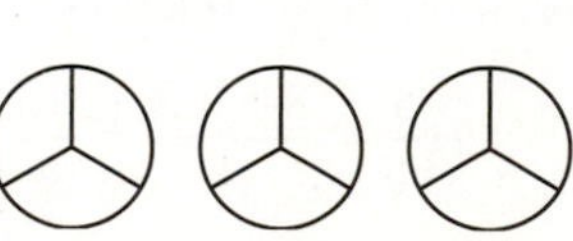

25. True or false? The fractional part of a mixed number is an improper fraction.

OBJECTIVE B To write an improper fraction as a mixed number or a whole number, and a mixed number as an improper fraction

For Exercises 26 to 49, write the improper fraction as a mixed number or a whole number.

26. $\frac{11}{4}$ **27.** $\frac{16}{3}$ **28.** $\frac{20}{4}$ **29.** $\frac{18}{9}$ **30.** $\frac{9}{8}$ **31.** $\frac{13}{4}$

32. $\frac{23}{10}$ **33.** $\frac{29}{2}$ **34.** $\frac{48}{16}$ **35.** $\frac{51}{3}$ **36.** $\frac{8}{7}$ **37.** $\frac{16}{9}$

38. $\frac{7}{3}$ **39.** $\frac{9}{5}$ **40.** $\frac{16}{1}$ **41.** $\frac{23}{1}$ **42.** $\frac{17}{8}$ **43.** $\frac{31}{16}$

44. $\frac{12}{5}$ **45.** $\frac{19}{3}$ **46.** $\frac{9}{9}$ **47.** $\frac{40}{8}$ **48.** $\frac{72}{8}$ **49.** $\frac{3}{3}$

For Exercises 50 to 73, write the mixed number as an improper fraction.

50. $2\frac{1}{3}$ **51.** $4\frac{2}{3}$ **52.** $6\frac{1}{2}$ **53.** $8\frac{2}{3}$ **54.** $6\frac{5}{6}$ **55.** $7\frac{3}{8}$

56. $9\frac{1}{4}$ **57.** $6\frac{1}{4}$ **58.** $10\frac{1}{2}$ **59.** $15\frac{1}{8}$ **60.** $8\frac{1}{9}$ **61.** $3\frac{5}{12}$

62. $5\frac{3}{11}$ **63.** $3\frac{7}{9}$ **64.** $2\frac{5}{8}$ **65.** $12\frac{2}{3}$ **66.** $1\frac{5}{8}$ **67.** $5\frac{3}{7}$

68. $11\frac{1}{9}$ **69.** $12\frac{3}{5}$ **70.** $3\frac{3}{8}$ **71.** $4\frac{5}{9}$ **72.** $6\frac{7}{13}$ **73.** $8\frac{5}{14}$

74. True or false? If an improper fraction is equivalent to 1, then the numerator and the denominator are the same number.

Applying the Concepts

75. Name three situations in which fractions are used. Provide an example of a fraction that is used in each situation.

SECTION

2.3 Writing Equivalent Fractions

OBJECTIVE A To find equivalent fractions by raising to higher terms

Equal fractions with different denominators are called **equivalent fractions.**

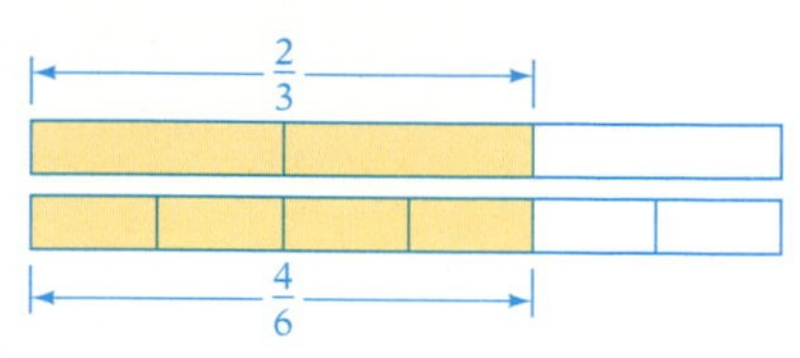

$\frac{4}{6}$ is equivalent to $\frac{2}{3}$.

Remember that the Multiplication Property of One states that the product of a number and one is the number. This is true for fractions as well as whole numbers. This property can be used to write equivalent fractions.

$$\frac{2}{3} \times 1 = \frac{2}{3} \times \frac{1}{1} = \frac{2 \cdot 1}{3 \cdot 1} = \frac{2}{3}$$

$$\frac{2}{3} \times 1 = \frac{2}{3} \times \boxed{\frac{2}{2}} = \frac{2 \cdot 2}{3 \cdot 2} = \frac{4}{6} \qquad \frac{4}{6} \text{ is equivalent to } \frac{2}{3}.$$

$$\frac{2}{3} \times 1 = \frac{2}{3} \times \boxed{\frac{4}{4}} = \frac{2 \cdot 4}{3 \cdot 4} = \frac{8}{12} \qquad \frac{8}{12} \text{ is equivalent to } \frac{2}{3}.$$

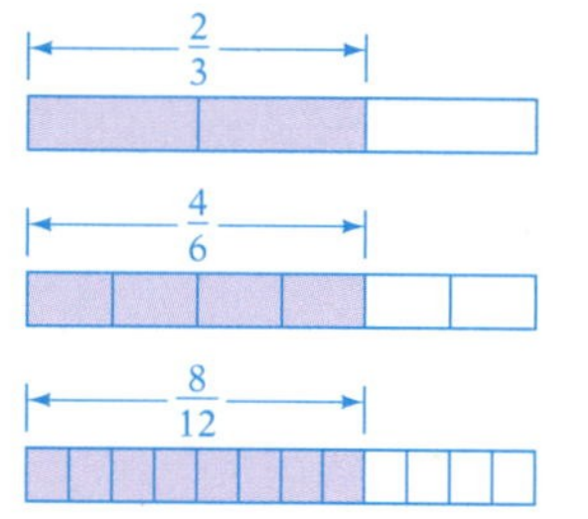

$\frac{2}{3}$ was rewritten as the equivalent fractions $\frac{4}{6}$ and $\frac{8}{12}$.

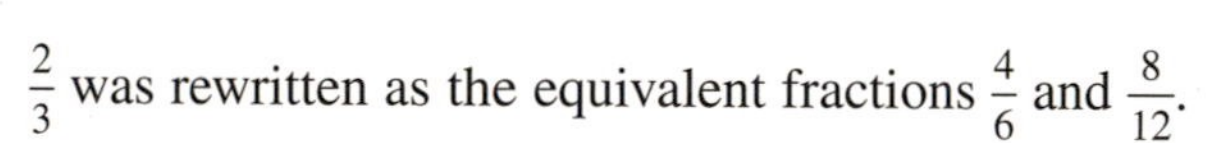

HOW TO • 1 Write a fraction that is equivalent to $\frac{5}{8}$ and has a denominator of 32.

$32 \div 8 = 4$
• Divide the larger denominator by the smaller.

$\frac{5}{8} = \frac{5 \cdot 4}{8 \cdot 4} = \frac{20}{32}$
• Multiply the numerator and denominator of the given fraction by the quotient (4).

$\frac{20}{32}$ is equivalent to $\frac{5}{8}$.

EXAMPLE • 1

Write $\frac{2}{3}$ as an equivalent fraction that has a denominator of 42.

Solution $42 \div 3 = 14 \quad \frac{2}{3} = \frac{2 \cdot 14}{3 \cdot 14} = \frac{28}{42}$

$\frac{28}{42}$ is equivalent to $\frac{2}{3}$.

YOU TRY IT • 1

Write $\frac{3}{5}$ as an equivalent fraction that has a denominator of 45.

Your solution

EXAMPLE • 2

Write 4 as a fraction that has a denominator of 12.

Solution Write 4 as $\frac{4}{1}$.

$12 \div 1 = 12 \quad 4 = \frac{4 \cdot 12}{1 \cdot 12} = \frac{48}{12}$

$\frac{48}{12}$ is equivalent to 4.

YOU TRY IT • 2

Write 6 as a fraction that has a denominator of 18.

Your solution

Solutions on p. S4

OBJECTIVE B To write a fraction in simplest form

Writing the **simplest form of a fraction** means writing it so that the numerator and denominator have no common factors other than 1.

The fractions $\frac{4}{6}$ and $\frac{2}{3}$ are equivalent fractions.

$\frac{4}{6}$ has been written in simplest form as $\frac{2}{3}$.

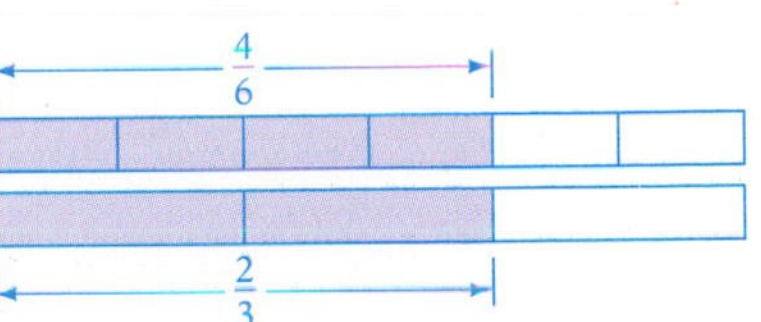

The Multiplication Property of One can be used to write fractions in simplest form. Write the numerator and denominator of the given fraction as a product of factors. Write factors common to both the numerator and denominator as an improper fraction equivalent to 1.

$$\frac{4}{6} = \frac{2 \cdot 2}{2 \cdot 3} = \frac{\boxed{2} \cdot 2}{\boxed{2} \cdot 3} = \boxed{\frac{2}{2}} \cdot \frac{2}{3} = 1 \cdot \frac{2}{3} = \frac{2}{3}$$

The process of eliminating common factors is displayed with slashes through the common factors as shown at the right.

$$\frac{4}{6} = \frac{\overset{1}{\cancel{2}} \cdot 2}{\underset{1}{\cancel{2}} \cdot 3} = \frac{2}{3}$$

To write a fraction in simplest form, eliminate the common factors.

$$\frac{18}{30} = \frac{\overset{1}{\cancel{2}} \cdot \overset{1}{\cancel{3}} \cdot 3}{\underset{1}{\cancel{2}} \cdot \underset{1}{\cancel{3}} \cdot 5} = \frac{3}{5}$$

An improper fraction can be changed to a mixed number.

$$\frac{22}{6} = \frac{\overset{1}{\cancel{2}} \cdot 11}{\underset{1}{\cancel{2}} \cdot 3} = \frac{11}{3} = 3\frac{2}{3}$$

EXAMPLE 3

Write $\frac{15}{40}$ in simplest form.

Solution

$$\frac{15}{40} = \frac{3 \cdot \overset{1}{\cancel{5}}}{2 \cdot 2 \cdot 2 \cdot \underset{1}{\cancel{5}}} = \frac{3}{8}$$

YOU TRY IT 3

Write $\frac{16}{24}$ in simplest form.

Your solution

EXAMPLE 4

Write $\frac{6}{42}$ in simplest form.

Solution

$$\frac{6}{42} = \frac{\overset{1}{\cancel{2}} \cdot \overset{1}{\cancel{3}}}{\underset{1}{\cancel{2}} \cdot \underset{1}{\cancel{3}} \cdot 7} = \frac{1}{7}$$

YOU TRY IT 4

Write $\frac{8}{56}$ in simplest form.

Your solution

EXAMPLE 5

Write $\frac{8}{9}$ in simplest form.

Solution $\frac{8}{9} = \frac{2 \cdot 2 \cdot 2}{3 \cdot 3} = \frac{8}{9}$

$\frac{8}{9}$ is already in simplest form because there are no common factors in the numerator and denominator.

YOU TRY IT 5

Write $\frac{15}{32}$ in simplest form.

Your solution

EXAMPLE 6

Write $\frac{30}{12}$ in simplest form.

Solution

$$\frac{30}{12} = \frac{\overset{1}{\cancel{2}} \cdot \overset{1}{\cancel{3}} \cdot 5}{\underset{1}{\cancel{2}} \cdot 2 \cdot \underset{1}{\cancel{3}}} = \frac{5}{2} = 2\frac{1}{2}$$

YOU TRY IT 6

Write $\frac{48}{36}$ in simplest form.

Your solution

Solutions on p. S4

2.3 EXERCISES

OBJECTIVE A To find equivalent fractions by raising to higher terms

For Exercises 1 to 35, write an equivalent fraction with the given denominator.

1. $\frac{1}{2} = \frac{}{10}$ **2.** $\frac{1}{4} = \frac{}{16}$ **3.** $\frac{3}{16} = \frac{}{48}$ **4.** $\frac{5}{9} = \frac{}{81}$ **5.** $\frac{3}{8} = \frac{}{32}$

6. $\frac{7}{11} = \frac{}{33}$ **7.** $\frac{3}{17} = \frac{}{51}$ **8.** $\frac{7}{10} = \frac{}{90}$ **9.** $\frac{3}{4} = \frac{}{16}$ **10.** $\frac{5}{8} = \frac{}{32}$

11. $3 = \frac{}{9}$ **12.** $5 = \frac{}{25}$ **13.** $\frac{1}{3} = \frac{}{60}$ **14.** $\frac{1}{16} = \frac{}{48}$ **15.** $\frac{11}{15} = \frac{}{60}$

16. $\frac{3}{50} = \frac{}{300}$ **17.** $\frac{2}{3} = \frac{}{18}$ **18.** $\frac{5}{9} = \frac{}{36}$ **19.** $\frac{5}{7} = \frac{}{49}$ **20.** $\frac{7}{8} = \frac{}{32}$

21. $\frac{5}{9} = \frac{}{18}$ **22.** $\frac{11}{12} = \frac{}{36}$ **23.** $7 = \frac{}{3}$ **24.** $9 = \frac{}{4}$ **25.** $\frac{7}{9} = \frac{}{45}$

26. $\frac{5}{6} = \frac{}{42}$ **27.** $\frac{15}{16} = \frac{}{64}$ **28.** $\frac{11}{18} = \frac{}{54}$ **29.** $\frac{3}{14} = \frac{}{98}$ **30.** $\frac{5}{6} = \frac{}{144}$

31. $\frac{5}{8} = \frac{}{48}$ **32.** $\frac{7}{12} = \frac{}{96}$ **33.** $\frac{5}{14} = \frac{}{42}$ **34.** $\frac{2}{3} = \frac{}{42}$ **35.** $\frac{17}{24} = \frac{}{144}$

36. When you multiply the numerator and denominator of a fraction by the same number, you are actually multiplying the fraction by the number _____.

OBJECTIVE B To write a fraction in simplest form

For Exercises 37 to 71, write the fraction in simplest form.

37. $\frac{4}{12}$ **38.** $\frac{8}{22}$ **39.** $\frac{22}{44}$ **40.** $\frac{2}{14}$ **41.** $\frac{2}{12}$

42. $\frac{50}{75}$ 43. $\frac{40}{36}$ 44. $\frac{12}{8}$ 45. $\frac{0}{30}$ 46. $\frac{10}{10}$

47. $\frac{9}{22}$ 48. $\frac{14}{35}$ 49. $\frac{75}{25}$ 50. $\frac{8}{60}$ 51. $\frac{16}{84}$

52. $\frac{20}{44}$ 53. $\frac{12}{35}$ 54. $\frac{8}{36}$ 55. $\frac{28}{44}$ 56. $\frac{12}{16}$

57. $\frac{16}{12}$ 58. $\frac{24}{18}$ 59. $\frac{24}{40}$ 60. $\frac{44}{60}$ 61. $\frac{8}{88}$

62. $\frac{9}{90}$ 63. $\frac{144}{36}$ 64. $\frac{140}{297}$ 65. $\frac{48}{144}$ 66. $\frac{32}{120}$

67. $\frac{60}{100}$ 68. $\frac{33}{110}$ 69. $\frac{36}{16}$ 70. $\frac{80}{45}$ 71. $\frac{32}{160}$

72. Suppose the denominator of a fraction is a multiple of the numerator. When the fraction is written in simplest form, what number is its numerator?

Applying the Concepts

73. Make a list of five different fractions that are equivalent to $\frac{2}{3}$.

74. Show that $\frac{15}{24} = \frac{5}{8}$ by using a diagram.

75. **Births** Generally, the Centers for Disease Control and Prevention (CDC) studies show that there are more boys born than girls. In a recent group of 500 births, there were 255 boys born.
   a. What fraction of the babies born were boys?
   b. What fraction of the babies born were girls?

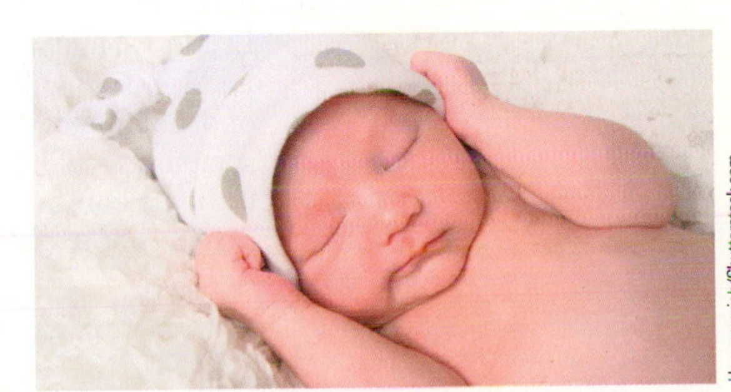

SECTION

2.4 Addition of Fractions and Mixed Numbers

OBJECTIVE A To add fractions with the same denominator

Fractions with the same denominator are added by adding the numerators and placing the sum over the common denominator. After adding, write the sum in simplest form.

HOW TO • 1 Add: $\frac{2}{7}+\frac{4}{7}$

$$\begin{array}{r} \frac{2}{7} \\ +\frac{4}{7} \\ \hline \frac{6}{7} \end{array}$$

• Add the numerators and place the sum over the common denominator.

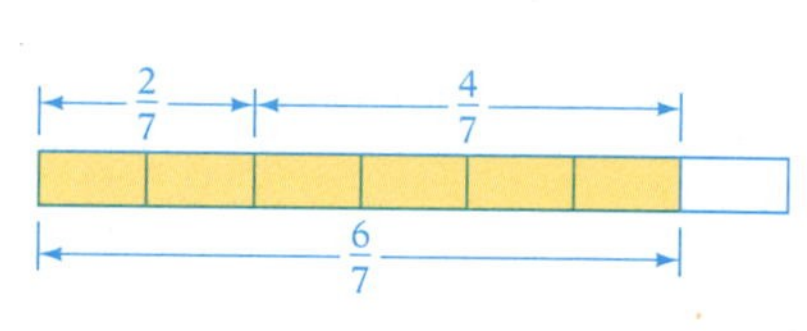

$$\frac{2}{7}+\frac{4}{7}=\frac{2+4}{7}=\frac{6}{7}$$

EXAMPLE • 1

Add: $\frac{5}{12}+\frac{11}{12}$

Solution

$$\begin{array}{r} \frac{5}{12} \\ +\frac{11}{12} \\ \hline \frac{16}{12}=\frac{4}{3}=1\frac{1}{3} \end{array}$$

• The denominators are the same. Add the numerators. Place the sum over the common denominator.

YOU TRY IT • 1

Add: $\frac{3}{8}+\frac{7}{8}$

Your solution

Solution on p. S5

OBJECTIVE B To add fractions with different denominators

Integrating Technology

Some scientific calculators have a fraction key, a b/c. It is used to perform operations on fractions. To use this key to simplify the expression at the right, enter

1 a b/c 2 + 1 a b/c 3 =

$\frac{1}{2}$ $\frac{1}{3}$

To add fractions with different denominators, first rewrite the fractions as equivalent fractions with a common denominator. The common denominator is the LCM of the denominators of the fractions.

HOW TO • 2 Find the total of $\frac{1}{2}$ and $\frac{1}{3}$.

The common denominator is the LCM of 2 and 3. The LCM = 6. The LCM of denominators is sometimes called the **least common denominator (LCD).**

Write equivalent fractions using the LCM.

$$\begin{array}{r} \frac{1}{2}=\frac{3}{6} \\ +\frac{1}{3}=\frac{2}{6} \\ \hline \end{array}$$

Add the fractions.

$$\begin{array}{r} \frac{1}{2}=\frac{3}{6} \\ +\frac{1}{3}=\frac{2}{6} \\ \hline \frac{5}{6} \end{array}$$

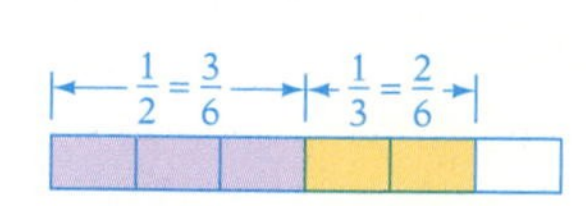

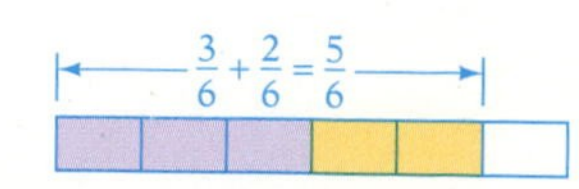

EXAMPLE 2

Find $\frac{7}{12}$ more than $\frac{3}{8}$.

Solution

$$\begin{aligned} \frac{3}{8} &= \frac{9}{24} \\ +\frac{7}{12} &= \frac{14}{24} \\ \hline & \frac{23}{24} \end{aligned}$$

• The LCM of 8 and 12 is 24.

YOU TRY IT 2

Find the sum of $\frac{5}{12}$ and $\frac{9}{16}$.

Your solution

EXAMPLE 3

Add: $\frac{5}{8} + \frac{7}{9}$

Solution

$$\begin{aligned} \frac{5}{8} &= \frac{45}{72} \\ +\frac{7}{9} &= \frac{56}{72} \\ \hline & \frac{101}{72} = 1\frac{29}{72} \end{aligned}$$

YOU TRY IT 3

Add: $\frac{7}{8} + \frac{11}{15}$

Your solution

EXAMPLE 4

Add: $\frac{2}{3} + \frac{3}{5} + \frac{5}{6}$

Solution

$$\begin{aligned} \frac{2}{3} &= \frac{20}{30} \\ \frac{3}{5} &= \frac{18}{30} \\ +\frac{5}{6} &= \frac{25}{30} \\ \hline & \frac{63}{30} = 2\frac{3}{30} = 2\frac{1}{10} \end{aligned}$$

• The LCM of 3, 5, and 6 is 30.

YOU TRY IT 4

Add: $\frac{3}{4} + \frac{4}{5} + \frac{5}{8}$

Your solution

Solutions on p. S5

OBJECTIVE C To add whole numbers, mixed numbers, and fractions

Take Note

The procedure at the right illustrates why $2 + \frac{2}{3} = 2\frac{2}{3}$. You do not need to show these steps when adding a whole number and a fraction. Here are two more examples:

$7 + \frac{1}{5} = 7\frac{1}{5}$

$6 + \frac{3}{4} = 6\frac{3}{4}$

The sum of a whole number and a fraction is a mixed number.

HOW TO 3 Add: $2 + \frac{2}{3}$

$$\boxed{2} + \frac{2}{3} = \boxed{\frac{6}{3}} + \frac{2}{3} = \frac{8}{3} = 2\frac{2}{3}$$

To add a whole number and a mixed number, write the fraction and then add the whole numbers.

HOW TO 4 Add: $7\frac{2}{5} + 4$

Write the fraction.

$$\begin{array}{r} 7\frac{2}{5} \\ +\ 4 \\ \hline \frac{2}{5} \end{array}$$

Add the whole numbers.

$$\begin{array}{r} 7\frac{2}{5} \\ +\ 4 \\ \hline 11\frac{2}{5} \end{array}$$

Integrating Technology

Use the fraction key on a calculator to enter mixed numbers. For the example at the right, enter

5 a b/c 4 a b/c 9 + → $5\frac{4}{9}$

6 a b/c 14 a b/c 15 = → $6\frac{14}{15}$

To add two mixed numbers, add the fractional parts and then add the whole numbers. Remember to reduce the sum to simplest form.

HOW TO • 5 What is $6\frac{14}{15}$ added to $5\frac{4}{9}$?

The LCM of 9 and 15 is 45.

Add the fractional parts.

$$\begin{aligned} 5\frac{4}{9} &= 5\frac{20}{45} \\ +\,6\frac{14}{15} &= 6\frac{42}{45} \\ \hline &\frac{62}{45} \end{aligned}$$

Add the whole numbers.

$$\begin{aligned} 5\frac{4}{9} &= 5\frac{20}{45} \\ +\,6\frac{14}{15} &= 6\frac{42}{45} \\ \hline \end{aligned}$$

$$11\frac{62}{45} = 11 + 1\frac{17}{45} = 12\frac{17}{45}$$

EXAMPLE • 5

Add: $5 + \frac{3}{8}$

Solution $5 + \frac{3}{8} = 5\frac{3}{8}$

YOU TRY IT • 5

What is 7 added to $\frac{6}{11}$?

Your solution

EXAMPLE • 6

Find 17 increased by $3\frac{3}{8}$.

Solution $17 + 3\frac{3}{8} = 20\frac{3}{8}$

YOU TRY IT • 6

Find the sum of 29 and $17\frac{5}{12}$.

Your solution

EXAMPLE • 7

Add: $5\frac{2}{3} + 11\frac{5}{6} + 12\frac{7}{9}$

Solution

$$\begin{aligned} 5\frac{2}{3} &= 5\frac{12}{18} \quad \text{• LCM = 18} \\ 11\frac{5}{6} &= 11\frac{15}{18} \\ +\,12\frac{7}{9} &= 12\frac{14}{18} \\ \hline &28\frac{41}{18} = 30\frac{5}{18} \end{aligned}$$

YOU TRY IT • 7

Add: $7\frac{4}{5} + 6\frac{7}{10} + 13\frac{11}{15}$

Your solution

EXAMPLE • 8

Add: $11\frac{5}{8} + 7\frac{5}{9} + 8\frac{7}{15}$

Solution

$$\begin{aligned} 11\frac{5}{8} &= 11\frac{225}{360} \quad \text{• LCM = 360} \\ 7\frac{5}{9} &= 7\frac{200}{360} \\ +\,8\frac{7}{15} &= 8\frac{168}{360} \\ \hline &26\frac{593}{360} = 27\frac{233}{360} \end{aligned}$$

YOU TRY IT • 8

Add: $9\frac{3}{8} + 17\frac{7}{12} + 10\frac{14}{15}$

Your solution

Solutions on p. S5

OBJECTIVE D To solve application problems

EXAMPLE • 9

A certified nurse assistant records the monthly weight gain of patients in a nursing home. If a patient gains $1\frac{1}{2}$ pounds in January, $\frac{3}{4}$ pound in February, and $2\frac{1}{4}$ pounds in March, find the patient's total weight gain.

Strategy

To find the total weight gain for the three months, add the three amounts gained $\left(1\frac{1}{2}, \frac{3}{4}, \text{ and } 2\frac{1}{4}\right)$.

Solution

$$\begin{aligned} 1\frac{1}{2} &= 1\frac{2}{4} \\ \frac{3}{4} &= \frac{3}{4} \\ +\,2\frac{1}{4} &= 2\frac{1}{4} \\ \hline & \quad 3\frac{6}{4} = 4\frac{2}{4} = 4\frac{1}{2} \end{aligned}$$

The patient's total weight gain was $4\frac{1}{2}$ pounds.

YOU TRY IT • 9

On Monday, you spent $4\frac{1}{2}$ hours in class, $3\frac{3}{4}$ hours studying, and $1\frac{1}{3}$ hours driving. Find the total number of hours spent on these three activities.

Your strategy

Your solution

EXAMPLE • 10

Barbara Walsh worked 4 hours, $2\frac{1}{3}$ hours, and $5\frac{2}{3}$ hours this week at a part-time job. Barbara is paid \$9 an hour. How much did she earn this week?

Strategy

To find how much Barbara earned:

- Find the total number of hours worked.
- Multiply the total number of hours worked by the hourly wage (9).

Solution

$$\begin{array}{r} 4 \\ 2\frac{1}{3} \\ +\,5\frac{2}{3} \\ \hline 11\frac{3}{3} \end{array} = 12 \text{ hours worked} \qquad \begin{array}{r} 12 \\ \times\ 9 \\ \hline 108 \end{array}$$

Barbara earned \$108 this week.

YOU TRY IT • 10

Jeff Sapone, an x-ray technician, worked $1\frac{2}{3}$ hours of overtime on Monday, $3\frac{1}{3}$ hours of overtime on Tuesday, and 2 hours of overtime on Wednesday. At an overtime hourly rate of \$36, find Jeff's overtime pay for these 3 days.

Your strategy

Your solution

Solutions on p. S5

2.4 EXERCISES

OBJECTIVE A To add fractions with the same denominator

For Exercises 1 to 16, add.

1. $\frac{2}{7} + \frac{1}{7}$

2. $\frac{3}{11} + \frac{5}{11}$

3. $\frac{1}{2} + \frac{1}{2}$

4. $\frac{1}{3} + \frac{2}{3}$

5. $\frac{8}{11} + \frac{7}{11}$

6. $\frac{9}{13} + \frac{7}{13}$

7. $\frac{8}{5} + \frac{9}{5}$

8. $\frac{5}{3} + \frac{7}{3}$

9. $\frac{3}{5} + \frac{8}{5} + \frac{3}{5}$

10. $\frac{3}{8} + \frac{5}{8} + \frac{7}{8}$

11. $\frac{3}{4} + \frac{1}{4} + \frac{5}{4}$

12. $\frac{2}{7} + \frac{4}{7} + \frac{5}{7}$

13. $\frac{3}{8} + \frac{7}{8} + \frac{1}{8}$

14. $\frac{5}{12} + \frac{7}{12} + \frac{1}{12}$

15. $\frac{4}{15} + \frac{7}{15} + \frac{11}{15}$

16. $\frac{5}{7} + \frac{4}{7} + \frac{5}{7}$

17. Find the sum of $\frac{5}{12}$, $\frac{1}{12}$, and $\frac{11}{12}$.

18. Find the total of $\frac{5}{8}$, $\frac{3}{8}$, and $\frac{7}{8}$.

For Exercises 19 to 22, each statement concerns a pair of fractions that have the same denominator. State whether the sum of the fractions is a proper fraction, the number 1, a mixed number, or a whole number other than 1.

19. The sum of the numerators is a multiple of the denominator.

20. The sum of the numerators is one more than the denominator.

21. The sum of the numerators is the denominator.

22. The sum of the numerators is smaller than the denominator.

OBJECTIVE B To add fractions with different denominators

For Exercises 23 to 42, add.

23. $\frac{1}{2} + \frac{2}{3}$

24. $\frac{2}{3} + \frac{1}{4}$

25. $\frac{3}{14} + \frac{5}{7}$

26. $\frac{3}{5} + \frac{7}{10}$

27. $\frac{8}{15} + \frac{7}{20}$

28. $\frac{1}{6} + \frac{7}{9}$

29. $\frac{3}{8} + \frac{9}{14}$

30. $\frac{5}{12} + \frac{5}{16}$

31. $\frac{3}{20} + \frac{7}{30}$

32. $\frac{5}{12} + \frac{7}{30}$

33. $\frac{1}{3} + \frac{5}{6} + \frac{7}{9}$

34. $\frac{2}{3} + \frac{5}{6} + \frac{7}{12}$

35. $\frac{5}{6} + \frac{1}{12} + \frac{5}{16}$

36. $\frac{2}{9} + \frac{7}{15} + \frac{4}{21}$

37. $\frac{2}{3} + \frac{1}{5} + \frac{7}{12}$

38. $\frac{3}{4} + \frac{4}{5} + \frac{7}{12}$

39. $\frac{2}{3} + \frac{3}{5} + \frac{7}{8}$

40. $\frac{3}{10} + \frac{14}{15} + \frac{9}{25}$

41. $\frac{2}{3} + \frac{5}{8} + \frac{7}{9}$

42. $\frac{1}{3} + \frac{2}{9} + \frac{7}{8}$

43. What is $\frac{3}{8}$ added to $\frac{3}{5}$?

44. What is $\frac{5}{9}$ added to $\frac{7}{12}$?

45. Find the sum of $\frac{3}{8}$, $\frac{5}{6}$, and $\frac{7}{12}$.

46. Find the total of $\frac{1}{2}$, $\frac{5}{8}$, and $\frac{7}{9}$.

47. Which statement describes a pair of fractions for which the least common denominator is the product of the denominators?
(i) The denominator of one fraction is a multiple of the denominator of the second fraction.
(ii) The denominators of the two fractions have no common factors.

OBJECTIVE C To add whole numbers, mixed numbers, and fractions

For Exercises 48 to 69, add.

48. $2\frac{2}{5} + 3\frac{3}{10}$

49. $4\frac{1}{2} + 5\frac{7}{12}$

50. $3\frac{3}{8} + 2\frac{5}{16}$

51. $4 + 5\frac{2}{7}$

52. $6\frac{8}{9} + 12$

53. $7\frac{5}{12} + 2\frac{9}{16}$

54. $9\frac{1}{2} + 3\frac{3}{11}$

55. $6 + 2\frac{3}{13}$

56. $8\frac{21}{40} + 6$

57. $8\frac{29}{30} + 7\frac{11}{40}$

58. $17\frac{5}{16} + 3\frac{11}{24}$

59. $17\frac{3}{8} + 7\frac{7}{20}$

60. $14\frac{7}{12} + 29\frac{13}{21}$

61. $5\frac{7}{8} + 27\frac{5}{12}$

62. $7\frac{5}{6} + 3\frac{5}{9}$

63. $7\frac{5}{9} + 2\frac{7}{12}$

64. $3\frac{1}{2} + 2\frac{3}{4} + 1\frac{5}{6}$

65. $2\frac{1}{2} + 3\frac{2}{3} + 4\frac{1}{4}$

66. $3\frac{1}{3} + 7\frac{1}{5} + 2\frac{1}{7}$

67. $3\frac{1}{2} + 3\frac{1}{5} + 8\frac{1}{9}$

68. $6\frac{5}{9} + 6\frac{5}{12} + 2\frac{5}{18}$

69. $2\frac{3}{8} + 4\frac{7}{12} + 3\frac{5}{16}$

70. Find the sum of $2\frac{4}{9}$ and $5\frac{7}{12}$.

71. Find $5\frac{5}{6}$ more than $3\frac{3}{8}$.

72. What is $4\frac{3}{4}$ added to $9\frac{1}{3}$?

73. What is $4\frac{8}{9}$ added to $9\frac{1}{6}$?

74. Find the total of 2, $4\frac{5}{8}$, and $2\frac{2}{9}$.

75. Find the total of $1\frac{5}{8}$, 3, and $7\frac{7}{24}$.

For Exercises 76 and 77, state whether the given sum can be a whole number. Answer *yes* or *no*.

76. The sum of two mixed numbers

77. The sum of a mixed number and a whole number

OBJECTIVE D To solve application problems

78. **Fluid Intake** A patient drinks $\frac{1}{4}$ cup of coffee, $\frac{3}{4}$ cup of fruit juice, and $\frac{1}{2}$ cup of water for breakfast. What is the patient's total fluid intake at breakfast?

79. **Weight** A pediatric nurse records the birth weight of a newborn as $7\frac{1}{4}$ pounds. On the first and second monthly visits, the infant gains $1\frac{3}{8}$ pounds and $1\frac{5}{8}$ pounds. How much does the baby weigh at the end of two months?

80. **Weight** A certified nurse assistant weighs a patient who is admitted to a nursing facility. The patient's initial weight is 118 pounds, and she gains $1\frac{3}{4}$ pounds, $\frac{1}{2}$ pound, and $\frac{3}{4}$ pound during the next three months. What is the patient's total weight at the end of three months?

81. **Length** A new baby measured $20\frac{1}{2}$ inches at birth. The baby grew $\frac{7}{16}$ inch in August and $\frac{3}{4}$ inch in September. How long is the baby at its two-month checkup?

82. Wages You are working a part-time job that pays \$11 an hour. You worked 5, $3\frac{3}{4}$, $2\frac{1}{3}$, $1\frac{1}{4}$, and $7\frac{2}{3}$ hours during the last five days.
a. Find the total number of hours you worked during the last five days.
b. Find your total wages for the five days.

83. Wages Elizabeth is a physical therapist's assistant who worked 6 hours, $3\frac{1}{4}$ hours, and $7\frac{3}{4}$ hours at a part-time job. She earns \$24 an hour. Find her earnings for the week.

84. Wages Jeff, a hospital engineer who is responsible for maintaining the hospital generators, worked $2\frac{1}{2}$ hours of overtime on Monday, $4\frac{1}{4}$ hours of overtime on Thursday, and $3\frac{1}{4}$ hours of overtime on Friday. His hourly overtime rate is \$38. Find Jeff's overtime pay for the three days.

85. Clinicals As a dental assistant student, you spend many hours a week in clinicals. You spent $4\frac{2}{3}$ hours in clinicals on Monday, $2\frac{1}{2}$ hours on Wednesday, and $3\frac{3}{4}$ hours on Friday. Find the total number of hours spent in clinicals each week.

86. Nutrition A cook for the nursing home is preparing 100 yeast rolls for dinner. The dry ingredients needed include $23\frac{1}{2}$ cups of flour, $2\frac{1}{16}$ cups of sugar, and $\frac{1}{8}$ cup of salt. How many cups of dry ingredients are in this recipe?

87. Nutrition The hospital cafeteria staff is preparing 150 servings of chocolate cake for lunch. Dry ingredients include $1\frac{1}{4}$ cups of instant coffee, $18\frac{3}{4}$ cups of white sugar, $9\frac{1}{3}$ cups of all-purpose flour, and $\frac{1}{2}$ cup of confectioner's sugar. Find the total amount of dry ingredients needed for this recipe.

Mi. Ti./Shutterstock.com

Applying the Concepts

88. What is a unit fraction? Find the sum of the three largest unit fractions. Is there a smallest unit fraction? If so, write it down. If not, explain why.

89. A survey was conducted to determine people's favorite color from among blue, green, red, purple, and other. The surveyor claims that $\frac{1}{3}$ of the people responded blue, $\frac{1}{6}$ responded green, $\frac{1}{8}$ responded red, $\frac{1}{12}$ responded purple, and $\frac{2}{5}$ responded some other color. Is this possible? Explain your answer.

SECTION

2.5 Subtraction of Fractions and Mixed Numbers

OBJECTIVE A To subtract fractions with the same denominator

Fractions with the same denominator are subtracted by subtracting the numerators and placing the difference over the common denominator. After subtracting, write the fraction in simplest form.

HOW TO 1 Subtract: $\frac{5}{7} - \frac{3}{7}$

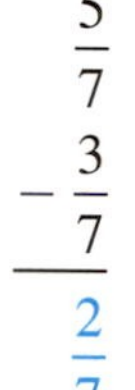

$$\begin{array}{r} \frac{5}{7} \\ -\frac{3}{7} \\ \hline \frac{2}{7} \end{array}$$

• Subtract the numerators and place the difference over the common denominator.

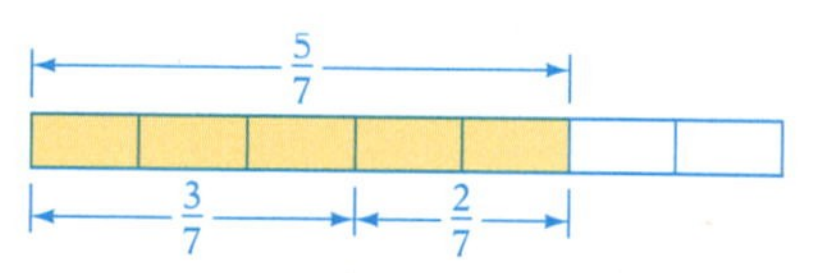

$$\frac{5}{7} - \frac{3}{7} = \frac{5-3}{7} = \frac{2}{7}$$

EXAMPLE 1

Find $\frac{17}{30}$ less $\frac{11}{30}$.

Solution

$$\begin{array}{r} \frac{17}{30} \\ -\frac{11}{30} \\ \hline \frac{6}{30} = \frac{1}{5} \end{array}$$

• The denominators are the same. Subtract the numerators. Place the difference over the common denominator.

YOU TRY IT 1

Subtract: $\frac{16}{27} - \frac{7}{27}$

Your solution

Solution on p. S5

OBJECTIVE B To subtract fractions with different denominators

To subtract fractions with different denominators, first rewrite the fractions as equivalent fractions with a common denominator. As with adding fractions, the common denominator is the LCM of the denominators of the fractions.

HOW TO 2 Subtract: $\frac{5}{6} - \frac{1}{4}$

The common denominator is the LCM of 6 and 4. The LCM = 12.

Write equivalent fractions using the LCM.

$$\begin{array}{r} \frac{5}{6} = \frac{10}{12} \\ -\frac{1}{4} = \frac{3}{12} \\ \hline \end{array}$$

Subtract the fractions.

$$\begin{array}{r} \frac{5}{6} = \frac{10}{12} \\ -\frac{1}{4} = \frac{3}{12} \\ \hline \frac{7}{12} \end{array}$$

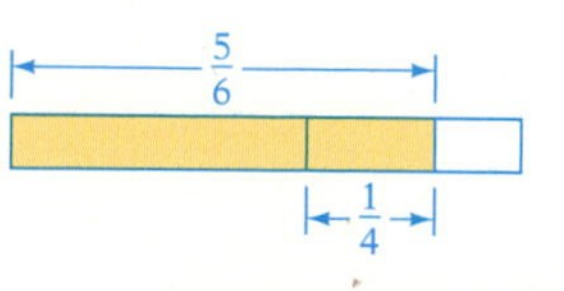

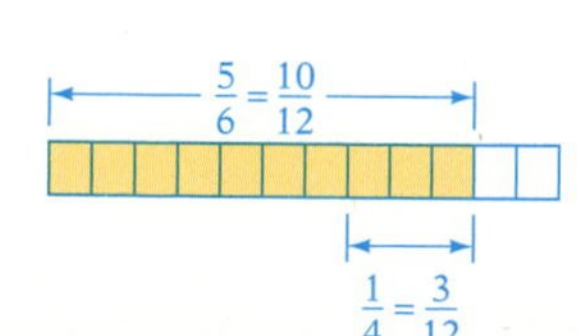

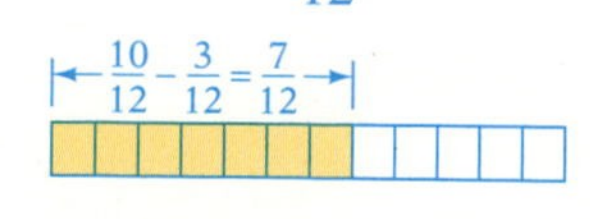

EXAMPLE • 2

Subtract: $\frac{11}{16} - \frac{5}{12}$

Solution

$$\begin{array}{r} \frac{11}{16} = \frac{33}{48} \\ -\frac{5}{12} = \frac{20}{48} \\ \hline \frac{13}{48} \end{array}$$

• LCM = 48

YOU TRY IT • 2

Subtract: $\frac{13}{18} - \frac{7}{24}$

Your solution

Solution on p. S5

OBJECTIVE C To subtract whole numbers, mixed numbers, and fractions

To subtract mixed numbers without borrowing, subtract the fractional parts and then subtract the whole numbers.

HOW TO • 3 Subtract: $5\frac{5}{6} - 2\frac{3}{4}$

Subtract the fractional parts.

$$\begin{array}{r} 5\frac{5}{6} = 5\frac{10}{12} \\ -2\frac{3}{4} = 2\frac{9}{12} \\ \hline \frac{1}{12} \end{array}$$

• The LCM of 6 and 4 is 12.

Subtract the whole numbers.

$$\begin{array}{r} 5\frac{5}{6} = 5\frac{10}{12} \\ -2\frac{3}{4} = 2\frac{9}{12} \\ \hline 3\frac{1}{12} \end{array}$$

Subtraction of mixed numbers sometimes involves borrowing.

HOW TO • 4 Subtract: $5 - 2\frac{5}{8}$

Borrow 1 from 5.

$$\begin{array}{r} 5 = \overset{4}{\cancel{5}}\ 1 \\ -2\frac{5}{8} = 2\frac{5}{8} \\ \hline \end{array}$$

Write 1 as a fraction so that the fractions have the same denominators.

$$\begin{array}{r} 5 = 4\frac{8}{8} \\ -2\frac{5}{8} = 2\frac{5}{8} \\ \hline \end{array}$$

Subtract the mixed numbers.

$$\begin{array}{r} 5 = 4\frac{8}{8} \\ -2\frac{5}{8} = 2\frac{5}{8} \\ \hline 2\frac{3}{8} \end{array}$$

HOW TO • 5 Subtract: $7\frac{1}{6} - 2\frac{5}{8}$

Write equivalent fractions using the LCM.

$$\begin{array}{r} 7\frac{1}{6} = 7\frac{4}{24} \\ -2\frac{5}{8} = 2\frac{15}{24} \\ \hline \end{array}$$

Borrow 1 from 7. Add the 1 to $\frac{4}{24}$. Write $1\frac{4}{24}$ as $\frac{28}{24}$.

$$\begin{array}{r} 7\frac{1}{6} = \overset{6}{\cancel{7}}1\frac{4}{24} = 6\frac{28}{24} \\ -2\frac{5}{8} = 2\frac{15}{24} = 2\frac{15}{24} \\ \hline \end{array}$$

Subtract the mixed numbers.

$$\begin{array}{r} 7\frac{1}{6} = 6\frac{28}{24} \\ -2\frac{5}{8} = 2\frac{15}{24} \\ \hline 4\frac{13}{24} \end{array}$$

EXAMPLE • 3

Subtract: $15\frac{7}{8} - 12\frac{2}{3}$

Solution

$$\begin{array}{r} 15\frac{7}{8} = 15\frac{21}{24} \\ -\ 12\frac{2}{3} = 12\frac{16}{24} \\ \hline 3\frac{5}{24} \end{array}$$

• LCM = 24

YOU TRY IT • 3

Subtract: $17\frac{5}{9} - 11\frac{5}{12}$

Your solution

EXAMPLE • 4

Subtract: $9 - 4\frac{3}{11}$

Solution

$$\begin{array}{r} 9 \ = 8\frac{11}{11} \\ -\ 4\frac{3}{11} = 4\frac{3}{11} \\ \hline 4\frac{8}{11} \end{array}$$

• LCM = 11

YOU TRY IT • 4

Subtract: $8 - 2\frac{4}{13}$

Your solution

EXAMPLE • 5

Find $11\frac{5}{12}$ decreased by $2\frac{11}{16}$.

Solution

$$\begin{array}{r} 11\frac{5}{12} = 11\frac{20}{48} = 10\frac{68}{48} \\ -\ 2\frac{11}{16} = 2\frac{33}{48} = 2\frac{33}{48} \\ \hline 8\frac{35}{48} \end{array}$$

• LCM = 48

YOU TRY IT • 5

What is $21\frac{7}{9}$ minus $7\frac{11}{12}$?

Your solution

Solutions on p. S6

OBJECTIVE D To solve application problems

HOW TO • 6 The outside diameter of a laboratory beaker is $5\frac{1}{8}$ inches and the thickness of the glass is $\frac{1}{8}$ inch. Find the inside diameter of the beaker.

Katrina Leigh/Shutterstock.com

$$\frac{1}{8} + \frac{1}{8} = \frac{2}{8} = \frac{1}{4}$$

• Add $\frac{1}{8}$ and $\frac{1}{8}$ to find the total thickness of the two walls of the beaker.

$$\begin{array}{r} 5\frac{1}{8} = 5\frac{1}{8} = 4\frac{9}{8} \\ -\ \frac{1}{4} = \frac{2}{8} = \frac{2}{8} \\ \hline 4\frac{7}{8} \end{array}$$

• Subtract the total thickness of the two walls from the outside diameter to find the inside diameter.

The inside diameter of the beaker is $4\frac{7}{8}$ inches.

EXAMPLE 6

As an LPN, you will be required to clean and dress the wounds of patients. You need to cut a $2\frac{2}{3}$-inch length of gauze from a $6\frac{5}{8}$-inch gauze pad. How much of the gauze pad remains?

Strategy
To find the remaining length, subtract the length of the piece cut from the total length of the gauze pad.

Solution

$$\begin{array}{r} 6\frac{5}{8} = 6\frac{15}{24} = 5\frac{39}{24} \\ -\ 2\frac{2}{3} = 2\frac{16}{24} = 2\frac{16}{24} \\ \hline 3\frac{23}{24} \end{array}$$

$3\frac{23}{24}$ inches of the gauze pad remain.

YOU TRY IT 6

A baby weighed $21\frac{1}{2}$ pounds but lost $1\frac{3}{4}$ pounds during an extended illness. What was the baby's weight after the illness?

Your strategy

Your solution

EXAMPLE 7

A new roll of gauze is 10 feet long. A medical assistant cuts off $1\frac{3}{4}$ feet of gauze to wrap a wound on one patient and later cuts another $2\frac{1}{3}$ feet to dress a wound on a second patient. How much gauze remains on the roll?

Strategy
To find out how much of the gauze remains:

- Find the total amount of the gauze used for dressing the wounds $\left(1\frac{3}{4} + 2\frac{1}{3}\right)$.
- Subtract the total amount from the original length of the roll (10).

Solution

$$\begin{array}{r} 1\frac{3}{4} = 1\frac{9}{12} \\ +\ 2\frac{1}{3} = 2\frac{4}{12} \\ \hline 3\frac{13}{12} = 4\frac{1}{12} \end{array} \qquad \begin{array}{r} 10 = 9\frac{12}{12} \\ -\ 4\frac{1}{12} = 4\frac{1}{12} \\ \hline 5\frac{11}{12} \end{array}$$

There are $5\frac{11}{12}$ feet of gauze left on the roll.

YOU TRY IT 7

A patient is put on a diet to lose 24 pounds in 3 months. The patient lost $7\frac{1}{2}$ pounds the first month and $5\frac{3}{4}$ pounds the second month. How much weight must be lost the third month to achieve the goal?

Your strategy

Your solution

Solutions on p. S6

2.5 EXERCISES

OBJECTIVE A To subtract fractions with the same denominator

For Exercises 1 to 10, subtract.

1. $\frac{9}{17} - \frac{7}{17}$

2. $\frac{11}{15} - \frac{3}{15}$

3. $\frac{11}{12} - \frac{7}{12}$

4. $\frac{13}{15} - \frac{4}{15}$

5. $\frac{9}{20} - \frac{7}{20}$

6. $\frac{48}{55} - \frac{13}{55}$

7. $\frac{42}{65} - \frac{17}{65}$

8. $\frac{11}{24} - \frac{5}{24}$

9. $\frac{23}{30} - \frac{13}{30}$

10. $\frac{17}{42} - \frac{5}{42}$

11. What is $\frac{5}{14}$ less than $\frac{13}{14}$?

12. Find the difference between $\frac{7}{8}$ and $\frac{5}{8}$.

13. Find $\frac{17}{24}$ decreased by $\frac{11}{24}$.

14. What is $\frac{19}{30}$ minus $\frac{11}{30}$?

For Exercises 15 and 16, each statement describes the difference between a pair of fractions that have the same denominator. State whether the difference of the fractions will need to be rewritten in order to be in simplest form. Answer *yes* or *no*.

15. The difference between the numerators is a factor of the denominator.

16. The difference between the numerators is 1.

OBJECTIVE B To subtract fractions with different denominators

For Exercises 17 to 26, subtract.

17. $\frac{2}{3} - \frac{1}{6}$

18. $\frac{7}{8} - \frac{5}{16}$

19. $\frac{5}{8} - \frac{2}{7}$

20. $\frac{5}{6} - \frac{3}{7}$

21. $\frac{5}{7} - \frac{3}{14}$

22. $\frac{5}{9} - \frac{7}{15}$

23. $\frac{8}{15} - \frac{7}{20}$

24. $\frac{7}{9} - \frac{1}{6}$

25. $\frac{9}{16} - \frac{17}{32}$

26. $\frac{29}{60} - \frac{3}{40}$

27. What is $\frac{3}{5}$ less than $\frac{11}{12}$?

28. What is $\frac{5}{9}$ less than $\frac{11}{15}$?

29. Find the difference between $\frac{11}{24}$ and $\frac{7}{18}$.

30. Find the difference between $\frac{9}{14}$ and $\frac{5}{42}$.

31. Find $\frac{11}{12}$ decreased by $\frac{11}{15}$.

32. Find $\frac{17}{20}$ decreased by $\frac{7}{15}$.

33. What is $\frac{13}{20}$ minus $\frac{1}{6}$?

34. What is $\frac{5}{6}$ minus $\frac{7}{9}$?

35. Which statement describes a pair of fractions for which the least common denominator is one of the denominators?

(i) The denominator of one fraction is a factor of the denominator of the second fraction.

(ii) The denominators of the two fractions have no common factors.

OBJECTIVE C To subtract whole numbers, mixed numbers, and fractions

For Exercises 36 to 50, subtract.

36. $5\frac{7}{12} - 2\frac{5}{12}$

37. $16\frac{11}{15} - 11\frac{8}{15}$

38. $6\frac{1}{3} - 2$

39. $5\frac{7}{8} - 1$

40. $10 - 6\frac{1}{3}$

41. $3 - 2\frac{5}{21}$

42. $6\frac{2}{5} - 4\frac{4}{5}$

43. $16\frac{3}{8} - 10\frac{7}{8}$

44. $25\frac{4}{9} - 16\frac{7}{9}$

45. $8\frac{3}{7} - 2\frac{6}{7}$

46. $16\frac{2}{5} - 8\frac{4}{9}$

47. $23\frac{7}{8} - 16\frac{2}{3}$

48. $82\frac{4}{33} - 16\frac{5}{22}$

49. $6 - 4\frac{3}{5}$

50. $17 - 7\frac{8}{13}$

51. What is $7\frac{3}{5}$ less than $23\frac{3}{20}$?

52. Find the difference between $12\frac{3}{8}$ and $7\frac{5}{12}$.

53. What is $10\frac{5}{9}$ minus $5\frac{11}{15}$?

54. Find $6\frac{1}{3}$ decreased by $3\frac{3}{5}$.

55. Can the difference between a whole number and a mixed number ever be a whole number?

OBJECTIVE D To solve application problems

56. **Nutrition** A hospital dietician had a 100-pound bag of sugar at the beginning of the month. When taking inventory at the end of the month, the dietician notes that $82\frac{1}{2}$ pounds of sugar have been used. How much sugar remains?

57. **Nutrition** A 5-pound bag of sugar contains about 12 cups. During the week, you use $\frac{1}{4}$ cup of sugar to make pancakes, $1\frac{1}{2}$ cups of sugar to make cookies, and $1\frac{3}{4}$ cups of sugar to make a cake. How may cups of sugar are left in your 5-pound bag?

58. **Health** There are two surgical technicians responsible for cleaning and maintaining the sterility of the surgical instruments utilized during surgery. Prior to lunch, one surgical technician cleaned $\frac{1}{6}$ of the instruments and the other cleaned $\frac{2}{3}$ of the instruments. How much of the cleaning remains?

59. **Fundraising** The community hospital is sponsoring a 10-mile run to raise money for a new pediatric family room. The first checkpoint is $2\frac{2}{5}$ miles from the starting point. The second checkpoint is $5\frac{1}{3}$ miles from the first checkpoint.

a. How many miles is it from the starting point to the second checkpoint?
b. How many miles is it from the first checkpoint to the finish line?

60. **Walkathon** A hospital is sponsoring a "Legs for Life Walkathon," which is a 12-mile event with three checkpoints. The first checkpoint is $3\frac{3}{8}$ miles from the starting point. The second checkpoint is $4\frac{1}{3}$ miles from the first.

a. How many miles is it from the starting point to the second checkpoint?
b. How many miles is it from the second checkpoint to the finish line?

61. Hiking Two U.S. Air Force medics are on a three-day, $27\frac{1}{2}$-mile training mission while carrying a total of 80 pounds. The medics plan to travel $7\frac{3}{8}$ miles the first day and $10\frac{1}{3}$ miles the second day.

a. How many total miles do the medics plan to travel the first two days?
b. How many miles will be left to travel on the third day?

Patrick Barth/Getty Images

For Exercises 62 and 63, refer to Exercise 61. Describe what each difference represents.

62. $27\frac{1}{2} - 7\frac{3}{8}$

63. $10\frac{1}{3} - 7\frac{3}{8}$

64. Health A patient with high blood pressure who weighs 225 pounds is put on a diet to lose 25 pounds in 3 months. The patient loses $8\frac{3}{4}$ pounds the first month and $11\frac{5}{8}$ pounds the second month. How much weight must be lost the third month for the goal to be achieved?

65. Health A patient needs to lose an additional $12\frac{3}{4}$ pounds to reach the goal weight set for him by his doctor. The patient loses $5\frac{1}{4}$ pounds the first week and $4\frac{1}{4}$ pounds the second week.

a. Without doing the calculations, determine whether the patient can reach his goal weight by losing less in the third week than was lost in the second week.
b. How many pounds must be lost in the third week for the desired weight to be reached?

Gerald Bernard/Shutterstock.com

66. Health A patient with type 2 diabetes weighs 325 pounds and is advised to follow a prescribed nutrition plan in order to lose weight. During the first month of the diet, weekly losses were recorded as follows: $4\frac{1}{4}$ pounds, $3\frac{3}{4}$ pounds, 2 pounds, and $2\frac{1}{2}$ pounds. What is the patient's weight at the end of the month?

67. Finances If $\frac{4}{15}$ of a nurse's income is spent for housing, what fraction of the nurse's income is not spent for housing?

Applying the Concepts

68. Fill in the square to produce a true statement: $5\frac{1}{3} - \square = 2\frac{1}{2}$

69. Fill in the square to produce a true statement: $\square - 4\frac{1}{2} = 1\frac{5}{8}$

70. Fill in the blank squares at the right so that the sum of the numbers is the same along any row, column, or diagonal. The resulting square is called a magic square.

		$\frac{3}{4}$
1	$\frac{5}{8}$	
$\frac{1}{2}$		$\frac{7}{8}$

SECTION

2.6 Multiplication of Fractions and Mixed Numbers

OBJECTIVE A To multiply fractions

Tips for Success

Before the class meeting in which your professor begins a new section, you should read each objective statement for that section. Next, browse through the material in that objective. The purpose of browsing through the material is to prepare your brain to accept and organize the new information when it is presented to you. See *AIM for Success* at the front of the book.

The product of two fractions is the product of the numerators over the product of the denominators.

HOW TO • 1 Multiply: $\frac{2}{3} \times \frac{4}{5}$

$$\frac{2}{3} \times \frac{4}{5} = \frac{2 \cdot 4}{3 \cdot 5} = \frac{8}{15}$$

- Multiply the numerators.
- Multiply the denominators.

The product $\frac{2}{3} \times \frac{4}{5}$ can be read "$\frac{2}{3}$ times $\frac{4}{5}$" or "$\frac{2}{3}$ of $\frac{4}{5}$."

Reading the times sign as "of" is useful in application problems.

$\frac{4}{5}$ of the bar is shaded.

Shade $\frac{2}{3}$ of the $\frac{4}{5}$ already shaded.

$\frac{8}{15}$ of the bar is then shaded light yellow.

$\frac{2}{3}$ of $\frac{4}{5} = \frac{2}{3} \times \frac{4}{5} = \frac{8}{15}$

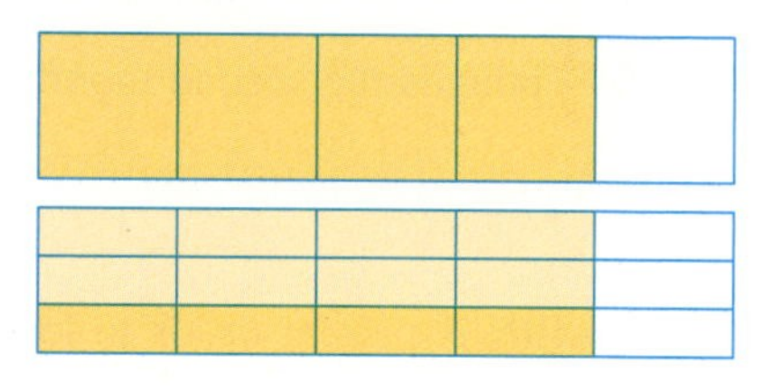

After multiplying two fractions, write the product in simplest form.

HOW TO • 2 Multiply: $\frac{3}{4} \times \frac{14}{15}$

$$\frac{3}{4} \times \frac{14}{15} = \frac{3 \cdot 14}{4 \cdot 15}$$

- Multiply the numerators.
- Multiply the denominators.

$$= \frac{3 \cdot 2 \cdot 7}{2 \cdot 2 \cdot 3 \cdot 5}$$

- Write the prime factorization of each number.

$$= \frac{\overset{1}{\cancel{3}} \cdot \overset{1}{\cancel{2}} \cdot 7}{\underset{1}{\cancel{2}} \cdot 2 \cdot \underset{1}{\cancel{3}} \cdot 5} = \frac{7}{10}$$

- Eliminate the common factors. Then multiply the remaining factors in the numerator and denominator.

This example could also be worked by using the GCF.

$$\frac{3}{4} \times \frac{14}{15} = \frac{42}{60}$$

- Multiply the numerators.
- Multiply the denominators.

$$= \frac{6 \cdot 7}{6 \cdot 10}$$

- The GCF of 42 and 60 is 6. Factor 6 from 42 and 60.

$$= \frac{\overset{1}{\cancel{6}} \cdot 7}{\underset{1}{\cancel{6}} \cdot 10} = \frac{7}{10}$$

- Eliminate the GCF.

EXAMPLE 1

Multiply $\frac{4}{15}$ and $\frac{5}{28}$.

Solution

$$\frac{4}{15} \times \frac{5}{28} = \frac{4 \cdot 5}{15 \cdot 28} = \frac{\overset{1}{\cancel{2}} \cdot \overset{1}{\cancel{2}} \cdot \overset{1}{\cancel{5}}}{3 \cdot \underset{1}{\cancel{5}} \cdot \underset{1}{\cancel{2}} \cdot \underset{1}{\cancel{2}} \cdot 7} = \frac{1}{21}$$

YOU TRY IT 1

Multiply $\frac{4}{21}$ and $\frac{7}{44}$.

Your solution

EXAMPLE 2

Find the product of $\frac{9}{20}$ and $\frac{33}{35}$.

Solution

$$\frac{9}{20} \times \frac{33}{35} = \frac{9 \cdot 33}{20 \cdot 35} = \frac{3 \cdot 3 \cdot 3 \cdot 11}{2 \cdot 2 \cdot 5 \cdot 5 \cdot 7} = \frac{297}{700}$$

YOU TRY IT 2

Find the product of $\frac{2}{21}$ and $\frac{10}{33}$.

Your solution

EXAMPLE 3

What is $\frac{14}{9}$ times $\frac{12}{7}$?

Solution

$$\frac{14}{9} \times \frac{12}{7} = \frac{14 \cdot 12}{9 \cdot 7} = \frac{2 \cdot \overset{1}{\cancel{7}} \cdot 2 \cdot 2 \cdot \overset{1}{\cancel{3}}}{3 \cdot \underset{1}{\cancel{3}} \cdot \underset{1}{\cancel{7}}} = \frac{8}{3} = 2\frac{2}{3}$$

YOU TRY IT 3

What is $\frac{16}{5}$ times $\frac{15}{24}$?

Your solution

Solutions on p. S6

OBJECTIVE B To multiply whole numbers, mixed numbers, and fractions

To multiply a whole number by a fraction or a mixed number, first write the whole number as a fraction with a denominator of 1.

HOW TO 3 Multiply: $4 \times \frac{3}{7}$

$$4 \times \frac{3}{7} = \frac{4}{1} \times \frac{3}{7} = \frac{4 \cdot 3}{1 \cdot 7} = \frac{2 \cdot 2 \cdot 3}{7} = \frac{12}{7} = 1\frac{5}{7}$$

- **Write 4 with a denominator of 1; then multiply the fractions.**

When one or more of the factors in a product is a mixed number, write the mixed number as an improper fraction before multiplying.

HOW TO 4 Multiply: $2\frac{1}{3} \times \frac{3}{14}$

$$2\frac{1}{3} \times \frac{3}{14} = \frac{7}{3} \times \frac{3}{14} = \frac{7 \cdot 3}{3 \cdot 14} = \frac{\overset{1}{\cancel{7}} \cdot \overset{1}{\cancel{3}}}{\underset{1}{\cancel{3}} \cdot 2 \cdot \underset{1}{\cancel{7}}} = \frac{1}{2}$$

- **Write $2\frac{1}{3}$ as an improper fraction; then multiply the fractions.**

EXAMPLE 4

Multiply: $4\frac{5}{6} \times \frac{12}{13}$

Solution

$$4\frac{5}{6} \times \frac{12}{13} = \frac{29}{6} \times \frac{12}{13} = \frac{29 \cdot 12}{6 \cdot 13}$$

$$= \frac{29 \cdot \overset{1}{\cancel{2}} \cdot 2 \cdot \overset{1}{\cancel{3}}}{\underset{1}{\cancel{2}} \cdot \underset{1}{\cancel{3}} \cdot 13} = \frac{58}{13} = 4\frac{6}{13}$$

YOU TRY IT 4

Multiply: $5\frac{2}{5} \times \frac{5}{9}$

Your solution

EXAMPLE 5

Find $5\frac{2}{3}$ times $4\frac{1}{2}$.

Solution

$$5\frac{2}{3} \times 4\frac{1}{2} = \frac{17}{3} \times \frac{9}{2} = \frac{17 \cdot 9}{3 \cdot 2}$$

$$= \frac{17 \cdot \overset{1}{\cancel{3}} \cdot 3}{\underset{1}{\cancel{3}} \cdot 2} = \frac{51}{2} = 25\frac{1}{2}$$

YOU TRY IT 5

Multiply: $3\frac{2}{5} \times 6\frac{1}{4}$

Your solution

EXAMPLE 6

Multiply: $4\frac{2}{5} \times 7$

Solution

$$4\frac{2}{5} \times 7 = \frac{22}{5} \times \frac{7}{1} = \frac{22 \cdot 7}{5 \cdot 1}$$

$$= \frac{2 \cdot 11 \cdot 7}{5} = \frac{154}{5} = 30\frac{4}{5}$$

YOU TRY IT 6

Multiply: $3\frac{2}{7} \times 6$

Your solution

Solutions on p. S6

OBJECTIVE C To solve application problems

HOW TO 5 A baby weighed $6\frac{5}{8}$ pounds at birth. At the baby's six-month checkup, the baby's weight was $2\frac{1}{2}$ times its original weight. Find the baby's weight in pounds at its six-month checkup.

Strategy
Multiply the weight of the newborn $\left(6\frac{5}{8}\right)$ by $2\frac{1}{2}$.

Solution

$$6\frac{5}{8} \times 2\frac{1}{2} = \frac{53}{8} \times \frac{5}{2} = \frac{53 \cdot 5}{8 \cdot 2} = \frac{265}{16} = 16\frac{9}{16}$$

The baby weighed $16\frac{9}{16}$ pounds at its six-month checkup.

EXAMPLE • 7

A dental hygienist earns $206 for each day worked. What are the hygienist's earnings for working $4\frac{1}{2}$ days?

Strategy
To find the hygienist's total earnings, multiply the daily earnings (206) by the number of days worked $\left(4\frac{1}{2}\right)$.

Solution

$$206 \times 4\frac{1}{2} = \frac{206}{1} \times \frac{9}{2} = \frac{206 \cdot 9}{1 \cdot 2} = 927$$

The dental hygienist's earnings are $927.

YOU TRY IT • 7

A dialysis technician earns a starting salary of $37,000 per year. After 10 years of experience, the technician earns $1\frac{1}{2}$ times the starting salary. What is the dialysis technician's salary after 10 years of experience?

Your strategy

Your solution

EXAMPLE • 8

The value of a small medical office building and the land on which it is built is $290,000. The value of the land is $\frac{1}{4}$ the total value. What is the dollar value of the building?

Strategy
To find the value of the building:
- Find the value of the land $\left(\frac{1}{4} \times 290{,}000\right)$.
- Subtract the value of the land from the total value (290,000).

Solution

$$\frac{1}{4} \times 290{,}000 = \frac{290{,}000}{4} = 72{,}500 \quad \text{• Value of the land}$$

$$290{,}000 - 72{,}500 = 217{,}500$$

The value of the building is $217,500.

YOU TRY IT • 8

A radiologist purchased a portable digital radiography x-ray machine for $125,000. The portable x-ray cassette costs $\frac{1}{5}$ of the total cost. What is the cost of the x-ray machine?

Your strategy

Your solution

Solutions on pp. S6–S7

2.6 EXERCISES

OBJECTIVE A To multiply fractions

For Exercises 1 to 32, multiply.

1. $\frac{2}{3} \times \frac{7}{8}$

2. $\frac{1}{2} \times \frac{2}{3}$

3. $\frac{5}{16} \times \frac{7}{15}$

4. $\frac{3}{8} \times \frac{6}{7}$

5. $\frac{1}{6} \times \frac{1}{8}$

6. $\frac{2}{5} \times \frac{5}{6}$

7. $\frac{11}{12} \times \frac{6}{7}$

8. $\frac{11}{12} \times \frac{3}{5}$

9. $\frac{8}{9} \times \frac{27}{4}$

10. $\frac{3}{5} \times \frac{3}{10}$

11. $\frac{5}{6} \times \frac{1}{2}$

12. $\frac{3}{8} \times \frac{5}{12}$

13. $\frac{16}{9} \times \frac{27}{8}$

14. $\frac{5}{8} \times \frac{16}{15}$

15. $\frac{3}{2} \times \frac{4}{9}$

16. $\frac{5}{3} \times \frac{3}{7}$

17. $\frac{7}{8} \times \frac{3}{14}$

18. $\frac{2}{9} \times \frac{1}{5}$

19. $\frac{1}{10} \times \frac{3}{8}$

20. $\frac{5}{12} \times \frac{6}{7}$

21. $\frac{15}{8} \times \frac{16}{3}$

22. $\frac{5}{6} \times \frac{4}{15}$

23. $\frac{1}{2} \times \frac{2}{15}$

24. $\frac{3}{8} \times \frac{5}{16}$

25. $\frac{5}{7} \times \frac{14}{15}$

26. $\frac{3}{8} \times \frac{15}{41}$

27. $\frac{5}{12} \times \frac{42}{65}$

28. $\frac{16}{33} \times \frac{55}{72}$

29. $\frac{12}{5} \times \frac{5}{3}$

30. $\frac{17}{9} \times \frac{81}{17}$

31. $\frac{16}{85} \times \frac{125}{84}$

32. $\frac{19}{64} \times \frac{48}{95}$

33. Give an example of a proper and an improper fraction whose product is 1.

34. Multiply $\frac{7}{12}$ and $\frac{15}{42}$.

35. Multiply $\frac{32}{9}$ and $\frac{3}{8}$.

36. Find the product of $\frac{5}{9}$ and $\frac{3}{20}$.

37. Find the product of $\frac{7}{3}$ and $\frac{15}{14}$.

38. What is $\frac{1}{2}$ times $\frac{8}{15}$?

39. What is $\frac{3}{8}$ times $\frac{12}{17}$?

OBJECTIVE B To multiply whole numbers, mixed numbers, and fractions

For Exercises 40 to 71, multiply.

40. $4 \times \frac{3}{8}$

41. $14 \times \frac{5}{7}$

42. $\frac{2}{3} \times 6$

43. $\frac{5}{12} \times 40$

44. $\frac{1}{3} \times 1\frac{1}{3}$

45. $\frac{2}{5} \times 2\frac{1}{2}$

46. $1\frac{7}{8} \times \frac{4}{15}$

47. $2\frac{1}{5} \times \frac{5}{22}$

48. $4 \times 2\frac{1}{2}$

49. $9 \times 3\frac{1}{3}$

50. $2\frac{1}{7} \times 3$

51. $5\frac{1}{4} \times 8$

52. $3\frac{2}{3} \times 5$

53. $4\frac{2}{9} \times 3$

54. $\frac{1}{2} \times 3\frac{3}{7}$

55. $\frac{3}{8} \times 4\frac{4}{5}$

56. $6\frac{1}{8} \times \frac{4}{7}$

57. $5\frac{1}{3} \times \frac{5}{16}$

58. $\frac{3}{8} \times 4\frac{1}{2}$

59. $\frac{5}{7} \times 2\frac{1}{3}$

60. $0 \times 2\frac{2}{3}$

61. $6\frac{1}{8} \times 0$

62. $2\frac{5}{8} \times 3\frac{2}{5}$

63. $5\frac{3}{16} \times 5\frac{1}{3}$

64. $3\frac{1}{7} \times 2\frac{1}{8}$

65. $16\frac{5}{8} \times 1\frac{1}{16}$

66. $2\frac{2}{5} \times 3\frac{1}{12}$

67. $2\frac{2}{3} \times \frac{3}{20}$

68. $5\frac{1}{5} \times 3\frac{1}{13}$

69. $3\frac{3}{4} \times 2\frac{3}{20}$

70. $12\frac{3}{5} \times 1\frac{3}{7}$

71. $6\frac{1}{2} \times 1\frac{3}{13}$

72. True or false? If the product of a whole number and a fraction is a whole number, then the denominator of the fraction is a factor of the original whole number.

73. Multiply $2\frac{1}{2}$ and $3\frac{3}{5}$.

74. Multiply $4\frac{3}{8}$ and $3\frac{3}{5}$.

75. Find the product of $2\frac{1}{8}$ and $\frac{5}{17}$.

76. Find the product of $12\frac{2}{5}$ and $3\frac{7}{31}$.

77. What is $1\frac{3}{8}$ times $2\frac{1}{5}$?

78. What is $3\frac{1}{8}$ times $2\frac{4}{7}$?

OBJECTIVE C To solve application problems

For Exercises 79 and 80, give your answer without actually doing a calculation.

79. Read Exercise 81. Will the requested cost be greater than or less than \$400?

80. Read Exercise 83. Will the remaining length be greater than or less than 200 feet?

81. **Purchasing** A massage therapist purchases all-natural massage oil at \$190 per case of 10 bottles. Find the cost of $2\frac{1}{2}$ cases.

82. **Exercise** Maria Rivera can walk $3\frac{1}{2}$ miles in 1 hour. At this rate, how far can Maria walk in $\frac{1}{3}$ hour?

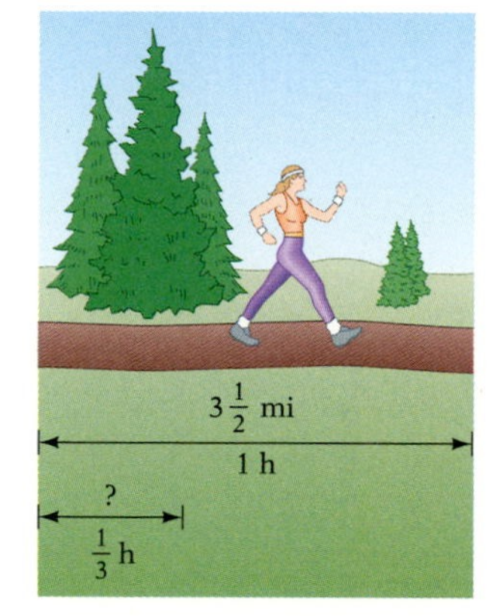

83. **Supplies** Examination table paper costs \$4 for 225 feet. One-third of the roll is used in a week. How many feet of the paper have been used in one week? How many feet remain on the roll after one week?

84. **Real Estate** An oral surgeon purchases a medical office building and the dental practice patient files valued at \$650,000. The value of the dental practice files is $\frac{2}{5}$ of the total value. What is the dollar value for the building?

85. **X-Rays** As a radiology technician, it takes you 10 minutes or $\frac{1}{6}$ hour to take a chest x-ray. Your schedule requires that you complete 27 x-rays today. How many hours will it take to complete your scheduled x-rays?

86. **IV Infusion** A patient is to receive an IV infusion of 1000 milliliters of normal saline over a period of 8 hours. A nurse checks on the patient after $\frac{3}{4}$ of the bag has been infused. How many milliliters of fluid has the patient received? How many hours have passed since the IV was started?

Nutrition The hospital dietician has directed the kitchen staff to use a low-fat recipe for banana bread. The ingredients are listed at the right. Use the recipe for Exercises 87 to 90.

Low-Fat Banana Bread

- 2 cups all-purpose flour
- $\frac{3}{4}$ teaspoon baking soda
- $\frac{1}{2}$ teaspoon salt
- 1 cup sugar
- $\frac{1}{4}$ cup trans fat-free shortening
- $\frac{1}{2}$ cup egg substitute
- $1\frac{1}{2}$ cups mashed ripe banana

Yield : 14 slices

87. A dietician notes that 40 patients have requested banana bread for breakfast. The staff will need to triple the recipe in order to have the requested amount. How many cups of mashed ripe bananas will be required in order to fill the patients' requests?

88. Find the total amount of baking soda and salt needed for one recipe of banana bread. If the recipe is quadrupled, how much soda and salt will be needed?

89. If 70 patients requested banana bread for their meals, by what number will the kitchen staff have to multiply the ingredients in the recipe to prepare enough slices? How many cups of trans fat-free shortening will be needed for a recipe to feed 70 patients?

90. You have a small loaf pan and want to make only half a recipe.
a. How many cups of egg substitute will you need?
b. How many cups of mashed bananas will you need?

91. **Wages** A clinical laboratory technician earns $168 for each day worked. What are the technician's earning for working $5\frac{1}{2}$ days?

92. **Wages** The initial salary of a laboratory supervisor 15 years ago was $38,000. Today the supervisor's salary is $1\frac{1}{2}$ times that amount. What is the supervisor's current salary?

Applying the Concepts

93. The product of 1 and a number is $\frac{1}{2}$. Find the number.

94. **Time** Our calendar is based on the solar year, which is $365\frac{1}{4}$ days. Use this fact to explain leap years.

95. Which of the labeled points on the number line at the right could be the graph of the product of B and C?

0 *A* *B* *C* 1 *D* 2 *E* 3

96. Fill in the circles on the square at the right with the fractions $\frac{1}{6}$, $\frac{5}{18}$, $\frac{4}{9}$, $\frac{5}{9}$, $\frac{2}{3}$, $\frac{3}{4}$, $1\frac{1}{9}$, $1\frac{1}{2}$, and $2\frac{1}{4}$ so that the product of any row is equal to $\frac{5}{18}$. (*Note:* There is more than one possible answer.)

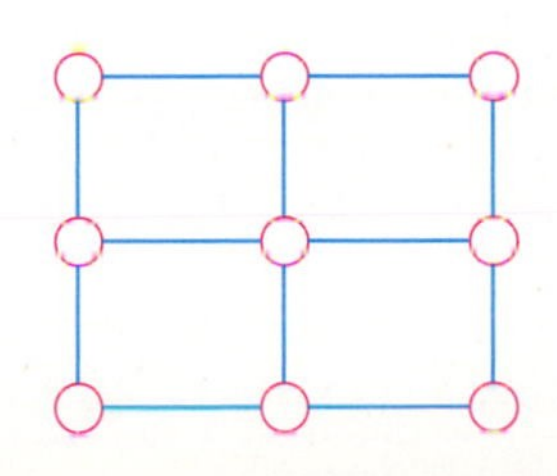

SECTION

2.7 Division of Fractions and Mixed Numbers

OBJECTIVE A To divide fractions

The **reciprocal of a fraction** is the fraction with the numerator and denominator interchanged.

The reciprocal of $\frac{2}{3}$ is $\frac{3}{2}$.

The process of interchanging the numerator and denominator is called **inverting a fraction.**

To find the reciprocal of a whole number, first write the whole number as a fraction with a denominator of 1. Then find the reciprocal of that fraction.

The reciprocal of 5 is $\frac{1}{5}$. $\left(\text{Think } 5 = \frac{5}{1}.\right)$

Reciprocals are used to rewrite division problems as related multiplication problems. Look at the following two problems:

$8 \div 2 = 4$ $\qquad$ $8 \times \frac{1}{2} = 4$

8 divided by 2 is 4. $\qquad$ 8 times the reciprocal of 2 is 4.

"Divided by" means the same as "times the reciprocal of." Thus "$\div$ 2" can be replaced with "$\times \frac{1}{2}$," and the answer will be the same. Fractions are divided by making this replacement.

HOW TO • 1 Divide: $\frac{2}{3} \div \frac{3}{4}$

$$\frac{2}{3} \div \frac{3}{4} = \frac{2}{3} \times \frac{4}{3} = \frac{2 \cdot 4}{3 \cdot 3} = \frac{2 \cdot 2 \cdot 2}{3 \cdot 3} = \frac{8}{9}$$

• Multiply the first fraction by the reciprocal of the second fraction.

EXAMPLE • 1

Divide: $\frac{5}{8} \div \frac{4}{9}$

Solution

$$\frac{5}{8} \div \frac{4}{9} = \frac{5}{8} \times \frac{9}{4} = \frac{5 \cdot 9}{8 \cdot 4}$$

$$= \frac{5 \cdot 3 \cdot 3}{2 \cdot 2 \cdot 2 \cdot 2 \cdot 2} = \frac{45}{32} = 1\frac{13}{32}$$

YOU TRY IT • 1

Divide: $\frac{3}{7} \div \frac{2}{3}$

Your solution

EXAMPLE • 2

Divide: $\frac{3}{5} \div \frac{12}{25}$

Solution

$$\frac{3}{5} \div \frac{12}{25} = \frac{3}{5} \times \frac{25}{12} = \frac{3 \cdot 25}{5 \cdot 12}$$

$$= \frac{\overset{1}{\cancel{3}} \cdot \overset{1}{\cancel{5}} \cdot 5}{\underset{1}{\cancel{5}} \cdot 2 \cdot 2 \cdot \underset{1}{\cancel{3}}} = \frac{5}{4} = 1\frac{1}{4}$$

YOU TRY IT • 2

Divide: $\frac{3}{4} \div \frac{9}{10}$

Your solution

Solutions on p. S7

OBJECTIVE B To divide whole numbers, mixed numbers, and fractions

To divide a fraction and a whole number, first write the whole number as a fraction with a denominator of 1.

HOW TO • 2 Divide: $\frac{3}{7} \div 5$

$$\frac{3}{7} \div \boxed{5} = \frac{3}{7} \div \boxed{\frac{5}{1}} = \frac{3}{7} \times \frac{1}{5} = \frac{3 \cdot 1}{7 \cdot 5} = \frac{3}{35}$$

• Write 5 with a denominator of 1. Then divide the fractions.

When a number in a quotient is a mixed number, write the mixed number as an improper fraction before dividing.

HOW TO • 3 Divide: $1\frac{13}{15} \div 4\frac{4}{5}$

Write the mixed numbers as improper fractions. Then divide the fractions.

$$1\frac{13}{15} \div 4\frac{4}{5} = \frac{28}{15} \div \frac{24}{5} = \frac{28}{15} \times \frac{5}{24} = \frac{28 \cdot 5}{15 \cdot 24} = \frac{\overset{1}{\cancel{2}} \cdot \overset{1}{\cancel{2}} \cdot 7 \cdot \overset{1}{\cancel{5}}}{3 \cdot \underset{1}{\cancel{5}} \cdot \underset{1}{\cancel{2}} \cdot \underset{1}{\cancel{2}} \cdot 2 \cdot 3} = \frac{7}{18}$$

EXAMPLE • 3

Divide $\frac{4}{9}$ by 5.

Solution

$$\frac{4}{9} \div 5 = \frac{4}{9} \div \frac{5}{1} = \frac{4}{9} \times \frac{1}{5}$$

$$= \frac{4 \cdot 1}{9 \cdot 5} = \frac{2 \cdot 2}{3 \cdot 3 \cdot 5} = \frac{4}{45}$$

• $5 = \frac{5}{1}$. The reciprocal of $\frac{5}{1}$ is $\frac{1}{5}$.

YOU TRY IT • 3

Divide $\frac{5}{7}$ by 6.

Your solution

EXAMPLE • 4

Find the quotient of $\frac{3}{8}$ and $2\frac{1}{10}$.

Solution

$$\frac{3}{8} \div 2\frac{1}{10} = \frac{3}{8} \div \frac{21}{10} = \frac{3}{8} \times \frac{10}{21}$$

$$= \frac{3 \cdot 10}{8 \cdot 21} = \frac{\overset{1}{\cancel{3}} \cdot \overset{1}{\cancel{2}} \cdot 5}{\underset{1}{\cancel{2}} \cdot 2 \cdot 2 \cdot \underset{1}{\cancel{3}} \cdot 7} = \frac{5}{28}$$

YOU TRY IT • 4

Find the quotient of $12\frac{3}{5}$ and 7.

Your solution

EXAMPLE • 5

Divide: $2\frac{3}{4} \div 1\frac{5}{7}$

Solution

$$2\frac{3}{4} \div 1\frac{5}{7} = \frac{11}{4} \div \frac{12}{7} = \frac{11}{4} \times \frac{7}{12} = \frac{11 \cdot 7}{4 \cdot 12}$$

$$= \frac{11 \cdot 7}{2 \cdot 2 \cdot 2 \cdot 2 \cdot 3} = \frac{77}{48} = 1\frac{29}{48}$$

YOU TRY IT • 5

Divide: $3\frac{2}{3} \div 2\frac{2}{5}$

Your solution

Solutions on p. S7

EXAMPLE • 6

Divide: $1\frac{13}{15} \div 4\frac{1}{5}$

Solution

$$1\frac{13}{15} \div 4\frac{1}{5} = \frac{28}{15} \div \frac{21}{5} = \frac{28}{15} \times \frac{5}{21} = \frac{28 \cdot 5}{15 \cdot 21}$$

$$= \frac{2 \cdot 2 \cdot \overset{1}{\cancel{7}} \cdot \overset{1}{\cancel{5}}}{3 \cdot \underset{1}{\cancel{5}} \cdot 3 \cdot \underset{1}{\cancel{7}}} = \frac{4}{9}$$

YOU TRY IT • 6

Divide: $2\frac{5}{6} \div 8\frac{1}{2}$

Your solution

EXAMPLE • 7

Divide: $4\frac{3}{8} \div 7$

Solution

$$4\frac{3}{8} \div 7 = \frac{35}{8} \div \frac{7}{1} = \frac{35}{8} \times \frac{1}{7}$$

$$= \frac{35 \cdot 1}{8 \cdot 7} = \frac{5 \cdot \overset{1}{\cancel{7}}}{2 \cdot 2 \cdot 2 \cdot \underset{1}{\cancel{7}}} = \frac{5}{8}$$

YOU TRY IT • 7

Divide: $6\frac{2}{5} \div 4$

Your solution

Solutions on p. S7

OBJECTIVE C To solve application problems

EXAMPLE • 8

A hospital ambulance used $17\frac{1}{2}$ gallons of gasoline transporting a post-surgical hip replacement patient 210 miles to a physical rehabilitation center. How many miles can the ambulance travel on 1 gallon of gasoline?

Strategy

To find the number of miles, divide the number of miles traveled by the number of gallons of gasoline used.

Solution

$$210 \div 17\frac{1}{2} = \frac{210}{1} \div \frac{35}{2}$$

$$= \frac{210}{1} \times \frac{2}{35}$$

$$= \frac{5 \times 7 \times 6 \times 2}{1 \times 5 \times 7}$$

$$= \frac{12}{1} = 12$$

The ambulance travels 12 miles on 1 gallon of gasoline.

YOU TRY IT • 8

A hospital engineering technician spends $2\frac{1}{2}$ hours assembling a hospital bed. If the technician worked a 10-hour day assembling new beds last week, how many beds were assembled?

Your strategy

Your solution

Solutions on p. S7

EXAMPLE • 9

A 15-foot roll of gauze bandage wrap is cut into lengths of $3\frac{1}{3}$ feet to be used as wound coverings. What is the length of the remaining piece of gauze after cutting as many pieces of wound covering gauze as possible?

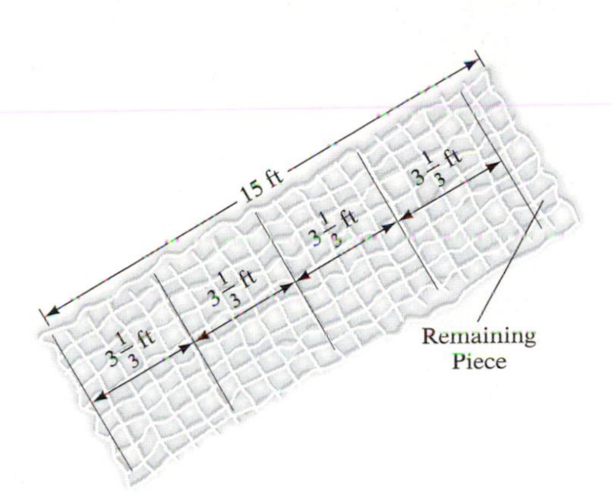

Strategy

To find the length of the remaining gauze:

- Divide the total length of the gauze (15) by the length of each wound covering $\left(3\frac{1}{3}\right)$. This will give you the number of wound coverings with a certain fraction of gauze left over.
- Multiply the fractional part of the result in step 1 by the length of one wound covering piece of gauze to determine the length of the remaining gauze.

Solution

$$15 \div 3\frac{1}{3} = \frac{15}{1} \div \frac{10}{3} = \frac{15}{1} \times \frac{3}{10}$$

$$= \frac{15 \times 3}{1 \times 10} = \frac{45}{10} = 4\frac{1}{2}$$

There are four pieces that are each $3\frac{1}{3}$-feet long.

There is one piece that is $\frac{1}{2}$ of $3\frac{1}{3}$-feet long.

$$\frac{1}{2} \times 3\frac{1}{3} = \frac{1}{2} \times \frac{10}{3} = \frac{10}{6} = 1\frac{2}{3}$$

The length of the gauze piece remaining is $1\frac{2}{3}$ feet.

YOU TRY IT • 9

A 12-foot piece of nasal cannula tubing is cut into pieces $2\frac{1}{6}$-feet long. What is the length of the remaining piece of tubing after cutting as many as possible?

Your strategy

Your solution

Solution on p. S7

2.7 EXERCISES

OBJECTIVE A To divide fractions

For Exercises 1 to 28, divide.

1. $\frac{1}{3} \div \frac{2}{5}$

2. $\frac{3}{7} \div \frac{3}{2}$

3. $\frac{3}{7} \div \frac{3}{7}$

4. $0 \div \frac{1}{2}$

5. $0 \div \frac{3}{4}$

6. $\frac{16}{33} \div \frac{4}{11}$

7. $\frac{5}{24} \div \frac{15}{36}$

8. $\frac{11}{15} \div \frac{1}{12}$

9. $\frac{1}{9} \div \frac{2}{3}$

10. $\frac{10}{21} \div \frac{5}{7}$

11. $\frac{2}{5} \div \frac{4}{7}$

12. $\frac{3}{8} \div \frac{5}{12}$

13. $\frac{1}{2} \div \frac{1}{4}$

14. $\frac{1}{3} \div \frac{1}{9}$

15. $\frac{1}{5} \div \frac{1}{10}$

16. $\frac{4}{15} \div \frac{2}{5}$

17. $\frac{7}{15} \div \frac{14}{5}$

18. $\frac{5}{8} \div \frac{15}{2}$

19. $\frac{14}{3} \div \frac{7}{9}$

20. $\frac{7}{4} \div \frac{9}{2}$

21. $\frac{5}{9} \div \frac{25}{3}$

22. $\frac{5}{16} \div \frac{3}{8}$

23. $\frac{2}{3} \div \frac{1}{3}$

24. $\frac{4}{9} \div \frac{1}{9}$

25. $\frac{5}{7} \div \frac{2}{7}$

26. $\frac{5}{6} \div \frac{1}{9}$

27. $\frac{2}{3} \div \frac{2}{9}$

28. $\frac{5}{12} \div \frac{5}{6}$

29. Divide $\frac{7}{8}$ by $\frac{3}{4}$.

30. Divide $\frac{7}{12}$ by $\frac{3}{4}$.

31. Find the quotient of $\frac{5}{7}$ and $\frac{3}{14}$.

32. Find the quotient of $\frac{6}{11}$ and $\frac{9}{32}$.

33. True or false? If a fraction has a numerator of 1, then the reciprocal of the fraction is a whole number.

34. True or false? The reciprocal of an improper fraction that is not equal to 1 is a proper fraction.

OBJECTIVE B **To divide whole numbers, mixed numbers, and fractions**

For Exercises 35 to 73, divide.

35. $4 \div \frac{2}{3}$

36. $\frac{2}{3} \div 4$

37. $\frac{3}{2} \div 3$

38. $3 \div \frac{3}{2}$

39. $\frac{5}{6} \div 25$

40. $22 \div \frac{3}{11}$

41. $6 \div 3\frac{1}{3}$

42. $5\frac{1}{2} \div 11$

43. $6\frac{1}{2} \div \frac{1}{2}$

44. $\frac{3}{8} \div 2\frac{1}{4}$

45. $8\frac{1}{4} \div 2\frac{3}{4}$

46. $3\frac{5}{9} \div 32$

47. $4\frac{1}{5} \div 21$

48. $6\frac{8}{9} \div \frac{31}{36}$

49. $\frac{11}{12} \div 2\frac{1}{3}$

50. $\frac{7}{8} \div 3\frac{1}{4}$

51. $35 \div \frac{7}{24}$

52. $\frac{3}{8} \div 2\frac{3}{4}$

53. $\frac{11}{18} \div 2\frac{2}{9}$

54. $\frac{21}{40} \div 3\frac{3}{10}$

55. $2\frac{1}{16} \div 2\frac{1}{2}$

56. $7\frac{3}{5} \div 1\frac{7}{12}$

57. $1\frac{2}{3} \div \frac{3}{8}$

58. $16 \div \frac{2}{3}$

59. $1\frac{5}{8} \div 4$

60. $13\frac{3}{8} \div \frac{1}{4}$

61. $16 \div 1\frac{1}{2}$

62. $9 \div \frac{7}{8}$

63. $1\frac{1}{3} \div 5\frac{8}{9}$

64. $13\frac{2}{3} \div 0$

65. $82\frac{3}{5} \div 19\frac{1}{10}$

66. $45\frac{3}{5} \div 15$

67. $102 \div 1\frac{1}{2}$

68. $0 \div 3\frac{1}{2}$

69. $8\frac{2}{7} \div 1$

70. $6\frac{9}{16} \div 1\frac{3}{32}$

71. $8\frac{8}{9} \div 2\frac{13}{18}$

72. $10\frac{1}{5} \div 1\frac{7}{10}$

73. $7\frac{3}{8} \div 1\frac{27}{32}$

74. Divide $7\frac{7}{9}$ by $5\frac{5}{6}$.

75. Divide $2\frac{3}{4}$ by $1\frac{23}{32}$.

76. Find the quotient of $8\frac{1}{4}$ and $1\frac{5}{11}$.

77. Find the quotient of $\frac{14}{17}$ and $3\frac{1}{9}$.

78. True or false? The reciprocal of a mixed number is an improper fraction.

79. True or false? A fraction divided by its reciprocal is 1.

OBJECTIVE C To solve application problems

For Exercises 80 and 81, give your answer without actually doing a calculation.

80. Read Exercise 82. Will the requested number of boxes be greater than or less than 600?

81. Read Exercise 83. Will the requested number of servings be greater than or less than 16?

82. **Consumerism** Individual cereal boxes contain $\frac{3}{4}$ ounce of cereal. How many boxes can be filled with 600 ounces of cereal?

David Young-Wolff/PhotoEdit, Inc.

83. **Consumerism** A box of Post's Great Grains cereal costing \$4 contains 16 ounces of cereal. How many $1\frac{1}{3}$-ounce servings are in this box?

84. **Consumerism** The cost of capsaicin arthritis rub is \$21 for a $1\frac{1}{2}$-ounce tube. As a physical therapist who works with chronic arthritis patients, you need to buy 42 ounces of capsaicin. How many tubes will you need to purchase?

85. **Unit Price** Refer to the cost of capsaicin arthritis rub in problem 84. What is the price per ounce of this arthritis rub?

86. **Real Estate** A doctor's Investment Coalition purchased $6\frac{3}{4}$ acres of land on which to build a hospital. The cost is \$810,000. What is the cost per acre of land?

87. **Fuel Efficiency** During one week, an EMT vehicle traveled 357 miles and used $25\frac{1}{2}$ gallons of gasoline responding to emergencies. How many miles can the vehicle travel on 1 gallon of gasoline?

88. **Bandages** A doctor uses $1\frac{1}{2}$ feet of a self-adherent gauze wrap to bandage a patient's foot after minor foot surgery. If a new roll of this gauze wrap is 18 feet long, how many $1\frac{1}{2}$-foot-long bandages could be cut from a roll?

89. **The Food Industry** A hospital chef purchased a roast that weighed $10\frac{3}{4}$ pounds. After the fat was trimmed and the bone removed, the roast weighed $9\frac{1}{3}$ pounds.

a. What was the total weight of the fat and bone?

b. How many $\frac{1}{3}$-pound servings can be cut from the trimmed roast?

Tom McCarthy/PhotoEdit, Inc.

90. **Nutrition** A nursing home cafeteria director purchases white sugar in 50-pound bags. Each bag of sugar contains about 120 cups. If it takes $1\frac{1}{3}$ cups of sugar for every gallon of sweet tea served, how many gallons of tea can be made with 50 pounds of sugar?

91. **Medication** How many $\frac{1}{2}$-ounce doses of cough syrup can be given from a $10\frac{1}{2}$-ounce bottle?

92. **Medication** There are 15 patients on the second floor at a nursing home. Each patient has been prescribed $1\frac{1}{2}$ ounces of a certain medication.

a. If one bottle of this medication contains 18 ounces of medication, will one bottle be enough to fill this order for all 15 patients?

b. How many doses can be given from one bottle?

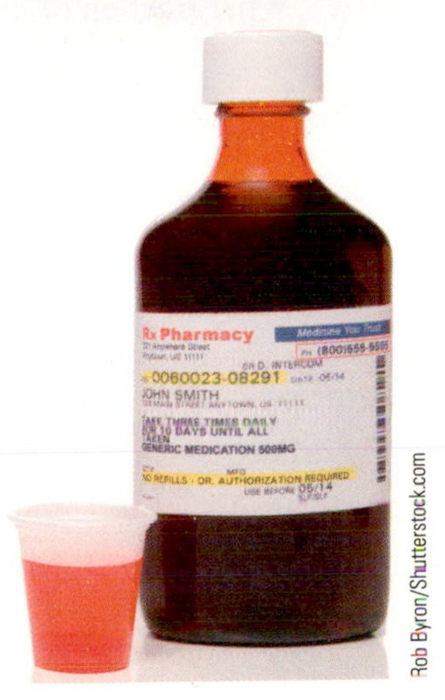

Rob Byron/Shutterstock.com

Applying the Concepts

Loans The figure below at the right shows how the money borrowed by a doctor to open a new office is spent. Use this graph for Exercises 93 and 94.

93. What fractional part of the money borrowed is spent on office renovations and equipment purchases?

94. What fractional part of the money borrowed is spent on equipment purchases, office insurance, and office overhead?

Medical Supplies $\frac{1}{25}$
Office Insurance $\frac{1}{20}$
Office Overhead Expenses $\frac{1}{20}$
Equipment Purchase $\frac{6}{25}$
Malpractice Insurance $\frac{6}{25}$
Office Renovations $\frac{19}{50}$

How Money Is Borrowed and Spent on New Office

95. **Health Coverage** Your health insurance reimbursed your doctor $\frac{1}{2}$ of the billed amount for an office visit. You paid $\frac{1}{6}$ of the billed amount when you saw the doctor. What fractional amount of the bill remains to be paid?

96. **Finances** A bank recommends that the maximum monthly payment for a home be $\frac{1}{3}$ of your total monthly income. Your monthly income is $4500. What would the bank recommend as your maximum monthly house payment?

Nutrition The dietician at a nursing home monitors the dietary restrictions of the 145 residents as their meals are prepared. Use the table for Exercises 97 to 99.

Dietary Restriction	*Number of Residents*
Low sodium	40
Low fat	25
Low sodium and low fat	30
No restrictions	50

97. What fraction of the residents require a low-sodium diet?

98. What fraction of the residents have no restrictions on their diets?

99. What fraction of the residents have some type of dietary restriction on their meals?

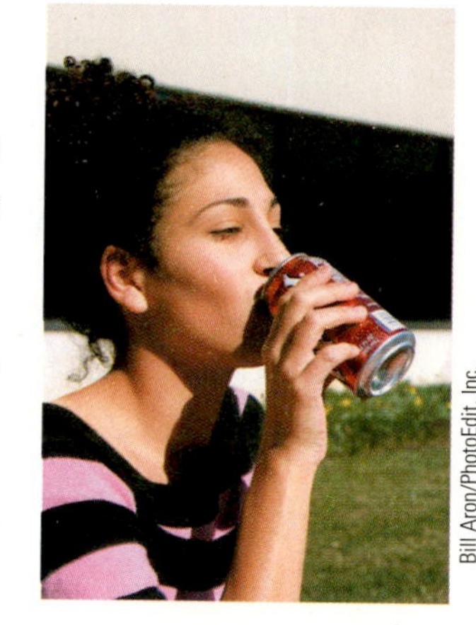
Bill Aron/PhotoEdit, Inc.

Nutrition According to the Center for Science in the Public Interest, the average teenage boy drinks $3\frac{1}{3}$ cans of soda per day. The average teenage girl drinks $2\frac{1}{3}$ cans of soda per day. Use this information for Exercises 100 and 101.

100. If a can of soda contains 150 calories, how many calories does the average teenage boy consume each week in soda?

101. How many more cans of soda per week does the average teenage boy drink than the average teenage girl?

102. **Wages** As a home health aide, you assist family caregivers by sitting with elderly relatives. The job pays $9 per hour, and last week you worked 5 hours on Monday, $3\frac{3}{4}$ hours on Tuesday, $1\frac{1}{4}$ hours on Wednesday, and $2\frac{1}{3}$ hours on Thursday. Find your total earnings for last week's work.

103. Fill in the box to make a true statement.

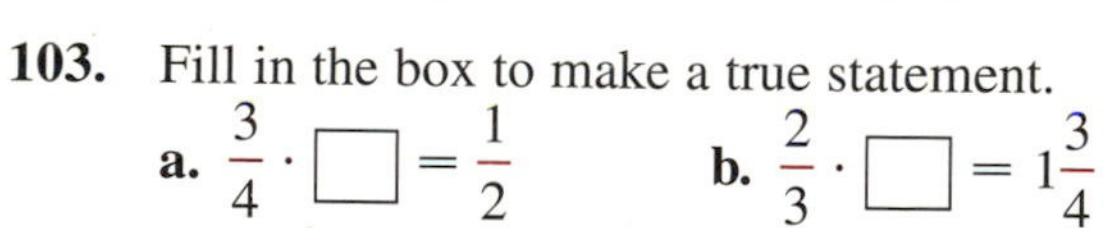

a. $\frac{3}{4} \cdot \square = \frac{1}{2}$ **b.** $\frac{2}{3} \cdot \square = 1\frac{3}{4}$

104. **Pain Relief** A large hydrocollator hot pack for the back measures 24 inches by 24 inches by $\frac{1}{3}$ inch. What are the new dimensions of the pack if it is folded in half to place over the thoracic spinal column?

SECTION 2.8 Order, Exponents, and the Order of Operations Agreement

OBJECTIVE A To identify the order relation between two fractions

Point of Interest

Leonardo of Pisa, who was also called Fibonacci (c. 1175–1250), is credited with bringing the Hindu-Arabic number system to the Western world and promoting its use in place of the cumbersome Roman numeral system.

He was also influential in promoting the idea of the fraction bar. His notation, however, was very different from what we use today.

For instance, he wrote $\frac{3}{4}\frac{5}{7}$ to mean $\frac{5}{7}+\frac{3}{7\cdot 4}$, which equals $\frac{23}{28}$.

Recall that whole numbers can be graphed as points on the number line. Fractions can also be graphed as points on the number line.

The graph of $\frac{3}{4}$ on the number line

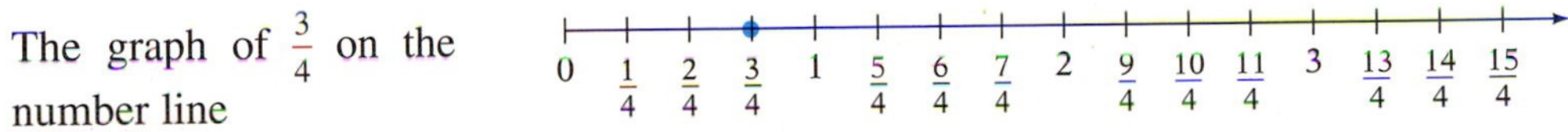

The number line can be used to determine the order relation between two fractions. A fraction that appears to the left of a given fraction is less than the given fraction. A fraction that appears to the right of a given fraction is greater than the given fraction.

$\frac{1}{8} < \frac{3}{8}$ $\quad \frac{6}{8} > \frac{3}{8}$

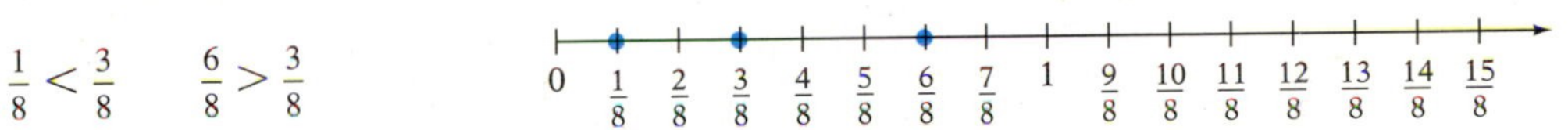

To find the order relation between two fractions with the same denominator, compare the numerators. The fraction that has the smaller numerator is the smaller fraction. When the denominators are different, begin by writing equivalent fractions with a common denominator; then compare the numerators.

HOW TO • 1 Find the order relation between $\frac{11}{18}$ and $\frac{5}{8}$.

The LCM of 18 and 8 is 72.

$\frac{11}{18} = \frac{44}{72}$ ← Smaller numerator

$\frac{5}{8} = \frac{45}{72}$ ← Larger numerator

$\frac{11}{18} < \frac{5}{8}$ or $\frac{5}{8} > \frac{11}{18}$

EXAMPLE • 1

Place the correct symbol, < or >, between the two numbers.

$\frac{5}{12} \quad \frac{7}{18}$

Solution $\frac{5}{12} = \frac{15}{36}$ $\quad \frac{7}{18} = \frac{14}{36}$

$\frac{5}{12} > \frac{7}{18}$

YOU TRY IT • 1

Place the correct symbol, < or >, between the two numbers.

$\frac{9}{14} \quad \frac{13}{21}$

Your solution

Solution on p. S8

OBJECTIVE B To simplify expressions containing exponents

Repeated multiplication of the same fraction can be written in two ways:

$$\frac{1}{2}\cdot\frac{1}{2}\cdot\frac{1}{2}\cdot\frac{1}{2} \text{ or } \left(\frac{1}{2}\right)^4 \leftarrow \text{Exponent}$$

The exponent indicates how many times the fraction occurs as a factor in the multiplication. The expression $\left(\frac{1}{2}\right)^4$ is in exponential notation.

EXAMPLE • 2

Simplify: $\left(\frac{5}{6}\right)^3 \cdot \left(\frac{3}{5}\right)^2$

Solution $\left(\frac{5}{6}\right)^3 \cdot \left(\frac{3}{5}\right)^2 = \left(\frac{5}{6} \cdot \frac{5}{6} \cdot \frac{5}{6}\right) \cdot \left(\frac{3}{5} \cdot \frac{3}{5}\right)$

$= \frac{\overset{1}{\cancel{5}} \cdot \overset{1}{\cancel{5}} \cdot 5 \cdot \overset{1}{\cancel{3}} \cdot \overset{1}{\cancel{3}}}{2 \cdot \underset{1}{\cancel{3}} \cdot 2 \cdot \underset{1}{\cancel{3}} \cdot 2 \cdot 3 \cdot \underset{1}{\cancel{5}} \cdot \underset{1}{\cancel{5}}} = \frac{5}{24}$

YOU TRY IT • 2

Simplify: $\left(\frac{7}{11}\right)^2 \cdot \left(\frac{2}{7}\right)$

Your solution

Solution on p. S8

OBJECTIVE C **To use the Order of Operations Agreement to simplify expressions**

The Order of Operations Agreement is used for fractions as well as whole numbers.

The Order of Operations Agreement

Step 1. Do all the operations inside parentheses.

Step 2. Simplify any number expressions containing exponents.

Step 3. Do multiplications and divisions as they occur from left to right.

Step 4. Do additions and subtractions as they occur from left to right.

HOW TO • 2 Simplify $\frac{14}{15} - \left(\frac{1}{2}\right)^2 \times \left(\frac{2}{3} + \frac{4}{5}\right)$.

$\frac{14}{15} - \left(\frac{1}{2}\right)^2 \times \left(\frac{2}{3} + \frac{4}{5}\right)$ 1. Perform operations in parentheses.

$\frac{14}{15} - \left(\frac{1}{2}\right)^2 \times \frac{22}{15}$ 2. Simplify expressions with exponents.

$\frac{14}{15} - \frac{1}{4} \times \frac{22}{15}$ 3. Do multiplication and division as they occur from left to right.

$\frac{14}{15} - \frac{11}{30}$ 4. Do addition and subtraction as they occur from left to right.

$\frac{17}{30}$

One or more of the above steps may not be needed to simplify an expression. In that case, proceed to the next step in the Order of Operations Agreement.

EXAMPLE • 3

Simplify: $\left(\frac{3}{4}\right)^2 \div \left(\frac{3}{8} - \frac{1}{12}\right)$

Solution $\left(\frac{3}{4}\right)^2 \div \left(\frac{3}{8} - \frac{1}{12}\right)$

$= \left(\frac{3}{4}\right)^2 \div \left(\frac{7}{24}\right) = \frac{9}{16} \div \frac{7}{24}$

$= \frac{9}{16} \cdot \frac{24}{7} = \frac{27}{14} = 1\frac{13}{14}$

YOU TRY IT • 3

Simplify: $\left(\frac{1}{13}\right)^2 \cdot \left(\frac{1}{4} + \frac{1}{6}\right) \div \frac{5}{13}$

Your solution

Solution on p. S8

2.8 EXERCISES

OBJECTIVE A To identify the order relation between two fractions

For Exercises 1 to 12, place the correct symbol, < or >, between the two numbers.

1. $\frac{11}{40} \quad \frac{19}{40}$
2. $\frac{92}{103} \quad \frac{19}{103}$
3. $\frac{2}{3} \quad \frac{5}{7}$
4. $\frac{2}{5} \quad \frac{3}{8}$
5. $\frac{5}{8} \quad \frac{7}{12}$
6. $\frac{11}{16} \quad \frac{17}{24}$
7. $\frac{7}{9} \quad \frac{11}{12}$
8. $\frac{5}{12} \quad \frac{7}{15}$
9. $\frac{13}{14} \quad \frac{19}{21}$
10. $\frac{13}{18} \quad \frac{7}{12}$
11. $\frac{7}{24} \quad \frac{11}{30}$
12. $\frac{13}{36} \quad \frac{19}{48}$

13. Without writing the fractions $\frac{4}{5}$ and $\frac{1}{7}$ with a common denominator, decide which fraction is larger.

OBJECTIVE B To simplify expressions containing exponents

For Exercises 14 to 29, simplify.

14. $\left(\frac{3}{8}\right)^2$
15. $\left(\frac{5}{12}\right)^2$
16. $\left(\frac{2}{9}\right)^3$
17. $\left(\frac{1}{2}\right) \cdot \left(\frac{2}{3}\right)^2$
18. $\left(\frac{2}{3}\right) \cdot \left(\frac{1}{2}\right)^4$
19. $\left(\frac{1}{3}\right)^2 \cdot \left(\frac{3}{5}\right)^3$
20. $\left(\frac{2}{5}\right)^3 \cdot \left(\frac{5}{7}\right)^2$
21. $\left(\frac{5}{9}\right)^3 \cdot \left(\frac{18}{25}\right)^2$
22. $\left(\frac{1}{3}\right)^4 \cdot \left(\frac{9}{11}\right)^2$
23. $\left(\frac{1}{2}\right)^6 \cdot \left(\frac{32}{35}\right)^2$
24. $\left(\frac{2}{3}\right)^4 \cdot \left(\frac{81}{100}\right)^2$
25. $\left(\frac{1}{6}\right) \cdot \left(\frac{6}{7}\right)^2 \cdot \left(\frac{2}{3}\right)$
26. $\left(\frac{2}{7}\right) \cdot \left(\frac{7}{8}\right)^2 \cdot \left(\frac{8}{9}\right)$
27. $3 \cdot \left(\frac{3}{5}\right)^3 \cdot \left(\frac{1}{3}\right)^2$
28. $4 \cdot \left(\frac{3}{4}\right)^3 \cdot \left(\frac{4}{7}\right)^2$
29. $11 \cdot \left(\frac{3}{8}\right)^3 \cdot \left(\frac{8}{11}\right)^2$

30. True or false? When simplified, the expression $\left(\frac{1}{2}\right)^{24} \cdot \left(\frac{1}{3}\right)^{35}$ is a fraction with a numerator of 1.

OBJECTIVE C To use the Order of Operations Agreement to simplify expressions

For Exercises 31 to 49, simplify.

31. $\frac{1}{2} - \frac{1}{3} + \frac{2}{3}$

32. $\frac{2}{5} + \frac{3}{10} - \frac{2}{3}$

33. $\frac{1}{3} \div \frac{1}{2} + \frac{3}{4}$

34. $\frac{4}{5} + \frac{3}{7} \cdot \frac{14}{15}$

35. $\left(\frac{3}{4}\right)^2 - \frac{5}{12}$

36. $\left(\frac{3}{5}\right)^3 - \frac{3}{25}$

37. $\frac{5}{6} \cdot \left(\frac{2}{3} - \frac{1}{6}\right) + \frac{7}{18}$

38. $\frac{3}{4} \cdot \left(\frac{11}{12} - \frac{7}{8}\right) + \frac{5}{16}$

39. $\frac{7}{12} - \left(\frac{2}{3}\right)^2 + \frac{5}{8}$

40. $\frac{11}{16} - \left(\frac{3}{4}\right)^2 + \frac{7}{12}$

41. $\frac{3}{4} \cdot \left(\frac{4}{9}\right)^2 + \frac{1}{2}$

42. $\frac{9}{10} \cdot \left(\frac{2}{3}\right)^3 + \frac{2}{3}$

43. $\left(\frac{1}{2} + \frac{3}{4}\right) \div \frac{5}{8}$

44. $\left(\frac{2}{3} + \frac{5}{6}\right) \div \frac{5}{9}$

45. $\frac{3}{8} \div \left(\frac{5}{12} + \frac{3}{8}\right)$

46. $\frac{7}{12} \div \left(\frac{2}{3} + \frac{5}{9}\right)$

47. $\left(\frac{3}{8}\right)^2 \div \left(\frac{3}{7} + \frac{3}{14}\right)$

48. $\left(\frac{5}{6}\right)^2 \div \left(\frac{5}{12} + \frac{2}{3}\right)$

49. $\frac{2}{5} \div \frac{3}{8} \cdot \frac{4}{5}$

50. Insert parentheses into the expression $\frac{2}{9} \cdot \frac{5}{6} + \frac{3}{4} \div \frac{3}{5}$ so that **a.** the first operation to be performed is addition and **b.** the first operation to be performed is division.

Applying the Concepts

51. **The Food Industry** The table at the right shows the results of a survey that asked fast-food patrons their criteria for choosing where to go for fast food. For example, 3 out of every 25 people surveyed said that the speed of the service was most important.

a. According to the survey, do more people choose a fast-food restaurant on the basis of its location or the quality of the food?

b. Which criterion was cited by the most people?

Fast-Food Patrons' Top Criteria for Fast-Food Restaurants	
Food quality	$\frac{1}{4}$
Location	$\frac{13}{50}$
Menu	$\frac{4}{25}$
Price	$\frac{2}{25}$
Speed	$\frac{3}{25}$
Other	$\frac{13}{100}$

Source: Maritz Marketing Research, Inc.

FOCUS ON PROBLEM SOLVING

Common Knowledge An application problem may not provide all the information that is needed to solve the problem. Sometimes, however, the necessary information is common knowledge.

HOW TO • 1 You are placing a call to New York City to obtain a patient's insurance coverage benefits. What time will it be in New York City if you are calling from Rapid City, South Dakota at 10 A.M.?

What other information do you need to solve this problem?

You will need to know that Rapid City is in the Mountain Standard Time zone and New York City is in the Eastern Standard Time zone. This represents a 2-hour time difference between these two cities. Using a 12-hour clock, the hours run

10 A.M.
11 A.M.
12 P.M.

Two hours after 10 A.M. is 12 P.M.

The time in New York City will be 12 P.M.

HOW TO • 2 A pharmacy technician sells you a bottle of multivitamins for $13. You hand the technician a 20-dollar bill. How much change do you receive?

What information do you need to solve this problem?

You need to know that there are 20 dollars in a $20 bill.

Your change is $20 – $13.

20 – 13 = 7

You receive $7 in change.

What information do you need to know to solve each of the following problems?

1. You sell a dozen tickets to a fundraiser. Each ticket costs $10. How much money do you collect?
2. The weekly lab period for your science course is 1 hour and 20 minutes long. Find the length of the science lab period in minutes.
3. An employee's monthly salary is $3750. Find the employee's annual salary.
4. A survey revealed that eighth graders spend an average of 3 hours each day watching television. Find the total time an eighth grader spends watching TV each week.
5. You want to buy a carpet for a room that is 15 feet wide and 18 feet long. Find the amount of carpet that you need.

PROJECTS AND GROUP ACTIVITIES

Music

In musical notation, notes are printed on a **staff,** which is a set of five horizontal lines and the spaces between them. The notes of a musical composition are grouped into **measures,** or **bars.** Vertical lines separate measures on a staff. The shape of a note indicates how long it should be held. The whole note has the longest time value of any note. Each time value is divided by 2 in order to find the next smallest time value.

The **time signature** is a fraction that appears at the beginning of a piece of music. The numerator of the fraction indicates the number of beats in a measure. The denominator indicates what kind of note receives 1 beat. For example, music written in $\frac{2}{4}$ time has 2 beats to a measure, and a quarter note receives 1 beat. One measure in $\frac{2}{4}$ time may have 1 half note, 2 quarter notes, 4 eighth notes, or any other combination of notes totaling 2 beats. Other common time signatures are $\frac{4}{4}$, $\frac{3}{4}$, and $\frac{6}{8}$.

1. Explain the meaning of the 6 and the 8 in the time signature $\frac{6}{8}$.
2. Give some possible combinations of notes in one measure of a piece written in $\frac{4}{4}$ time.
3. What does a dot at the right of a note indicate? What is the effect of a dot at the right of a half note? At the right of a quarter note? At the right of an eighth note?
4. Symbols called rests are used to indicate periods of silence in a piece of music. What symbols are used to indicate the different time values of rests?
5. Find some examples of musical compositions written in different time signatures. Use a few measures from each to show that the sum of the time values of the notes and rests in each measure equals the numerator of the time signature.

Construction

Suppose you are involved in building your own home. Design a stairway from the first floor of the house to the second floor. Some of the questions you will need to answer follow.

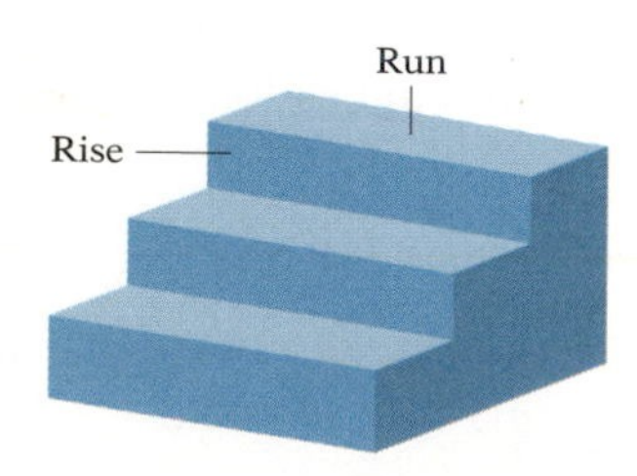

What is the distance from the floor of the first story to the floor of the second story?

Typically, what is the number of steps in a stairway?

What is a reasonable length for the run of each step?

What is the width of the wood being used to build the staircase?

In designing the stairway, remember that each riser should be the same height, that each run should be the same length, and that the width of the wood used for the steps will have to be incorporated into the calculation.

Fractions of Diagrams The diagram that follows has been broken up into nine areas separated by heavy lines. Eight of the areas have been labeled *A* through *H*. The ninth area is shaded. Determine which lettered areas would have to be shaded so that half of the entire diagram is shaded and half is not shaded. Write down the strategy that you or your group used to arrive at the solution. Compare your strategy with that of other individual students or groups.

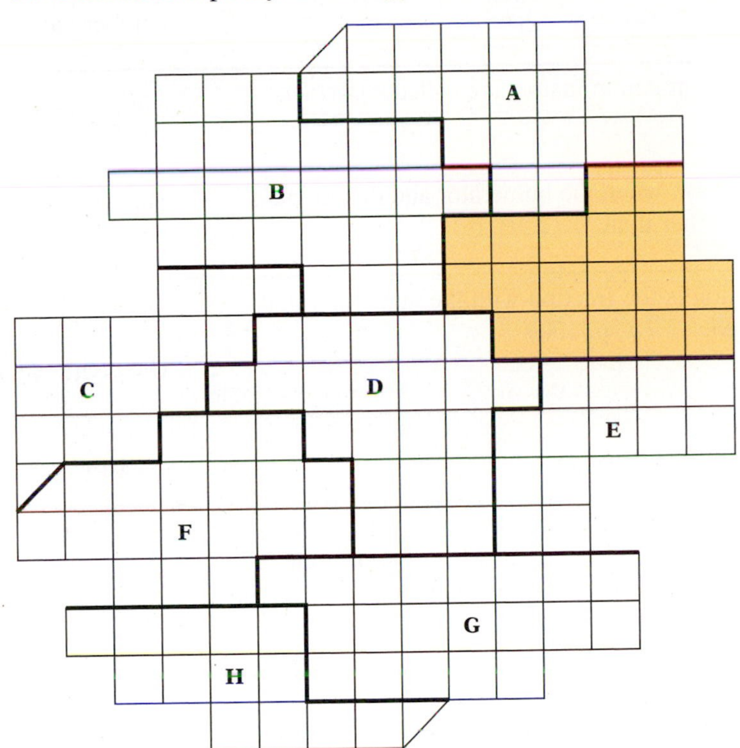

Tips for Success

Three important features of this text that can be used to prepare for a test are the

- Chapter Summary
- Chapter Review Exercises
- Chapter Test

See *AIM for Success* at the front of the book.

CHAPTER 2

SUMMARY

KEY WORDS	EXAMPLES
A number that is a multiple of two or more numbers is a *common multiple* of those numbers. The *least common multiple (LCM)* is the smallest common multiple of two or more numbers. [2.1A, p. 64]	12, 24, 36, 48, . . . are common multiples of 4 and 6. The LCM of 4 and 6 is 12.
A number that is a factor of two or more numbers is a *common factor* of those numbers. The *greatest common factor (GCF)* is the largest common factor of two or more numbers. [2.1B, p. 65]	The common factors of 12 and 16 are 1, 2, and 4. The GCF of 12 and 16 is 4.
A *fraction* can represent the number of equal parts of a whole. In a fraction, the *fraction bar* separates the *numerator* and the *denominator*. [2.2A, p. 68]	In the fraction $\frac{3}{4}$, the numerator is 3 and the denominator is 4.

In a *proper fraction,* the numerator is smaller than the denominator; a proper fraction is a number less than 1. In an *improper fraction,* the numerator is greater than or equal to the denominator; an improper fraction is a number greater than or equal to 1. A *mixed number* is a number greater than 1 with a whole-number part and a fractional part. [2.2A, p. 68]

$\frac{2}{5}$ is proper fraction.

$\frac{7}{6}$ is an improper fraction.

$4\frac{1}{10}$ is a mixed number; 4 is the whole-number part and $\frac{1}{10}$ is the fractional part.

Equal fractions with different denominators are called *equivalent fractions.* [2.3A, p. 72]

$\frac{3}{4}$ and $\frac{6}{8}$ are equivalent fractions.

A fraction is in *simplest form* when the numerator and denominator have no common factors other than 1. [2.3B, p. 73]

The fraction $\frac{11}{12}$ is in simplest form.

The *reciprocal* of a fraction is the fraction with the numerator and denominator interchanged. [2.7A, p. 100]

The reciprocal of $\frac{3}{8}$ is $\frac{8}{3}$.

The reciprocal of 5 is $\frac{1}{5}$.

ESSENTIAL RULES AND PROCEDURES

EXAMPLES

To find the LCM of two or more numbers, find the prime factorization of each number and write the factorization of each number in a table. Circle the greatest product in each column. The LCM is the product of the circled numbers. [2.1A, p. 64]

	2	3
12 =	(2 · 2)	3
18 =	2	(3 · 3)

The LCM of 12 and 18 is $2 \cdot 2 \cdot 3 \cdot 3 = 36$.

To find the GCF of two or more numbers, find the prime factorization of each number and write the factorization of each number in a table. Circle the least product in each column that does not have a blank. The GCF is the product of the circled numbers. [2.1B, p. 65]

	2	3
12 =	2 · 2	(3)
18 =	(2)	3 · 3

The GCF of 12 and 18 is $2 \cdot 3 = 6$.

To write an improper fraction as a mixed number or a whole number, divide the numerator by the denominator. [2.2B, p. 69]

$$\frac{29}{6} = 29 \div 6 = 4\frac{5}{6}$$

To write a mixed number as an improper fraction, multiply the denominator of the fractional part of the mixed number by the whole-number part. Add this product and the numerator of the fractional part. The sum is the numerator of the improper fraction. The denominator remains the same. [2.2B, p. 69]

$$3\frac{2}{5} = \frac{5 \times 3 + 2}{5} = \frac{17}{5}$$

To find equivalent fractions by raising to higher terms, multiply the numerator and denominator of the fraction by the same number. [2.3A, p. 72]

$$\frac{3}{4} = \frac{3 \cdot 5}{4 \cdot 5} = \frac{15}{20}$$

$\frac{3}{4}$ and $\frac{15}{20}$ are equivalent fractions.

To write a fraction in simplest form, factor the numerator and denominator of the fraction; then eliminate the common factors. [2.3B, p. 73]

$$\frac{30}{45} = \frac{2 \cdot \overset{1}{\cancel{3}} \cdot \overset{1}{\cancel{5}}}{\underset{1}{\cancel{3}} \cdot 3 \cdot \underset{1}{\cancel{5}}} = \frac{2}{3}$$

To add fractions with the same denominator, add the numerators and place the sum over the common denominator. [2.4A, p. 76]

$$\frac{5}{12} + \frac{11}{12} = \frac{16}{12} = 1\frac{4}{12} = 1\frac{1}{3}$$

To add fractions with different denominators, first rewrite the fractions as equivalent fractions with a common denominator. (The common denominator is the LCM of the denominators of the fractions.) Then add the fractions. [2.4B, p. 76]

$$\frac{1}{4} + \frac{2}{5} = \frac{5}{20} + \frac{8}{20} = \frac{13}{20}$$

To subtract fractions with the same denominator, subtract the numerators and place the difference over the common denominator. [2.5A, p. 84]

$$\frac{9}{16} - \frac{5}{16} = \frac{4}{16} = \frac{1}{4}$$

To subtract fractions with different denominators, first rewrite the fractions as equivalent fractions with a common denominator. (The common denominator is the LCM of the denominators of the fractions.) Then subtract the fractions. [2.5B, p. 84]

$$\frac{2}{3} - \frac{7}{16} = \frac{32}{48} - \frac{21}{48} = \frac{11}{48}$$

To multiply two fractions, multiply the numerators; this is the numerator of the product. Multiply the denominators; this is the denominator of the product. [2.6A, p. 92]

$$\frac{3}{4} \cdot \frac{2}{9} = \frac{3 \cdot 2}{4 \cdot 9} = \frac{\overset{1}{\cancel{3}} \cdot \overset{1}{\cancel{2}}}{\underset{1}{\cancel{2}} \cdot 2 \cdot \underset{1}{\cancel{3}} \cdot 3} = \frac{1}{6}$$

To divide two fractions, multiply the first fraction by the reciprocal of the second fraction. [2.7A, p. 100]

$$\frac{8}{15} \div \frac{4}{5} = \frac{8}{15} \cdot \frac{5}{4} = \frac{8 \cdot 5}{15 \cdot 4}$$

$$= \frac{\overset{1}{\cancel{2}} \cdot \overset{1}{\cancel{2}} \cdot 2 \cdot \overset{1}{\cancel{5}}}{3 \cdot \underset{1}{\cancel{5}} \cdot \underset{1}{\cancel{2}} \cdot \underset{1}{\cancel{2}}} = \frac{2}{3}$$

The find the order relation between two fractions with the same denominator, compare the numerators. The fraction that has the smaller numerator is the smaller fraction. [2.8A, p. 109]

$\frac{17}{25}$ ← Smaller numerator

$\frac{19}{25}$ ← Larger numerator

$$\frac{17}{25} < \frac{19}{25}$$

To find the order relation between two fractions with different denominators, first rewrite the fractions with a common denominator. The fraction that has the smaller numerator is the smaller fraction. [2.8A, p. 109]

$$\frac{3}{5} = \frac{24}{40} \qquad \frac{5}{8} = \frac{25}{40}$$

$$\frac{24}{40} < \frac{25}{40}$$

$$\frac{3}{5} < \frac{5}{8}$$

Order of Operations Agreement [2.8C, p. 110]

Step 1 Do all the operations inside parentheses.

Step 2 Simplify any numerical expressions containing exponents.

Step 3 Do multiplication and division as they occur from left to right.

Step 4 Do addition and subtraction as they occur from left to right.

$$\left(\frac{1}{3}\right)^2 + \left(\frac{5}{6} - \frac{7}{12}\right) \cdot (4)$$

$$= \left(\frac{1}{3}\right)^2 + \left(\frac{1}{4}\right) \cdot (4)$$

$$= \frac{1}{9} + \left(\frac{1}{4}\right) \cdot (4)$$

$$= \frac{1}{9} + 1 = 1\frac{1}{9}$$

CHAPTER 2

CONCEPT REVIEW

Test your knowledge of the concepts presented in this chapter. Answer each question. Then check your answers against the ones provided in the Answer Section.

1. How do you find the LCM of 75, 30, and 50?

2. How do you find the GCF of 42, 14, and 21?

3. How do you write an improper fraction as a mixed number?

4. When is a fraction in simplest form?

5. When adding fractions, why do you have to convert to equivalent fractions with a common denominator?

6. How do you add mixed numbers?

7. If you are subtracting a mixed number from a whole number, why do you need to borrow?

8. When multiplying two fractions, why is it better to eliminate the common factors before multiplying the remaining factors in the numerator and denominator?

9. When multiplying two fractions that are less than 1, will the product be greater than 1, less than the smaller number, or between the smaller number and the bigger number?

10. How are reciprocals used when dividing fractions?

11. When a fraction is divided by a whole number, why do we write the whole number as a fraction before dividing?

12. When comparing two fractions, why is it important to look at both the numerators and denominators to determine which is larger?

13. In the expression $\left(\frac{5}{6}\right)^2 - \left(\frac{3}{4} - \frac{2}{3}\right) \div \frac{1}{2}$, in what order should the operations be performed?

CHAPTER 2

REVIEW EXERCISES

1. Write $\frac{30}{45}$ in simplest form.

2. Simplify: $\left(\frac{3}{4}\right)^3 \cdot \frac{20}{27}$

3. Express the shaded portion of the circles as an improper fraction.

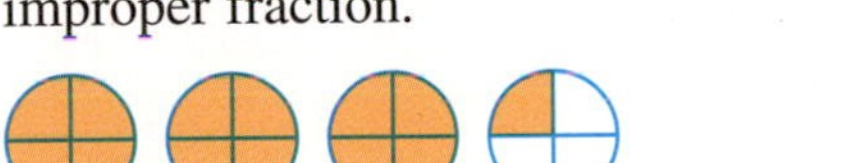

4. Find the total of $\frac{2}{3}$, $\frac{5}{6}$, and $\frac{2}{9}$.

5. Place the correct symbol, < or >, between the two numbers.

$\frac{11}{18} \quad \frac{17}{24}$

6. Subtract: $\begin{array}{r} 18\frac{1}{6} \\ -3\frac{5}{7} \\ \hline \end{array}$

7. Simplify: $\frac{2}{7}\left(\frac{5}{8} - \frac{1}{3}\right) \div \frac{3}{5}$

8. Multiply: $2\frac{1}{3} \times 3\frac{7}{8}$

9. Divide: $1\frac{1}{3} \div \frac{2}{3}$

10. Find $\frac{17}{24}$ decreased by $\frac{3}{16}$.

11. Divide: $8\frac{2}{3} \div 2\frac{3}{5}$

12. Find the GCF of 20 and 48.

13. Write an equivalent fraction with the given denominator.

$\frac{2}{3} = \frac{\quad}{36}$

14. What is $\frac{15}{28}$ divided by $\frac{5}{7}$?

15. Write an equivalent fraction with the given denominator.

$\frac{8}{11} = \frac{\quad}{44}$

16. Multiply: $2\frac{1}{4} \times 7\frac{1}{3}$

17. Find the LCM of 18 and 12.

18. Write $\frac{16}{44}$ in simplest form.

19. Add: $\frac{3}{8} + \frac{5}{8} + \frac{1}{8}$

20. Subtract: $\begin{array}{r} 16 \\ -\ 5\frac{7}{8} \\ \hline \end{array}$

21. Add: $4\frac{4}{9} + 2\frac{1}{6} + 11\frac{17}{27}$

22. Find the GCF of 15 and 25.

23. Write $\frac{17}{5}$ as a mixed number.

24. Simplify: $\left(\frac{4}{5} - \frac{2}{3}\right)^2 \div \frac{4}{15}$

25. Add: $\frac{3}{8} + 1\frac{2}{3} + 3\frac{5}{6}$

26. Find the LCM of 18 and 27.

27. Subtract: $\frac{11}{18} - \frac{5}{18}$

28. Write $2\frac{5}{7}$ as an improper fraction.

29. Divide: $\frac{5}{6} \div \frac{5}{12}$

30. Multiply: $\frac{5}{12} \times \frac{4}{25}$

31. What is $\frac{11}{50}$ multiplied by $\frac{25}{44}$?

32. Express the shaded portion of the circles as a mixed number.

33. **Weight Loss** A nurse at a medical spa records the monthly weight loss of clients. If a client loses $4\frac{1}{2}$ pounds in May, $3\frac{3}{4}$ pounds in June, and $4\frac{1}{4}$ pounds in July, find the client's total weight loss for these three months.

34. **Real Estate** The Community Care Clinic purchased and remodeled a building for $324,000, dividing the property into eight offices of equal size. If the value of each office is equal to $\frac{1}{8}$ of the original cost, what is the value of each office?

35. **Fundraiser** A 15-mile "Race for the Cure" event has three checkpoints. The first checkpoint is $4\frac{1}{2}$ miles from the starting point. The second checkpoint is $5\frac{3}{4}$ miles from the first checkpoint. How many miles is the second checkpoint from the finish line?

Holly Kuchera/Shutterstock.com

36. **Fuel Efficiency** An ambulance gets 12 miles on each gallon of gasoline. How many miles can the ambulance travel on $34\frac{1}{2}$ gallons of gasoline?

CHAPTER 2

TEST

1. Multiply: $\frac{9}{11} \times \frac{44}{81}$

2. Find the GCF of 24 and 80.

3. Divide: $\frac{5}{9} \div \frac{7}{18}$

4. Simplify: $\left(\frac{3}{4}\right)^2 \div \left(\frac{2}{3} + \frac{5}{6}\right) - \frac{1}{12}$

5. Write $9\frac{4}{5}$ as an improper fraction.

6. What is $5\frac{2}{3}$ multiplied by $1\frac{7}{17}$?

7. Write $\frac{40}{64}$ in simplest form.

8. Place the correct symbol, < or >, between the two numbers.
   $\frac{3}{8} \quad \frac{5}{12}$

9. Simplify: $\left(\frac{1}{4}\right)^3 \div \left(\frac{1}{8}\right)^2 - \frac{1}{6}$

10. Find the LCM of 24 and 40.

11. Subtract: $\frac{17}{24} - \frac{11}{24}$

12. Write $\frac{18}{5}$ as a mixed number.

13. Find the quotient of $6\frac{2}{3}$ and $3\frac{1}{6}$.

14. Write an equivalent fraction with the given denominator: $\frac{5}{8} = \frac{}{72}$

15. Add:

$$\begin{array}{r} \frac{5}{6} \\ \frac{7}{9} \\ +\frac{1}{15} \\ \hline \end{array}$$

16. Subtract:

$$\begin{array}{r} 23 \\ -9\frac{9}{44} \\ \hline \end{array}$$

17. What is $\frac{9}{16}$ minus $\frac{5}{12}$?

18. Simplify: $\left(\frac{2}{3}\right)^4 \cdot \frac{27}{32}$

19. Add: $\frac{7}{12} + \frac{11}{12} + \frac{5}{12}$

20. What is $12\frac{5}{12}$ more than $9\frac{17}{20}$?

21. Express the shaded portion of the circles as an improper fraction.

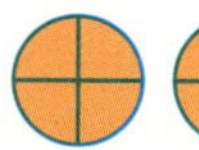  

22. Compensation A radiologist earns $240 for each day worked. What is the total of the radiologist's earnings for working $3\frac{1}{2}$ days?

23. Measurement A podiatrist uses plaster wraps to mold and produce orthotics for patients. The wraps used measured $14\frac{1}{4}$ inches, $12\frac{5}{8}$ inches, and $15\frac{4}{6}$ inches. Find the total length of material used.

24. Measurement A roll of adhesive tape is 36 feet long. How many $\frac{1}{2}$-foot sections can be cut from this roll?

25. Weight Loss A nurse's aide is monitoring closely the weight loss of a geriatric patient who weighed 122 pounds when admitted to the floor. The patient lost $1\frac{1}{2}$ pounds the first week and $\frac{3}{4}$ pound the second week. How much does the patient weigh at the end of the second week?

CUMULATIVE REVIEW EXERCISES

1. Round 290,496 to the nearest thousand.

2. Subtract: $\begin{array}{r} 390{,}047 \\ -\ 98{,}769 \\ \hline \end{array}$

3. Find the product of 926 and 79.

4. Divide: $57\overline{)30{,}792}$

5. Simplify: $4 \cdot (6 - 3) \div 6 - 1$

6. Find the prime factorization of 44.

7. Find the LCM of 30 and 42.

8. Find the GCF of 60 and 80.

9. Write $7\frac{2}{3}$ as an improper fraction.

10. Write $\frac{25}{4}$ as a mixed number.

11. Write an equivalent fraction with the given denominator.

$$\frac{5}{16} = \frac{}{48}$$

12. Write $\frac{24}{60}$ in simplest form.

13. What is $\frac{9}{16}$ more than $\frac{7}{12}$?

14. Add: $\begin{array}{r} 3\frac{7}{8} \\ 7\frac{5}{12} \\ +\ 2\frac{15}{16} \\ \hline \end{array}$

15. Find $\frac{3}{8}$ less than $\frac{11}{12}$.

16. Subtract: $\begin{array}{r} 5\frac{1}{6} \\ -\ 3\frac{7}{18} \\ \hline \end{array}$

17. Multiply: $\frac{3}{8} \times \frac{14}{15}$

18. Multiply: $3\frac{1}{8} \times 2\frac{2}{5}$

19. Divide: $\frac{7}{16} \div \frac{5}{12}$

20. Find the quotient of $6\frac{1}{8}$ and $2\frac{1}{3}$.

21. Simplify: $\left(\frac{1}{2}\right)^3 \cdot \frac{8}{9}$

22. Simplify: $\left(\frac{1}{2} + \frac{1}{3}\right) \div \left(\frac{2}{5}\right)^2$

23. Banking Molly O'Brien had \$1359 in a checking account. During the week, Molly wrote checks for \$128, \$54, and \$315. Find the amount in the checking account at the end of the week.

24. Purchasing The office manager of a medical practice orders 10 boxes of latex safety gloves for \$17 per box and 20 boxes of tongue depressors for \$4 per box. Find the total cost of this order.

25. Growth A new baby measures $19\frac{7}{8}$ inches at its birth on July 31. It grew $\frac{3}{4}$ inch in August, $\frac{5}{8}$ inch in September, and $\frac{1}{2}$ inch in October. How long is the baby at the end of three months?

Gaby Kooijman/Shutterstock.com

26. IV Infusion A patient is receiving 500 milliliters of a solution containing an antibiotic through an IV infusion. The nurse estimates that $\frac{2}{5}$ of the fluid remains in the bag after an hour has passed. How much fluid remains to be infused?

27. Dosage A pediatric dosage of a cough medicine is $\frac{2}{3}$ of an adult dose. If an adult dose of this medicine is $1\frac{1}{2}$ ounces every 4 hours, how much should you give a child?

28. Dosage If a bottle of cough syrup contains 12 ounces of medication, how many doses are in the bottle if an adult dose is $\frac{3}{4}$ ounce every 4 hours?

Decimals

CHAPTER 3

OBJECTIVES

SECTION 3.1

A To write decimals in standard form and in words

B To round a decimal to a given place value

SECTION 3.2

A To add decimals

B To solve application problems

SECTION 3.3

A To subtract decimals

B To solve application problems

SECTION 3.4

A To multiply decimals

B To solve application problems

SECTION 3.5

A To divide decimals

B To solve application problems

SECTION 3.6

A To convert fractions to decimals

B To convert decimals to fractions

C To identify the order relation between two decimals or between a decimal and a fraction

AIM for the Future

NURSING AIDES (CNAs) help care for physically ill, injured, disabled, or infirm individuals in hospitals and nursing care facilities. Nursing aides and home health aides are among the occupations commonly referred to as direct-care workers due to their roles in working with patients who need long-term care. Employment for nursing aides, orderlies, and attendants will grow 19% from 2008 to 2018, faster than the average for all occupations, predominantly in response to the long-term care needs of an increasing elderly population. Median hourly wages of nursing aides, orderlies, and attendants were $11.46 in May 2008. (*Source: Bureau of Labor Statistics:* http://www.bls.gov/oco/ocos327.htm)

PREP TEST

Are you ready to succeed in this chapter?

Take the Prep Test below to find out if you are ready to learn the new material.

1. Express the shaded portion of the rectangle as a fraction.

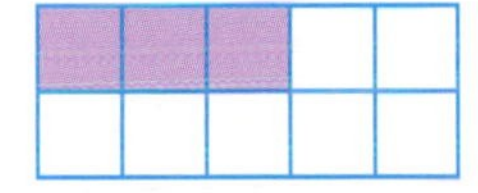

2. Round 36,852 to the nearest hundred.

3. Write 4791 in words.

4. Write six thousand eight hundred forty-two in standard form.

For Exercises 5 to 8, add, subtract, multiply, or divide.

5. $37 + 8892 + 465$

6. $2403 - 765$

7. 844×91

8. $23\overline{)6412}$

SECTION

3.1 Introduction to Decimals

OBJECTIVE A To write decimals in standard form and in words

Take Note

In decimal notation, the part of the number that appears to the left of the decimal point is the **whole-number part.** The part of the number that appears to the right of the decimal point is the **decimal part.** The **decimal point** separates the whole-number part from the decimal part.

The price tag on a sweater reads $61.88. The number 61.88 is in **decimal notation.** A number written in decimal notation is often called simply a **decimal.**

A number written in decimal notation has three parts.

61	.	88
Whole-number part	**Decimal point**	**Decimal part**

The decimal part of the number represents a number less than 1. For example, $0.88 is less than $1. The decimal point (.) separates the whole-number part from the decimal part.

The position of a digit in a decimal determines the digit's place value. The place-value chart is extended to the right to show the place value of digits to the right of a decimal point.

Point of Interest

The idea that all fractions should be represented in tenths, hundredths, and thousandths was presented in 1585 in Simon Stevin's publication *De Thiende* and its French translation, *La Disme,* which was widely read and accepted by the French. This may help to explain why the French accepted the metric system so easily two hundred years later.

In *De Thiende*, Stevin argued in favor of his notation by including examples for astronomers, tapestry makers, surveyors, tailors, and the like. He stated that using decimals would enable calculations to be "performed . . . with as much ease as counterreckoning."

In the decimal 458.302719, the position of the digit 7 determines that its place value is ten-thousandths.

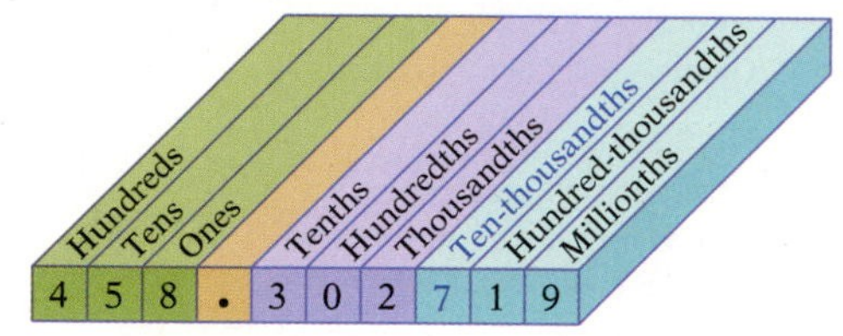

Note the relationship between fractions and numbers written in decimal notation.

Seven tenths	Seven hundredths	Seven thousandths
$\frac{7}{10} = 0.7$	$\frac{7}{100} = 0.07$	$\frac{7}{1000} = 0.007$
1 zero in 10	2 zeros in 100	3 zeros in 1000
1 decimal place in 0.7	2 decimal places in 0.07	3 decimal places in 0.007

To write a decimal in words, write the decimal part of the number as though it were a whole number, and then name the place value of the last digit.

0.9684 Nine thousand six hundred eighty-four ten-thousandths

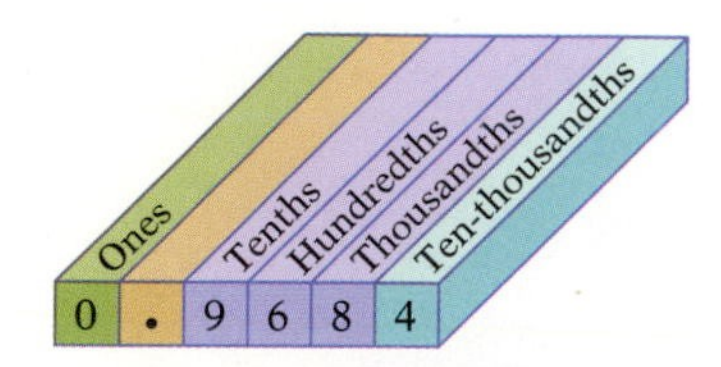

The decimal point in a decimal is read as "and."

372.516 Three hundred seventy-two and five hundred sixteen thousandths

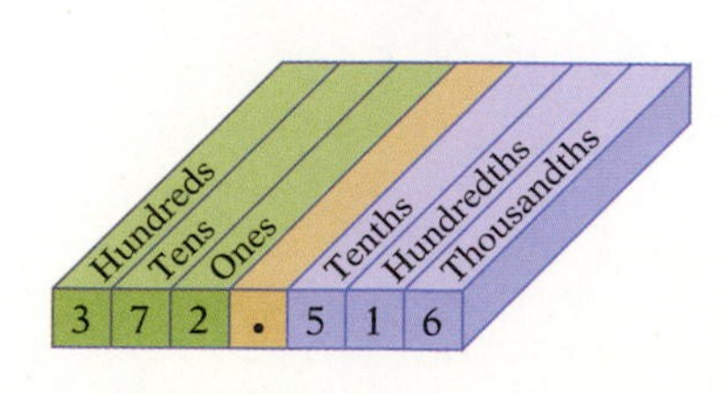

Point of Interest

The decimal point did not make its appearance until the early 1600s. Stevin's notation used subscripts with circles around them after each digit: 0 for ones, 1 for tenths (which he called "primes"), 2 for hundredths (called "seconds"), 3 for thousandths ("thirds"), and so on. For example, 1.375 would have been written

1 3 7 5
⓪ ① ② ③

To write a decimal in standard form when it is written in words, write the whole-number part, replace the word *and* with a decimal point, and write the decimal part so that the last digit is in the given place-value position.

Four and twenty-three <u>hundredths</u>

3 is in the hundredths place.

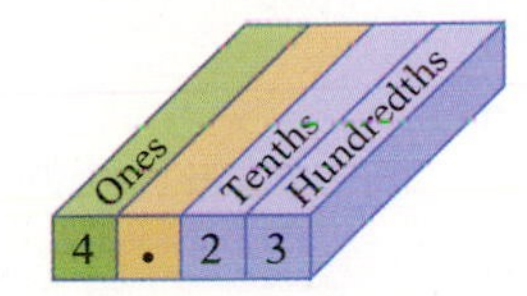

When writing a decimal in standard form, you may need to insert zeros after the decimal point so that the last digit is in the given place-value position.

Ninety-one and eight <u>thousandths</u>

8 is in the thousandths place.

Insert two zeros so that the 8 is in the thousandths place.

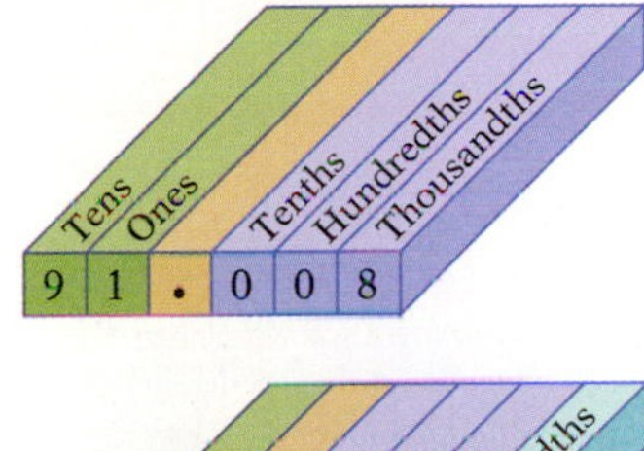

Sixty-five <u>ten-thousandths</u>

5 is in the ten-thousandths place.

Insert two zeros so that the 5 is in the ten-thousandths place.

Ones | Tenths | Hundredths | Thousandths | Ten-thousandths

0 . 0 0 6 5

EXAMPLE • 1

Name the place value of the digit 8 in the number 45.687.

Solution

The digit 8 is in the hundredths place.

YOU TRY IT • 1

Name the place value of the digit 4 in the number 907.1342.

Your solution

EXAMPLE • 2

Write $\frac{43}{100}$ as a decimal.

Solution

$\frac{43}{100} = 0.43$ • **Forty-three hundredths**

YOU TRY IT • 2

Write $\frac{501}{1000}$ as a decimal.

Your solution

EXAMPLE • 3

Write 0.289 as a fraction.

Solution

$0.289 = \frac{289}{1000}$ • **289 thousandths**

YOU TRY IT • 3

Write 0.67 as a fraction.

Your solution

EXAMPLE • 4

Write 293.50816 in words.

Solution

Two hundred ninety-three and fifty thousand eight hundred sixteen hundred-thousandths

YOU TRY IT • 4

Write 55.6083 in words.

Your solution

Solutions on p. S8

EXAMPLE • 5

Write twenty-three and two hundred forty-seven millionths in standard form.

Solution

23.000247 • 7 is in the millionths place.

YOU TRY IT • 5

Write eight hundred six and four hundred ninety-one hundred-thousandths in standard form.

Your solution

Solution on p. S8

OBJECTIVE B To round a decimal to a given place value

Tips for Success

Have you considered joining a study group? Getting together regularly with other students in the class to go over material and quiz each other can be very beneficial. See *AIM for Success* at the front of the book.

In general, rounding decimals is similar to rounding whole numbers except that the digits to the right of the given place value are dropped instead of being replaced by zeros.

If the digit to the right of the given place value is less than 5, that digit and all digits to the right are dropped.

Round 6.9237 to the nearest hundredth.

6.9237 — Given place value (hundredths): 2

$3 < 5$ Drop the digits 3 and 7.

6.9237 rounded to the nearest hundredth is 6.92.

If the digit to the right of the given place value is greater than or equal to 5, increase the digit in the given place value by 1, and drop all digits to its right.

Round 12.385 to the nearest tenth.

12.385 — Given place value (tenths): 3

$8 > 5$ Increase 3 by 1 and drop all digits to the right of 3.

12.385 rounded to the nearest tenth is 12.4.

Take Note

In the example at the right, the zero in the given place value is not dropped. This indicates that the number is rounded to the nearest thousandth. If we dropped the zero and wrote 0.47, it would indicate that the number was rounded to the nearest hundredth.

HOW TO • 1 Round 0.46972 to the nearest thousandth.

0.46972 — Given place value (thousandths): 9

$7 > 5$ Round up by adding 1 to the 9 $(9 + 1 = 10)$. Carry the 1 to the hundredths place $(6 + 1 = 7)$.

0.46972 rounded to the nearest thousandth is 0.470.

EXAMPLE • 6

Round 0.9375 to the nearest thousandth.

Solution

Given place value

0.9375

$5 = 5$

0.9375 rounded to the nearest thousandth is 0.938.

YOU TRY IT • 6

Round 3.675849 to the nearest ten-thousandth.

Your solution

EXAMPLE • 7

Round 2.5963 to the nearest hundredth.

Solution

Given place value

2.5963

$6 > 5$

2.5963 rounded to the nearest hundredth is 2.60.

YOU TRY IT • 7

Round 48.907 to the nearest tenth.

Your solution

EXAMPLE • 8

Round 72.416 to the nearest whole number.

Solution

Given place value

72.416

$4 < 5$

72.416 rounded to the nearest whole number is 72.

YOU TRY IT • 8

Round 31.8652 to the nearest whole number.

Your solution

EXAMPLE • 9

The average fetal length of a baby at 40 weeks (full-term) is 20.16 inches. To the nearest whole number, how long is the average baby at 40 weeks?

Solution

20.16 rounded to the nearest whole number is 20.

The average length of a baby at 40 weeks is about 20 inches.

YOU TRY IT • 9

The average fetal weight of a baby at 40 weeks is 7.63 pounds. To the nearest pound, what is the average weight of a newborn at 40 weeks?

Your solution

Solutions on p. S8

3.1 EXERCISES

OBJECTIVE A **To write decimals in standard form and in words**

For Exercises 1 to 6, name the place value of the digit 5.

1. 76.31587

2. 291.508

3. 432.09157

4. 0.0006512

5. 38.2591

6. 0.0000853

For Exercises 7 to 12, write the fraction as a decimal.

7. $\frac{3}{10}$

8. $\frac{9}{10}$

9. $\frac{21}{100}$

10. $\frac{87}{100}$

11. $\frac{461}{1000}$

12. $\frac{853}{1000}$

For Exercises 13 to 18, write the decimal as a fraction.

13. 0.1

14. 0.3

15. 0.47

16. 0.59

17. 0.289

18. 0.601

For Exercises 19 to 27, write the number in words.

19. 0.37

20. 25.6

21. 9.4

22. 1.004

23. 0.0053

24. 41.108

25. 0.045

26. 3.157

27. 26.04

For Exercises 28 to 35, write the number in standard form.

28. Six hundred seventy-two thousandths

29. Three and eight hundred six ten-thousandths

30. Nine and four hundred seven ten-thousandths

31. Four hundred seven and three hundredths

32. Six hundred twelve and seven hundred four thousandths

33. Two hundred forty-six and twenty-four thousandths

34. Two thousand sixty-seven and nine thousand two ten-thousandths

35. Seventy-three and two thousand six hundred eighty-four hundred-thousandths

36. Suppose the first nonzero digit to the right of the decimal point in a decimal number is in the hundredths place. If the number has three consecutive nonzero digits to the right of the decimal point, and all other digits are zero, what place value names the number?

OBJECTIVE B To round a decimal to a given place value

For Exercises 37 to 51, round the number to the given place value.

37. 6.249 Tenths

38. 5.398 Tenths

39. 21.007 Tenths

40. 30.0092 Tenths

41. 18.40937 Hundredths

42. 413.5972 Hundredths

43. 72.4983 Hundredths

44. 6.061745 Thousandths

45. 936.2905 Thousandths

46. 96.8027 Whole number

47. 47.3192 Whole number

48. 5439.83 Whole number

49. 7014.96 Whole number

50. 0.023591 Ten-thousandths

51. 2.975268 Hundred-thousandths

52. **Atomic Mass** The atomic mass of oxygen given in the Periodic Table of Elements is 15.9994. Round this number to the nearest tenth.

53. **Measurement** The length of the small intestine is approximately 6.1 meters. Round this measurement to the nearest whole number.

Cogipix/Shutterstock.com

For Exercises 54 and 55, give an example of a decimal number that satisfies the given condition.

54. The number rounded to the nearest tenth is greater than the number rounded to the nearest hundredth.

55. The number rounded to the nearest hundredth is equal to the number rounded to the nearest thousandth.

Applying the Concepts

56. Indicate which digits of the number, if any, need not be entered on a calculator.
a. 1.500 **b.** 0.908 **c.** 60.07 **d.** 0.0032

57. **a.** Find a number between 0.1 and 0.2. **b.** Find a number between 1 and 1.1.
c. Find a number between 0 and 0.005.

SECTION

3.2 Addition of Decimals

OBJECTIVE A To add decimals

To add decimals, write the numbers so that the decimal points are on a vertical line. Add as for whole numbers, and write the decimal point in the sum directly below the decimal points in the addends.

HOW TO 1 Add: 0.237 + 4.9 + 27.32

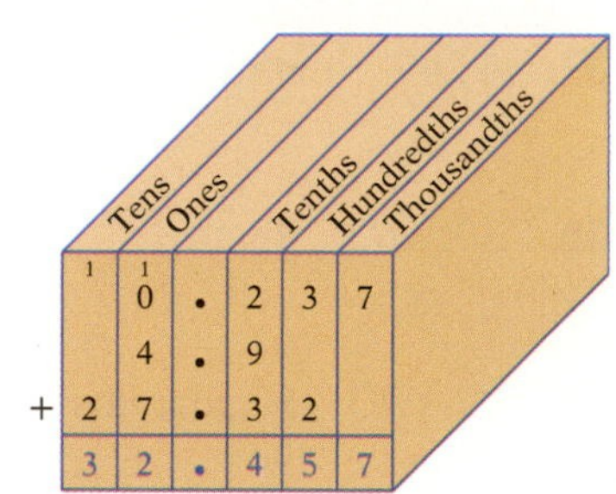

Note that by placing the decimal points on a vertical line, we make sure that digits of the same place value are added.

EXAMPLE 1

Find the sum of 42.3, 162.903, and 65.0729.

Solution

$$\begin{array}{r} \scriptstyle 1\,1\,1 \\ 42.3 \\ 162.903 \\ +\ 65.0729 \\ \hline 270.2759 \end{array}$$

• **Place the decimal points on a vertical line.**

YOU TRY IT 1

Find the sum of 4.62, 27.9, and 0.62054.

Your solution

EXAMPLE 2

Add: 0.83 + 7.942 + 15

Solution

$$\begin{array}{r} \scriptstyle 1\,1 \\ 0.83 \\ 7.942 \\ +15. \\ \hline 23.772 \end{array}$$

YOU TRY IT 2

Add: 6.05 + 12 + 0.374

Your solution

Solutions on p. S8

ESTIMATION

Estimating the Sum of Two or More Decimals

Calculate 23.037 + 16.7892. Then use estimation to determine whether the sum is reasonable.

Add to find the exact sum. 23.037 [+] 16.7892 [=] 39.8262

To estimate the sum, round each number to the same place value. Here we have rounded to the nearest whole number. Then add. The estimated answer is 40, which is very close to the exact sum, 39.8262.

$$\begin{array}{rcr} 23.037 & \approx & 23 \\ +16.7892 & \approx & +17 \\ \hline & & 40 \end{array}$$

OBJECTIVE B To solve application problems

The graph at the right shows the breakdown by age group of Americans who are hearing-impaired. Use this graph for Example 3 and You Try It 3.

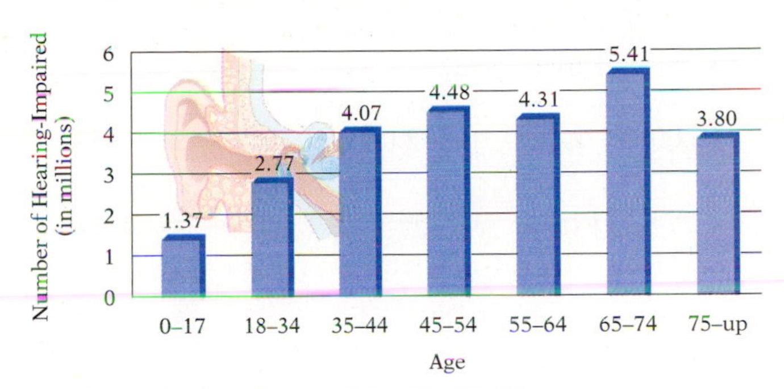

Breakdown by Age Group of Americans Who Are Hearing-Impaired

Source: American Speech-Language-Hearing Association

EXAMPLE 3

Determine the number of Americans under the age of 45 who are hearing-impaired.

Strategy

To determine the number, add the numbers of hearing impaired ages 0 to 17, 18 to 34, and 35 to 44.

Solution

$$\begin{array}{r} 1.37 \\ 2.77 \\ +4.07 \\ \hline 8.21 \end{array}$$

8.21 million Americans under the age of 45 are hearing-impaired.

YOU TRY IT 3

Determine the number of Americans ages 45 and older who are hearing-impaired.

Your strategy

Your solution

EXAMPLE 4

A pharmaceutical representative earns $2685.52 a month in salary. The pharmaceutical representative also earned $348.95, $287.34, $411.26, and $167.44 in commissions during the four weeks of this month. Find the total income for the month.

Strategy

To find the total income, add the commissions (348.95, 287.34, 411.26, 167.44) to the salary (2685.52).

Solution

$2685.52 + 348.95 + 287.34 + 411.26 + 167.44$
$= 3900.51$

The pharmaceutical representative's total income for the month was $3900.51.

YOU TRY IT 4

A nurse's aide works a second job sitting for an elderly patient. Her regular weekly pay for working at a nursing home is $480.15. She also worked four nights at her second job sitting with the elderly patient and earned $81.20, $60.90, $40.60, and $81.20. Find her total income for this week.

Your strategy

Your solution

Solutions on p. S8

3.2 EXERCISES

OBJECTIVE A To add decimals

For Exercises 1 to 17, add.

1. 16.008 + 2.0385 + 132.06
2. 17.32 + 1.0579 + 16.5
3. 1.792 + 67 + 27.0526
4. 8.772 + 1.09 + 26.5027
5. 3.02 + 62.7 + 3.924
6. 9.06 + 4.976 + 59.6
7. 82.006 + 9.95 + 0.927
8. 0.826 + 8.76 + 79.005
9. 4.307 + 99.82 + 9.078

10. 0.3 + 0.07

11. 0.29 + 0.4

12. 1.007 + 2.1

13. 7.3 + 9.005

14. 4.9257 + 27.05 + 9.0063

15. 8.72 + 99.073 + 2.9736

16. 62.4 + 9.827 + 692.44

17. 8 + 89.43 + 7.0659

For Exercises 18 to 21, use a calculator to add. Then round the numbers to the nearest whole number and use estimation to determine whether the sum you calculated is reasonable.

18. 342.42 + 89.625 + 176.2

19. 219.9 + 0.872 + 13.42

20. 823.9 + 82.65 + 46.923

21. 678.92 + 97.6 + 5.423

22. For a certain decimal addition problem, each addend rounded to the nearest whole number is greater than the addend itself. Must the sum of the rounded numbers be greater than the exact sum?

23. If none of the addends of a decimal addition problem is a whole number, is it possible for the sum to be a whole number?

OBJECTIVE B To solve application problems

24. **Temperature** In the morning, a patient's temperature was 97.9°F. By midafternoon, the patient's temperature had risen 1.3°, and by 5 P.M, the patient's temperature had risen another 0.9°. What was the patient's temperature at 5 P.M.?

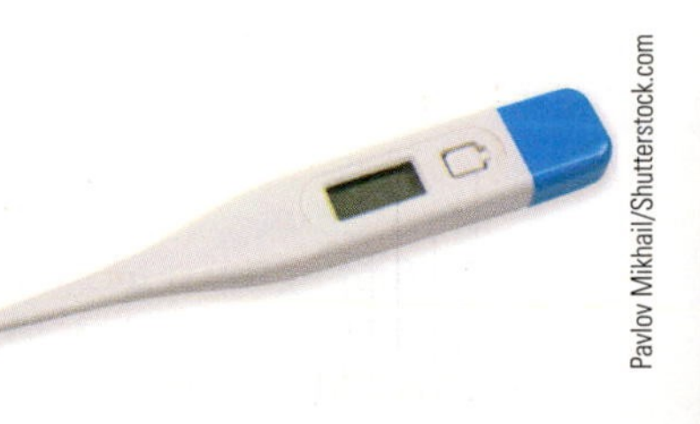

Pavlov Mikhail/Shutterstock.com

25. **Smoking Seminars** A smoking cessation hypnotist holds three online seminars in a weekend. The online seminars were viewed by 1.2 million, 0.8 million, and 1.4 million people. Calculate the total number of viewers.

26. **Banking** You have \$2143.57 in your checking account. You make deposits of \$210.98, \$45.32, \$1236.34, and \$27.99. Find the amount in your checking account after you have made the deposits if no money has been withdrawn.

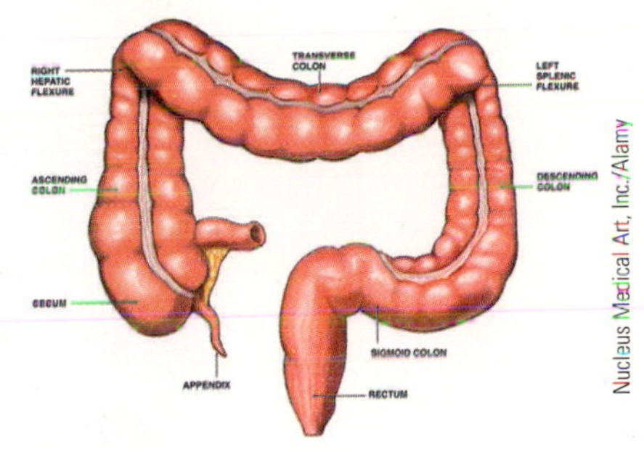

Nucleus Medical Art, Inc./Alamy

27. **Measurement** There are four sections to the large intestine or colon. The approximate lengths of the four sections are: the ascending colon, approximately 20.32 centimeters; the transverse colon, approximately 45.72 centimeters; the descending colon, approximately 30.48 centimeters; and the sigmoid colon, approximately 45.72 centimeters. Calculate the approximate total length of the large intestine.

28. **Demography** The world's population in 2050 is expected to be 8.9 billion people. It is projected that in that year, Asia's population will be 5.3 billion and Africa's population will be 1.8 billion. What are the combined populations of Asia and Africa expected to be in 2050? (*Source:* United Nations Population Division, World Population Prospects)

29. **Weight** Some medications, especially for children, have dosages based on the patient's weight in kilograms. A child weighs 15.9 kilograms and over the next two months gains 1.5 kilograms and 0.8 kilograms. How much does the child weigh now?

Fiber Intake Nutritionists recommend that someone on a 2000-calorie diet eat about 28 grams of fiber. The given table lists the grams of fiber in selected food items. Use this table for Exercises 30 and 31.

Food	*Grams of Fiber*
1 medium apple	3.5
$\frac{1}{2}$ grapefruit	0.9
$\frac{1}{2}$ cup bran cereal	10.4
$\frac{1}{2}$ cup green beans	2.1
1 cup whole wheat spaghetti	5.6
1 small baked potato	4.2
$\frac{1}{2}$ cup pinto beans	18.8
1 oat bran muffin	5.2

30. If Maria eats $\frac{1}{2}$ grapefruit and $\frac{1}{2}$ cup of bran cereal for breakfast, 1 small baked potato at lunch, and 1 cup of whole wheat spaghetti for supper, how much fiber has she consumed?

31. John is on a 2000-calorie diet and eats the following food in one day: $\frac{1}{2}$ cup of bran cereal, $\frac{1}{2}$ cup green beans, 1 medium apple, and $\frac{1}{2}$ cup of pinto beans. Has John met his requirement of 28 grams of fiber per day?

Applying the Concepts

Consumerism The table at the right gives the prices for selected products in a grocery store. Use this table for Exercises 32 and 33.

Product	*Cost*
Raisin bran	\$3.29
Butter	\$2.79
Bread	\$1.99
Popcorn	\$2.19
Potatoes	\$3.49
Cola (6-pack)	\$2.99
Mayonnaise	\$3.99
Lunch meat	\$3.39
Milk	\$2.59
Toothpaste	\$2.69

32. Does a customer with \$10 have enough money to purchase raisin bran, bread, milk, and butter?

33. Name three items that would cost more than \$8 but less than \$9. (There is more than one answer.)

SECTION

3.3 Subtraction of Decimals

OBJECTIVE A To subtract decimals

To subtract decimals, write the numbers so that the decimal points are on a vertical line. Subtract as for whole numbers, and write the decimal point in the difference directly below the decimal point in the subtrahend.

HOW TO • 1 Subtract 21.532 − 9.875 and check.

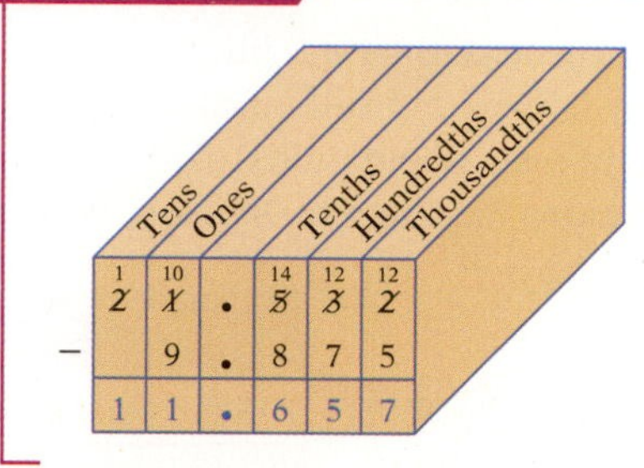

Placing the decimal points on a vertical line ensures that digits of the same place value are subtracted.

Check:

Subtrahend	9.875
+ Difference	+ 11.657
= Minuend	21.532

HOW TO • 2 Subtract 4.3 − 1.7942 and check.

```
  4.3000
- 1.7942
  2.5058
```

If necessary, insert zeros in the minuend before subtracting.

Check:

```
  1.7942
+ 2.5058
  4.3000
```

EXAMPLE • 1

Subtract 39.047 − 7.96 and check.

Solution

```
  39.047
-  7.96
  31.087
```

Check:

```
   7.96
+ 31.087
  39.047
```

YOU TRY IT • 1

Subtract 72.039 − 8.47 and check.

Your solution

EXAMPLE • 2

Find 9.23 less than 29 and check.

Solution

```
  29.00
-  9.23
  19.77
```

Check:

```
   9.23
+ 19.77
  29.00
```

YOU TRY IT • 2

Subtract 35 − 9.67 and check.

Your solution

EXAMPLE • 3

Subtract 1.2 − 0.8235 and check.

Solution

```
  1.2000
- 0.8235
  0.3765
```

Check:

```
  0.8235
+ 0.3765
  1.2000
```

YOU TRY IT • 3

Subtract 3.7 − 1.9715 and check.

Your solution

Solutions on pp. S8–S9

ESTIMATION

Estimating the Difference Between Two Decimals

Calculate 820.23 − 475.748. Then use estimation to determine whether the difference is reasonable.

Subtract to find the exact difference. 820.23 [−] 475.748 [=] 344.482

To estimate the difference, round each number to the same place value. Here we have rounded to the nearest ten. Then subtract. The estimated answer is 340, which is very close to the exact difference, 344.482.

$$\begin{array}{rcr} 820.23 & \approx & 820 \\ -475.748 & \approx & -480 \\ \hline & & 340 \end{array}$$

OBJECTIVE B To solve application problems

EXAMPLE 4

You bought a book for $15.87. How much change did you receive from a $20.00 bill?

Strategy

To find the amount of change, subtract the cost of the book (15.87) from $20.00.

Solution

$$\begin{array}{r} 20.00 \\ -15.87 \\ \hline 4.13 \end{array}$$

You received $4.13 in change.

YOU TRY IT 4

Your breakfast cost $6.85. How much change did you receive from a $10.00 bill?

Your strategy

Your solution

EXAMPLE 5

You had a balance of $87.93 on your student debit card. You then used the card, deducting $15.99 for a calculator, $6.85 for a nursing assistant workbook, and $28.50 for a lab manual. What is your new student debit card balance?

Strategy

To find your new debit card balance:

- Add to find the total of the three deductions (15.99 + 6.85 + 28.50).
- Subtract the total of the three deductions from the old balance (87.93).

Solution

$$\begin{array}{r} 15.99 \\ 6.85 \\ +28.50 \\ \hline 51.34 \end{array} \text{ total of deductions} \qquad \begin{array}{r} 87.93 \\ -51.34 \\ \hline 36.59 \end{array}$$

Your new debit card balance is $36.59.

YOU TRY IT 5

You had a balance of $2472.69 in your checking account. You then wrote checks for $1025.60, $79.85, and $162.47. Find the new balance in your checking account.

Your strategy

Your solution

Solutions on p. S9

3.3 EXERCISES

OBJECTIVE A To subtract decimals

For Exercises 1 to 24, subtract and check.

1. 24.037 − 18.41
2. 26.029 − 19.31
3. 123.07 − 9.4273
4. 214 − 7.143
5. 16.5 − 9.7902
6. 13.2 − 8.6205
7. 235.79 − 20.093
8. 463.27 − 40.095
9. 63.005 − 9.1274
10. 23.004 − 7.2175
11. 92 − 19.2909
12. 41.2405 − 25.2709
13. $\begin{array}{r} 0.32 \\ -\ 0.0058 \\ \hline \end{array}$
14. $\begin{array}{r} 0.78 \\ -\ 0.0073 \\ \hline \end{array}$
15. $\begin{array}{r} 3.005 \\ -\ 1.982 \\ \hline \end{array}$
16. $\begin{array}{r} 6.007 \\ -\ 2.734 \\ \hline \end{array}$
17. $\begin{array}{r} 352.16 \\ -\ \ 90.994 \\ \hline \end{array}$
18. $\begin{array}{r} 872 \\ -\ \ 80.753 \\ \hline \end{array}$
19. $\begin{array}{r} 724.32 \\ -\ \ 69 \\ \hline \end{array}$
20. $\begin{array}{r} 625.46 \\ -\ \ 77.509 \\ \hline \end{array}$
21. $\begin{array}{r} 362.394 \\ -\ \ 19.4672 \\ \hline \end{array}$
22. $\begin{array}{r} 421.385 \\ -\ \ 17.5293 \\ \hline \end{array}$
23. $\begin{array}{r} 19 \\ -\ 10.372 \\ \hline \end{array}$
24. $\begin{array}{r} 23.4 \\ -\ \ 0.921 \\ \hline \end{array}$

For Exercises 25 to 27, use the relationship between addition and subtraction to write the subtraction problem you would use to find the missing addend.

25. ______ + 2.325 = 7.01
26. 5.392 + ______ = 8.07
27. ______ + 8.967 = 19.35

For Exercises 28 to 31, use a calculator to subtract. Then round the numbers to the nearest whole number and use estimation to determine whether the difference you calculated is reasonable.

28. $\begin{array}{r} 93.079256 \\ -\ 66.09249 \\ \hline \end{array}$
29. $\begin{array}{r} 3.7529 \\ -\ 1.00784 \\ \hline \end{array}$
30. $\begin{array}{r} 76.53902 \\ -\ 45.73005 \\ \hline \end{array}$
31. $\begin{array}{r} 9.07325 \\ -\ 1.924 \\ \hline \end{array}$

OBJECTIVE B To solve application problems

32. **Purchases** A home-health nurse purchased medical supplies for her patient at a cost of $80.43. How much change is returned if the bill is paid with a $100 bill?

33. **Purchases** Shao Vang is a third-year medical residency student and just received her weekly stipend salary of $350 on a debit card. She purchases a medical laboratory diagnostic textbook for $89.65, lunch for $6.73, and groceries for $134.22. What is the remaining balance on the debit card?

34. **Business** The manager of MedSupply takes a reading of the cash register tape each hour. At 1:00 P.M. the tape read $967.54. At 2:00 P.M. the tape read $1437.15. Find the amount of sales between 1:00 P.M. and 2:00 P.M.

35. **Nutrition** Nutritionists recommend that women eat about 25 grams of fiber per day. If Sheila eats one medium pear with 5.5 grams of fiber, one oat bran muffin with 5.2 grams of fiber, and one cup of baked beans with 10.4 grams of fiber, how many additional grams of fiber should she consume today in order to meet the nutritionists' recommendation?

36. **Infant Mortality** The graph at the right shows the estimated number of deaths per 1000 live births in 2010 of infants younger than 1 year old. Find the difference between the number of infant deaths in Mexico per 1000 live births and the number in Canada.

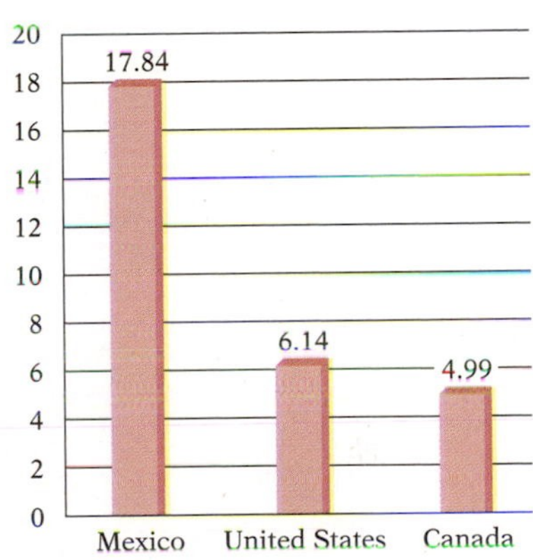

Infant Mortality Rates per 1000 Live Births

Source: Central Intelligence Agency, *The World Factbook,* www.cia.gov

37. **Temperature** A child had a fever of 102.1°. After being given acetaminophen, the child's fever came down 1.4° the first hour and another 0.8° the second hour. What was the child's temperature two hours after taking acetaminophen?

38. You have $30 to spend, and you make purchases that cost $6.74 and $13.68. Which expressions correctly represent the amount of money you have left?

(i) $30 - 6.74 + 13.68$ (ii) $(6.74 + 13.68) - 30$
(iii) $30 - (6.74 + 13.68)$ (iv) $30 - 6.74 - 13.68$

Applying the Concepts

39. Find the largest amount by which the estimate of the sum of two decimals rounded to the given place value could differ from the exact sum.
a. Tenths **b.** Hundredths **c.** Thousandths

SECTION

3.4 Multiplication of Decimals

OBJECTIVE A To multiply decimals

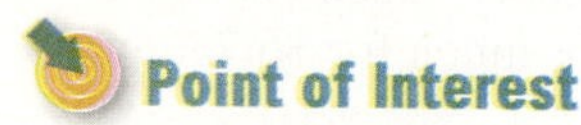

Point of Interest

Benjamin Banneker (1731–1806) was the first African American to earn distinction as a mathematician and scientist. He was on the survey team that determined the boundaries of Washington, D.C. The mathematics of surveying requires extensive use of decimals.

Decimals are multiplied as though they were whole numbers. Then the decimal point is placed in the product. Writing the decimals as fractions shows where to write the decimal point in the product.

$$0.\underline{3} \times 5 = \frac{3}{10} \times \frac{5}{1} = \frac{15}{10} = 1.\underline{5}$$

1 decimal place — 1 decimal place

$$0.\underline{3} \times 0.\underline{5} = \frac{3}{10} \times \frac{5}{10} = \frac{15}{100} = 0.\underline{15}$$

1 decimal place — 1 decimal place — 2 decimal places

$$0.\underline{3} \times 0.\underline{05} = \frac{3}{10} \times \frac{5}{100} = \frac{15}{1000} = 0.\underline{015}$$

1 decimal place — 2 decimal places — 3 decimal places

To multiply decimals, multiply the numbers as with whole numbers. Write the decimal point in the product so that the number of decimal places in the product is the sum of the decimal places in the factors.

Integrating Technology

Scientific calculators have a floating decimal point. This means that the decimal point is automatically placed in the answer. For example, for the product at the right, enter

21 . 4 x 0 . 36 =

The display reads 7.704, with the decimal point in the correct position.

HOW TO 1 Multiply: 21.4×0.36

$$\begin{array}{r} 21.4 \\ \times\ 0.36 \\ \hline 1284 \\ 642 \\ \hline 7.704 \end{array}$$

21.4 — 1 decimal place
0.36 — 2 decimal places
7.704 — 3 decimal places

HOW TO 2 Multiply: 0.037×0.08

$$\begin{array}{r} 0.037 \\ \times\ 0.08 \\ \hline 0.00296 \end{array}$$

0.037 — 3 decimal places
0.08 — 2 decimal places
0.00296 — 5 decimal places

• Two zeros must be inserted between the 2 and the decimal point so that there are 5 decimal places in the product.

To multiply a decimal by a power of 10 (10, 100, 1000, . . .), move the decimal point to the right the same number of places as there are zeros in the power of 10.

$3.8925 \times 1\underline{0} = 38.925$
1 zero — 1 decimal place

$3.8925 \times 1\underline{00} = 389.25$
2 zeros — 2 decimal places

$3.8925 \times 1\underline{000} = 3892.5$
3 zeros — 3 decimal places

$3.8925 \times 1\underline{0,000} = 38,925.$
4 zeros — 4 decimal places

$3.8925 \times 1\underline{00,000} = 389,250.$
5 zeros — 5 decimal places

• Note that a zero must be inserted before the decimal point.

Note that if the power of 10 is written in exponential notation, the exponent indicates how many places to move the decimal point.

$3.8925 \times 10^1 = 38.925$
1 decimal place

$3.8925 \times 10^2 = 389.25$
2 decimal places

$3.8925 \times 10^3 = 3892.5$
3 decimal places

$3.8925 \times 10^4 = 38{,}925.$
4 decimal places

$3.8925 \times 10^5 = 389{,}250.$
5 decimal places

EXAMPLE • 1

Multiply: 920×3.7

Solution

```
    920
×   3.7     • 1 decimal place
  644 0
  2760
 3404.0     • 1 decimal place
```

YOU TRY IT • 1

Multiply: 870×4.6

Your solution

EXAMPLE • 2

Find 0.00079 multiplied by 0.025.

Solution

```
   0.00079     • 5 decimal places
×    0.025     • 3 decimal places
       395
       158
0.00001975     • 8 decimal places
```

YOU TRY IT • 2

Find 0.000086 multiplied by 0.057.

Your solution

EXAMPLE • 3

Find the product of 3.69 and 2.07.

Solution

```
   3.69     • 2 decimal places
×  2.07     • 2 decimal places
   2583
  7380
 7.6383     • 4 decimal places
```

YOU TRY IT • 3

Find the product of 4.68 and 6.03.

Your solution

EXAMPLE • 4

Multiply: $42.07 \times 10{,}000$

Solution

$42.07 \times 10{,}000 = 420{,}700$

YOU TRY IT • 4

Multiply: 6.9×1000

Your solution

EXAMPLE • 5

Find 3.01 times 10^3.

Solution

$3.01 \times 10^3 = 3010$

YOU TRY IT • 5

Find 4.0273 times 10^2.

Your solution

Solutions on p. S9

ESTIMATION

Estimating the Product of Two Decimals

Calculate 28.259×0.029. Then use estimation to determine whether the product is reasonable.

Multiply to find the exact product.

28.259 **x** 0.029 **=** 0.819511

To estimate the product, round each number so that it contains one nonzero digit. Then multiply. The estimated answer is 0.90, which is very close to the exact product, 0.819511.

$$\begin{array}{rcr} 28.259 & \approx & 30 \\ \times 0.029 & \approx & \times 0.03 \\ \hline & & 0.90 \end{array}$$

OBJECTIVE B To solve application problems

The tables that follow list water rates and meter fees for a city. These tables are used for Example 6 and You Try It 6.

Water Charges	
Commercial	$1.39/1000 gal
Comm Restaurant	$1.39/1000 gal
Industrial	$1.39/1000 gal
Institutional	$1.39/1000 gal
Res—No Sewer	
Residential—SF	
>0 <200 gal per day	$1.15/1000 gal
>200 <1500 gal per day	$1.39/1000 gal
>1500 gal per day	$1.54/1000 gal

Meter Charges	
Meter	*Meter Fee*
5/8" & 3/4"	$13.50
1"	$21.80
1-1/2"	$42.50
2"	$67.20
3"	$133.70
4"	$208.20
6"	$415.10
8"	$663.70

EXAMPLE • 6

Find the total bill for an industrial water user with a 6-inch meter that used 152,000 gallons of water for July and August.

Strategy

To find the total cost of water:

- Find the cost of water by multiplying the cost per 1000 gallons (1.39) by the number of 1000-gallon units used.
- Add the cost of the water to the meter fee (415.10).

Solution

$$\text{Cost of water} = \frac{152{,}000}{1000} \cdot 1.39 = 211.28$$

$$\text{Total cost} = 211.28 + 415.10 = 626.38$$

The total cost is $626.38.

YOU TRY IT • 6

Find the total bill for a hospital that used 5000 gallons of water per day for July and August. The user has a 3-inch meter.

Your strategy

Your solution

Solution on p. S9

EXAMPLE 7

A pediatrician prescribes 1 teaspoon of amoxicillin every 4 hours. The patient is required to take this 1 teaspoon (0.167 ounce/teaspoon) of medication four times a day for 10 days. Find the total number of ounces of amoxicillin ingested during the 10 days. Round your answer to the nearest tenth.

Strategy

- To find the ounces ingested per day, multiply 1 teaspoon in ounces (0.167 ounce) by the times per day (4).
- To find the total ounces ingested for the prescription, multiply the ounces ingested per day (answer from step 1) by the total number of days (10).
- Round the final answer to one decimal place.

Solution

$$\begin{array}{r} 0.167 \\ \times \quad 4 \\ \hline 0.668 \end{array}$$ ounces per day

$$\begin{array}{r} 0.668 \\ \times \quad 10 \\ \hline 6.668 \end{array}$$ total ounces per 10-day prescription

Round 6.668 to 6.7.

The patient has ingested 6.7 ounces of amoxicillin for the 10 days.

YOU TRY IT 7

The cost to operate an x-ray processor in July 2010 was $0.1201 per kilowatt hour. The x-ray processor ran for 56 hours during the first week of July. Find the total cost to operate the x-ray processor during that week. Round to the nearest cent.

Your strategy

Your solution

EXAMPLE 8

Jason Ng earns a salary of $440 for a 40-hour workweek. This week he worked 12 hours of overtime at a rate of $16.50 for each hour of overtime worked. Find his total income for the week.

Strategy

To find Jason's total income for the week:

- Find the overtime pay by multiplying the hourly overtime rate (16.50) by the number of hours of overtime worked (12).
- Add the overtime pay to the weekly salary (440).

Solution

$$\begin{array}{r} 16.50 \\ \times \quad 12 \\ \hline 33\ 00 \\ 165\ 0 \\ \hline 198.00 \end{array}$$ Overtime pay

$$\begin{array}{r} 440.00 \\ +\ 198.00 \\ \hline 638.00 \end{array}$$

Jason's total income for the week is $638.00.

YOU TRY IT 8

As the athletic trainer for a college hockey team, you ordered an electrical muscle stimulation machine for knee rehabilitation. A deposit of $277.31 was made and the college agreed to make monthly payments of $111.20 for 20 months to repay the remaining balance. Find the total cost of the electrical muscle stimulation machine.

Your strategy

Your solution

Solutions on p. S9

3.4 EXERCISES

OBJECTIVE A To multiply decimals

For Exercises 1 to 73, multiply.

1. 0.9×0.4
2. 0.7×0.9
3. 0.5×0.5
4. 0.7×0.7
5. 7.7×0.9
6. 3.4×0.4
7. 9.2×0.2
8. 2.6×0.7
9. 7.4×0.1
10. 3.8×0.1
11. 7.9×5
12. 9.3×7
13. 0.68×4
14. 0.83×9
15. 0.67×0.9
16. 0.84×0.3
17. 2.5×5.4
18. 3.9×1.9
19. 0.83×5.2
20. 0.24×2.7
21. 1.47×0.09
22. 6.37×0.05
23. 8.92×0.004
24. 6.75×0.007
25. 0.49×0.16
26. 0.38×0.21
27. 7.6×0.01
28. 5.1×0.01
29. 8.62×4
30. 5.83×7
31. 64.5×9
32. 37.8×8
33. 2.19×9.2
34. 1.25×5.6
35. 1.85×0.023
36. 37.8×0.052
37. 0.478×0.37
38. 0.526×0.22
39. 48.3×0.0041
40. 67.2×0.0086

41. $\begin{array}{r} 0.413 \\ \times\ 0.0016 \\ \hline \end{array}$

42. $\begin{array}{r} 0.517 \\ \times\ 0.0029 \\ \hline \end{array}$

43. $\begin{array}{r} 8.005 \\ \times\ 0.067 \\ \hline \end{array}$

44. $\begin{array}{r} 9.032 \\ \times\ 0.019 \\ \hline \end{array}$

45. 4.29×0.1

46. 6.78×0.1

47. 5.29×0.4

48. 6.78×0.5

49. 0.68×0.7

50. 0.56×0.9

51. 1.4×0.73

52. 6.3×0.37

53. 5.2×7.3

54. 7.4×2.9

55. 3.8×0.61

56. 7.2×0.72

57. 0.32×10

58. 6.93×10

59. 0.065×100

60. 0.039×100

61. 6.2856×1000

62. 3.2954×1000

63. 3.2×1000

64. $0.006 \times 10{,}000$

65. $3.57 \times 10{,}000$

66. 8.52×10^1

67. 0.63×10^1

68. 82.9×10^2

69. 0.039×10^2

70. 6.8×10^3

71. 4.9×10^4

72. 6.83×10^4

73. 0.067×10^2

74. A number is rounded to the nearest thousandth. What is the smallest power of 10 the number must be multiplied by to give a product that is a whole number?

75. Find the product of 0.0035 and 3.45.

76. Find the product of 237 and 0.34.

77. Multiply 3.005 by 0.00392.

78. Multiply 20.34 by 1.008.

79. Multiply 1.348 by 0.23.

80. Multiply 0.000358 by 3.56.

81. Find the product of 23.67 and 0.0035.

82. Find the product of 0.00346 and 23.1.

83. Find the product of 5, 0.45, and 2.3.

84. Find the product of 0.03, 23, and 9.45.

For Exercises 85 to 96, use a calculator to multiply. Then use estimation to determine whether the product you calculated is reasonable.

85. 28.5×3.2

86. 86.3×4.4

87. 2.38×0.44

88. 9.82×0.77

89. 0.866×4.5

90. 0.239×8.2

91. 4.34×2.59

92. 6.87×9.98

93. 8.434×0.044

94. 7.037×0.094

95. 28.44×1.12

96. 86.57×7.33

OBJECTIVE B To solve application problems

97. Medication Mariel is diabetic and uses U-100 syringes to administer insulin daily. Each unit of insulin on the syringe is equal to 0.01 milliliter of insulin. If she averages 48 units of insulin per day, how many milliliters of insulin does she use in a week?

98. Transportation A hospital waste management company charges $0.24 per pound to dispose of regulated medical waste. Find the total charge to dispose of 1752 pounds of regulated medical waste.

99. Recycling Four hundred empty soft drink cans weigh 18.75 pounds. A recycling center pays $0.75 per pound for the cans. Find the amount received for the 400 cans. Round to the nearest cent.

100. Recycling A recycling center pays $0.045 per pound for newspapers.
a. Estimate the payment for recycling 520 pounds of newspapers.
b. Find the actual amount received from recycling the newspapers.

Marcio Jose Bastos Silva/Shutterstock.com

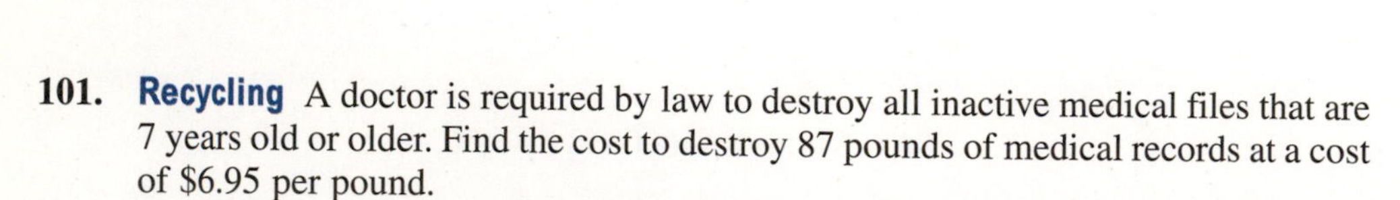

101. Recycling A doctor is required by law to destroy all inactive medical files that are 7 years old or older. Find the cost to destroy 87 pounds of medical records at a cost of $6.95 per pound.

102. **Vaccinations** A local family medical clinic offers pediatric flu vaccinations to its patients 6 months old and older. The clinic pays a federal excise tax of $0.75 per vaccination. (*Source:* CDC Vaccine Price List) Find the total tax if 198 vaccines were given this month.

103. **Weight** A baby weighs 7.16 pounds at birth. By the time the baby is 1 year old, its weight has tripled. How much does the baby weigh at its 1-year checkup?

104. **Finance** You bought a car for $5000 down and made payments of $499.50 each month for 36 months.
a. Find the amount of the payments over the 36 months.
b. Find the total cost of the car.

105. **Compensation** A nurse earns a salary of $1396 for a 40-hour workweek. This week the nurse worked 15 hours of overtime at a rate of $52.35 for each hour of overtime worked.
a. Find the nurse's overtime pay.
b. Find the nurse's total income for the week.

106. **Compensation** A radiologic technologist earns an average salary of $25.11 per hour for a regular 40-hour workweek. This rate increases to $37.67 per hour on Saturday. Find the total wages earned by the technologist for a week if he works a 40-hour week and 7 hours on Saturday.

107. **Medication** A popular sleeping pill contains 0.025 gram of diphenhydramine hydrochloride. If a patient takes a sleeping pill every day for a week, how much diphenhydramine hydrochloride does the patient ingest during the week?

Fiber Intake For Exercises 108 to 110, use the table of fiber content on page 135.

108. Calculate the total amount of fiber consumed by a patient who eats two medium apples, one oat bran muffin, and 1 cup of green beans during one day.

109. Calculate the total amount of fiber consumed by a patient who eats $\frac{1}{2}$ grapefruit, $1\frac{1}{2}$ cups of green beans, and 1 cup of bran cereal.

110. Write a verbal description of what this expression represents: $(3 \times 3.5) + (2 \times 18.8)$.

111. **Electronic Checks** As a general practitioner, you receive electronic transfer payments from some patients. These electronic transfer payments are processed by a consulting company for you at a cost of $0.55 each. This month the office received 87 payments by electronic transfer. Find the total cost of the electronic payment transfers to the medical office.

Business The table at the right lists the pay rates for employees in a medical clinic. Employees who work more than 40 hours per week are paid overtime rates for all hours in excess of 40. Use this table for Exercises 112 and 113.

Employee Classification	*Regular Pay Rate*	*Overtime Pay Rate*
RN (registered nurse)	$28.50/hour	$42.75/hour
LPN (licensed practical nurse)	$16.30/hour	$24.45/hour
CMA (certified medical assistant)	$13.80/hour	$20.70/hour

112. **a.** Find the weekly pay for an RN who works 40 hours in a week.
b. Find the weekly pay for an LPN who works 42 hours in a week.
c. Find the weekly pay for a CMA who works 45 hours in a week.

113. Other than the physicians and office staff, this medical clinic has 12 employees, including 4 RNs, 3 LPNs, and 5 CMAs. If everyone works a 40-hour week, what is the weekly payroll amount for these 12 employees?

Code	*Description*	*Price*
112	Almonds 16 oz	$6.75
116	Cashews 8 oz	$5.90
117	Cashews 16 oz	$8.50
130	Macadamias 7 oz	$7.25
131	Macadamias 16 oz	$11.95
149	Pecan halves 8 oz	$8.25
155	Mixed nuts 8 oz	$6.80
160	Cashew brittle 8 oz	$5.95
182	Pecan roll 8 oz	$6.70
199	Chocolate peanuts 8 oz	$5.90

114. **Business** A confectioner ships holiday packs of candy and nuts anywhere in the United States. Above is a price list for nuts and candy, and below is a table of shipping charges to zones in the United States. For any fraction of a pound, use the next higher weight. Sixteen ounces (16 oz) is equal to 1 pound.

Pounds	*Zone 1*	*Zone 2*	*Zone 3*	*Zone 4*
1–3	$7.55	$7.85	$8.25	$8.75
4–6	$8.10	$8.40	$8.80	$9.30
7–9	$8.50	$8.80	$9.20	$9.70
10–12	$8.90	$9.20	$9.60	$10.10

Find the cost of sending the following orders to the given mail zone.

a. Code	*Quantity*	*b. Code*	*Quantity*	*c. Code*	*Quantity*
116	2	112	1	117	3
130	1	117	4	131	1
149	3	131	2	155	2
182	4	160	3	160	4
Mail to zone 4.		182	5	182	1
		Mail to zone 3.		199	3
				Mail to zone 2.	

© Steve Lupton/Corbis

Applying the Concepts

115. Show how the decimal is placed in the product of 1.3×2.31 by first writing each number as a fraction and then multiplying. Then change the product back to decimal notation.

116. **Automotive Repair** Chris works at B & W Garage as an auto mechanic and has just completed an engine overhaul for a customer. To determine the cost of the repair job, Chris keeps a list of times worked and parts used. A parts list and a list of the times worked are shown below.

Parts Used		*Time Spent*	
Item	*Quantity*	*Day*	*Hours*
Gasket set	1	Monday	7.0
Ring set	1	Tuesday	7.5
Valves	8	Wednesday	6.5
Wrist pins	8	Thursday	8.5
Valve springs	16	Friday	9.0
Rod bearings	8		
Main bearings	5		
Valve seals	16		
Timing chain	1		

Price List		
Item Number	*Description*	*Unit Price*
27345	Valve spring	$9.25
41257	Main bearing	$17.49
54678	Valve	$16.99
29753	Ring set	$169.99
45837	Gasket set	$174.90
23751	Timing chain	$50.49
23765	Fuel pump	$229.99
28632	Wrist pin	$23.55
34922	Rod bearing	$13.69
2871	Valve seal	$1.69

a. Organize a table of data showing the parts used, the unit price for each part, and the price of the quantity used. *Hint:* Use the following headings for the table.

Quantity   Item Number   Description   Unit Price   Total

b. Add up the numbers in the "Total" column to find the total cost of the parts.

c. If the charge for labor is $46.75 per hour, compute the cost of labor.

d. What is the total cost for parts and labor?

117. Explain how the decimal point is placed when a number is multiplied by 10, 100, 1000, 10,000, etc.

118. Explain how the decimal point is placed in the product of two decimals.

SECTION

3.5 Division of Decimals

OBJECTIVE A To divide decimals

To divide decimals, move the decimal point in the divisor to the right to make the divisor a whole number. Move the decimal point in the dividend the same number of places to the right. Place the decimal point in the quotient directly over the decimal point in the dividend, and then divide as with whole numbers.

HOW TO • 1 Divide: $3.25\overline{)15.275}$

$3.25.\overline{)15.27.5}$

- Move the decimal point 2 places to the right in the divisor and then in the dividend. Place the decimal point in the quotient.

$$\begin{array}{r} 4.7 \\ 325.\overline{)\ 1527.5} \\ -1300 \\ \hline 227\ 5 \\ -227\ 5 \\ \hline 0 \end{array}$$

- Divide as with whole numbers.

Tips for Success

To learn mathematics, you must be an active participant. Listening and watching your professor do mathematics are not enough. Take notes in class, mentally think through every question your instructor asks, and try to answer it even if you are not called on to answer it verbally. Ask questions when you have them. See *AIM for Success* at the front of the book for other ways to be an active learner.

Moving the decimal point the same number of decimal places in the divisor and dividend does not change the value of the quotient, because this process is the same as multiplying the numerator and denominator of a fraction by the same number. In the example above,

$$3.25\overline{)15.275} = \frac{15.275}{3.25} = \frac{15.275 \times 100}{3.25 \times 100} = \frac{1527.5}{325} = 325\overline{)1527.5}$$

When dividing decimals, we usually round the quotient off to a specified place value, rather than writing the quotient with a remainder.

HOW TO • 2 Divide: $0.3\overline{)0.56}$

Round to the nearest hundredth.

$$\begin{array}{r} 1.866 \approx 1.87 \\ 0.3.\overline{)\ 0.5.600} \\ -\ 3 \\ \hline 2\ 6 \\ -2\ 4 \\ \hline 20 \\ -18 \\ \hline 20 \\ -18 \\ \hline \end{array}$$

We must carry the division to the thousandths place to round the quotient to the nearest hundredth. Therefore, zeros must be inserted in the dividend so that the quotient has a digit in the thousandths place.

Integrating Technology

A calculator displays the quotient to the limit of the calculator's display. Enter

=

to determine the number of places your calculator displays.

HOW TO • 3 Divide 57.93 by 3.24. Round to the nearest thousandth.

```
            17.8796 ≈ 17.880
3.24.) 57.93.0000
      −32 4
       25 53
      −22 68
        2 85 0
       −2 59 2
          25 80
         −22 68
           3 120
          −2 916
             2040
            −1944
```

Zeros must be inserted in the dividend so that the quotient has a digit in the ten-thousandths place.

To divide a decimal by a power of 10 (10, 100, 1000, . . .), move the decimal point to the left the same number of places as there are zeros in the power of 10.

$34.65 \div 10 = 3.465$ — 1 zero, 1 decimal place

$34.65 \div 100 = 0.3465$ — 2 zeros, 2 decimal places

$34.65 \div 1000 = 0.03465$ — 3 zeros, 3 decimal places

- **Note that a zero must be inserted between the 3 and the decimal point.**

$34.65 \div 10{,}000 = 0.003465$ — 4 zeros, 4 decimal places

- **Note that two zeros must be inserted between the 3 and the decimal point.**

If the power of 10 is written in exponential notation, the exponent indicates how many places to move the decimal point.

$34.65 \div 10^1 = 3.465$ — 1 decimal place

$34.65 \div 10^2 = 0.3465$ — 2 decimal places

$34.65 \div 10^3 = 0.03465$ — 3 decimal places

$34.65 \div 10^4 = 0.003465$ — 4 decimal places

EXAMPLE • 1

Divide: $0.1344 \div 0.032$

Solution

```
             4.2
0.032.)0.134.4
        −128
           6 4
          −6 4
             0
```

- **Move the decimal point 3 places to the right in the divisor and the dividend.**

YOU TRY IT • 1

Divide: $0.1404 \div 0.052$

Your solution

Solution on p. S9

EXAMPLE • 2

Divide: $58.092 \div 82$
Round to the nearest thousandth.

Solution

$$\begin{array}{r} 0.7084 \approx 0.708 \\ 82\overline{)\ 58.0920} \\ -57\ 4 \\ \hline 69 \\ -\ \ 0 \\ \hline 692 \\ -656 \\ \hline 360 \\ -328 \\ \hline \end{array}$$

YOU TRY IT • 2

Divide: $37.042 \div 76$
Round to the nearest thousandth.

Your solution

EXAMPLE • 3

Divide: $420.9 \div 7.06$
Round to the nearest tenth.

Solution

$$\begin{array}{r} 59.61 \approx 59.6 \\ 7.06.\overline{)\ 420.90.00} \\ -353\ 0 \\ \hline 67\ 90 \\ -63\ 54 \\ \hline 4\ 36\ 0 \\ -4\ 23\ 6 \\ \hline 12\ 40 \\ -\ \ 7\ 06 \\ \hline \end{array}$$

YOU TRY IT • 3

Divide: $370.2 \div 5.09$
Round to the nearest tenth.

Your solution

EXAMPLE • 4

Divide: $402.75 \div 1000$

Solution
$402.75 \div 1000 = 0.40275$

YOU TRY IT • 4

Divide: $309.21 \div 10{,}000$

Your solution

EXAMPLE • 5

What is 0.625 divided by 10^2?

Solution
$0.625 \div 10^2 = 0.00625$

YOU TRY IT • 5

What is 42.93 divided by 10^4?

Your solution

Solutions on p. S10

ESTIMATION

Estimating the Quotient of Two Decimals

Calculate $282.18 \div 0.48$. Then use estimation to determine whether the quotient is reasonable.

Divide to find the exact quotient.

282.18 [÷] 0.48 [=] 587.875

To estimate the quotient, round each number so that it contains one nonzero digit. Then divide. The estimated answer is 600, which is very close to the exact quotient, 587.875.

$$\begin{aligned} 282.18 \div 0.48 &\approx 300 \div 0.5 \\ &= 600 \end{aligned}$$

OBJECTIVE B To solve application problems 

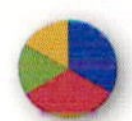 The graph at the right shows average hourly earnings in the United States for selected years. Use this table for Example 6 and You Try It 6.

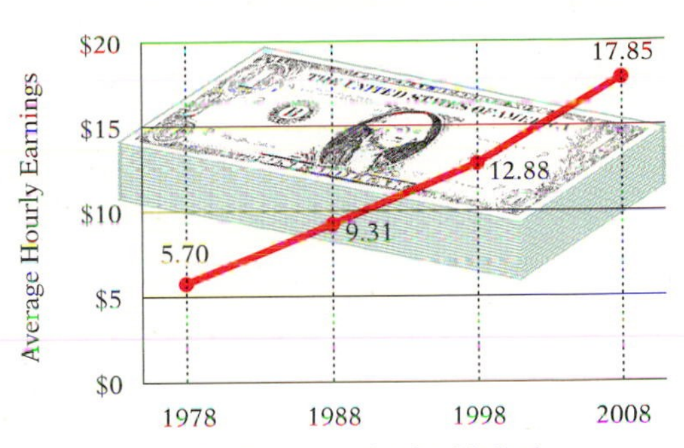

Average Hourly Earnings in the United States *Source:* Bureau of Labor Statistics

EXAMPLE 6

How many times greater were the average hourly earnings in 2008 than in 1978? Round to the nearest whole number.

Strategy
To find how many times greater the average hourly earnings were, divide the 2008 average hourly earnings (17.85) by the 1978 average hourly earnings (5.70).

Solution
$17.85 \div 5.70 \approx 3$

The average hourly earnings in 2008 were about 3 times greater than in 1978.

YOU TRY IT 6

How many times greater were the average hourly earnings in 1998 than in 1978? Round to the nearest tenth.

Your strategy

Your solution

EXAMPLE 7

A 1-year subscription to a monthly medical magazine costs $90. The price of each issue at the newsstand is $9.80. How much would you save per issue by buying a year's subscription rather than buying each issue at the newsstand?

Strategy
To find the amount saved:

- Find the subscription price per issue by dividing the cost of the subscription (90) by the number of issues (12).
- Subtract the subscription price per issue from the newsstand price (9.80).

Solution
$90 \div 12 = 7.50$
$9.80 - 7.50 = 2.30$

The savings would be $2.30 per issue.

YOU TRY IT 7

The CIA's *The World Factbook* (2009 estimates) lists the life expectancy at birth of the citizens in four large European countries as follows: France, 80.98 years; Spain, 80.05 years; Italy, 80.2 years; and Germany, 79.4 years. Find the average life expectancy for these countries.

Your strategy

Your solution

Solutions on p. S10

3.5 EXERCISES

OBJECTIVE A To divide decimals

For Exercises 1 to 20, divide.

1. $3\overline{)2.46}$
2. $7\overline{)3.71}$
3. $0.8\overline{)3.84}$
4. $0.9\overline{)6.93}$
5. $0.7\overline{)62.3}$
6. $0.4\overline{)52.8}$
7. $0.4\overline{)24}$
8. $0.5\overline{)65}$
9. $0.7\overline{)59.01}$
10. $0.9\overline{)8.721}$
11. $0.5\overline{)16.15}$
12. $0.8\overline{)77.6}$
13. $0.7\overline{)3.542}$
14. $0.6\overline{)2.436}$
15. $6.3\overline{)8.19}$
16. $3.2\overline{)7.04}$
17. $3.6\overline{)0.396}$
18. $2.7\overline{)0.648}$
19. $6.9\overline{)26.22}$
20. $1.7\overline{)84.66}$

For Exercises 21 to 29, divide. Round to the nearest tenth.

21. $55.62 \div 8.8$
22. $25.43 \div 5.4$
23. $5.427 \div 9.5$
24. $1.837 \div 1.4$
25. $18.4 \div 7.3$
26. $52.9 \div 8.1$
27. $0.183 \div 0.17$
28. $0.381 \div 0.47$
29. $6.924 \div 0.053$

For Exercises 30 to 38, divide. Round to the nearest hundredth.

30. $4.817 \div 16$
31. $6.467 \div 8$
32. $0.0418 \div 0.53$
33. $0.0647 \div 0.72$
34. $7 \div 0.55$
35. $38.665 \div 0.95$
36. $13.97 \div 25.4$
37. $27.738 \div 60.8$
38. $3.171 \div 45.6$

For Exercises 39 to 47, divide. Round to the nearest thousandth.

39. $1.028 \div 54$

40. $6.729 \div 27$

41. $0.0437 \div 0.5$

42. $75.469 \div 77.8$

43. $34.31 \div 95.3$

44. $0.2695 \div 2.67$

45. $0.4871 \div 4.72$

46. $0.1142 \div 17.2$

47. $0.2307 \div 26.7$

For Exercises 48 to 56, divide. Round to the nearest whole number.

48. $16.5 \div 4$

49. $89.76 \div 90$

50. $1.94 \div 0.3$

51. $1.0478 \div 0.413$

52. $2.148 \div 0.519$

53. $0.79 \div 0.778$

54. $3.092 \div 0.075$

55. $392 \div 6.9$

56. $8.729 \div 0.075$

For Exercises 57 to 74, divide.

57. $4.07 \div 10$

58. $0.039 \div 10$

59. $42.67 \div 10$

60. $389.7 \div 100$

61. $1.037 \div 100$

62. $237.835 \div 100$

63. $8.295 \div 1000$

64. $82{,}547 \div 1000$

65. $825.37 \div 1000$

66. $8.35 \div 10^1$

67. $0.32 \div 10^1$

68. $87.65 \div 10^1$

69. $23.627 \div 10^2$

70. $2.954 \div 10^2$

71. $0.0053 \div 10^2$

72. $289.32 \div 10^3$

73. $1.8932 \div 10^3$

74. $0.139 \div 10^3$

75. Divide 44.208 by 2.4.

76. Divide 0.04664 by 0.44.

77. Find the quotient of 723.15 and 45.

78. Find the quotient of 3.3463 and 3.07.

79. Divide 13.5 by 10^3.

80. Divide 0.045 by 10^5.

81. Find the quotient of 23.678 and 1000.

82. Find the quotient of 7.005 and 10,000.

83. What is 0.0056 divided by 0.05?

84. What is 123.8 divided by 0.02?

For Exercises 85 to 93, use a calculator to divide. Round to the nearest ten-thousandth. Then use estimation to determine whether the quotient you calculated is reasonable.

85. $42.42 \div 3.8$

86. $69.8 \div 7.2$

87. $389 \div 0.44$

88. $642 \div 0.83$

89. $6.394 \div 3.5$

90. $8.429 \div 4.2$

91. $1.235 \div 0.021$

92. $7.456 \div 0.072$

93. $95.443 \div 1.32$

94. A four-digit whole number is divided by 1000. Is the quotient less than 1 or greater than 1?

OBJECTIVE B To solve application problems

95. A 12-pack of bottled spring water sells for $3.85. State whether to use *multiplication* or *division* to find the specified amount.
a. The cost of one bottle of spring water
b. The cost of four 12-packs of spring water

96. **Tuition** The tuition at a massage therapy school in Provo, Utah, is $19,568, whereas the tuition at a massage therapy school in Salt Lake City, Utah, is $7500. Which of the following represents the number of times greater the tuition in Provo, Utah, is compared to the Salt Lake City, Utah, school?

a. $19,568 \times 7500$ **b.** $19,568 + 7500$ **c.** $19,568 - 7500$ **d.** $19,568 \div 7500$

97. **Consumerism** An OB/GYN office purchases a box of 50 paper medical exam gowns for $21.35. Find the cost per paper gown. Round to the nearest cent.

98. **Insurance** An orthodontist charges $5000 for a complete set of braces. The insurance company will reimburse the orthodontist in 12 equal payments. Find the amount of each payment. Round to the nearest cent.

99. **Weight Loss** A physician advises an overweight patient to lose $\frac{1}{5}$ of her current weight, or 49.8 pounds, over the next 12 months. How many pounds should the patient try to lose each month in order to reach this goal?

Point of Interest

Not all tablets of medication can be safely divided in half. If a tablet is scored with a line down the middle, it is probably safe to cut into two pieces. However, always check with your doctor or pharmacist.

100. **Medication** A doctor prescribes a dose of 1.25 milligrams of a medication for a patient. The medication is in liquid form and each milliliter of liquid contains 0.25 milligram of medication. How many milliliters of the liquid should a nurse give the patient?

101. **Medication** A certain pain medication is in tablet form and contains 0.15 gram of medication per tablet. If the doctor has ordered a patient to have 0.3 gram of pain medication, how many tablets should a nurse give the patient?

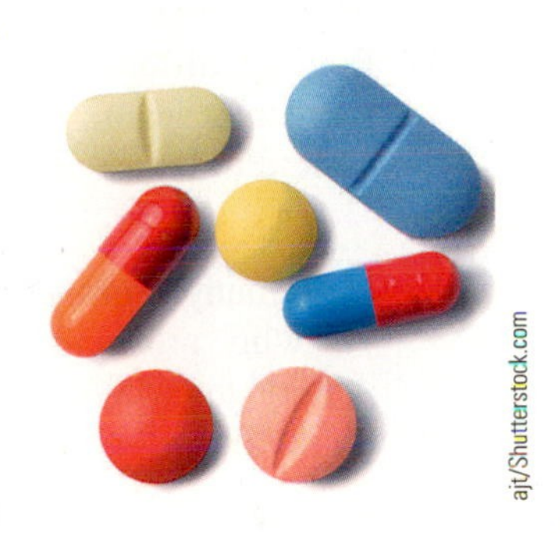

ajt/Shutterstock.com

102. **Medication** A doctor orders 0.5 gram of a medication for a patient. The pharmacy has scored tablets on hand and each one contains 0.2 gram of medication. How many tablets will be required for one dose of medication?

103. **Investments** A pharmaceutical company has issued 3,541,221,500 shares of stock. The company paid $6,090,990,120 in dividends. Find the dividend for each share of stock. Round to the nearest cent.

104. **Insurance** Earl is 52 years old and is buying $70,000 of life insurance for an annual premium of $703.80. If he pays the annual premium in 12 equal installments, how much is each monthly payment?

105. **Fuel Efficiency** A car with an odometer reading of 17,814.2 is filled with 9.4 gallons of gas. At an odometer reading of 18,130.4, the tank is empty and the car is filled with 12.4 gallons of gas. How many miles does the car travel on 1 gallon of gasoline?

106. **Salaries** A survey of four regional hospitals revealed different annual salaries for operating room nurses with less than 5 years of experience. Find the average salary for an operating room nurse with less than 5 years of experience in this region if the four annual salaries are $52,505, $55,250, $52,355, and $53,265.

107. **Hemoglobin Levels** A patient with a low hemoglobin count is anemic. A normal hemoglobin range for a woman after middle age is 11.7 to 13.8 grams of hemoglobin per 100 milliliters of blood. Sonja's hemoglobin numbers for the past 6 months are 10.9, 11.5, 12.1, 11.8, 13.1, and 11.0. Find her average hemoglobin level for the 6-month period. Round to the nearest tenth.

108. **Going Green** See the news clipping at the right.
   a. Find the reduction in solid waste per month if every U.S. household viewed and paid its bills online.
   b. Find the reduction in greenhouse gas emissions per month if every household viewed and paid its bills online.

Write your answers in standard form, rounded to the nearest whole number.

In the News

"Green" Banking Has Far-Reaching Effects

Banking and paying bills online not only saves trees; it cuts down on the amount of fuel used by vehicles that transport paper checks. According to Javelin Strategy and Research, if every household in the United States paid its bills online, solid waste would be reduced by 1.6 billion tons a year and greenhouse-gas emissions would be cut by 2.1 million tons a year.

Source: Time, April 9, 2008

Applying the Concepts

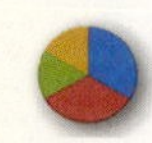

109. **Education** According to the National Center for Education Statistics, 10.03 million women and 7.46 million men were enrolled at institutions of higher learning in a recent year. How many more women than men were attending institutions of higher learning in that year?

Health Insurance The table at the right shows the number of Americans without health insurance by age in 2009. Use this table for Exercises 110 to 112.

Americans Without Health Insurance by Age in 2009	
Age	***Number Uninsured (in thousands)***
Under 18 years	75.04
18 to 24 years	29.31
25 to 34 years	41.09
35 to 44 years	40.45
45 to 64 years	79.78
65 years and older	38.61

Source: U.S. Census Bureau, Current Population Survey, 2009 and 2010 Annual Social and Economic Supplements

110. Find the difference between the number of Americans without health insurance who are aged under 18 and those aged 65 and older.

111. How many times greater is the number of Americans under 18 who are uninsured than the number of uninsured Americans who are 18 to 24 years old? Round to the nearest hundredth.

112. What was the total number of Americans without health insurance in 2009?

113. **Population Growth** The U.S. population of people ages 85 and over is expected to grow from 4.2 million in 2000 to 8.9 million in 2030. How many times greater is the population of this segment expected to be in 2030 than in 2000? Round to the nearest tenth.

114. Explain how the decimal point is moved when a number is divided by 10, 100, 1000, 10,000, etc.

115. **Grade Point Average** Explain how to calculate a Grade Point Average (GPA) for a semester. Then calculate the GPA for a student taking four 3-hour courses and earning 2 Cs, 1 B, and 1 A.

116. Explain how the decimal point is placed in the quotient when a number is divided by a decimal.

For Exercises 117 to 122, insert $+$, $-$, $\times$, or $\div$ into the square so that the statement is true.

117. $3.45 \square 0.5 = 6.9$

118. $3.46 \square 0.24 = 0.8304$

119. $6.009 \square 4.68 = 1.329$

120. $0.064 \square 1.6 = 0.1024$

121. $9.876 \square 23.12 = 32.996$

122. $3.0381 \square 1.23 = 2.47$

For Exercises 123 to 125, fill in the square to make a true statement.

123. $6.47 - \square = 1.253$

124. $6.47 + \square = 9$

125. $0.009 \div \square = 0.36$

SECTION

3.6 Comparing and Converting Fractions and Decimals

OBJECTIVE A To convert fractions to decimals

Every fraction can be written as a decimal. To write a fraction as a decimal, divide the numerator of the fraction by the denominator. The quotient can be rounded to the desired place value.

HOW TO • 1 Convert $\frac{3}{7}$ to a decimal.

$$7\overline{)3.00000} = 0.42857$$

$\frac{3}{7}$ rounded to the nearest hundredth is 0.43.

$\frac{3}{7}$ rounded to the nearest thousandth is 0.429.

$\frac{3}{7}$ rounded to the nearest ten-thousandth is 0.4286.

HOW TO • 2 Convert $3\frac{2}{9}$ to a decimal. Round to the nearest thousandth.

$$3\frac{2}{9} = \frac{29}{9} \qquad 9\overline{)29.0000} = 3.2222$$

$3\frac{2}{9}$ rounded to the nearest thousandth is 3.222.

EXAMPLE • 1

Convert $\frac{3}{8}$ to a decimal.
Round to the nearest hundredth.

Solution $8\overline{)3.000} = 0.375 \approx 0.38$

YOU TRY IT • 1

Convert $\frac{9}{16}$ to a decimal.
Round to the nearest tenth.

Your solution

EXAMPLE • 2

Convert $2\frac{3}{4}$ to a decimal.
Round to the nearest tenth.

Solution $2\frac{3}{4} = \frac{11}{4}$ $\qquad 4\overline{)11.00} = 2.75 \approx 2.8$

YOU TRY IT • 2

Convert $4\frac{1}{6}$ to a decimal.
Round to the nearest hundredth.

Your solution

Solutions on p. S10

OBJECTIVE B To convert decimals to fractions

To convert a decimal to a fraction, remove the decimal point and place the decimal part over a denominator equal to the place value of the last digit in the decimal.

$0.47 = \frac{47}{100}$ (hundredths) $\qquad 7.45 = 7\frac{45}{100} = 7\frac{9}{20}$ (hundredths)

$0.275 = \frac{275}{1000} = \frac{11}{40}$ (thousandths) $\qquad 0.16\frac{2}{3} = \frac{16\frac{2}{3}}{100} = 16\frac{2}{3} \div 100 = \frac{50}{3} \times \frac{1}{100} = \frac{1}{6}$ (hundredths)

EXAMPLE • 3

Convert 0.82 and 4.75 to fractions.

Solution $0.82 = \frac{82}{100} = \frac{41}{50}$

$4.75 = 4\frac{75}{100} = 4\frac{3}{4}$

YOU TRY IT • 3

Convert 0.56 and 5.35 to fractions.

Your solution

EXAMPLE • 4

Convert $0.15\frac{2}{3}$ to a fraction.

Solution $0.15\frac{2}{3} = \frac{15\frac{2}{3}}{100} = 15\frac{2}{3} \div 100$

$= \frac{47}{3} \times \frac{1}{100} = \frac{47}{300}$

YOU TRY IT • 4

Convert $0.12\frac{7}{8}$ to a fraction.

Your solution

Solutions on p. S10

OBJECTIVE C To identify the order relation between two decimals or between a decimal and a fraction

Decimals, like whole numbers and fractions, can be graphed as points on the number line. The number line can be used to show the order of decimals. A decimal that appears to the right of a given number is greater than the given number. A decimal that appears to the left of a given number is less than the given number.

3.00 3.05 3.10 3.15 3.20 3.25 3.30 3.35 3.40

Note that 3, 3.0, and 3.00 represent the same number.

HOW TO • 3 Find the order relation between $\frac{3}{8}$ and 0.38.

$\frac{3}{8} = 0.375$ $0.38 = 0.380$ • **Convert the fraction $\frac{3}{8}$ to a decimal.**

$0.375 < 0.380$ • **Compare the two decimals.**

$\frac{3}{8} < 0.38$ • **Convert 0.375 back to a fraction.**

EXAMPLE • 5

Place the correct symbol, < or >, between the numbers.

$\frac{5}{16}$ 0.32

Solution $\frac{5}{16} \approx 0.313$ • **Convert $\frac{5}{16}$ to a decimal.**

$0.313 < 0.32$ • **Compare the two decimals.**

$\frac{5}{16} < 0.32$ • **Convert 0.313 back to a fraction.**

YOU TRY IT • 5

Place the correct symbol, < or >, between the numbers.

0.63 $\frac{5}{8}$

Your solution

Solution on p. S10

3.6 EXERCISES

OBJECTIVE A To convert fractions to decimals

For Exercises 1 to 24, convert the fraction to a decimal. Round to the nearest thousandth.

1. $\frac{5}{8}$ **2.** $\frac{7}{12}$ **3.** $\frac{2}{3}$ **4.** $\frac{5}{6}$ **5.** $\frac{1}{6}$ **6.** $\frac{7}{8}$

7. $\frac{5}{12}$ **8.** $\frac{9}{16}$ **9.** $\frac{7}{4}$ **10.** $\frac{5}{3}$ **11.** $1\frac{1}{2}$ **12.** $2\frac{1}{3}$

13. $\frac{16}{4}$ **14.** $\frac{36}{9}$ **15.** $\frac{3}{1000}$ **16.** $\frac{5}{10}$ **17.** $7\frac{2}{25}$ **18.** $16\frac{7}{9}$

19. $37\frac{1}{2}$ **20.** $\frac{5}{24}$ **21.** $\frac{4}{25}$ **22.** $3\frac{1}{3}$ **23.** $8\frac{2}{5}$ **24.** $5\frac{4}{9}$

For Exercises 25 to 28, without actually doing any division, state whether the decimal equivalent of the given fraction is greater than 1 or less than 1.

25. $\frac{54}{57}$ **26.** $\frac{176}{129}$ **27.** $\frac{88}{80}$ **28.** $\frac{2007}{2008}$

OBJECTIVE B To convert decimals to fractions

For Exercises 29 to 53, convert the decimal to a fraction.

29. 0.8 **30.** 0.4 **31.** 0.32 **32.** 0.48 **33.** 0.125

34. 0.485 **35.** 1.25 **36.** 3.75 **37.** 16.9 **38.** 17.5

39. 8.4 **40.** 10.7 **41.** 8.437 **42.** 9.279 **43.** 2.25

44. 7.75 **45.** $0.15\frac{1}{3}$ **46.** $0.17\frac{2}{3}$ **47.** $0.87\frac{7}{8}$ **48.** $0.12\frac{5}{9}$

49. 7.38 **50.** 0.33 **51.** 0.57 **52.** $0.33\frac{1}{3}$ **53.** $0.66\frac{2}{3}$

54. Is $0.27\frac{4}{9}$ greater than 0.27 or less than 0.27?

OBJECTIVE C To identify the order relation between two decimals or between a decimal and a fraction

For Exercises 55 to 74, place the correct symbol, $<$ or $>$, between the numbers.

55. 0.15 0.5 **56.** 0.6 0.45 **57.** 6.65 6.56 **58.** 3.89 3.98

59. 2.504 2.054 **60.** 0.025 0.105 **61.** $\frac{3}{8}$ 0.365 **62.** $\frac{4}{5}$ 0.802

63. $\frac{2}{3}$ 0.65 **64.** 0.85 $\frac{7}{8}$ **65.** $\frac{5}{9}$ 0.55 **66.** $\frac{7}{12}$ 0.58

67. 0.62 $\frac{7}{15}$ **68.** $\frac{11}{12}$ 0.92 **69.** 0.161 $\frac{1}{7}$ **70.** 0.623 0.6023

71. 0.86 0.855 **72.** 0.87 0.087 **73.** 1.005 0.5 **74.** 0.033 0.3

75. Use the inequality symbol $<$ to rewrite the order relation expressed by the inequality $17.2 > 0.172$.

76. Use the inequality symbol $>$ to rewrite the order relation expressed by the inequality $0.0098 < 0.98$.

Applying the Concepts

77. Air Pollution An emissions test for cars requires that of the total engine exhaust, less than 1 part per thousand $\left(\frac{1}{1000} = 0.001\right)$ be hydrocarbon emissions. Using this figure, determine which of the cars in the table at the right would fail the emissions test.

Car	*Total Engine Exhaust*	*Hydrocarbon Emission*
1	367,921	360
2	401,346	420
3	298,773	210
4	330,045	320
5	432,989	450

78. Explain how terminating, repeating, and nonrepeating decimals differ. Give an example of each kind of decimal.

FOCUS ON PROBLEM SOLVING

Relevant Information

Problems in mathematics or real life involve a question or a need and information or circumstances related to that question or need. Solving problems in the sciences usually involves a question, an observation, and measurements of some kind.

Tony Freeman/PhotoEdit, Inc.

One of the challenges of problem solving in the sciences is to separate the information that is relevant to the problem from other information. Following is an example from the physical sciences in which some relevant information was omitted.

Hooke's Law states that the distance that a weight will stretch a spring is directly proportional to the weight on the spring. That is, $d = kF$, where d is the distance the spring is stretched and F is the force. In an experiment to verify this law, some physics students were continually getting inconsistent results. Finally, the instructor discovered that the heat produced when the lights were turned on was affecting the experiment. In this case, relevant information was omitted—namely, that the temperature of the spring can affect the distance it will stretch.

A physician drove 8 miles to the train station. After a 35-minute ride of 18 miles, the physician walked 10 minutes to the office. Find the total time it took the physician to get to work.

From this situation, answer the following before reading on.

a. What is asked for?

b. Is there enough information to answer the question?

c. Is information given that is not needed?

Here are the answers.

a. We want the total time for the physician to get to work.

b. No. We do not know the time it takes the physician to get to the train station.

c. Yes. Neither the distance to the train station nor the distance of the train ride is necessary to answer the question.

For each of the following problems, answer the questions printed in red above.

1. An emergency room doctor treats 48 patients in 8 hours. How many patients were treated per hour?

2. Hospital beds are 3-feet wide by 8-feet long and require a distance of 6 feet between each bed. If the hospital room is 17-feet long by 12-feet wide, can three hospital beds be placed in this room?

3. A family rented a car for their vacation and drove 680 miles. The cost of the rental car was \$21 per day with 150 free miles per day and \$0.15 for each mile driven above the number of free miles allowed. How many miles did the family drive per day?

4. An investor bought 8 acres of land for \$80,000. One and one-half acres were set aside for a park, and the remaining land was developed into one-half-acre lots. How many lots were available for sale?

5. You wrote checks of \$43.67, \$122.88, and \$432.22 after making a deposit of \$768.55. How much do you have left in your checking account?

PROJECTS AND GROUP ACTIVITIES

Fractions as Terminating or Repeating Decimals

The fraction $\frac{3}{4}$ is equivalent to 0.75. The decimal 0.75 is a **terminating decimal** because there is a remainder of zero when 3 is divided by 4. The fraction $\frac{1}{3}$ is equivalent to 0.333 . . . . The three dots mean the pattern continues on and on. 0.333 . . . is a **repeating decimal.** To determine whether a fraction can be written as a terminating decimal, first write the fraction in simplest form. Then look at the denominator of the fraction. If it contains prime factors of only 2s and/or 5s, then it can be expressed as a terminating decimal. If it contains prime factors other than 2s or 5s, it represents a repeating decimal.

Take Note

If the denominator of a fraction in simplest form is 20, then it can be written as a terminating decimal because $20 = 2 \cdot 2 \cdot 5$ (only prime factors of 2 and 5). If the denominator of a fraction in simplest form is 6, it represents a repeating decimal because it contains the prime factor 3 (a number other than 2 or 5).

1. Assume that each of the following numbers is the denominator of a fraction written in simplest form. Does the fraction represent a terminating or repeating decimal?

   a. 4 **b.** 5 **c.** 7 **d.** 9 **e.** 10 **f.** 12 **g.** 15
   h. 16 **i.** 18 **j.** 21 **k.** 24 **l.** 25 **m.** 28 **n.** 40

2. Write two other numbers that, as denominators of fractions in simplest form, represent terminating decimals, and write two other numbers that, as denominators of fractions in simplest form, represent repeating decimals.

CHAPTER 3

SUMMARY

KEY WORDS	EXAMPLES
A number written in *decimal notation* has three parts: a *whole-number part,* a *decimal point,* and a *decimal part.* The decimal part of a number represents a number less than 1. A number written in decimal notation is often simply called a *decimal.* [3.1A, p. 126]	For the decimal 31.25, 31 is the whole-number part and 25 is the decimal part.

ESSENTIAL RULES AND PROCEDURES	EXAMPLES
To write a decimal in words, write the decimal part as if it were a whole number. Then name the place value of the last digit. The decimal point is read as "and." [3.1A, p. 126]	The decimal 12.875 is written in words as twelve and eight hundred seventy-five thousandths.
To write a decimal in standard form when it is written in words, write the whole-number part, replace the word *and* with a decimal point, and write the decimal part so that the last digit is in the given place-value position. [3.1A, p. 127]	The decimal forty-nine and sixty-three thousandths is written in standard form as 49.063.
To round a decimal to a given place value, use the same rules used with whole numbers, except drop the digits to the right of the given place value instead of replacing them with zeros. [3.1B, p. 128]	2.7134 rounded to the nearest tenth is 2.7. 0.4687 rounded to the nearest hundredth is 0.47.

To add decimals, write the decimals so that the decimal points are on a vertical line. Add as you would with whole numbers. Then write the decimal point in the sum directly below the decimal points in the addends. [3.2A, p. 132]

$$\begin{array}{r} \scriptstyle 1\ 1 \\ 1.35 \\ 20.8 \\ +\ 0.76 \\ \hline 22.91 \end{array}$$

To subtract decimals, write the decimals so that the decimal points are on a vertical line. Subtract as you would with whole numbers. Then write the decimal point in the difference directly below the decimal point in the subtrahend. [3.3A, p. 136]

$$\begin{array}{r} \scriptstyle 2\ 15\qquad 6\ 10 \\ \not{3}\not{5}.8\not{7}\not{0} \\ -\ \ 9.641 \\ \hline 26.229 \end{array}$$

To multiply decimals, multiply the numbers as you would whole numbers. Then write the decimal point in the product so that the number of decimal places in the product is the sum of the decimal places in the factors. [3.4A, p. 140]

$$\begin{array}{rl} 26.83 & \text{2 decimal places} \\ \times\quad 0.45 & \text{2 decimal places} \\ \hline 13415 & \\ 10732 & \\ \hline 12.0735 & \text{4 decimal places} \end{array}$$

To multiply a decimal by a power of 10, move the decimal point to the right the same number of places as there are zeros in the power of 10. If the power of 10 is written in exponential notation, the exponent indicates how many places to move the decimal point. [3.4A, pp. 140, 141]

$3.97 \cdot 10{,}000 = 39{,}700$

$0.641 \cdot 10^5 = 64{,}100$

To divide decimals, move the decimal point in the divisor to the right so that it is a whole number. Move the decimal point in the dividend the same number of places to the right. Place the decimal point in the quotient directly above the decimal point in the dividend. Then divide as you would with whole numbers. [3.5A, p. 150]

$$\begin{array}{r} 6.2 \\ 0.39.\overline{)2.41.8} \\ -2\,34 \\ \hline 7\,8 \\ -7\,8 \\ \hline 0 \end{array}$$

To divide a decimal by a power of 10, move the decimal point to the left the same number of places as there are zeros in the power of 10. If the power of 10 is written in exponential notation, the exponent indicates how many places to move the decimal point. [3.5A, p. 151]

$972.8 \div 1000 = 0.9728$

$61.305 \div 10^4 = 0.0061305$

To convert a fraction to a decimal, divide the numerator of the fraction by the denominator. [3.6A, p. 159]

$\frac{7}{8} = 7 \div 8 = 0.875$

To convert a decimal to a fraction, remove the decimal point and place the decimal part over a denominator equal to the place value of the last digit in the decimal. [3.6B, p. 159]

0.85 is eighty-five <u>hundredths</u>.

$0.85 = \frac{85}{100} = \frac{17}{20}$

To find the order relation between a decimal and a fraction, first rewrite the fraction as a decimal. Then compare the two decimals. [3.6C, p. 160]

Because $\frac{3}{11} \approx 0.273$, and $0.273 > 0.26$, $\frac{3}{11} > 0.26$.

CHAPTER 3

CONCEPT REVIEW

Test your knowledge of the concepts presented in this chapter. Answer each question. Then check your answers against the ones provided in the Answer Section.

1. How do you round a decimal to the nearest tenth?

2. How do you write the decimal 0.37 as a fraction?

3. How do you write the fraction $\frac{173}{10{,}000}$ as a decimal?

4. When adding decimals of different place values, what do you do with the decimal points?

5. Where do you put the decimal point in the product of two decimals?

6. How do you estimate the product of two decimals?

7. What do you do with the decimal point when dividing decimals?

8. Which is greater, the decimal 0.63 or the fraction $\frac{5}{8}$?

9. How many zeros must be inserted when dividing 0.763 by 0.6 and rounding to the nearest hundredth?

10. How do you subtract a decimal from a whole number that has no decimal point?

CHAPTER 3

REVIEW EXERCISES

1. Find the quotient of 3.6515 and 0.067.

2. Find the sum of 369.41, 88.3, 9.774, and 366.474.

3. Place the correct symbol, < or >, between the two numbers.
0.055   0.1

4. Write 22.0092 in words.

5. Round 0.05678235 to the nearest hundred-thousandth.

6. Convert $2\frac{1}{3}$ to a decimal. Round to the nearest hundredth.

7. Convert 0.375 to a fraction.

8. Add: $3.42 + 0.794 + 32.5$

9. Write thirty-four and twenty-five thousandths in standard form.

10. Place the correct symbol, < or >, between the two numbers.
$\frac{5}{8}$ 0.62

11. Convert $\frac{7}{9}$ to a decimal. Round to the nearest thousandth.

12. Convert 0.66 to a fraction.

13. Subtract: $27.31 - 4.4465$

14. Round 7.93704 to the nearest hundredth.

15. Find the product of 3.08 and 2.9.

16. Write 342.37 in words.

17. Write three and six thousand seven hundred fifty-three hundred-thousandths in standard form.

18. Multiply: $\begin{array}{r} 34.79 \\ \times\ 0.74 \\ \hline \end{array}$

19. Divide: $0.053\overline{)0.349482}$

20. What is 7.796 decreased by 2.9175?

21. **Banking** You had a balance of $895.68 in your checking account. You then wrote checks for $145.72 and $88.45. Find the new balance in your checking account.

22. **Fuel Consumption** A county ambulance company has 15 ambulances in their fleet. Each ambulance travels an average of 760 miles per week at a cost of $118. Find the cost per mile traveled. Round to the nearest cent.

23. **Salary.** Full-time paramedics are paid for a 40-hour workweek or 2080 hours per year. The average salary for a paramedic in a certain town is $40,040. Find the hourly wage for these paramedics.

24. **Nutrition** Nutritionists recommend that adults over 50 years old eat about 30 grams of fiber per day. If Mario is 65 years old and eats one medium orange with 3.1 grams of fiber, one oat bran muffin with 5.2 grams of fiber, and 1 cup of lima beans with 13.2 grams of fiber, how many additional grams of fiber should he consume today in order to meet the nutritionists' recommendation?

25. **Nutrition** According to the American School Food Service Association, 1.9 million gallons of milk are served in school cafeterias every day. How many gallons of milk are served in school cafeterias during a 5-day school week?

CHAPTER 3

TEST

1. Place the correct symbol, < or >, between the two numbers.
0.66 0.666

2. Subtract:
$$\begin{array}{r} 13.027 \\ -\ 8.94 \\ \hline \end{array}$$

3. Write 45.0302 in words.

4. Convert $\frac{9}{13}$ to a decimal. Round to the nearest thousandth.

5. Convert 0.825 to a fraction.

6. Round 0.07395 to the nearest ten-thousandth.

7. Find 0.0569 divided by 0.037. Round to the nearest thousandth.

8. Find 9.23674 less than 37.003.

9. Round 7.0954625 to the nearest thousandth.

10. Divide: $0.006\overline{)1.392}$

11. Add:
$$\begin{array}{r} 270.93 \\ 97. \\ 1.976 \\ +\ 88.675 \\ \hline \end{array}$$

12. **Purchases** A home health nurse purchased over-the-counter medications at a pharmacy for her patient at a cost of \$12.43. How much change is returned if the bill is paid with a \$20 bill?

13. Multiply: $\begin{array}{r} 1.37 \\ \times\ 0.004 \\ \hline \end{array}$

14. What is the total of 62.3, 4.007, and 189.65?

15. Write two hundred nine and seven thousand eighty-six hundred-thousandths in standard form.

16. **Finances** A car was bought for $16,734.40, with a down payment of $2500. The balance was paid in 36 monthly payments. Find the amount of each monthly payment.

17. **Infant Mortality** Refer to the Infant Mortality graph on page 139. How many times greater is the infant mortality rate in Mexico than the infant mortality rate in the United States? Round to the nearest tenth.

18. **Physical Therapy** A physical therapist working with a patient following back surgery recommends that the patient walk five days each week, increasing the distance walked each week. The patient is told to walk the following distances: 0.25 mile per day during week 1, 0.5 mile per day during week 2, and 0.75 mile per day during week 3. How many miles has the patient walked at the end of the three-week period?

Computers The table at the right shows the average number of hours per week that students use a computer. Use this table for Exercises 19 and 20.

Grade Level	*Average Number of Hours of Computer Use per Week*
Prekindergarten–kindergarten	3.9
1st – 3rd	4.9
4th – 6th	4.2
7th – 8th	6.9
9th – 12th	6.7

Source: Find/SVP American Learning Household Survey

19. On average, how many hours per year does a 10th-grade student use a computer? Use a 52-week year.

20. On average, how many more hours per year does a 2nd-grade student use a computer than a 5th-grade student? Use a 52-week year.

CUMULATIVE REVIEW EXERCISES

1. Divide: $89\overline{)20{,}932}$

2. Simplify: $2^3 \cdot 4^2$

3. Simplify: $2^2 - (7 - 3) \div 2 + 1$

4. Find the LCM of 9, 12, and 24.

5. Write $\frac{22}{5}$ as a mixed number.

6. Write $4\frac{5}{8}$ as an improper fraction.

7. Write an equivalent fraction with the given denominator.

$$\frac{5}{12} = \frac{}{60}$$

8. Add: $\frac{3}{8} + \frac{5}{12} + \frac{9}{16}$

9. What is $5\frac{7}{12}$ increased by $3\frac{7}{18}$?

10. Subtract: $9\frac{5}{9} - 3\frac{11}{12}$

11. Multiply: $\frac{9}{16} \times \frac{4}{27}$

12. Find the product of $2\frac{1}{8}$ and $4\frac{5}{17}$.

13. Divide: $\frac{11}{12} \div \frac{3}{4}$

14. What is $2\frac{3}{8}$ divided by $2\frac{1}{2}$?

15. Simplify: $\left(\frac{2}{3}\right)^2 \cdot \left(\frac{3}{4}\right)^3$

16. Simplify: $\left(\frac{2}{3}\right)^2 - \left(\frac{2}{3} - \frac{1}{2}\right) + 2$

17. Write 65.0309 in words.

18. Add:

$$\begin{array}{r} 379.006 \\ 27.523 \\ 9.8707 \\ +\ 88.2994 \\ \hline \end{array}$$

19. What is 29.005 decreased by 7.9286?

20. Multiply: $\begin{array}{r} 9.074 \\ \times\ 6.09 \\ \hline \end{array}$

21. Divide: $8.09\overline{)17.42963}$
Round to the nearest thousandth

22. Convert $\frac{11}{15}$ to a decimal. Round to the nearest thousandth.

23. Convert $0.16\frac{2}{3}$ to a fraction.

24. Place the correct symbol, $<$ or $>$, between the two numbers.

$\frac{8}{9}$ 0.98

25. **Calories** Using a calorie calculator, we determine that a 50-year-old, 6-foot tall male who leads a sedentary lifestyle burns only 2747 calories per day. The same male who leads an active lifestyle burns 3387 calories per day. How many more calories per day does the active male burn than the sedentary male?

26. **Health** A patient is put on a diet to lose 24 pounds in 3 months. The patient loses $9\frac{1}{2}$ pounds the first month and $6\frac{3}{4}$ pounds the second month. How much weight must this patient lose the third month to achieve the goal?

27. **Banking** You have a checking account balance of $814.35. You then write checks for $42.98, $16.43, and $137.56. Find your checking account balance after you write the checks.

28. **Medication** A doctor has ordered that a patient be given 1.25 milligrams of medication. The only tablets available are scored tablets with a strength of 0.5 milligram. How many tablets should this patient be given in order to receive the proper dosage?

29. **Taxes** The state income tax on your business is $820 plus 0.08 times your profit. You made a profit of $64,860 last year. Find the amount of income tax you paid last year.

30. **Weight Gain** After an extended illness, a patient is advised to gain 10 pounds in order to reach a healthy weight. During the next month, the patient gains $2\frac{1}{2}$ pounds the first week, $3\frac{3}{4}$ pounds the second week, 2 pounds the third week, and $1\frac{3}{4}$ pounds the fourth week. How many more pounds does the patient need to gain in order to reach the goal of a 10-pound increase?

CHAPTER

4

Ratio and Proportion

OBJECTIVES

SECTION 4.1

- **A** To write the ratio of two quantities in simplest form
- **B** To solve application problems

SECTION 4.2

- **A** To write rates
- **B** To write unit rates
- **C** To solve application problems

SECTION 4.3

- **A** To determine whether a proportion is true
- **B** To solve proportions
- **C** To solve application problems

AIM for the Future

MEDICAL ASSISTANTS perform administrative and clinical tasks to keep the offices of physicians running smoothly. Medical assistants have various duties, which may include taking medical histories and recording vital signs, explaining treatment procedures to patients, preparing patients for examinations, and assisting physicians during examinations. They may also update and file patients' medical records, fill out insurance forms, and arrange for hospital admissions and laboratory services. Employment of medical assistants is expected to grow 34% from 2008 to 2018, much faster than the average for all occupations. Median annual wages of wage-and-salary medical assistants were $28,300 in May 2008. (*Source: Bureau of Labor Statistics:* http://www.bls.gov/oco/ocos164.htm)

PREP TEST

Are you ready to succeed in this chapter?

Take the Prep Test below to find out if you are ready to learn the new material.

1. Simplify: $\dfrac{8}{10}$

2. Simplify: $\dfrac{450}{650 + 250}$

3. Write as a decimal: $\dfrac{372}{15}$

4. Which is greater, 4×33 or 62×2?

5. Complete: $? \times 5 = 20$

SECTION

4.1 Ratio

OBJECTIVE A To write the ratio of two quantities in simplest form

Point of Interest

In the 1990s, the major-league pitchers with the best strikeout-to-walk ratios (having pitched a minimum of 100 innings) were

Dennis Eckersley	6.46:1
Shane Reynolds	4.13:1
Greg Maddux	4:1
Bret Saberhagen	3.92:1
Rod Beck	3.81:1

The best single-season strikeout-to-walk ratio for starting pitchers in the same period was that of Bret Saberhagen, 11:1. (*Source:* Elias Sports Bureau)

Quantities such as 3 feet, 12 cents, and 9 cars are number quantities written with units.

3 feet
12 cents
9 cars
↑
units

These are some examples of units. Shirts, dollars, trees, miles, and gallons are further examples.

A **ratio** is a comparison of two quantities that have the *same* units. This comparison can be written three different ways:

1. As a fraction
2. As two numbers separated by a colon (:)
3. As two numbers separated by the word *to*

The ratio of the lengths of two boards, one 8 feet long and the other 10 feet long, can be written as

1. $\frac{8 \text{ feet}}{10 \text{ feet}} = \frac{8}{10} = \frac{4}{5}$
2. 8 feet:10 feet = 8:10 = 4:5
3. 8 feet to 10 feet = 8 to 10 = 4 to 5

Writing the **simplest form of a ratio** means writing it so that the two numbers have no common factor other than 1.

This ratio means that the smaller board is $\frac{4}{5}$ the length of the longer board.

EXAMPLE 1

Write the comparison \$6 to \$8 as a ratio in simplest form using a fraction, a colon, and the word *to*.

Solution $\frac{\$6}{\$8} = \frac{6}{8} = \frac{3}{4}$

\$6:\$8 = 6:8 = 3:4

\$6 to \$8 = 6 to 8 = 3 to 4

YOU TRY IT 1

Write the comparison 20 pounds to 24 pounds as a ratio in simplest form using a fraction, a colon, and the word *to*.

Your solution

EXAMPLE 2

Write the comparison 18 quarts to 6 quarts as a ratio in simplest form using a fraction, a colon, and the word *to*.

Solution $\frac{18 \text{ quarts}}{6 \text{ quarts}} = \frac{18}{6} = \frac{3}{1}$

18 quarts:6 quarts = 18:6 = 3:1

18 quarts to 6 quarts = 18 to 6 = 3 to 1

YOU TRY IT 2

Write the comparison 64 miles to 8 miles as a ratio in simplest form using a fraction, a colon, and the word *to*.

Your solution

Solutions on p. S10

OBJECTIVE B To solve application problems

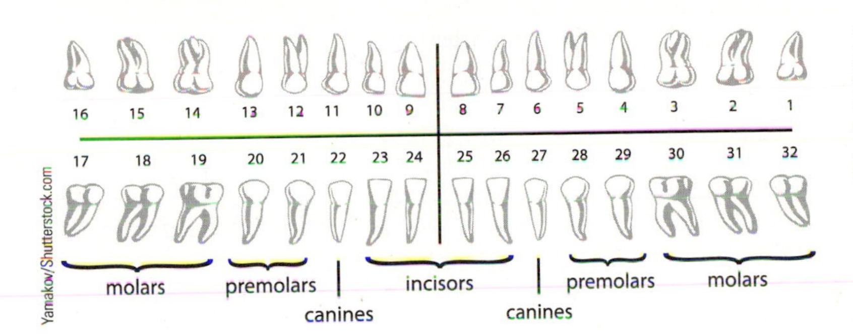

The universal tooth numbering system shown in the table is a method of identifying teeth that is approved and adopted by the American Dental Association. Use this information for Example 3 and You Try It 3.

EXAMPLE • 3

Find, as a fraction in simplest form, the ratio of the number of molars in an adult mouth to the total number of teeth in an adult mouth.

Strategy
To find the ratio, write the ratio of molars (12) to the total number of teeth (32) in simplest form.

Solution

$$\frac{12}{32} = \frac{3}{8}$$

The ratio is $\frac{3}{8}$.

YOU TRY IT • 3

Find, as a fraction in simplest form, the ratio of canines in an adult mouth to the number of incisors in an adult mouth.

Your strategy

Your solution

EXAMPLE • 4

A **dilution ratio** as used in a laboratory is a very specific ratio. The numerator is the amount of substance being diluted and the denominator is the total amount of the mixture, including the diluting agent and the substance being diluted. A dilution of alcohol was done in the lab using 10 milliliters of alcohol and 40 milliliters of saline. What, as a fraction in simplest form, is the dilution ratio for this solution?

Strategy
To find the ratio, write the ratio of the amount of alcohol (10 milliliters) to the total amount of solution (10 milliliters + 40 milliliters) in simplest form.

Solution

$$\frac{10 \text{ milliliters}}{10 \text{ millilters} + 40 \text{ milliliters}} = \frac{10}{50} = \frac{1}{5}$$

The dilution ratio is $\frac{1}{5}$.

YOU TRY IT • 4

For medical testing, a dilution of 2 milliliters of blood with 98 milliliters of water and reagents is required. What, as a fraction in simplest form, is the dilution ratio for this solution? See Example 4.

Your strategy

Your solution

Solutions on p. S10

4.1 EXERCISES

OBJECTIVE A To write the ratio of two quantities in simplest form

For Exercises 1 to 18, write the comparison as a ratio in simplest form using a fraction, a colon (:), and the word *to*.

1. 3 pints to 15 pints

2. 6 pounds to 8 pounds

3. $40 to $20

4. 10 feet to 2 feet

5. 3 miles to 8 miles

6. 2 hours to 3 hours

7. 6 minutes to 6 minutes

8. 8 days to 12 days

9. 35 cents to 50 cents

10. 28 inches to 36 inches

11. 30 minutes to 60 minutes

12. 25 cents to 100 cents

13. 32 ounces to 16 ounces

14. 12 quarts to 4 quarts

15. 30 yards to 12 yards

16. 12 quarts to 18 quarts

17. 20 gallons to 28 gallons

18. 14 days to 7 days

19. To write a ratio that compares 3 days to 3 weeks, change 3 weeks into an equivalent number of __________.

20. Is the ratio 3 : 4 the same as the ratio 4 : 3?

OBJECTIVE B To solve application problems

For Exercises 21 to 23, write ratios in simplest form using a fraction.

Family Budget						
Housing	*Food*	*Transportation*	*Taxes*	*Utilities*	*Miscellaneous*	*Total*
$1600	$800	$600	$700	$300	$800	$4800

21. **Budgets** Use the table to find the ratio of housing costs to total expenses.

22. **Budgets** Use the table to find the ratio of food costs to total expenses.

23. **Budgets** Use the table to find the ratio of utilities costs to food costs.

24. Refer to the table above. Write a verbal description of the ratio represented by 1 : 2. (*Hint:* There is more than one answer.)

25. **Facial Hair** Using the data in the news clipping at the right and the figure 50 million for the number of adult males in the United States, write the ratio of the number of men who participated in Movember to the number of adult males in the U.S. Write the ratio as a fraction in simplest form.

In the News

Grow a Mustache, Save a Life

Last fall, in an effort to raise money for the Prostate Cancer Foundation, approximately 65,000 men participated in a month-long mustache-growing competition. The event was dubbed Movember.

Source: http://us.movember.com/momoney

Dilution Ratio Refer to the definition of a dilution ratio as explained in Example 4 on page 175 to determine the ratios in Exercises 26 to 28.

26. Calculate the dilution ratio of a solution of 3 milliliters of urine diluted with 447 milliliters of distilled water. Write the ratio in simplest form using a fraction.

27. Calculate the dilution ratio of a solution of 20 microliters of blood diluted with 1980 microliters of diluting agents. Write the ratio in simplest form using a fraction.

corepics/Shutterstock.com

28. Calculate the dilution ratio of a solution of 100 microliters of serum diluted with 900 microliters of saline. Write the ratio in simplest form using a fraction.

29. **Business Expenses** A medical practice has $1,850,000 in business expenses during a year. The salaries of the physicians and staff total $1,036,000. Write the ratio as a fraction in simplest form of the salaries of those in the practice to the total business expenses of the practice.

Malpractice Insurance The bar graph at the right shows the approximate costs in 2009 of medical malpractice insurance in Florida for certain specialties. Use this information for Exercises 30 to 32. Write the ratios in simplest form using a fraction.

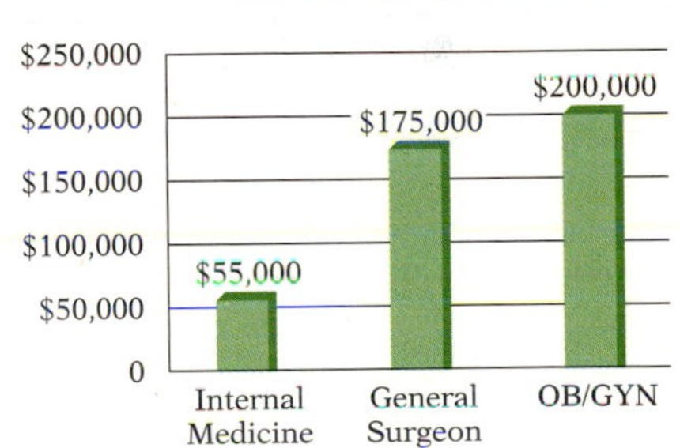

2009 Medical Malpractice Insurance Rates in Florida

30. Find the ratio of yearly malpractice rates paid by general surgeons to the total of the three specialties.

31. Find the ratio in yearly malpractice rates paid by doctors of internal medicine to those paid by OB/GYN physicians.

32. Find the ratio in yearly malpractice rates paid by OB/GYN physicians to the rates paid by general surgeons.

Applying the Concepts

33. Is the value of a ratio always less than 1? Explain.

SECTION

4.2 Rates

OBJECTIVE A To write rates

Point of Interest

Listed below are rates at which various crimes are committed in our nation.

Crime	*Every*
Larceny	4 seconds
Burglary	14 seconds
Robbery	60 seconds
Rape	6 minutes
Murder	31 minutes

A **rate** is a comparison of two quantities that have *different* units. A rate is written as a fraction.

A distance runner ran 26 miles in 4 hours. The distance-to-time rate is written

$$\frac{26 \text{ miles}}{4 \text{ hours}} = \frac{13 \text{ miles}}{2 \text{ hours}}$$

Writing the **simplest form of a rate** means writing it so that the two numbers that form the rate have no common factor other than 1.

EXAMPLE • 1

Write "4 nurses for every 24 patients" as a rate in simplest form.

Solution

$$\frac{4 \text{ nurses}}{24 \text{ patients}} = \frac{1 \text{ nurse}}{6 \text{ patients}}$$

YOU TRY IT • 1

Write "9 milligrams of medication in 3 tablets" as a rate in its simplest form.

Your solution

Solution on p. S11

OBJECTIVE B To write unit rates

Point of Interest

According to a Gallup Poll, women see doctors more often than men do. On average, men visit the doctor 3.8 times per year, whereas women go to the doctor 5.8 times per year.

A **unit rate** is a rate in which the number in the denominator is 1.

$\frac{\$3.25}{1 \text{ pound}}$ or \$3.25/pound is read "\$3.25 per pound."

To find a unit rate, divide the number in the numerator of the rate by the number in the denominator of the rate.

A car traveled 344 miles on 16 gallons of gasoline. To find the miles per gallon (unit rate), divide the numerator of the rate by the denominator of the rate.

$\frac{344 \text{ miles}}{16 \text{ gallons}}$ is the rate. $\quad 16\overline{)344.0} = 21.5 \quad$ 21.5 miles/gallon is the unit rate.

EXAMPLE • 2

Write "2 grams of medication per 5 milliliters of solution" as a unit rate.

Solution

$$\frac{2 \text{ grams}}{5 \text{ milliliters}} = 5\overline{)2.0} = 0.4$$

0.4 gram/milliliter

YOU TRY IT • 2

Write "150 milligrams of medication in 3 tablets" as a unit rate.

Your solution

Solution on p. S11

OBJECTIVE C To solve application problems

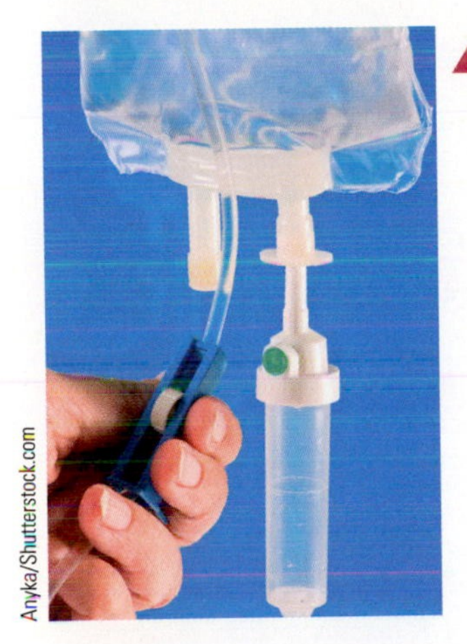
Anyka/Shutterstock.com

Some Common IV Fluids	
5% dextrose in water	D5W
Normal saline	NS
Ringer's lactate	RL
Dextrose/ normal saline	D5NS

HOW TO • 1 Intravenous fluids are usually ordered on the basis of milliliters of solution per hour to be infused. Today, many IV fluids are administered with volumetric pumps that must be set to correctly administer the medications at the proper unit rate. Calculate the unit rates for the following orders.

a. A doctor orders that a patient receive 1000 milliliters of normal saline (NS) solution over a 10-hour period. Find the unit rate for this infusion.

b. A doctor orders that a patient receive 2000 milliliters of dextrose in water (D5W) in 16 hours. Find the unit rate for the infusion.

Strategy

To find the rate per hour, divide the total volume to be given to the patient by the number of hours of the infusion.

Solution **a.** $\frac{1000 \text{ milliliters}}{10 \text{ hours}} = 100 \text{ milliliters/hour}$

The patient will receive 100 milliliters of normal saline per hour.

b. $\frac{2000 \text{ milliliters}}{16 \text{ hours}} = 125 \text{ milliliters/hour}$

The patient will receive 125 milliliters of D5W solution per hour.

EXAMPLE • 3

A pediatric dose of an IV medication for a particular patient is 250 units per hour. The prescribed dosage is based on the weight of this child, which is 12.5 kilograms. Calculate the recommended dosage rate for this medication in units per kilograms per hour.

Strategy

To find the recommended dosage rate for this medication, divide the prescribed units per hour (250) by the weight in kilograms of the child (12.5). The dosage rate will be in units per kilogram per hour.

Solution

$250 \div 12.5 = 20$

The recommended dosage rate for this medication is 20 units per kilogram per hour.

YOU TRY IT • 3

A certain solution contains 5 milligrams of sodium chloride (NaCl) in 100 milliliters of solution. Find the strength of the solution as a unit rate.

Your strategy

Your solution

Solution on p. S11

4.2 EXERCISES

OBJECTIVE A To write rates

For Exercises 1 to 8, write each phrase as a rate in simplest form.

1. 3 pounds of meat for 4 people
2. 30 ounces in 24 glasses
3. $80 for 12 thermometers
4. 500 tongue depressors for $5
5. 15 patients to 5 ICU nurses
6. 4 tablets in 24 hours
7. 16 gallons in 2 hours
8. 25 ounces in 5 minutes

9. For television advertising rates, what units are **a.** in the numerator and **b.** in the denominator?

OBJECTIVE B To write unit rates

For Exercises 10 to 12, complete the unit rate.

10. 5 miles in ___ hour
11. 15 feet in ___ second
12. 5 grams of fat in ___ serving

For Exercises 13 to 22, write each phrase as a unit rate.

13. 10 feet in 4 seconds
14. 816 miles in 6 days
15. $3900 earned in 4 weeks
16. $51,000 earned in 12 months
17. 125 milligrams of medication in 100 milliliters of solution
18. 300 milliliters infused in 4 hours
19. $131.88 earned in 7 hours
20. 628.8 miles in 12 hours
21. 409.4 miles on 11.5 gallons of gasoline
22. $11.05 for 3.4 pounds

OBJECTIVE C To solve application problems

Miles per Dollar One measure of how expensive it is to drive your car is calculated as miles per dollar, which is the number of miles you drive on 1 dollar's worth of gasoline.

23. Suppose you get 26 miles per gallon of gasoline and gasoline costs $3.49 per gallon. Calculate your miles per dollar. Round to the nearest tenth.
24. Suppose you get 23 miles per gallon of gasoline and gasoline costs $3.15 per gallon. It costs you $44.10 to fill the tank. Calculate your miles per dollar. Round to the nearest tenth.

Consumerism The table below is a price list of some items from MedPro Supply Company. Use this table for Exercises 25 to 27.

Medical Supply Item	*Cost*
Sterile strip skin closures, 3/pack, 50 packs/box	$66 per box
Butterfly wound closures, 100/box, 24 boxes/case	$55 per case
Sterile latex exam gloves, 100/box, 8 boxes/case	$153 per case
Cotton swabs, 265/pack, 24 packs/case	$90 per case

25. A medical office manager orders a box of sterile strip skin closures. What is the cost per pack of these skin closures?

26. When comparison shopping, the medical office manager needs to find the cost per glove for the sterile latex exam gloves. Use the information in the table to calculate this cost. Round to the nearest cent.

27. The office manager is planning to place a large order for butterfly wound closures. MediSupp Store sells the same product as MedPro for $3.50 per box of 100 closures. Which store has the better deal?

28. **Medication Dosage** The dosage of a certain antibiotic prescribed by a doctor is determined by the weight of the patient in kilograms. If a patient who weighs 25 kilograms is given a shot containing 250 milligrams of this antibiotic, find the dosage rate for this medication in milligrams per kilogram.

29. **Medication Dosage** A certain medication is administered through injection, and the label on the vial states the strength of the medication is 75 milligrams per 37.5 milliliters. How many milligrams of medication are in 1 milliliter?

30. **IV Infusions** A doctor orders an IV of 800 milliliters of Ringer's lactate to be infused over 5 hours. Find the unit rate for the infusion.

31. **IV Infusions** A doctor orders an IV of 1500 milliliters of D5W to be infused over a 12-hour period. Find the unit rate for the infusion.

Applying the Concepts

32. **Compensation** You have a choice of receiving a wage of $34,000 per year, $2840 per month, $650 per week, or $18 per hour. Which pay choice would you take? Assume a 40-hour workweek with 52 weeks per year.

33. The price–earnings ratio of a company's stock is one measure used by stock market analysts to assess the financial well-being of the company. Explain the meaning of the price–earnings ratio.

SECTION

4.3 Proportions

OBJECTIVE A To determine whether a proportion is true

Point of Interest

Proportions were studied by the earliest mathematicians. Clay tablets uncovered by archaeologists show evidence of proportions in Egyptian and Babylonian cultures dating from 1800 B.C.

A **proportion** is an expression of the equality of two ratios or rates.

$\frac{50 \text{ miles}}{4 \text{ gallons}} = \frac{25 \text{ miles}}{2 \text{ gallons}}$ Note that the units of the numerators are the same and the units of the denominators are the same.

$\frac{3}{6} = \frac{1}{2}$ This is the equality of two ratios.

A proportion is **true** if the fractions are equal when written in lowest terms.

In any true proportion, the **cross products** are equal.

HOW TO • 1 Is $\frac{2}{3} = \frac{8}{12}$ a true proportion?

$\frac{2}{3} \quad \frac{8}{12}$ → $3 \times 8 = 24$, $2 \times 12 = 24$

The cross products *are* equal.

$\frac{2}{3} = \frac{8}{12}$ is a true proportion.

A proportion is **not true** if the fractions are not equal when reduced to lowest terms.

If the cross products are not equal, then the proportion is not true.

HOW TO • 2 Is $\frac{4}{5} = \frac{8}{9}$ a true proportion?

$\frac{4}{5} \quad \frac{8}{9}$ → $5 \times 8 = 40$, $4 \times 9 = 36$

The cross products *are not* equal.

$\frac{4}{5} = \frac{8}{9}$ is not a true proportion.

EXAMPLE • 1

Is $\frac{5}{8} = \frac{10}{16}$ a true proportion?

Solution

$\frac{5}{8} \quad \frac{10}{16}$ → $8 \times 10 = 80$, $5 \times 16 = 80$

The cross products are equal.

The proportion is true.

YOU TRY IT • 1

Is $\frac{6}{10} = \frac{9}{15}$ a true proportion?

Your solution

EXAMPLE • 2

Is $\frac{62 \text{ miles}}{4 \text{ gallons}} = \frac{33 \text{ miles}}{2 \text{ gallons}}$ a true proportion?

Solution

$\frac{62}{4} \quad \frac{33}{2}$ → $4 \times 33 = 132$, $62 \times 2 = 124$

The cross products are not equal.

The proportion is not true.

YOU TRY IT • 2

Is $\frac{\$32}{6 \text{ hours}} = \frac{\$90}{8 \text{ hours}}$ a true proportion?

Your solution

Solutions on p. S11

OBJECTIVE B To solve proportions

Tips for Success

An important element of success is practice. We cannot do anything well if we do not practice it repeatedly. Practice is crucial to success in mathematics. In this objective you are learning a new skill: how to solve a proportion. You will need to practice this skill over and over again in order to be successful at it.

Sometimes one of the numbers in a proportion is unknown. In this case, it is necessary to *solve* the proportion.

To **solve a proportion,** find a number to replace the unknown so that the proportion is true.

HOW TO • 3 Solve: $\frac{9}{6} = \frac{3}{n}$

$$\frac{9}{6} = \frac{3}{n}$$

$9 \times n = 6 \times 3$ • **Find the cross products.**

$9 \times n = 18$

$n = 18 \div 9$ • **Think of $9 \times n = 18$ as $9\overline{)18}$.**

$n = 2$

Check:

$\frac{9}{6}$ $\frac{3}{2}$ → $6 \times 3 = 18$

→ $9 \times 2 = 18$

EXAMPLE • 3

Solve $\frac{n}{12} = \frac{25}{60}$ and check.

Solution

$n \times 60 = 12 \times 25$ • **Find the cross products. Then solve for n.**

$n \times 60 = 300$

$n = 300 \div 60$

$n = 5$

Check:

$\frac{5}{12}$ $\frac{25}{60}$ → $12 \times 25 = 300$

→ $5 \times 60 = 300$

YOU TRY IT • 3

Solve $\frac{n}{14} = \frac{3}{7}$ and check.

Your solution

EXAMPLE • 4

Solve $\frac{4}{9} = \frac{n}{16}$. Round to the nearest tenth.

Solution

$4 \times 16 = 9 \times n$ • **Find the cross products. Then solve for n.**

$64 = 9 \times n$

$64 \div 9 = n$

$7.1 \approx n$

Note: A rounded answer is an approximation. Therefore, the answer to a check will not be exact.

YOU TRY IT • 4

Solve $\frac{5}{7} = \frac{n}{20}$. Round to the nearest tenth.

Your solution

Solutions on p. S11

EXAMPLE • 5

Solve $\frac{28}{52} = \frac{7}{n}$ and check.

Solution

$28 \times n = 52 \times 7$
$28 \times n = 364$
$n = 364 \div 28$
$n = 13$

• Find the cross products. Then solve for n.

Check:

$\frac{28}{52} = \frac{7}{13}$ → $52 \times 7 = 364$; $28 \times 13 = 364$

YOU TRY IT • 5

Solve $\frac{15}{20} = \frac{12}{n}$ and check.

Your solution

EXAMPLE • 6

Solve $\frac{15}{n} = \frac{8}{3}$. Round to the nearest hundredth.

Solution

$15 \times 3 = n \times 8$
$45 = n \times 8$
$45 \div 8 = n$
$5.63 \approx n$

YOU TRY IT • 6

Solve $\frac{12}{n} = \frac{7}{4}$. Round to the nearest hundredth.

Your solution

EXAMPLE • 7

Solve $\frac{n}{9} = \frac{3}{1}$ and check.

Solution

$n \times 1 = 9 \times 3$
$n \times 1 = 27$
$n = 27 \div 1$
$n = 27$

Check:

$\frac{27}{9} = \frac{3}{1}$ → $9 \times 3 = 27$; $27 \times 1 = 27$

YOU TRY IT • 7

Solve $\frac{n}{12} = \frac{4}{1}$ and check.

Your solution

Solutions on p. S11

OBJECTIVE C To solve application problems

The application problems in this objective require you to write and solve a proportion. When setting up a proportion, remember to keep the same units in the numerators and the same units in the denominators.

EXAMPLE • 8

The dosage of a certain medication is 2 ounces for every 50 pounds of body weight. How many ounces of this medication are required for a person who weighs 175 pounds?

Strategy

To find the number of ounces of medication for a person weighing 175 pounds, write and solve a proportion using n to represent the number of ounces of medication for a 175-pound person.

Solution

$$\frac{2 \text{ ounces}}{50 \text{ pounds}} = \frac{n \text{ ounces}}{175 \text{ pounds}}$$

• The unit "ounces" is in the numerator. The unit "pounds" is in the denominator.

$$2 \times 175 = 50 \times n$$
$$350 = 50 \times n$$
$$350 \div 50 = n$$
$$7 = n$$

A 175-pound person requires 7 ounces of medication.

YOU TRY IT • 8

To clean the floors in a lab, a technician adds 3 tablespoons of disinfectant to every 4 gallons of water. How many tablespoons of disinfectant are required for 10 gallons of water?

Your strategy

Your solution

EXAMPLE • 9

A pediatrician prescribes amoxicillin to a patient. The dosage is 1 teaspoon three times a day for 10 days. If the patient ingests a total of 150 milliliters of amoxicillin, find the amount of each individual dose in milliliters.

Strategy

To find the amount per dose in milliliters, determine the total number of doses the patient will take and then solve a proportion using n to represent the number of milliliters per dose.

Solution

- Three doses per day for 10 days : $3 \times 10 = 30$ doses
- $\frac{150 \text{ milliliters}}{30 \text{ doses}} = \frac{n \text{ milliliters}}{1 \text{ dose}}$

$$150 \times 1 = 30 \times n$$
$$150 = 30n$$
$$150 \div 30 = n$$
$$5 = n$$

Each individual dose is 5 milliliters (the equivalent of 1 teaspoon).

YOU TRY IT • 9

An IV infusion of 1000 milliliters of D5W is ordered for a patient and is to be infused at a rate of 75 milliliters per hour. Calculate the number of hours it will take for this infusion to be complete.

Your strategy

Your solution

Solutions on p. S11

4.3 EXERCISES

OBJECTIVE A To determine whether a proportion is true

For Exercises 1 to 18, determine whether the proportion is true or not true.

1. $\frac{4}{8} = \frac{10}{20}$

2. $\frac{39}{48} = \frac{13}{16}$

3. $\frac{7}{8} = \frac{11}{12}$

4. $\frac{15}{7} = \frac{17}{8}$

5. $\frac{27}{8} = \frac{9}{4}$

6. $\frac{3}{18} = \frac{4}{19}$

7. $\frac{45}{135} = \frac{3}{9}$

8. $\frac{3}{4} = \frac{54}{72}$

9. $\frac{50 \text{ miles}}{2 \text{ gallons}} = \frac{25 \text{ miles}}{1 \text{ gallon}}$

10. $\frac{16 \text{ feet}}{10 \text{ seconds}} = \frac{24 \text{ feet}}{15 \text{ seconds}}$

11. $\frac{6 \text{ minutes}}{5 \text{ cents}} = \frac{30 \text{ minutes}}{25 \text{ cents}}$

12. $\frac{16 \text{ pounds}}{12 \text{ days}} = \frac{20 \text{ pounds}}{14 \text{ days}}$

13. $\frac{125 \text{ milligrams}}{5 \text{ milliliters}} = \frac{250 \text{ milligrams}}{10 \text{ milliliters}}$

14. $\frac{250 \text{ milligrams}}{5 \text{ milliliters}} = \frac{1250 \text{ milligrams}}{75 \text{ milliliters}}$

15. $\frac{300 \text{ feet}}{4 \text{ rolls}} = \frac{450 \text{ feet}}{7 \text{ rolls}}$

16. $\frac{1 \text{ gallon}}{4 \text{ quarts}} = \frac{7 \text{ gallons}}{28 \text{ quarts}}$

17. $\frac{\$65}{5 \text{ days}} = \frac{\$26}{2 \text{ days}}$

18. $\frac{125 \text{ milliliters}}{1 \text{ hour}} = \frac{1000 \text{ milliliters}}{8 \text{ hours}}$

19. Suppose that in a true proportion you switch the numerator of the first fraction with the denominator of the second fraction. Must the result be another true proportion?

20. Write a true proportion in which the cross products are equal to 36.

OBJECTIVE B To solve proportions

21. Consider the proportion $\frac{n}{7} = \frac{9}{21}$ in Exercise 23. In lowest terms, $\frac{9}{21} = \frac{3}{7}$. Will solving the proportion $\frac{n}{7} = \frac{3}{7}$ give the same result for n as found in Exercise 23?

For Exercises 22 to 41, solve. Round to the nearest hundredth, if necessary.

22. $\frac{n}{4} = \frac{6}{8}$

23. $\frac{n}{7} = \frac{9}{21}$

24. $\frac{12}{18} = \frac{n}{9}$

25. $\frac{7}{21} = \frac{35}{n}$

26. $\frac{6}{n} = \frac{24}{36}$

27. $\frac{3}{n} = \frac{15}{10}$

28. $\frac{n}{6} = \frac{2}{3}$

29. $\frac{5}{12} = \frac{n}{144}$

30. $\frac{n}{5} = \frac{7}{8}$

31. $\frac{4}{n} = \frac{9}{5}$

32. $\frac{5}{12} = \frac{n}{8}$

33. $\frac{36}{20} = \frac{12}{n}$

34. $\frac{n}{15} = \frac{21}{12}$

35. $\frac{40}{n} = \frac{15}{8}$

36. $\frac{28}{8} = \frac{12}{n}$

37. $\frac{n}{30} = \frac{65}{120}$

38. $\frac{0.3}{5.6} = \frac{n}{25}$

39. $\frac{1.3}{16} = \frac{n}{30}$

40. $\frac{0.7}{9.8} = \frac{3.6}{n}$

41. $\frac{1.9}{7} = \frac{13}{n}$

OBJECTIVE C To solve application problems

42. Jesse walked 3 miles in 40 minutes. Let n be the number of miles Jesse can walk in 60 minutes at the same rate. To determine how many miles Jesse can walk in 60 minutes, a student used the proportion $\frac{40}{3} = \frac{60}{n}$. Is this a valid proportion to use in solving this problem?

For Exercises 43 to 61, solve. Round to the nearest hundredth.

43. **Nutrition** A 6-ounce package of Puffed Wheat contains 600 calories. How many calories are in a 0.5-ounce serving of the cereal?

44. Health Using the data at the right and a figure of 300 million for the number of Americans, determine the number of morbidly obese Americans.

> **In the News**
>
> **Number of Obese Americans Increasing**
>
> In the past 20 years, the number of obese Americans (those at least 30 pounds overweight) has doubled. The number of morbidly obese (those at least 100 pounds overweight) has quadrupled to 1 in 50.
>
> *Source: Time,* July 9, 2006

45. Fuel Efficiency A car travels 70.5 miles on 3 gallons of gas. Find the distance the car can travel on 14 gallons of gas.

46. Nurses According to the National Sample Survey of Registered Nurses, approximately one in five American Registered Nurses is male. If there are a total of 300 nurses working at a local hospital, how many of them would you expect to be male?

47. Solutions The concentration of an alcohol solution is 25 milliliters of alcohol in 100 milliliters of solution. If you have 175 milliliters of alcohol, how many milliliters of solution can be prepared?

48. Solutions A saline solution has a concentration ratio of 2.5 milligrams of salt in 100 milliliters of solution. How many milligrams of salt will be needed to produce 500 milliliters of a saline solution having this same concentration?

49. Solutions You need to prepare a sucrose solution with a concentration of 1.5 grams of sucrose in 100 milliliters of solution. How much solution can you produce with 3 grams of sucrose?

50. Disinfectant Solution In many medical offices, a 10% bleach solution is used as a disinfectant for cleaning the exam rooms. If $\frac{1}{4}$ cup of bleach is mixed with $2\frac{1}{4}$ cups of water, how much bleach should be mixed with 9 cups of water in order to make a larger volume of this bleach solution?

51. Medication The dosage for a medication is $\frac{1}{3}$ ounce for every 40 pounds of body weight. At this rate, how many ounces of medication should a physician prescribe for a patient who weighs 150 pounds? Write the answer as a decimal.

52. Medication A pharmacist suspends a mixture of cefaclor using 100 milliliters of fluid. Each dose of cefaclor is 250 milligrams per 5 milliliters. If the patient is told to take all of the medication in the bottle during the next week, find the total number of milligrams of cefaclor the patient will consume.

53. Medication A prescription-strength medication contains 500 milligrams of acetaminophen in each tablet. If a patient takes one tablet four times a day for three days, how many milligrams of acetaminophen has the patient ingested at the end of the three-day period?

54. Patient Care One state's requirement for minimum staffing levels mandates a ratio of one nurse for every three patients in a pediatric unit. Under these guidelines, how many nurses are required to be on duty if there are 14 patients in the pediatric unit?

55. Patient Care The staffing requirements of a local nursing home include a ratio of 1:12 for nurse's aides to residents. How many nurse's aides should be on duty when there are 52 residents in the nursing home?

56. **Food Waste** At the rate given in the news clipping, find the cost of food wasted yearly by **a.** the average family of three and **b.** the average family of five.

> **In the News**
>
> **How Much Food Do You Waste?**
>
> In the United States, the estimated cost of food wasted each year by the average family of four is $590.
>
> *Source:* University of Arizona

57. **Insurance** A 60-year-old male can obtain $10,000 of life insurance for $35.35 per month. At this rate, what is the monthly cost for $50,000 of life insurance?

58. **Investments** You own 240 shares of stock in a pharmaceutical company. The company declares a stock split of 5 shares for every 3 owned. How many shares of stock will you own after the stock split?

59. **Investments** Carlos Capasso owns 50 shares of Medical Supply Inc. that pays dividends of $153. At this rate, what dividend would Carlos receive after buying 300 additional shares of Medical Supply Inc.?

60. **Nutrition** There are 200 calories in $\frac{1}{2}$ cup of cake flour. How many calories are in $2\frac{1}{4}$ cups of cake flour?

61. **Nutrition** There are 6.6 grams of carbohydrates in 2 tablespoons of smooth peanut butter. How many grams are there in 8 tablespoons $\left(\frac{1}{2}\text{ cup}\right)$?

Applying the Concepts

62. **Applications** According to the Association of American Medical Colleges, one out of every three applicants to medical school is a first-time applicant (http://www.aamc.org/facts). Explain how a proportion can be used to determine the number of first-time applicants in 2011 if you are given the total number of applicants for that year.

63. **Social Security** According to the Social Security Administration, the number of workers per retiree in the future are expected to be as given in the table below.

Year	2020	2030	2040
Number of workers per retiree	2.5	2.1	2.0

Why is the shrinking number of workers per retiree of importance to the Social Security Administration?

64. **Elections** A survey of voters in a city claimed that 2 people of every 5 who voted cast a ballot in favor of city amendment A and that 3 people of every 4 who voted cast a ballot against amendment A. Is this possible? Explain your answer.

65. Write a word problem that requires solving a proportion to find the answer.

FOCUS ON PROBLEM SOLVING

Looking for a Pattern

A very useful problem-solving strategy is looking for a pattern.

Reproduced by Permission of the State Hermitage Museum, St. Petersburg, Russia/Corbis

Problem A legend says that a peasant invented the game of chess and gave it to a very rich king as a present. The king so enjoyed the game that he gave the peasant the choice of anything in the kingdom. The peasant's request was simple: "Place one grain of wheat on the first square, 2 grains on the second square, 4 grains on the third square, 8 on the fourth square, and continue doubling the number of grains until the last square of the chessboard is reached." How many grains of wheat must the king give the peasant?

Solution A chessboard consists of 64 squares. To find the total number of grains of wheat on the 64 squares, we begin by looking at the amount of wheat on the first few squares.

Square 1	Square 2	Square 3	Square 4	Square 5	Square 6	Square 7	Square 8
1	2	4	8	16	32	64	128
1	3	7	15	31	63	127	255

The bottom row of numbers represents the sum of the number of grains of wheat up to and including that square. For instance, the number of grains of wheat on the first 7 squares is $1 + 2 + 4 + 8 + 16 + 32 + 64 = 127$.

Notice that the number of grains of wheat on a square can be expressed as a power of 2.

The number of grains on square $n = 2^{n-1}$.

For example, the number of grains on square $7 = 2^{7-1} = 2^6 = 64$.

A second pattern of interest is that **the number *below* a square** (the total number of grains up to and including that square) **is 1 less than the number of grains of wheat *on the next square.*** For example, the number *below* square 7 is 1 less than the number on square 8 ($128 - 1 = 127$). From this observation, the number of grains of wheat on the first 8 squares is the number on square 8 (128) plus 1 less than the number on square 8 (127): The total number of grains of wheat on the first 8 squares is $128 + 127 = 255$.

From this observation,

$$\text{Number of grains of wheat on the chessboard} = \text{number of grains on square 64} + \text{1 less than the number of grains on square 64}$$

$$= 2^{64-1} + (2^{64-1} - 1)$$

$$= 2^{63} + 2^{63} - 1 \approx 18{,}000{,}000{,}000{,}000{,}000{,}000$$

To give you an idea of the magnitude of this number, this is more wheat than has been produced in the world since chess was invented.

The same king decided to have a banquet in the long banquet room of the palace to celebrate the invention of chess. The king had 50 square tables, and each table could seat only one person on each side. The king pushed the tables together to form one long banquet table. How many people could sit at this table? *Hint:* Try constructing a pattern by using 2 tables, 3 tables, and 4 tables.

PROJECTS AND GROUP ACTIVITIES

The Golden Ratio

There are certain designs that have been repeated over and over in both art and architecture. One of these involves the **golden rectangle.**

A golden rectangle is drawn at the right. Begin with a square that measures, say, 2 inches on a side. Let A be the midpoint of a side (halfway between two corners). Now measure the distance from A to B. Place this length along the bottom of the square, starting at A. The resulting rectangle is a golden rectangle.

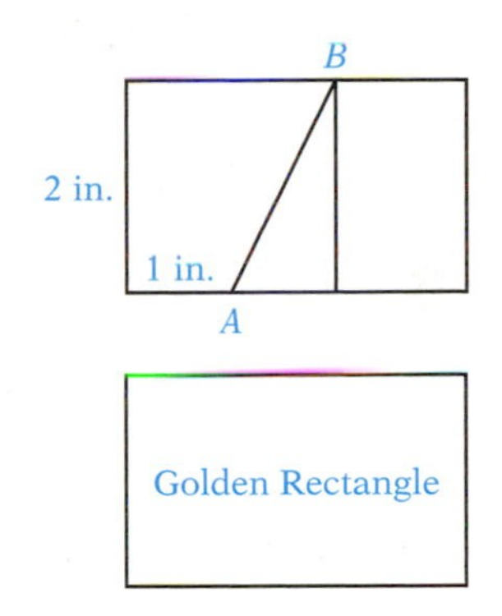

The **golden ratio** is the ratio of the length of the golden rectangle to its width. If you have drawn the rectangle following the procedure above, you will find that the golden ratio is approximately 1.6 to 1.

The golden ratio appears in many different situations. Some historians claim that some of the great pyramids of Egypt are based on the golden ratio. The drawing at the right shows the Pyramid of Giza, which dates from approximately 2600 B.C. The ratio of the height to a side of the base is approximately 1.6 to 1.

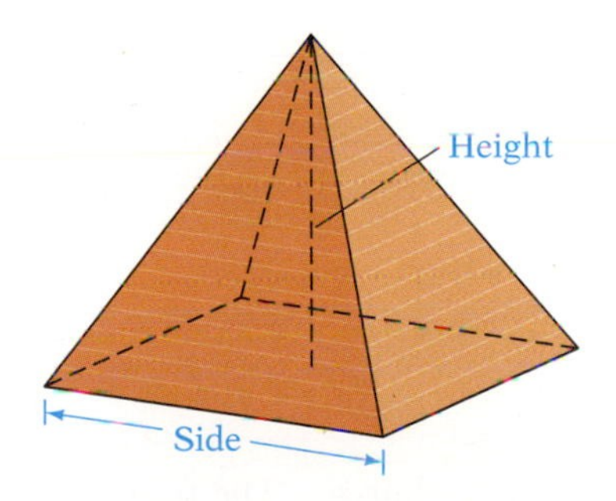

Dallas & John Heaton/Corbis

1. There are instances of the golden rectangle in the Mona Lisa painted by Leonardo da Vinci. Do some research on this painting and write a few paragraphs summarizing your findings.
2. What do 3×5 and 5×8 index cards have to do with the golden rectangle?
3. What does the United Nations Building in New York City have to do with the golden rectangle?
4. When was the Parthenon in Athens, Greece, built? What does the front of that building have to do with the golden rectangle?

Drawing the Floor Plans for a Building

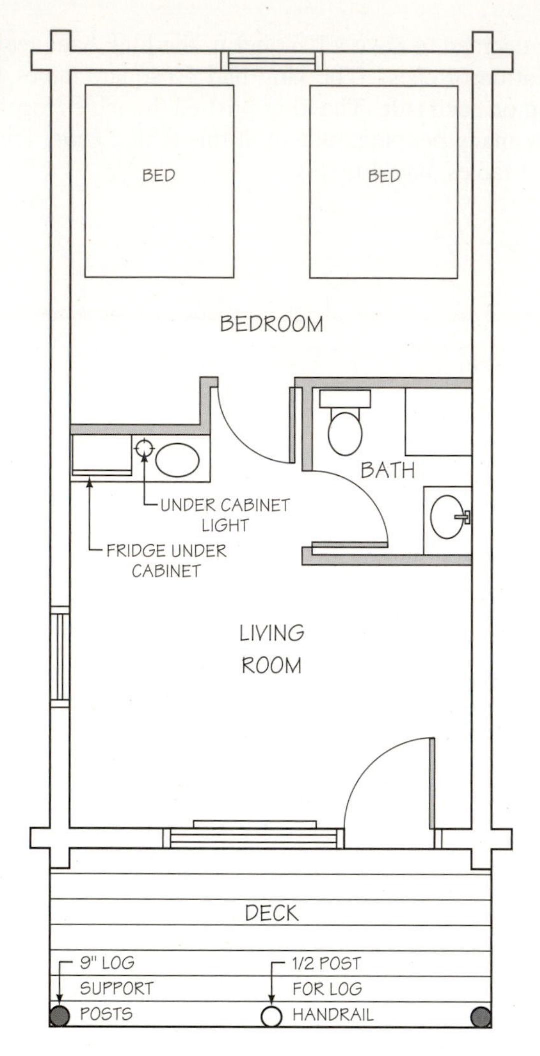

The drawing at the left is a sketch of the floor plan for a cabin at a resort in the mountains of Utah. The measurements are missing. Assume that you are the architect and will finish the drawing. You will have to decide the size of the rooms and put in the measurements to scale.

Design a cabin that you would like to own. Select a scale and draw all the rooms to scale.

If you are interested in architecture, visit an architect who is using CAD (computer-aided design) software to create a floor plan. Computer technology has revolutionized the field of architectural design.

The U.S. House of Representatives

The framers of the Constitution decided to use a ratio to determine the number of representatives from each state. It was determined that each state would have one representative for every 30,000 citizens, with a minimum of one representative. Congress has changed this ratio over the years, and we now have 435 representatives.

Find the number of representatives from your state. Determine the ratio of citizens to representatives. Also do this for the most populous state and for the least populous state.

You might consider getting information on the number of representatives for each state and the populations of different states via the Internet.

CHAPTER 4

SUMMARY

KEY WORDS	EXAMPLES
A *ratio* is the comparison of two quantities with the same units. A ratio can be written in three ways: as a fraction, as two numbers separated by a colon (:), or as two numbers separated by the word *to*. A ratio is in *simplest form* when the two numbers do not have a common factor. [4.1A, p. 174]	The comparison 16 to 24 ounces can be written as a ratio in simplest form as $\frac{2}{3}$, 2:3, or 2 to 3.
A *rate* is the comparison of two quantities with different units. A rate is written as a fraction. A rate is in *simplest form* when the numbers that form the rate do not have a common factor. [4.2A, p. 178]	You earned \$63 for working 6 hours. The rate is written in simplest form as $\frac{\$21}{2 \text{ hours}}$.
A *unit rate* is a rate in which the number in the denominator is 1. [4.2B, p. 178]	You traveled 144 miles in 3 hours. The unit rate is 48 miles per hour.
A *proportion* is an expression of the equality of two ratios or rates. A proportion is true if the fractions are equal when written in lowest terms; in any true proportion, the *cross products* are equal. A proportion is not true if the fractions are not equal when written in lowest terms; if the cross products are not equal, the proportion is not true. [4.3A, p. 182]	The proportion $\frac{3}{5} = \frac{12}{20}$ is true because the cross products are equal: $3 \times 20 = 5 \times 12$. The proportion $\frac{3}{4} = \frac{12}{20}$ is not true because the cross products are not equal: $3 \times 20 \neq 4 \times 12$.

ESSENTIAL RULES AND PROCEDURES	EXAMPLES
To find a unit rate, divide the number in the numerator of the rate by the number in the denominator of the rate. [4.2B, p. 178]	You earned \$41 for working 4 hours. $41 \div 4 = 10.25$ The unit rate is \$10.25/hour.
To solve a proportion, find a number to replace the unknown so that the proportion is true. [4.3B, p. 183]	$\frac{6}{24} = \frac{9}{n}$ $6 \times n = 24 \times 9$ • **Find the cross products.** $6 \times n = 216$ $n = 216 \div 6$ $n = 36$
To set up a proportion, keep the same units in the numerator and the same units in the denominator. [4.3C, p. 184]	Three machines fill 5 cereal boxes per minute. How many boxes can 8 machines fill per minute? $\frac{3 \text{ machines}}{5 \text{ cereal boxes}} = \frac{8 \text{ machines}}{n \text{ cereal boxes}}$

CHAPTER 4

CONCEPT REVIEW

Test your knowledge of the concepts presented in this chapter. Answer each question. Then check your answers against the ones provided in the Answer Section.

1. If the units in a comparison are different, is it a ratio or a rate?

2. How do you find a unit rate?

3. How do you write the ratio $\frac{6}{7}$ using a colon?

4. How do you write the ratio 12 : 15 in simplest form?

5. How do you write the rate $\frac{342 \text{ miles}}{9.5 \text{ gallons}}$ as a unit rate?

6. When is a proportion true?

7. How do you solve a proportion?

8. How do the units help you to set up a proportion?

9. How do you check the solution of a proportion?

10. How do you write the ratio 19 : 6 as a fraction?

CHAPTER 4

REVIEW EXERCISES

1. Determine whether the proportion is true or not true.
$\frac{2}{9} = \frac{10}{45}$

2. Write the comparison 32 dollars to 80 dollars as a ratio in simplest form using a fraction, a colon (:), and the word *to*.

3. Write "250 miles in 4 hours" as a unit rate.

4. Determine whether the proportion is true or not true.
$\frac{8}{15} = \frac{32}{60}$

5. Solve the proportion.
$\frac{16}{n} = \frac{4}{17}$

6. Write "$500 earned in 40 hours" as a unit rate.

7. Write "$8.75 for 5 pounds" as a unit rate.

8. Write the comparison 8 feet to 28 feet as a ratio in simplest form using a fraction, a colon (:), and the word *to*.

9. Solve the proportion.
$\frac{n}{8} = \frac{9}{2}$

10. Solve the proportion. Round to the nearest hundredth.
$\frac{18}{35} = \frac{10}{n}$

11. Write the comparison 6 inches to 15 inches as a ratio in simplest form using a fraction, a colon (:), and the word *to*.

12. Determine whether the proportion is true or not true.
$\frac{3}{8} = \frac{10}{24}$

13. Write "$35 in 4 hours" as a rate in simplest form.

14. Write "326.4 miles on 12 gallons" as a unit rate.

15. Write the comparison 12 days to 12 days as a ratio in simplest form using a fraction, a colon (:), and the word *to*.

16. Determine whether the proportion is true or not true.
$\frac{5}{7} = \frac{25}{35}$

17. Solve the proportion. Round to the nearest hundredth.

$$\frac{24}{11} = \frac{n}{30}$$

18. Write "250 milligrams in 5 milliliters" as a rate in simplest form.

19. **Insurance Increases** In the past year, the cost of a doctor visit co-pay on a health insurance policy was increased to $25 from $20. What is the ratio, as a fraction in simplest form, of the increase in cost to the original co-pay?

20. **Staffing Requirements** The total nurse staffing hours per resident day at a local nursing home averaged 3.62 hours per resident day. If there are 50 residents in this nursing home, how many total nurse staffing hours are needed per resident day?

21. **Staffing Requirements** An intensive care unit has 15 patients and 5 nurses on duty. What is the staffing rate in the ICU? Is this rate higher or lower than the $\frac{1}{4}$ rate required by the hospital regulations?

22. **Working Hours** You are employed by two different nursing homes. One requires that you work 8 hours per day and the other requires an additional 4 hours a day from you. Write as a ratio, in simplest form, the total number of hours you work in one day to the total number of hours in a day.

23. **Business** Dr. Ned Kelly and Dr. Sadat Khan establish a new internal medicine practice, and their office and professional expenses are $15,500 the first month. The collections for the first month total $9750. Write, in simplest form, the ratio of collections compared to expenses for the month.

24. **Consumerism** The cost of a 128-ounce container of massage therapy gel is $59.95. Determine the cost per ounce. Round to the nearest cent.

25. **Medication** A pediatrician prescribes an antibiotic to a patient. The dosage is 1 teaspoon four times a day for seven days. If the patient ingests a total of 140 milliliters of this antibiotic, find the amount of each individual dose in milliliters.

26. **Medication** A prescription-strength medication contains 250 milligrams of acetaminophen in each tablet. If a patient takes two tablets three times a day for four days, how many milligrams of acetaminophen has the patient ingested?

27. **Dilution Ratio** A dilution was done in the lab using 100 microliters of blood and 4900 microliters of a diluting agent. What, as a fraction in simplest form, is the dilution ratio for this solution?

28. **Dilution Ratio** If 50 milliliters of alcohol is mixed with 100 milliliters of water, what is the dilution ratio, in simplest form, of this solution?

29. **Solutions** The dilution ratio of an alcohol solution is $\frac{3}{20}$. If you need to prepare 250 milliliters of this solution, how many milliliters of alcohol will you need?

30. **Solution** As a lab technician, you need to make an alcohol solution that has a dilution ratio of $\frac{1}{100}$. If you have 2.5 milliliters of alcohol, how much of this solution can you make?

CHAPTER 4

TEST

1. Write "$46,036.80 earned in 12 months" as a unit rate.

2. Write the comparison 40 miles to 240 miles as a ratio in simplest form using a fraction, a colon (:), and the word *to*.

3. Write "125 milligrams for every 5 milliliters" as a rate in simplest form.

4. Determine whether the proportion is true or not true.
$$\frac{40}{125} = \frac{5}{25}$$

5. Write the comparison 12 days to 8 days as a ratio in simplest form using a fraction, a colon (:), and the word *to*.

6. Solve the proportion.
$$\frac{5}{12} = \frac{60}{n}$$

7. Write "256.2 miles on 8.4 gallons of gas" as a unit rate.

8. Write the comparison 27 dollars to 81 dollars as a ratio in simplest form using a fraction, a colon (:), and the word *to*.

9. Determine whether the proportion is true or not true.
$$\frac{5}{14} = \frac{25}{70}$$

10. Solve the proportion.
$$\frac{n}{18} = \frac{9}{4}$$

11. Write "$72 for 16 thermometers" as a rate in simplest form.

12. Write the comparison 18 feet to 30 feet as a ratio in simplest form using a fraction, a colon (:), and the word *to*.

13. **Investments** Fifty shares of a healthcare stock pay a dividend of $62.50. At the same rate, what is the dividend paid on 500 shares of the healthcare stock?

14. **Anatomy** Write, as a fraction in simplest form, the ratio of 60,000 miles of blood vessels in the human body to 45 miles of nerve cells (Vogel, Steven. *Vital Circuits*, pp.15–16).

15. **Infusion Rates** A unit of blood contains 450 milliliters. If a unit of blood is infused over a 3-hour period, find the infusion rate in milliliters per hour.

16. **Physiology** A research scientist estimates that the human body contains 88 pounds of water for every 100 pounds of body weight. At this rate, estimate the number of pounds of water in a college student who weighs 150 pounds.

17. **Nutrition** There are 108 calories in $\frac{1}{2}$ cup of long-grain brown rice. How many calories are in 2 cups of long-grain brown rice?

18. **Medicine** The dosage of a certain medication is $\frac{1}{4}$ ounce for every 50 pounds of body weight. How many ounces of this medication are required for a person who weighs 175 pounds? Write the answer as a decimal.

19. **Medicine** A doctor prescribe 350 milligrams of ibuprofen four times a day for a patient with arthritis. The strength of the liquid form of ibuprofen is 100 milligrams/5 milliliters of liquid. How many milliliters of liquid should be consumed by the patient for a single dose? How many milliliters will be consumed in a single day?

20. **Solutions** You need to prepare a saline solution with a concentration of 2.25 grams of salt in 250 milliliters of distilled water. How much solution can you produce with 4.5 grams of salt?

CUMULATIVE REVIEW EXERCISES

1. Subtract: $\begin{array}{r} 20{,}095 \\ -\ 10{,}937 \\ \hline \end{array}$

2. Write $2 \cdot 2 \cdot 2 \cdot 2 \cdot 3 \cdot 3 \cdot 3$ in exponential notation.

3. Simplify: $4 - (5 - 2)^2 \div 3 + 2$

4. Find the prime factorization of 160.

5. Find the LCM of 9, 12, and 18.

6. Find the GCF of 28 and 42.

7. Write $\frac{40}{64}$ in simplest form.

8. Find $4\frac{7}{15}$ more than $3\frac{5}{6}$.

9. What is $4\frac{5}{9}$ less than $10\frac{1}{6}$?

10. Multiply: $\frac{11}{12} \times 3\frac{1}{11}$

11. Find the quotient of $3\frac{1}{3}$ and $\frac{5}{7}$.

12. Simplify: $\left(\frac{2}{5} + \frac{3}{4}\right) \div \frac{3}{2}$

13. Write 4.0709 in words.

14. Round 2.09762 to the nearest hundredth.

15. Divide: $8.09\overline{)16.0976}$
    Round to the nearest thousandth.

16. Convert $0.06\frac{2}{3}$ to a fraction.

17. Write the comparison 25 miles to 200 miles as a ratio in simplest form using a fraction.

18. Write "87 cents for 6 pencils" as a rate in simplest form.

19. Write "250.5 miles on 7.5 gallons of gas" as a unit rate.

20. Solve $\frac{40}{n} = \frac{160}{17}$.

21. **Travel** A car traveled 457.6 miles in 8 hours. Find the car's speed in miles per hour.

22. Solve the proportion.
    $$\frac{12}{5} = \frac{n}{15}$$

23. **Banking** As a home health aide, you need to purchase scrubs to wear to work. The scrubs cost $25.95 per set. You purchase a stethoscope and blood pressure cuff for $75.45, and three sets of scrubs. You currently have $485.80 in your checking account. Find the total cost of your purchases and the balance in your checking account after the purchases are made.

24. **Consumerism** An acupuncture student purchases a small acupuncture model for $89.95 and three books at a cost of $205. The student pays with two $100 bills and two $50 bills. Determine the change the student should receive.

Demographics for the United States The table provides the age distribution of males and females in the United States in 2010. Use this table for Exercises 25 to 27.

Age Distribution of Males and Females in the United States			
Males	80 years and older = 4,000,000	*Females*	80 years and older = 7,900,000
	55 to 59 years = 10,500,000		55 to 59 years = 10,000,000
	40 to 44 years = 10,200,000		40 to 44 years = 10,200,000
	35 to 39 years = 9,900,000		35 to 39 years = 10,100,000

Source: www.nationmaster.com/country/us/Age_distribution

25. Which sex has the greater population?

26. What is the total difference in population between males aged 55 to 59 and aged 80 and older, as compared to females of the same ages?

27. Find the difference between the male and female populations aged 35 to 39.

28. **Weight Loss** A patient is advised to lose 20 pounds in order to reach a healthy weight. During the next month, the patient loses $2\frac{1}{2}$ pounds the first week, $3\frac{3}{4}$ pounds the second week, 2 pounds the third week, and $1\frac{3}{4}$ pounds the fourth week. How many more pounds does the patient need to lose in order to reach the goal of a 20-pound loss?

29. **Infusion Rate** A doctor orders an IV infusion of 500 milliliters of D5W solution over a 4-hour period for a patient. Find the unit rate for the infusion.

30. **Medicine** The dosage of a certain medication is $\frac{1}{2}$ ounce for every 50 pounds of body weight. How many ounces of this medication are required for a person who weighs 160 pounds? Write the answer as a decimal.

CHAPTER

5

Percents

OBJECTIVES

SECTION 5.1

A To write a percent as a fraction or a decimal

B To write a fraction or a decimal as a percent

SECTION 5.2

A To find the amount when the percent and the base are given

B To solve application problems

SECTION 5.3

A To find the percent when the base and amount are given

B To solve application problems

SECTION 5.4

A To find the base when the percent and amount are given

B To solve application problems

SECTION 5.5

A To solve percent problems using proportions

B To solve application problems

AIM for the Future

PHYSICAL THERAPIST AIDES help make therapy sessions productive. They work under the direct supervision of a physical therapist or physical therapist assistant. They usually are responsible for keeping the treatment area clean and organized and for preparing for each patient's therapy. When patients need assistance moving to or from a treatment area, aides assist in their transport. Employment of physical therapist aides is expected to grow by 35% from 2008 through 2018, much faster than the average for all occupations. Median annual wages of physical therapist aides were $23,760 in May 2008. (*Source: Bureau of Labor Statistics:* http://www.bls.gov/oco/ocos167.htm)

PREP TEST

Are you ready to succeed in this chapter?

Take the Prep Test below to find out if you are ready to learn the new material.

For Exercises 1 to 6, multiply or divide.

1. $19 \times \frac{1}{100}$

2. 23×0.01

3. 0.47×100

4. $0.06 \times 47{,}500$

5. $60 \div 0.015$

6. $8 \div \frac{1}{4}$

7. Multiply $\frac{5}{8} \times 100$. Write the answer as a decimal.

8. Write $\frac{200}{3}$ as a mixed number.

9. Divide $28 \div 16$. Write the answer as a decimal.

SECTION

5.1 Introduction to Percents

OBJECTIVE A To write a percent as a fraction or a decimal

Percent means "parts of 100." In the figure at the right, there are 100 parts. Because 13 of the 100 parts are shaded, 13% of the figure is shaded. The symbol % is the **percent sign.**

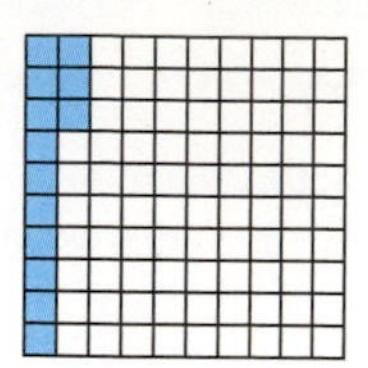

In most applied problems involving percents, it is necessary either to rewrite a percent as a fraction or a decimal or to rewrite a fraction or a decimal as a percent.

To write a percent as a fraction, remove the percent sign and multiply by $\frac{1}{100}$.

$$13\% = 13 \times \frac{1}{100} = \frac{13}{100}$$

Take Note

Recall that division is defined as multiplication by the reciprocal. Therefore, multiplying by $\frac{1}{100}$ is equivalent to dividing by 100.

To write a percent as a decimal, remove the percent sign and multiply by 0.01.

$$13\% = 13 \times 0.01 = 0.13$$

Move the decimal point two places to the left. Then remove the percent sign.

EXAMPLE 1

a. Write 120% as a fraction.
b. Write 120% as a decimal.

Solution a. $120\% = 120 \times \frac{1}{100} = \frac{120}{100}$

$= 1\frac{1}{5}$

b. $120\% = 120 \times 0.01 = 1.2$

Note that percents larger than 100 are greater than 1.

YOU TRY IT 1

a. Write 125% as a fraction.
b. Write 125% as a decimal.

Your solution

EXAMPLE 2

Write $16\frac{2}{3}\%$ as a fraction.

Solution $16\frac{2}{3}\% = 16\frac{2}{3} \times \frac{1}{100}$

$= \frac{50}{3} \times \frac{1}{100} = \frac{50}{300} = \frac{1}{6}$

YOU TRY IT 2

Write $33\frac{1}{3}\%$ as a fraction.

Your solution

EXAMPLE 3

Write 0.5% as a decimal.

Solution $0.5\% = 0.5 \times 0.01 = 0.005$

YOU TRY IT 3

Write 0.25% as a decimal.

Your solution

Solutions on pp. S11–S12

OBJECTIVE B To write a fraction or a decimal as a percent

A fraction or a decimal can be written as a percent by multiplying by 100%.

HOW TO • 1 Write $\frac{3}{8}$ as a percent.

$$\frac{3}{8} = \frac{3}{8} \times 100\% = \frac{3}{8} \times \frac{100}{1}\% = \frac{300}{8}\% = 37\frac{1}{2}\% \text{ or } 37.5\%$$

HOW TO • 2 Write 0.37 as a percent.

$$0.37 = 0.37 \times 100\% = 37\%$$

Move the decimal point two places to the right. Then write the percent sign.

EXAMPLE • 4

Write 0.015, 2.15, and $0.33\frac{1}{3}$ as percents.

Solution

$0.015 = 0.015 \times 100\%$
$= 1.5\%$

$2.15 = 2.15 \times 100\% = 215\%$

$0.33\frac{1}{3} = 0.33\frac{1}{3} \times 100\%$
$= 33\frac{1}{3}\%$

YOU TRY IT • 4

Write 0.048, 3.67, and $0.62\frac{1}{2}$ as percents.

Your solution

EXAMPLE • 5

Write $\frac{2}{3}$ as a percent.
Write the remainder in fractional form.

Solution $\frac{2}{3} = \frac{2}{3} \times 100\% = \frac{200}{3}\%$
$= 66\frac{2}{3}\%$

YOU TRY IT • 5

Write $\frac{5}{6}$ as a percent.
Write the remainder in fractional form.

Your solution

EXAMPLE • 6

Write $2\frac{2}{7}$ as a percent.
Round to the nearest tenth.

Solution $2\frac{2}{7} = \frac{16}{7} = \frac{16}{7} \times 100\%$
$= \frac{1600}{7}\% \approx 228.6\%$

YOU TRY IT • 6

Write $1\frac{4}{9}$ as a percent.
Round to the nearest tenth.

Your solution

Solutions on p. S12

5.1 EXERCISES

OBJECTIVE A To write a percent as a fraction or a decimal

For Exercises 1 to 16, write as a fraction and as a decimal.

1. 25% **2.** 40% **3.** 130% **4.** 150%

5. 100% **6.** 87% **7.** 73% **8.** 45%

9. 383% **10.** 425% **11.** 70% **12.** 55%

13. 88% **14.** 64% **15.** 32% **16.** 18%

For Exercises 17 to 28, write as a fraction.

17. $66\frac{2}{3}\%$ **18.** $12\frac{1}{2}\%$ **19.** $83\frac{1}{3}\%$ **20.** $3\frac{1}{8}\%$ **21.** $11\frac{1}{9}\%$ **22.** $\frac{3}{8}\%$

23. $45\frac{5}{11}\%$ **24.** $15\frac{3}{8}\%$ **25.** $4\frac{2}{7}\%$ **26.** $5\frac{3}{4}\%$ **27.** $6\frac{2}{3}\%$ **28.** $8\frac{2}{3}\%$

For Exercises 29 to 40, write as a decimal.

29. 6.5% **30.** 9.4% **31.** 12.3% **32.** 16.7% **33.** 0.55% **34.** 0.45%

35. 8.25% **36.** 6.75% **37.** 5.05% **38.** 3.08% **39.** 2% **40.** 7%

41. When a certain percent is written as a fraction, the result is an improper fraction. Is the percent less than, equal to, or greater than 100%?

OBJECTIVE B To write a fraction or a decimal as a percent

For Exercises 42 to 53, write as a percent.

42. 0.16 **43.** 0.73 **44.** 0.05 **45.** 0.01 **46.** 1.07 **47.** 2.94

48. 0.004 **49.** 0.006 **50.** 1.012 **51.** 3.106 **52.** 0.8 **53.** 0.7

For Exercises 54 to 65, write as a percent. If necessary, round to the nearest tenth of a percent.

54. $\frac{27}{50}$ **55.** $\frac{37}{100}$ **56.** $\frac{1}{3}$ **57.** $\frac{2}{5}$ **58.** $\frac{5}{8}$ **59.** $\frac{1}{8}$

60. $\frac{1}{6}$ **61.** $1\frac{1}{2}$ **62.** $\frac{7}{40}$ **63.** $1\frac{2}{3}$ **64.** $1\frac{7}{9}$ **65.** $\frac{7}{8}$

For Exercises 66 to 73, write as a percent. Write the remainder in fractional form.

66. $\frac{15}{50}$ **67.** $\frac{12}{25}$ **68.** $\frac{7}{30}$ **69.** $\frac{1}{3}$

70. $2\frac{3}{8}$ **71.** $1\frac{2}{3}$ **72.** $2\frac{1}{6}$ **73.** $\frac{7}{8}$

74. Does a mixed number represent a percent greater than 100% or less than 100%?

75. A decimal number less than 0 has zeros in the tenths and hundredths places. Does the decimal represent a percent greater than 1% or less than 1%?

76. Write the part of the square that is shaded as a fraction, as a decimal, and as a percent. Write the part of the square that is not shaded as a fraction, as a decimal, and as a percent.

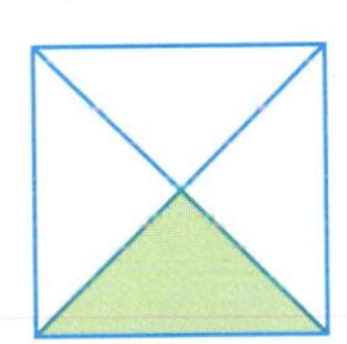

Applying the Concepts

77. Blood Types Among Caucasians, the American Red Cross lists the following blood types for the U.S. population: 37% have O+, 8% have O−, 33% have A+, 7% have A−, 9% have B+, 2% have B−, and 3% have AB+ (*Source:* http://www.redcrossblood.org/learn-about-blood/blood-types). Based on these percents, what percent of the U.S. population have blood type AB−?

78. Dilution Ratio A dilution ratio can indicate the concentration of a solution and many times is expressed as a percent. What is the percent concentration of a solution having a dilution ratio of $\frac{1}{20}$?

Chepko Danil Vitalevich/Shutterstock.com

79. Solutions An alcohol solution is a mixture of alcohol and water. If an alcohol solution contains 15% alcohol, what fractional part of the solution is alcohol?

80. Solutions If $\frac{1}{10}$ of an alcohol solution is alcohol, what percent of the solution is water?

SECTION

5.2 Percent Equations: Part 1

OBJECTIVE A To find the amount when the percent and the base are given

A real estate broker receives a payment that is 4% of a $285,000 sale. To find the amount the broker receives requires answering the question "4% of $285,000 is what?"

This sentence can be written using mathematical symbols and then solved for the unknown number.

4%	of	$285,000	is	what?
↓	↓	↓	↓	↓
Percent 4%	×	base 285,000	=	amount n
0.04	×	285,000	=	n
		11,400	=	n

of is written as × (times)
is is written as = (equals)
what is written as n (the unknown number)

Note that the percent is written as a decimal.

The broker receives a payment of $11,400.

The solution was found by solving the **basic percent equation** for amount.

The Basic Percent Equation

Percent × base = amount

In most cases, the percent is written as a decimal before the basic percent equation is solved. However, some percents are more easily written as a fraction than as a decimal. For example,

$$33\frac{1}{3}\% = \frac{1}{3} \qquad 66\frac{2}{3}\% = \frac{2}{3} \qquad 16\frac{2}{3}\% = \frac{1}{6} \qquad 83\frac{1}{3}\% = \frac{5}{6}$$

EXAMPLE • 1

Find 5.7% of 160.

Solution

Percent × base = amount
$0.057 \times 160 = n$
$9.12 = n$

• The word *Find* is used instead of the words *what is*.

YOU TRY IT • 1

Find 6.3% of 150.

Your solution

EXAMPLE • 2

What is $33\frac{1}{3}\%$ of 90?

Solution

Percent × base = amount
$\frac{1}{3} \times 90 = n$
$30 = n$

• $33\frac{1}{3}\% = \frac{1}{3}$

YOU TRY IT • 2

What is $16\frac{2}{3}\%$ of 66?

Your solution

Solutions on p. S12

OBJECTIVE B To solve application problems

Solving percent problems requires identifying the three elements of the basic percent equation. Recall that these three parts are the *percent,* the *base,* and the *amount.* Usually the base follows the phrase "percent of."

During a recent year, Americans gave $212 billion to charities. The circle graph at the right shows where that money came from. Use these data for Example 3 and You Try It 3.

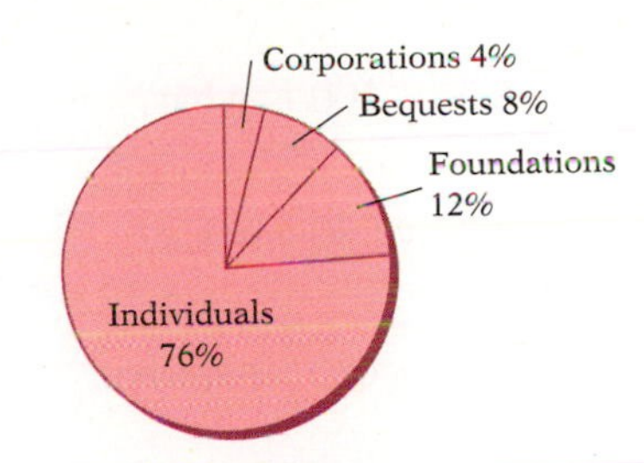

Charitable Giving
Sources: American Association of Fundraising Counsel; AP

EXAMPLE 3

How much of the amount given to charities came from individuals?

Strategy
To determine the amount that came from individuals, write and solve the basic percent equation using n to represent the amount. The percent is 76%. The base is $212 billion.

Solution

$$\text{Percent} \times \text{base} = \text{amount}$$
$$76\% \times 212 = n$$
$$0.76 \times 212 = n$$
$$161.12 = n$$

Individuals gave $161.12 billion to charities.

YOU TRY IT 3

How much of the amount given to charities was given by corporations?

Your strategy

Your solution

EXAMPLE 4

Figures from the American Heart Association state that approximately 23.1% of American men smoke. (*Source:* National Health Interview Survey [NHIS], 2008, National Center for Health Statistics) If there are approximately 138 million men in the United States, how many men don't smoke?

Strategy
To find the number of men who don't smoke:

- Find the number of men who do smoke. Write and solve the basic percent equation using n to represent the number of males who smoke (amount). The percent is 23.1% and the base is 138 million.
- Subtract the number of men who smoke from the total number of men in the United States.

Solution

$$23.1\% \times 138 \text{ million} = n$$
$$0.231 \times 138 \text{ million} = n$$
$$31.878 \text{ million} = n \text{ smokers}$$

$138 \text{ million} - 31.878 \text{ million} = 106.122 \text{ million}$ or 106,122,000

106,122,000 American men do not smoke.

YOU TRY IT 4

A physical therapist's hourly wage was $33.50 before an 8% raise. What is the new hourly wage?

Your strategy

Your solution

Solutions on p. S12

5.2 EXERCISES

OBJECTIVE A To find the amount when the percent and the base are given

1. 8% of 100 is what?

2. 16% of 50 is what?

3. 27% of 40 is what?

4. 52% of 95 is what?

5. 0.05% of 150 is what?

6. 0.075% of 625 is what?

7. 125% of 64 is what?

8. 210% of 12 is what?

9. Find 10.7% of 485.

10. Find 12.8% of 625.

11. What is 0.25% of 3000?

12. What is 0.06% of 250?

13. 80% of 16.25 is what?

14. 26% of 19.5 is what?

15. What is $1\frac{1}{2}$% of 250?

16. What is $5\frac{3}{4}$% of 65?

17. $16\frac{2}{3}$% of 120 is what?

18. What is $66\frac{2}{3}$% of 891?

19. Which is larger: 5% of 95, or 75% of 6?

20. Which is larger: 112% of 5, or 0.45% of 800?

21. Which is smaller: 79% of 16, or 20% of 65?

22. Which is smaller: 15% of 80, or 95% of 15?

23. Is 15% of a number greater than or less than the number?

24. Is 150% of a number greater than or less than the number?

OBJECTIVE B To solve application problems

25. Read Exercise 26. Without doing any calculations, determine whether the number of people in the United States aged 18 to 24 who do not have health insurance is *less than, equal to,* or *greater than* 44 million.

26. **Health Insurance** Approximately 30% of the 44 million people in the United States who do not have health insurance are between the ages of 18 and 24. (*Source:* U.S. Census Bureau) About how many people in the United States aged 18 to 24 do not have health insurance?

27. **Blood Donors** According to the American Red Cross, 16 million blood donations were collected in 2006. These donations were made by occasional donors (19%), first-time donors (31%), and regular donors (50%). Of the donations in 2006, how many donations were made by occasional donors?

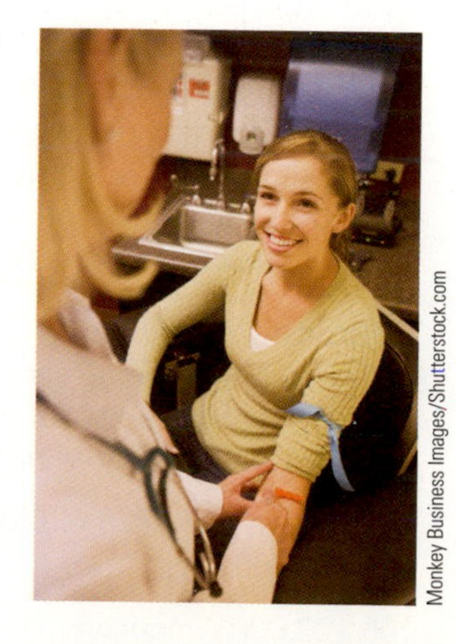

Monkey Business Images/Shutterstock.com

Alternative Medicine The statistics shown in the graph at the right are from the 2007 National Health Interview Survey (NHIS), an annual in-person survey of Americans regarding their health- and illness-related experiences. The Complementary and Alternative Medicine (CAM) section gathered information on 23,393 adults aged 18 and older. (*Source*: http://nccam.nih.gov/news/camstats/2007/) Use this graph for Exercises 28 to 30.

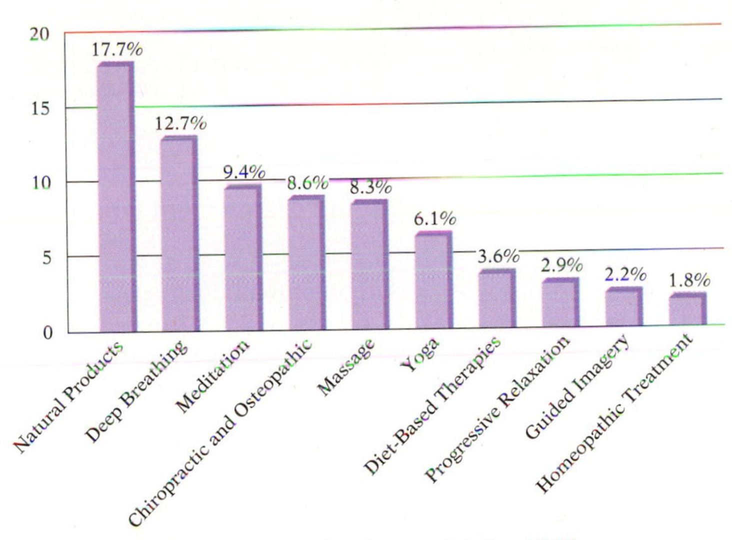

10 Most Common CAM Therapies Among Adults—2007

Source: Barners PM, Bloom B, Nahin R, *CDC National Health Statistics Report #12.* Complementary and Alternative Medicines Use Among Adults and Children: United States, 2007. December 2008.

28. In the United States, according to this survey, approximately 38% of adults are using some form of complementary and alternative medicine. If the adult population of the United States is approximately 210,000,000 people, how may adults are estimated to use some form of complementary and alternative medicine?

29. Of the 23,393 adults included in the survey, how many use natural products as part of CAM therapy? Round to the nearest whole number.

30. Of the 23,393 adults included in the survey, how many used both deep breathing and meditation as part of CAM therapy? Round to the nearest whole number.

31. According to the NHIS survey, approximately 12% of children aged 17 and younger use some form of complementary and alternative medicine. If there are approximately 73,000,000 children aged 17 and younger in the United States, how many do not use CAM?

32. **Cancer** Based upon a report from the Oral Cancer Foundation, a total of 37,000 Americans will be diagnosed with oral cancer this year. Approximately 21.6% of that number will die this year, and only slightly more than half of those 37,000 will be alive in 5 years. Determine the approximate number of deaths from oral cancer this year.

33. **Hip Replacement** A commonly used artificial hip joint replacement consists of a cobalt–chrome alloy combination. The cobalt–chrome replacement consists of 34% cobalt and 19% chrome with the balance made up of a mix of other metals. If the hip joint replacement weighs 18 ounces, how many ounces of cobalt and chrome are in the replacement? Round to the nearest tenth.

34. **Taxes** A sales tax of 6% of the cost of a car is added to the purchase price of $29,500. What is the total cost of the car, including sales tax?

35. **Wages** A registered dental hygienist's hourly wage was $32.40 before receiving a 2.5% raise. What is the new hourly wage?

36. **Human body** The human body is primarily made up of water. The components that contribute to the mass or weight of a human body are water (65% of the mass), fats or fatty acids called lipids (12% of the mass), protein (20% of the mass), and other inorganics and organics (3%). If you weigh 250 pounds, how much of your body weight is water?

SECTION

5.3 Percent Equations: Part II

OBJECTIVE A To find the percent when the base and amount are given

A recent promotional game at a grocery store listed the probability of winning a prize as "1 chance in 2." A percent can be used to describe the chance of winning. This requires answering the question "What percent of 2 is 1?"

The chance of winning can be found by solving the basic percent equation for *percent*.

Integrating Technology

The percent key % on a scientific calculator moves the decimal point to the right two places when pressed after a multiplication or division computation. For the example at the right, enter

The display reads 50.

What percent of 2 is 1?

$$\boxed{\text{Percent } n} \times \boxed{\text{base } 2} = \boxed{\text{amount } 1}$$

$$n \times 2 = 1$$
$$n = 1 \div 2$$
$$n = 0.5$$
$$n = 50\%$$

• **The solution must be written as a percent in order to answer the question.**

There is a 50% chance of winning a prize.

EXAMPLE 1

What percent of 40 is 30?

Solution

$$\text{Percent} \times \text{base} = \text{amount}$$
$$n \times 40 = 30$$
$$n = 30 \div 40$$
$$n = 0.75$$
$$n = 75\%$$

YOU TRY IT 1

What percent of 32 is 16?

Your solution

EXAMPLE 2

What percent of 12 is 27?

Solution

$$\text{Percent} \times \text{base} = \text{amount}$$
$$n \times 12 = 27$$
$$n = 27 \div 12$$
$$n = 2.25$$
$$n = 225\%$$

YOU TRY IT 2

What percent of 15 is 48?

Your solution

EXAMPLE 3

25 is what percent of 75?

Solution

$$\text{Percent} \times \text{base} = \text{amount}$$
$$n \times 75 = 25$$
$$n = 25 \div 75$$
$$n = \frac{1}{3} = 33\frac{1}{3}\%$$

YOU TRY IT 3

30 is what percent of 45?

Your solution

Solutions on p. S12

OBJECTIVE B To solve application problems

To solve percent problems, remember that it is necessary to identify the percent, base, and amount. Usually the base follows the phrase "percent of."

EXAMPLE • 4

The monthly house payment for the Kaminski family is $787.50. What percent of the Kaminskis' monthly income of $3750 is the house payment?

Strategy

To find what percent of the income the house payment is, write and solve the basic percent equation using n to represent the percent. The base is $3750 and the amount is $787.50.

Solution

$$n \times 3750 = 787.50$$
$$n = 787.50 \div 3750$$
$$n = 0.21 = 21\%$$

The house payment is 21% of the monthly income.

YOU TRY IT • 4

Tomo Nagata had an income of $33,500 and paid $5025 in income tax. What percent of the income is the income tax?

Your strategy

Your solution

EXAMPLE • 5

The Centers for Disease Control and Prevention (CDC) reports that 1.7 million people sustain a traumatic brain injury annually. Of that number, 1,640,000 are seen in the emergency room and 275,000 of those require hospitalization. What percent of those seen in the emergency room do not require hospitalization? Round the nearest percent. (*Source:* http://www.cdc.gov/TraumaticBrainInjury/statistics.html)

Strategy

To find the percent of patients who are not hospitalized:

- Subtract to find the number of patients who are not hospitalized. (1,640,000 – 275,000)
- Write and solve the basic percent equation using n to represent the percent. The base is 1,640,000, and the amount is the number of patients who are not hospitalized.

Solution

$$1{,}640{,}000 - 275{,}000 = 1{,}365{,}000$$

1,365,000 patients with traumatic brain injuries seen in the emergency room are not hospitalized.

$$n \times 1{,}640{,}000 = 1{,}365{,}000$$
$$n = 1{,}365{,}000 \div 1{,}640{,}000$$
$$n \approx 0.83$$

Approximately 83% of patients with a traumatic brain injury seen in the emergency room do not require hospitalization.

YOU TRY IT • 5

Of the 1.7 million traumatic brain injuries that occur annually, about 1,275,000 are concussions or other forms of mild traumatic brain injuries. What percent of traumatic brain injuries are considered to be serious injuries, such as those requiring hospitalization or those causing death?

Your strategy

Your solution

Solutions on p. S12

5.3 EXERCISES

OBJECTIVE A To find the percent when the base and amount are given

1. What percent of 75 is 24?
2. What percent of 80 is 20?
3. 15 is what percent of 90?
4. 24 is what percent of 60?
5. What percent of 12 is 24?
6. What percent of 6 is 9?
7. What percent of 16 is 6?
8. What percent of 24 is 18?
9. 18 is what percent of 100?
10. 54 is what percent of 100?
11. 5 is what percent of 2000?
12. 8 is what percent of 2500?
13. What percent of 6 is 1.2?
14. What percent of 2.4 is 0.6?
15. 16.4 is what percent of 4.1?
16. 5.3 is what percent of 50?
17. 1 is what percent of 40?
18. 0.3 is what percent of 20?
19. What percent of 48 is 18?
20. What percent of 11 is 88?
21. What percent of 2800 is 7?
22. What percent of 400 is 12?

23. True or false? If the *base* is larger than the *amount* in the basic percent equation, then the *percent* is larger than 100.

OBJECTIVE B To solve application problems

24. Read Exercise 25. Without doing any calculations, determine whether the percent of expenses the technicians' salaries represent is *less than* or *greater than* 50%.

25. **Office Expenses** The average monthly expenses for a veterinary practice are \$12,300. The total monthly cost in salaries for the veterinary technicians is \$5800. Determine the percent of expenses that the technicians' salaries represent. Round to the nearest percent.

26. **Office Expenses** The same veterinarian office mentioned in Exercise 25 pays \$2150 a month for the office mortgage. This payment is a portion of the \$12,300 monthly expenses. What percent of the office expenses is the monthly mortgage payment?

27. **Agriculture** According to the U.S. Department of Agriculture, of the 63 billion pounds of vegetables produced in the United States in 1 year, 16 billion pounds were wasted. What percent of the vegetables produced were wasted? Round to the nearest tenth of a percent.

28. **Injuries** In 2008, the injury treated in the emergency room most frequently was an injury resulting from an unintentional fall. Of the 8,551,037 patients treated for unintentional falls, 2,114,113 were patients aged 65 and older. (*Source:* NEISS All Injury Program operated by the Consumer Product Safety Commission) What percent of the patients treated for unintentional falls were aged 65 and older? Round to the nearest tenth of a percent.

29. **Diabetes** Approximately 7% of the American population has diabetes. Within this group, 14.6 million are diagnosed, while 6.2 million are undiagnosed. (*Source:* The National Diabetes Education Program) What percent of Americans with diabetes have not been diagnosed with the disease? Round to the nearest tenth of a percent.

30. **Solutions** An alcohol solution (alcohol and water) has a total volume of 250 milliliters. 25 milliliters of that solution is alcohol. What percent of the solution is water?

Nutrition Labels The nutrition labels on food provide the consumer with information about the serving size, number of calories, fat content, cholesterol, sodium, carbohydrates, and protein in a particular food. The % daily values on most labels are based on a 2000-calorie diet. The footnote at the bottom of the label shows recommended dietary advice for all Americans and is the same on all food labels. Use the cereal label at the right for Exercises 31 to 34.

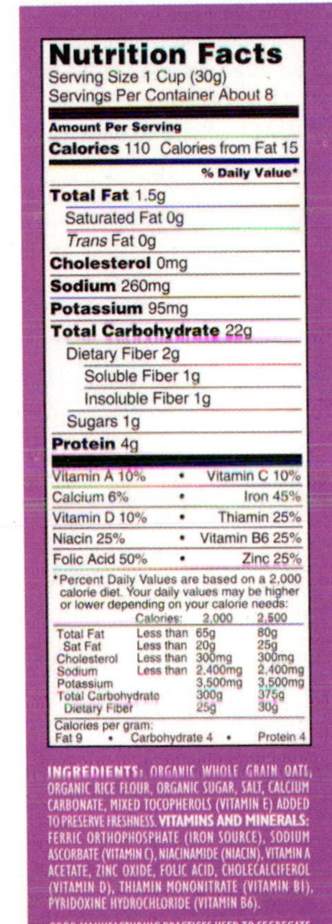

31. If you are on a 2000-calorie diet, what percent of the recommended amount of total carbohydrates is in one serving of this cereal? Round to the nearest tenth of a percent.

32. If you are on a 2000-calorie diet, what percent of the total recommended amount of sodium is in one serving of this cereal? Round to the nearest tenth of a percent.

33. What percent of the total calories per serving of this cereal come from fat? Round to the nearest tenth of a percent.

34. You are on a 2500-calorie per day diet. If you eat two cups of this cereal, what percent of your daily calorie allowance have you consumed? Round to the nearest percent.

35. **Weight Loss** A weight loss clinic weighs their 150 clients at the end of each month to determine how many pounds each has lost during the month. The results are given in the table at the right. Explain why these results are not possible.

Amount of Weight Lost	*Percent*
Gained weight	5%
0	28%
1 to 2	35%
3 to 4	32%
5 to 6	13%

SECTION

5.4 Percent Equations: Part III

OBJECTIVE A To find the base when the percent and amount are given

In 1780, the population of Virginia was 538,000. This was 19% of the total population of the United States at that time. To find the total population at that time, you must answer the question "19% of what number is 538,000?"

Tips for Success

After completing this objective, you will have learned to solve the basic percent equation for each of the three elements: percent, base, and amount. You will need to be able to recognize these three different types of problems. To test yourself, try the Chapter 5 Review Exercises.

19%	of	what	is	538,000?
↓	↓	↓	↓	↓
Percent 19%		base n		amount 538,000

$$0.19 \times n = 538{,}000$$
$$n = 538{,}000 \div 0.19$$
$$n \approx 2{,}832{,}000$$

• The population of the United States in 1780 can be found by solving the basic percent equation for the base.

The population of the United States in 1780 was approximately 2,832,000.

EXAMPLE • 1

18% of what is 900?

Solution

Percent × base = amount

$$0.18 \times n = 900$$
$$n = 900 \div 0.18$$
$$n = 5000$$

YOU TRY IT • 1

86% of what is 215?

Your solution

EXAMPLE • 2

30 is 1.5% of what?

Solution

Percent × base = amount

$$0.015 \times n = 30$$
$$n = 30 \div 0.015$$
$$n = 2000$$

YOU TRY IT • 2

15 is 2.5% of what?

Your solution

EXAMPLE • 3

$33\frac{1}{3}\%$ of what is 7?

Solution

Percent × base = amount

$$\frac{1}{3} \times n = 7$$
$$n = 7 \div \frac{1}{3}$$
$$n = 21$$

• Note that the percent is written as a fraction.

YOU TRY IT • 3

$16\frac{2}{3}\%$ of what is 5?

Your solution

Solutions on p. S13

OBJECTIVE B To solve application problems

To solve percent problems, it is necessary to identify the percent, the base, and the amount. Usually the base follows the phrase "percent of."

EXAMPLE • 4

A medical office bought a used copy machine for \$900, which was 75% of the original cost. What was the original cost of the copier?

Strategy
To find the original cost of the copier, write and solve the basic percent equation using n to represent the original cost (base). The percent is 75% and the amount is \$900.

Solution

$75\% \times n = 900$
$0.75 \times n = 900$
$n = 900 \div 0.75$
$n = 1200$

The original cost of the copier was \$1200.

YOU TRY IT • 4

A used car has a value of \$10,458, which is 42% of the car's original value. What was the car's original value?

Your strategy

Your solution

EXAMPLE • 5

A licensed massage therapist's wage this year is \$26.40 per hour, which is 110% of last year's wage. What was the increase in the hourly wage over last year?

Strategy
To find the increase in the hourly wage over last year:

- Find last year's wage. Write and solve the basic percent equation using n to represent last year's wage (base). The percent is 110% and the amount is \$26.40.
- Subtract last year's wage from this year's wage (26.40).

Solution

$110\% \times n = 26.40$
$1.10 \times n = 26.40$
$n = 26.40 \div 1.10$
$n = 24.00$ • **Last year's wage**

$26.40 - 24.00 = 2.40$

The increase in the hourly wage was \$2.40.

YOU TRY IT • 5

A medical supply warehouse was offering a sale on in-office hematology analyzers. The purchase price was \$4999, which was 63% of the original price. Determine the difference between the original price and sale price of the hematology analyzer.

Your strategy

Your solution

Solutions on p. S13

5.4 EXERCISES

OBJECTIVE A To find the base when the percent and amount are given

1. 12% of what is 9?

2. 38% of what is 171?

3. 8 is 16% of what?

4. 54 is 90% of what?

5. 10 is 10% of what?

6. 37 is 37% of what?

7. 30% of what is 25.5?

8. 25% of what is 21.5?

9. 2.5% of what is 30?

10. 10.4% of what is 52?

11. 125% of what is 24?

12. 180% of what is 21.6?

13. 18 is 240% of what?

14. 24 is 320% of what?

15. 4.8 is 15% of what?

16. 87.5 is 50% of what?

17. 25.6 is 12.8% of what?

18. 45.014 is 63.4% of what?

19. 30% of what is 2.7?

20. 78% of what is 3.9?

21. 84 is $16\frac{2}{3}\%$ of what?

22. 120 is $33\frac{1}{3}\%$ of what?

23. Consider the question "*P*% of what number is 50?" If the percent *P* is greater than 100%, is the unknown number greater than 50 or less than 50?

OBJECTIVE B To solve application problems

24. **Salary** A physical therapist was given a raise after a performance review. The new salary was $62,475, which was 105% of the previous salary. Calculate the salary prior to the raise.

25. **Office Expenses** A podiatry group purchased a telephone system for the office at a discount. The purchase price was $1500, which was 70% of the original cost. Determine the original cost of the telephone system. Round to the nearest cent.

26. **Sleep** In a recent survey, 30% of those polled sleep 7 hours every night. If 72 people responded that they sleep 7 hours each night, how many people participated in the survey?

27. **Treatment** In one month, a chiropractor treated 51 patients with disc-related complaints. This number represents 17% of all patients treated during the month. Find the number of patients that were treated this month.

28. **Education** In the United States today, 23.1% of women and 27.5% of men have earned a bachelor's or graduate degree. (*Source:* Census Bureau) How many women in the United States have earned a bachelor's or graduate degree?

29. **Exam** On the medical laboratory national board exam, a student only completed 120 questions, which equaled 75% of the questions. How many questions were on the national board exam?

30. **Nutrition** A one-serving size bag of potato chips contains 90 calories from fat, which represents 60% of the total calories of the chips in the bag. How many calories are there in this bag of chips?

Marc Dietrich/Shutterstock.com

31. **Blood Sugar** Tyler is diabetic and checks his blood sugar on a regular basis using a glucometer. His blood sugar was higher than normal six times during one week. This result was 15% of the number of times he checked it that week.
    a. How many times did he check his blood sugar that week?
    b. How many times was his blood sugar in the normal range?

32. **Weight Loss** A doctor advises his patient to lose 10% of his current weight in order to improve his health and avoid the onset of adult diabetes. The patient reaches this goal in four months by losing 21 pounds.
    a. How much did the patient weigh at the beginning of the diet?
    b. How much does the patient weigh at the end of the four-month period?

NUTRITION INFORMATION

SERVING SIZE: 1.4 OZ WHEAT FLAKES WITH 0.4 OZ. RAISINS: 39.4 g. ABOUT 1/2 CUP

SERVINGS PER PACKAGE: 14

PERCENTAGE OF U.S. RECOMMENDED DAILY ALLOWANCES (U.S. RDA)

	CEREAL & RAISINS	WITH 1/2 CUP VITAMINS A & D SKIM MILK
PROTEIN	4	15
VITAMIN A	15	20
VITAMIN C	**	2
THIAMIN	25	30
RIBOFLAVIN	25	35
NIACIN	25	35
CALCIUM	**	15
IRON	100	100
VITAMIN D	10	25
VITAMIN B_6	25	25
FOLIC ACID	25	25
VITAMIN B_{12}	25	30
PHOSPHOROUS	10	15
MAGNESIUM	10	20
ZINC	25	30
COPPER	2	4

* 2% MILK SUPPLIES AN ADDITIONAL 20 CALORIES. 2 g FAT, AND 10 mg CHOLESTEROL.

** CONTAINS LESS THAN 2% OF THE U.S. RDA OF THIS NUTRIENT

Applying the Concepts

33. **Nutrition** The table at the right contains nutrition information about a breakfast cereal. The amount of thiamin in one serving of this cereal with skim milk is 0.45 milligram. Find the recommended daily allowance of thiamin for an adult.

34. Increase a number by 10%. Now decrease the number by 10%. Is the result the original number? Explain.

SECTION

5.5 Percent Problems: Proportion Method

OBJECTIVE A To solve percent problems using proportions

Problems that can be solved using the basic percent equation can also be solved using proportions.

The proportion method is based on writing two ratios. One ratio is the percent ratio, written as $\frac{\text{percent}}{100}$. The second ratio is the amount-to-base ratio, written as $\frac{\text{amount}}{\text{base}}$. These two ratios form the proportion

$$\frac{\textbf{percent}}{\textbf{100}} = \frac{\textbf{amount}}{\textbf{base}}$$

To use the proportion method, first identify the percent, the amount, and the base (the base usually follows the phrase "percent of").

What is 23% of 45?

$$\frac{23}{100} = \frac{n}{45}$$
$$23 \times 45 = 100 \times n$$
$$1035 = 100 \times n$$
$$1035 \div 100 = n$$
$$10.35 = n$$

What percent of 25 is 4?

$$\frac{n}{100} = \frac{4}{25}$$
$$n \times 25 = 100 \times 4$$
$$n \times 25 = 400$$
$$n = 400 \div 25$$
$$n = 16$$

12 is 60% of what number?

$$\frac{60}{100} = \frac{12}{n}$$
$$60 \times n = 100 \times 12$$
$$60 \times n = 1200$$
$$n = 1200 \div 60$$
$$n = 20$$

Integrating Technology

To use a calculator to solve the proportions at the right for *n*, enter

23 × 45 ÷ 100 =

100 × 4 ÷ 25 =

100 × 12 ÷ 60 =

EXAMPLE 1

15% of what is 7? Round to the nearest hundredth.

Solution

$$\frac{15}{100} = \frac{7}{n}$$
$$15 \times n = 100 \times 7$$
$$15 \times n = 700$$
$$n = 700 \div 15$$
$$n \approx 46.67$$

YOU TRY IT 1

26% of what is 22? Round to the nearest hundredth.

Your solution

EXAMPLE 2

30% of 63 is what?

Solution

$$\frac{30}{100} = \frac{n}{63}$$
$$30 \times 63 = 100 \times n$$
$$1890 = 100 \times n$$
$$1890 \div 100 = n$$
$$18.90 = n$$

YOU TRY IT 2

16% of 132 is what?

Your solution

Solutions on p. S13

OBJECTIVE B To solve application problems

EXAMPLE • 3

An IV solution of normal saline has a 0.9% concentration, indicating that there is 0.9 gram of sodium chloride (NaCl) in every 100 milliliters of water. If a patient is given an IV solution of 1000 milliliters, how many grams of salt has the patient received when the infusion is complete?

Strategy

To find the amount of salt in 1000 milliliters of IV solution, set up a proportion using the percent concentration (0.9%) as the percent, 1000 milliliters as the base, and n as the amount of salt in grams.

Solution

$$\frac{0.9}{100} = \frac{n}{1000}$$
$$0.9 \times 1000 = 100 \times n$$
$$900 = 100 \times n$$
$$900 \div 100 = n$$
$$9 = n$$

The patient received 9 grams of salt from the IV solution.

YOU TRY IT • 3

A 5% IV dextrose solution (D5W) indicates that 5 grams of dextrose (sugar) are dissolved in 100 milliliters of solution. If a patient receives 250 milliliters of a 5% dextrose solution in an IV, how many grams of dextrose does the patient receive when the infusion is complete?

Your strategy

Your solution

EXAMPLE • 4

Last year at the local eye clinic, doctors performed cataract surgery on a total of 1240 patients. Of those patients, 1178 recovered without complications. What is the success rate for this eye clinic?

Strategy

A success rate is the percent of successful surgeries out of the total number of surgeries done. Write and solve a proportion using n to represent the percent of successful surgeries. The base is 1240, the total number of patients, and the amount is the number of patients without complications (1178).

Solution

$$\frac{n}{100} = \frac{1178}{1240}$$
$$n \times 1240 = 100 \times 1178$$
$$n \times 1240 = 117{,}800$$
$$n = 117{,}800 \div 1240$$
$$n = 95$$

The success rate for the clinic is 95%.

YOU TRY IT • 4

When companies test new medications, they try to determine the percent of patients who will experience certain side effects such as nausea. If 60 patients are given an experimental medication and three experience nausea, what percent experienced nausea?

Your strategy

Your solution

Solutions on p. S13

5.5 EXERCISES

OBJECTIVE A To solve percent problems using proportions

1. 26% of 250 is what?

2. What is 18% of 150?

3. 37 is what percent of 148?

4. What percent of 150 is 33?

5. 68% of what is 51?

6. 126 is 84% of what?

7. What percent of 344 is 43?

8. 750 is what percent of 50?

9. 82 is 20.5% of what?

10. 2.4% of what is 21?

11. What is 6.5% of 300?

12. 96% of 75 is what?

13. 7.4 is what percent of 50?

14. What percent of 1500 is 693?

15. 50.5% of 124 is what?

16. What is 87.4% of 255?

17. 33 is 220% of what?

18. 160% of what is 40?

19. **a.** Which equation(s) below can be used to answer the question "What is 12% of 75?"
b. Which equation(s) below can be used to answer the question "75 is 12% of what?"

(i) $\frac{12}{100} = \frac{75}{n}$ (ii) $0.12 \times 75 = n$ (iii) $\frac{12}{100} = \frac{n}{75}$ (iv) $0.12 \times n = 75$

OBJECTIVE B To solve application problems

20. Read Exercise 21. Without doing any calculations, determine whether the length of time the drug will be effective is *less than* or *greater than* 6 hours.

21. **Medicine** A manufacturer of an anti-inflammatory drug claims that the drug will be effective for 6 hours. An independent testing service determined that the drug was effective for only 80% of the length of time claimed by the manufacturer. Find the length of time the drug will be effective as determined by the testing service.

22. **Smoking** According to the American Heart Association 2010 update, *Heart Disease and Stroke Statistics*, approximately 21% of Americans smoke cigarettes. Based on the latest census, there are approximately 308,700,000 people in the United States. Approximately how many Americans smoke?

23. IV Solution If 350 milliliters of an IV 10% dextrose solution are infused, how many grams of dextrose does the patient receive when the infusion is complete?

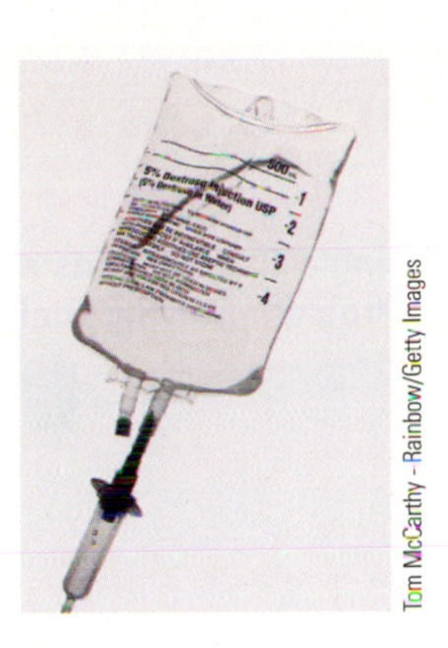

Tom McCarthy - Rainbow/Getty Images

24. Charities The American Red Cross spent $185,048,179 for administrative expenses. This amount was 3.16% of its total revenue. Find the American Red Cross's total revenue. Round to the nearest hundred million.

25. Blood Types During a recent blood drive, 18 donors had type O blood. This was 45% of the total number of donors who gave blood that day. How many donors participated in the blood drive?

26. Blood pressure In a recent clinical study, 85 adults were found to have high blood pressure. This represents 34% of those who were screened. How many patients were screened?

In the News

Over Half of Baby Boomers Have College Experience

Of the 78 million baby boomers living in the United States, 45 million have some college experience but no college degree. Twenty million baby boomers have one or more college degrees.

Sources: The National Center for Education Statistics; U.S. Census Bureau; *McCook Daily Gazette*

27. Education See the news clipping at the right. What percent of the baby boomers living in the United States have some college experience but have not earned a college degree? Round to the nearest tenth of a percent.

28. Demography According to a 25-city survey of the status of hunger and homelessness by the U.S. Conference of Mayors, 41% of the homeless in the United States are single men, 41% are families with children, 13% are single women, and 5% are unaccompanied minors. How many homeless people in the United States are single men?

29. Morbidity The graph at the right from the Centers for Disease Control and Prevention (CDC) shows the estimated annual number of smoking-attributable deaths in the United States by specific causes. What percent of the deaths attributable to smoking result from lung cancer? Round to the nearest tenth of a percent.

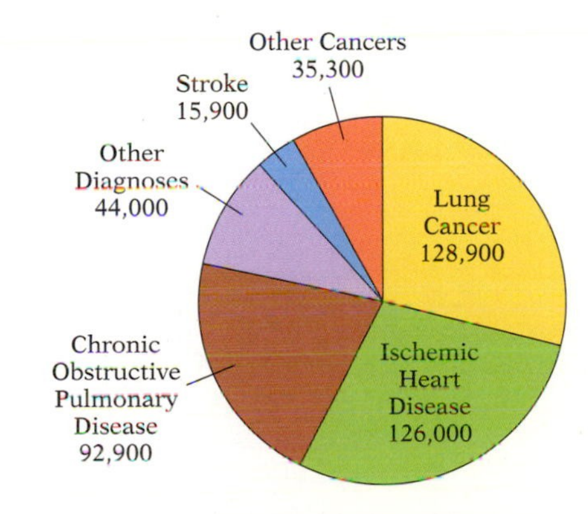

About 443,000 U.S. Deaths Attributable Each Year to Cigarette Smoking*

*Average annual number of deaths, 2000–2004.

(*Source:* http://www.cdc.gov/tobaccodata_statistics/tables/health/attrdeaths/index.htm)

Applying the Concepts

30. Side Effects of Medication In a recent clinical trial, 250 patients received Medication A while 180 patients received Medication B. 15 of the patients receiving Medication A reported dizziness after taking the medication, whereas only 12 of those on Medication B reported dizziness. Which medication had the higher percentage of patients reporting dizziness?

FOCUS ON PROBLEM SOLVING

Using a Calculator as a Problem-Solving Tool

A calculator is an important tool for problem solving. Here are a few problems to solve with a calculator. You may need to research some of the questions to find information you do not know.

1. Choose any single-digit positive number. Multiply the number by 1507 and 7373. What is the answer? Choose another positive single-digit number and again multiply by 1507 and 7373. What is the answer? What pattern do you see? Why does this work?

2. The gross domestic product in 2007 was \$13,841,300,000. Is this more or less than the amount of money that would be placed on the last square of a standard checkerboard if 1 cent were placed on the first square, 2 cents were placed on the second square, 4 cents were placed on the third square, 8 cents were placed on the fourth square, and so on, until the 64th square was reached?

3. Which of the reciprocals of the first 16 natural numbers have a terminating-decimal representation and which have a repeating-decimal representation?

4. What is the largest natural number n for which $4^n > 1 \cdot 2 \cdot 3 \cdot 4 \cdot 5 \cdot \cdots \cdot n$?

5. If \$1000 bills are stacked one on top of another, is the height of \$1 billion less than or greater than the height of the Washington Monument?

6. What is the value of $1 + \cfrac{1}{1 + \cfrac{1}{1 + \cfrac{1}{1 + \cfrac{1}{1 + 1}}}}$?

7. Calculate 15^2, 35^2, 65^2, and 85^2. Study the results. Make a conjecture about a relationship between a number ending in 5 and its square. Use your conjecture to find 75^2 and 95^2. Does your conjecture work for 125^2?

8. Find the sum of the first 1000 natural numbers. (*Hint:* You could just start adding $1 + 2 + 3 + \cdots$, but even if you performed one operation every 3 seconds, it would take you an hour to find the sum. Instead, try pairing the numbers and then adding the pairs. Pair 1 and 1000, 2 and 999, 3 and 998, and so on. What is the sum of each pair? How many pairs are there? Use this information to answer the original question.)

9. For a borrower to qualify for a home loan, a bank requires that the monthly mortgage payment be less than 25% of the borrower's monthly take-home income. A laboratory technician has deductions for taxes, insurance, and retirement that amount to 25% of the technician's monthly gross income. What minimum monthly income must this technician earn to receive a bank loan that has a mortgage payment of \$1200 per month?

Using Estimation as a Problem-Solving Tool

You can use your knowledge of rounding, your understanding of percent, and your experience with the basic percent equation to quickly estimate the answer to a percent problem. Here is an example.

HOW TO 1 What is 11.2% of 978?

Round the given numbers. $11.2\% \approx 10\%$
$978 \approx 1000$

Mentally calculate with the rounded numbers. $10\% \text{ of } 1000 = \frac{1}{10} \text{ of } 1000 = 100$

11.2% of 978 is approximately 100.

Take Note
The exact answer is $0.112 \times 978 = 109.536$. The exact answer 109.536 is close to the approximation of 100.

For Exercises 1 to 8, state which quantity is greater.

1. 49% of 51, or 201% of 15

2. 99% of 19, or 22% of 55

3. 8% of 31, or 78% of 10

4. 24% of 402, or 76% of 205

5. 10.2% of 51, or 20.9% of 41

6. 51.8% of 804, or 25.3% of 1223

7. 26% of 39.217, or 9% of 85.601

8. 66% of 31.807, or 33% of 58.203

© Ariel Skelly/Corbis

For Exercises 9 to 12, use estimation to provide an approximate number.

9. A company found that 24% of its 2096 employees favored a new dental plan. How many employees favored the new dental plan?

10. A local newspaper reported that 52.3% of the 29,875 eligible voters in the town voted in the last election. How many people voted in the last election?

11. 19.8% of the 2135 first-year students at a community college have part-time jobs. How many of the first-year students at the college have part-time jobs?

12. A couple made a down payment of 33% of the $310,000 cost of a home. Find the down payment.

PROJECTS AND GROUP ACTIVITIES

Health

The American College of Sports Medicine (ACSM) recommends that you know how to determine your target heart rate in order to get the full benefit of exercise. Your **target heart rate** is the rate at which your heart should beat during any aerobic exercise such as running, cycling, fast walking, or participating in an aerobics class. According to the ACSM, you should reach your target rate and then maintain it for 20 minutes or more to achieve cardiovascular fitness. The intensity level varies for different individuals. A sedentary person might begin at the 60% level and gradually work up to 70%, whereas athletes and very fit individuals might work at the 85% level. The ACSM suggests that you calculate both 50% and 85% of your maximum heart rate. This will give you the low and high ends of the range within which your heart rate should stay.

To calculate your target heart rate:

	Example
Subtract your age from 220. This is your maximum heart rate.	$220 - 20 = 200$
Multiply your maximum heart rate by 50%. This is the low end of your range.	$200(0.50) = 100$
Divide the low end by 6. This is your low 10-second heart rate.	$100 \div 6 \approx 17$
Multiply your maximum heart rate by 85%. This is the high end of your range.	$200(0.85) = 170$
Divide the high end by 6. This is your high 10-second heart rate.	$170 \div 6 \approx 28$

1. Why are the low end and high end divided by 6 in order to determine the low and high 10-second heart rates?
2. Calculate your target heart rate, both the low and high end of your range.

Consumer Price Index

The consumer price index (CPI) is a percent that is written without the percent sign. For instance, a CPI of 160.1 means 160.1%. This number means that an item that cost \$100 between 1982 and 1984 (the base years) would cost \$160.10 today. Determining the cost is an application of the basic percent equation.

$$\text{Percent} \times \text{base} = \text{amount}$$
$$\text{CPI} \times \text{cost in base year} = \text{cost today}$$
$$1.601 \times 100 = 160.1 \qquad \bullet\ 160.1\% = 1.601$$

The table below gives the CPI for various products in March of 2008. If you have Internet access, you can obtain current data for the items below, as well as other items not on this list, by visiting the website of the Bureau of Labor Statistics.

Product	*CPI*
All items	213.5
Food and beverages	209.7
Housing	214.4
Clothes	120.9
Transportation	195.2
Medical care	363.0
Entertainment[1]	112.7
Education[1]	121.8

[1]Indexes on December 1997 = 100

1. Of the items listed, are there any items that in 2008 cost more than twice as much as they cost during the base year? If so, which items?
2. Of the items listed, are there any items that in 2008 cost more than one-and-one-half times as much as they cost during the base years but less than twice as much as they cost during the base years? If so, which items?
3. If the cost for textbooks for one semester was \$120 in the base years, how much did similar textbooks cost in 2008? Use the "Education" category.
4. If a new car cost \$40,000 in 2008, what would a comparable new car have cost during the base years? Use the "Transportation" category.

5. If a movie ticket cost \$10 in 2008, what would a comparable movie ticket have cost during the base years? Use the "Entertainment" category.

6. The base year for the CPI was 1967 before the change to 1982–1984. If 1967 were still used as the base year, the CPI for all items in 2008 (not just those listed above) would be 639.6.

a. Using the base year of 1967, explain the meaning of a CPI of 639.6.

b. Using the base year of 1967 and a CPI of 639.6, if textbooks cost \$75 for one semester in 1967, how much did similar textbooks cost in 2008?

c. Using the base year of 1967 and a CPI of 639.6, if a family's food budget in 2008 is \$1000 per month, what would a comparable family budget have been in 1967?

CHAPTER 5

SUMMARY

KEY WORDS / EXAMPLES

Percent means "parts of 100." [5.1A, p. 202]

23% means 23 of 100 equal parts.

ESSENTIAL RULES AND PROCEDURES / EXAMPLES

To write a percent as a fraction, drop the percent sign and multiply by $\frac{1}{100}$. [5.1A, p. 202]

$$56\% = 56\left(\frac{1}{100}\right) = \frac{56}{100} = \frac{14}{25}$$

To write a percent as a decimal, drop the percent sign and multiply by 0.01. [5.1A, p. 202]

$$87\% = 87(0.01) = 0.87$$

To write a fraction as a percent, multiply by 100%. [5.1B, p. 203]

$$\frac{7}{20} = \frac{7}{20}(100\%) = \frac{700}{20}\% = 35\%$$

To write a decimal as a percent, multiply by 100%. [5.1B, p. 203]

$$0.325 = 0.325(100\%) = 32.5\%$$

The Basic Percent Equation [5.2A, p. 206]

The basic percent equation is

$$\text{Percent} \times \text{base} = \text{amount}$$

Solving percent problems requires identifying the three elements of this equation. Usually the base follows the phrase "percent of."

8% of 250 is what number?

$$\text{Percent} \times \text{base} = \text{amount}$$
$$0.08 \times 250 = n$$
$$20 = n$$

Proportion Method of Solving a Percent Problem [5.5A, p. 218]

The following proportion can be used to solve percent problems.

$$\frac{\text{percent}}{100} = \frac{\text{amount}}{\text{base}}$$

To use the proportion method, first identify the percent, the amount, and the base. The base usually follows the phrase "percent of."

8% of 250 is what number?

$$\frac{\text{percent}}{100} = \frac{\text{amount}}{\text{base}}$$
$$\frac{8}{100} = \frac{n}{250}$$
$$8 \times 250 = 100 \times n$$
$$2000 = 100 \times n$$
$$2000 \div 100 = n$$
$$20 = n$$

CHAPTER 5

CONCEPT REVIEW

Test your knowledge of the concepts presented in this chapter. Answer each question. Then check your answers against the ones provided in the Answer Section.

1. How do you write 197% as a fraction?

2. How do you write 6.7% as a decimal?

3. How do you write $\frac{9}{5}$ as a percent?

4. How do you write 56.3 as a percent?

5. What is the basic percent equation?

6. What percent of 40 is 30? Did you multiply or divide?

7. Find 11.7% of 532. Did you multiply or divide?

8. 36 is 240% of what number? Did you multiply or divide?

9. How do you use the proportion method to solve a percent problem?

10. What percent of 1400 is 763? Use the proportion method to solve.

CHAPTER 5

REVIEW EXERCISES

1. What is 30% of 200?

2. 16 is what percent of 80?

3. Write $1\frac{3}{4}$ as a percent.

4. 20% of what is 15?

5. Write 12% as a fraction.

6. Find 22% of 88.

7. What percent of 20 is 30?

8. $16\frac{2}{3}\%$ of what is 84?

9. Write 42% as a decimal.

10. What is 7.5% of 72?

11. $66\frac{2}{3}\%$ of what is 105?

12. Write 7.6% as a decimal.

13. Find 125% of 62.

14. Write $16\frac{2}{3}\%$ as a fraction.

15. Use the proportion method to find what percent of 25 is 40.

16. 20% of what number is 15? Use the proportion method.

17. Write 0.38 as a percent.

18. 78% of what is 8.5? Round to the nearest tenth.

19. What percent of 30 is 2.2? Round to the nearest tenth of a percent.

20. What percent of 15 is 92? Round to the nearest tenth of a percent.

21. Education Trent missed 9 out of 60 questions on an anatomy exam. What percent of the questions did he answer correctly? Use the proportion method.

22. Nutrition Dieticians recommend that a person get a maximum of 35% of their calories each day from fat. Based on an 1850-calorie daily diet, what is the maximum number of calories that should come from fat? Round to the nearest whole number.

23. Health Expenditures The graph at the right shows the Johnson family's annual budget. If their total income is $45,000, what percent of the budget is allocated for health care?

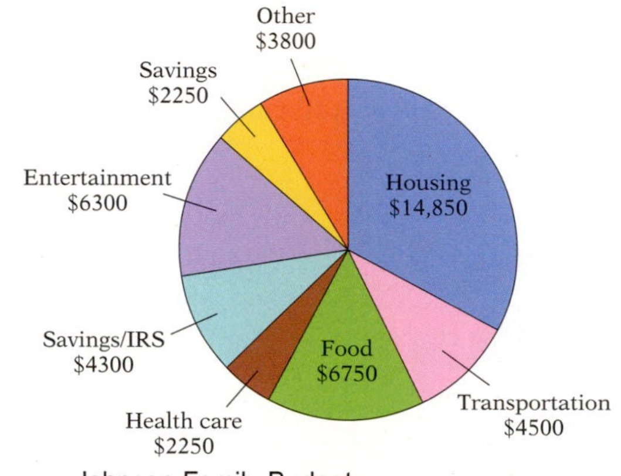

Johnson Family Budget

24. Equipment Costs A college bought an $1850 automated defibrillator for its first aid station and paid a sales tax of 6.5% of the cost. What was the total cost of the defibrillator?

25. Health In a survey of 350 women and 420 men, 275 of the women and 300 of the men reported that they wore sunscreen often. To the nearest tenth of a percent, what percent of the women wore sunscreen often?

26. Demography It is estimated that the world's population will be 9,100,000,000 by the year 2050. This is 149% of the population in 2000. (*Source:* U.S. Census Bureau). What was the world's population in 2000? Round to the nearest hundred million.

27. Medical Procedures During a recent year, 636,000 colonoscopies were performed in the United States. Of these procedures, 60% were performed on women. Find the number of women who had colonoscopies.

28. Health Care Spending According to the Centers for Medicare and Medicaid Services, Office of the Actuary, health spending in 2008 exceeded $2.3 trillion. This amount was approximately 16.2% of the United States gross domestic product (GDP). Find the approximate amount of the GDP in 2008. Round to the nearest trillion.

CHAPTER 5

TEST

1. Write 97.3% as a decimal.

2. Write $83\frac{1}{3}\%$ as a fraction.

3. Write 0.3 as a percent.

4. Write 1.63 as a percent.

5. Write $\frac{3}{2}$ as a percent.

6. Write $\frac{2}{3}$ as a percent.

7. What is 77% of 65?

8. 47.2% of 130 is what?

9. Which is larger:
7% of 120, or 76% of 13?

10. Which is smaller:
13% of 200, or 212% of 12?

11. **Advertising** A large hospital uses 6% of its \$750,000 budget for advertising. What amount of the budget is spent on advertising?

12. **Nutrition** According to USDA dietary guidelines for Americans, diets that receive fewer than 45% of calories from carbohydrates are difficult to adhere to and are not recommended for weight loss. If you are on a 2000-calorie diet, what is the minimum number of calories you should get daily from carbohydrates?

Nutrition The table at the right contains nutrition information about a breakfast cereal. Solve Exercises 13 and 14 with information taken from this table.

NUTRITION INFORMATION

SERVING SIZE: 1.4 OZ WHEAT FLAKES WITH 0.4 OZ. RAISINS: 39.4 g. ABOUT 1/2 CUP
SERVINGS PER PACKAGE: 14

		CEREAL & RAISINS	WITH 1/2 CUP VITAMINS A & D SKIM MILK
CALORIES		120	180
PROTEIN, g		3	7
CARBOHYDRATE, g		28	34
FAT, TOTAL, g		1	1*
UNSATURATED, g	1		
SATURATED, g	0		
CHOLESTEROL, mg		0	0*
SODIUM, mg		125	190
POTASSIUM, mg		240	440

* 2% MILK SUPPLIES AN ADDITIONAL 20 CALORIES. 2 g FAT, AND 10 mg CHOLESTEROL.
** CONTAINS LESS THAN 2% OF THE U.S. RDA OF THIS NUTRIENT

13. The recommended amount of potassium per day for an adult is 3000 milligrams (mg). What percent, to the nearest tenth of a percent, of the daily recommended amount of potassium is provided by one serving of this cereal with skim milk?

14. The daily recommended number of calories for a 190-pound man is 2200 calories. What percent, to the nearest tenth of a percent, of the daily recommended number of calories is provided by one serving of this cereal with 2% milk?

15. **Surgical Complications** Although surgical complications of tonsillectomies are rare, last year, 18 out of Dr. Juarez's 540 tonsillectomy patients had complications after surgery. What percent of Dr. Juarez's patients did not have complications after surgery? Round to the nearest tenth of a percent.

16. **Education** Conchita missed 7 out of 80 questions on a medical terminology exam. What percent of the questions did she answer correctly? Round to the nearest tenth of a percent.

17. 12 is 15% of what?

18. 42.5 is 150% of what? Round to the nearest tenth.

19. **IV Solution** If a patient receives 500 milliliters of a 5% dextrose solution in an IV, how many grams of dextrose does the patient receive when the infusion is complete?

20. **Real Estate** A dental office was bought for $285,000. Five years later the office was sold for $456,000. The increase was what percent of the original price?

21. 123 is 86% of what number? Use the proportion method. Round to the nearest tenth.

22. What percent of 12 is 120? Use the proportion method.

23. **Wages** An administrative assistant receives a wage of $16.24 per hour. This amount is 112% of last year's wage. What is the dollar increase in the hourly wage over last year? Use the proportion method.

24. **Nutrition** One slice of deep pan pepperoni pizza has approximately 300 calories. If you are on a diet restricting you to 1800 calories daily, what percent of your total daily calorie allowance is represented by this slice of pizza? Round to the nearest tenth of a percent.

25. **Weight Loss** Jeanne started a diet on New Year's Day and by Valentine's Day, she had lost 9 pounds. This amount represents a loss of 3.6% of her original weight. How much did she weigh when she began the diet?

CUMULATIVE REVIEW EXERCISES

1. Simplify: $18 \div (7 - 4)^2 + 2$

2. Find the LCM of 16, 24, and 30.

3. Find the sum of $2\frac{1}{3}$, $3\frac{1}{2}$, and $4\frac{5}{8}$.

4. Subtract: $27\frac{5}{12} - 14\frac{9}{16}$

5. Multiply: $7\frac{1}{3} \times 1\frac{5}{7}$

6. What is $\frac{14}{27}$ divided by $1\frac{7}{9}$?

7. Simplify: $\left(\frac{3}{4}\right)^3 \cdot \left(\frac{8}{9}\right)^2$

8. Simplify: $\left(\frac{2}{3}\right)^2 - \left(\frac{3}{8} - \frac{1}{3}\right) \div \frac{1}{2}$

9. Round 3.07973 to the nearest hundredth.

10. Subtract: $\begin{array}{r} 3.0902 \\ -\ 1.9706 \\ \hline \end{array}$

11. Divide: $0.032\overline{)1.097}$
Round to the nearest ten-thousandth.

12. Convert $3\frac{5}{8}$ to a decimal.

13. Convert 1.75 to a fraction.

14. Place the correct symbol, < or >, between the two numbers.
$\frac{3}{8}$ 0.87

15. Solve the proportion $\frac{3}{8} = \frac{20}{n}$.
Round to the nearest tenth.

16. Write "$153.60 earned in 8 hours" as a unit rate.

17. Write $18\frac{1}{3}\%$ as a fraction.

18. Write $\frac{5}{6}$ as a percent.

19. 16.3% of 120 is what?

20. 24 is what percent of 18?

21. 12.4 is 125% of what?

22. What percent of 35 is 120? Round to the nearest tenth.

23. **Taxes** Sergio has an income of $740 per week. One-fifth of his income is deducted for income tax payments. Find his take-home pay.

24. **Finance** Eunice bought a used car for $12,530, with a down payment of $2000. The balance was paid in 36 equal monthly payments. Find the monthly payment.

25. **Medication** A certain pain medication is in tablet form and contains 0.25 gram of medication per tablet. If the doctor has ordered a patient to have 0.5 gram of pain medication, how many tablets should a nurse give the patient?

26. **Property Taxes** The property tax on a $344,000 dental office is $6880. At the same rate, find the property tax on a medical office valued at $500,000.

27. **Blood Types** The graph at the right gives the percentages of the African-American population in the United States with each blood type. (*Source:* http://www.redcrossblood.org/learn-about-blood/blood-types) The Census Bureau estimated in 2009 that there were approximately 37,300,000 African-Americans in the United States. How many would you expect to have type A+ blood?

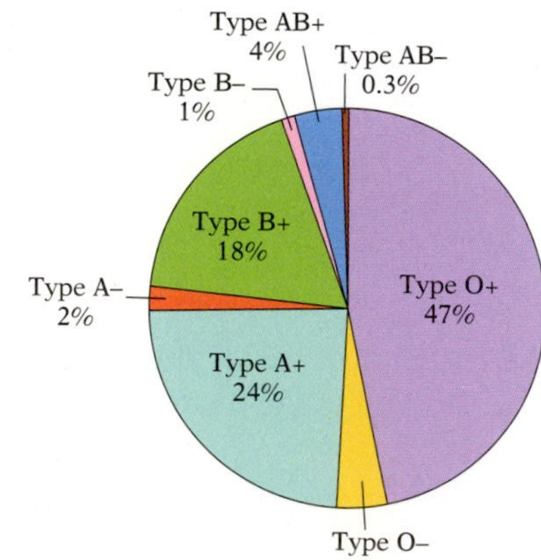

Blood Types of African-Americans in the United States
Source: American Red Cross

28. **Surveys** A group of consumers was asked to taste two cough syrups and rate the one with the best flavor. The results show that of 300 people, 165 favored Cough Syrup A. What percent of those surveyed favored Cough Syrup B?

29. **Nutrition** The Mayo Clinic recommends that you limit your daily intake of saturated fat to a maximum of 7% of your total calories. If you are on a 2000-calorie diet, what is the maximum number of calories that you should get from saturated fat in one week?

30. **Health** The Environmental Protection Agency found that 990 out of 5500 children tested had levels of lead in their blood that exceeded federal guidelines. What percent of the children tested had levels of lead in the blood that exceeded federal standards?

Applications for Business and Consumers

CHAPTER 6

OBJECTIVES

SECTION 6.1

- **A** To find unit cost
- **B** To find the most economical purchase
- **C** To find total cost

SECTION 6.2

- **A** To find percent increase
- **B** To apply percent increase to business—markup
- **C** To find percent decrease
- **D** To apply percent decrease to business—discount

SECTION 6.3

- **A** To calculate simple interest
- **B** To calculate finance charges on a credit card bill
- **C** To calculate compound interest

SECTION 6.4

- **A** To calculate the initial expenses of buying a home
- **B** To calculate the ongoing expenses of owning a home

SECTION 6.5

- **A** To calculate the initial expenses of buying a car
- **B** To calculate the ongoing expenses of owning a car

SECTION 6.6

- **A** To calculate commissions, total hourly wages, and salaries

SECTION 6.7

- **A** To calculate checkbook balances
- **B** To balance a checkbook

AIM for the Future

MEDICAL TRANSCRIPTIONISTS listen to dictated recordings made by physicians and other healthcare professionals and transcribe them into medical reports, correspondence, and other administrative materials. To understand and accurately transcribe dictated reports, medical transcriptionists must understand medical terminology, anatomy and physiology, diagnostic procedures, pharmacology, and treatment assessments. Employment of medical transcriptionists is projected to grow by 11% from 2008 to 2018, about as fast as the average for all occupations. Wage-and-salary medical transcriptionists had median hourly wages of $15.41 in May 2008. (*Source: Bureau of Labor Statistics:* http://www.bls.gov/oco/ocos271.htm)

PREP TEST

Are you ready to succeed in this chapter?

Take the Prep Test below to find out if you are ready to learn the new material.

For Exercises 1 to 6, add, subtract, multiply, or divide.

1. Divide: $3.75 \div 5$
2. Multiply: 3.47×15
3. Subtract: $874.50 - 369.99$
4. Multiply: $0.065 \times 150{,}000$
5. Multiply: $1500 \times 0.06 \times 0.5$
6. Add: $1372.47 + 36.91 + 5.00 + 2.86$
7. Divide $10 \div 3$. Round to the nearest hundredth.
8. Divide $345 \div 570$. Round to the nearest thousandth.
9. Place the correct symbol, $<$ or $>$, between the two numbers.
   0.379   0.397

SECTION

6.1 Applications to Purchasing

OBJECTIVE A To find unit cost

Frequently, stores promote items for purchase by advertising, say, 2 Red Baron Bake to Rise Pizzas for \$10.50 or 5 cans of StarKist tuna for \$4.25. The **unit cost** is the cost of *one* Red Baron Pizza or of *one* can of StarKist tuna. To find the unit cost, divide the total cost by the number of units.

2 pizzas for \$10.50

$10.50 \div 2 = 5.25$

\$5.25 is the cost of one pizza.

Unit cost: \$5.25 per pizza

5 cans for \$4.25

$4.25 \div 5 = 0.85$

\$0.85 is the cost of one can.

Unit cost: \$0.85 per can

EXAMPLE • 1

Find the unit cost. Round to the nearest tenth of a cent.

a. A box of 100 insulin syringes for \$53.82
b. 4 ounces of Crest toothpaste for \$2.29

Strategy
To find the unit cost, divide the total cost by the number of units.

Solution

a. $53.82 \div 100 \approx 0.538$
\$0.538 per syringe

b. $2.29 \div 4 = 0.5725$
\$0.573 per ounce

YOU TRY IT • 1

Find the unit cost. Round to the nearest tenth of a cent.

a. A box of 50 disposable latex safety gloves for \$16.80
b. 15 ounces of Suave shampoo for \$2.29

Your strategy

Your solution

Solution on p. S13

OBJECTIVE B To find the most economical purchase

Comparison shoppers often find the most economical buy by comparing unit costs.

One store is selling 6 twelve-ounce cans of ginger ale for \$2.99, and a second store is selling 24 twelve-ounce cans of ginger ale for \$11.79. To find the better buy, compare the unit costs.

$2.99 \div 6 \approx 0.498$

Unit cost: \$0.498 per can

$11.79 \div 24 \approx 0.491$

Unit cost: \$0.491 per can

Because $\$0.491 < \$0.498$, the better buy is 24 cans for \$11.79.

EXAMPLE 2

Find the more economical purchase: a box of 250 tongue depressors for $14.75 or a box of 100 tongue depressors for $6.20.

Strategy
To find the more economical purchase, compare the unit costs.

Solution
$14.75 \div 250 = 0.059$
$6.20 \div 100 = 0.062$
$0.059 < 0.062$

The more economical purchase is a box of 250 for $14.75.

YOU TRY IT 2

Find the more economical purchase: 6 cans of fruit for $8.70, or 4 cans of fruit for $6.96.

Your strategy

Your solution

Solution on p. S14

OBJECTIVE C To find total cost

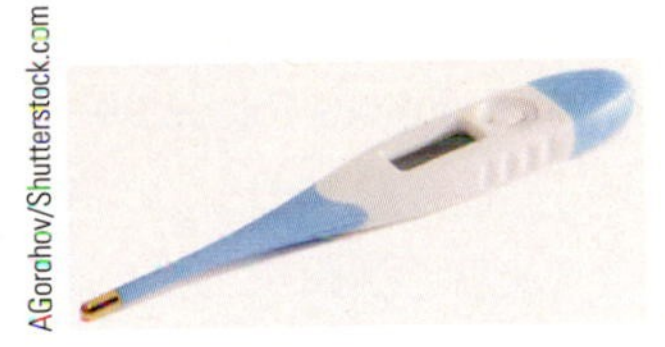
AGorohov/Shutterstock.com

The office manager of a nursing home needs to place an order for a dozen digital thermometers. The unit cost for identical thermometers is listed below.

Store 1	*Store 2*	*Store 3*
$6.75 per thermometer	$6.39 per thermometer	$7.22 per thermometer

By comparing the unit costs, the office manager determined that store 2 would provide the most economical purchase.

The office manager also uses the unit cost to find the total cost of purchasing two dozen thermometers at store 2. The **total cost** is found by multiplying the unit cost by the number of units purchased.

Unit cost	×	number of units	=	total cost
6.39	×	24	=	153.36

The total cost is $153.36.

EXAMPLE 3

Reusable penlights for patient examinations are $4.40 each. How much would a dozen penlights cost?

Strategy
To find the total cost, multiply the unit cost (4.40) by the number of units (one dozen = 12 units).

Solution

Unit cost	×	number of units	=	total cost
4.40	×	12	=	52.80

The total cost is $52.80.

YOU TRY IT 3

First aid kits are $14.99 each. If the county purchases 28 new kits for its EMS units, how much will they cost?

Your strategy

Your solution

Solution on p. S14

6.1 EXERCISES

OBJECTIVE A To find unit cost

For Exercises 1 to 10, find the unit cost. Round to the nearest tenth of a cent.

1. Heinz B·B·Q sauce, 18 ounces for $0.99
2. Birds-eye maple, 6 feet for $18.75
3. Diamond walnuts, $2.99 for 8 ounces
4. A&W root beer, 6 cans for $2.99
5. Ibuprofen, 50 tablets for $3.99
6. Visine eye drops, 0.5 ounce for $3.89
7. Dry skin therapy lotion, $5.35 for 18 ounces
8. Corn, 6 ears for $2.85
9. Cheerios cereal, 15 ounces for $2.99
10. Doritos Cool Ranch chips, 14.5 ounces for $2.99

11. A store advertises a "buy one, get one free" sale on pint containers of ice cream. How would you find the unit cost of one pint of ice cream?

OBJECTIVE B To find the most economical purchase

For Exercises 12 to 21, suppose your local supermarket offers the following products at the given prices. Find the more economical purchase.

12. Sutter Home pasta sauce, 25.5 ounces for $3.29, or Muir Glen Organic pasta sauce, 26 ounces for $3.79
13. Kraft mayonnaise, 40 ounces for $3.98, or Springfield mayonnaise, 32 ounces for $3.39
14. Ortega salsa, 20 ounces for $3.29 or 12 ounces for $1.99
15. L'Oréal shampoo, 13 ounces for $4.69, or Cortexx shampoo, 12 ounces for $3.99
16. Golden Sun vitamin E, 200 tablets for $12.99 or 400 tablets for $18.69
17. Ultra Mr. Clean, 20 ounces for $2.67, or Ultra Spic and Span, 14 ounces for $2.19
18. 16 ounces of Kraft cheddar cheese for $4.37, or 9 ounces of Land O'Lakes cheddar cheese for $2.29
19. Bertolli olive oil, 34 ounces for $9.49, or Pompeian olive oil, 8 ounces for $2.39

20. Maxwell House coffee, 4 ounces for $3.99, or Sanka coffee, 2 ounces for $2.39

21. Wagner's vanilla extract, $3.95 for 1.5 ounces, or Durkee vanilla extract, 1 ounce for $2.84

For Exercises 22 and 23, suppose a box of Tea A contains twice as many tea bags as a box of Tea B. Decide which box of tea is the more economical purchase.

22. The price of a box of Tea A is less than twice the price of a box of Tea B.

23. The price of a box of Tea B is greater than half the price of a box of Tea A.

OBJECTIVE C To find total cost

24. If sliced bacon costs $4.59 per pound, find the total cost of 3 pounds.

25. Healing lip balm costs $0.98 per stick. Find the total cost of a box of 75 sticks.

26. Kiwi fruit cost $0.43 each. Find the total cost of 8 kiwi.

27. Boneless chicken filets cost $4.69 per pound. Find the cost of 3.6 pounds. Round to the nearest cent.

28. Herbal tea costs $0.98 per ounce. Find the total cost of 6.5 ounces.

29. If Stella Swiss Lorraine cheese costs $5.99 per pound, find the total cost of 0.65 pound. Round to the nearest cent.

30. Red Delicious apples cost $1.29 per pound. Find the total cost of 2.1 pounds. Round to the nearest cent.

31. Choice rib eye steak costs $9.49 per pound. Find the total cost of 2.8 pounds. Round to the nearest cent.

32. Suppose a store flyer advertises cantaloupes as "buy one, get one free." True or false? The total cost of 6 cantaloupes at the sale price is the same as the total cost of 3 cantaloupes at the regular price.

Applying the Concepts

33. Explain in your own words the meaning of unit pricing.

34. What is the UPC (Universal Product Code) and how is it used?

SECTION

6.2 Percent Increase and Percent Decrease

OBJECTIVE A To find percent increase

Point of Interest

According to the U.S. Census Bureau, the number of persons aged 65 and over in the United States will increase to about 82.0 million by 2050, a 136% increase from 2000.

Percent increase is used to show how much a quantity has increased over its original value. The statements "Food prices increased by 2.3% last year" and "City council members received a 4% pay increase" are examples of percent increase.

HOW TO • 1 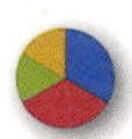 According to the Energy Information Administration, the number of alternative-fuel vehicles increased from approximately 277,000 to 352,000 in four years. Find the percent increase in alternative-fuel vehicles. Round to the nearest percent.

$$\boxed{\text{New value}} - \boxed{\text{original value}} = \boxed{\text{amount of increase}}$$

$$352{,}000 - 277{,}000 = 75{,}000$$

Amount of increase (75,000)
Original value (277,000)
New value (352,000)

Now solve the basic percent equation for percent.

$$\text{Percent} \times \text{base} = \text{amount}$$

$$\boxed{\text{Percent increase}} \times \boxed{\text{original value}} = \boxed{\text{amount of increase}}$$

$$n \times 277{,}000 = 75{,}000$$
$$n = 75{,}000 \div 277{,}000$$
$$n \approx 0.27$$

The number of alternative-fuel vehicles increased by approximately 27%.

EXAMPLE • 1

The average wholesale price of coffee increased from $2 per pound to $3 per pound in one year. What was the percent increase in the price of 1 pound of coffee?

Strategy

To find the percent increase:

- Find the amount of the increase.
- Solve the basic percent equation for *percent*.

Solution

$$\boxed{\text{New value}} - \boxed{\text{original value}} = \boxed{\text{amount of increase}}$$

$$3 - 2 = 1$$

$$\text{Percent} \times \text{base} = \text{amount}$$
$$n \times 2 = 1$$
$$n = 1 \div 2$$
$$n = 0.5 = 50\%$$

The percent increase was 50%.

YOU TRY IT • 1

The average price of gasoline rose from $3.46 to $3.83 in 5 months. What was the percent increase in the price of gasoline? Round to the nearest percent.

Your strategy

Your solution

Solution on p. S14

EXAMPLE • 2

Chris Carley was earning \$13.50 an hour as a nursing assistant before receiving a 10% increase in pay. What is Chris's new hourly pay?

Strategy

To find the new hourly wage:

- Solve the basic percent equation for *amount.*
- Add the amount of the increase to the original wage.

Solution

Percent × base = amount

$0.10 \times 13.50 = n$

$1.35 = n$

The amount of the increase was \$1.35.

$13.50 + 1.35 = 14.85$

The new hourly wage is \$14.85.

YOU TRY IT • 2

Yolanda Liyama was making a wage of \$12.50 an hour as a pharmacy tech before receiving a 14% increase in hourly pay. What is Yolanda's new hourly wage?

Your strategy

Your solution

Solution on p. S14

OBJECTIVE B To apply percent increase to business—markup

Some of the expenses involved in operating a business are salaries, rent, equipment, and utilities. To pay these expenses and earn a profit, a business must sell a product at a higher price than it paid for the product.

Cost is the price a business pays for a product, and **selling price** is the price at which a business sells a product to a customer. The difference between selling price and cost is called **markup.**

Selling price − cost = markup

or

Cost + markup = selling price

Markup

Cost

Selling price

Markup is frequently expressed as a percent of a product's cost. This percent is called the **markup rate.**

Markup rate × cost = markup

Point of Interest

According to *Managing a Small Business*, from Liraz Publishing Company, goods in a store are often marked up 50% to 100% of the cost. This allows a business to make a profit of 5% to 10%.

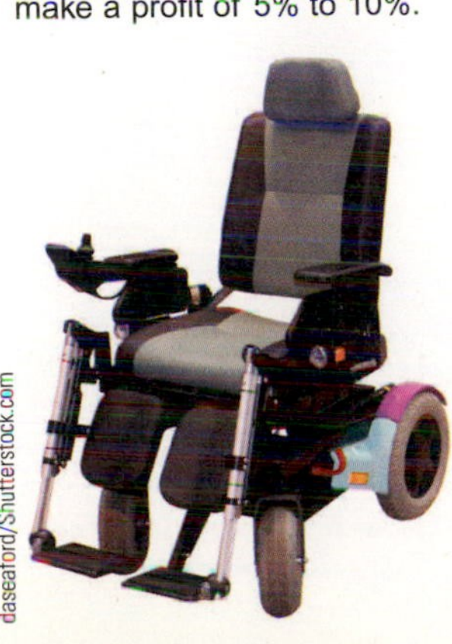

daseaford/Shutterstock.com

HOW TO • 2 Suppose MedSupplies, Inc. purchases an ACME power wheelchair for \$2119.20 and sells it for \$2649. What markup rate does MedSupplies, Inc. use?

Selling price − cost = markup

$2649.00 - 2119.20 = 529.80$

- **First find the markup.**

Percent × base = amount

Markup rate × cost = markup

$n \times 2119.20 = 529.80$

- **Then solve the basic percent equation for *percent.***

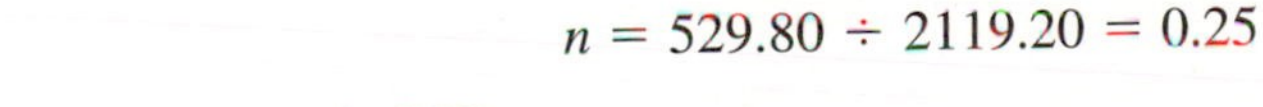

$n = 529.80 \div 2119.20 = 0.25$

The markup rate is 25%.

EXAMPLE • 3

The manager of a medical supply store determines that a markup rate of 36% is necessary to make a profit. What is the markup on an LCD Deluxe Digital Microscope that costs the store $225?

Strategy

To find the markup, solve the basic percent equation for *amount*.

Solution

Percent	×	base	=	amount
Markup rate	×	cost	=	markup
0.36	×	225	=	n
		81	=	n

The markup is $81.

YOU TRY IT • 3

A drugstore manager determines that a markup rate of 20% is necessary to make a profit. What is the markup on a large bottle of joint flex vitamins that costs the drugstore $32?

Your strategy

Your solution

EXAMPLE • 4

A big-box store bought a new brand of blood glucose monitor for $9.50 and used a markup rate of 46%. What is the selling price of this brand of blood glucose monitor?

Strategy

To find the selling price:

- Find the markup by solving the basic percent equation for *amount*.
- Add the markup to the cost.

Solution

Percent	×	base	=	amount
Markup rate	×	cost	=	markup
0.46	×	9.50	=	n
		4.37	=	n

Cost	+	markup	=	selling price
9.50	+	4.37	=	13.87

The selling price is $13.87.

YOU TRY IT • 4

A clothing store bought a leather jacket for $72 and used a markup rate of 55%. What is the selling price?

Your strategy

Your solution

Solutions on p. S14

OBJECTIVE C To find percent decrease

Percent decrease is used to show how much a quantity has decreased from its original value. The statements "The number of family farms decreased by 2% last year" and "There has been a 50% decrease in the cost of a Pentium chip" are examples of percent decrease.

HOW TO • 3 During a 2-year period, the value of U.S. agricultural products exported decreased from approximately \$60.6 billion to \$52.0 billion. Find the percent decrease in the value of U.S. agricultural exports. Round to the nearest tenth of a percent.

Amount of decrease (8.6)
New value (52.0)
Original value (60.6)

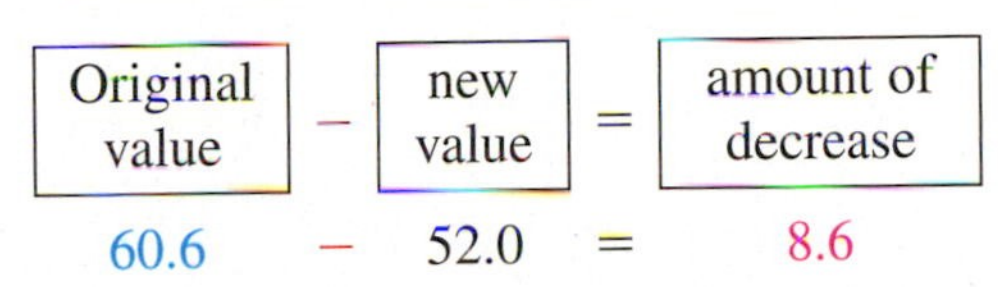

$60.6 - 52.0 = 8.6$

Now solve the basic percent equation for percent.

Percent × base = amount

Percent decrease × original value = amount of decrease

$n \times 60.6 = 8.6$

$n = 8.6 \div 60.6$

$n \approx 0.142$

The value of agricultural exports decreased approximately 14.2%.

Tips for Success

Note in the example below that solving a word problem involves stating a strategy and using the strategy to find a solution. If you have difficulty with a word problem, write down the known information. Be very specific. Write out a phrase or sentence that states what you are trying to find. See *AIM for Success* at the front of the book.

EXAMPLE • 5

During an 8-year period, the population of Baltimore, Maryland, decreased from approximately 736,000 to 646,000. Find the percent decrease in Baltimore's population. Round to the nearest tenth of a percent.

Strategy

To find the percent decrease:

- Find the amount of the decrease.
- Solve the basic percent equation for *percent.*

Solution

Original value − new value = amount of decrease

$736{,}000 - 646{,}000 = 90{,}000$

Percent × base = amount

$n \times 736{,}000 = 90{,}000$

$n = 90{,}000 \div 736{,}000$

$n \approx 0.122$

Baltimore's population decreased approximately 12.2%.

YOU TRY IT • 5

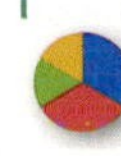

During an 8-year period, the population of Norfolk, Virginia, decreased from approximately 261,000 to 215,000. Find the percent decrease in Norfolk's population. Round to the nearest tenth of a percent.

Your strategy

Your solution

Solution on p. S14

EXAMPLE • 6

The total sales for December for a medical supply store were $96,000. For January, total sales showed an 8% decrease from December's sales. What were the total sales for January?

Strategy

To find the total sales for January:

- Find the amount of decrease by solving the basic percent equation for *amount.*
- Subtract the amount of decrease from the December sales.

Solution

$$\text{Percent} \times \text{base} = \text{amount}$$
$$0.08 \times 96{,}000 = n$$
$$7680 = n$$

The decrease in sales was $7680.

$96{,}000 - 7680 = 88{,}320$

The total sales for January were $88,320.

YOU TRY IT • 6

Maria weighed 220 pounds when she started her diet on January 1. After a month, she had decreased her weight by 5%. What was her weight at the end of the first month?

Your strategy

Your solution

Solution on p. S14

OBJECTIVE D To apply percent decrease to business—discount 

To promote sales, a store may reduce the regular price of some of its products temporarily. The reduced price is called the **sale price.** The difference between the regular price and the sale price is called the **discount.**

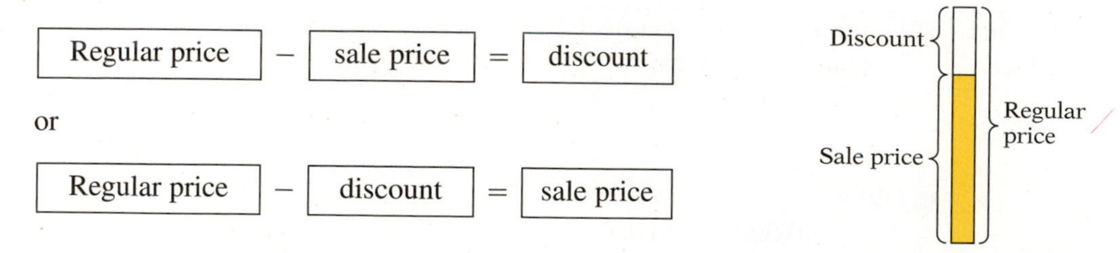

Discount is frequently stated as a percent of a product's regular price. This percent is called the **discount rate.**

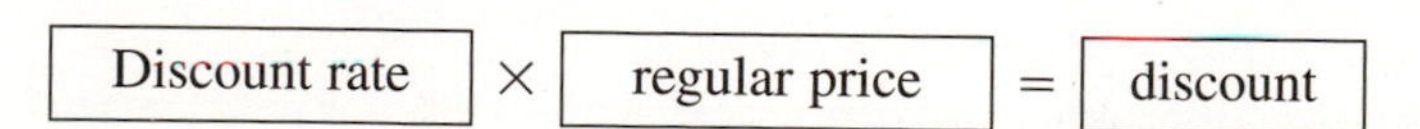

EXAMPLE • 7

A 28-inch black stethoscope that regularly sells for \$72.99 is on sale for \$68.99. Find the discount rate. Round to the nearest tenth of a percent.

Strategy

To find the discount rate:

- Find the discount.
- Solve the basic percent equation for *percent*.

Solution

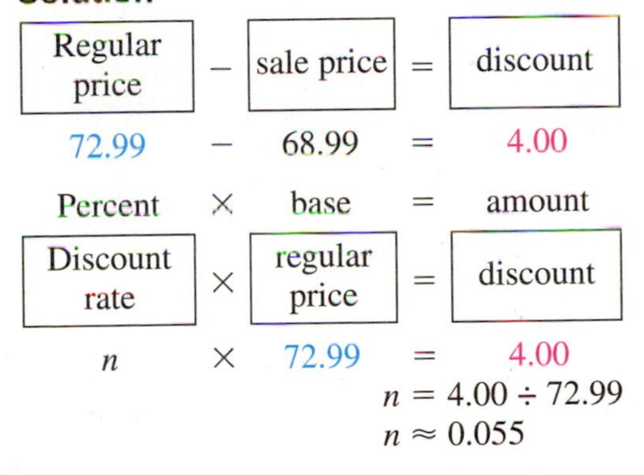

Regular price − sale price = discount

72.99 − 68.99 = 4.00

Percent × base = amount

Discount rate × regular price = discount

$n \times 72.99 = 4.00$

$n = 4.00 \div 72.99$

$n \approx 0.055$

The discount rate is 5.5%.

YOU TRY IT • 7

A case of 24 bottles of instant hand sanitizer with aloe regularly sells for \$44.99. It is currently on sale for \$24.99. Find the discount rate. Round to the nearest tenth of a percent.

Your strategy

Your solution

EXAMPLE • 8

A 4-wheel travel scooter is on sale for 25% off the regular price of \$1525. Find the sale price.

Strategy

To find the sale price:

- Find the discount by solving the basic percent equation for *amount*.
- Subtract to find the sale price.

Solution

Percent × base = amount

Discount rate × regular price = discount

$0.25 \times 1525 = n$

$381.25 = n$

Regular price − discount = sale price

1525 − 381.25 = 1143.75

The sale price is \$1143.75.

YOU TRY IT • 8

A drugstore is selling a manual blood pressure monitor kit for 15% off the regular price of \$17.85. Find the sale price.

Your strategy

Your solution

Solutions on p. S15

6.2 EXERCISES

OBJECTIVE A To find percent increase

Solve. If necessary, round percents to the nearest tenth of a percent.

In the News

A Taste for Bison

In 2005, the meat of 17,674 bison was consumed in the United States. This year, that number will reach 50,000. However, the consumption of bison is still a small fraction of beef consumption. Every day, the meat of 90,000 cattle is consumed in this country.

Source: Time, March 26, 2007

1. **Bison** See the news clipping at the right. Find the percent increase in human consumption of bison from 2005 to the date of this news article.

2. **Fuel Efficiency** An automobile manufacturer increased the average mileage on a car from 17.5 miles per gallon to 18.2 miles per gallon. Find the percent increase in mileage.

3. **Business** In the 1990s, the number of Target stores increased from 420 stores to 914 stores. (*Source:* Target) What was the percent increase in the number of Target stores in the 1990s?

4. **Demography** The graph at the right shows the number of unmarried American couples living together. (*Source:* U.S. Census Bureau) Find the percent increase in the number of unmarried couples living together from 1980 to 2000.

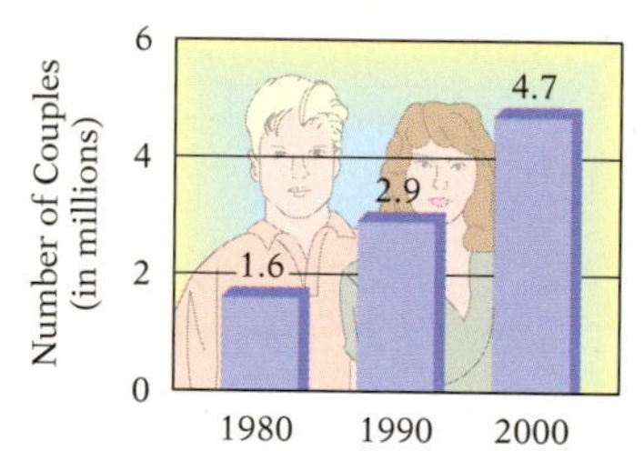

Unmarried U.S. Couples Living Together

5. **Health Insurance** Based on data from the Agency for Healthcare Research and Quality, the average cost of health insurance for a family in the United States in 2003 was $9249. By 2008, the average cost for a family had risen to $12,298. Find the percent increase in health care cost for this 5-year period. Round to the nearest tenth of a percent.

6. **Cancer** In 2009, the American Cancer Society estimated the number of new cancer cases in the United States would be 1,479,350. The 2010 prediction reflected a 3.4% increase in the estimate. Find the estimate of new cancer cases in 2010. Round to the nearest whole number.

© iStockphoto.com/Mariya Bibikova

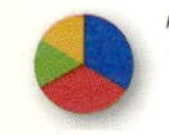

7. **Pets** In a recent year, Americans spent $35.9 billion on their pets. This was up from $17 billion a decade earlier. (*Source: Time,* February 4, 2008) Find the percent increase in the amount Americans spent on their pets during the 10-year period.

8. **Demography** From 1970 to 2000, the average age of American mothers giving birth to their first child rose 16.4%. (*Source:* Centers for Disease Control and Prevention) If the average age in 1970 was 21.4 years, what was the average age in 2000? Round to the nearest tenth.

9. **Compensation** A certified nursing assistant (CNA) earning $12 per hour is given a 10% raise. To find the new wage, we can multiply $12 by 0.10 and add the product to $12. Can the new wage be found by multiplying $12 by 1.10?

OBJECTIVE B To apply percent increase to business—markup

The three important markup equations are:

(1) Selling price − cost = markup
(2) Cost + markup = selling price
(3) Markup rate × cost = markup

For Exercises 10 and 11, list, in the order in which they will be used, the equations needed to solve each problem.

10. A book that cost the seller $17 is sold for $23. Find the markup rate.

11. A DVD that cost the seller $12 has a markup rate of 55%. Find the selling price.

12. A dental chair costs Dental Supply Discount Inc. $1850. Find the markup on the dental chair if the markup rate is 75% of the cost.

13. Adam's Grocery uses a markup rate of 40% on its pre-cut vegetables and fruits in the produce section. What is the markup on a tray of pre-cut fruit that costs $8.75?

14. Computer Inc. uses a markup of $975 on a computer system that costs $3250. What is the markup rate on this system?

15. Saizon Pen & Office Supply uses a markup of $12 on a calculator that costs $20. What markup rate does this amount represent?

16. Giant Photo Service uses a markup rate of 48% on its Model ZA cameras, which cost the shop $162. What is the selling price?

17. A medical device costs $12 to produce. If the retail store uses a 45% markup rate on the device, what is its selling price?

18. Harris Drugstore uses a 52% markup on a cervical collar that costs $9. What is the selling price of the cervical collar?

19. Seniors Medical Supply uses a markup rate of 48%. What is the selling price of a pair of aluminum crutches that costs $15?

OBJECTIVE C To find percent decrease

Solve. If necessary, round to the nearest tenth of a percent.

20. **Law School** Use the news clipping at the right to find the percent decrease in the number of people who took the LSATs in the last three years.

21. **Travel** A new bridge reduced the normal 45-minute travel time between two cities by 18 minutes. What percent decrease does this represent?

In the News

Fewer Students Take LSATs

This year 137,444 people took the Law School Admission Test (LSATs). Three years ago, the LSATs were administered to 148,014 people.

Source: Law School Admission Council

22. **Emergency Room Visits** A local hospital implemented a program to attempt to reduce the number of unnecessary visits to the emergency room. Prior to the implementation, there was an average of 593 visits and after the program began, there was an average of 479 visits per month. What percent decrease does this represent? Round to the nearest tenth.

	1990 Census	*2000 Census*	*2005 Population Estimate*
Chicago	1,783,726	2,896,016	2,842,518
Detroit	1,027,974	951,270	886,671
Phildelphia	1,585,577	1,517,550	1,463,281

Source: Census Bureau

23. **Urban Populations** The table above shows the populations of three cities in the United States.
    a. Find the percent decrease in the population of Detroit from 1990 to 2005.
    b. Find the percent decrease in the population of Philadelphia from 1990 to 2005.
    c. Find the percent decrease in the population of Chicago from 2000 to 2005.

24. **Missing Persons** See the news clipping at the right. Find the percent decrease over the last 10 years in the number of people entered into the National Crime Information Center's Missing Person File.

In the News

Missing-Person Cases Decrease

This year, 834,536 missing-person cases were entered into the National Crime Information Center's Missing Person File. Ten years ago, the number was 969,264.

Source: National Crime Information Center

25. **Depreciation** It is estimated that the value of a new car is reduced 30% after 1 year of ownership. Using this estimate, find how much value a $28,200 new car loses after 1 year.

26. **Weight Loss** Bob's weight is currently 280 pounds. After a visit with his doctor, Bob is advised that based on his height and age, he should lose 15% of his weight over the next 12 months. How much weight does he need to lose?

27. **Finance** Juanita's average monthly expense for gasoline was $176. After joining a car pool, she was able to reduce the expense by 20%.
    a. What was the amount of the decrease?
    b. What is the average monthly gasoline bill now?

28. **Investments** A pharmaceutical company paid a dividend of $1.60 per share. After a reorganization, the company reduced the dividend by 37.5%.
    a. What was the amount of the decrease?
    b. What is the new dividend?

29. **Smoking** A local business initiated a "Stop Smoking Today" incentive plan for its employees. Of the 45 smokers who started the program in January, only 36 were smoking one year later. What was the percent decrease in the smokers who participated in the program?

Kaspri/Shutterstock.com

30. In a math class, the average grade on the second test was 5% lower than the average grade on the first test. What should you multiply the first test average by to find the difference between the average grades on the two tests?

OBJECTIVE D To apply percent decrease to business—discount

The three important discount equations are:

(1) Regular price − sale price = discount
(2) Regular price − discount = sale price
(3) Discount rate × regular price = discount

For Exercises 31 and 32, list, in the order in which they will be used, the equations needed to solve each problem.

31. Shoes that regularly sell for $65 are on sale for 15% off the regular price. Find the sale price.

32. A radio with a regular price of $89 is on sale for $59. Find the discount rate.

33. The Austin College Bookstore is giving a discount of $8 on calculators that normally sell for $24. What is the discount rate?

34. A discount medical supply store is selling a $72 set of scrubs for $24 off the regular price. What is the discount rate?

35. A disc player that regularly sells for $400 is selling for 20% off the regular price. What is the discount?

36. A $45 blood glucose monitoring set is on sale at Discount Drugs for 15% off the regular price. What is the discount?

37. A deluxe folding aluminum walker is regularly $60 but is on sale for $18 off the regular price. What is the discount rate?

38. Quick Service Gas Station has its regularly priced $125 tune-up on sale for 16% off the regular price.
a. What is the discount?
b. What is the sale price?

39. Tomatoes that regularly sell for $1.25 per pound are on sale for 20% off the regular price.
a. What is the discount?
b. What is the sale price?

40. A box of 50 diabetic test strips normally sells for $35 at LowCost Drugs. This week only, the same box is on sale for $21. What is the discount rate?

41. Aromatic Massage Therapy sells its 8-ounce bottle of unscented massage oil for $15.50. During sale week, the price is only $12.40. What is the discount rate?

Applying the Concepts

42. **Business** A promotional sale at a department store offers 25% off the sale price. The sale price itself is 25% off the regular price. Is this the same as a sale that offers 50% off the regular price? If not, which sale gives the better price? Explain your answer.

SECTION

6.3 Interest

OBJECTIVE A To calculate simple interest

When you deposit money in a bank—for example, in a savings account—you are permitting the bank to use your money. The bank may use the deposited money to lend customers the money to buy cars or make renovations on their homes. The bank pays you for the privilege of using your money. The amount paid to you is called **interest.** If you are the one borrowing money from the bank, the amount you pay for the privilege of using that money is also called interest.

Take Note

If you deposit $1000 in a savings account paying 5% interest, the $1000 is the principal and 5% is the interest rate.

The original amount deposited or borrowed is called the **principal.** The amount of interest paid is usually given as a percent of the principal. The percent used to determine the amount of interest is the **interest rate.**

Interest paid on the original principal is called **simple interest.** To calculate simple interest, multiply the principal by the interest rate per period by the number of time periods. In this objective, we are working with annual interest rates, so the time periods are years. The simple interest formula for an annual interest rate is given below.

Simple Interest Formula for Annual Interest Rates

Principal × annual interest rate × time (in years) = interest

Interest rates are generally given as percents. Before performing calculations involving an interest rate, write the interest rate as a decimal.

HOW TO • 1 Calculate the simple interest due on a 2-year loan of $1500 that has an annual interest rate of 7.5%.

Principal	×	annual interest rate	×	time (in years)	=	interest
1500	×	0.075	×	2	=	225

The simple interest due is $225.

When we borrow money, the total amount to be repaid to the lender is the sum of the principal and the interest. This amount is called the **maturity value of a loan.**

Maturity Value Formula for Simple Interest Loans

Principal + interest = maturity value

In the example above, the simple interest due on the loan of $1500 was $225. The maturity value of the loan is therefore $1500 + $225 = $1725.

Take Note

The time of the loan must be in years. Eight months is $\frac{8}{12}$ of a year.

See Example 1. The time of the loan must be in years. 180 days is $\frac{180}{365}$ of a year.

HOW TO • 2 Calculate the maturity value of a simple interest, 8-month loan of \$8000 if the annual interest rate is 9.75%.

First find the interest due on the loan.

Principal	×	annual interest rate	×	time (in years)	=	interest
8000	×	0.0975	×	$\frac{8}{12}$	=	520

Find the maturity value.

Principal	+	interest	=	maturity value
8000	+	520	=	8520

The maturity value of the loan is \$8520.

The monthly payment on a loan can be calculated by dividing the maturity value by the length of the loan in months.

Monthly Payment on a Simple Interest Loan

Maturity value ÷ length of the loan in months = monthly payment

In the example above, the maturity value of the loan is \$8520. To find the monthly payment on the 8-month loan, divide 8520 by 8.

Maturity value	÷	length of the loan in months	=	monthly payment
8520	÷	8	=	1065

The monthly payment on the loan is \$1065.

EXAMPLE • 1

Kamal borrowed \$500 from a savings and loan association for 180 days at an annual interest rate of 7%. What is the simple interest due on the loan?

Strategy

To find the simple interest due, multiply the principal (500) times the annual interest rate (7% = 0.07) times the time in years (180 days = $\frac{180}{365}$ year).

Solution

Principal	×	annual interest rate	×	time (in years)	=	interest
500	×	0.07	×	$\frac{180}{365}$	≈	17.26

The simple interest due is \$17.26.

YOU TRY IT • 1

A company borrowed \$15,000 from a bank for 18 months at an annual interest rate of 8%. What is the simple interest due on the loan?

Your strategy

Your solution

Solution on p. S15

EXAMPLE • 2

Calculate the maturity value of a simple interest, 9-month loan of $4000 if the annual interest rate is 8.75%.

Strategy

To find the maturity value:

- Use the simple interest formula to find the simple interest due.
- Find the maturity value by adding the principal and the interest.

Solution

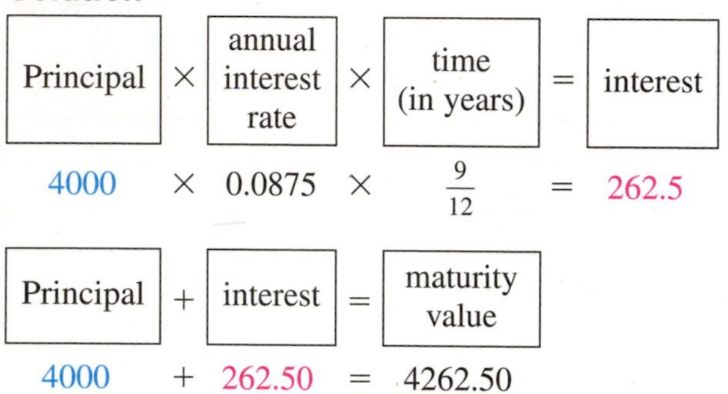

The maturity value is $4262.50.

YOU TRY IT • 2

Calculate the maturity value of a simple interest, 90-day loan of $3800. The annual interest rate is 6%.

Your strategy

Your solution

EXAMPLE • 3

The simple interest due on a 3-month loan of $1400 is $26.25. Find the monthly payment on the loan.

Strategy

To find the monthly payment:

- Find the maturity value by adding the principal and the interest.
- Divide the maturity value by the length of the loan in months (3).

Solution

Principal + interest = maturity value
1400 + 26.25 = 1426.25

Maturity value ÷ length of the loan = payment
1426.25 ÷ 3 ≈ 475.42

The monthly payment is $475.42.

YOU TRY IT • 3

The simple interest due on a 1-year loan of $1900 is $152. Find the monthly payment on the loan.

Your strategy

Your solution

Solutions on p. S15

OBJECTIVE B To calculate finance charges on a credit card bill

When a customer uses a credit card to make a purchase, the customer is actually receiving a loan. Therefore, there is frequently an added cost to the consumer who purchases on credit. This may be in the form of an annual fee and interest charges on purchases. The interest charges on purchases are called **finance charges.**

The finance charge on a credit card bill is calculated using the simple interest formula. In the last objective, the interest rates were annual interest rates. However, credit card companies generally issue *monthly* bills and express interest rates on credit card purchases as *monthly* interest rates. Therefore, when using the simple interest formula to calculate finance charges on credit card purchases, use a monthly interest rate and express the time in months.

Note: In the simple interest formula, the time must be expressed in the same period as the rate. For an *annual* interest rate, the time must be expressed in years. For a *monthly* interest rate, the time must be expressed in months.

EXAMPLE 4

A credit card company charges a customer 1.5% per month on the unpaid balance of charges on the credit card. What is the finance charge in a month in which the customer has an unpaid balance of $254?

Strategy

To find the finance charge, multiply the principal, or unpaid balance (254), times the monthly interest rate (1.5%) times the number of months (1).

Solution

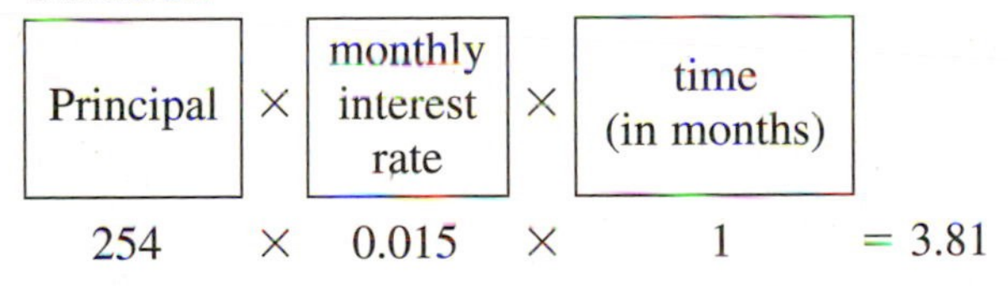

254 × 0.015 × 1 = 3.81

The finance charge is $3.81.

YOU TRY IT 4

The credit card that Francesca uses charges her 1.6% per month on her unpaid balance. Find the finance charge when her unpaid balance for the month is $1250.

Your strategy

Your solution

Solution on p. S15

OBJECTIVE C To calculate compound interest

Usually, the interest paid on money deposited or borrowed is compound interest. **Compound interest** is computed not only on the original principal but also on interest already earned. Here is an illustration.

Suppose $1000 is invested for 3 years at an annual interest rate of 9% compounded annually. Because this is an *annual* interest rate, we will calculate the interest earned each year.

During the first year, the interest earned is calculated as follows:

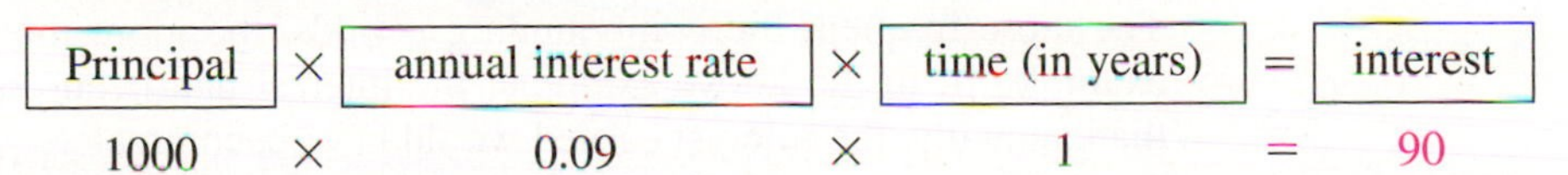

1000 × 0.09 × 1 = 90

At the end of the first year, the total amount in the account is

$$1000 + 90 = 1090$$

During the second year, the interest earned is calculated on the amount in the account at the end of the first year.

Principal	×	annual interest rate	×	time (in years)	=	interest
1090	×	0.09	×	1	=	98.10

Note that the interest earned during the second year ($98.10) is greater than the interest earned during the first year ($90). This is because the interest earned during the first year was added to the original principal, and the interest for the second year was calculated using this sum. If the account earned simple interest, the interest earned would be the same every year ($90).

At the end of the second year, the total amount in the account is the sum of the amount in the account at the end of the first year and the interest earned during the second year.

$$1090 + 98.10 = 1188.10$$

The interest earned during the third year is calculated using the amount in the account at the end of the second year ($1188.10).

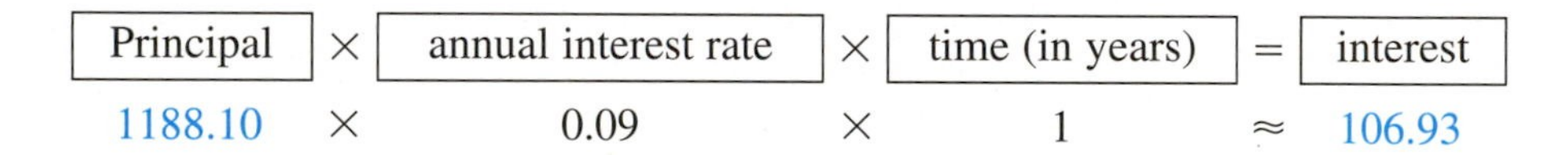

Principal	×	annual interest rate	×	time (in years)	=	interest
1188.10	×	0.09	×	1	≈	106.93

Take Note

The interest earned each year keeps increasing. This is the effect of compound interest.

The amount in the account at the end of the third year is

$$1188.10 + 106.93 = 1295.03$$

To find the interest earned for the three years, subtract the original principal from the new principal.

New principal	−	original principal	=	interest earned
1295.03	−	1000	=	295.03

Note that the compound interest earned is $295.03. The simple interest earned on the investment would have been only $1000 × 0.09 × 3 = $270.

In this example, the interest was compounded annually. However, interest can be compounded

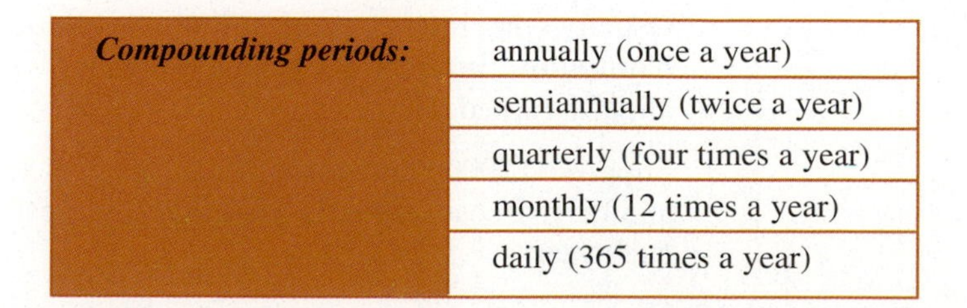

Compounding periods:	annually (once a year)
	semiannually (twice a year)
	quarterly (four times a year)
	monthly (12 times a year)
	daily (365 times a year)

The more frequent the compounding periods, the more interest the account earns. For example, if, in the above example, the interest had been compounded quarterly rather than annually, the interest earned would have been greater.

Calculating compound interest can be very tedious, so there are tables that can be used to simplify these calculations. A portion of a Compound Interest Table is given in the Appendix.

HOW TO • 3 What is the value after 5 years of $1000 invested at 7% annual interest, compounded quarterly?

To find the interest earned, multiply the original principal (1000) by the factor found in the Compound Interest Table. To find the factor, first find the table headed "Compounded Quarterly" in the Compound Interest Table in the Appendix. Then look at the number where the 7% column and the 5-year row meet.

Compounded Quarterly							
	4%	*5%*	*6%*	*7%*	*8%*	*9%*	*10%*
1 year	1.04060	1.05094	1.06136	1.07186	1.08243	1.09308	1.10381
5 years	1.22019	1.28204	1.34686	**1.41478**	1.48595	1.56051	1.63862
10 years	1.48886	1.64362	1.81402	2.00160	2.20804	2.43519	2.68506
15 years	1.81670	2.10718	2.44322	2.83182	3.28103	3.80013	4.39979
20 years	2.21672	2.70148	3.29066	4.00639	4.87544	5.93015	7.20957

The factor is 1.41478.

$$1000 \times 1.41478 = 1414.78$$

The value of the investment after 5 years is $1414.78.

EXAMPLE • 5

An investment of $650 pays 8% annual interest, compounded semiannually. What is the interest earned in 5 years?

Strategy

To find the interest earned:

- Find the new principal by multiplying the original principal (650) by the factor found in the Compound Interest Table (1.48024).
- Subtract the original principal from the new principal.

Solution

$650 \times 1.48024 \approx 962.16$

The new principal is $962.16.

$962.16 - 650 = 312.16$

The interest earned is $312.16.

YOU TRY IT • 5

An investment of $1000 pays 6% annual interest, compounded quarterly. What is the interest earned in 20 years?

Your strategy

Your solution

Solution on pp. S15–S16

6.3 EXERCISES

OBJECTIVE A To calculate simple interest

1. A 2-year student loan of $10,000 is made at an annual simple interest rate of 4.25%. The simple interest on the loan is $850. Identify **a.** the principal, **b.** the interest, **c.** the interest rate, and **d.** the time period of the loan.

2. A physician obtained a 9-month loan for $80,000 to buy office equipment at an annual simple interest rate of 9.75%. The simple interest on the loan is $5850. Identify **a.** the principal, **b.** the interest, **c.** the interest rate, and **d.** the time period of the loan.

3. Find the simple interest Jacob Zucker owes on a 2-year student loan of $8000 at an annual interest rate of 6%.

4. Find the simple interest Kara Tanamachi owes on a $1\frac{1}{2}$-year loan of $1500 at an annual interest rate of 7.5%.

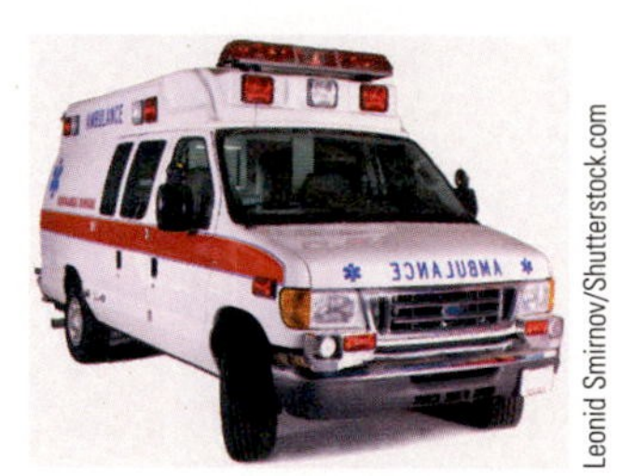
Leonid Smirnov/Shutterstock.com

5. To finance the purchase of two new ambulances, a local hospital borrowed $100,000 for 9 months at an annual interest rate of 4.5%. What is the simple interest due on the loan?

6. A builder obtained a preconstruction loan of $50,000 for 8 months at an annual interest rate of 9.5% to build a new medical facility near the hospital. What is the simple interest due on the loan?

7. A bank lent Gloria Masters $20,000 at an annual interest rate of 8.8%. The period of the loan was 9 months. Find the simple interest due on the loan.

8. Eugene Madison obtained an 8-month loan of $4500 at an annual interest rate of 6.2%. Find the simple interest Eugene owes on the loan.

9. Jorge Elizondo took out a 75-day loan of $7500 at an annual interest rate of 5.5%. Find the simple interest due on the loan.

10. Kristi Yang borrowed $15,000. The term of the loan was 90 days, and the annual simple interest rate was 7.4%. Find the simple interest due on the loan.

11. The simple interest due on a 4-month loan of $4800 is $320. What is the maturity value of the loan?

12. The simple interest due on a 60-day loan of $6500 is $80.14. Find the maturity value of the loan.

13. William Carey borrowed $12,500 for 8 months at an annual simple interest rate of 4.5%. Find the total amount due on the loan.

14. You arrange for a 9-month bank loan of $9000 at an annual simple interest rate of 8.5%. Find the total amount you must repay to the bank.

15. Capital City Bank approves a home-improvement loan application for $14,000 at an annual simple interest rate of 5.25% for 270 days. What is the maturity value of the loan?

16. A credit union lends a member $5000 for college tuition. The loan is made for 18 months at an annual simple interest rate of 6.9%. What is the maturity value of this loan?

17. Grace Hospital purchased some imaging equipment for $225,000 and financed the full amount at 8% annual simple interest for 4 years. The simple interest on the loan is $72,000. Find the monthly payment.

18. For the purchase of an entertainment center, a $1900 loan is obtained for 2 years at an annual simple interest rate of 9.4%. The simple interest due on the loan is $357.20. What is the monthly payment on the loan?

19. To attract new customers, Heller Ford is offering car loans at an annual simple interest rate of 4.5%.

a. Find the interest charged to a customer who finances a car loan of $12,000 for 2 years.

b. Find the monthly payment.

20. Bertie EMS purchased an ambulance for $57,000 and financed the full amount for 5 years at an annual simple interest rate of 9%.

a. Find the interest due on the loan.

b. Find the monthly payment.

Knut Platon/STONE/Getty Images

21. Dennis Pappas decided to build onto his present home instead of buying a new, larger house. He borrowed $142,000 for $5\frac{1}{2}$ years at an annual simple interest rate of 7.5%. Find the monthly payment.

22. Rosalinda Johnson took out a 6-month, $12,000 loan. The annual simple interest rate on the loan was 8.5%. Find the monthly payment.

23. Student A and Student B borrow the same amount of money at the same annual interest rate. Student A has a 2-year loan and Student B has a 1-year loan. In each case, state whether the first quantity is *less than, equal to,* or *greater than* the second quantity.

a. Student A's principal; Student B's principal

b. Student A's maturity value; Student B's maturity value

c. Student A's monthly payment; Student B's monthly payment

OBJECTIVE B **To calculate finance charges on a credit card bill**

24. A credit card company charges a customer 1.25% per month on the unpaid balance of charges on the credit card. What is the finance charge in a month in which the customer has an unpaid balance of $118.72?

25. The credit card that Dee Brown uses charges her 1.75% per month on her unpaid balance. Find the finance charge when her unpaid balance for the month is $391.64.

26. What is the finance charge on an unpaid balance of $12,368.92 on a credit card that charges 1.5% per month on any unpaid balance?

27. Suppose you have an unpaid balance of $995.04 on a credit card that charges 1.2% per month on any unpaid balance. What finance charge do you owe the company?

28. A credit card customer has an unpaid balance of $1438.20. What is the difference between monthly finance charges of 1.15% per month on the unpaid balance and monthly finance charges of 1.85% per month?

29. One credit card company charges 1.25% per month on any unpaid balance, and a second company charges 1.75%. What is the difference between the finance charges that these two companies assess on an unpaid balance of $687.45?

Your credit card company requires a minimum monthly payment of $10. You plan to pay off the balance on your credit card by paying the minimum amount each month and making no further purchases using this credit card. For Exercises 30 and 31, state whether the finance charge for the second month will be *less than, equal to,* or *greater than* the finance charge for the first month, and state whether you will eventually be able to pay off the balance.

30. The finance charge for the first month was less than $10.

31. The finance charge for the first month was exactly $10.

OBJECTIVE C **To calculate compound interest**

32. North Island Federal Credit Union pays 4% annual interest, compounded daily, on time savings deposits. Find the value after 1 year of $750 deposited in this account.

33. Tanya invested $2500 in a tax-sheltered annuity that pays 8% annual interest, compounded daily. Find the value of her investment after 20 years.

34. Sal Travato invested $3000 in a corporate retirement account that pays 6% annual interest, compounded semiannually. Find the value of his investment after 15 years.

35. To replace equipment, a pediatrician invested \$20,000 in an account that pays 7% annual interest, compounded monthly. What is the value of the investment after 5 years?

36. Green River Lodge invests \$75,000 in a trust account that pays 8% interest, compounded quarterly.
a. What will the value of the investment be in 5 years?
b. How much interest will be earned in the 5 years?

37. To save for retirement, a couple deposited \$3000 in an account that pays 7% annual interest, compounded daily.
a. What will the value of the investment be in 10 years?
b. How much interest will be earned in the 10 years?

38. To save for a child's education, the Petersens deposited \$2500 into an account that pays 6% annual interest, compounded daily. Find the amount of interest earned on this account over a 20-year period.

39. How much interest is earned in 2 years on \$4000 deposited in an account that pays 6% interest, compounded quarterly?

40. The compound interest factor for a 5-year investment at an annual interest rate of 6%, compounded semiannually, is 1.34392. What does the expression $3500 - (3500 \times 1.34392)$ represent?

Applying the Concepts

41. **Banking** At 4 P.M. on July 31, you open a savings account that pays 5% annual interest and you deposit \$500 in the account. Your deposit is credited as of August 1. At the beginning of September, you receive a statement from the bank that shows that during the month of August, you received \$2.12 in interest. The interest has been added to your account, bringing the total on deposit to \$502.12. At the beginning of October, you receive a statement from the bank that shows that during the month of September, you received \$2.06 in interest on the \$502.12 on deposit. Explain why you received less interest during the second month when there was more money on deposit.

42. **Banking** Suppose you have a savings account that earns interest at the rate of 6% per year, compounded monthly. On January 1, you open this account with a deposit of \$100.
a. On February 1, you deposit an additional \$100 into the account. What is the value of the account after the deposit?
b. On March 1, you deposit an additional \$100 into the account. What is the value of the account after the deposit?
Note: This type of savings plan, wherein equal amounts (\$100) are saved at equal time intervals (every month), is called an annuity.

SECTION

6.4 Real Estate Expenses

OBJECTIVE A To calculate the initial expenses of buying a home

One of the largest investments most people ever make is the purchase of a home. The major initial expense in the purchase is the **down payment,** which is normally a percent of the purchase price. This percent varies among banks, but it usually ranges from 5% to 25%.

The **mortgage** is the amount that is borrowed to buy real estate. The mortgage amount is the difference between the purchase price and the down payment.

HOW TO • 1 A home is purchased for $140,000, and a down payment of $21,000 is made. Find the mortgage.

Purchase price	−	down payment	=	mortgage
140,000	−	21,000	=	119,000

The mortgage is $119,000.

Take Note

Because *points* means percent, a loan origination fee of $2\frac{1}{2}$ points = $2\frac{1}{2}\%$ = 2.5% = 0.025.

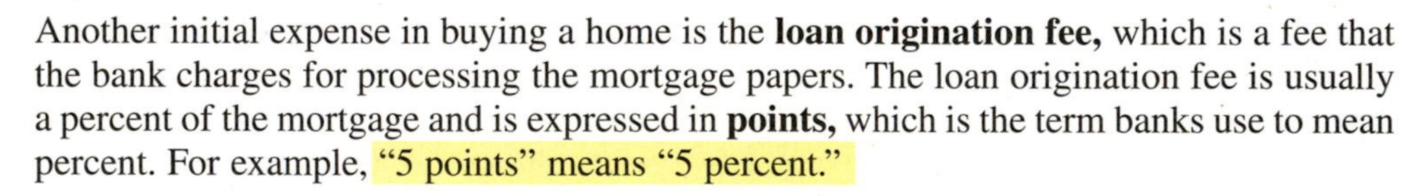

Another initial expense in buying a home is the **loan origination fee,** which is a fee that the bank charges for processing the mortgage papers. The loan origination fee is usually a percent of the mortgage and is expressed in **points,** which is the term banks use to mean percent. For example, "5 points" means "5 percent."

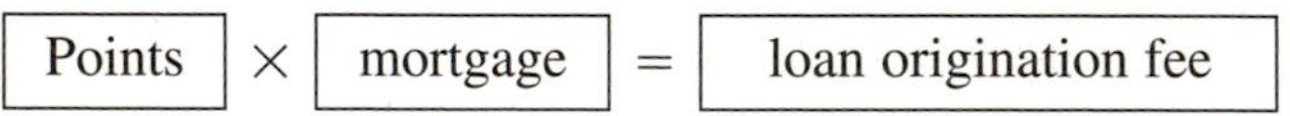

Points × mortgage = loan origination fee

EXAMPLE • 1

A house is purchased for $250,000, and a down payment, which is 20% of the purchase price, is made. Find the mortgage.

Strategy

To find the mortgage:

- Find the down payment by solving the basic percent equation for *amount.*
- Subtract the down payment from the purchase price.

Solution

Percent	×	base	=	amount
Percent	×	purchase price	=	down payment
0.20	×	250,000	=	n
		50,000	=	n

Purchase price	−	down payment	=	mortgage
250,000	−	50,000	=	200,000

The mortgage is $200,000.

YOU TRY IT • 1

An office building is purchased for $1,500,000, and a down payment, which is 25% of the purchase price, is made. Find the mortgage.

Your strategy

Your solution

Solution on p. S16

EXAMPLE 2

A home is purchased with a mortgage of \$165,000. The buyer pays a loan origination fee of $3\frac{1}{2}$ points. How much is the loan origination fee?

Strategy
To find the loan origination fee, solve the basic percent equation for *amount*.

Solution

Percent	×	base	=	amount
Points	×	mortgage	=	fee
0.035	×	165,000	=	n

$$5775 = n$$

The loan origination fee is \$5775.

YOU TRY IT 2

The mortgage on a real estate investment is \$180,000. The buyer paid a loan origination fee of $4\frac{1}{2}$ points. How much was the loan origination fee?

Your strategy

Your solution

Solution on p. S16

OBJECTIVE B To calculate the ongoing expenses of owning a home

Point of Interest

The number-one response of adults when asked what they would spend money on first if they suddenly became wealthy (for example, by winning the lottery) was a house; 31% gave this response. (*Source:* Yankelovich Partners for Lutheran Brotherhood)

Besides the initial expenses of buying a house, there are continuing monthly expenses involved in owning a home. The **monthly mortgage payment** (one of 12 payments due each year to the lender of money to buy real estate), utilities, insurance, and **property tax** (a tax based on the value of real estate) are some of these ongoing expenses. Of these expenses, the largest one is normally the monthly mortgage payment.

For a **fixed-rate mortgage,** the monthly mortgage payment remains the same throughout the life of the loan. The calculation of the monthly mortgage payment is based on the amount of the loan, the interest rate on the loan, and the number of years required to pay back the loan. Calculating the monthly mortgage payment is fairly difficult, so tables such as the one in the Appendix are used to simplify these calculations.

Integrating Technology

In general, when a problem requests a monetary payment, the answer is rounded to the nearest cent. For the example at the right, enter

160000 **x** 0.0080462 **=**

The display reads 1287.392. Round this number to the nearest hundredth: 1287.39. The answer is \$1287.39.

HOW TO 2 Find the monthly mortgage payment on a 30-year, \$160,000 mortgage at an interest rate of 9%. Use the Monthly Payment Table in the Appendix.

$$160{,}000 \times \underset{\substack{\uparrow \\ \text{From the table}}}{\underline{0.0080462}} \approx 1287.39$$

The monthly mortgage payment is \$1287.39.

The monthly mortgage payment includes the payment of both principal and interest on the mortgage. The interest charged during any one month is charged on the unpaid balance of the loan. Therefore, during the early years of the mortgage, when the unpaid balance is high, most of the monthly mortgage payment is interest charged on the loan. During the last few years of a mortgage, when the unpaid balance is low, most of the monthly mortgage payment goes toward paying off the loan.

Point of Interest

Home buyers rated the following characteristics "extremely important" in their purchase decision.

Natural, open space: 77%
Walking and biking paths: 74%
Gardens with native plants: 56%
Clustered retail stores: 55%
Wilderness area: 52%
Outdoor pool: 52%
Community recreation center: 52%
Interesting little parks: 50%

(*Sources:* American Lives, Inc; Intercommunications, Inc.)

HOW TO • 3 Find the interest paid on a mortgage during a month in which the monthly mortgage payment is $886.26 and $358.08 of that amount goes toward paying off the principal.

Monthly mortgage payment	−	principal	=	interest
886.26	−	358.08	=	528.18

The interest paid on the mortgage is $528.18.

Property tax is another ongoing expense of owning a house. Property tax is normally an annual expense that may be paid on a monthly basis. The monthly property tax, which is determined by dividing the annual property tax by 12, is usually added to the monthly mortgage payment.

HOW TO • 4 A homeowner must pay $3120 in property tax annually. Find the property tax that must be added each month to the homeowner's monthly mortgage payment.

$3120 \div 12 = 260$

Each month, $260 must be added to the monthly mortgage payment for property tax.

EXAMPLE • 3

Serge purchased some land for $120,000 and made a down payment of $25,000. The savings and loan association charges an annual interest rate of 8% on Serge's 25-year mortgage. Find the monthly mortgage payment.

Strategy

To find the monthly mortgage payment:

- Subtract the down payment from the purchase price to find the mortgage.
- Multiply the mortgage by the factor found in the Monthly Payment Table in the Appendix.

Solution

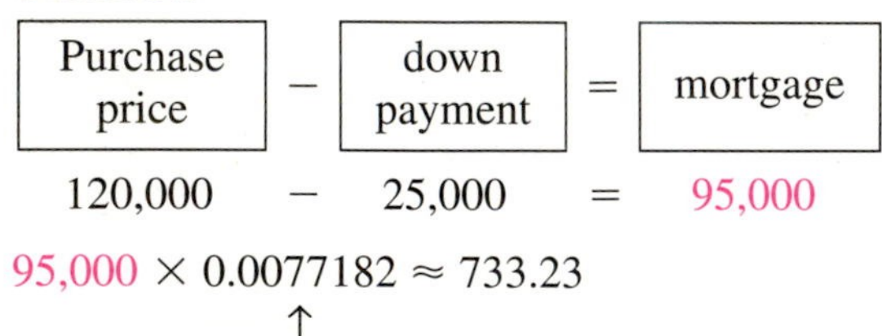

Purchase price	−	down payment	=	mortgage
120,000	−	25,000	=	95,000

$95{,}000 \times 0.0077182 \approx 733.23$

↑ From the table

The monthly mortgage payment is $733.23.

YOU TRY IT • 3

A new condominium project is selling townhouses for $175,000. A down payment of $17,500 is required, and a 20-year mortgage at an annual interest rate of 9% is available. Find the monthly mortgage payment.

Your strategy

Your solution

Solution on p. S16

EXAMPLE • 4

A home has a mortgage of $134,000 for 25 years at an annual interest rate of 7%. During a month in which $375.88 of the monthly mortgage payment is principal, how much of the payment is interest?

Strategy

To find the interest:

- Multiply the mortgage by the factor found in the Monthly Payment Table in the Appendix to find the monthly mortgage payment.
- Subtract the principal from the monthly mortgage payment.

Solution

$134{,}000 \times 0.0070678 \approx 947.09$

↑ From the table (0.0070678) ↑ Monthly mortgage payment (947.09)

Monthly mortgage payment	−	principal	=	interest
947.09	−	375.88	=	571.21

$571.21 of the payment is interest on the mortgage.

YOU TRY IT • 4

An office building has a mortgage of $625,000 for 25 years at an annual interest rate of 7%. During a month in which $2516.08 of the monthly mortgage payment is principal, how much of the payment is interest?

Your strategy

Your solution

EXAMPLE • 5

The monthly mortgage payment for a home is $998.75. The annual property tax is $4020. Find the total monthly payment for the mortgage and property tax.

Strategy

To find the monthly payment:

- Divide the annual property tax by 12 to find the monthly property tax.
- Add the monthly property tax to the monthly mortgage payment.

Solution

$4020 \div 12 = 335$ • **Monthly property tax**

$998.75 + 335 = 1333.75$

The total monthly payment is $1333.75.

YOU TRY IT • 5

The monthly mortgage payment for a home is $815.20. The annual property tax is $3000. Find the total monthly payment for the mortgage and property tax.

Your strategy

Your solution

Solutions on p. S16

6.4 EXERCISES

OBJECTIVE A To calculate the initial expenses of buying a home

Chad McDermott/Shutterstock.com

1. A condominium at Twin Lakes Retirement Community was purchased for $197,000, and a down payment of $24,550 was made. Find the mortgage.

2. An insurance business was purchased for $173,000, and a down payment of $34,600 was made. Find the mortgage.

3. Brian Stedman made a down payment of 25% of the $850,000 purchase price of an apartment building. How much was the down payment?

4. A medical equipment store was purchased for $625,000, and a down payment that was 25% of the purchase price was made. Find the down payment.

5. A loan of $150,000 is obtained to purchase a home. The loan origination fee is $2\frac{1}{2}$ points. Find the amount of the loan origination fee.

6. Security Savings & Loan requires a borrower to pay $3\frac{1}{2}$ points for a loan. Find the amount of the loan origination fee for a loan of $90,000.

7. Baja Construction Inc. is selling homes for $350,000. A down payment of 10% is required. Find the mortgage.

8. A nurse purchased a new home for $240,000. The bank requires a down payment of 15% of the purchase price. Find the mortgage.

9. Vivian Tom purchased a home for $210,000. Find the mortgage if the down payment Vivian made is 10% of the purchase price.

10. A mortgage lender requires a down payment of 5% of the $180,000 purchase price of a condominium. How much is the mortgage?

11. A home is purchased for $435,000. The mortgage lender requires a 10% down payment. Which expression below represents the mortgage?
    (i) $0.10 \times 435{,}000$
    (ii) $0.10 \times 435{,}000 - 435{,}000$
    (iii) $435{,}000 - 0.10 \times 435{,}000$
    (iv) $435{,}000 + 0.10 \times 435{,}000$

OBJECTIVE B To calculate the ongoing expenses of owning a home

For Exercises 12 to 23, solve. Use the Monthly Payment Table in the Appendix. Round to the nearest cent.

12. An investor obtained a loan of $850,000 to buy a medical spa. The monthly mortgage payment was based on 25 years at 8%. Find the monthly mortgage payment.

13. A physical therapist obtained a 20-year mortgage of $90,000 to expand the business. The credit union charges an annual interest rate of 6%. Find the monthly mortgage payment.

14. A couple interested in buying a home determines that they can afford a monthly mortgage payment of $800. Can they afford to buy a home with a 30-year, $110,000 mortgage at 8% interest?

15. A podiatrist is considering purchasing a new office building with a 15-year, $400,000 mortgage at 6% interest. The podiatrist can afford a monthly mortgage payment of $3500. Can the podiatrist afford the monthly mortgage payment on the new office building?

16. The county tax assessor has determined that the annual property tax on a $325,000 house is $3032.40. Find the monthly property tax.

17. The annual property tax on a $155,000 home is $1992. Find the monthly property tax.

18. Midtown Medical Supply has a warehouse with a 25-year mortgage of $200,000 at an annual interest rate of 9%.
  a. Find the monthly mortgage payment.
  b. During a month in which $941.72 of the monthly mortgage payment is principal, how much of the payment is interest?

19. A vacation home has a mortgage of $135,000 for 30 years at an annual interest rate of 7%.
  a. Find the monthly mortgage payment.
  b. During a month in which $392.47 of the monthly mortgage payment is principal, how much of the payment is interest?

20. The annual mortgage payment on a duplex is $20,844.40. The owner must pay an annual property tax of $1944. Find the total monthly payment for the mortgage and property tax.

21. The monthly mortgage payment on a home is $716.40, and the homeowner pays an annual property tax of $1512. Find the total monthly payment for the mortgage and property tax.

22. Maria Hernandez purchased a home for $210,000 and made a down payment of $15,000. The balance was financed for 15 years at an annual interest rate of 6%. Find the monthly mortgage payment.

23. A customer of a savings and loan purchased a $385,000 home and made a down payment of $40,000. The savings and loan charges its customers an annual interest rate of 7% for 30 years for a home mortgage. Find the monthly mortgage payment.

24. The monthly mortgage payment for a home is $623.57. The annual property tax is $1400. Which expression below represents the total monthly payment for the mortgage and property tax? Which expression represents the total amount of money the owner will spend on the mortgage and property tax in one year?
  (i) $623.57 + 1400$ (ii) $12 \times 623.57 + 1400$
  (iii) $\frac{623.57 + 1400}{12}$ (iv) $623.57 + \frac{1400}{12}$

Applying the Concepts

25. **Mortgages** A couple considering a mortgage of $100,000 have a choice of loans. One loan is an 8% loan for 20 years, and the other loan is at 8% for 30 years. Find the amount of interest that the couple can save by choosing the 20-year loan.

SECTION 6.5 Car Expenses

OBJECTIVE A To calculate the initial expenses of buying a car

The initial expenses in the purchase of a car usually include the down payment, the **license fees** (fees charged for authorization to operate a vehicle), and the **sales tax** (a tax levied by a state or municipality on purchases). The down payment may be very small or as much as 25% or 30% of the purchase price of the car, depending on the lending institution. License fees and sales tax are regulated by each state, so these expenses vary from state to state.

EXAMPLE 1

A car is purchased for \$38,500, and the lender requires a down payment of 15% of the purchase price. Find the amount financed.

Strategy
To find the amount financed:

- Find the down payment by solving the basic percent equation for *amount*.
- Subtract the down payment from the purchase price.

Solution

Percent	×	base	=	amount
Percent	×	purchase price	=	down payment
0.15	×	38,500	=	n

$5775 = n$

$38{,}500 - 5775 = 32{,}725$

The amount financed is \$32,725.

YOU TRY IT 1

A down payment of 20% of the \$19,200 purchase price of a new car is made. Find the amount financed.

Your strategy

Your solution

EXAMPLE 2

A nurse's aide purchases a used car for \$16,500 and pays a sales tax that is 5% of the purchase price. How much is the sales tax?

Strategy
To find the sales tax, solve the basic percent equation for *amount*.

Solution

Percent	×	base	=	amount
Percent	×	purchase price	=	sales tax
0.05	×	16,500	=	n

$825 = n$

The sales tax is \$825.

YOU TRY IT 2

A car is purchased for \$27,350. The car license fee is 1.5% of the purchase price. How much is the license fee?

Your strategy

Your solution

Solutions on pp. S16–S17

OBJECTIVE B To calculate the ongoing expenses of owning a car

 **Take Note**

The same formula that is used to calculate a monthly mortgage payment is used to calculate a monthly car payment.

Besides the initial expenses of buying a car, there are continuing expenses involved in owning a car. These ongoing expenses include car insurance, gas and oil, general maintenance, and the monthly car payment. The monthly car payment is calculated in the same manner as the monthly mortgage payment on a home loan. A monthly payment table, such as the one in the Appendix, is used to simplify the calculation of monthly car payments.

EXAMPLE • 3

At a cost of \$0.38 per mile, how much does it cost to operate a car during a year in which the car is driven 15,000 miles?

Strategy

To find the cost, multiply the cost per mile (0.38) by the number of miles driven (15,000).

Solution

$15{,}000 \times 0.38 = 5700$

The cost is \$5700.

YOU TRY IT • 3

At a cost of \$0.41 per mile, how much does it cost to operate a car during a year in which the car is driven 23,000 miles?

Your strategy

Your solution

EXAMPLE • 4

During one month, your total gasoline bill was \$252 and the car was driven 1200 miles. What was the cost per mile for gasoline?

Strategy

To find the cost per mile, divide the cost for gasoline (252) by the number of miles driven (1200).

Solution

$252 \div 1200 = 0.21$

The cost per mile was \$0.21.

YOU TRY IT • 4

In a year in which your total car insurance bill was \$360 and the car was driven 15,000 miles, what was the cost per mile for car insurance?

Your strategy

Your solution

EXAMPLE • 5

A car is purchased for \$18,500 with a down payment of \$3700. The balance is financed for 3 years at an annual interest rate of 6%. Find the monthly car payment.

Strategy

To find the monthly payment:

- Subtract the down payment from the purchase price to find the amount financed.
- Multiply the amount financed by the factor found in the Monthly Payment Table in the Appendix.

Solution

$18{,}500 - 3700 = 14{,}800$

$14{,}800 \times 0.0304219 \approx 450.24$

The monthly payment is \$450.24.

YOU TRY IT • 5

A truck is purchased for \$25,900 with a down payment of \$6475. The balance is financed for 4 years at an annual interest rate of 8%. Find the monthly car payment.

Your strategy

Your solution

Solutions on p. S17

6.5 EXERCISES

OBJECTIVE A To calculate the initial expenses of buying a car

1. Amanda has saved $780 to make a down payment on a used minivan that costs $7100. The car dealer requires a down payment of 12% of the purchase price. Has she saved enough money to make the down payment?

2. A sedan was purchased for $23,500. A down payment of 15% of the purchase price was required. How much was the down payment?

3. A medical testing facility purchased a van to pick up and deliver blood samples to the lab. The purchase price of the van was $26,500, and a 4.5% sales tax was paid. How much was the sales tax?

4. A small community purchased a used ambulance for $28,500. The sales tax on the ambulance was 4%. Find the sales tax.

5. A license fee of 2% of the purchase price is paid on a pickup truck costing $32,500. Find the license fee for the truck.

6. Your state charges a license fee of 1.5% on the purchase price of a car. How much is the license fee for a car that costs $16,998?

7. A physician's assistant bought a $32,000 new SUV. A state license fee of $275 and a sales tax of 3.5% of the purchase price are required.
   a. Find the sales tax.
   b. Find the total cost of the sales tax and the license fee.

8. A physical therapist bought a used car for $9375 and made a down payment of $1875. The sales tax is 5% of the purchase price.
   a. Find the sales tax.
   b. Find the total cost of the sales tax and the down payment.

9. Martin bought a motorcycle for $16,200 and made a down payment of 25% of the purchase price. Find the amount financed.

10. Olivia bought a minivan for $24,900 and made a down payment of 15% of the purchase price. Find the amount financed.

11. An author bought a sports car for $45,000 and made a down payment of 20% of the purchase price. Find the amount financed.

12. Tania purchased a used car for $13,500 and made a down payment of 25% of the cost. Find the amount financed.

13. The purchase price of a car is $25,700. The car dealer requires a down payment of 15% of the purchase price. There is a license fee of 2.5% of the purchase price and sales tax of 6% of the purchase price. What does the following expression represent?
    $25{,}700 + 0.025 \times 25{,}700 + 0.06 \times 25{,}700$

OBJECTIVE B To calculate the ongoing expenses of owning a car

14. A driver had $1100 in car expenses and drove his car 8500 miles. Would you use *multiplication* or *division* to find the cost per mile to operate the car?

15. A car costs $0.36 per mile to operate. Would you use *multiplication* or *division* to find the cost of driving the car 23,000 miles?

For Exercises 16 to 25, solve. Use the Monthly Payment Table in the Appendix. Round to the nearest cent.

16. A radiologist financed $24,000 for the purchase of a truck through a credit union at 5% interest for 4 years. Find the monthly truck payment.

17. A car loan of $18,000 is financed for 3 years at an annual interest rate of 4%. Find the monthly car payment.

18. An estimate of the cost of owning a compact car is $0.38 per mile. Using this estimate, find how much it costs to operate a car during a year in which the car is driven 16,000 miles.

19. An estimate of the cost of care and maintenance of automobile tires is $0.018 per mile. Using this estimate, find how much it costs for care and maintenance of tires during a year in which the car is driven 14,000 miles.

20. A family spent $2600 on gas, oil, and car insurance during a period in which the car was driven 14,000 miles. Find the cost per mile for gas, oil, and car insurance.

21. Last year you spent $2400 for gasoline for your car. The car was driven 15,000 miles. What was your cost per mile for gasoline?

22. The city of Colton purchased a fire truck for $164,000 and made a down payment of $10,800. The balance is financed for 5 years at an annual interest rate of 6%.
a. Find the amount financed.
b. Find the monthly truck payment.

23. A used car is purchased for $14,999, and a down payment of $2999 is made. The balance is financed for 3 years at an annual interest rate of 5%.
a. Find the amount financed.
b. Find the monthly car payment.

24. A lab technician purchased a new car costing $27,500 and made a down payment of $5500. The balance is financed for 3 years at an annual interest rate of 4%. Find the monthly car payment.

25. A camper is purchased for $39,500, and a down payment of $5000 is made. The balance is financed for 4 years at an annual interest rate of 6%. Find the monthly payment.

Ulrich Mueller/Flickr/Getty Images

Applying the Concepts

26. Car Loans One bank offers a 4-year car loan at an annual interest rate of 7% plus a loan application fee of $45. A second bank offers 4-year car loans at an annual interest rate of 8% but charges no loan application fee. If you need to borrow $5800 to purchase a car, which of the two bank loans has the lesser loan costs? Assume you keep the car for 4 years.

27. Car Loans How much interest is paid on a 5-year car loan of $19,000 if the interest rate is 9%? Round to the nearest dollar.

SECTION

6.6 Wages

OBJECTIVE A To calculate commissions, total hourly wages, and salaries

Commissions, hourly wage, and salary are three ways to receive payment for doing work.

Commissions are usually paid to salespersons and are calculated as a percent of total sales.

HOW TO • 1 As a real estate broker, Emma Smith receives a commission of 4.5% of the selling price of a house. Find the commission she earned for selling a home for $275,000.

To find the commission Emma earned, solve the basic percent equation for *amount*.

Percent	×	base	=	amount
Commission rate	×	total sales	=	commission
0.045	×	275,000	=	12,375

The commission is $12,375.

An employee who receives an **hourly wage** is paid a certain amount for each hour worked.

HOW TO • 2 A licensed physical therapy assistant receives an hourly wage of $25.50. Find the therapy assistant's total wages for working 30 hours.

To find the therapy assistant's total wages, multiply the hourly wage by the number of hours worked.

Hourly wage	×	number of hours worked	=	total wages
25.50	×	30	=	765

The physical therapy assistant's total wages for working 30 hours are $765.

An employee who is paid a **salary** receives payment based on a weekly, biweekly (every other week), monthly, or annual time schedule. Unlike the employee who receives an hourly wage, the salaried worker does not receive additional pay for working more than the regularly scheduled workday.

HOW TO • 3 Ravi Basar is a physician's assistant who receives a weekly salary of $1360. Find his salary for 1 month (4 weeks).

To find Ravi's salary for 1 month, multiply the salary per pay period by the number of pay periods.

Salary per pay period	×	number of pay periods	=	total salary
1360	×	4	=	5440

Ravi's total salary for 1 month is $5440.

EXAMPLE • 1

A pharmacist's hourly wage is $48. On Saturday, the pharmacist earns time and a half (1.5 times the regular hourly wage). How much does the pharmacist earn for working 6 hours on Saturday?

Strategy

To find the pharmacist's earnings:

- Find the hourly wage for working on Saturday by multiplying the hourly wage by 1.5.
- Multiply the hourly wage by the number of hours worked.

Solution

$48 \times 1.5 = 72$ $72 \times 6 = 432$

The pharmacist earns $432.

YOU TRY IT • 1

A respiratory therapist, whose hourly wage is $24.50, earns double time (two times the regular hourly wage) for working overtime. Find the therapist's wages for working 8 hours of overtime.

Your strategy

Your solution

EXAMPLE • 2

An efficiency expert received a contract for $3000. The expert spent 75 hours on the project. Find the consultant's hourly wage.

Strategy

To find the hourly wage, divide the total earnings by the number of hours worked.

Solution

$3000 \div 75 = 40$

The hourly wage was $40.

YOU TRY IT • 2

A radiologist receives an annual salary of $70,980. What is the radiologist's salary per month?

Your strategy

Your solution

EXAMPLE • 3

Dani Greene earns $38,500 per year plus a 5.5% commission on pharmaceutical sales over $100,000. During one year, Dani sold $150,000 worth of pharmaceuticals. Find Dani's total earnings for the year.

Strategy

To find the total earnings:

- Find the sales over $100,000.
- Multiply the commission rate by sales over $100,000.
- Add the commission to the annual pay.

Solution

$150{,}000 - 100{,}000 = 50{,}000$

$50{,}000 \times 0.055 = 2750$ • Commission

$38{,}500 + 2750 = 41{,}250$

Dani earned $41,250.

YOU TRY IT • 3

A medical equipment salesperson earns $37,000 per year plus a 9.5% commission on sales over $50,000. During one year, the salesperson's sales totaled $175,000. Find the salesperson's total earnings for the year.

Your strategy

Your solution

Solutions on p. S17

6.6 EXERCISES

OBJECTIVE A To calculate commissions, total hourly wages, and salaries

1. Lewis works in a hospital cafeteria and earns $11.50 per hour. How much does he earn in a 40-hour work week?

2. Sasha pays a nurse's aide an hourly wage of $11. How much does she pay the nurse's aide for working 25 hours?

3. A real estate agent receives a 3% commission for selling a house. Find the commission that the agent earned for selling a house for $131,000.

4. Ron Caruso works as an insurance agent and receives a commission of 40% of the first year's premium. Find Ron's commission for selling a life insurance policy with a first-year premium of $1050.

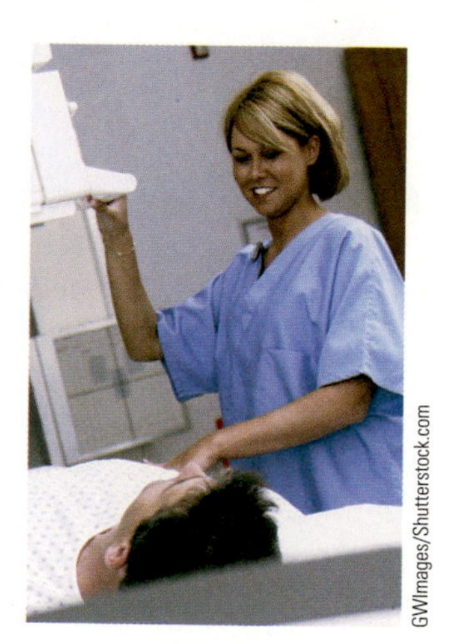
GWImages/Shutterstock.com

5. A stockbroker receives a commission of 1.5% of the price of stock that is bought or sold. Find the commission on 100 shares of stock that were bought for $5600.

6. A pharmaceutical salesman receives a commission of 20% of the sales that he makes to physicians and medical groups. Find the commission on sales of $22,500 of pharmaceuticals he made this month.

7. Keisha Brown receives an annual salary of $38,928 as a certified radiology x-ray technician. How much does Keisha receive each month?

8. A physician's assistant receives an annual salary of $78,000. How much does he receive per month?

9. Carlos receives a commission of 12% of his weekly sales as a sales representative for a medical supply company. Find the commission he earned during a week in which sales were $4500.

10. A sales clerk receives a 25% commission on all durable medical equipment he sells. Find the commission earned by the sales clerk if he sells a wheelchair for $450.

11. A medical transcriptionist makes $15.20 per hour transcribing surgical pathology reports. If she works 25 hours this week completing reports for one surgeon, how much will she be paid?

12. A typist charges $3.75 per page for typing technical material. How much does the typist earn for typing a 225-page book?

13. A nuclear chemist received $15,000 in consulting fees while working on a nuclear power plant. The chemist worked 120 hours on the project. Find the chemist's hourly wage.

14. Maxine received \$3400 for working 40 hours as a computer consultant training office staff to use a new dental software package. Find her hourly wage.

15. Gil Stratton's hourly wage is \$10.78. For working overtime, he receives double time.
a. What is Gil's hourly wage for working overtime?
b. How much does he earn for working 16 hours of overtime?

16. Mark is a medical laboratory technician and receives an hourly wage of \$15.90. When working on Saturday, he receives time and a half.
a. What is Mark's hourly wage on Saturday?
b. How much does he earn for working 8 hours on Saturday?

17. A home health aide earns \$8.20 an hour. For working the night shift, the health aide's wage increases by 15%.
a. What is the increase in hourly pay for working the night shift?
b. What is the aide's hourly wage for working the night shift?

18. A nurse earns \$31.50 an hour. For working the night shift, the nurse receives a 10% increase in pay.
a. What is the increase in hourly pay for working the night shift?
b. What is the hourly pay for working the night shift?

19. Nicole Tobin, a salesperson, receives a salary of \$250 per week plus a commission of 15% on all sales over \$1500. Find her earnings during a week in which sales totaled \$3000.

20. A veterinarian's assistant works 35 hours a week at \$20 an hour. The assistant is paid time and a half for overtime hours. Which expression represents the assistant's earnings for a week in which the assistant worked 41 hours?
(i) 41×20 (ii) $(35 \times 20) + (41 \times 30)$
(iii) $(35 \times 20) + (6 \times 30)$ (iv) 41×30

Applying the Concepts

Compensation The table at the right shows the average starting salaries for recent college graduates. Use this table for Exercises 21 to 24. Round to the nearest dollar.

Average Starting Salaries		
Bachelor's Degree	*Average Starting Salary*	*Change from Previous Year*
Chemical Engineering	\$52,169	1.8% increase
Electrical Engineering	\$50,566	0.4% increase
Computer Science	\$46,536	7.6% decrease
Accounting	\$41,360	2.6% increase
Business	\$36,515	3.7% increase
Biology	\$29,554	1.0% decrease
Political Science	\$28,546	12.6% decrease
Psychology	\$26,738	10.7% decrease

Source: National Association of Colleges

21. What was the starting salary in the previous year for an accountant?

22. How much did the starting salary for a chemical engineer increase over that of the previous year?

23. What was the starting salary in the previous year for a computer science major?

24. How much did the starting salary for a biology major decrease from that of the previous year?

SECTION

6.7 Bank Statements

OBJECTIVE A To calculate checkbook balances

Take Note

A **checking account** is a bank account that enables you to withdraw money or make payments to other people using checks. A **check** is a printed form that, when filled out and signed, instructs a bank to pay a specified sum of money to the person named on it.

A **deposit slip** is a form for depositing money in a checking account.

A checking account can be opened at most banks and savings and loan associations by depositing an amount of money in the bank. A checkbook contains checks and deposit slips and a checkbook register in which to record checks written and amounts deposited in the checking account. A sample check is shown below.

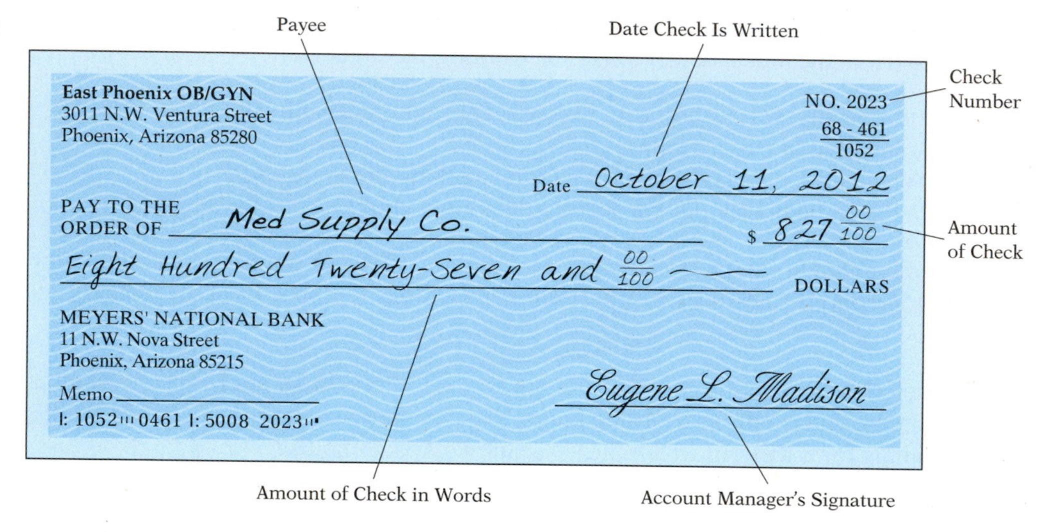

Each time a check is written, the amount of the check is subtracted from the amount in the account. When a deposit is made, the amount deposited is added to the amount in the account.

Point of Interest

There are a number of computer programs that serve as "electronic" checkbooks. With these programs, you can pay your bills by using a computer to write the check and then transmit the check over telephone lines using a modem.

A portion of a checkbook register is shown below. The account holder had a balance of $587.93 before writing two checks, one for $286.87 and the other for $202.38, and making one deposit of $345.00.

RECORD ALL CHARGES OR CREDITS THAT AFFECT YOUR ACCOUNT

NUMBER	DATE	DESCRIPTION OF TRANSACTION	PAYMENT/DEBIT (−)		√ T	FEE (IF ANY) (−)	DEPOSIT/CREDIT (+)		BALANCE	
									$ 587	93
108	8/4	Plumber	$286	87		$	$		301	06
109	8/10	Car Payment	202	38					98	68
	8/14	Deposit					345	00	443	68

To find the current checking account balance, subtract the amount of each check from the previous balance. Then add the amount of the deposit.

The current checking account balance is $443.68.

EXAMPLE 1

A nurse's aide had a checking account balance of $485.93 before writing two checks, one for $18.98 and another for $35.72, and making a deposit of $250. Find the current checking account balance.

Strategy
To find the current balance:

- Subtract the amount of each check from the old balance.
- Add the amount of the deposit.

Solution

485.93	
− 18.98	first check
466.95	
− 35.72	second check
431.23	
+ 250.00	deposit
681.23	

The current checking account balance is $681.23.

YOU TRY IT 1

A dietician had a checking account balance of $302.46 before writing a check for $20.59 and making two deposits, one in the amount of $176.86 and another in the amount of $94.73. Find the current checking account balance.

Your strategy

Your solution

Solution on p. S17

OBJECTIVE B To balance a checkbook

Each month a bank statement is sent to the account holder. A **bank statement** is a document showing all the transactions in a bank account during the month. It shows the checks that the bank has paid, the deposits received, and the current bank balance.

A bank statement and checkbook register are shown on the next page.

Balancing a checkbook, or determining whether the checking account balance is accurate, requires a number of steps.

1. In the checkbook register, put a check mark (✓) by each check paid by the bank and by each deposit recorded by the bank.

RECORD ALL CHARGES OR CREDITS THAT AFFECT YOUR ACCOUNT

NUMBER	DATE	DESCRIPTION OF TRANSACTION	PAYMENT/DEBIT (−)	√ T	FEE (IF ANY) (−)	DEPOSIT/CREDIT (+)	BALANCE
							$ 840 27
263	5/20	Dentist	$ 75 00	√	$	$	765 27
264	5/22	Post Office	33 61	√			731 66
265	5/22	Gas Company	67 14				664 52
	5/29	Deposit		√		192 00	856 52
266	5/29	Pharmacy	38 95	√			817 57
267	5/30	Telephone	63 85				753 72
268	6/2	Groceries	73 19	√			680 53
	6/3	Deposit		√		215 00	895 53
269	6/7	Insurance	103 00	√			792 53
	6/10	Deposit				225 00	1017 53
270	6/15	Photo Shop	16 63	√			1000 90
271	6/18	Newspaper	27 00				973 90

CHECKING ACCOUNT Monthly Statement Account Number: 924-297-8

Date	Transaction	Amount	Balance
5/20	OPENING BALANCE		840.27
5/21	CHECK	75.00	765.27
5/23	CHECK	33.61	731.66
5/29	DEPOSIT	192.00	923.66
6/1	CHECK	38.95	884.71
6/1	INTEREST	4.47	889.18
6/3	CHECK	73.19	815.99
6/3	DEPOSIT	215.00	1030.99
6/9	CHECK	103.00	927.99
6/16	CHECK	16.63	911.36
6/20	SERVICE CHARGE	3.00	908.36
6/20	CLOSING BALANCE		908.36

2. Add to the current checkbook balance all checks that have been written but have not yet been paid by the bank and any interest paid on the account.

Current checkbook balance:	973.90
Checks: 265	67.14
267	63.85
271	27.00
Interest:	+ 4.47
	1136.36

3. Subtract any service charges and any deposits not yet recorded by the bank. This is the checkbook balance.

Service charge:	− 3.00
	1133.36
Deposit:	− 225.00
Checkbook balance:	908.36

Take Note

A **service charge** is an amount of money charged by a bank for handling a transaction.

4. Compare the balance with the bank balance listed on the bank statement. If the two numbers are equal, the bank statement and the checkbook "balance."

Closing bank balance from bank statement $908.36 = Checkbook balance $908.36

The bank statement and checkbook balance.

HOW TO 1

RECORD ALL CHARGES OR CREDITS THAT AFFECT YOUR ACCOUNT

NUMBER	DATE	DESCRIPTION OF TRANSACTION	PAYMENT/DEBIT (−)		√ T	FEE (IF ANY) (−)	DEPOSIT/CREDIT (+)		BALANCE	
									$ 1620	42
413	3/2	Car Payment	$232	15	√	$	$		1388	27
414	3/2	Utilities	67	14	√				1321	13
415	3/5	Restaurant	78	14					1242	99
	3/8	Deposit			√		1842	66	3085	65
416	3/10	House Payment	672	14	√				2413	51
417	3/14	Insurance	177	10					2236	41

CHECKING ACCOUNT Monthly Statement — Account Number: 924-297-8

Date	Transaction	Amount	Balance
3/1	OPENING BALANCE		1620.42
3/4	CHECK	232.15	1388.27
3/5	CHECK	67.14	1321.13
3/8	DEPOSIT	1842.66	3163.79
3/10	INTEREST	6.77	3170.56
3/12	CHECK	672.14	2498.42
3/25	SERVICE CHARGE	2.00	2496.42
3/30	CLOSING BALANCE		2496.42

Balance the checkbook shown above.

1. In the checkbook register, put a check mark (✓) by each check paid by the bank and by each deposit recorded by the bank.

2. Add to the current checkbook balance all checks that have been written but have not yet been paid by the bank and any interest paid on the account.

Current checkbook balance:	2236.41
Checks: 415	78.14
417	177.10
Interest:	+ 6.77
	2498.42

3. Subtract any service charges and any deposits not yet recorded by the bank. This is the checkbook balance.

Service charge:	− 2.00
Checkbook balance:	2496.42

4. Compare the balance with the bank balance listed on the bank statement. If the two numbers are equal, the bank statement and the checkbook "balance."

Closing bank balance from bank statement $2496.42 = Checkbook balance $2496.42

The bank statement and checkbook balance.

EXAMPLE 2

Balance the checkbook shown below.

RECORD ALL CHARGES OR CREDITS THAT AFFECT YOUR ACCOUNT

NUMBER	DATE	DESCRIPTION OF TRANSACTION	PAYMENT/DEBIT (−)		√ T	FEE (IF ANY) (−)	DEPOSIT/CREDIT (+)		BALANCE	
									$ 412	64
345	1/14	Phone Bill	$ 54	75	√	$	$		357	89
346	1/19	News Shop	18	98	√				338	91
347	1/23	Theater Tickets	95	00					243	91
	1/31	Deposit			√		947	00	1190	91
348	2/5	Cash	250	00	√				940	91
349	2/12	Rent	840	00					100	91

CHECKING ACCOUNT Monthly Statement — Account Number: 924-297-8

Date	Transaction	Amount	Balance
1/10	OPENING BALANCE		412.64
1/18	CHECK	54.75	357.89
1/23	CHECK	18.98	338.91
1/31	DEPOSIT	947.00	1285.91
2/1	INTEREST	4.52	1290.43
2/10	CHECK	250.00	1040.43
2/10	CLOSING BALANCE		1040.43

Solution

Current checkbook balance:	100.91
Checks: 347	95.00
349	840.00
Interest:	+ 4.52
	1040.43
Service charge:	− 0.00
	1040.43
Deposit:	− 0.00
Checkbook balance:	1040.43

Closing bank balance from bank statement: $1040.43

Checkbook balance: $1040.43

The bank statement and the checkbook balance.

YOU TRY IT 2

Balance the checkbook shown below.

RECORD ALL CHARGES OR CREDITS THAT AFFECT YOUR ACCOUNT

NUMBER	DATE	DESCRIPTION OF TRANSACTION	PAYMENT/DEBIT (−)		√ T	FEE (IF ANY) (−)	DEPOSIT/CREDIT (+)		BALANCE	
									$ 903	17
	2/15	Deposit	$			$	$ 523	84	1427	01
234	2/20	Mortgage	773	21					653	80
235	2/27	Cash	200	00					453	80
	3/1	Deposit					523	84	977	64
236	3/12	Insurance	275	50					702	14
237	3/12	Telephone	78	73					623	41

CHECKING ACCOUNT Monthly Statement   Account Number: 314-271-4

Date	Transaction	Amount	Balance
2/14	OPENING BALANCE		903.17
2/15	DEPOSIT	523.84	1427.01
2/21	CHECK	773.21	653.80
2/28	CHECK	200.00	453.80
3/1	INTEREST	2.11	455.91
3/14	CHECK	275.50	180.41
3/14	CLOSING BALANCE		180.41

Your solution

Solution on p. S17

6.7 EXERCISES

OBJECTIVE A To calculate checkbook balances

1. You had a checking account balance of $342.51 before making a deposit of $143.81. What is your new checking account balance?

2. The business checking account for R and R Medispa showed a balance of $1536.97. What is the balance in this account after a deposit of $439.21 has been made?

3. A nutritionist had a checking account balance of $1204.63 before writing one check for $119.27 and another check for $260.09. Find the current checkbook balance.

4. Sam had a checking account balance of $3046.93 before writing a check for $1027.33 and making a deposit of $150. Find the current checkbook balance.

5. The business checking account for Rachael's Drugstore had a balance of $3476.85 before a deposit of $1048.53 was made. The store manager then wrote two checks, one for $848.37 and another for $676.19. Find the current checkbook balance.

6. Joel had a checking account balance of $427.38 before a deposit of $127.29 was made. Joel then wrote two checks, one for $43.52 and one for $249.78. Find the current checkbook balance.

7. A nurse had a checkbook balance of $404.96 before making a deposit of $350 and writing a check for $71.29. Is there enough money in the account to purchase a refrigerator for $675?

8. A paramedic had a checkbook balance of $149.85 before making a deposit of $245 and writing a check for $387.68. Is there enough money in the account for the bank to pay the check?

9. A medical supply outlet has the opportunity to buy power wheelchairs at a manufacturer's closeout sale for $2215 each. The manager wants to purchase two chairs at this low price. If the store's checking account currently has $5127.25 in it, is there enough money to buy two power wheelchairs by check?

10. A small health clinic's current checkbook balance is $2848.25. The office manager wants to purchase a new examination table for $2175, a mobile medical cabinet for $629, and a rolling stool for $125. Is there enough money in the account to make these purchases?

For Exercises 11 and 12, suppose the given transactions take place on an account in one day. State whether the account's ending balance on that day *must be less than, might be less than,* or *cannot be less than* its starting balance on that day.

11. Two deposits and one check written

12. Three checks written

OBJECTIVE B To balance a checkbook

13. Balance the checkbook.

RECORD ALL CHARGES OR CREDITS THAT AFFECT YOUR ACCOUNT

NUMBER	DATE	DESCRIPTION OF TRANSACTION	PAYMENT/DEBIT (−)		√ T	FEE (IF ANY) (−)	DEPOSIT/CREDIT (+)		BALANCE $ 2466	79
223	3/2	Groceries	$ 167	32		$	$		2299	47
	3/5	Deposit					960	70	3260	17
224	3/5	Rent	860	00					2400	17
225	3/7	Gas & Electric	142	35					2257	82
226	3/7	Cash	300	00					1957	82
227	3/7	Insurance	218	44					1739	38
228	3/7	Credit Card	419	32					1320	06
229	3/12	Dentist	92	00					1228	06
230	3/13	Drug Store	47	03					1181	03
	3/19	Deposit					960	70	2141	73
231	3/22	Car Payment	241	35					1900	38
232	3/25	Cash	300	00					1600	38
233	3/25	Oil Company	166	40					1433	98
234	3/28	Plumber	155	73					1278	25
235	3/29	Department Store	288	39					989	86

CHECKING ACCOUNT Monthly Statement — Account Number: 122-345-1

Date	Transaction	Amount	Balance
3/1	OPENING BALANCE		2466.79
3/5	DEPOSIT	960.70	3427.49
3/7	CHECK	167.32	3260.17
3/8	CHECK	860.00	2400.17
3/8	CHECK	300.00	2100.17
3/9	CHECK	142.35	1957.82
3/12	CHECK	218.44	1739.38
3/14	CHECK	92.00	1647.38
3/18	CHECK	47.03	1600.35
3/19	DEPOSIT	960.70	2561.05
3/25	CHECK	241.35	2319.70
3/27	CHECK	300.00	2019.70
3/29	CHECK	155.73	1863.97
3/30	INTEREST	13.22	1877.19
4/1	CLOSING BALANCE		1877.19

14. Balance the checkbook.

RECORD ALL CHARGES OR CREDITS THAT AFFECT YOUR ACCOUNT

NUMBER	DATE	DESCRIPTION OF TRANSACTION	PAYMENT/DEBIT (−)		√ T	FEE (IF ANY) (−)	DEPOSIT/CREDIT (+)		BALANCE	
									$ 1219	43
	5/1	Deposit	$			$	$ 619	14	1838	57
515	5/2	Electric Bill	42	35					1796	22
516	5/2	Groceries	95	14					1701	08
517	5/4	Insurance	122	17					1578	91
518	5/5	Theatre Tickets	84	50					1494	41
	5/8	Deposit					619	14	2113	55
519	5/10	Telephone	37	39					2076	16
520	5/12	Newspaper	22	50					2053	66
	5/15	Deposit					619	14	2672	80
521	5/20	Computer Store	172	90					2499	90
522	5/21	Credit Card	313	44					2186	46
523	5/22	Eye Exam	82	00					2104	46
524	5/24	Groceries	107	14					1997	32
525	5/24	Deposit					619	14	2616	46
526	5/25	Oil Company	144	16					2472	30
527	5/30	Car Payment	288	62					2183	68
528	5/30	Mortgage Payment	877	42					1306	26

CHECKING ACCOUNT Monthly Statement — Account Number: 122-345-1

Date	Transaction	Amount	Balance
5/1	OPENING BALANCE		1219.43
5/1	DEPOSIT	619.14	1838.57
5/3	CHECK	95.14	1743.43
5/4	CHECK	42.35	1701.08
5/6	CHECK	84.50	1616.58
5/8	CHECK	122.17	1494.41
5/8	DEPOSIT	619.14	2113.55
5/15	INTEREST	7.82	2121.37
5/15	CHECK	37.39	2083.98
5/15	DEPOSIT	619.14	2703.12
5/23	CHECK	82.00	2621.12
5/23	CHECK	172.90	2448.22
5/24	CHECK	107.14	2341.08
5/24	DEPOSIT	619.14	2960.22
5/30	CHECK	288.62	2671.60
6/1	CLOSING BALANCE		2671.60

15. Balance the checkbook.

RECORD ALL CHARGES OR CREDITS THAT AFFECT YOUR ACCOUNT

NUMBER	DATE	DESCRIPTION OF TRANSACTION	PAYMENT/DEBIT (−)		√ T	FEE (IF ANY) (−)	DEPOSIT/CREDIT (+)		BALANCE	
									$ 2035	18
218	7/2	Mortgage	$ 984	60		$	$		1050	58
219	7/4	Telephone	63	36					987	22
220	7/7	Cash	200	00					787	22
	7/12	Deposit					792	60	1579	82
221	7/15	Insurance	292	30					1287	52
222	7/18	Investment	500	00					787	52
223	7/20	Credit Card	414	83					372	69
	7/26	Deposit					792	60	1165	29
224	7/27	Department Store	113	37					1051	92

CHECKING ACCOUNT Monthly Statement — Account Number: 122-345-1

Date	Transaction	Amount	Balance
7/1	OPENING BALANCE		2035.18
7/1	INTEREST	5.15	2040.33
7/4	CHECK	984.60	1055.73
7/6	CHECK	63.36	992.37
7/12	DEPOSIT	792.60	1784.97
7/20	CHECK	292.30	1492.67
7/24	CHECK	500.00	992.67
7/26	DEPOSIT	792.60	1785.27
7/28	CHECK	200.00	1585.27
7/30	CLOSING BALANCE		1585.27

16. The ending balance on a monthly bank statement is greater than the beginning balance, and the bank did not include a service charge. Was the total of all deposits recorded *less than* or *greater than* the total of all checks paid?

17. When balancing your checkbook, you find that all the deposits in your checkbook register have been recorded by the bank, four checks in the register have not yet been paid by the bank, and the bank did not include a service charge. Is the ending balance on the monthly bank statement *less than* or *greater than* the ending balance on the check register?

Applying the Concepts

18. Define the words *credit* and *debit* as they apply to checkbooks.

FOCUS ON PROBLEM SOLVING

Counterexamples

An example that is given to show that a statement is not true is called a **counterexample.** For instance, suppose someone makes the statement "All colors are red." A counterexample to that statement would be to show someone the color blue or some other color.

If a statement is *always* true, there are no counterexamples. The statement "All even numbers are divisible by 2" is always true. It is not possible to give an example of an even number that is not divisible by 2.

Take Note

Recall that a prime number is a natural number greater than 1 that can be divided by only itself and 1. For instance, 17 is a prime number. 12 is not a prime number because 12 is divisible by numbers other than 1 and 12—for example, 4.

In mathematics, statements that are always true are called *theorems,* and mathematicians are always searching for theorems. Sometimes a conjecture by a mathematician appears to be a theorem. That is, the statement appears to be always true, but later on someone finds a counterexample.

One example of this occurred when the French mathematician Pierre de Fermat (1601–1665) conjectured that $2^{(2^n)} + 1$ is always a prime number for any natural number n. For instance, when $n = 3$, we have $2^{(2^3)} + 1 = 2^8 + 1 = 257$, and 257 is a prime number. However, in 1732 Leonhard Euler (1707–1783) showed that when $n = 5$, $2^{(2^5)} + 1 = 4{,}294{,}967{,}297$, and that $4{,}294{,}967{,}297 = 641 \cdot 6{,}700{,}417$—without a calculator! Because 4,294,967,297 is the product of two numbers (other than itself and 1), it is not a prime number. This counterexample showed that Fermat's conjecture is not a theorem.

For Exercises 1 to 5, find at least one counterexample.

1. All composite numbers are divisible by 2.

2. All prime numbers are odd numbers.

3. The square of any number is always bigger than the number.

4. The reciprocal of a number is always less than 1.

5. A number ending in 9 is always larger than a number ending in 3.

When a problem is posed, it may not be known whether the problem statement is true or false. For instance, Christian Goldbach (1690–1764) stated that every even number greater than 2 can be written as the sum of two prime numbers. For example,

$$12 = 5 + 7 \qquad 32 = 3 + 29$$

Although this problem is approximately 250 years old, mathematicians have not been able to prove it is a theorem, nor have they been able to find a counterexample.

For Exercises 6 to 9, answer true if the statement is always true. If there is an instance in which the statement is false, give a counterexample.

6. The sum of two positive numbers is always larger than either of the two numbers.

7. The product of two positive numbers is always larger than either of the two numbers.

8. Percents always represent a number less than or equal to 1.

9. It is never possible to divide by zero.

PROJECTS AND GROUP ACTIVITIES

Buying a Car Suppose a student has an after-school job to earn money to buy and maintain a car. We will make assumptions about the monthly costs in several categories in order to determine how many hours per week the student must work to support the car. Assume the student earns $10.50 per hour.

1. Monthly payment

   Assume that the car cost $18,500 with a down payment of $2220. The remainder is financed for 3 years at an annual simple interest rate of 9%.

   Monthly payment = ____________

2. Insurance

   Assume that insurance costs $3000 per year.

   Monthly insurance payment = ____________

3. Gasoline

   Assume that the student travels 750 miles per month, that the car travels 25 miles per gallon of gasoline, and that gasoline costs $3.50 per gallon.

   Number of gallons of gasoline purchased per month = ____________

   Monthly cost for gasoline = ____________

4. Miscellaneous

   Assume $0.42 per mile for upkeep.

   Monthly expense for upkeep = ____________

5. Total monthly expenses for the monthly payment, insurance, gasoline, and miscellaneous = ____________

6. To find the number of hours per month that the student must work to finance the car, divide the total monthly expenses by the hourly rate.

   Number of hours per month = ____________

7. To find the number of hours per week that the student must work, divide the number of hours per month by 4.

   Number of hours per week = ____________

   The student has to work ____________ hours per week to pay the monthly car expenses.

If you own a car, make out your own expense record. If you do not own a car, make assumptions about the kind of car that you would like to purchase, and calculate the total monthly expenses that you would have. An insurance company will give you rates on different kinds of insurance. An automobile club can give you approximations of miscellaneous expenses.

CHAPTER 6

SUMMARY

KEY WORDS	EXAMPLES
The *unit cost* is the cost of one item. [6.1A, p. 234]	Three paperback books cost $36. The unit cost is the cost of one paperback book, $12.
Percent increase is used to show how much a quantity has increased over its original value. [6.2A, p. 238]	The city's population increased 5%, from 10,000 people to 10,500 people.
Cost is the price a business pays for a product. *Selling price* is the price at which a business sells a product to a customer. *Markup* is the difference between selling price and cost. *Markup rate* is the markup expressed as a percent of the product's cost. [6.2B, p. 239]	A business pays $90 for a pair of cross trainers; the cost is $90. The business sells the cross trainers for $135; the selling price is $135. The markup is $135 − $90 = $45.
Percent decrease is used to show how much a quantity has decreased from its original value. [6.2C, p. 241]	Sales decreased 10%, from 10,000 units in the third quarter to 9000 units in the fourth quarter.
Sale price is the price after a reduction from the regular price. *Discount* is the difference between the regular price and the sale price. *Discount rate* is the discount as a percent of the product's regular price. [6.2D, p. 242]	A glucose monitor that regularly sells for $50 is on sale for $40. The regular price is $50. The sale price is $40. The discount is $50 − $40 = $10.
Interest is the amount paid for the privilege of using someone else's money. *Principal* is the amount of money originally deposited or borrowed. The percent used to determine the amount of interest is the *interest rate.* Interest computed on the original amount is called *simple interest.* The principal plus the interest owed on a loan is called the *maturity value.* [6.3A, p. 248]	Consider a 1-year loan of $5000 at an annual simple interest rate of 8%. The principal is $5000. The interest rate is 8%. The interest paid on the loan is $5000 × 0.08 = $400. The maturity value is $5000 + $400 = $5400.
The interest charged on purchases made with a credit card is called a *finance charge.* [6.3B, p. 250]	A credit card company charges 1.5% per month on any unpaid balance. The finance charge on an unpaid balance of $1000 is $1000 × 0.015 × 1 = $15.
Compound interest is computed not only on the original principal but also on the interest already earned. [6.3C, p. 251]	$10,000 is invested at 5% annual interest, compounded monthly. The value of the investment after 5 years can be found by multiplying 10,000 by the factor found in the Compound Interest Table in the Appendix. $10,000 × 1.283359 = $12,833.59
A *mortgage* is an amount that is borrowed to buy real estate. The *loan origination fee* is usually a percent of the mortgage and is expressed as *points.* [6.4A, p. 258]	The loan origination fee of 3 points paid on a mortgage of $200,000 is 0.03 × $200,000 = $6000.
A *commission* is usually paid to a salesperson and is calculated as a percent of sales. [6.6A, p. 268]	A commission of 5% on sales of $50,000 is 0.05 × $50,000 = $2500.

An employee who receives an *hourly wage* is paid a certain amount for each hour worked. [6.6A, p. 268]	An employee is paid an hourly wage of \$15. The employee's wages for working 10 hours are $\$15 \times 10 = \$150$.
An employee who is paid a *salary* receives payment based on a weekly, biweekly, monthly, or annual time schedule. [6.6A, p. 268]	An employee paid an annual salary of \$60,000 is paid $\$60{,}000 \div 12 = \$5000$ per month.
Balancing a checkbook is determining whether the checkbook balance is accurate. [6.7B, pp. 273–274]	To balance a checkbook: (1) Put a check mark in the checkbook register by each check paid by the bank and by each deposit recorded by the bank. (2) Add to the current checkbook balance all checks that have been written but have not yet been paid by the bank and any interest paid on the account. (3) Subtract any charges and any deposits not yet recorded by the bank. This is the checkbook balance. (4) Compare the balance with the bank balance listed on the bank statement. If the two numbers are equal, the bank statement and the checkbook "balance."

ESSENTIAL RULES AND PROCEDURES	EXAMPLES
To find unit cost, divide the total cost by the number of units. [6.1A, p. 234]	Three paperback books cost \$36. The unit cost is $\$36 \div 3 = \$12$ per book.
To find total cost, multiply the unit cost by the number of units purchased. [6.1C, p. 235]	One melon costs \$3. The total cost for 5 melons is $\$3 \times 5 = \$15$.
Basic Markup Equations [6.2B, p. 239] Selling price − cost = markup Cost + markup = selling price Markup rate × cost = markup	A pair of cross trainers that cost a business \$90 has a 50% markup rate. The markup is $0.50 \times \$90 = \$45$. The selling price is $\$90 + \$45 = \$135$.
Basic Discount Equations [6.2D, p. 242] Regular price − sale price = discount Regular price − discount = sale price Discount rate × regular price = discount	A first aid kit is on sale for 20% off the regular price of \$50. The discount is $0.20 \times \$50 = \$10$. The sale price is $\$50 - \$10 = \$40$.
Simple Interest Formula for Annual Interest Rates [6.3A, p. 248] Principal × annual interest rate × time (in years) = interest	The simple interest due on a 2-year loan of \$5000 that has an annual interest rate of 5% is $\$5000 \times 0.05 \times 2 = \$500$.
Maturity Value Formula for a Simple Interest Loan [6.3A, p. 248] Principal + interest = maturity value	The interest to be paid on a 2-year loan of \$5000 is \$500. The maturity value of the loan is $\$5000 + \$500 = \$5500$.
Monthly Payment on a Simple Interest Loan [6.3A, p. 249] Maturity value ÷ length of the loan in months = monthly payment	The maturity value of a simple interest 8-month loan is \$8000. The monthly payment is $\$8000 \div 8 = \$1000$.

CHAPTER 6

CONCEPT REVIEW

Test your knowledge of the concepts presented in this chapter. Answer each question. Then check your answers against the ones provided in the Answer Section.

1. Find the unit cost if 4 cans cost $2.96.

2. Find the total cost of 3.4 pounds of apples if apples cost $0.85 per pound.

3. How do you find the selling price if you know the cost and the markup?

4. How do you use the markup rate to find the markup?

5. How do you find the amount of decrease if you know the percent decrease?

6. How do you find the discount if you know the regular price and the sale price?

7. How do you find the discount rate?

8. How do you find simple interest?

9. How do you find the maturity value for a simple interest loan?

10. What is the principal?

11. How do you find the monthly payment for a loan of 18 months if you know the maturity value of the loan?

12. What is compound interest?

13. What is a fixed-rate mortgage?

14. What expenses are involved in owning a car?

15. How do you balance a checkbook?

CHAPTER 6

REVIEW EXERCISES

1. **Consumerism** A 20-ounce box of cereal costs $3.90. Find the unit cost.

2. **Car Expenses** An account executive had car expenses of $1025.58 for insurance, $1805.82 for gas, $37.92 for oil, and $288.27 for maintenance during a year in which 11,320 miles were driven. Find the cost per mile for these four items taken as a group. Round to the nearest tenth of a cent.

3. **Investments** A health care company's stock was bought for $42.375 per share. Six months later, the stock was selling for $55.25 per share. Find the percent increase in the price of the stock over the 6 months. Round to the nearest tenth of a percent.

4. **Markup** A medical uniform store uses a markup rate of 40%. What is the markup on a set of scrubs that costs the store $21?

5. **Simple Interest** A physician borrowed $100,000 from a credit union for 9 months at an annual interest rate of 4%. What is the simple interest due on the loan?

6. **Compound Interest** A computer programmer invested $25,000 in a retirement account that pays 6% interest, compounded daily. What is the value of the investment in 10 years? Use the Compound Interest Table in the Appendix. Round to the nearest cent.

7. **Investments** Last year a pharmaceutical company had earnings of $4.12 per share. This year the earnings are $4.73 per share. What is the percent increase in earnings per share? Round to the nearest percent.

8. **Real Estate** The monthly mortgage payment for a condominium is $923.67. The owner must pay an annual property tax of $2582.76. Find the total monthly payment for the mortgage and property tax.

Car Culture/Getty Images

9. **Car Expenses** A used pickup truck is purchased for $24,450. A down payment of 8% is made, and the remaining cost is financed for 4 years at an annual interest rate of 5%. Find the monthly payment. Use the Monthly Payment Table in the Appendix. Round to the nearest cent.

10. **Compound Interest** A medical practice invested $50,000 in an account that pays 7% annual interest, compounded quarterly. What is the value of the investment in 1 year? Use the Compound Interest Table in the Appendix.

11. **Real Estate** Paula Mason purchased a home for $195,000. The lender requires a down payment of 15%. Find the amount of the down payment.

12. **Car Expenses** A nurse bought a hybrid car for $28,500. A state license fee of $315 and a sales tax of 6.25% of the purchase price are required. Find the total cost of the sales tax and the license fee.

13. **Markup** Techno-Center uses a markup rate of 35% on all computer systems. Find the selling price of a computer system that costs the store $1540.

14. **Car Expenses** Mien pays a monthly car payment of $222.78. During a month in which $65.45 is principal, how much of the payment is interest?

15. **Compensation** A salesperson for a medical equipment company receives a commission of 3% on all sales he makes. Find the total commission received during a month in which he had $108,000 in sales.

16. **Discount** A suit that regularly costs $235 is on sale for 40% off the regular price. Find the sale price.

17. **Banking** Luke had a checking account balance of $1568.45 before writing checks for $123.76, $756.45, and $88.77. He then deposited a check for $344.21. Find Luke's current checkbook balance.

18. **Simple Interest** Metro Medical Supply borrowed $30,000 at an annual interest rate of 8% for 6 months. Find the maturity value of the loan.

19. **Real Estate** A credit union requires a borrower to pay $2\frac{1}{2}$ points for a loan. Find the origination fee for a $75,000 loan.

20. **Consumerism** Sixteen ounces of mouthwash cost $3.49. A 33-ounce container of the same brand of mouthwash costs $6.99. Which is the better buy?

21. **Real Estate** The Sweeneys bought a home for $356,000. The family made a 10% down payment and financed the remainder with a 30-year loan at an annual interest rate of 7%. Find the monthly mortgage payment. Use the Monthly Payment Table in the Appendix. Round to the nearest cent.

22. **Compensation** Richard Valdez receives $12.60 per hour for working 40 hours a week and time and a half for working over 40 hours. Find his total income during a week in which he worked 48 hours.

23. **Banking** The business checking account of a medical spa showed a balance of $9567.44 before checks of $1023.55, $345.44, and $23.67 were written and checks of $555.89 and $135.91 were deposited. Find the current checkbook balance.

24. **Simple Interest** The simple interest due on a 4-month loan of $55,000 is $1375. Find the monthly payment on the loan.

25. **Simple Interest** A credit card company charges a customer 1.25% per month on the unpaid balance of charges on the card. What is the finance charge in a month in which the customer has an unpaid balance of $576?

CHAPTER 6

TEST

1. **Consumerism** A case of 24 bottles of instant hand sanitizer costs $44.88. What is the cost per bottle?

2. **Consumerism** Which is the more economical purchase: 3 pounds of tomatoes for $7.49 or 5 pounds of tomatoes for $12.59?

3. **Consumerism** Red snapper costs $4.15 per pound. Find the cost of $3\frac{1}{2}$ pounds. Round to the nearest cent.

4. **Business** An exercise bicycle increased in price from $415 to $498. Find the percent increase in the cost of the exercise bicycle.

5. **Markup** A department store uses a 40% markup rate. Find the selling price of a blu-ray disc player that the store purchased for $315.

6. **Inflation** The price of a certain liquid medication rose from $25.99 per bottle to $29.99 per bottle. What percent increase does this amount represent? Round to the nearest tenth of a percent.

7. **Consumerism** The price of a wheelchair dropped from $420 to $357. What percent decrease does this price drop represent?

8. **Discount** A rolling medical cabinet with a regular price of $299 is on sale for 30% off the regular price. Find the sale price.

9. **Discount** A box of stationery that regularly sells for $9.50 is on sale for $5.70. Find the discount rate.

10. **Simple Interest** A construction company borrowed $75,000 for 4 months at an annual interest rate of 8%. Find the simple interest due on the loan.

11. **Simple Interest** Craig Allen borrowed $25,000 at an annual interest rate of 9.2% for 9 months. Find the maturity value of the loan.

12. **Simple Interest** A credit card company charges a customer 1.2% per month on the unpaid balance of charges on the card. What is the finance charge in a month in which the customer has an unpaid balance of $374.95?

13. **Compound Interest** Jorge, who is self-employed, placed $30,000 in an account that pays 6% annual interest, compounded quarterly. How much interest was earned in 10 years? Use the Compound Interest Table in the Appendix.

14. **Real Estate** A savings and loan institution is offering mortgage loans that have a loan origination fee of $2\frac{1}{2}$ points. Find the loan origination fee when a home is purchased with a loan of $134,000.

15. **Real Estate** A new housing development offers homes with a mortgage of $222,000 for 25 years at an annual interest rate of 8%. Find the monthly mortgage payment. Use the Monthly Payment Table in the Appendix.

16. **Car Expenses** A Chevrolet was purchased for $23,750, and a 20% down payment was made. Find the amount financed.

17. **Car Expenses** A physician's assistant purchased an SUV for $33,714 and made a down payment of 15% of the cost. The balance was financed for 4 years at an annual interest rate of 7%. Find the monthly truck payment. Use the Monthly Payment Table in the Appendix.

18. **Compensation** Shaney receives an hourly wage of $30.40 an hour as an emergency room nurse. When called in at night, she receives time and a half. How much does Shaney earn in a week when she works 30 hours at normal rates and 15 hours during the night?

19. **Banking** The business checking account for a dental office had a balance of $7349.44 before checks for $1349.67 and $344.12 were written. The office manager then made a deposit of $956.60. Find the current checkbook balance.

20. **Banking** Balance the checkbook shown.

RECORD ALL CHARGES OR CREDITS THAT AFFECT YOUR ACCOUNT

NUMBER	DATE	DESCRIPTION OF TRANSACTION	PAYMENT/DEBIT (−)		√ T	FEE (IF ANY) (−)	DEPOSIT/CREDIT (+)		BALANCE $ 1422	13
843	8/1	House Payment	$ 713	72		$	$		708	41
	8/4	Deposit					852	60	1561	01
844	8/5	Loan Payment	162	40					1398	61
845	8/6	Groceries	166	44					1232	17
846	8/10	Car Payment	322	37					909	80
	8/15	Deposit					852	60	1762	40
847	8/16	Credit Card	413	45					1348	95
848	8/18	Pharmacy	92	14					1256	81
849	8/22	Utilities	72	30					1184	51
850	8/28	Telephone	78	20					1106	31

CHECKING ACCOUNT Monthly Statement — Account Number: 122-345-1

Date	Transaction	Amount	Balance
8/1	OPENING BALANCE		1422.13
8/3	CHECK	713.72	708.41
8/4	DEPOSIT	852.60	1561.01
8/8	CHECK	166.44	1394.57
8/8	CHECK	162.40	1232.17
8/15	DEPOSIT	852.60	2084.77
8/23	CHECK	72.30	2012.47
8/24	CHECK	92.14	1920.33
9/1	CLOSING BALANCE		1920.33

CUMULATIVE REVIEW EXERCISES

1. Simplify: $12 - (10 - 8)^2 \div 2 + 3$

2. Add: $3\frac{1}{3} + 4\frac{1}{8} + 1\frac{1}{12}$

3. Find the difference between $12\frac{3}{16}$ and $9\frac{5}{12}$.

4. Find the product of $5\frac{5}{8}$ and $1\frac{9}{15}$.

5. Divide: $3\frac{1}{2} \div 1\frac{3}{4}$

6. Simplify: $\left(\frac{3}{4}\right)^2 \div \left(\frac{3}{8} - \frac{1}{4}\right) + \frac{1}{2}$

7. Divide: $0.059\overline{)3.0792}$
Round to the nearest tenth.

8. Convert $\frac{17}{12}$ to a decimal. Round to the nearest thousandth.

9. Write "$410 in 8 hours" as a unit rate.

10. Solve the proportion $\frac{5}{n} = \frac{16}{35}$.
Round to the nearest hundredth.

11. Write $\frac{5}{8}$ as a percent.

12. Find 6.5% of 420.

13. Write 18.2% as a decimal.

14. What percent of 20 is 8.4?

15. 30 is 12% of what?

16. 65 is 42% of what? Round to the nearest hundredth.

17. **Meteorology** A series of late-summer storms produced rainfall of $3\frac{3}{4}$, $8\frac{1}{2}$, and $1\frac{2}{3}$ inches during a 3-week period. Find the total rainfall during the 3 weeks.

18. **Taxes** The Homer family pays $\frac{1}{5}$ of its total monthly income for taxes. The family has a total monthly income of $4850. Find the amount of their monthly income that the Homers pay in taxes.

19. **Consumerism** In 5 years, the cost of a scientific calculator went from $75 to $30. What is the ratio of the decrease in price to the original price?

20. **Fuel Efficiencies** A compact car was driven 417.5 miles on 12.5 gallons of gasoline. Find the number of miles driven per gallon of gasoline.

21. **Consumerism** A 14-pound turkey costs $15.40. Find the unit cost. Round to the nearest cent.

22. **Investments** Eighty shares of a stock paid a dividend of $112. At the same rate, find the dividend on 200 shares of the stock.

23. **Discount** A laptop computer that regularly sells for $900 is on sale for 20% off the regular price. What is the sale price?

24. **Markup** A medical equipment store bought an EKG machine for $1600 and used a markup rate of 40%. Find the selling price of the EKG machine.

25. **Compensation** Sook Kim, a physical therapist assistant, received an increase in salary from $2800 per month to $3024 per month. Find the percent increase in her salary.

26. **Simple Interest** A contractor borrowed $120,000 for 6 months at an annual interest rate of 4.5%. How much simple interest is due on the loan?

27. **Car Expenses** A red Ford Mustang was purchased for $26,900, and a down payment of $2000 was made. The balance is financed for 3 years at an annual interest rate of 9%. Find the monthly payment. Use the Monthly Payment Table in the Appendix. Round to the nearest cent.

28. **Banking** A family had a checking account balance of $1846.78. A check of $568.30 was deposited into the account, and checks of $123.98 and $47.33 were written. Find the new checking account balance.

29. **Car Expenses** During 1 year, Anna Gonzalez spent $1840 on gasoline and oil, $820 on insurance, $185 on tires, and $432 on repairs. Find the cost per mile to drive the car 10,000 miles during the year. Round to the nearest cent.

30. **Real Estate** A house has a mortgage of $172,000 for 20 years at an annual interest rate of 6%. Find the monthly mortgage payment. Use the Monthly Payment Table in the Appendix. Round to the nearest cent.

CHAPTER 7

Statistics and Probability

OBJECTIVES

SECTION 7.1
- A To read a pictograph
- B To read a circle graph

SECTION 7.2
- A To read a bar graph
- B To read a broken-line graph

SECTION 7.3
- A To read a histogram
- B To read a frequency polygon

SECTION 7.4
- A To find the mean, median, and mode of a distribution
- B To draw a box-and-whiskers plot

SECTION 7.5
- A To calculate the probability of simple events

AIM for the Future

DENTAL ASSISTANTS perform a variety of patient care, as well as office and laboratory duties. They sterilize and disinfect instruments and equipment, prepare and lay out the instruments and materials required to treat each patient, and obtain and update patients' dental records. During dental procedures, assistants work alongside the dentist to provide assistance. Employment is expected to grow 36% from 2008 to 2018, which is much faster than the average for all occupations. In fact, dental assistants are expected to be among the fastest growing occupations over the 2008–2018 projection period. Median annual wages of dental assistants were $32,380 in May 2008. (*Source: Bureau of Labor Statistics: http://www.bls.gov/oco/ocos163.htm*)

PREP TEST

Are you ready to succeed in this chapter?

Take the Prep Test below to find out if you are ready to learn the new material.

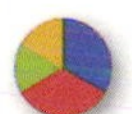

1. **Mail** Bill-related mail accounted for 49 billion of the 102 billion pieces of first-class mail handled by the U.S. Postal Service during a recent year. (*Source:* U.S. Postal Service) What percent of the pieces of first-class mail handled by the U.S. Postal Service was bill-related mail? Round to the nearest tenth of a percent.

2. **Education** The table at the right shows the estimated costs of funding an education at a public college.
   a. Between which two enrollment years is the increase in cost greatest?
   b. What is the increase between these two years?

Enrollment Year	*Cost of Public College*
2005	$70,206
2006	$74,418
2007	$78,883
2008	$83,616
2009	$88,633
2010	$93,951

Source: The College Board's Annual Survey of Colleges

3. **Sports** During the 1924 Summer Olympics in Paris, France, the United States won 45 gold medals, 27 silver medals, and 27 bronze medals. (*Source: The Ultimate Book of Sports Lists*) Find the ratio of gold medals won by the United States to silver medals won by the United States during the 1924 Summer Olympics. Write the ratio as a fraction in simplest form.

4. **The Military** Approximately 198,000 women serve in the U.S. military. Six percent of these women serve in the Marine Corps. (*Source:* www.fedstats.gov) What fractional amount of women in the military are in the Marine Corps?

SECTION

7.1 Pictographs and Circle Graphs

OBJECTIVE A To read a pictograph

Statistics is the branch of mathematics concerned with **data,** or numerical information. **Graphs** are displays that provide a pictorial representation of data. The advantage of graphs is that they present information in a way that is easily read.

A **pictograph** uses symbols to represent information. The pictograph in Figure 1 represents the net worth of America's richest billionaires. Each symbol represents 10 billion dollars.

Bill Gates

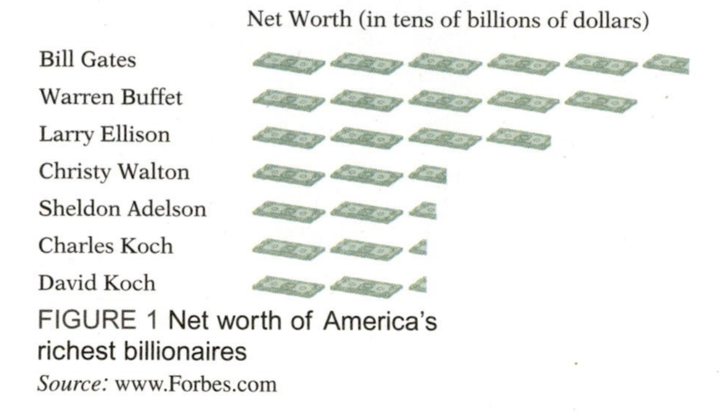

FIGURE 1 Net worth of America's richest billionaires

Source: www.Forbes.com

From the pictograph, we can determine that Bill Gates has the greatest net worth. His net worth is about $6 billion more than Warren Buffet's.

HOW TO • 1 The pictograph in Figure 2 represents the number of patients having their teeth cleaned at a local dentist's office during the first half of the year. What percent of the patients seen during the first six months of the year had their teeth cleaned in June?

January
February
March
April
May
June

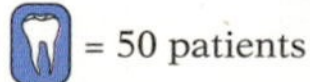
= 50 patients

FIGURE 2 Dental Cleanings at Dr. Mabry's Dental Clinic (January–June)

Strategy

Use the basic percent equation. The base is 1200 (the total number of patients during the six-month period), and the amount is 300 (the number served in June).

Solution

Percent $\times$ base $=$ amount

$$n \times 1200 = 300$$

$$n = 300 \div 1200$$

$$n = 0.25$$

25% of the patients had their teeth cleaned in June.

Tips for Success

Remember that the How To feature indicates a worked-out example. Using paper and pencil, work through the example. See *AIM for Success* at the front of the book.

The pictograph in Figure 3 shows the number of pints of blood donated at the quarterly blood drives conducted by the local community college SGA last year.

The ratio of the number of pints collected on July 20 to the number collected on October 31 is

$$\frac{15}{40} = \frac{3}{8}$$

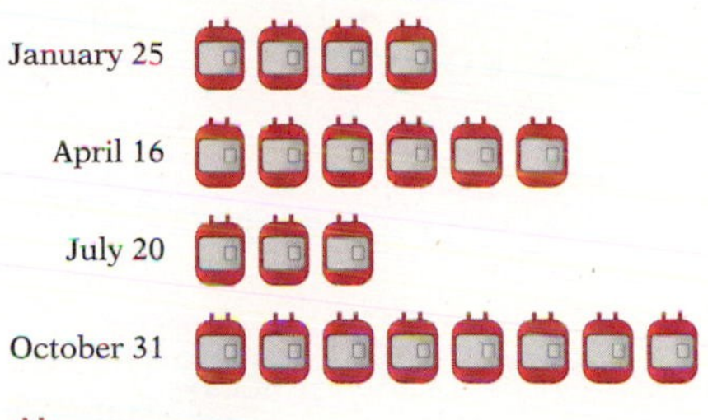

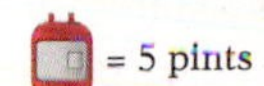

FIGURE 3 Results of SGA Blood Drives

EXAMPLE • 1

Use Figure 3 to find the total number of pints of blood collected during the community college blood drives last year.

Strategy

To find the number of pints of blood given during the year:

- Read the pictograph to determine the number of pints of blood given each of the four days.
- Add the four numbers.

Solution

Blood donations on January 25: 20 pints
Blood donations on April 16: 30 pints
Blood donations on July 20: 15 pints
Blood donations on October 31: 40 pints

Total number of pints for the year:

$$\begin{array}{r} 20 \\ 30 \\ 15 \\ +40 \\ \hline 105 \end{array}$$

There were 105 pints of blood collected last year.

YOU TRY IT • 1

According to Figure 3, the number of pints of blood collected on January 25 represents what percent of the total number of pints collected last year? Round to the nearest whole percent.

Your strategy

Your solution

Solution on p. S18

OBJECTIVE B To read a circle graph

A **circle graph** represents data by the size of the sectors. A **sector of a circle** is one of the "pieces of the pie" into which a circle graph is divided.

 **Take Note**

One quadrillion is 1,000,000,000,000,000.

Point of Interest

Fossil fuels include coal, natural gas, and petroleum. Renewable energy includes hydroelectric power, solar energy, wood burning, and wind energy.

The circle graph in Figure 4 shows the consumption of energy sources in the United States during a recent year. The complete circle graph represents the total amount of energy consumed, 96.6 quadrillion Btu. Each sector of the circle represents the consumption of energy from a different source.

To find the percent of the total energy consumed that originated from nuclear power, solve the basic percent equation for percent. The base is 96.6 quadrillion Btu, and the amount is 8.2 quadrillion Btu.

$$\text{Percent} \times \text{base} = \text{amount}$$

$$n \times 96.6 = 8.2$$

$$n = 8.2 \div 96.6$$

$$n \approx 0.085$$

To the nearest tenth of a percent, 8.5% of the energy consumed originated from nuclear power.

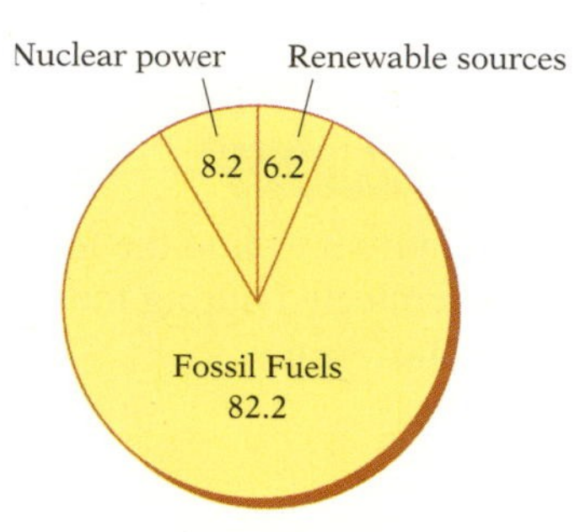

FIGURE 4 Annual energy consumption in quadrillion Btu in the United States
Source: The World Almanac and Book of Facts 2003

The U.S. Department of Health and Human Services reports that the national health expenditure in the United States grew 4% to \$2.5 trillion in 2009. The circle graph in Figure 5 shows what percents of the \$2.5 trillion were paid by different agencies. The complete circle represents 100% of all health expenditures in 2009. Each sector of the graph represents the percent of the total spent by specific entities.

HOW TO • 2 According to Figure 5, how much of the health expenditure was paid by private health insurance?

Strategy
Use the basic percent equation. The base is \$2.5 trillion, and the percent is 32%.

Solution

$$\text{Percent} \times \text{base} = \text{amount}$$

$$0.32 \times 2.5 = n$$

$$0.8 = n$$

0.8 trillion = 800,000,000,000 or 800 billion

The amount spent by private health insurance in 2009 was \$800 billion.

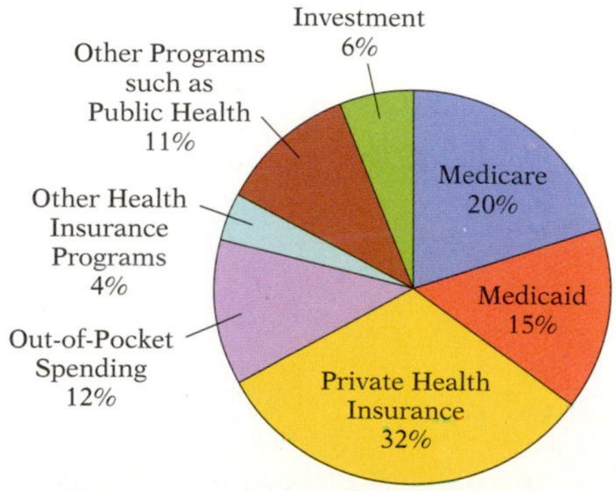

Investment 6%
Other Programs such as Public Health 11%
Other Health Insurance Programs 4%
Out-of-Pocket Spending 12%
Medicare 20%
Medicaid 15%
Private Health Insurance 32%

FIGURE 5 National Health Expenditure Data 2009
Source: Centers for Medicare & Medicaid Services, Office of the Actuary, National Health Statistics Group

The circle graph in Figure 6 shows typical annual expenses of owning, operating, and financing a car. Use this figure for Example 2 and You Try It 2.

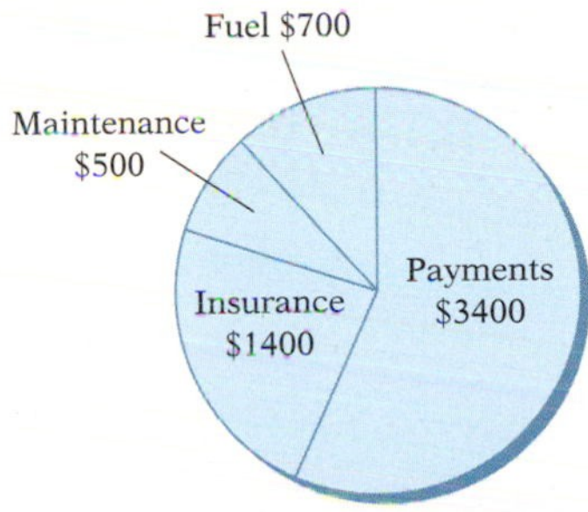

FIGURE 6 Annual expenses of $6000 for owning, operating, and financing a car
Source: Based on data from IntelliChoice

The circle graph in Figure 7 shows the distribution of an employee's gross monthly income. Use this figure for Example 3 and You Try It 3.

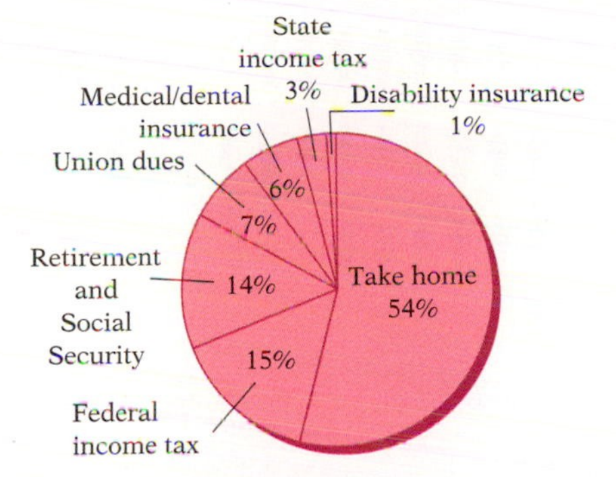

FIGURE 7 Distribution of gross monthly income of $2900

EXAMPLE 2

Use Figure 6 to find the ratio of the annual insurance expense to the total annual cost of the car.

Strategy
To find the ratio:

- Locate the annual insurance expense in the circle graph.
- Write in simplest form the ratio of the annual insurance expense to the total annual cost of operating the car.

Solution
Annual insurance expense: $1400

$$\frac{1400}{6000} = \frac{7}{30}$$

The ratio is $\frac{7}{30}$.

YOU TRY IT 2

Use Figure 6 to find the ratio of the annual cost of fuel to the annual cost of maintenance.

Your strategy

Your solution

EXAMPLE 3

Use Figure 7 to find the employee's take-home pay.

Strategy
To find the take-home pay:

- Locate the percent of the distribution that is take-home pay.
- Solve the basic percent equation for amount.

Solution
Take-home pay: 54%

$$\begin{aligned} \text{Percent} \times \text{base} &= \text{amount} \\ 0.54 \times 2900 &= n \\ 1566 &= n \end{aligned}$$

The employee's take-home pay is $1566.

YOU TRY IT 3

Use Figure 7 to find the amount paid for medical/dental insurance.

Your strategy

Your solution

Solutions on p. S18

7.1 EXERCISES

OBJECTIVE A To read a pictograph

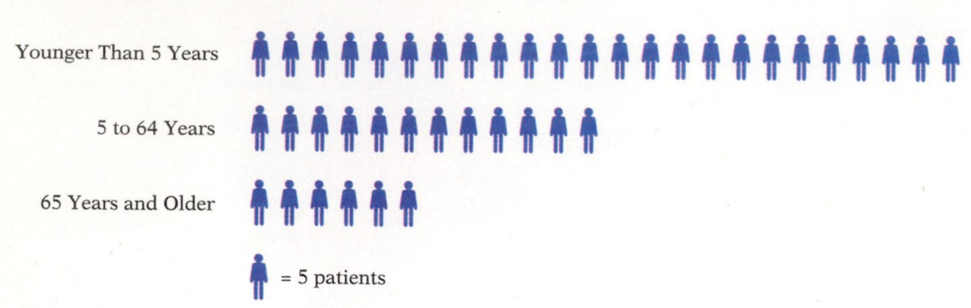

FIGURE 8 Emergency Room Visits for Patients Diagnosed with Asthma

Emergency Room Visits The pictograph in Figure 8 shows the number of patients seen in the emergency room at Memorial Hospital last year who received a diagnosis of asthma. Use this graph for Exercise 1 to 3.

1. Find the total number of patients who visited the emergency room and received a diagnosis of asthma.

2. Find the ratio of the number of patients aged 65 and older to those younger than 5 years old.

3. Find the percent of patients seen in the emergency room diagnosed with asthma aged 5 and younger. Round to the nearest percent.

Wages The pictograph in Figure 9 gives typical hourly rates for these health care providers. Use this graph for Exercises 4 to 6.

4. Find the ratio of the hourly rate for a physical therapist assistant to the hourly rate for a certified medical assistant.

5. How much more is the hourly rate for a certified dental assistant than for a certified nurse assistant?

6. Is the hourly rate of the physical therapist assistant more than twice the hourly rate of the certified nurse assistant?

Certified Dental Assistant $ $ $ $ $ $ $
Certified Medical Assistant $ $ $ $ $ $
Physical Therapist Assistant $ $ $ $ $ $ $ $ $
Certified Nurse Assistant $ $ $ $ $
$ = $2/hour

FIGURE 9 Hourly Rates for Health Care Providers
Source: http://www.payscale.com

Children's Behavior The pictograph in Figure 10 is based on a survey of children aged 7 through 12. The percents of children's responses to the survey are shown. Assume that 500 children were surveyed. Use this graph for Exercises 7 to 9.

7. Find the number of children who said they hid vegetables under a napkin.

8. What is the difference between the number of children who fed vegetables to the dog and the number who dropped them on the floor?

9. Were the responses given in the graph the only responses given by the children? Explain your answer.

FIGURE 10 How children try to hide vegetables
Source: Strategic Consulting and Research for Del Monte

For Exercises 10 and 11, refer to the pictograph in Figure 1 on page 294.

10. Which statement in blue below the pictograph does not depend upon knowing what each small dollar bill symbol represents?

11. Write down two more facts you can determine from the pictograph without knowing what each small dollar bill symbol represents.

OBJECTIVE B To read a circle graph

Education A nursing student at a community college recorded the number of units required in each discipline to graduate with an Associate Degree Nursing. The results are shown in the circle graph in Figure 11. Use this graph for Exercises 12 to 14.

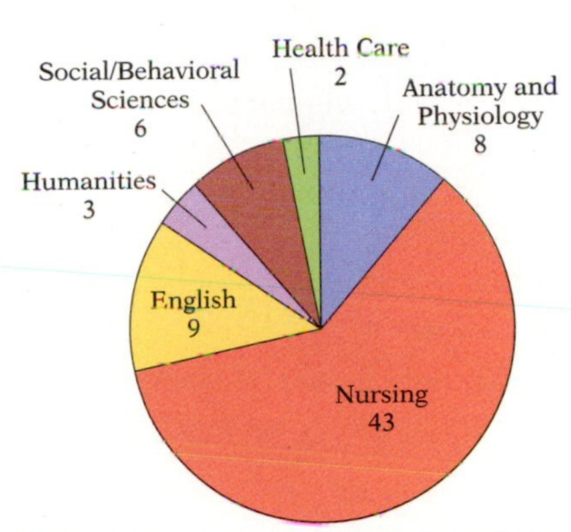

FIGURE 11 Number of Units Required to Graduate with an Associate Degree Nursing

12. How many units are required to graduate with an Associate Degree Nursing?

13. Is the number of units required in anatomy & physiology *less than* or *greater than* twice the number of units required in social/behavioral sciences?

14. Is the ratio of the number of units required in nursing to the number of units required in English *less than, equal to, or greater than* 5?

Blood Types The circle graph in Figure 12 shows the number of blood donors of each blood type who participated in the previous blood drive at a community college. Use this graph for Exercises 15 to 18.

15. Find the total number of donors.

16. What is the ratio of donors with type O− blood to those with type O+ blood?

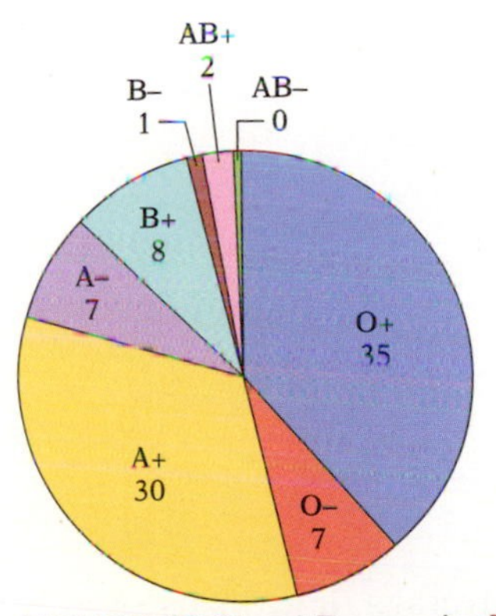

FIGURE 12 Blood Donors by Type at the Community College Blood Drive

17. What percent of donors had blood type A+? Round to the nearest tenth of a percent.

18. What percent of donors had blood type AB+? Round to the nearest tenth of a percent.

Budget The circle graph in Figure 13 shows the budget allocation for the Griner family's income. The annual family income is $125,000. Use this graph for Exercises 19 to 22.

19. Find the amount of money budgeted for housing and utilities annually. What is the monthly allocation?

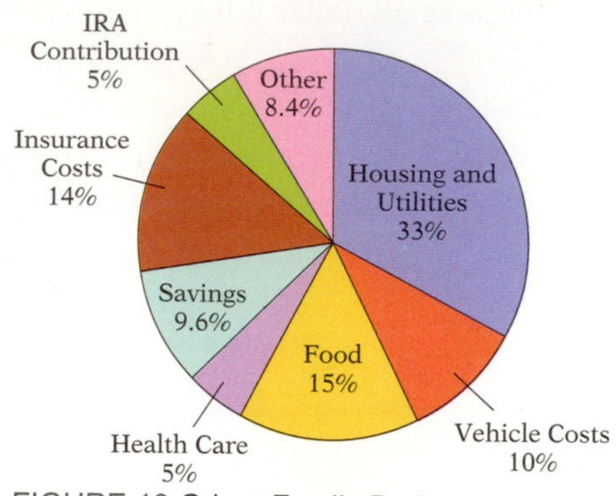

FIGURE 13 Griner Family Budget

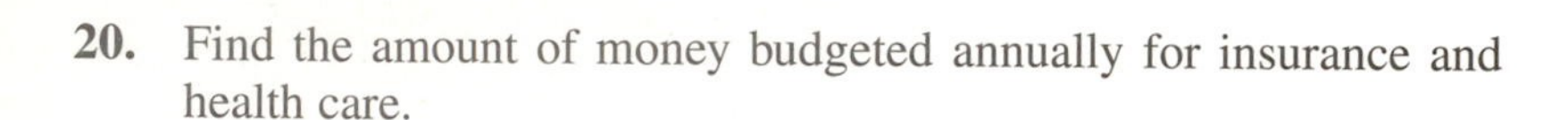

20. Find the amount of money budgeted annually for insurance and health care.

21. What fractional amount of the budget is designated for food?

22. Is the amount budgeted for housing and utilities more than three times the amount budgeted for savings?

Demographics The circle graph in Figure 14 shows a breakdown, according to age, of the homeless in America. Use this graph for Exercises 23 to 26.

23. What age group represents the largest segment of the homeless population?

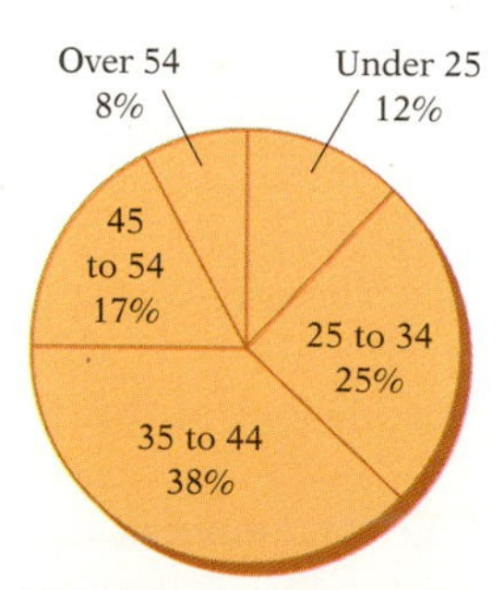

FIGURE 14 Ages of the homeless in America
Source: The Department of Housing and Urban Development

24. Is the number of homeless who are aged 25 to 34 more or less than twice the number who are under the age of 25?

25. What percent of the homeless population is under the age of 35?

26. On average, how many of every 100,000 homeless people in America are over the age of 54?

Health Programs The circle graph in Figure 15 shows the number of students enrolled in the various health programs offered at Southeastern Community College. Use this graph for Exercises 27 to 30.

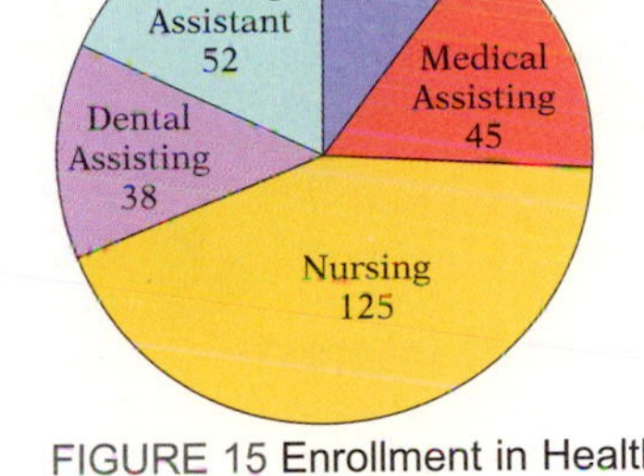

FIGURE 15 Enrollment in Health Programs at Southeastern Community College

27. Find the total number of students enrolled in health programs at Southeastern Community College.

28. How many more students are in the nursing program than in the nursing assistant program?

29. What percent of students enrolled in health programs are enrolled in the medical assisting program? Round to the nearest tenth of a percent.

30. What percent of the students enrolled in health programs are enrolled in the dental assisting program? Round to the nearest tenth of a percent.

Demographics There are approximately 300,000,000 people living in the United States. The circle graph in Figure 16 shows the breakdown of the U.S. population by ethnic origin. Use this graph for Exercises 31 to 33.

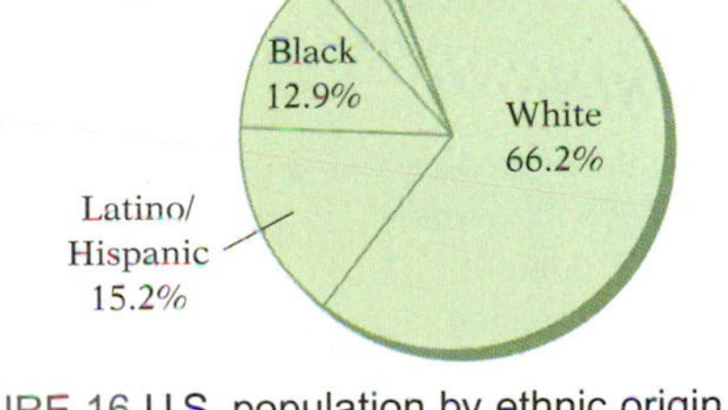

FIGURE 16 U.S. population by ethnic origin
Source: Entertainment Weekly, June 20, 2008

31. Approximately how many people living in the United States are of Asian ethnic origin?

32. Approximately how many more people of American Indian/Alaska Native ethnic origin live in the United States than people of Hawaiian/Pacific Islander ethnic origin?

33. On average, how many people of black ethnic origin would there be in a random sample of 500,000 people living in the United States?

Applying the Concepts

34. **a.** What are the advantages of presenting data in the form of a pictograph?
b. What are the disadvantages?

35. The circle graph at the right shows a couple's expenditures last month. Write two observations about this couple's expenses.

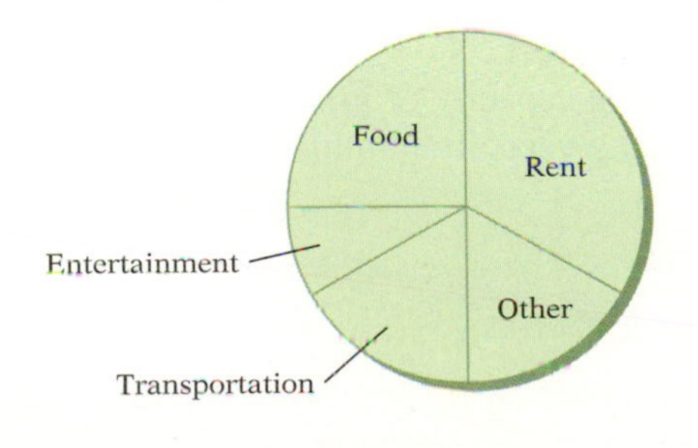

SECTION

7.2 Bar Graphs and Broken-Line Graphs

OBJECTIVE A To read a bar graph

Point of Interest

The first bar graph appeared in 1786 in the book *The Commercial and Political Atlas.* The author, William Playfair (1759–1823), was a pioneer in the use of graphical displays.

A **bar graph** represents data by the height of the bars. The bar graph in Figure 17 shows temperature data recorded for Cincinnati, Ohio, for the months March through November. For each month, the height of the bar indicates the average daily high temperature during that month. The jagged line near the bottom of the graph indicates that the vertical scale is missing the numbers between 0 and 50.

The daily high temperature in September was 78°F. Because the bar for July is the tallest, the daily high temperature was highest in July.

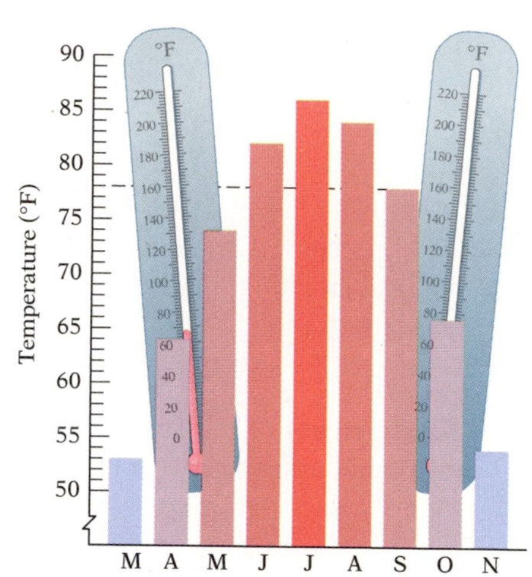

FIGURE 17 Daily high temperatures in Cincinnati, Ohio

Source: U.S. Weather Bureau

A **double-bar graph** is used to display data for purposes of comparison. The double-bar graph in Figure 18 compares the lung capacities of inactive and athletic 45-year-olds.

The lung capacity of an athletic female is 55 milliliters of oxygen per kilogram of body weight per minute.

Take Note

The bar for athletic females is halfway between the marks for 50 and 60. Therefore, we estimate that the lung capacity is halfway between these two numbers, at 55.

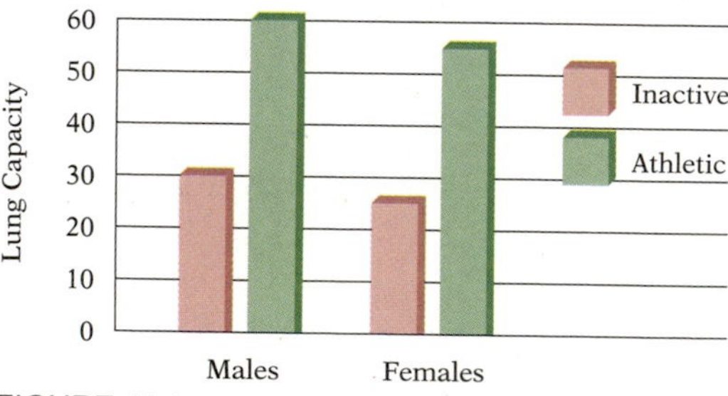

FIGURE 18 Lung capacity (in milliliters of oxygen per kilogram of body weight per minute)

EXAMPLE • 1

What is the ratio of the lung capacity of an inactive male to that of an athletic male?

Strategy

To write the ratio:

- Read the graph to find the lung capacity of an inactive male and of an athletic male.
- Write the ratio in simplest form.

Solution

Lung capacity of inactive male: 30

Lung capacity of athletic male: 60

$$\frac{30}{60} = \frac{1}{2}$$

The ratio is $\frac{1}{2}$.

YOU TRY IT • 1

What is the ratio of the lung capacity of an inactive female to that of an athletic female?

Your strategy

Your solution

Solution on p. S18

OBJECTIVE B To read a broken-line graph

A **broken-line graph** represents data by the positions of the lines. It is used to show trends.

The broken-line graph in Figure 19 shows the effect of inflation on the value of a $100,000 life insurance policy. The height of each dot indicates the value of the policy.

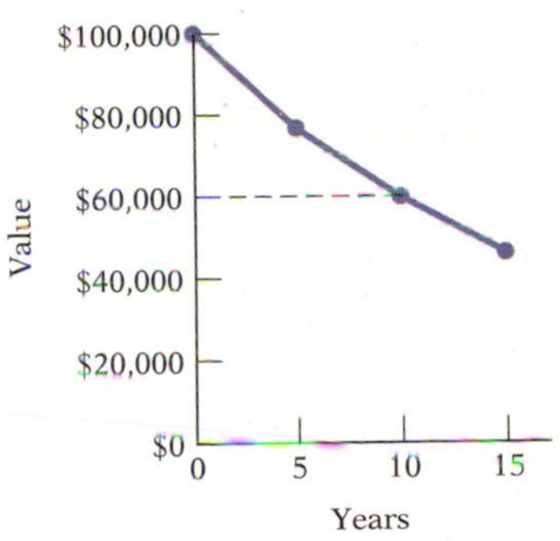

FIGURE 19 Effect of inflation on the value of a $100,000 life insurance policy

After 10 years, the purchasing power of the $100,000 has decreased to approximately $60,000.

Broken-line graphs are often shown in the same figure for comparison. Figure 20 shows a comparison of cigarette smoking among males of different ages.

Several things can be determined from the graph:

The percent of males aged 65 and older who smoke cigarettes has remained fairly constant throughout the 10-year period at approximately 10%.

The percent of males in grades 9 to 12 who smoke declined sharply from 1999 until 2003, at which time it rose slightly and then declined to approximately 20% by 2009.

In general, cigarette smoking among males appears to be declining.

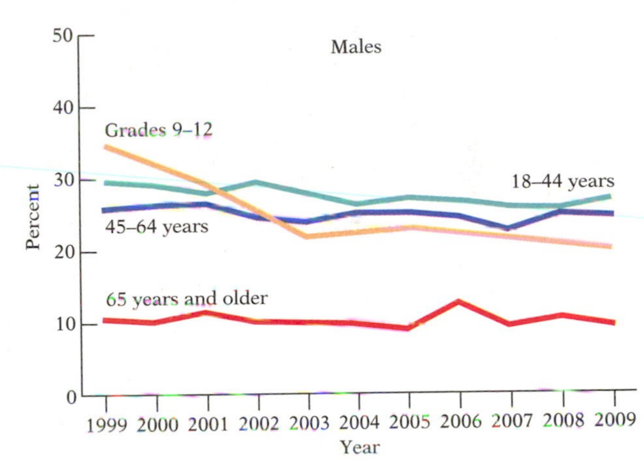

FIGURE 20 Cigarette Smoking Among Male Students in Grades 9 to 12 and Males Aged 18 and Older in the United States, 1999–2009

Source: CCDC/NCHS, National Health Interview Survey and CDC, Youth Risk Behavior Survey

EXAMPLE 2

Use Figure 20 to approximate the difference between the percent of males in grades 9 to 12 who smoked in 2009 and the percent of males aged 65 and older who smoked in 2009.

Strategy

To write the difference:

- Read the line graph to determine the percent of males in grades 9 to 12 who smoked in 2009 and the percent of males aged 65 and older who smoked in 2009.
- Subtract to find the difference.

Solution

Percent of males in grades 9 to 12 who smoked in 2009: 20%

Percent of males aged 65 and older who smoked in 2009: 10%

20% − 10% = 10%

The difference between the percents in 2009 was 10%.

YOU TRY IT 2

Use Figure 20 to determine in which year the percent of males smoking in grades 9 to 12 fell below the percent of males aged 18 to 44 and those aged 45 to 64 who smoked.

Your strategy

Your solution

Solution on p. S18

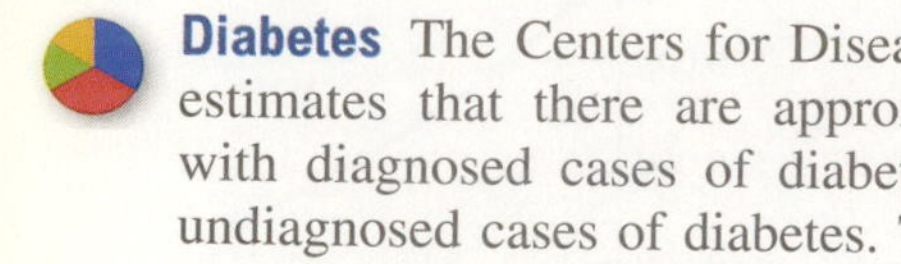

7.2 EXERCISES

OBJECTIVE A To read a bar graph

Diabetes The Centers for Disease Control and Prevention (CDC) estimates that there are approximately 18.8 million Americans with diagnosed cases of diabetes and 7 million American with undiagnosed cases of diabetes. The bar graph in Figure 21 shows the percentages of Americans in different age groups with diabetes, both diagnosed and undiagnosed.

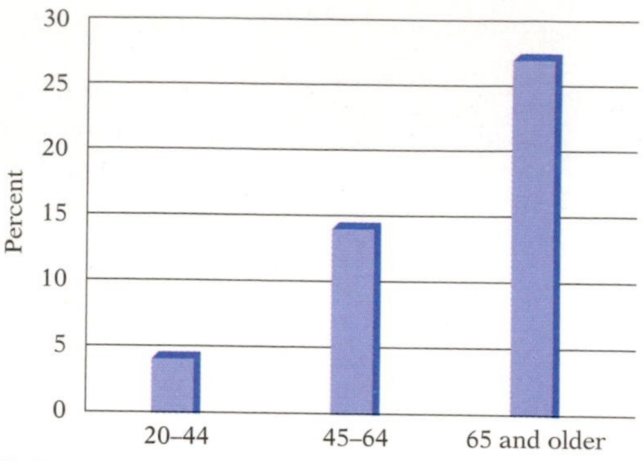

FIGURE 21 Estimated Percentage of People Aged 20 and Older with Diagnosed and Undiagnosed Diabetes, by Age Group, United States, 2005-2008

Source: Centers for Disease Control and Prevention. National diabetes fact sheet: national estimates and general information on diabetes and prediabetes in the United States, 2011. Atlanta, GA: U.S. Department of Health and Human Services, Centers for Disease Control and Prevention, 2011

1. Approximately what percent of Americans aged 65 and older have diabetes?
2. According to a recent estimate by the Census Bureau, there are approximately 38.9 million Americans 65 years old and over. Using the percents shown in the graph, approximately how many of those 65 years and older have diabetes?
3. Find the ratio of the percent of Americans with diabetes aged 45 to 64 to the percent of Americans with diabetes aged 20 to 44. Write the ratio in simplest form using the word *to*.

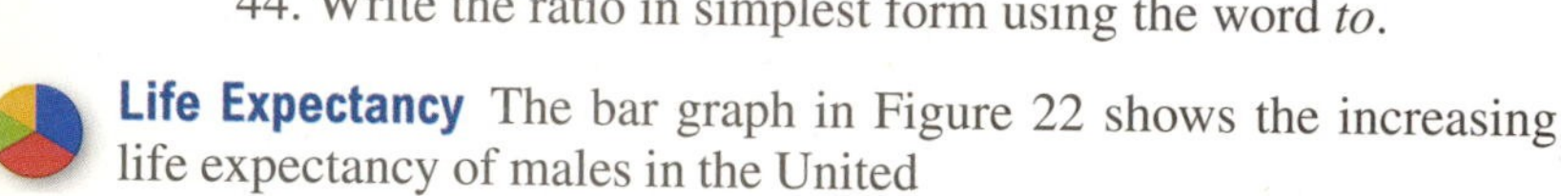

Life Expectancy The bar graph in Figure 22 shows the increasing life expectancy of males in the United States from 1930 to 2010. Use this graph for Exercises 4 to 6.

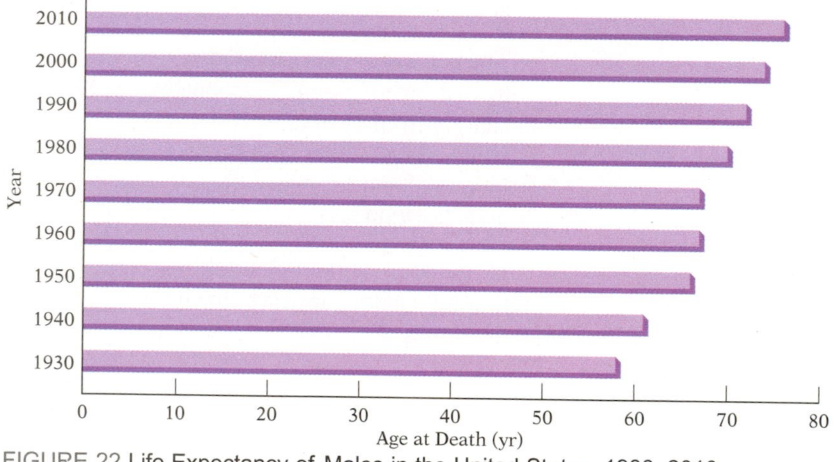

FIGURE 22 Life Expectancy of Males in the United States, 1930–2010

Source: Health, United States, 2010 http://www.cdc.gov/nchs/hus.htm

4. In which of the years shown was the life expectancy of males in the United States approximately the same?
5. How much greater is the life expectancy for American males in 2010 than it was in 1930?
6. Between what two decades shown on the graph did the life expectancy increase by approximately five years?

Salaries The double-bar graph in Figure 23 shows the range of annual salaries for health care workers in various jobs. Use this graph for Exercises 7 to 10.

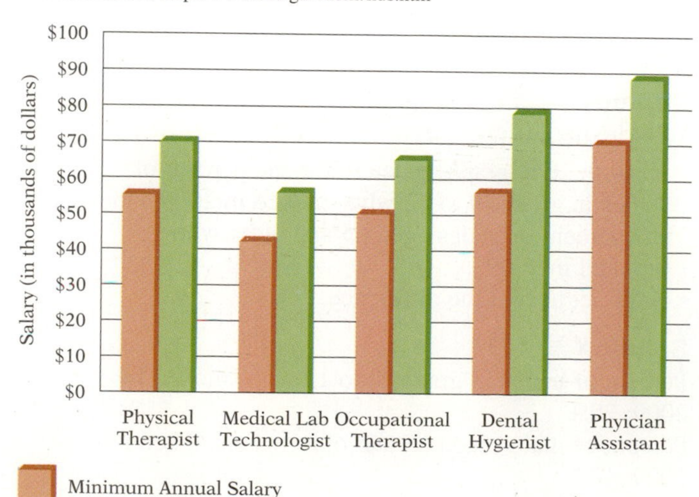

FIGURE 23 Salary Ranges for Health Care Jobs

Source: http://www.payscale.com

7. Estimate the difference in the maximum and minimum annual salaries for a physical therapist.
8. Which occupation has the highest minimum salary? Which occupation has the lowest minimum salary?
9. Which occupation has the greatest difference between the minimum and maximum salaries?
10. Which occupations have maximum salaries greater than $70,000?

For Exercises 11 and 12, refer to Figure 18 on page 302. Match the given statement about the double-bar graph to one of the following statements.

(i) The lung capacity of inactive males is less than the lung capacity of athletic males.
(ii) The lung capacity of inactive males is less than the lung capacity of inactive females.
(iii) The lung capacity of inactive males is greater than the lung capacity of inactive females.

11. The brown bar for males is longer than the brown bar for females.

12. The brown bar for males is shorter than the green bar for males.

OBJECTIVE B To read a broken-line graph

Blood Sugar A nurse monitors the blood glucose levels of a patient during the patient's hospital stay. The broken-line graph in Figure 24 shows the readings during the three-day period. Use this graph for Exercises 13 to 16.

13. What was the patient's blood sugar level at 10 P.M. on Tuesday?

14. What was the highest blood sugar reading recorded? When was it recorded?

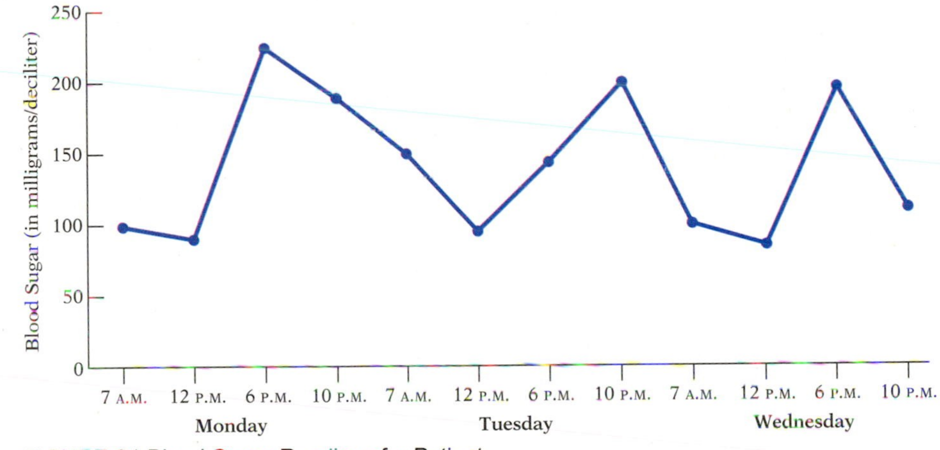

FIGURE 24 Blood Sugar Readings for Patient

15. If a normal blood sugar reading is between 70 and 125 milligrams/deciliter, how many readings are outside the normal range?

16. Find the difference between the blood sugar reading at 7 A.M. Monday and the reading at 7 A.M. Tuesday.

Influenza Vaccine The broken-line graph in Figure 25 shows the percent of adults in the United States over 18 years old who took the influenza vaccine during recent years. Use this graph for Exercises 17 to 19.

17. What percent of U.S. adults aged 18 and older took the influenza vaccine in 2007?

18. Between what two consecutive years was there a decrease in the percent of U.S. adults aged 18 and older taking the influenza vaccine?

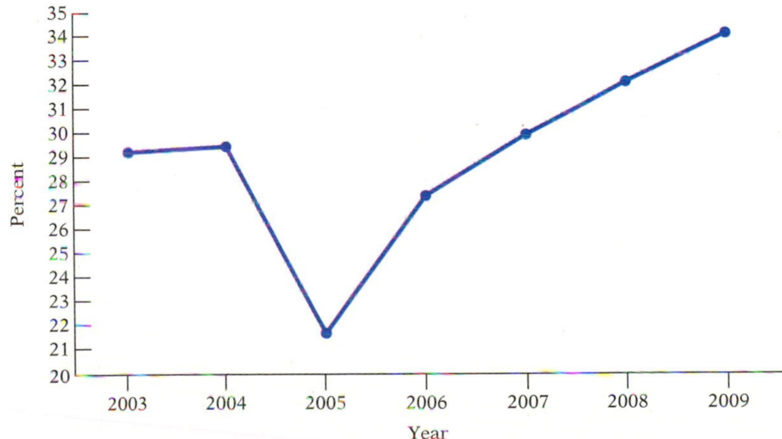

FIGURE 25 Percent of American Adults Aged 18 and Older Taking the Influenza Vaccine

Source: CDC/NCHS, National Health Interview Survey. Health, United States 2010

19. What was the percent increase in U.S. adults aged 18 and older taking the influenza vaccine from 2008 to 2009?

Health The double-broken-line graph in Figure 26 shows the number of Calories per day that should be consumed by women and men in various age groups. Use this graph for Exercises 20 to 22.

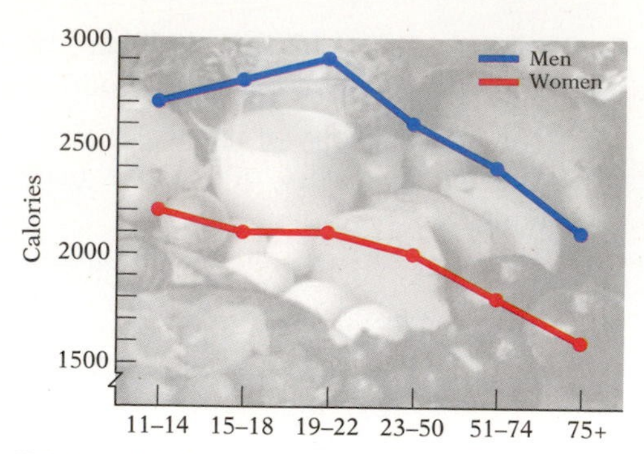

FIGURE 26 Recommended number of Calories per day for women and men
Source: Numbers, by Andrea Sutcliffe (HarperCollins)

20. What is the difference between the number of Calories recommended for men and the number recommended for women 19 to 22 years of age?

21. People of what age and gender have the lowest recommended number of Calories?

22. Find the ratio of the number of Calories recommended for women 15 to 18 years old to the number recommended for women 51 to 74 years old.

For Exercises 23 and 24, each statement refers to a line graph (not shown) that displays the population of a particular state every 10 years between 1950 and 2000. Determine whether the statement is *true* or *false.*

23. If the population decreased between 1990 and 2000, then the segment joining the point for 1990 and the point for 2000 slants down from left to right.

24. If the points for 1960 and 1970 are connected by a horizontal line, the population in 1970 was the same as the population in 1960.

Applying the Concepts

Wind Power The graph in Figure 27 shows how wind power capacity increased from 2000 to 2007 for the states with the largest wind energy capacity, Texas and California. Use this graph for Exercises 25 and 26.

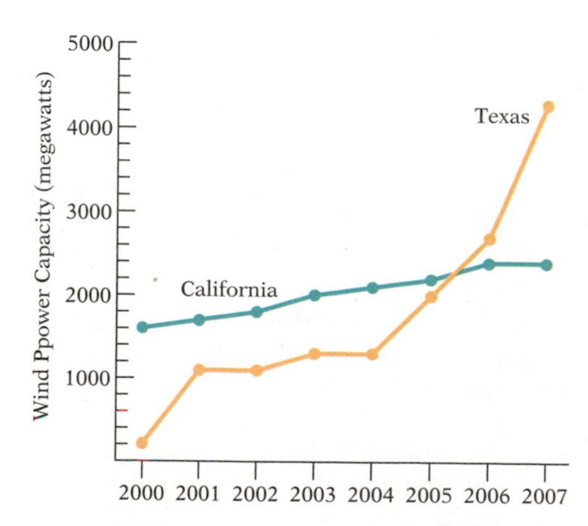

FIGURE 27 Wind power capacity in Texas and California
Source: www.eere.energy.gov

25. Create a table that shows the wind power capacity of each state for each of the years 2000 through 2007.

26. Create a table that shows the difference in the wind power capacities of Texas and California for each year from 2000 to 2007, and indicate which state had the greater wind power capacity. During which years did the wind power capacity of Texas exceed that of California?

SECTION

7.3 Histograms and Frequency Polygons

OBJECTIVE A To read a histogram

A research group measured the fuel usage of 92 cars. The results are recorded in the histogram in Figure 28. A **histogram** is a special type of bar graph. The width of each bar corresponds to a range of numbers called a **class interval.** The height of each bar corresponds to the number of occurrences of data in each class interval and is called the **class frequency.**

Class Intervals (miles per gallon)	*Class Frequencies (number of cars)*
18–19.9	12
20–21.9	19
22–23.9	24
24–25.9	17
26–27.9	15
28–29.9	5

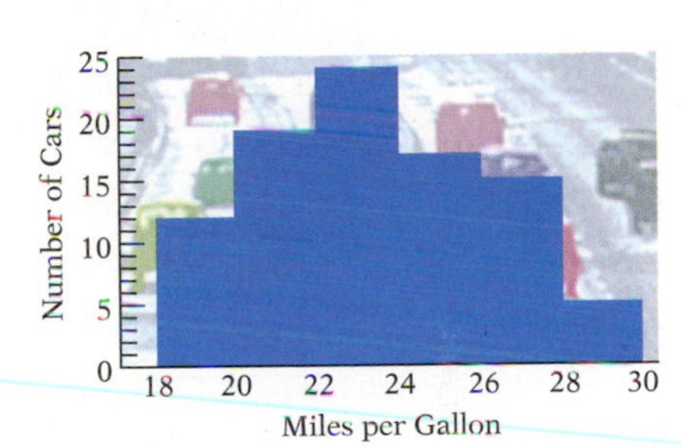

FIGURE 28

Twenty-four cars get between 22 and 24 miles per gallon.

A walk-in clinic has 85 employees. Their hourly wages are recorded in the histogram in Figure 29.

The ratio of the number of employees whose hourly wage is between \$14 and \$16 to the total number of employees is $\frac{17 \text{ employees}}{85 \text{ employees}} = \frac{1}{5}$.

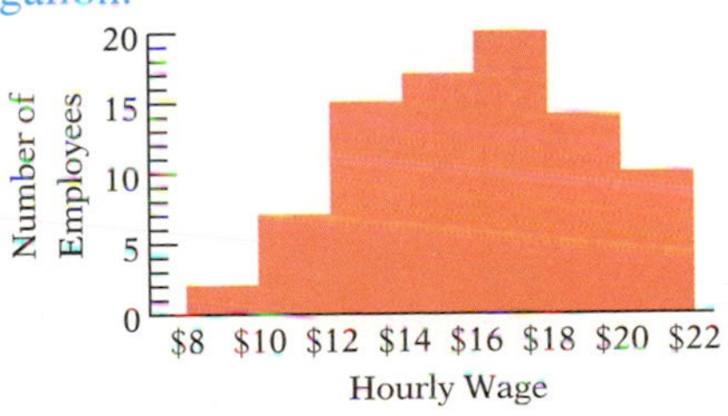

FIGURE 29

EXAMPLE • 1

Use Figure 29 to find the number of employees whose hourly wage is between \$16 and \$20.

Strategy

To find the number of employees:

- Read the histogram to find the number of employees whose hourly wage is between \$16 and \$18 and the number whose hourly wage is between \$18 and \$20.
- Add the two numbers.

Solution

Number with wages between \$16 and \$18: 20
Number with wages between \$18 and \$20: 14

20 + 14 = 34

34 employees have an hourly wage between \$16 and \$20.

YOU TRY IT • 1

Use Figure 29 to find the number of employees whose hourly wage is between \$10 and \$14.

Your strategy

Your solution

Solution on p. S19

OBJECTIVE B To read a frequency polygon

Take Note

The green portion of the graph at the right is a histogram. The red portion of the graph is a frequency polygon

The third grade boys at Newlin Elementary School were weighed during a recent school health fair. The results are recorded in the frequency polygon in Figure 30. A **frequency polygon** is a graph that displays information in a manner similar to a histogram. A dot is placed above the center of each class interval at a height corresponding to that class's frequency. The dots are then connected to form a broken-line graph. The center of a class interval is called the **class midpoint.**

Weights in Pounds	*Class Midpoint*	*Class Frequency*
40–49	45	3
50–59	55	12
60–69	65	26
70–79	75	15
80–89	85	1

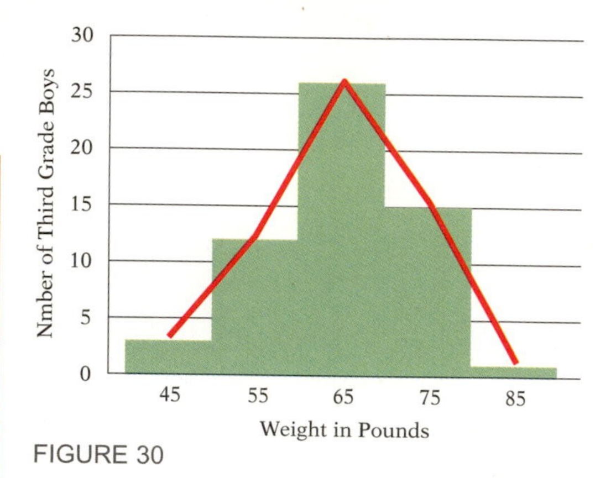

FIGURE 30

There are 26 third grade boys whose weights ranged from 60 to 69 pounds.

The per capita incomes in a recent year for the 50 states are recorded in the frequency polygon in Figure 31.

The number of states with a per capita income between $28,000 and $32,000 is 17.

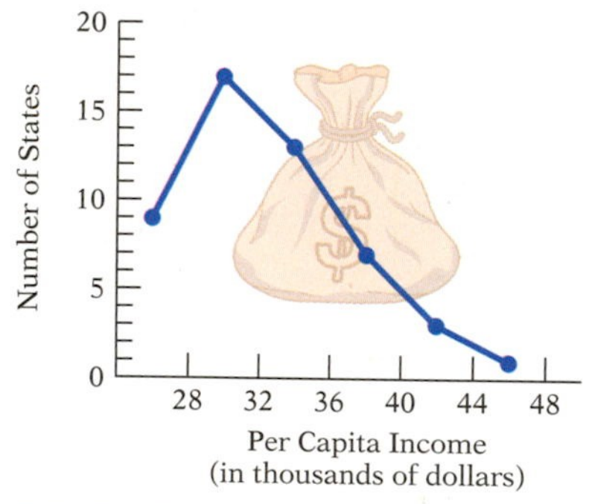

FIGURE 31

Source: Bureau of Economic Analysis

EXAMPLE • 2

According to Figure 31, what percent of the states have a per capita income between $28,000 and $32,000?

Strategy

To find the percent, solve the basic percent equation for percent. The base is 50. The amount is 17.

Solution

Percent × base = amount

$n \times 50 = 17$

$n = 17 \div 50$

$n = 0.34$

34% of the states have a per capita income between $28,000 and $32,000.

YOU TRY IT • 2

Use Figure 31 to find the ratio of the number of states with a per capita income between $36,000 and $40,000 to the number with a per capita income between $44,000 and $48,000.

Your strategy

Your solution

Solution on p. S19

7.3 EXERCISES

OBJECTIVE A To read a histogram

Customer Credit A total of 50 monthly credit account balances were recorded. Figure 32 is a histogram of these data. Use this figure for Exercises 1 to 4.

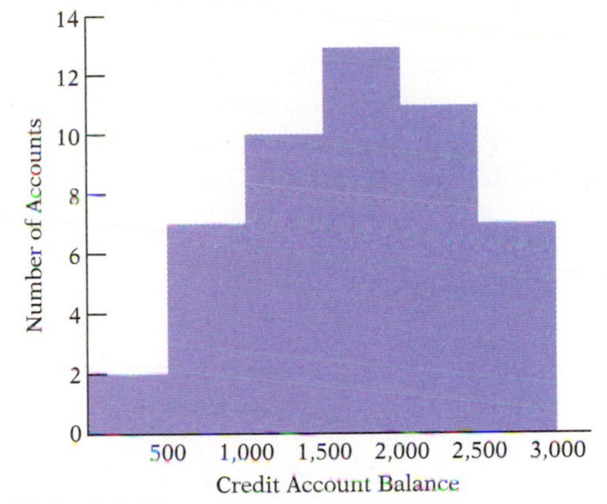

FIGURE 32

1. How many account balances were between $1500 and $2000?

2. How many account balances were less than $2000?

3. What percent of the account balances were between $2000 and $2500?

4. What percent of the account balances were greater than $1500?

Body Temperatures Most health statistics list normal body temperature as 98.6°. A recent study of 30 healthy adults recorded their body temperatures at 8 A.M. and then displayed the results in the histogram in Figure 33. Use this figure for Exercises 5 to 9.

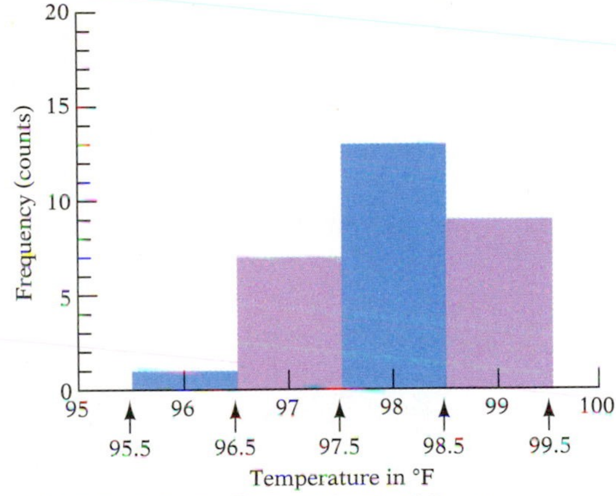

FIGURE 33 Body Temperatures at 8 A.M.

5. How many body temperatures were lower than 98.5°?

6. Find the percent of body temperatures that were higher than 98.5°.

7. Find the ratio of the number of adults with temperatures between 96.5° and 97.5° to the total number of adults in the study. Write the ratio as a fraction in simplest form.

8. Which class has the lowest frequency?

9. If you had to draw a conclusion about the normal body temperature for an adult based on just this histogram, what would you report?

Cholesterol Readings The total cholesterol readings for 40 female patients seen this week at a doctor's office were recorded. Figure 34 is a histogram of these data. Use this figure for Exercises 10 to 12.

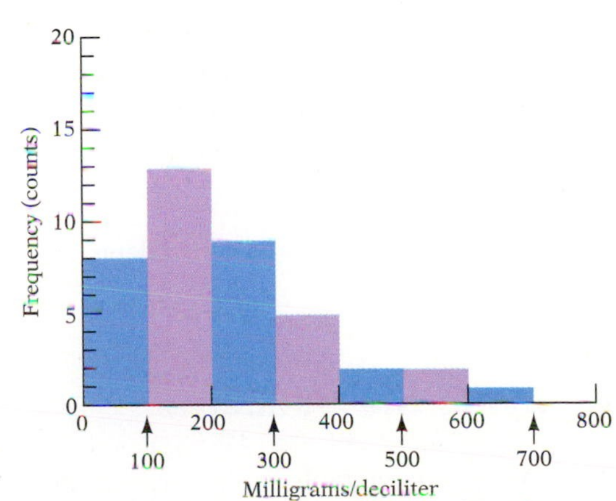

FIGURE 34 Cholesterol Readings for Female Patients

10. How many patients had total cholesterol readings between 200 and 300?

11. Total cholesterol readings lower than 200 are considered normal. How many patients had readings lower than 200?

12. If a patient has a total cholesterol reading higher than 300, a doctor may prescribe a cholesterol-lowering medication as well as a change in diet. What percent of these patients had total cholesterol readings higher than 300?

OBJECTIVE B To read a frequency polygon

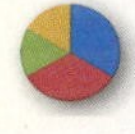

Marathons The frequency polygon in Figure 35 shows the approximate numbers of runners in the 2008 Boston Marathon to finish in each of the given time slots (times are given in hours and minutes). Use this figure for Exercises 13 to 15.

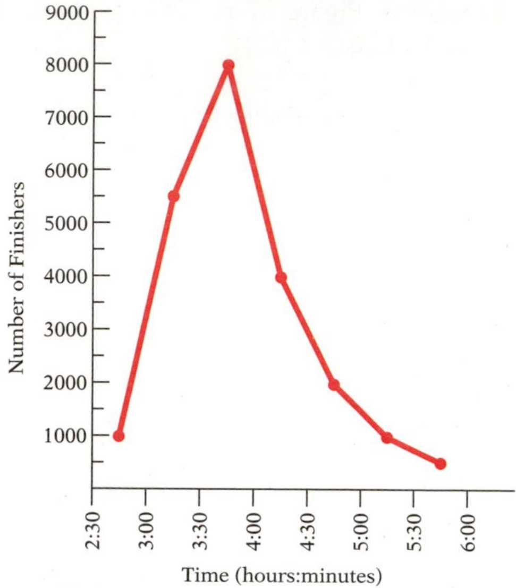

FIGURE 35

Source: www.marathonguide.com

13. Determine the approximate number of runners who finished with times between $2\frac{1}{2}$ hours and 6 hours.

14. Find the approximate number of marathoners who finished with times of more than 4 hours.

15. State whether the frequency polygon can be used to draw the following conclusion:

More runners had finishing times between 4 and $4\frac{1}{2}$ hours than had finishing times between $4\frac{1}{2}$ and 5 hours.

Heights The heights, in inches, of the women in a nursing class were recorded. The results are displayed in the frequency polygon in Figure 36. Use this figure for Exercises 16 to 18.

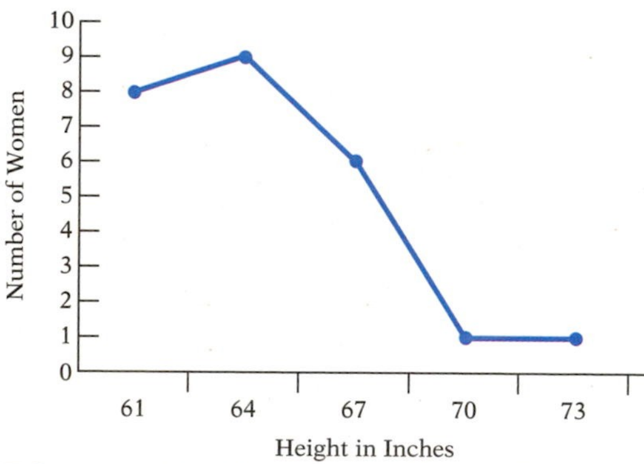

FIGURE 36 Heights of Female Nursing Students

16. How many women are between 63 inches and 65 inches tall?

17. What percent of the women are taller than 68 inches?

18. Is it possible to know how many of the women were 67 inches tall? Explain.

Education The frequency polygon in Figure 37 shows the distribution of scores of the approximately 1,548,000 students who took the SAT Math exam in 2010. Use this figure for Exercises 19 to 21.

19. Approximately how many students scored between 700 and 800 on the SAT Math exam?

20. Approximately what percent of these students scored between 400 and 500 on the SAT Math exam? Round to the nearest tenth of a percent.

21. Approximately how many students scored lower than 500?

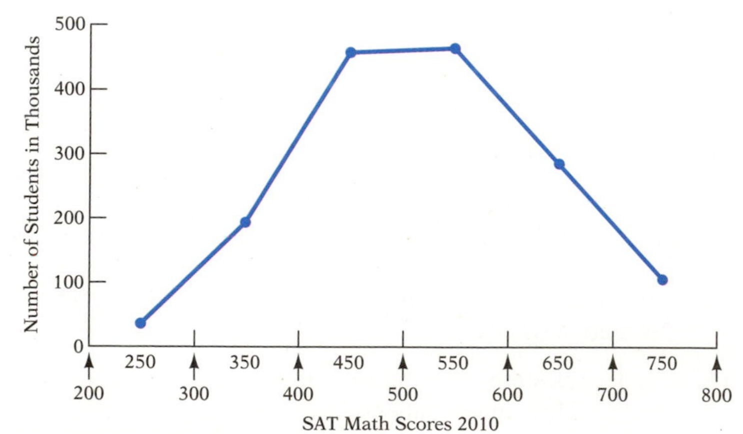

FIGURE 37 SAT Math Scores 2010

Source: Educational Testing Service

Applying the Concepts

22. Write a paragraph explaining the difference between a histogram and a bar graph.

SECTION 7.4 Statistical Measures

OBJECTIVE A To find the mean, median, and mode of a distribution

The average score on the math portion of the SAT was 432. The EPA estimates that a 2007 Toyota Camry Hybrid averages 35 miles per gallon on the highway. The average rainfall for portions of Kauai is 350 inches per year. Each of these statements uses one number to describe an entire collection of numbers. Such a number is called an *average.*

In statistics there are various ways to calculate an average. Three of the most common—*mean, median,* and *mode*—are discussed here.

An automotive engineer tests the miles-per-gallon ratings of 15 cars and records the results as follows:

Miles-per-Gallon Ratings of 15 Cars														
25	22	21	27	25	35	29	31	25	26	21	39	34	32	28

The **mean** of the data is the sum of the measurements divided by the number of measurements. The symbol for the mean is $\bar{x}$.

Formula for the Mean

$$\bar{x} = \frac{\text{sum of the data values}}{\text{number of data values}}$$

To find the mean for the data above, add the numbers and then divide by 15.

$$\bar{x} = \frac{25 + 22 + 21 + 27 + 25 + 35 + 29 + 31 + 25 + 26 + 21 + 39 + 34 + 32 + 28}{15}$$

$$= \frac{420}{15} = 28$$

The mean number of miles per gallon for the 15 cars tested was 28 miles per gallon.

The mean is one of the most frequently computed averages. It is the one that is commonly used to calculate a student's performance in a class.

Integrating Technology

When using a calculator to calculate the mean, use parentheses to group the sum in the numerator.

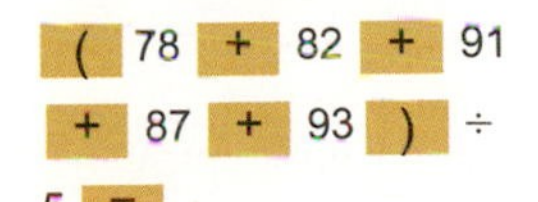

HOW TO 1 The test scores for a student taking American history were 78, 82, 91, 87, and 93. What was the mean score for this student?

Strategy
To find the mean, divide the sum of the test scores by 5, the number of scores.

Solution

$$\bar{x} = \frac{78 + 82 + 91 + 87 + 93}{5} = \frac{431}{5} = 86.2$$

The mean score for the history student was 86.2.

The **median** of the data is the number that separates the data into two equal parts when the numbers are arranged from least to greatest (or from greatest to least). There is an equal number of values above the median and below the median.

To find the median of a set of numbers, first arrange the numbers from least to greatest. The median is the number in the middle.

The result of arranging the miles-per-gallon ratings given on the previous page from least to greatest is shown below.

21 21 22 25 25 25 26 27 28 29 31 32 34 35 39

7 values below the median — Middle number **Median** — 7 values above the median

The median is 27 miles per gallon.

If data contain an *even* number of values, the median is the mean of the two middle numbers.

Tips for Success

Word problems are difficult because we must read the problem, determine the quantity we must find, think of a method to find it, actually solve the problem, and then check the answer. In short, we must devise a *strategy* and then use that strategy to find the *solution*. See *AIM for Success* at the front of the book.

HOW TO • 2 The selling prices of the last six homes sold by a real estate agent were $275,000, $250,000, $350,000, $230,000, $345,000, and $290,000. Find the median selling price of these homes.

Strategy

To find the median, arrange the numbers from least to greatest. Because there is an even number of values, the median is the mean of the two middle numbers.

Solution

230,000 250,000 275,000 290,000 345,000 350,000

Middle 2 numbers: 275,000 290,000

$$\text{Median} = \frac{275{,}000 + 290{,}000}{2} = 282{,}500$$

The median selling price was $282,500.

The **mode** of a set of numbers is the value that occurs most frequently. If a set of numbers has no number occurring more than once, then the data have no mode.

Here again are the data for the gasoline mileage ratings of 15 cars.

Miles-per-Gallon Ratings of 15 Cars														
25	22	21	27	25	35	29	31	25	26	21	39	34	32	28

25 is the number that occurs most frequently.

The mode is 25 miles per gallon.

Note from the miles-per-gallon example that the mean, median, and mode may be different.

EXAMPLE • 1

Twenty students were asked the number of units in which they were enrolled. The responses were as follows:

15	12	13	15	17	18	13	20	9	16
14	10	15	12	17	16	6	14	15	12

Find the mean number of units taken by these students.

Strategy

To find the mean number of units:

- Find the sum of the 20 numbers.
- Divide the sum by 20.

Solution

$15 + 12 + 13 + 15 + 17 + 18 + 13 + 20 + 9 + 16 + 14 + 10 + 15 + 12 + 17 + 16 + 6 + 14 + 15 + 12 = 279$

$$\bar{x} = \frac{279}{20} = 13.95$$

The mean is 13.95 units.

YOU TRY IT • 1

The blood sugar readings in milligrams/deciliter for a patient over a two-day period were as follows:

85	120	105	155	204	260
180	95	135	150	180	165

Find the mean blood sugar reading for this patient. Round to the nearest whole number.

Your strategy

Your solution

EXAMPLE • 2

The starting hourly wages for a nursing assistant for six different work locations are \$12.50, \$11.25, \$10.90, \$11.56, \$13.75, and \$14.55. Find the median starting hourly wage.

Strategy

To find the median starting hourly wage:

- Arrange the numbers from least to greatest.
- Because there is an even number of values, the median is the mean of the two middle numbers.

Solution

10.90, 11.25, 11.56, 12.50, 13.75, 14.55

$$\text{Median} = \frac{11.56 + 12.50}{2} = 12.03$$

The median starting hourly wage is \$12.03.

YOU TRY IT • 2

The amounts of weight lost, in pounds, by 10 participants in a 6-month weight-reduction program were 22, 16, 31, 14, 27, 16, 29, 31, 40, and 10. Find the median weight loss for these participants.

Your strategy

Your solution

Solutions on p. S19

OBJECTIVE B To draw a box-and-whiskers plot

Recall from the last objective that an average is one number that helps to describe all the numbers in a set of data. For example, we know from the following statement that Erie gets a lot of snow each winter.

The average annual snowfall in Erie, Pennsylvania, is 85 inches.

Now look at these two statements.

The average annual temperature in San Francisco, California, is 57°F.

The average annual temperature in St. Louis, Missouri, is 57°F.

San Francisco

The average annual temperature in both cities is the same. However, we do not expect the climate in St. Louis to be like San Francisco's climate. Although both cities have the same average annual temperature, their temperature ranges differ. In fact, the difference between the average monthly high temperatures in July and January in San Francisco is 14°F, whereas the difference between the average monthly high temperatures in July and January in St. Louis is 50°F.

Note that for this example, a single number (the average annual temperature) does not provide us with a very comprehensive picture of the climate of either of these two cities.

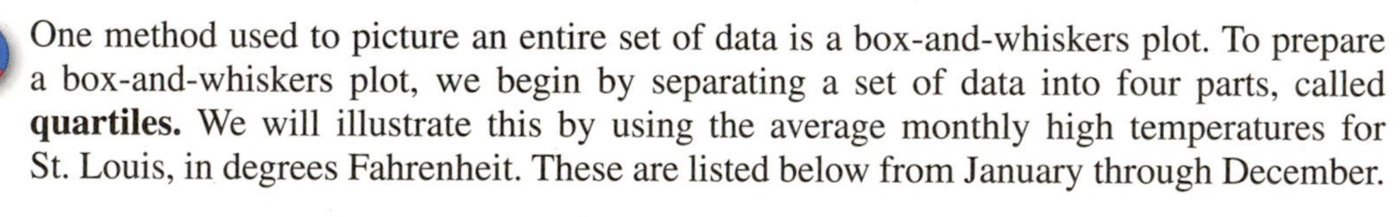

One method used to picture an entire set of data is a box-and-whiskers plot. To prepare a box-and-whiskers plot, we begin by separating a set of data into four parts, called **quartiles.** We will illustrate this by using the average monthly high temperatures for St. Louis, in degrees Fahrenheit. These are listed below from January through December.

39	47	58	72	81	88	89	89	85	76	49	47

Source: The Weather Channel

St. Louis

First list the numbers in order from least to greatest and determine the median.

39	47	47	49	58	72	76	81	85	88	89	89

↑
Median = 74

Now find the median of the data values below the median. The median of the data values below the median is called the **first quartile,** symbolized by Q_1. Also find the median of the data values above the median. The median of the data values above the median is called the **third quartile,** symbolized by Q_3.

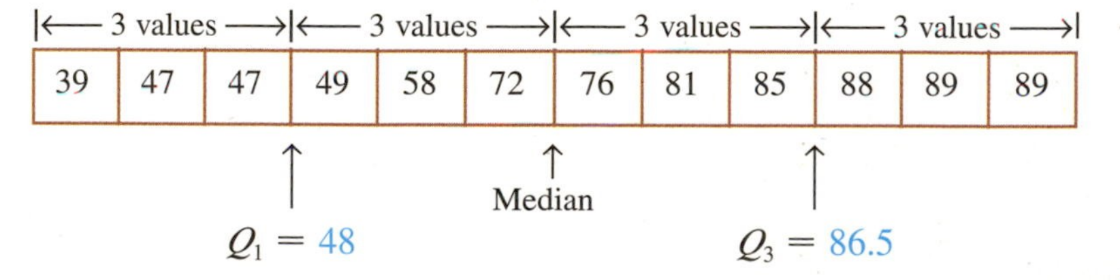

39	47	47	49	58	72	76	81	85	88	89	89

The first quartile, Q_1, is the number that one-quarter of the data lie below. This means that 25% of the data lie below the first quartile. The third quartile, Q_3, is the number that one-quarter of the data lie above. This means that 25% of the data lie above the third quartile.

The **range** of a set of numbers is the difference between the greatest number and the least number in the set. The range describes the spread of the data. For the data above,

$$\text{Range} = \text{greatest value} - \text{least value} = 89 - 39 = 50$$

Take Note

50% of the data in a distribution lie in the interquartile range.

The **interquartile range** is the difference between the third quartile, Q_3, and the first quartile, Q_1. For the data above,

$$\text{Interquartile range} = Q_3 - Q_1 = 86.5 - 48 = 38.5$$

The interquartile range is the distance that spans the "middle" 50% of the data values. Because it excludes the bottom fourth of the data values and the top fourth of the data values, it excludes any extremes in the numbers of the set.

A **box-and-whiskers plot,** or **boxplot,** is a graph that shows five numbers: the least value, the first quartile, the median, the third quartile, and the greatest value. Here are these five values for the data on St. Louis temperatures.

The least number	39
The first quartile, Q_1	48
The median	74
The third quartile, Q_3	86.5
The greatest number	89

Take Note

The number line at the right is shown as a reference only. It is not a part of the boxplot. It shows that the numbers labeled on the boxplot are plotted according to the distances on the number line above it.

Think of a number line that includes the five values listed above. With this in mind, mark off the five values. Draw a box that spans the distance from Q_1 to Q_3. Draw a vertical line the height of the box at the median.

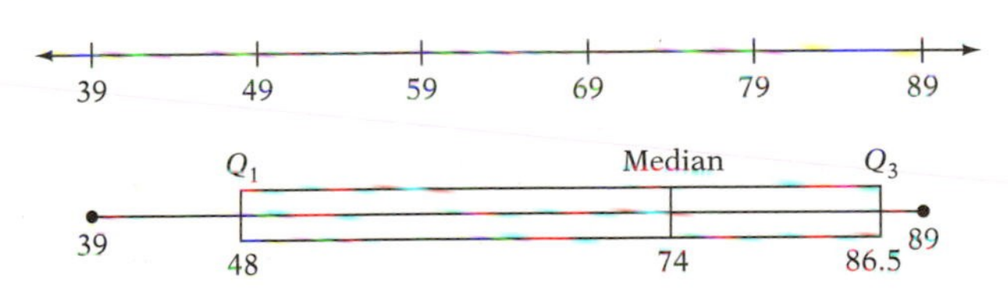

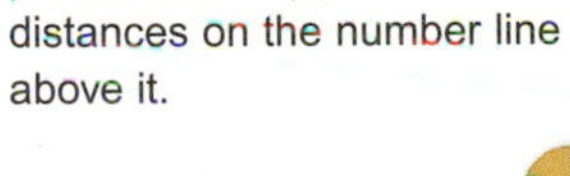

Listed below are the average monthly high temperatures for San Francisco.

57	60	61	64	68	71	71	73	74	73	60	59

Source: The Weather Channel

We can perform the same calculations on these data to determine the five values needed for the box-and-whiskers plot.

The least number	57
The first quartile, Q_1	60
The median	66
The third quartile, Q_3	72
The greatest number	74

Take Note

It is the "whiskers" on the box-and-whiskers plot that show the range of the data. The "box" on the box-and-whiskers plot shows the interquartile range of the data.

The box-and-whiskers plot is shown at the right with the same scale used for the data on the St. Louis temperatures.

Note that by comparing the two boxplots, we can see that the range of temperatures in St. Louis is greater than the range of temperatures in San Francisco. For the St. Louis temperatures, there is a greater spread of the data below the median than above the median, whereas the data above and below the median of the San Francisco boxplot are spread nearly equally.

HOW TO • 3 The weights of babies born on New Year's Day in a local hospital were 88 ounces, 92 ounces, 140 ounces, 105 ounces, 99 ounces, 120 ounces, 112 ounces, 95 ounces, and 110 ounces. Draw a box-and-whiskers plot of the data, and determine the interquartile range.

Strategy

To draw the box-and-whiskers plot, arrange the data from least to greatest. Then find the median, Q_1, and Q_3. Use the least value, Q_1, the median, Q_3, and the greatest value to draw the box-and-whiskers plot.

To find the interquartile range, find the difference between Q_3 and Q_1.

Solution

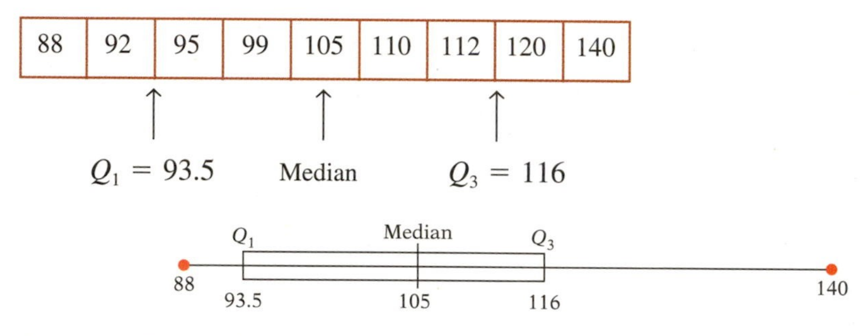

88	92	95	99	105	110	112	120	140

Interquartile range $= Q_3 - Q_1 = 116 - 93.5 = 22.5$

The interquartile range is 22.5 ounces.

✓ Take Note

Note that the left whisker in this box-and-whiskers plot is quite short. This illustrates a set of data in which the median is closer to the lowest data value. If the two whiskers are approximately the same length, then the least and greatest values are about the same distance from the median. See Example 3 below.

EXAMPLE • 3

The amount of weight loss for seven clients of a weight-loss clinic during the month of May was 5 pounds, 3 pounds, 4 pounds, 2 pounds, 3 pounds, 1 pound, and 4.5 pounds. Draw a box-and-whisker plot of the data.

Strategy

To draw the box-and-whiskers plot:

- Arrange the data from least to greatest.
- Find the median, Q_1, and Q_3.
- Use the least value, Q_1, the median, Q_3, and the greatest value to draw the box-and-whiskers plot.

Solution

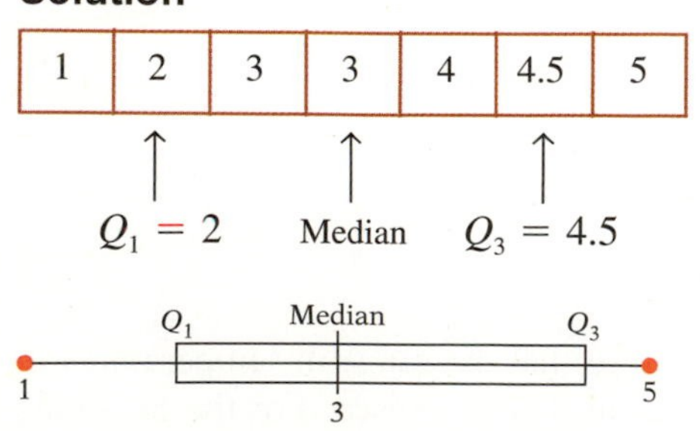

1	2	3	3	4	4.5	5

YOU TRY IT • 3

Seven different bran cereals have the following numbers of calories per serving: 92, 70, 120, 50, 75, 80 and 90. Draw a box-and-whiskers plot of the data.

Your strategy

Your solution

Solution on p. S19

7.4 EXERCISES

OBJECTIVE A **To find the mean, median, and mode of a distribution**

1. State whether the mean, median, or mode is being used.
   a. Half of the nurses at the hospital make less than $55,000 annually.

   b. The average bill for lunch at the college union is $11.95.
   c. The college bookstore sells more green college sweatshirts than any other color.

   d. In a recent year, there were as many people age 26 and younger in the world as there were people age 26 and older.
   e. The majority of full-time students carry a load of 12 credit hours per semester.

   f. The average waiting time for patients seeking treatment at the emergency room during the last seven days was 95 minutes.

2. **Heart Rates** The resting heart rates for a group of men involved in a clinical trial of a new medication were recorded at the beginning of the study. The results in beats per minute were 69, 84, 66, 68, 70, 78, 88, 81, 79, 73, 90, and 78. Calculate the mean, the median, and the mode of the resting heart rates of the group.

3. **Nutrition** A consumer research group recently reported the amount of sugar in 12-ounce servings of six popular soft drinks. The results in grams were 40.5, 40, 47, 40, 39, and 41. Calculate the mean, the median, and the mode of the amount of sugar in these soft drinks.

4. **Blood sugar** A patient's blood sugar readings were recorded during a recent stay in the hospital. The results in milligrams/deciliter were 216, 204, 165, 180, 192, 151, 134, 150, 120, 102, 100, and 98.
   a. Calculate the mean blood sugar reading for this patient.
   b. Calculate the median blood sugar reading for this patient.

5. **Consumerism** A consumer research group purchased identical items in eight drugstores. The costs for the purchased items were $85.89, $92.12, $81.43, $80.67, $88.73, $82.45, $87.81, and $85.82. Calculate the mean and the median costs of the purchased items.

6. **Ages** There are 15 students in the medical assisting class at the community college. Their ages are 21, 25, 28, 31, 40, 27, 23, 21, 42, 35, 36, 28, 21, 20, and 45. Find the mean and median ages of this class of students. Round to the nearest tenth.

7. **Health Plans** Eight health maintenance organizations (HMOs) presented group health insurance plans to a company. The monthly rates per employee were $423, $390, $405, $396, $426, $355, $404, and $430. Calculate the mean and the median monthly rates for these eight companies.

8. **Hospital Stay** The number of days that 12 patients stayed in a hospital after hip replacement was recorded during January and are shown in the table below. Find the mean and median length of hospital stay for these patients. Round to the nearest tenth.

5	3	4	4	3	6
4	5	5	3	5	4

9. **Life Expectancy** The life expectancies, in years, in ten selected Central and South American countries are given at the right.
   a. Find the mean life expectancy in this group of countries.
   b. Find the median life expectancy in this group of countries.

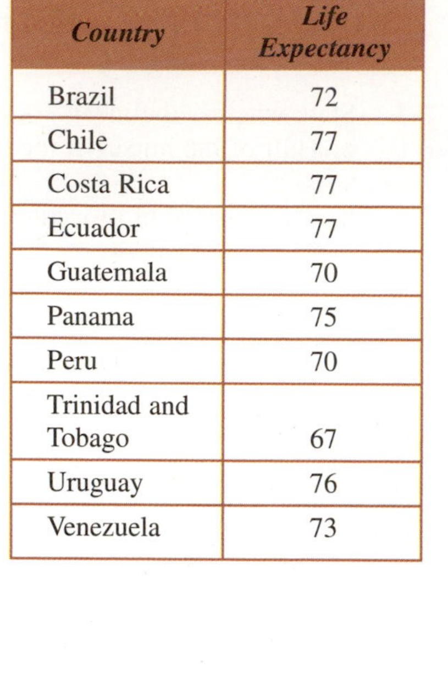

Country	*Life Expectancy*
Brazil	72
Chile	77
Costa Rica	77
Ecuador	77
Guatemala	70
Panama	75
Peru	70
Trinidad and Tobago	67
Uruguay	76
Venezuela	73

10. **Education** Your scores on six biology tests were 78, 92, 95, 77, 94, and 88. If an "average score" of 90 receives an A for the course, which average, the mean or the median, would you prefer that the instructor use?

11. **Education** One student received scores of 85, 92, 86, and 89. A second student received scores of 90, 97, 91, and 94 (exactly 5 points more on each exam). Are the means of the two students the same? If not, what is the relationship between the means of the two students?

12. **Prescription Drugs** The number of retail prescription drug sales from 2003 to 2009 are given in the table below. The numbers represent sales in the millions.

2003	*2004*	*2005*	*2006*	*2007*	*2008*	*2009*
3215	3274	3279	3419	3510	3535	3606

   a. Calculate the mean number of prescription drug sales for these years. Round to the nearest tenth of a million.
   b. Find the median number of drug sales.

   c. If the year 2003 were eliminated from the data, how would that affect the mean? The median?

OBJECTIVE B To draw a box-and-whiskers plot

13. a. What percent of the data in a set of numbers lie above Q_3?
    b. What percent of the data in a set of numbers lie above Q_1?
    c. What percent of the data in a set of numbers lie below Q_3?
    d. What percent of the data in a set of numbers lie below Q_1?

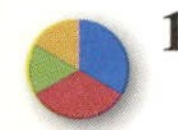

14. **U.S. Presidents** The box-and-whiskers plot below shows the distribution of the ages of presidents of the United States at the time of their inauguration.
    a. What is the youngest age in the set of data?
    b. What is the oldest age?
    c. What is the first quartile?
    d. What is the third quartile?
    e. What is the median?
    f. Find the range.
    g. Find the interquartile range.

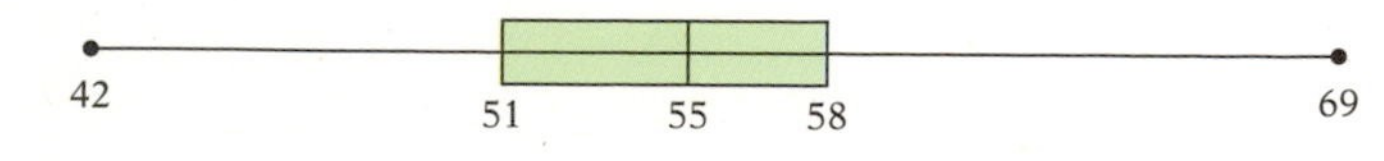

15. **Compensation** The box-and-whiskers plot below shows the distribution of median incomes for 50 states and the District of Columbia. What is the lowest value in the set of data? The highest value? The first quartile? The third quartile? The median? Find the range and the interquartile range.

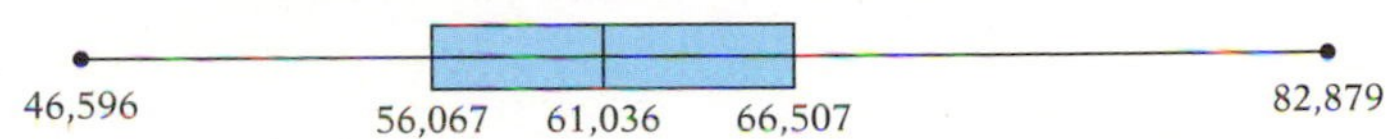

16. **Education** An aptitude test was taken by 200 pre-nursing students at Coastal Community College. The box-and-whiskers plot at the right shows the distribution of their scores.

a. How many students scored over 88?
b. How many students scored below 72?
c. How many scores are represented in each quartile?
d. What percent of the students had scores of at least 54?

17. **Health** The cholesterol levels for 80 adults were recorded and then displayed in the box-and-whiskers plot shown at the right.

a. How many adults had a cholesterol level above 217?
b. How many adults had a cholesterol level below 254?
c. How many cholesterol levels are represented in each quartile?
d. What percent of the adults had a cholesterol level of not more than 198?

18. **Body Temperatures** The body temperatures of 19 healthy adults in a clinical drug trial were measured and recorded in the table below.

Adult Body Temperatures									
98.7°	98.2°	98.9°	98.4°	98.0°	97.1°	97.0°	98.4°	96.5°	98.2°
98.6°	97.8°	97.6°	98.5°	99.0°	97.3°	98.0°	97.6°	99.2°	

a. Find the range, the first quartile, the third quartile, and the interquartile range.

b. Draw a box-and-whiskers plot of the data.
c. Is the value 97.9° in the interquartile range?

19. **Environment** Carbon dioxide is among the gases that contribute to global warming. The world's biggest emitters of carbon dioxide are listed below. The figures are emissions in millions of metric tons per year.
a. Find the range, the first quartile, the third quartile, and the interquartile range.

b. Draw a box-and-whiskers plot of the data.
c. What data value is responsible for the long whisker at the right?

Carbon Dioxide Emissions (in millions of metric tons per year)			
Canada	0.64	Japan	1.26
China	5.01	Russian Federation	1.52
Germany	0.81	South Korea	0.47
India	1.34	United Kingdom	0.59
Italy	0.45	United States	6.05

Source: U.S. Department of Energy

20. **Meteorology** The average monthly amounts of rainfall, in inches, from January through December for Seattle, Washington, and Houston, Texas, are listed below.

a. Is the difference between the means greater than 1 inch?

b. What is the difference between the medians?

c. Draw a box-and-whiskers plot of each set of data. Use the same scale.

d. Describe the difference between the distributions of the data for Seattle and Houston.

Seattle	6.0	4.2	3.6	2.4	1.6	1.4	0.7	1.3	2.0	3.4	5.6	6.3
Houston	3.2	3.3	2.7	4.2	4.7	4.1	3.3	3.7	4.9	3.7	3.4	3.7

Source: The Weather Channel

21. **Meteorology** The average monthly amounts of rainfall, in inches, from January through December for Orlando, Florida, and Portland, Oregon, are listed below.

a. Is the difference between the means greater than 1 inch?

b. What is the difference between the medians?

c. Draw a box-and-whiskers plot of each set of data. Use the same scale.

d. Describe the difference between the distributions of the data for Orlando and Portland.

Orlando	2.1	2.8	3.2	2.2	4.0	7.4	7.8	6.3	5.6	2.8	1.8	1.8
Portland	6.2	3.9	3.6	2.3	2.1	1.5	0.5	1.1	1.6	3.1	5.2	6.4

Source: The Weather Channel

22. Refer to the box-and-whiskers plot in Exercise 15. Which of the following fractions most accurately represents the fraction of states with median incomes less than $66,507?

(i) $\frac{1}{4}$ (ii) $\frac{1}{3}$ (iii) $\frac{1}{2}$ (iv) $\frac{3}{4}$

23. Write a set of data with five data values for which the mean, median, and mode are all 55.

Applying the Concepts

24. A set of data has a mean of 16, a median of 15, and a mode of 14. Which of these numbers must be a value in the data set? Explain your answer.

25. Explain each notation.

a. Q_1 **b.** Q_3 **c.** $\bar{x}$

26. The box in a box-and-whiskers plot represents 50%, or one-half, of the data in a set. Why is the box in Example 3 of this section not one-half of the entire length of the box-and-whiskers plot?

27. Create a set of data containing 25 numbers that would correspond to the box-and-whiskers plot shown at the right.

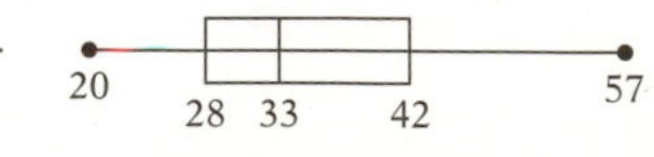

SECTION

7.5 Introduction to Probability

OBJECTIVE A To calculate the probability of simple events

Point of Interest

It was dice playing that led Antoine Gombaud, Chevalier de Mere, to ask Blaise Pascal, a French mathematician, to figure out the probability of throwing two sixes. Pascal and Pierre Fermat solved the problem, and their explorations led to the birth of probability theory.

A weather forecaster estimates that there is a 75% chance of rain. A state lottery director claims that there is a $\frac{1}{9}$ chance of winning a prize in a new game offered by the lottery. Each of these statements involves uncertainty to some extent. The degree of uncertainty is called **probability.** For the statements above, the probability of rain is 75%, and the probability of winning a prize in the new lottery game is $\frac{1}{9}$.

A probability is determined from an **experiment,** which is any activity that has an observable outcome. Examples of experiments include

Tossing a coin and observing whether it lands heads up or tails up
Interviewing voters to determine their preference for a political candidate
Drawing a card from a standard deck of 52 cards

All the possible outcomes of an experiment are called the **sample space** of the experiment. The outcomes are listed between braces. For example:

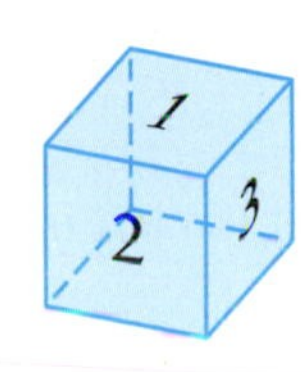

The number cube shown at the left is rolled once. Any of the numbers from 1 to 6 could show on the top of the cube. The sample space is

$$\{1, 2, 3, 4, 5, 6\}$$

A fair coin is tossed once. (A fair coin is one for which heads and tails have an equal chance of landing face up.) If H represents "heads up" and T represents "tails up," then the sample space is

$$\{H, T\}$$

An **event** is one or more outcomes of an experiment. For the experiment of rolling the six-sided cube described above, some possible events are

The number is even: $\{2, 4, 6\}$
The number is a multiple of 3: $\{3, 6\}$
The number is less than 10: $\{1, 2, 3, 4, 5, 6\}$

Note that in the last case, the event is the entire sample space.

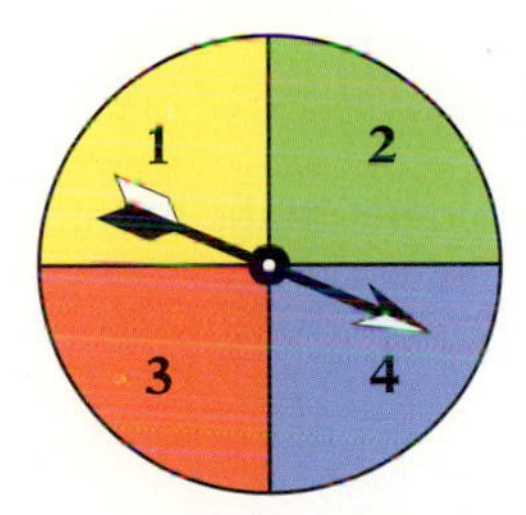

HOW TO • 1 The spinner at the left is spun once. Assume that the spinner does not come to rest on a line.

a. What is the sample space?

The arrow could come to rest on any one of the four sectors.
The sample space is $\{1, 2, 3, 4\}$.

b. List the outcomes in the event that the spinner points to an odd number.

$\{1, 3\}$

In discussing experiments and events, it is convenient to refer to the **favorable outcomes** of an experiment. These are the outcomes of an experiment that satisfy the requirements of a particular event.

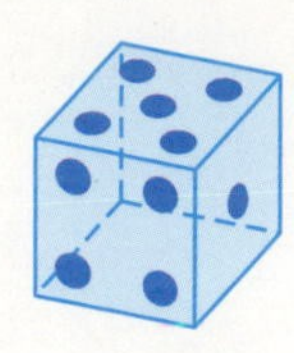

For instance, consider the experiment of rolling a fair die once. The sample space is

$$\{1, 2, 3, 4, 5, 6\}$$

and one possible event would be rolling a number that is divisible by 3. The outcomes of the experiment that are favorable to the event are 3 and 6:

$$\{3, 6\}$$

The outcomes of the experiment of tossing a fair coin are *equally likely.* Either one of the outcomes is just as likely as the other. If a fair coin is tossed once, the probability of a head is $\frac{1}{2}$, and the probability of a tail is $\frac{1}{2}$. Both events are equally likely. The theoretical probability formula, given below, applies to experiments for which the outcomes are equally likely.

Theoretical Probability Formula

The theoretical probability of an event is a fraction with the number of favorable outcomes of the experiment in the numerator and the total number of possible outcomes in the denominator.

$$\text{Probability of an event} = \frac{\text{number of favorable outcomes}}{\text{number of possible outcomes}}$$

A probability of an event is a number from 0 to 1 that tells us how likely it is that this outcome will happen.

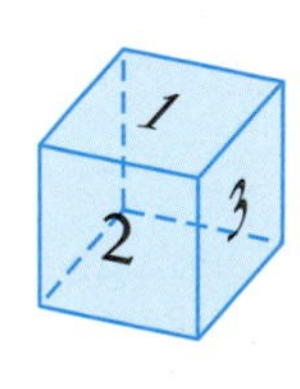

A probability of 0 means that the event is impossible.

The probability of getting a heads when rolling the die shown at the left is 0.

A probability of 1 means that the event must happen.

The probability of getting either heads or tails when tossing a coin is 1.

A probability of $\frac{1}{4}$ means that it is expected that the outcome will happen 1 in every 4 times the experiment is performed.

The phrase **at random** means that each card has an equal chance of being drawn.

HOW TO • 2 Each of the letters of the word *TENNESSEE* is written on a card, and the cards are placed in a hat. If one card is drawn at random from the hat, what is the probability that the card has the letter *E* on it?

Count the possible outcomes of the experiment.

There are 9 letters in *TENNESSEE.*

There are 9 possible outcomes of the experiment.

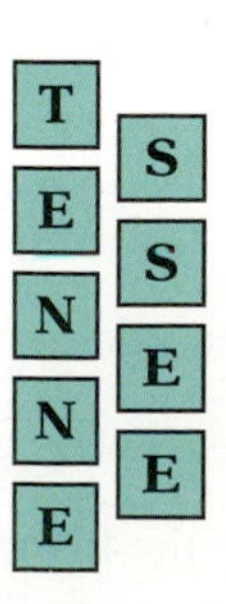

Count the number of outcomes of the experiment that are favorable to the event that a card with the letter *E* on it is drawn.

There are 4 cards with an *E* on them.

Use the probability formula.

$$\text{Probability of the event} = \frac{\text{number of favorable outcomes}}{\text{number of possible outcomes}} = \frac{4}{9}$$

The probability of drawing an *E* is $\frac{4}{9}$.

As just discussed, calculating the probability of an event requires counting the number of possible outcomes of an experiment and the number of outcomes that are favorable to the event. One way to do this is to list the outcomes of the experiment in a systematic way. Using a table is often helpful.

When two dice are rolled, the sample space for the experiment can be recorded systematically as in the following table.

Point of Interest

Romans called a die that was marked on four faces a *talus*, which meant "anklebone." The anklebone was considered an ideal die because it is roughly a rectangular solid and it has no marrow, so loose anklebones from sheep were more likely than other bones to be lying around after the wolves had left their prey.

Possible Outcomes from Rolling Two Dice

(1, 1)	(2, 1)	(3, 1)	(4, 1)	(5, 1)	(6, 1)
(1, 2)	(2, 2)	(3, 2)	(4, 2)	(5, 2)	(6, 2)
(1, 3)	(2, 3)	(3, 3)	(4, 3)	(5, 3)	(6, 3)
(1, 4)	(2, 4)	(3, 4)	(4, 4)	(5, 4)	(6, 4)
(1, 5)	(2, 5)	(3, 5)	(4, 5)	(5, 5)	(6, 5)
(1, 6)	(2, 6)	(3, 6)	(4, 6)	(5, 6)	(6, 6)

HOW TO • 3 Two dice are rolled once. Calculate the probability that the sum of the numbers on the two dice is 7.

Use the table above to count the number of possible outcomes of the experiment.

There are 36 possible outcomes.

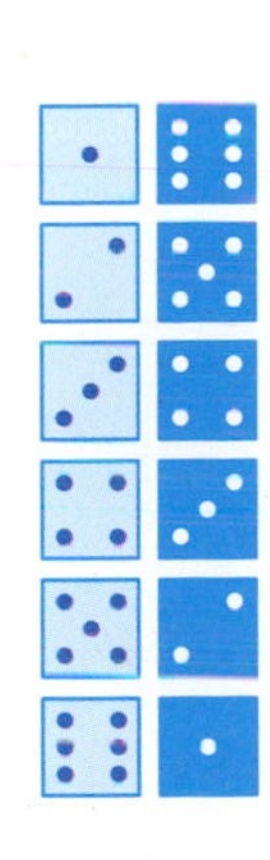

Count the number of outcomes of the experiment that are favorable to the event that a sum of 7 is rolled.

There are 6 favorable outcomes: (1, 6), (2, 5), (3, 4), (4, 3), (5, 2), and (6, 1).

Use the probability formula.

$$\text{Probability of the event} = \frac{\text{number of favorable outcomes}}{\text{number of possible outcomes}} = \frac{6}{36} = \frac{1}{6}$$

The probability of a sum of 7 is $\frac{1}{6}$.

The probabilities calculated above are theoretical probabilities. The calculation of a **theoretical probability** is based on theory—for example, that either side of a coin is equally likely to land face up or that each of the six sides of a fair die is equally likely to land face up. Not all probabilities arise from such assumptions.

An **empirical probability** is based on observations of certain events. For instance, a weather forecast of a 75% chance of rain is an empirical probability. From historical records kept by the weather bureau, when a similar weather pattern existed, rain occurred 75% of the time. It is theoretically impossible to predict the weather, and only observations of past weather patterns can be used to predict future weather conditions.

Empirical Probability Formula

The empirical probability of an event is the ratio of the number of observations of the event to the total number of observations.

$$\text{Probability of an event} = \frac{\text{number of observations of the event}}{\text{total number of observations}}$$

For example, a drug company is preparing to market a new medication. In running clinical trials, it is determined that 350 patients out of 2000 patients in the trial suffer from nausea while taking the medication. Therefore, the company determines the probability of someone having nausea as a side effect of this medication as the ratio of the 350 patients who experienced nausea over the total number of patients in the clinical trial, 2000. This number is usually written as a percent.

$$\frac{350}{2000} = \frac{7}{40} = 17.5\%$$

The probability of having nausea as a side effect of taking this medication is 17.5%.

EXAMPLE • 1

There are three choices, *a, b,* or *c,* for each of the two questions on a multiple-choice quiz. If the instructor randomly chooses which questions will have an answer of *a, b,* or *c,* what is the probability that the two correct answers on this quiz will be the same letter?

Strategy

To find the probability:

- List the outcomes of the experiment in a systematic way.
- Count the number of possible outcomes of the experiment.
- Count the number of outcomes of the experiment that are favorable to the event that the two correct answers on the quiz will be the same letter.
- Use the probability formula.

Solution

Possible outcomes: (a, a) (b, a) (c, a)
(a, b) (b, b) (c, b)
(a, c) (b, c) (c, c)

There are 9 possible outcomes.

There are 3 favorable outcomes:
(a, a), (b, b), (c, c)

$$\text{Probability} = \frac{\text{number of favorable outcomes}}{\text{number of possible outcomes}}$$

$$= \frac{3}{9} = \frac{1}{3}$$

The probability that the two correct answers will be the same letter is $\frac{1}{3}$.

YOU TRY IT • 1

A professor writes three true/false questions for a quiz. If the professor randomly chooses which questions will have a true answer and which will have a false answer, what is the probability that the test will have 2 true questions and 1 false question?

Your strategy

Your solution

Solution on pp. S19–S20

7.5 EXERCISES

OBJECTIVE A To calculate the probability of simple events

1. A coin is tossed four times. List all the possible outcomes of the experiment as a sample space.

2. Three cards—one red, one green, and one blue—are to be arranged in a stack. Using R for red, G for green, and B for blue, list all the different stacks that can be formed. (Some computer monitors are called RGB monitors for the colors red, green, and blue.)

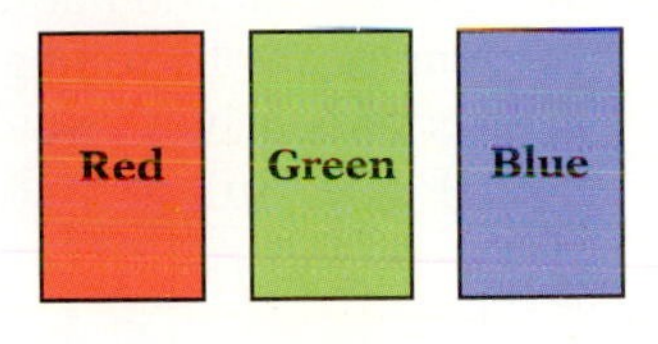

3. A tetrahedral die is one with four triangular sides. The sides show the numbers from 1 to 4. Say two tetrahedral dice are rolled. List all the possible outcomes of the experiment as a sample space.

Tetrahedral die

4. A coin is tossed and then a die is rolled. List all the possible outcomes of the experiment as a sample space. [To get you started, (H, 1) is one of the possible outcomes.]

5. The spinner at the right is spun once. Assume that the spinner does not come to rest on a line.
   a. What is the sample space?
   b. List the outcomes in the event that the number is less than 4.

6. A coin is tossed four times. Find the probability of the given event.
   a. The outcomes are exactly in the order HHTT. (See Exercise 1.)
   b. The outcomes consist of two heads and two tails.
   c. The outcomes consist of one head and three tails.

7. Two dice are rolled. Find the probability of the given outcome.
   a. The sum of the dots on the upward faces is 5.
   b. The sum of the dots on the upward faces is 15.
   c. The sum of the dots on the upward faces is less than 15.
   d. The sum of the dots on the upward faces is 2.

8. A dodecahedral die has 12 sides numbered from 1 to 12. The die is rolled once. Find the probability of the given outcome.
   a. The upward face shows an 11.
   b. The upward face shows a 5.

Dodecahedral die

9. A dodecahedral die has 12 sides numbered from 1 to 12. The die is rolled once. Find the probability of the given outcome.
   a. The upward face shows a number that is divisible by 4.
   b. The upward face shows a number that is a multiple of 3.

10. Two tetrahedral dice are rolled (see Exercise 3).
    a. What is the probability that the sum on the upward faces is 4?
    b. What is the probability that the sum on the upward faces is 6?

11. Two dice are rolled. Which has the greater probability, throwing a sum of 10 or throwing a sum of 5?

12. Two dice are rolled once. Calculate the probability that the two numbers on the dice are equal.

13. Each of the letters of the word *MISSISSIPPI* is written on a card, and the cards are placed in a hat. One card is drawn at random from the hat.

a. What is the probability that the card has the letter *I* on it?

b. Which is greater, the probability of choosing an *S* or that of choosing a *P*?

14. Use the situation described in Exercise 12. Suppose you decide to test your result empirically by rolling a pair of dice 30 times and recording the results. Which number of "doubles" would confirm the result found in Exercise 12?

(i) 1 (ii) 5 (iii) 6 (iv) 30

15. Use the situation described in Exercise 13. What probability does the fraction $\frac{1}{11}$ represent?

16. Three blue marbles, four green marbles, and five red marbles are placed in a bag. One marble is chosen at random.

a. What is the probability that the marble chosen is green?

b. Which is greater, the probability of choosing a blue marble or that of choosing a red marble?

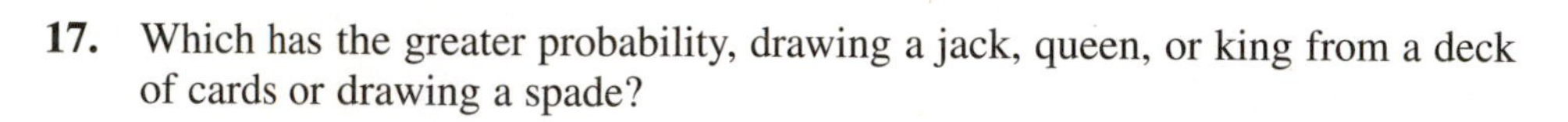

17. Which has the greater probability, drawing a jack, queen, or king from a deck of cards or drawing a spade?

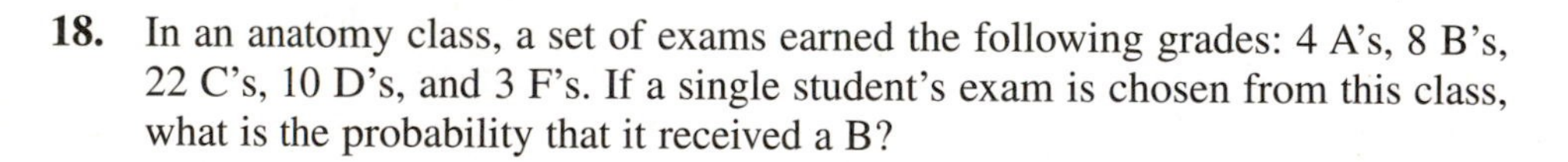

18. In an anatomy class, a set of exams earned the following grades: 4 A's, 8 B's, 22 C's, 10 D's, and 3 F's. If a single student's exam is chosen from this class, what is the probability that it received a B?

19. Of 3328 obese patients who had weight loss bariatric surgery including stomach stapling, stomach banding and gastric bypass, 63 died after surgery. Suppose these statistics are used to determine the probability of death following bariatric surgery. Calculate that probability in the form of a percent. Round to the nearest tenth of a percent.

20. A survey of 725 people showed that 587 had a group health insurance plan where they worked. On the basis of this survey, what is the empirical probability that an employee has a group health insurance plan? Write the answer as a decimal rounded to the nearest hundredth.

21. Cymbalta® is a prescription medication used to treat depression. The probabilities of some of the side effects from this medication are listed in the table at the right.

a. If 35 patients take this medication, how many would you expect to experience headaches as a side effect?

b. If 65 patients take this medication and five of them report dizziness as a side effect, is this more or less than the number you would expect based on the probabilities in the table?

Side Effect	*Probability*
Nausea	25%
Dry mouth	15%
Headaches	14%
Fatigue	11%
Dizziness	10%

Source: http://depression.emedtv.com/cymbalta/cymbalta-side-effects.html

Applying the Concepts

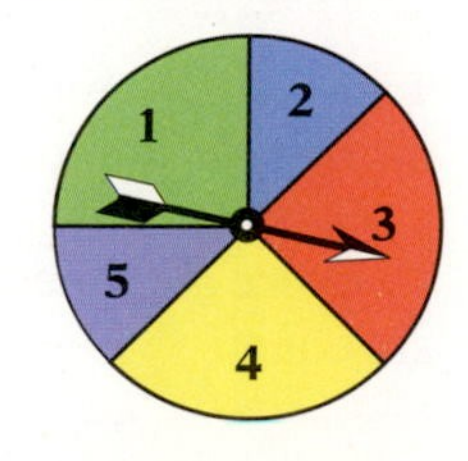

22. If the spinner at the right is spun once, is each of the numbers 1 through 5 equally likely? Why or why not?

23. Why can the probability of an event not be $\frac{5}{3}$?

FOCUS ON PROBLEM SOLVING

Growth Charts Clinical growth charts are published by the Centers for Disease Control and Prevention (CDC) and are used by pediatricians, nurses, and parents to monitor the growth of children in the United States. The charts have percentile curves that provide ranges of values illustrating the heights or weights that most children attain at certain ages. They are used by health care professionals to assist in monitoring the growth of children and adolescents.

In the growth chart shown for boys from birth to 36 months, you will see that most boys aged 18 months are approximately 30 inches to 34.5 inches long by reading the scale on the left side of the chart between the lower and upper curved lines. If you read the scale on the right side of the chart using the lower set of curves, you can determine that a normal weight for an 18-month-old boy is approximately 21 pounds to 31.5 pounds.

Find the normal ranges of weight and height for a 24-month-old boy.

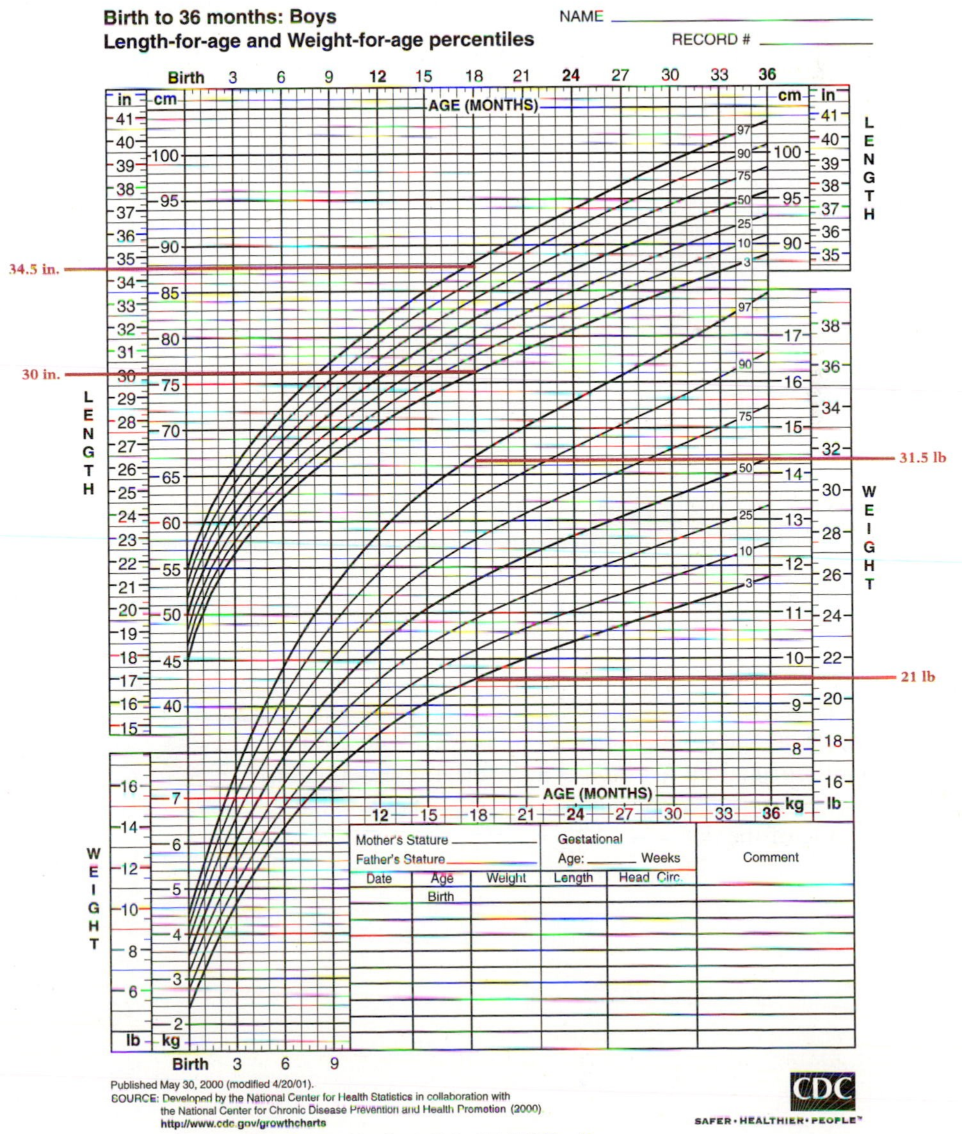

Source: http://www.cdc/gov/growthcharts/data/set2clinical/cj411067.pdf

PROJECTS AND GROUP ACTIVITIES

Collecting, Organizing, Displaying, and Analyzing Data

Before standardized units of measurement became commonplace, measurements were made in terms of the human body. For example, the cubit was the distance from the end of the elbow to the tips of the fingers. The yard was the distance from the tip of the nose to the tip of the fingers on an outstretched arm.

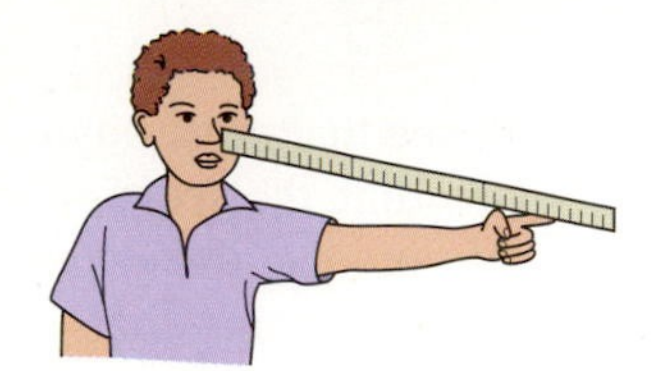

For each student in the class, find the measure from the tip of the nose to the tip of the fingers on an outstretched arm. Round each measure to the nearest centimeter. Record all the measurements on the board.

1. From the data collected, determine each of the following.

Mean __________

Median __________

Mode __________

Range __________

First quartile, Q_1 __________

Third quartile, Q_3 __________

Interquartile range __________

2. Prepare a box-and-whiskers plot of the data.

3. Write a description of the spread of the data.

4. Explain why we need standardized units of measurement.

CHAPTER 7

SUMMARY

KEY WORDS

Statistics is the branch of mathematics concerned with *data,* or numerical information. A *graph* is a pictorial representation of data. A *pictograph* represents data by using a symbol that is characteristic of the data. [7.1A, p. 294]

EXAMPLES

The pictograph shows the annual per capita turkey consumption in different countries.

Per Capita Turkey Consumption

Source: National Turkey Federation

A *circle graph* represents data by the sizes of the sectors. [7.1B, p. 296]

The circle graph shows the results of a survey of 300 people who were asked to name their favorite sport.

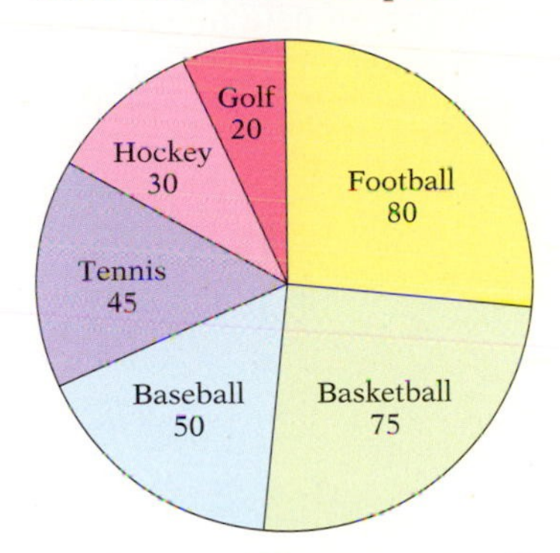

Distribution of Responses in a Survey

A *bar graph* represents data by the heights of the bars. [7.2A, p. 302]

The bar graph shows the expected U.S. population aged 100 and over.

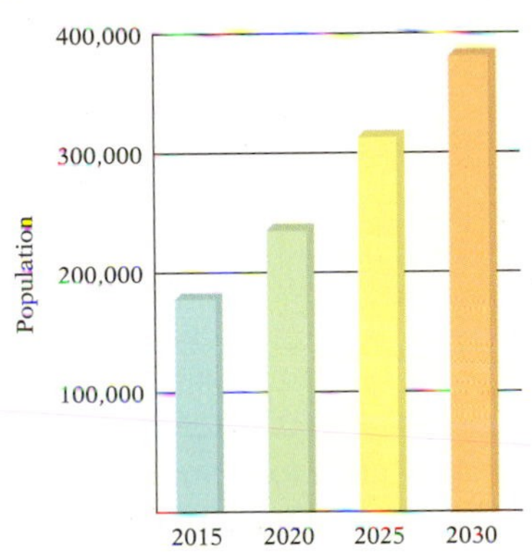

Expected U.S. Population Aged 100 and Over
Source: Census Bureau

A *broken-line graph* represents data by the positions of the lines and shows trends or comparisons. [7.2B, p. 303]

The line graph shows a recent graduate's cumulative debt in college loans at the end of each of the four years of college.

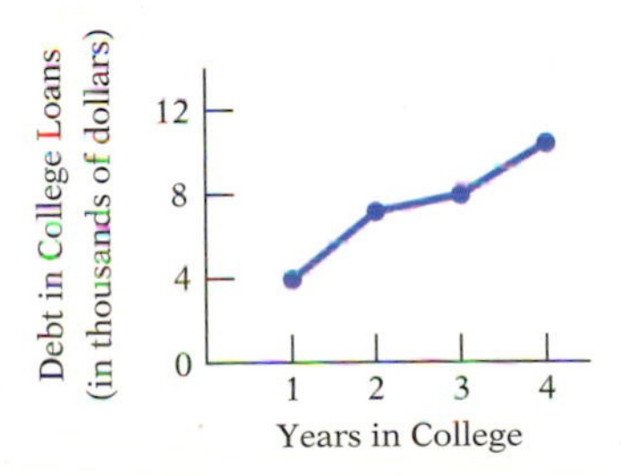

Cumulative Debt in College Loans

A *histogram* is a special kind of bar graph. In a histogram, the width of each bar corresponds to a range of numbers called a *class interval.* The height of each bar corresponds to the number of occurrences of data in each class interval and is called the *class frequency.* [7.3A, p. 307]

An Internet service provider (ISP) surveyed 1000 of its subscribers to determine the time required for each subscriber to download a particular file. The results of the survey are shown in the histogram below.

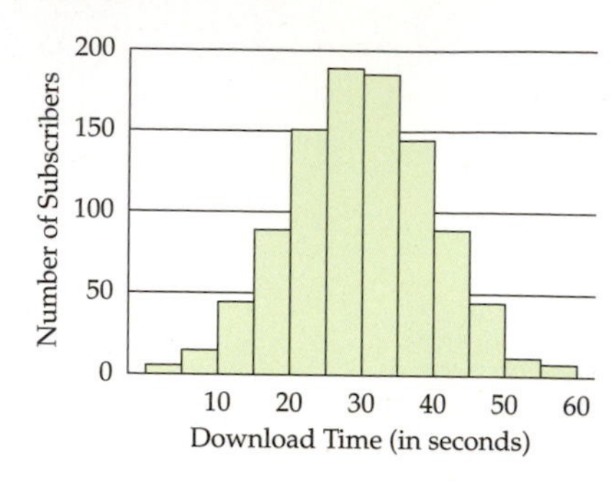

A *frequency polygon* is a graph that displays information in a manner similar to a histogram. A dot is placed above the center of each class interval at a height corresponding to that class's frequency. The dots are connected to form a broken-line graph. The center of a class interval is called the *class midpoint.* [7.3B, p. 308]

Below is a frequency polygon for the data in the histogram above.

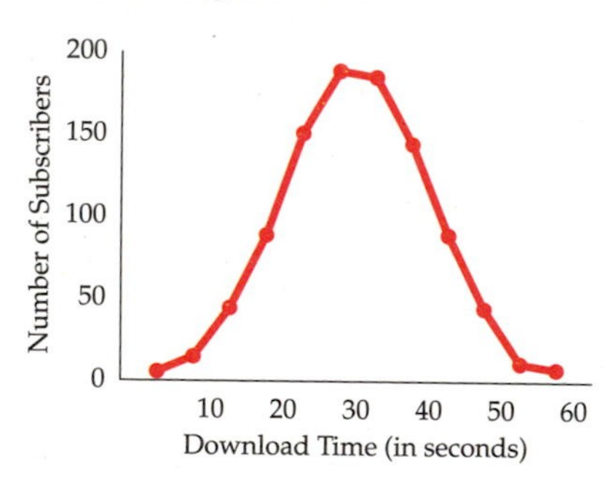

The *mean, median,* and *mode* are three types of averages used in statistics. The *mean* of a set of data is the sum of the data values divided by the number of values in the set. The *median* of a set of data is the number that separates the data into two equal parts when the data have been arranged from least to greatest (or greatest to least). There is an equal number of values above the median and below the median. The *mode* of a set of numbers is the value that occurs most frequently. [7.4A, pp. 311, 312]

Consider the following set of data.

24, 28, 33, 45, 45

The mean is 35.
The median is 33.
The mode is 45.

A *box-and-whiskers plot,* or *boxplot,* is a graph that shows five numbers: the least value, the first quartile, the median, the third quartile, and the greatest value. The *first quartile,* Q_1, is the number below which one-fourth of the data lie. The *third quartile,* Q_3, is the number above which one-fourth of the data lie. The box is placed around the values between the first quartile and the third quartile. The *range* is the difference between the greatest number and the least number in the set. The range describes the spread of the data. The *interquartile range* is the difference between Q_3 and Q_1. [7.4B, pp. 314–315]

The box-and-whiskers plot for a set of test scores is shown below.

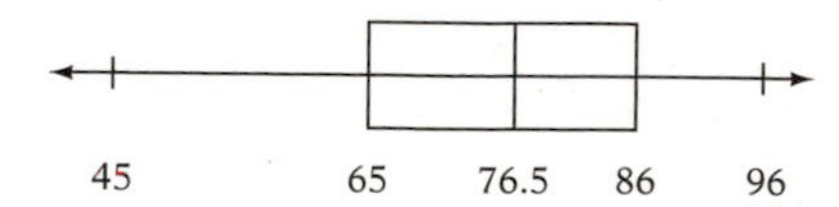

Range $= 96 - 45 = 51$

$Q_1 = 65$

$Q_3 = 86$

Interquartile range

$= Q_3 - Q_1 = 86 - 65 = 21$

Probability is a number from 0 to 1 that tells us how likely it is that a certain outcome of an experiment will happen. An *experiment* is an activity with an observable outcome. All the possible outcomes of an experiment are called the *sample space* of the experiment. An *event* is one or more outcomes of an experiment. The *favorable outcomes* of an experiment are the outcomes that satisfy the requirements of a particular event. [7.5A, p. 321]

Tossing a single die is an example of an experiment. The sample space for this experiment is the set of possible outcomes:

$$\{1, 2, 3, 4, 5, 6\}$$

The event that the number landing face up is an odd number is represented by

$$\{1, 3, 5\}$$

ESSENTIAL RULES AND PROCEDURES

EXAMPLES

To Find the Mean of a Set of Data [7.4A, p. 311]

Divide the sum of the numbers by the number of values in the set.

$$\bar{x} = \frac{\text{sum of the data values}}{\text{number of data values}}$$

Consider the following set of data.

24, 28, 33, 45, 45

$$\bar{x} = \frac{24 + 28 + 33 + 45 + 45}{5} = 35$$

To Find the Median [7.4A, p. 312]

1. Arrange the numbers from least to greatest.
2. If there is an *odd* number of values in the set of data, the median is the middle number. If there is an *even* number of values in the set of data, the median is the mean of the two middle numbers.

Consider the following set of data.

24, 28, 33, 35, 45, 45

The median is $\frac{33 + 35}{2} = 34$.

To Find Q_1 [7.4B, p. 314]

Arrange the numbers from least to greatest and locate the median. Q_1 is the median of the lower half of the data.

Consider the following data.

8 10 12 14 16 19 22

(10 ↑ Q_1; 14 ↑ Median)

To Find Q_3 [7.4B, p. 314]

Arrange the numbers from least to greatest and locate the median. Q_3 is the median of the upper half of the data.

Consider the following data.

8 10 12 14 16 19 22

(14 ↑ Median; 19 ↑ Q_3)

Theoretical Probability Formula [7.5A, p. 322]

$$\text{Probability of an event} = \frac{\text{number of favorable outcomes}}{\text{number of possible outcomes}}$$

A die is rolled. The probability of rolling a 2 or a 4 is $\frac{2}{6} = \frac{1}{3}$.

Empirical Probability Formula [7.5A, p. 324]

$$\text{Probability of an event} = \frac{\text{number of observations of the event}}{\text{total number of observations}}$$

A thumbtack is tossed 100 times. It lands point up 15 times and lands on its side 85 times. From this experiment, the empirical probability of "point up" is $\frac{15}{100} = \frac{3}{20}$.

CHAPTER 7

CONCEPT REVIEW

Test your knowledge of the concepts presented in this chapter. Answer each question. Then check your answers against the ones provided in the Answer Section.

1. What is a sector of a circle?

2. How does a pictograph give numerical information?

3. Why is a portion of the vertical axis jagged on some bar graphs?

4. How does a broken-line graph show changes over time?

5. What is class frequency in a histogram?

6. What is a class interval in a histogram?

7. What is a class midpoint?

8. What is the formula for the mean?

9. To find the median, why must the data be arranged in order from least to greatest?

10. When does a set of data have no mode?

11. What five values are shown in a box-and-whiskers plot?

12. How do you find the first quartile for a set of data values?

13. What is the empirical probability formula?

14. What is the theoretical probability formula?

CHAPTER 7

REVIEW EXERCISES

Medicare Benefits Payments The circle graph in Figure 38 shows the Medicare benefits payments by type of service in 2010. All numbers shown are in billions of dollars. Use this graph for Exercises 1 to 3.

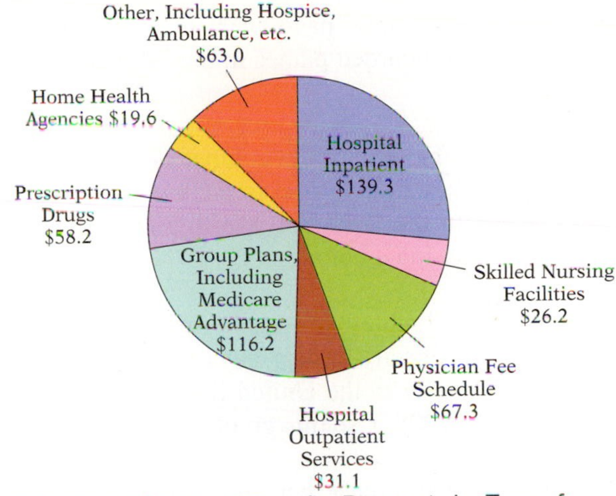

FIGURE 38 Medicare Benefits Payments by Type of Service, 2010, in Billions of Dollars
Source: Congressional Budget Office, August 2010 Baseline

1. Find the total amount of money that Medicare paid in benefits in 2010.

2. What is the ratio of the amount spent for skilled nursing facilities to the amount spent for prescription drugs? Write the ratio as a fraction in simplest form.

3. What percent of the total benefits payments was paid for hospital inpatient services? Round to the nearest tenth of a percent.

Demographics The double-line graph in Figure 39 shows the populations of California and Texas for selected years. Use this graph for Exercises 4 to 6.

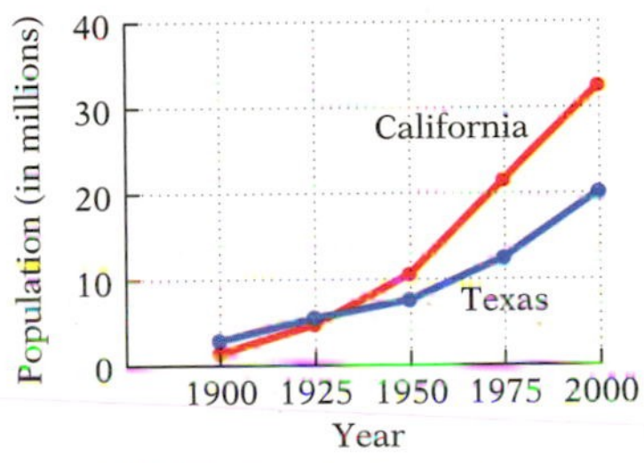

FIGURE 39 Populations of California and Texas

4. In 1900, which state had the larger population?

5. In 2000, approximately how much greater was the population of California than the population of Texas?

6. During which 25-year period did the population of Texas increase the least?

Heights The frequency polygon in Figure 40 shows the heights of 36 women involved in a clinical study of osteoporosis. All heights were recorded in inches. Use this figure for Exercises 7 to 9.

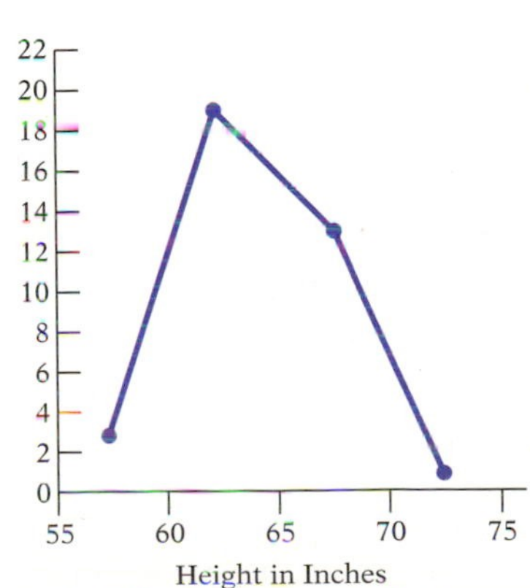

FIGURE 40 Heights of Females in Clinical Study of Osteoporosis

7. Find the number of women who are shorter than 65 inches.

8. What is the ratio of the number of women with heights between 55 inches and 60 inches to the number of women with heights between 70 and 75 inches? Write the ratio as a fraction in simplest form.

9. What percent of the women are taller than 65 inches? Round to the nearest tenth of a percent.

Airports The pictograph in Figure 41 shows the numbers of passengers that boarded planes in the five busiest U.S. airports in a recent year. Use this graph for Exercises 10 and 11.

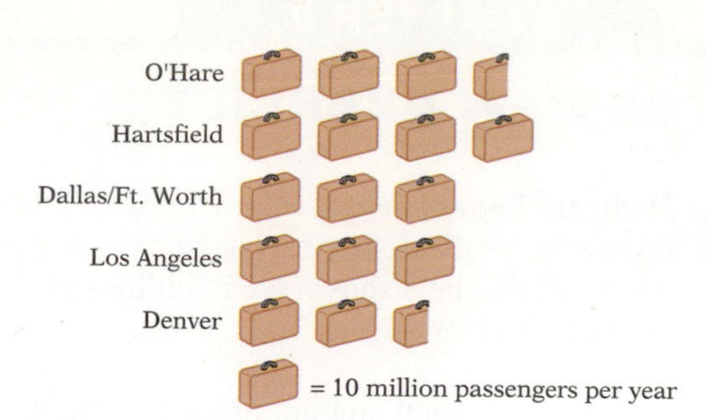

FIGURE 41 The busiest U.S. airports
Source: Federal Aviation Administration

10. How many more passengers boarded planes in O'Hare each year than boarded planes in the Denver airport each year?

11. What is the ratio of the number of passengers boarding planes in the Hartsfield airport to the number of passengers boarding planes in the Dallas/Ft. Worth airport each year? Write your answer using a colon.

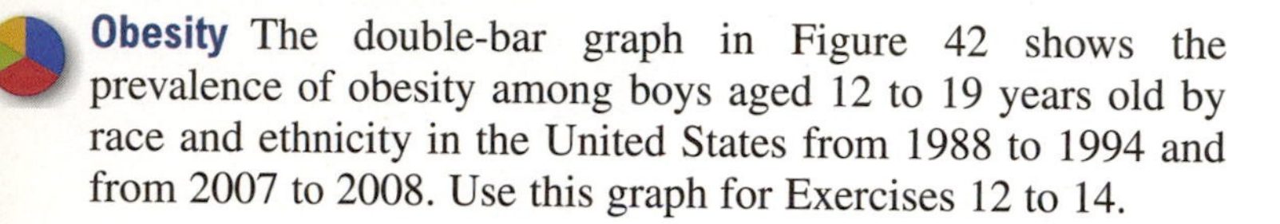

Obesity The double-bar graph in Figure 42 shows the prevalence of obesity among boys aged 12 to 19 years old by race and ethnicity in the United States from 1988 to 1994 and from 2007 to 2008. Use this graph for Exercises 12 to 14.

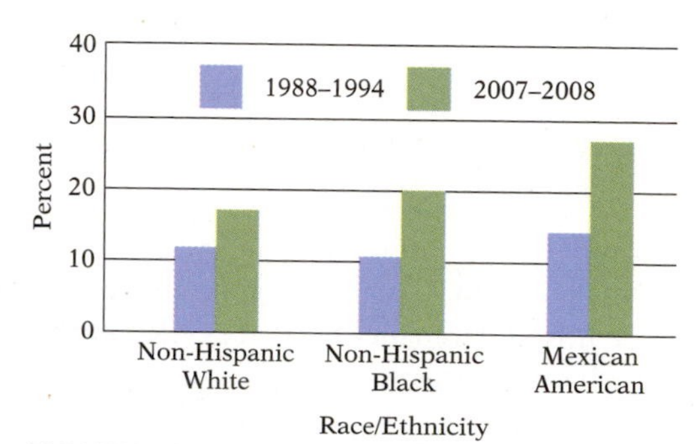

FIGURE 42 Prevalence of Obesity among Boys Aged 12 to 19, by Race/Ethnicity: United States 1988–1994, and 2007–2008

12. Find the difference between the percent of non-Hispanic white boys who were classified as obese from 1988 to 1994 and those who were classified as obese from 2007 to 2008.

13. In a study of 150 non-Hispanic black boys aged 12 to 19 during the years 1988 to 1994, how many of that group would you expect to be classified as obese?

14. Which group showed the largest increase in the percent of boys aged 12 to 19 classified as obese between 1988 to 1994 and 2007 to 2008? How large was the increase?

15. A coin is tossed four times. What is the probability that the outcomes of the tosses consist of one tail and three heads?

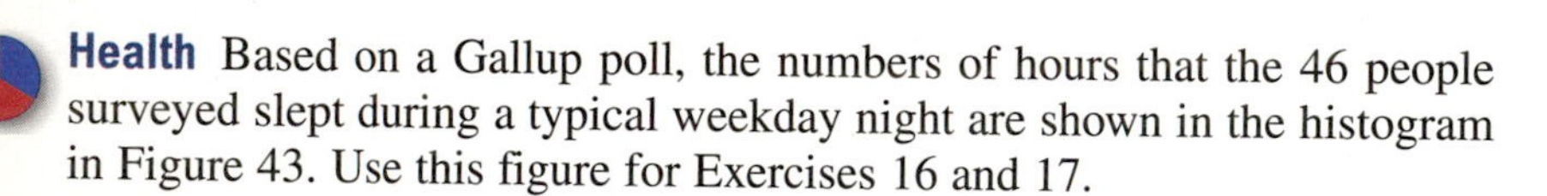

Health Based on a Gallup poll, the numbers of hours that the 46 people surveyed slept during a typical weekday night are shown in the histogram in Figure 43. Use this figure for Exercises 16 and 17.

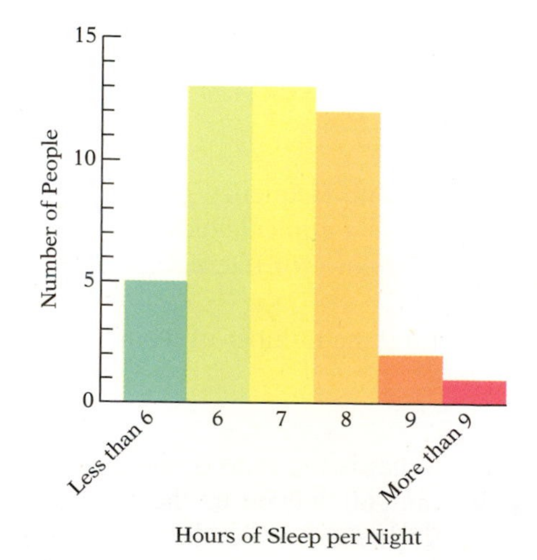

FIGURE 43

16. How many people slept 8 hours or more?

17. What percent of the people surveyed slept 7 hours? Round to the nearest tenth of a percent.

18. **Heart Rates** The heart rates of 24 women tennis players were measured after each of them had run one-quarter of a mile. The results are listed in the table below.

80	82	99	91	93	87	103	94	73	96	86	80
97	94	108	81	100	109	91	84	78	96	96	100

a. Find the mean, median, and mode for the data. Round to the nearest tenth.
b. Find the range and the interquartile range for the data.

CHAPTER 7

TEST

Consumerism Forty college students were surveyed to see how much money they spent each week on dining out in restaurants. The results are recorded in the frequency polygon shown in Figure 44. Use this figure for Exercises 1 to 3.

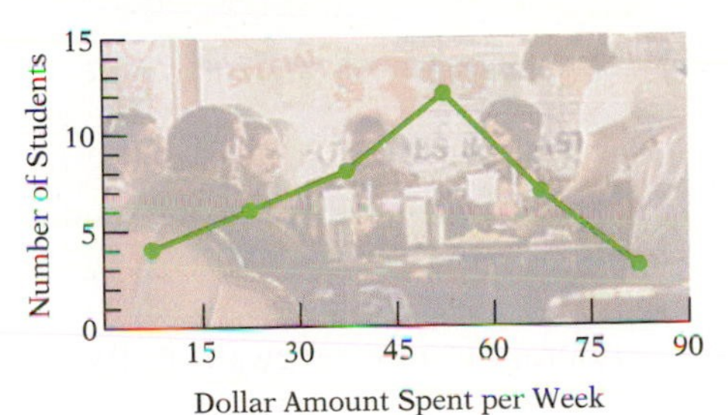

FIGURE 44

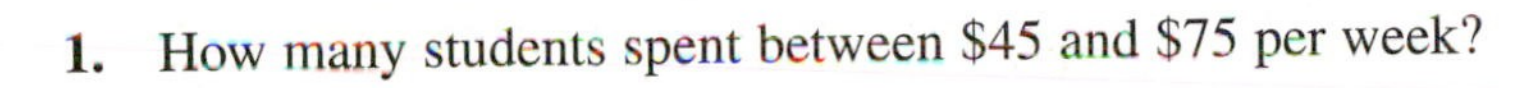

1. How many students spent between $45 and $75 per week?

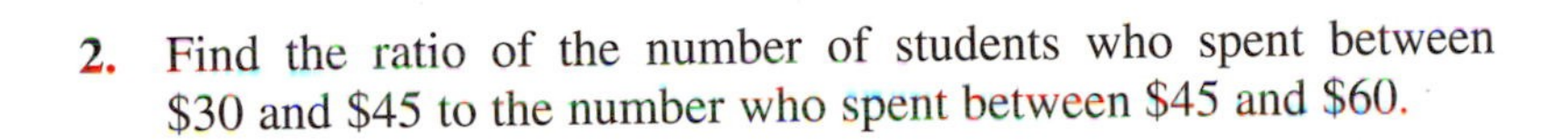

2. Find the ratio of the number of students who spent between $30 and $45 to the number who spent between $45 and $60.

3. What percent of the students surveyed spent less than $45 per week?

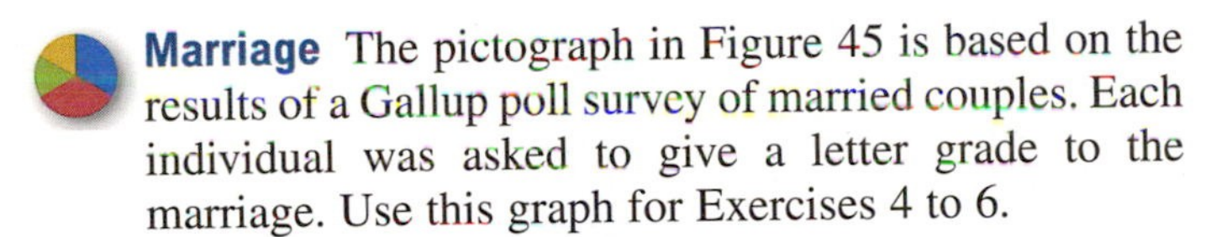

Marriage The pictograph in Figure 45 is based on the results of a Gallup poll survey of married couples. Each individual was asked to give a letter grade to the marriage. Use this graph for Exercises 4 to 6.

FIGURE 45 Survey of married couples rating their marriage

4. Find the total number of people who were surveyed.

5. Find the ratio of the number of people who gave their marriage a B to the number who gave it a C.

6. What percent of the total number of people surveyed gave their marriage an A? Round to the nearest tenth of a percent.

Health Occupations The bar graph in Figure 46 shows the health occupations with the largest job growth 2008, and projected 2018, as reported by the U.S. Bureau of Labor Statistics. Use this graph for Exercises 7 to 9.

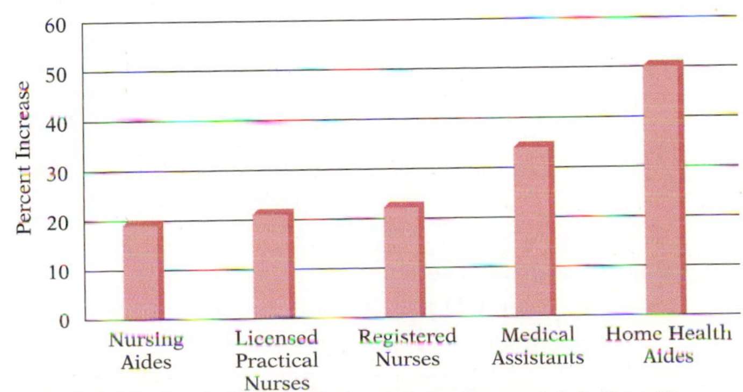

FIGURE 46 Health Occupations with the Largest Job Growth Rates, 2008 and Projected 2018

Source: Employment Projections Program, U.S. Department of Labor, U.S. Bureau of Labor Statistics

7. What is the job growth rate for medical assistants?

8. Which two occupations are expected to have job growth rates slightly higher than 20%?

9. How much higher is the predicted growth rate for home health aides than the predicted growth rate for nursing aides?

Grade Distribution The circle graph in Figure 47 gives information about the grade distribution for final grades during the fall semester for all nursing students enrolled in microbiology. Use this graph for Exercises 10 to 12.

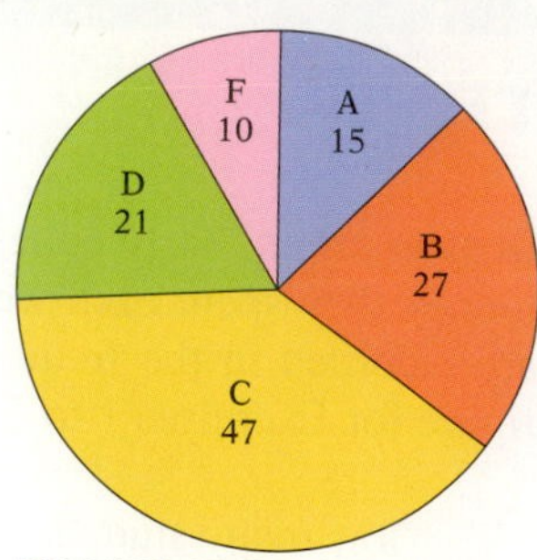

FIGURE 47 Final Grades in Microbiology Class

10. How many more students made Cs than made As?

11. What percent of the nursing students earned a B in microbiology?

12. In order to move forward in the nursing curriculum, students must earn a grade of C or better in all of their courses. What percent of students received grades lower than a C? Round to the nearest tenth of a percent.

Compensation The histogram in Figure 48 gives information about median incomes, by state, in the United States. Use this figure for Exercises 13 to 15.

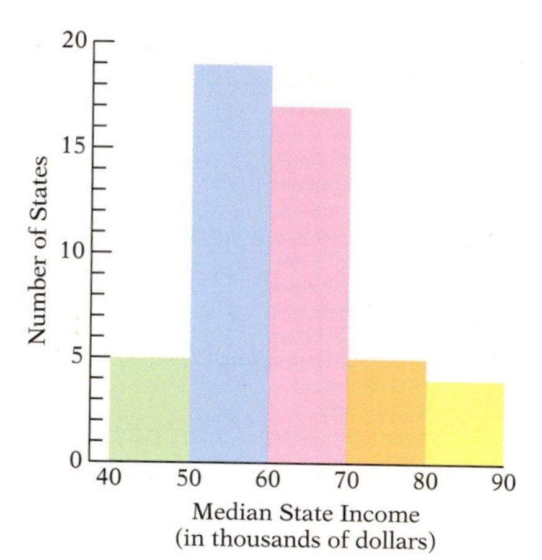

FIGURE 48
Source: U.S. Census Bureau

13. How many states have median incomes between \$40,000 and \$60,000?

14. What percent of the states have a median income that is between \$50,000 and \$70,000?

15. What percent of the states have a median income that is \$70,000 or more?

16. **Probability** A box contains 50 balls, of which 15 are red. If 1 ball is randomly selected from the box, what is the probability of the ball's being red?

Education The broken-line graph in Figure 49 shows the numbers of students enrolled in colleges for selected years. Use this figure for Exercises 17 and 18.

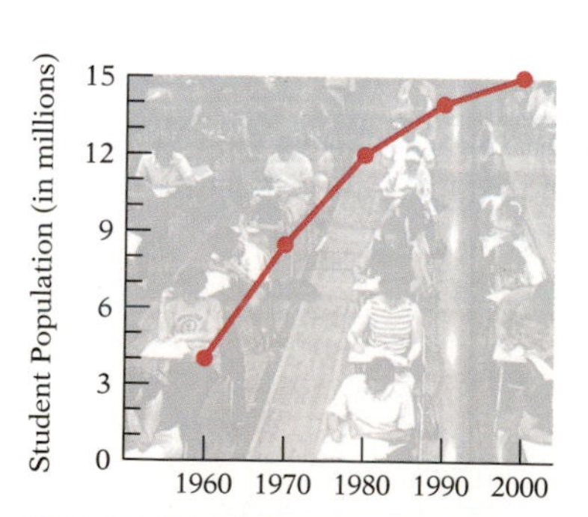

FIGURE 49 Student enrollment in public and private colleges
Source: National Center for Educational Statistics

17. During which decade did the student population increase the least?

18. Approximate the increase in college enrollment from 1960 to 2000.

19. **Quality Control** The lengths of time (in days) that various batteries operated a portable CD player continuously are given in the table below.

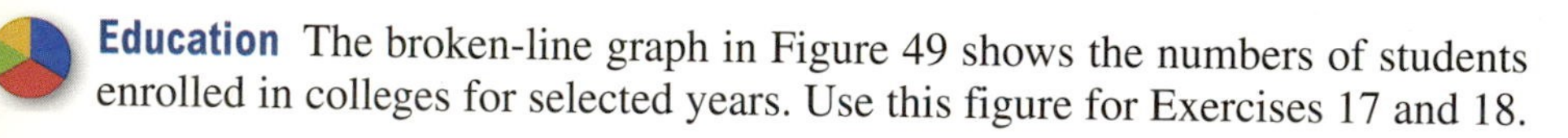

2.9	2.4	3.1	2.5	2.6	2.0	3.0	2.3	2.4	2.7
2.0	2.4	2.6	2.7	2.1	2.9	2.8	2.4	2.0	2.8

a. Find the mean for the data.
b. Find the median for the data.
c. Draw a box-and-whiskers plot for the data.

CUMULATIVE REVIEW EXERCISES

1. Simplify: $2^2 \cdot 3^3 \cdot 5$

2. Simplify: $3^2 \cdot (5 - 2) \div 3 + 5$

3. Find the LCM of 24 and 40.

4. Write $\frac{60}{144}$ in simplest form.

5. Find the total of $4\frac{1}{2}$, $2\frac{3}{8}$, and $5\frac{1}{5}$.

6. Subtract: $12\frac{5}{8} - 7\frac{11}{12}$

7. Multiply: $\frac{5}{8} \times 3\frac{1}{5}$

8. Find the quotient of $3\frac{1}{5}$ and $4\frac{1}{4}$.

9. Simplify: $\frac{5}{8} \div \left(\frac{3}{4} - \frac{2}{3}\right) + \frac{3}{4}$

10. Write two hundred nine and three hundred five thousandths in standard form.

11. Find the product of 4.092 and 0.69.

12. Convert $16\frac{2}{3}$ to a decimal. Round to the nearest hundredth.

13. Write "330 miles on 12.5 gallons of gas" as a unit rate.

14. Solve the proportion: $\frac{n}{5} = \frac{16}{25}$

15. Write $\frac{4}{5}$ as a percent.

16. 8 is 10% of what?

17. What is 38% of 43?

18. What percent of 75 is 30?

19. Compensation Tanim Kamal, a salesperson at a medical equipment supply company, receives \$100 per week plus 2% commission on sales. Find the income for a week in which Tanim had \$27,500 in sales.

20. Insurance A life insurance policy costs \$8.15 for every \$1000 of insurance. At this rate, what is the cost for \$50,000 of life insurance?

21. Simple Interest A medical practice borrowed \$125,000 for 6 months at an annual simple interest rate of 6%. Find the interest due on the loan.

22. Markup A wheelchair with a cost of \$180 is sold for \$279. Find the markup rate.

23. Finance The circle graph in Figure 50 shows how a family's monthly income of \$4500 is budgeted. How much is budgeted for food?

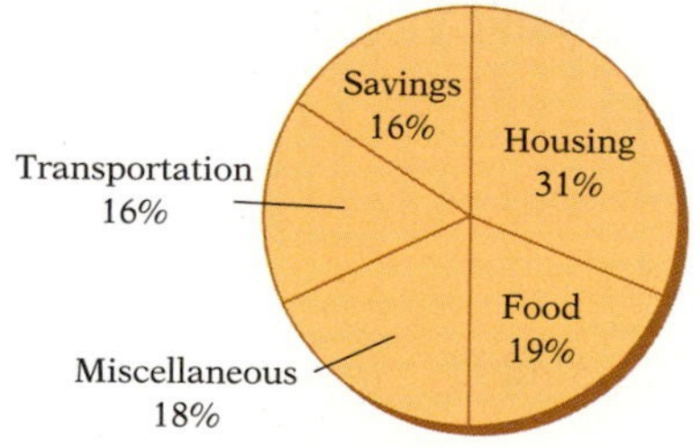

FIGURE 50 Budget for a monthly income of \$4500

24. Education The double-broken-line graph in Figure 51 shows two students' scores on 5 math tests of 30 problems each. Find the difference between the numbers of problems that the two students answered correctly on Test 1.

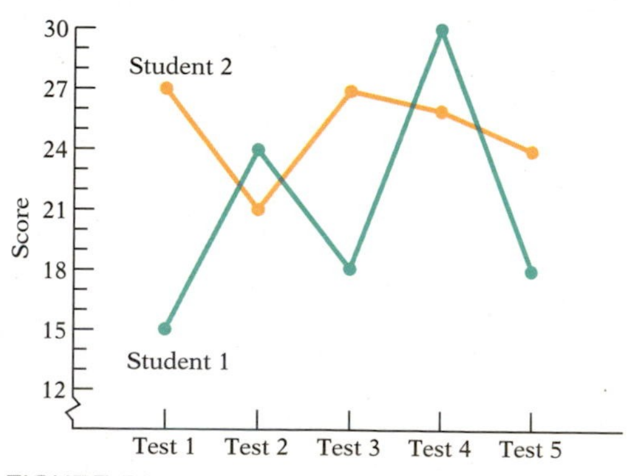

FIGURE 51

25. Hospital Stay The CDC reports that the average length of stay in a hospital for U.S. citizens is 4.8 days. (*Source*: http://www.cdc.gov/nchs/fastats/hospital.htm) The following data represent the length of hospital stays for 10 patients released recently from a local hospital: 6, 2, 1, 3, 4, 5, 6, 7, 5, 2. Find the mean of these numbers. Is the mean length of hospital stay for these patients higher or lower than the national average?

26. Probability Two dice are rolled. What is the probability that the sum of the dots on the upward faces is 8?

CHAPTER

8

U.S. Customary Units of Measurement

OBJECTIVES

SECTION 8.1

- **A** To convert measurements of length in the U.S. Customary System
- **B** To perform arithmetic operations with measurements of length
- **C** To solve application problems

SECTION 8.2

- **A** To convert measurements of weight in the U.S. Customary System
- **B** To perform arithmetic operations with measurements of weight
- **C** To solve application problems

SECTION 8.3

- **A** To convert measurements of capacity in the U.S. Customary System
- **B** To perform arithmetic operations with measurements of capacity
- **C** To solve application problems

SECTION 8.4

- **A** To convert units of time

SECTION 8.5

- **A** To use units of energy in the U.S. Customary System
- **B** To use units of power in the U.S. Customary System

AIM for the Future

PHYSICAL THERAPIST ASSISTANTS assist physical therapists in providing care to patients. Under the direction and supervision of physical therapists, they provide exercise and instruction; therapeutic methods like electrical stimulation, mechanical traction, and ultrasound; massage; and gait and balance training. Physical therapist assistants record the patient's responses to treatment and report the outcome of each treatment to the physical therapist. Employment of physical therapist assistants is expected to grow by 35% from 2008 through 2018, much faster than the average for all occupations. Median annual wages of physical therapist assistants were $46,140 in May 2008. (*Source: Bureau of Labor Statistics:* http://www.bls.gov/oco/ocos167.htm)

PREP TEST

Are you ready to succeed in this chapter?

Take the Prep Test below to find out if you are ready to learn the new material.

1. $\begin{array}{r} 485 \\ +\ 217 \\ \hline \end{array}$

2. $\begin{array}{r} 145 \\ -\ 87 \\ \hline \end{array}$

3. $36 \times \frac{1}{9}$

4. $\frac{5}{3} \times 6$

5. $400 \times \frac{1}{8} \times \frac{1}{2}$

6. $5\frac{3}{4} \times 8$

7. $3\overline{)714}$

8. $12\overline{)18}$

SECTION

8.1 Length

OBJECTIVE A To convert measurements of length in the U.S. Customary System

Point of Interest

The ancient Greeks devised the foot measurement, which they usually divided into 16 fingers. It was the Romans who subdivided the foot into 12 units called inches. The word *inch* is derived from the Latin word *uncia,* which means "a twelfth part."

The Romans also used a unit called *pace,* which equaled two steps. One thousand paces equaled 1 mile. The word *mile* is derived from the Latin word *mille,* which means "1000."

A **measurement** includes a number and a unit.

Number	Unit
3	feet
7	miles
12	yards

Standard units of measurement have been established to simplify trade and commerce.

The unit of **length,** or distance, that is called the yard was originally defined as the length of a specified bronze bar located in London.

The standard U.S. Customary System units of length are **inch, foot, yard,** and **mile.**

Equivalences Between Units of Length in the U.S. Customary System

12 inches (in.) = 1 foot (ft)

3 ft = 1 yard (yd)

36 in. = 1 yard (yd)

5280 ft = 1 mile (mi)

These equivalences can be used to form conversion rates; a **conversion rate** is a relationship used to change one unit of measurement to another. For example, because 3 ft = 1 yd, the conversion rates $\frac{3 \text{ ft}}{1 \text{ yd}}$ and $\frac{1 \text{ yd}}{3 \text{ ft}}$ are both equivalent to 1.

HOW TO • 1 Convert 27 ft to yards.

$$27 \text{ ft} = 27 \text{ ft} \times \boxed{\frac{1 \text{ yd}}{3 \text{ ft}}}$$

$$= 27 \cancel{\text{ft}} \times \frac{1 \text{ yd}}{3 \cancel{\text{ft}}}$$

$$= \frac{27 \text{ yd}}{3}$$

$$= 9 \text{ yd}$$

HOW TO • 2 Convert 5 yd to feet.

$$5 \text{ yd} = 5 \text{ yd} \times \boxed{\frac{3 \text{ ft}}{1 \text{ yd}}}$$

$$= 5 \cancel{\text{yd}} \times \frac{3 \text{ ft}}{1 \cancel{\text{yd}}}$$

$$= \frac{15 \text{ ft}}{1}$$

$$= 15 \text{ ft}$$

Note that in the conversion rate chosen, the unit in the numerator is the same as the unit desired in the answer. The unit in the denominator is the same as the unit in the given measurement.

EXAMPLE • 1

Convert 40 in. to feet.

Solution $40 \text{ in.} = 40 \text{ in.} \times \dfrac{1 \text{ ft}}{12 \text{ in.}}$

$= \dfrac{40 \text{ ft}}{12} = 3\dfrac{1}{3} \text{ ft}$

YOU TRY IT • 1

Convert 14 ft to yards.

Your solution

EXAMPLE • 2

Convert $3\frac{1}{4}$ yd to feet.

Solution $3\dfrac{1}{4} \text{ yd} = \dfrac{13}{4} \text{ yd} = \dfrac{13}{4} \text{ yd} \times \dfrac{3 \text{ ft}}{1 \text{ yd}}$

$= \dfrac{39 \text{ ft}}{4} = 9\dfrac{3}{4} \text{ ft}$

YOU TRY IT • 2

Convert 9240 ft to miles.

Your solution

Solutions on p. S20

OBJECTIVE B To perform arithmetic operations with measurements of length

When performing arithmetic operations with measurements of length, write the answer in simplest form. For example, 1 ft 14 in. should be written as 2 ft 2 in.

HOW TO • 3 Convert: 50 in. = ____ ft ____ in.

$$\begin{array}{r} 4 \text{ ft } 2 \text{ in.} \\ 12\overline{)\,50} \\ \underline{-48} \\ 2 \end{array}$$

- Because 12 in. = 1 ft, divide 50 in. by 12. The whole-number part of the quotient is the number of feet. The remainder is the number of inches.

50 in. = 4 ft 2 in.

EXAMPLE • 3

Convert: 17 in. = ____ ft ____ in.

Solution

$$\begin{array}{r} 1 \text{ ft } 5 \text{ in.} \\ 12\overline{)\,17} \\ \underline{-12} \\ 5 \end{array}$$

- 12 in. = 1 ft

17 in. = 1 ft 5 in.

YOU TRY IT • 3

Convert: 42 in. = ____ ft ____ in.

Your solution

EXAMPLE • 4

Convert: 31 ft = ____ yd ____ ft

Solution

$$\begin{array}{r} 10 \text{ yd } 1 \text{ ft} \\ 3\overline{)\,31} \\ \underline{-30} \\ 1 \end{array}$$

- 3 ft = 1 yd

31 ft = 10 yd 1 ft

YOU TRY IT • 4

Convert: 14 ft = ____ yd ____ ft

Your solution

Solutions on p. S20

EXAMPLE • 5

Find the sum of 4 ft 4 in. and 1 ft 11 in.

Solution

$$\begin{array}{r} 4\text{ ft }\ 4\text{ in.} \\ +\ 1\text{ ft }11\text{ in.} \\ \hline 5\text{ ft }15\text{ in.} \end{array}$$

• **15 in. = 1 ft 3 in.**

5 ft 15 in. = 6 ft 3 in.

YOU TRY IT • 5

Find the sum of 3 ft 5 in. and 4 ft 9 in.

Your solution

EXAMPLE • 6

Subtract: 9 ft 6 in. − 3 ft 8 in.

Solution

$$\begin{array}{r} \overset{8\text{ ft}}{\cancel{9\text{ ft}}}\ \ \overset{18\text{ in.}}{\cancel{6\text{ in.}}} \\ -\ 3\text{ ft }\ \ 8\text{ in.} \\ \hline 5\text{ ft }10\text{ in.} \end{array}$$

• **Borrow 1 ft (12 in.) from 9 ft and add to 6 in.**

YOU TRY IT • 6

Subtract: 4 ft 2 in. − 1 ft 8 in.

Your solution

EXAMPLE • 7

Multiply: 3 yd 2 ft × 4

Solution

$$\begin{array}{r} 3\text{ yd }2\text{ ft} \\ \times\qquad 4 \\ \hline 12\text{ yd }8\text{ ft} \end{array}$$

• **8 ft = 2 yd 2 ft**

12 yd 8 ft = 14 yd 2 ft

YOU TRY IT • 7

Multiply: 4 yd 1 ft × 8

Your solution

EXAMPLE • 8

Find the quotient of 4 ft 3 in. and 3.

Solution

$$\begin{array}{rll} & 1\text{ ft} & \ \ 5\text{ in.} \\ 3\overline{)} & 4\text{ ft} & \ \ 3\text{ in.} \\ & -3\text{ ft} & \\ \hline & 1\text{ ft} = & 12\text{ in.} \\ & & 15\text{ in.} \\ & & -15\text{ in.} \\ \hline & & 0 \end{array}$$

YOU TRY IT • 8

Find the quotient of 7 yd 1 ft and 2.

Your solution

EXAMPLE • 9

Multiply: $2\frac{3}{4}$ ft × 3

Solution

$$2\frac{3}{4}\text{ ft} \times 3 = \frac{11}{4}\text{ ft} \times 3$$
$$= \frac{33}{4}\text{ ft}$$
$$= 8\frac{1}{4}\text{ ft}$$

YOU TRY IT • 9

Subtract: $6\frac{1}{4}$ ft − $3\frac{2}{3}$ ft

Your solution

Solutions on p. S20

OBJECTIVE C To solve application problems

EXAMPLE • 10

Applicants seeking acceptance into the Coast Guard must be between 60 in. and 80 in. tall. (*Source:* http://www.military.com) Does an applicant who is 5 ft 2 in. tall meet this requirement?

Strategy

To determine if the applicant's height is within the acceptable range, convert the height in feet (5 ft) into inches and add this number to the number of inches over 5 ft (2 in.).

Solution

$$5 \text{ ft} = \frac{5 \text{ ft}}{1} \cdot \frac{12 \text{ in.}}{1 \text{ ft}} = 60 \text{ in.}$$

$$60 \text{ in.} + 2 \text{ in.} = 62 \text{ in.}$$

This applicant is 62 in. tall and, therefore, meets the height requirement.

YOU TRY IT • 10

A nurse measures a patient's height as 66 in. Express this measurement in feet and inches.

Your strategy

Your solution

EXAMPLE • 11

A roll of gauze bandage is 4 yd 1 ft long. If you cut this roll into three bandages of equal length, how long will each piece be?

Strategy

To find the length of each piece of gauze bandage, divide the total length (4 yd 1 ft) by 3.

Solution

$$\begin{array}{r l}
 & \text{1 yd 1 ft 4 in.} \\
3\,) & \text{4 yd 1 ft} \\
 & \underline{-\,3 \text{ yd}} \\
 & 1 \text{ yd} = 3 \text{ ft} \\
 & \qquad\quad \underline{4 \text{ ft}} \\
 & \qquad\quad \underline{-3 \text{ ft}} \\
 & \qquad\quad 1 \text{ ft} = 12 \text{ in.} \\
 & \qquad\qquad\quad \underline{-12 \text{ in.}} \\
 & \qquad\qquad\qquad 0
\end{array}$$

Each piece of gauze bandage will be 1 yd 1 ft 4 in. long.

YOU TRY IT • 11

A medical examination table must be covered with clean paper for each patient. A roll of exam table paper is 225 ft long and at the end of one week, 192 ft 3 in. of a new roll had been used. How much paper remained on the roll?

Your strategy

Your solution

Solutions on p. S20

8.1 EXERCISES

OBJECTIVE A To convert measurements of length in the U.S. Customary System

For Exercises 1 to 3, suppose you convert units of measurement as given. Will the number part of the converted measurement be *less than* or *greater than* the number part of the original measurement?

1. Convert feet to inches

2. Convert inches to miles

3. Convert yards to feet

For Exercises 4 to 15, convert.

4. 6 ft = _______ in.

5. 9 ft = _______ in.

6. 30 in. = _______ ft

7. 64 in. = _______ ft

8. 13 yd = _______ ft

9. $4\frac{1}{2}$ yd = _______ ft

10. 16 ft = _______ yd

11. $4\frac{1}{2}$ ft = _______ yd

12. $2\frac{1}{3}$ yd = _______ in.

13. 5 yd = _______ in.

14. 2 mi = _______ ft

15. $1\frac{1}{2}$ mi = _______ ft

OBJECTIVE B To perform arithmetic operations with measurements of length

For Exercises 16 and 17, look at the indicated exercise. The number that goes in the second blank must be less than what number?

16. Exercise 18

17. Exercise 19

For Exercises 18 to 29, perform the arithmetic operation.

18. 100 in. = ___ ft ___ in.

19. 6400 ft = ___ mi ___ ft

20. $\begin{array}{r} 6 \text{ ft } 7 \text{ in.} \\ + \; 3 \text{ ft } 4 \text{ in.} \\ \hline \end{array}$

21. $\begin{array}{r} 9 \text{ ft } 11 \text{ in.} \\ + \; 3 \text{ ft } \;\; 6 \text{ in.} \\ \hline \end{array}$

22. $\begin{array}{r} 1 \text{ mi } 4200 \text{ ft} \\ + \; 2 \text{ mi } 3600 \text{ ft} \\ \hline \end{array}$

23. $4\frac{2}{3}$ ft + $6\frac{1}{2}$ ft

24. $\begin{array}{r} 5 \text{ ft } 3 \text{ in.} \\ - \; 2 \text{ ft } 6 \text{ in.} \\ \hline \end{array}$

25. $\begin{array}{r} 9 \text{ yd } 1 \text{ ft} \\ - \; 3 \text{ yd } 2 \text{ ft} \\ \hline \end{array}$

26. $\begin{array}{r} 2 \text{ ft } 5 \text{ in.} \\ \times \quad 6 \\ \hline \end{array}$

27. $3\frac{2}{3}$ ft × 4

28. $2\overline{)5 \text{ ft } 4 \text{ in.}}$

29. $12\frac{1}{2}$ in. ÷ 3

OBJECTIVE C To solve application problems

30. **Length** The length of the examination table in a doctor's office is 57 in. long. Express this length in feet and inches.

31. **Height** A nurse measures the height of a patient as 71 in. Convert this measurement to feet and inches.

32. **Height** To enlist in the Marines, an applicant must be between 58 in. tall and 80 in. tall. (*Source:* http://www.military.com) An applicant lists his height as 5 ft 10 in. Calculate this height in inches. Does this applicant meet the requirement for enlistment?

33. **Height** The height requirement for students entering ROTC in order to become pilots in the Air Force is more restrictive because of the size of the airplane cockpit. Candidates must have a standing height between 5 ft 4 in. and 6 ft 5 in. tall. (*Source:* http://www.csus.edu/afrotc/pilot.html) If an applicant is 78 in. tall, will this applicant qualify?

34. **Height** A patient with osteoporosis was 5 ft tall at age 50. By age 60, the patient had lost $1\frac{1}{2}$ in. of height. How tall was the patient at age 60? Give your answer in feet and inches.

35. **Height** At their yearly checkup, the nurse measures Evan's height as 4 ft 3 in. and his younger sister Mary's height as 3 ft 8 in. How much taller is Evan than Mary?

36. **Growth** The height of a 12-year-old child at her yearly checkup was 4 ft 8 in. At the next checkup one year later, she had grown 7 in. Give her height at age 13 in feet and inches.

37. **Growth** A pediatrician uses a standard growth chart to monitor the growth of his patients. The minimum height shown on the growth chart for most 4-year-old girls is 3 ft 1 in. If a 4-year-old girl is 3 in. below this minimum height, what is her height? Give your answer in feet and inches.

38. **Medical Supplies** A roll of surgical tape is 1.5 yd long. How many pieces of tape 3 in. long can be cut from this roll?

39. **Medical Supplies** A package of waxed dental floss is 12 yd long. If a dental hygienist uses approximately 12 in. of floss per patient when cleaning one person's teeth, how many patients will one package of dental floss serve?

40. **Walking Distance** An orderly at a hospital wears a pedometer to measure the distance he walks during one day at work. At the end of the day, the pedometer indicates that he has walked 9240 ft. Express this distance in miles, and then express it in miles and feet.

41. **World Record** The Guinness Book of World Records lists Robert Wadlow as the tallest man for whom there is irrefutable evidence. (*Source:* http//www.guinnessworldrecords.com) He died at the age of 22 having reached a height of 8 ft $11\frac{1}{10}$ in. tall. His arm span was 9 ft $5\frac{3}{4}$ in. How much longer was his arm span than his height?

Imagno/Getty Images

For Exercises 42 and 43, use the following information. A roll of gauze is cut into five equal pieces. The length of each piece is 2 ft and a number of inches (the number of inches is less than 12). Determine whether each of the following statements is *true* or *false*.

42. The roll of gauze must be longer than 11 ft.

43. The roll of gauze must be shorter than 15 ft.

Applying the Concept

44. **Measurement** There are approximately 200,000,000 adults living in the United States. Assume that the average adult is 19 in. wide from shoulder to shoulder. If all the adults in the United States stood shoulder-to-shoulder, could they reach around Earth at the equator, a distance of approximately 25,000 mi?

Patty Orly/Shutterstock.com

SECTION

8.2 Weight

OBJECTIVE A To convert measurements of weight in the U.S. Customary System

Point of Interest

The Romans used two different systems of weights. In both systems, the smallest unit was the *uncia*, abbreviated to "oz," from which the term *ounce* is derived. In one system, there were 16 ounces to 1 pound. In the second system, a pound, which was called a libra, equaled 12 unciae. The abbreviation "lb" for pound comes from the word *libra*.

The avoirdupois system of measurement and the troy system of measurement have their heritage in the two Roman systems.

Weight is a measure of how strongly Earth is pulling on an object. The unit of weight called the pound is defined as the weight of a standard solid kept at the Bureau of Standards in Washington, D.C. The U.S. Customary System units of weight are **ounce, pound,** and **ton.**

Equivalences Between Units of Weight in the U.S. Customary System

16 ounces (oz) = 1 pound (lb)

2000 lb = 1 ton

These equivalences can be used to form conversion rates to change one unit of measurement to another. For example, because 16 oz = 1 lb, the conversion rates $\frac{16\text{ oz}}{1\text{ lb}}$ and $\frac{1\text{ lb}}{16\text{ oz}}$ are both equivalent to 1.

HOW TO • 1 Convert 62 oz to pounds.

$$62\text{ oz} = 62\text{ oz} \times \frac{1\text{ lb}}{16\text{ oz}}$$

- The conversion rate must contain lb (the unit desired in the answer) in the numerator and must contain oz (the original unit) in the denominator.

$$= \frac{62\text{ oz}}{1} \times \frac{1\text{ lb}}{16\text{ oz}}$$

$$= \frac{62\text{ lb}}{16}$$

$$= 3\frac{7}{8}\text{ lb}$$

EXAMPLE • 1

Convert $3\frac{1}{2}$ tons to pounds.

Solution

$$3\frac{1}{2}\text{ tons} = \frac{7}{2}\text{ tons} \times \frac{2000\text{ lb}}{1\text{ ton}}$$

$$= \frac{14{,}000\text{ lb}}{2} = 7000\text{ lb}$$

YOU TRY IT • 1

Convert 3 lb to ounces.

Your solution

EXAMPLE • 2

Convert 42 oz to pounds.

Solution

$$42\text{ oz} = 42\text{ oz} \times \frac{1\text{ lb}}{16\text{ oz}}$$

$$= \frac{42\text{ lb}}{16} = 2\frac{5}{8}\text{ lb}$$

YOU TRY IT • 2

Convert 4200 lb to tons.

Your solution

Solutions on p. S20

OBJECTIVE B To perform arithmetic operations with measurements of weight

When performing arithmetic operations with measurements of weight, write the answer in simplest form. For example, 1 lb 22 oz should be written 2 lb 6 oz.

EXAMPLE 3

Find the difference between 14 lb 5 oz and 8 lb 14 oz.

Solution

```
  13 lb   21 oz
  14 lb    5 oz
−  8 lb   14 oz
  5 lb     7 oz
```

- Borrow 1 lb (16 oz) from 14 lb and add it to 5 oz.

YOU TRY IT 3

Find the difference between 7 lb 1 oz and 3 lb 4 oz.

Your solution

EXAMPLE 4

Divide: 7 lb 14 oz ÷ 3

Solution

```
     2 lb     10 oz
3) 7 lb      14 oz
  −6 lb
    1 lb =   16 oz
             30 oz
            −30 oz
              0
```

YOU TRY IT 4

Multiply: 3 lb 6 oz × 4

Your solution

Solutions on p. S21

OBJECTIVE C To solve application problems

EXAMPLE 5

Sirina Jasper purchased 4 lb 8 oz of oat bran and 2 lb 11 oz of wheat bran. She plans to blend the two brans and then repackage the mixture in 3-ounce packages for a diet supplement. How many 3-ounce packages can she make?

Strategy

To find the number of 3-ounce packages:

- Add the amount of oat bran (4 lb 8 oz) to the amount of wheat bran (2 lb 11 oz).
- Convert the sum to ounces.
- Divide the total ounces by the weight of each package (3 oz).

Solution

```
  4 lb  8 oz
+ 2 lb 11 oz
  6 lb 19 oz = 7 lb 3 oz = 115 oz
```

$\frac{115 \text{ oz}}{3 \text{ oz}} \approx 38.3$

She can make 38 packages.

YOU TRY IT 5

Find the weight in pounds of 12 bars of soap. Each bar weighs 7 oz.

Your strategy

Your solution

Solution on p. S21

8.2 EXERCISES

OBJECTIVE A To convert measurements of weight in the U.S. Customary System

For Exercises 1 to 3, suppose you convert units of measurement as given. Will the number part of the converted measurement be *less than* or *greater than* the number part of the original measurement?

1. Convert pounds to tons

2. Convert pounds to ounces

3. Convert tons to pounds

For Exercises 4 to 18, convert.

4. 64 oz = _______ lb

5. 36 oz = _______ lb

6. 8 lb = _______ oz

7. 7 lb = _______ oz

8. 3200 lb = _______ tons

9. 9000 lb = _______ tons

10. 6 tons = _______ lb

11. $1\frac{1}{4}$ tons = _______ lb

12. 66 oz = _______ lb

13. 90 oz = _______ lb

14. $1\frac{1}{2}$ lb = _______ oz

15. $2\frac{5}{8}$ lb = _______ oz

16. $\frac{4}{5}$ ton = _______ lb

17. 5000 lb = _______ tons

18. 180 oz = _______ lb

OBJECTIVE B To perform arithmetic operations with measurements of weight

For Exercises 19 and 20, look at the indicated exercise. The number that goes in the second blank must be less than what number?

19. Exercise 21

20. Exercise 22

For Exercises 21 to 32, perform the arithmetic operation.

21. 9000 lb = _____ tons _____ lb

22. 85 oz = _____ lb _____ oz

23. $\begin{array}{r} 4\text{ lb }\ 7\text{ oz} \\ +\ 3\text{ lb }12\text{ oz} \\ \hline \end{array}$

24. $\begin{array}{r} 1\text{ ton }\ \ 800\text{ lb} \\ +\ 3\text{ tons }1600\text{ lb} \\ \hline \end{array}$

25. $\begin{array}{r} 7\text{ lb }5\text{ oz} \\ -\ 3\text{ lb }8\text{ oz} \\ \hline \end{array}$

26. $\begin{array}{r} 3\text{ tons }500\text{ lb} \\ -\ 1\text{ ton }\ 800\text{ lb} \\ \hline \end{array}$

27. $6\frac{3}{8}$ lb $-$ $2\frac{5}{6}$ lb

28. $\begin{array}{r} 3\text{ lb }6\text{ oz} \\ \times\qquad 4 \\ \hline \end{array}$

29. $5\frac{1}{2}$ lb $\times$ 6

30. $4\frac{2}{3}$ lb $\times$ 3

31. $2\overline{)3\text{ lb }8\text{ oz}}$

32. 5 lb 12 oz ÷ 4

OBJECTIVE C To solve application problems

33. Read Exercise 35. Without actually finding the total weight of the babies, determine whether the total weight will be *less than* or *greater than* 12 lb.

34. Read Exercise 36. Without actually finding the difference in the weights of the babies, determine whether the difference in their weights will be *greater than* or *less than* 12 lb.

35. **Triplets** A mother gives birth to triplets. The individual weights of the babies at birth are 4 lb 3 oz, 3 lb 14 oz, and 4 lb 1 oz. Find the total birth weight of the triplets.

36. **Birth Weights** The average weight for a newborn baby is 7 lb 8 oz. In September 2009, Baby Akbar was born in Indonesia weighing 19 lb 3 oz. (*Source:* http//www.baby2see.com/large_babies.html) How much more did Baby Akbar weigh than the average newborn baby?

37. **Weights** A college bookstore received 1200 textbooks, each weighing 9 oz. Find the total weight of the 1200 textbooks in pounds.

38. **Weights** The hospital kitchen has a 50-pound bag of flour. At the end of the week, $\frac{3}{4}$ of the bag has been used. How many ounces of flour remain in the bag at the end of the week?

39. **Weights** A case of soft drinks contains 24 cans, each weighing 6 oz. Find the weight, in pounds, of the case of soft drinks.

40. **Child Development** A baby weighed 7 lb 8 oz at birth. At 6 months of age, the baby weighed 15 lb 13 oz. Find the baby's increase in weight during the 6 months.

Tony Freeman/PhotoEdit, Inc.

41. **Child Development** A baby weighed 6 lb 11 oz at birth. By the time the baby was 12 months old, its weight was three times its birth weight. How much did the baby weigh when it was 12 months old?

42. **Packaging** Massage oil weighing 5 lb 4 oz is divided equally and poured into four containers. How much massage oil is in each container?

43. **Recycling** Use the news clipping at the right. How many tons of plastic bottles are *not* recycled each year?

In the News

Recycling Efforts Fall Short

Every year, 2 billion pounds of plastic bottles are dumped in landfills because consumers recycle only 25% of them.

Source: Time, August 20, 2007

44. **Markup** A candy store buys peppermint candy weighing 12 lb for \$14.40. The candy is repackaged and sold in 6-ounce packages for \$1.15 each. Find the markup on the 12 lb of candy.

45. **Shipping** A knee brace weighing 2 lb 3 oz is mailed at the rate of \$0.34 per ounce. Find the cost of mailing the knee brace.

Applying the Concepts

46. Estimate the weight of a nickel, a textbook, a friend, and a car. Then find the actual weights and compare them with your estimates.

SECTION

8.3 Capacity

OBJECTIVE A To convert measurements of capacity in the U.S. Customary System

Liquid substances are measured in units of **capacity.** The standard U.S. Customary units of capacity are the **fluid ounce, cup, pint, quart,** and **gallon.**

Equivalences Between Units of Capacity in the U.S. Customary System

8 fluid ounces (fl oz) = 1 cup (c)

2 c = 1 pint (pt)

2 pt = 1 quart (qt)

4 qt = 1 gallon (gal)

Point of Interest

The word *quart* has its root in the medieval Latin word *quartus,* which means "fourth." Thus a quart is $\frac{1}{4}$ of a gallon.

The same Latin word is the source of such other English words as *quarter, quartile, quadrilateral,* and *quartet.*

These equivalences can be used to form conversion rates to change one unit of measurement to another. For example, because 8 fl oz = 1 c, the conversion rates $\frac{8 \text{ fl oz}}{1 \text{ c}}$ and $\frac{1 \text{ c}}{8 \text{ fl oz}}$ are both equivalent to 1.

HOW TO • 1 Convert 36 fl oz to cups.

$$36 \text{ fl oz} = 36 \text{ fl oz} \times \boxed{\frac{1 \text{ c}}{8 \text{ fl oz}}}$$

$$= \frac{36 \cancel{\text{fl oz}}}{1} \times \frac{1 \text{ c}}{8 \cancel{\text{fl oz}}}$$

$$= \frac{36 \text{ c}}{8} = 4\frac{1}{2} \text{ c}$$

• The conversion rate must contain c in the numerator and fl oz in the denominator.

HOW TO • 2 Convert 3 qt to cups.

$$3 \text{ qt} = 3 \text{ qt} \times \boxed{\frac{2 \text{ pt}}{1 \text{ qt}}} \times \boxed{\frac{2 \text{ c}}{1 \text{ pt}}}$$

$$= \frac{3 \cancel{\text{qt}}}{1} \times \frac{2 \cancel{\text{pt}}}{1 \cancel{\text{qt}}} \times \frac{2 \text{ c}}{1 \cancel{\text{pt}}}$$

$$= \frac{12 \text{ c}}{1} = 12 \text{ c}$$

• The direct equivalence is not given above. Use two conversion rates. First convert quarts to pints, and then convert pints to cups. The unit in the denominator of the second conversion rate and the unit in the numerator of the first conversion rate must be the same in order to cancel.

EXAMPLE • 1

Convert 42 c to quarts.

Solution

$$42 \text{ c} = 42 \cancel{\text{c}} \times \frac{1 \cancel{\text{pt}}}{2 \cancel{\text{c}}} \times \frac{1 \text{ qt}}{2 \cancel{\text{pt}}}$$

$$= \frac{42 \text{ qt}}{4} = 10\frac{1}{2} \text{ qt}$$

YOU TRY IT • 1

Convert 18 pt to gallons.

Your solution

Solutions on p. S21

OBJECTIVE B To perform arithmetic operations with measurements of capacity

When performing arithmetic operations with measurements of capacity, write the answer in simplest form. For example, 1 c 12 fl oz should be written as 2 c 4 fl oz.

EXAMPLE • 2

What is 4 gal 1 qt decreased by 2 gal 3 qt?

Solution

$$\begin{array}{r} \overset{3\text{ gal}}{\cancel{4\text{ gal}}}\ \overset{5\text{ qt}}{\cancel{1\text{ qt}}} \\ -\ 2\text{ gal }3\text{ qt} \\ \hline 1\text{ gal }2\text{ qt} \end{array}$$

• Borrow 1 gal (4 qt) from 4 gal and add to 1 qt.

YOU TRY IT • 2

Find the quotient of 4 gal 2 qt and 3.

Your solution

Solution on p. S21

OBJECTIVE C To solve application problems

EXAMPLE • 3

A can of apple juice contains 25 fl oz. Find the number of quarts of apple juice in a case of 24 cans.

Strategy

To find the number of quarts of apple juice in one case:

- Multiply the number of cans (24) by the number of fluid ounces per can (25) to find the total number of fluid ounces in the case.
- Convert the number of fluid ounces in the case to quarts.

Solution

$24 \times 25 \text{ fl oz} = 600 \text{ fl oz}$

$$600 \text{ fl oz} = \frac{600 \cancel{\text{fl oz}}}{1} \cdot \frac{1 \cancel{\text{c}}}{8 \cancel{\text{fl oz}}} \cdot \frac{1 \cancel{\text{pt}}}{2 \cancel{\text{c}}} \cdot \frac{1 \text{ qt}}{2 \cancel{\text{pt}}}$$

$$= \frac{600 \text{ qt}}{32} = 18\frac{3}{4} \text{ qt}$$

One case of apple juice contains $18\frac{3}{4}$ qt.

YOU TRY IT • 3

Five students are going backpacking in the desert. Each student requires 5 qt of water per day. How many gallons of water should they take for a 3-day trip?

Your strategy

Your solution

Solution on p. S21

8.3 EXERCISES

OBJECTIVE A **To convert measurements of capacity in the U.S. Customary System**

For Exercises 1 to 3, suppose you convert units of measurement as given. Will the number part of the converted measurement be *less than* or *greater than* the number part of the original measurement?

1. Convert cups to fluid ounces

2. Convert quarts to gallons

3. Convert fluid ounces to pints

For Exercises 4 to 18, convert.

4. 60 fl oz = _______ c

5. 48 fl oz = _______ c

6. 3 c = _______ fl oz

7. $2\frac{1}{2}$ c = _______ fl oz

8. 8 c = _______ pt

9. 5 c = _______ pt

10. $7\frac{1}{2}$ pt = _______ qt

11. 12 pt = _______ qt

12. 22 qt = _______ gal

13. 10 qt = _______ gal

14. $2\frac{1}{4}$ gal = _______ qt

15. 7 gal = _______ qt

16. $1\frac{1}{2}$ pt = _______ fl oz

17. 17 c = _______ qt

18. $1\frac{1}{2}$ qt = _______ c

OBJECTIVE B **To perform arithmetic operations with measurements of capacity**

For Exercises 19 and 20, look at the indicated exercise. The number that goes in the second blank must be less than what number?

19. Exercise 21

20. Exercise 22

For Exercises 21 to 36, perform the arithmetic operation.

21. 14 qt = ___ gal ___ qt

22. 9 pt = ___ qt ___ pt

23. 3 gal 2 qt + 4 gal 3 qt

24. 5 c 3 fl oz + 3 c 6 fl oz

25. 3 gal 3 qt + 1 gal 2 qt

26. $1\frac{1}{2}$ pt + $2\frac{2}{3}$ pt

27. 3 gal 1 qt − 1 gal 2 qt

28. 3 c 3 fl oz − 2 c 5 fl oz

29. 4 c 6 fl oz − 2 c 7 fl oz

30. 3 gal − 1 gal 2 qt

31. $4\frac{1}{2}$ gal − $1\frac{3}{4}$ gal

32. 2 qt 1 pt × 5

33. $3\frac{1}{2}$ pt × 5

34. $5\overline{)6 \text{ gal } 1 \text{ qt}}$

35. $3\frac{1}{2}$ gal ÷ 4

36. $2\overline{)3 \text{ gal } 2 \text{ qt}}$

OBJECTIVE C To solve application problems

37. Nutrition The kitchen at Caring Hearts Assisted Living Facility prepares breakfast for 60 adult residents each morning. If 2 c of coffee are prepared per resident, how many gallons of coffee are prepared?

design56/Shutterstock.com

38. Nutrition The staff at a health care facility is planning a 4th of July party for the residents. They expect 200 residents to attend the party and drink 1 c of punch each. How many gallons of punch should be prepared?

39. Consumerism One brand of tomato juice costs $1.59 for 1 qt. Another brand costs $1.25 for 24 fl oz. Which is the more economical purchase?

40. Camping Mandy carried 12 qt of water for 3 days of desert camping. Water weighs $8\frac{1}{3}$ lb per gallon. Find the weight of water that she carried.

41. Bottled Water Use the news clipping at the right. On average, how many cups of bottled water does an American drink per month? Round to the nearest tenth.

In the News

Bottled Water Consumption Increasing

Americans drink 28.3 gal of bottled water per year, up from 18.8 gal per year five years ago.

Source: Beverage Marketing Corp.

42. Food Service A cafeteria sold 124 cartons of milk in 1 day. Each carton contained 1 c of milk. How many quarts of milk were sold that day?

43. Blood Volume The average human body contains 10 pt of blood. How many gallons of blood are in the average human body?

44. Medical Spa A spa owner bought hand lotion in a 5-quart container and then repackaged the lotion in 8-fluid-ounce bottles. The lotion and bottles cost $81.50, and each 8-fluid-ounce bottle was sold for $8.25. How much profit was made on the 5-quart package of lotion?

45. Repackaging A pharmacy technician's job includes repackaging bulk cough syrup into smaller bottles for dispensing. The pharmacy purchases a 500-ounce bottle of cough syrup for $120 and the technician repackages it into 4-ounce bottles, which sell for $2 each. Find the profit the pharmacy makes on the cough syrup.

For Exercises 46 to 48, use the following information. A punch is made from 3 qt of lemonade and 5 qt of sparkling water. What does each product represent?

46. $\frac{8\text{ qt}}{1} \cdot \frac{1\text{ gal}}{4\text{ qt}}$

47. $\frac{3\text{ qt}}{1} \cdot \frac{4\text{ c}}{1\text{ qt}}$

48. $\frac{8\text{ qt}}{1} \cdot \frac{4\text{ c}}{1\text{ qt}} \cdot \frac{8\text{ fl oz}}{1\text{ c}}$

Applying the Concepts

49. Assume that you want to invent a new measuring system. Discuss some of the features that would have to be incorporated into the system.

SECTION

8.4 Time

OBJECTIVE A To convert units of time

The units in which time is generally measured are the **second, minute, hour, day,** and **week**.

Equivalences Between Units of Time

60 seconds (s) = 1 minute (min)

60 min = 1 hour (h)

24 h = 1 day

7 days = 1 week

These equivalences can be used to form conversion rates to change one unit of time to another. For example, because 24 h = 1 day, the conversion rates $\frac{24\text{ h}}{1\text{ day}}$ and $\frac{1\text{ day}}{24\text{ h}}$ are both equivalent to 1. An example using each of these two rates is shown below.

HOW TO • 1 Convert $5\frac{1}{2}$ days to hours.

$$5\frac{1}{2}\text{ days} = 5\frac{1}{2}\text{ days} \times \frac{24\text{ h}}{1\text{ day}}$$

$$= \frac{11\text{ days}}{2} \times \frac{24\text{ h}}{1\text{ day}}$$

$$= \frac{264\text{ h}}{2} = 132\text{ h}$$

• **The conversion rate must contain h (the unit desired in the answer) in the numerator and must contain day (the original unit) in the denominator.**

HOW TO • 2 Convert 156 h to days.

$$156\text{ h} = 156\text{ h} \times \frac{1\text{ day}}{24\text{ h}}$$

$$= \frac{156\text{ h}}{1} \times \frac{1\text{ day}}{24\text{ h}}$$

$$= \frac{156\text{ days}}{24} = 6\frac{1}{2}\text{ days}$$

• **The conversion rate must contain day (the unit desired in the answer) in the numerator and must contain h (the original unit) in the denominator.**

EXAMPLE • 1

Convert 2880 min to days.

Solution

$$2880\text{ min} = 2880\text{ min} \times \frac{1\text{ h}}{60\text{ min}} \times \frac{1\text{ day}}{24\text{ h}}$$

$$= \frac{2880\text{ days}}{1440} = 2\text{ days}$$

YOU TRY IT • 1

Convert 18,000 s to hours.

Your solution

Solution on p. S21

8.4 EXERCISES

OBJECTIVE A To convert units of time

For Exercises 1 to 3, suppose you convert units of measurement as given. Will the number part of the converted measurement be *less than* or *greater than* the number part of the original measurement?

1. Convert days to minutes

2. Convert seconds to hours

3. Convert weeks to minutes

For Exercises 4 to 27, convert.

4. 98 days = ______ weeks

5. 12 weeks = ______ days

6. $6\frac{1}{4}$ days = ______ h

7. 114 h = ______ days

8. 555 min = ______ h

9. $7\frac{3}{4}$ h = ______ min

10. $18\frac{1}{2}$ min = ______ s

11. 750 s = ______ min

12. 12,600 s = ______ h

13. 15,300 s = ______ h

14. $6\frac{1}{2}$ h = ______ s

15. $5\frac{3}{4}$ h = ______ s

16. 5040 min = ______ days

17. 6840 min = ______ days

18. $2\frac{1}{2}$ days = ______ min

19. $6\frac{1}{4}$ days = ______ min

20. 672 h = ______ weeks

21. 588 h = ______ weeks

22. 3 weeks = ______ h

23. $5\frac{1}{2}$ weeks = ______ h

24. 172,800 s = ______ days

25. 20,160 min = ______ weeks

26. 3 days = ______ s

27. 3 weeks = ______ min

Applying the Concepts

28. An IV of 1000 milliliters of normal saline is set to infuse for $2\frac{1}{4}$ h.

a. How many minutes will it take to complete this infusion?
b. If the IV is begun at 8:00 A.M., at what time will the infusion be complete?

29. A nurse sets up an IV of D5W to infuse for 3 h 30 min. The nurse is required to check the IV after it is halfway infused. How long after beginning the infusion should the nurse check the IV?

SECTION

8.5 Energy and Power

OBJECTIVE A To use units of energy in the U.S. Customary System

Energy can be defined as the ability to do work. Energy is stored in coal, in gasoline, in water behind a dam, and in one's own body.

One **foot-pound** (ft · lb) of energy is the amount of energy necessary to lift 1 pound a distance of 1 foot.

To lift 50 lb a distance of 5 ft requires $50 \times 5 = 250$ ft · lb of energy.

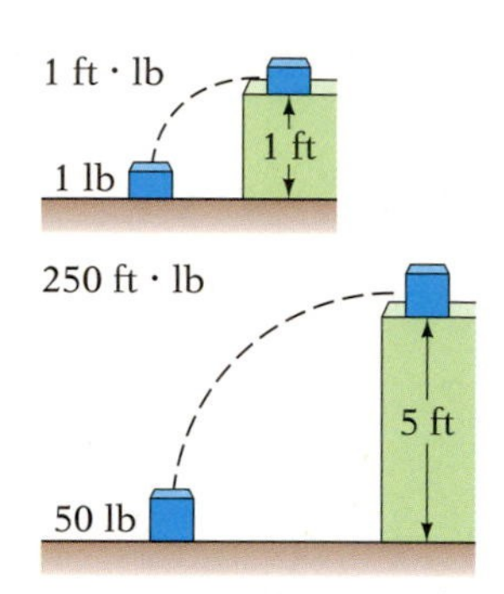

Consumer items that use energy, such as furnaces, stoves, and air conditioners, are rated in terms of the **British thermal unit** (Btu). For example, a furnace might have a rating of 35,000 Btu per hour, which means that it releases 35,000 Btu of energy in 1 hour (1 h).

Because 1 Btu = 778 ft · lb, the following conversion rate, equivalent to 1, can be written:

$$\frac{\textbf{778 ft} \cdot \textbf{lb}}{\textbf{1 Btu}} = \mathbf{1}$$

EXAMPLE 1

Convert 250 Btu to foot-pounds.

Solution

$$250 \text{ Btu} = 250 \cancel{\text{Btu}} \times \frac{778 \text{ ft} \cdot \text{lb}}{1 \cancel{\text{Btu}}}$$
$$= 194{,}500 \text{ ft} \cdot \text{lb}$$

YOU TRY IT 1

Convert 4.5 Btu to foot-pounds.

Your solution

EXAMPLE 2

Find the energy required for a 125-pound person to climb a mile-high mountain.

Solution

In climbing the mountain, the person is lifting 125 lb a distance of 5280 ft.

$$\text{Energy} = 125 \text{ lb} \times 5280 \text{ ft}$$
$$= 660{,}000 \text{ ft} \cdot \text{lb}$$

YOU TRY IT 2

Find the energy required for a motor to lift 800 lb a distance of 16 ft.

Your solution

Solutions on p. S21

EXAMPLE 3

A furnace is rated at 80,000 Btu per hour. How many foot-pounds of energy are released in 1 h?

Solution

$$80{,}000 \text{ Btu} = 80{,}000 \cancel{\text{Btu}} \times \frac{778 \text{ ft} \cdot \text{lb}}{1 \cancel{\text{Btu}}} = 62{,}240{,}000 \text{ ft} \cdot \text{lb}$$

YOU TRY IT 3

A furnace is rated at 56,000 Btu per hour. How many foot-pounds of energy are released in 1 h?

Your solution

Solutions on p. S21

OBJECTIVE B To use units of power in the U.S. Customary System

Power is the rate at which work is done or the rate at which energy is released. Power is measured in **foot-pounds per second** $\left(\frac{\text{ft} \cdot \text{lb}}{\text{s}}\right)$. In each of the following examples, the amount of energy released is the same, but the time taken to release the energy is different; thus the power is different.

100 lb is lifted 10 ft in 10 s.

$$\text{Power} = \frac{10 \text{ ft} \times 100 \text{ lb}}{10 \text{ s}} = 100 \frac{\text{ft} \cdot \text{lb}}{\text{s}}$$

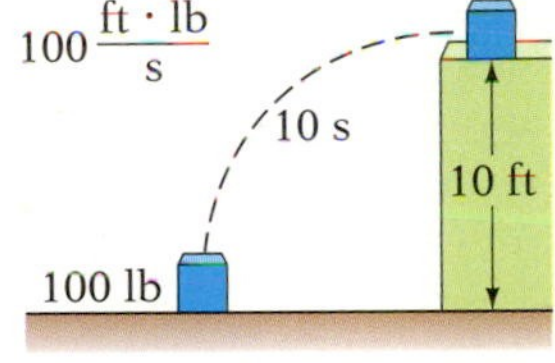

100 lb is lifted 10 ft in 5 s.

$$\text{Power} = \frac{10 \text{ ft} \times 100 \text{ lb}}{5 \text{ s}} = 200 \frac{\text{ft} \cdot \text{lb}}{\text{s}}$$

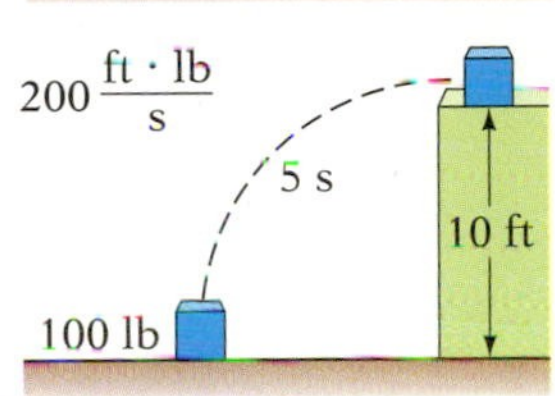

The U.S. Customary unit of power is the **horsepower.** A horse doing average work can pull 550 lb a distance of 1 ft in 1 s and can continue this work all day.

$$\textbf{1 horsepower (hp)} = \mathbf{550 \frac{ft \cdot lb}{s}}$$

EXAMPLE 4

Find the power needed to raise 300 lb a distance of 30 ft in 15 s.

Solution

$$\text{Power} = \frac{30 \text{ ft} \times 300 \text{ lb}}{15 \text{ s}} = 600 \frac{\text{ft} \cdot \text{lb}}{\text{s}}$$

YOU TRY IT 4

Find the power needed to raise 1200 lb a distance of 90 ft in 24 s.

Your solution

EXAMPLE 5

A motor has a power of 2750 $\frac{\text{ft} \cdot \text{lb}}{\text{s}}$. Find the horsepower of the motor.

Solution

$$\frac{2750}{550} = 5 \text{ hp}$$

YOU TRY IT 5

A motor has a power of 3300 $\frac{\text{ft} \cdot \text{lb}}{\text{s}}$. Find the horsepower of the motor.

Your solution

Solution on p. S21

8.5 EXERCISES

OBJECTIVE A **To use units of energy in the U.S. Customary System**

1. Convert 25 Btu to foot-pounds.

2. Convert 6000 Btu to foot-pounds.

3. Convert 25,000 Btu to foot-pounds.

4. Convert 40,000 Btu to foot-pounds.

5. Find the energy required to lift 150 lb a distance of 10 ft.

6. Find the energy required to lift 300 lb a distance of 16 ft.

7. Find the energy required to lift a 3300-pound car a distance of 9 ft.

8. Find the energy required to lift a 3680-pound elevator a distance of 325 ft.

9. Three tons are lifted 5 ft. Find the energy required in foot-pounds.

10. Seven tons are lifted 12 ft. Find the energy required in foot-pounds.

11. A construction worker carries 3-pound blocks up a 10-foot flight of stairs. How many foot-pounds of energy are required to carry 850 blocks up the stairs?

12. A crane lifts an 1800-pound steel beam to the roof of a building 36 ft high. Find the amount of energy the crane requires in lifting the beam.

13. A furnace is rated at 45,000 Btu per hour. How many foot-pounds of energy are released by the furnace in 1 h?

14. A furnace is rated at 22,500 Btu per hour. How many foot-pounds of energy does the furnace release in 1 h?

15. Find the amount of energy in foot-pounds given off when 1 lb of coal is burned. One pound of coal gives off 12,000 Btu of energy when burned.

16. Find the amount of energy in foot-pounds given off when 1 lb of gasoline is burned. One pound of gasoline gives off 21,000 Btu of energy when burned.

17. Without finding the equivalent number of foot-pounds, determine whether 360 Btu is *less than* or *greater than* 360,000 ft · lb.

OBJECTIVE B **To use units of power in the U.S. Customary System**

18. When you convert horsepower to foot-pounds per second, is the number part of the converted measurement *less than* or *greater than* the number part of the original measurement?

19. Convert 1100 $\frac{\text{ft} \cdot \text{lb}}{\text{s}}$ to horsepower.

20. Convert 6050 $\frac{\text{ft} \cdot \text{lb}}{\text{s}}$ to horsepower.

21. Convert 4400 $\frac{\text{ft} \cdot \text{lb}}{\text{s}}$ to horsepower.

22. Convert 1650 $\frac{\text{ft} \cdot \text{lb}}{\text{s}}$ to horsepower.

23. Convert 9 hp to foot-pounds per second.

24. Convert 4 hp to foot-pounds per second.

25. Convert 7 hp to foot-pounds per second.

26. Convert 8 hp to foot-pounds per second.

27. Find the power in foot-pounds per second needed to raise 125 lb a distance of 12 ft in 3 s.

28. Find the power in foot-pounds per second needed to raise 500 lb a distance of 60 ft in 8 s.

Christopher Gould/Photographer's Choice/Getty Images

29. Find the power in foot-pounds per second needed to raise 3000 lb a distance of 40 ft in 25 s.

30. Find the power in foot-pounds per second of an engine that can raise 180 lb to a height of 40 ft in 5 s.

31. Find the power in foot-pounds per second of an engine that can raise 1200 lb to a height of 18 ft in 30 s.

32. A motor has a power of 4950 $\frac{\text{ft} \cdot \text{lb}}{\text{s}}$. Find the horsepower of the motor.

33. A motor has a power of 16,500 $\frac{\text{ft} \cdot \text{lb}}{\text{s}}$. Find the horsepower of the motor.

34. A motor has a power of 6600 $\frac{\text{ft} \cdot \text{lb}}{\text{s}}$. Find the horsepower of the motor.

Applying the Concepts

35. Pick a source of energy and write an article about it. Include the source, possible pollution problems, and future prospects associated with this form of energy.

FOCUS ON PROBLEM SOLVING

Applying Solutions to Other Problems

Problem solving in the previous chapters concentrated on solving specific problems. After a problem is solved, however, there is an important question to be asked: "Does the solution to this problem apply to other types of problems?"

To illustrate this extension of problem solving, we will consider *triangular numbers,* which were studied by ancient Greek mathematicians. The numbers 1, 3, 6, 10, 15, and 21 are the first six triangular numbers. What is the next triangular number?

To answer this question, note in the diagram below that a triangle can be formed using the number of dots that correspond to a triangular number.

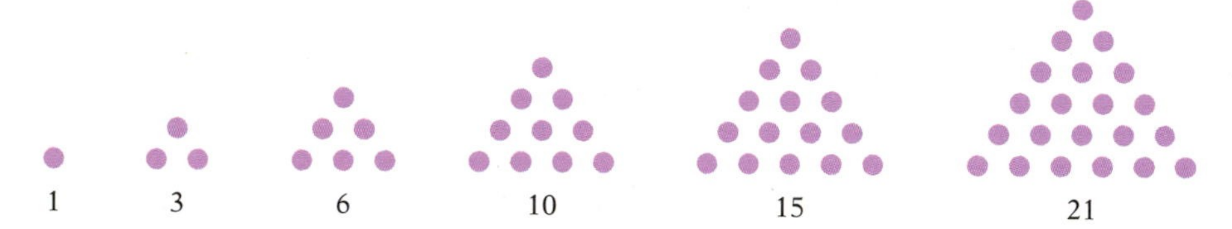

Observe that the number of dots in each row is one more than the number of dots in the row above. The total number of dots can be found by addition.

The pattern suggests that the next triangular number (the seventh one) is the sum of the first seven natural numbers. The seventh triangular number is 28. The diagram at the right shows the seventh triangular number.

Using the pattern for triangular numbers, it is easy to determine that the tenth triangular number is

$$1 + 2 + 3 + 4 + 5 + 6 + 7 + 8 + 9 + 10 = 55$$

Now consider a situation that may seem to be totally unrelated to triangular numbers. Suppose you are in charge of scheduling softball games for a league. There are seven teams in the league, and each team must play every other team once. How many games must be scheduled?

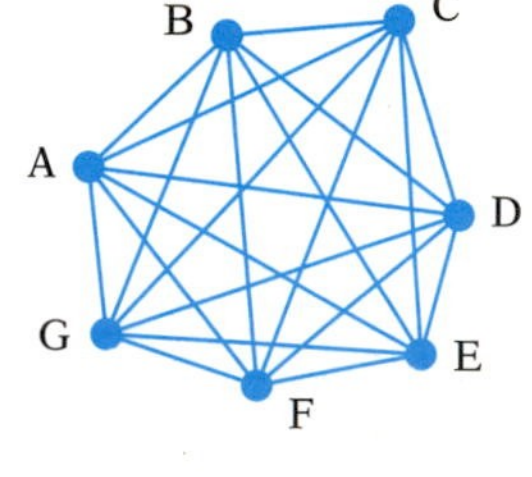

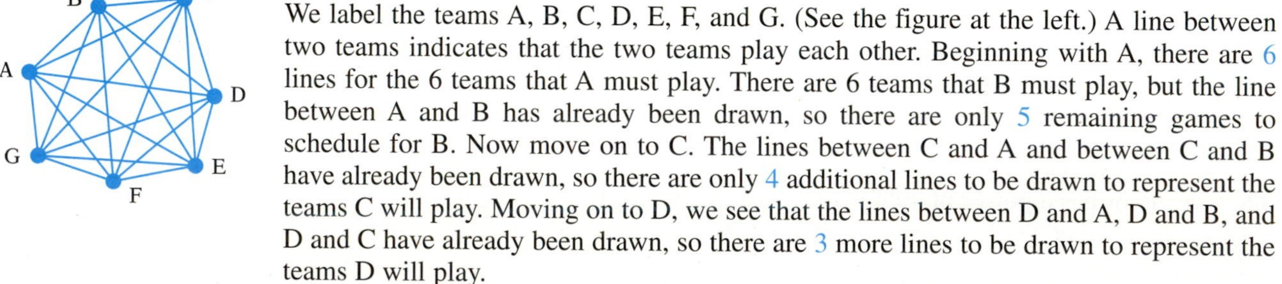

We label the teams A, B, C, D, E, F, and G. (See the figure at the left.) A line between two teams indicates that the two teams play each other. Beginning with A, there are 6 lines for the 6 teams that A must play. There are 6 teams that B must play, but the line between A and B has already been drawn, so there are only 5 remaining games to schedule for B. Now move on to C. The lines between C and A and between C and B have already been drawn, so there are only 4 additional lines to be drawn to represent the teams C will play. Moving on to D, we see that the lines between D and A, D and B, and D and C have already been drawn, so there are 3 more lines to be drawn to represent the teams D will play.

Note that each time we move from team to team, one fewer line needs to be drawn. When we reach F, there is only one line to be drawn, the one between F and G. The total number of lines drawn is $6 + 5 + 4 + 3 + 2 + 1 = 21$, the sixth triangular number. For a league with seven teams, the number of games that must be scheduled so that each team plays every other team once is the sixth triangular number. If there were ten teams in the league, the number of games that must be scheduled would be the ninth triangular number, which is 45.

A college chess team wants to schedule a match so that each of its 15 members plays each other member of the team once. How many matches must be scheduled?

PROJECTS AND GROUP ACTIVITIES

Nomographs

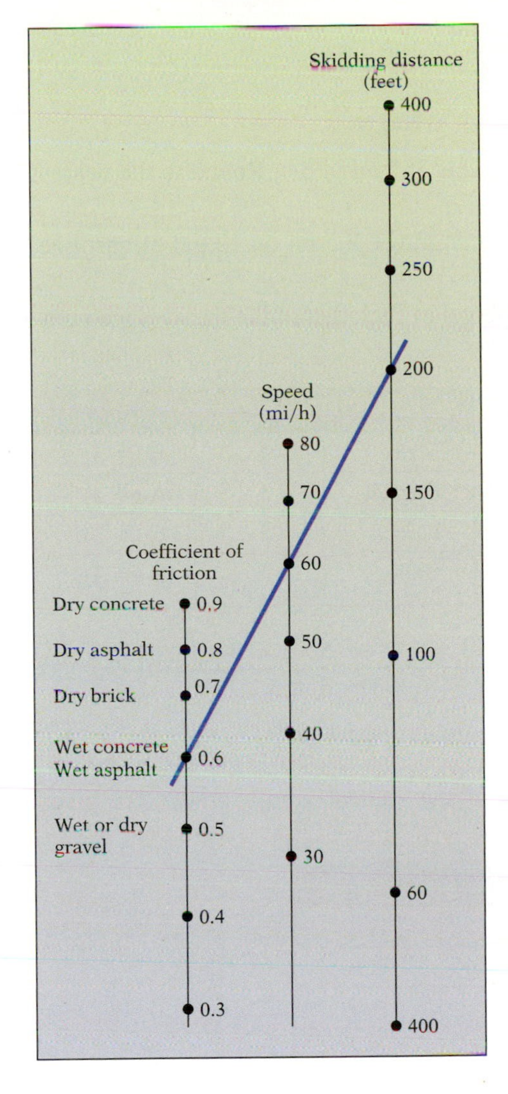

A chart is another tool that is used in problem solving. The chart at the left is a nomograph. A **nomograph** is a chart that represents numerical relationships among variables.

One of the details a traffic accident investigator checks when looking into a car accident is the length of the skid marks made by the car. This length can help the investigator determine the speed of the car when the brakes were applied.

The nomograph at the left can be used to determine the speed of a car under given conditions. It shows the relationship among the speed of the car, the skidding distance, and the *coefficient of friction.*

The coefficient of friction is an experimentally obtained value that reflects how easy or hard it is to drag one object over another. For instance, it is easier to drag a box across ice than to drag it across a carpet. The coefficient of friction is smaller for the box and ice than it is for the box and carpet.

To use the nomograph at the left, an investigator would draw a line from the coefficient of friction to the skidding distance. The point at which the line crosses the speed line shows how fast the car was going when the brakes were applied. The line from 0.6 to 200 intersects the speed line at 60. This indicates that a car that skidded 200 ft on wet asphalt or wet concrete was traveling at 60 mi/h.

1. Use the nomograph to determine the speed of a car when the brakes were applied for a car traveling on gravel and for skid marks of 100 ft.
2. Use the nomograph to determine the speed of a car when the brakes were applied for a car traveling on dry concrete and for skid marks of 150 ft.
3. Suppose a car is traveling 80 mi/h when the brakes are applied. Find the difference in skidding distance if the car is traveling on wet concrete rather than dry concrete.

Averages

If two towns are 150 mi apart and you drive between the two towns in 3 h, then your

$$\text{Average speed} = \frac{\text{total distance}}{\text{total time}} = \frac{150 \text{ mi}}{3 \text{ h}} = 50 \text{ mi/h}$$

It is highly unlikely that your speed was *exactly* 50 mi/h the entire time of the trip. Sometimes you will have traveled faster than 50 mi/h, and other times you will have traveled slower than 50 mi/h. Dividing the total distance you traveled by the total time it took to go that distance is an example of calculating an average.

There are many other averages that may be calculated. For instance, the Environmental Protection Agency calculates an estimated miles per gallon (mpg) for new cars. Miles per gallon is an average calculated from the formula

$$\frac{\text{Miles traveled}}{\text{Gallons of gasoline consumed}}$$

Digital Vision/Getty Images

For instance, the miles per gallon for a car that travels 308 mi on 11 gal of gas is $\frac{308 \text{ mi}}{11 \text{ gal}} = 28 \text{ mpg}$.

A pilot would not use miles per gallon as a measure of fuel efficiency. Rather, pilots use gallons per hour. A plane that travels 5 h and uses 400 gal of fuel has an average fuel efficiency of

$$\frac{\text{Gallons of fuel}}{\text{Hours flown}} = \frac{400 \text{ gal}}{5 \text{ h}} = 80 \text{ gal/h}$$

Using the examples above, calculate the following averages.

1. Determine the average speed of a car that travels 355 mi in 6 h. Round to the nearest tenth.
2. Determine the miles per gallon of a car that can travel 405 mi on 12 gal of gasoline. Round to the nearest tenth.
3. If a plane flew 2000 mi in 5 h and used 1000 gal of fuel, determine the average number of gallons per hour that the plane used.

Another type of average is grade-point average (GPA). It is calculated by multiplying the units for each class by the grade point for that class, adding the results, and dividing by the total number of units taken. Here is an example using the grading scale $A = 4$, $B = 3$, $C = 2$, $D = 1$, and $F = 0$.

Class	*Units*	*Grade*
Math	4	B (= 3)
English	3	A (= 4)
French	5	C (= 2)
Biology	3	B (= 3)

$$\text{GPA} = \frac{4 \cdot 3 + 3 \cdot 4 + 5 \cdot 2 + 3 \cdot 3}{4 + 3 + 5 + 3} = \frac{43}{15} \approx 2.87$$

4. A grading scale that provides for plus or minus grades uses $A = 4$, $A- = 3.7$, $B+ = 3.3$, $B = 3$, $B- = 2.7$, $C+ = 2.3$, $C = 2$, $C- = 1.7$, $D+ = 1.3$, $D = 1$, $D- = 0.7$, and $F = 0$. Calculate the GPA of the student whose grades are given below.

Class	*Units*	*Grade*
Math	5	B+
English	3	C+
Spanish	5	A−
Physical science	3	B−

CHAPTER 8

SUMMARY

KEY WORDS	EXAMPLES
A *measurement* includes a number and a unit. [8.1A, p. 340]	9 inches, 6 feet, 3 yards, and 50 miles are measurements.
Equivalent measures are used to form *conversion rates* to change one unit in the U.S. Customary System of measurement to another. In the conversion rate chosen, the unit in the numerator is the same as the unit desired in the answer. The unit in the denominator is the same as the unit in the given measurement. [8.1A, p. 340]	Because 12 in. = 1 ft, the conversion rate $\frac{12 \text{ in.}}{1 \text{ ft}}$ is used to convert feet to inches. The conversion rate $\frac{1 \text{ ft}}{12 \text{ in.}}$ is used to convert inches to feet.

Energy is the ability to do work. One *foot-pound* (ft · lb) of energy is the amount of energy necessary to lift 1 pound a distance of 1 foot. Consumer items that use energy are rated in *British thermal units* (Btu). [8.5A, p. 356]

Find the energy required for a 110-pound person to climb a set of stairs 12 ft high.

Energy = 110 lb × 12 ft = 1320 ft · lb

Power is the rate at which work is done or energy is released.

Power is measured in *foot-pounds per second* $\left(\frac{\text{ft}\cdot\text{lb}}{\text{s}}\right)$ and *horsepower* (hp). [8.5B, p. 357]

Find the power needed to raise 250 lb a distance of 20 ft in 10 s.

$$\text{Power} = \frac{20\text{ ft} \times 250\text{ lb}}{10\text{ s}} = 500\,\frac{\text{ft}\cdot\text{lb}}{\text{s}}$$

ESSENTIAL RULES AND PROCEDURES

EXAMPLES

Equivalences Between Units of Length [8.1A, p. 340]
The U.S. Customary units of length are inch (in.), foot (ft), yard (yd), and mile (mi).
12 in. = 1 ft
3 ft = 1 yd
36 in. = 1 yd
5280 ft = 1 mi

Convert 52 in. to ft.

$$52\text{ in.} = 52\text{ in.} \times \frac{1\text{ ft}}{12\text{ in.}} = \frac{52\text{ ft}}{12} = 4\frac{1}{3}\text{ ft}$$

Equivalences Between Units of Weight [8.2A, p. 346]
Weight is a measure of how strongly Earth is pulling on an object. The U.S. Customary units of weight are ounce (oz), pound (lb), and ton.
16 oz = 1 lb
2000 lb = 1 ton

Convert 9 lb to ounces.

$$9\text{ lb} = 9\text{ lb} \times \frac{16\text{ oz}}{1\text{ lb}} = 144\text{ oz}$$

Equivalences Between Units of Capacity [8.3A, p. 350]
Liquid substances are measured in units of *capacity*. The U.S. Customary units of capacity are fluid ounce (fl oz), cup (c), pint (pt), quart (qt), and gallon (gal).
8 fl oz = 1 c
2 c = 1 pt
2 pt = 1 qt
4 qt = 1 gal

Convert 14 qt to gallons.

$$14\text{ qt} = 14\text{ qt} \times \frac{1\text{ gal}}{4\text{ qt}} = \frac{14\text{ gal}}{4} = 3\frac{1}{2}\text{ gal}$$

Equivalences Between Units of Time [8.4A, p. 354]
Units of time are seconds (s), minutes (min), hours (h), days, and weeks.
60 s = 1 min
60 min = 1 h
24 h = 1 day
7 days = 1 week

Convert 8 days to hours.

$$8\text{ days} = 8\text{ days} \times \frac{24\text{ h}}{1\text{ day}} = 192\text{ h}$$

Equivalences Between Units of Energy [8.5A, p. 356]
1 Btu = 778 ft · lb

Convert 70 Btu to foot-pounds.

$$70\text{ Btu} = 70\text{ Btu} \times \frac{778\text{ ft}\cdot\text{lb}}{1\text{ Btu}} = 54{,}460\text{ ft}\cdot\text{lb}$$

Equivalences Between Units of Power [8.5B, p. 357]
The U.S. Customary unit of power is the horsepower (hp).

$$1\text{ hp} = 550\,\frac{\text{ft}\cdot\text{lb}}{\text{s}}$$

Convert 5 hp to foot-pounds per second.

$$5 \times 550 = 2750\,\frac{\text{ft}\cdot\text{lb}}{\text{s}}$$

CHAPTER 8

CONCEPT REVIEW

Test your knowledge of the concepts presented in this chapter. Answer each question. Then check your answers against the ones provided in the Answer Section.

1. What operation is used to convert from feet to inches?

2. How do you convert 5 ft 7 in. to all inches?

3. How do you divide 7 ft 8 in. by 2?

4. What conversion rate is used to convert 5240 lb to tons?

5. How do you multiply 6 lb 9 oz by 3?

6. Name five measures of capacity.

7. What conversion rate is used to convert 7 gal to quarts?

8. What conversion rate is used to convert 374 min to hours?

9. What is a foot-pound of energy?

10. What operations are needed to find the power to raise 200 lb a distance of 24 ft in 12 s?

CHAPTER 8

REVIEW EXERCISES

1. Convert 4 ft to inches.

2. What is 7 ft 6 in. divided by 3?

3. Convert 53 in. to feet and inches.

4. Convert $2\frac{1}{2}$ pt to fluid ounces.

5. Convert 14 ft to yards.

6. Convert 84 oz to pounds and ounces.

7. Find the quotient of 7 lb 5 oz and 3.

8. Convert $3\frac{3}{8}$ lb to ounces.

9. Add: $\begin{array}{r} 3 \text{ ft } 9 \text{ in.} \\ + \ 5 \text{ ft } 6 \text{ in.} \\ \hline \end{array}$

10. Subtract: $\begin{array}{r} 3 \text{ tons } \ 500 \text{ lb} \\ - \ 1 \text{ ton } \ 1500 \text{ lb} \\ \hline \end{array}$

11. Add: $\begin{array}{r} 4 \text{ c } 7 \text{ fl oz} \\ + \ 2 \text{ c } 3 \text{ fl oz} \\ \hline \end{array}$

12. Subtract: $\begin{array}{r} 5 \text{ yd } 1 \text{ ft} \\ - \ 3 \text{ yd } 2 \text{ ft} \\ \hline \end{array}$

13. Convert 12 c to quarts.

14. Convert 375 min to hours.

15. Convert 2.5 hp to foot-pounds per second. $\left(1 \text{ hp} = 550 \dfrac{\text{ft} \cdot \text{lb}}{\text{s}}\right)$

16. Multiply:

$$\begin{array}{r} 5 \text{ lb } 8 \text{ oz} \\ \times \quad 8 \\ \hline \end{array}$$

17. Height A patient's height is 67 in. Convert this measurement to feet and inches.

18. Birth Weight A premature baby weighs 70 oz. Convert this weight to pounds.

19. Medical Supplies Three pieces of gauze are cut from a roll. They measure 1 ft 3 in., 2 ft 3 in., and 11 in. Find the total amount of gauze that has been cut from the roll.

20. Birth Weight The Suleman octuplets were born in 2009 and had the following birth weights: 2 lb 8 oz, 2 lb 2 oz, 3 lb 1 oz, 2 lb 3 oz, 1 lb 12 oz, 2 lb 9 oz, 1 lb 13 oz, and 2 lb 7 oz. Find their total birth weight.

21. Capacity A can of pineapple juice contains 18 fl oz. Find the number of quarts in a case of 24 cans.

AP Photo/David Kohl

22. Food Service A cafeteria sold 256 cartons of milk in one school day. Each carton contains 1 c of milk. How many gallons of milk were sold that day?

23. Energy A furnace is rated at 35,000 Btu per hour. How many foot-pounds of energy does the furnace release in 1 h? (1 Btu = 778 ft · lb)

24. Power Find the power in foot-pounds per second of an engine that can raise 800 lb to a height of 15 ft in 25 s.

CHAPTER 8

TEST

1. Convert $2\frac{1}{2}$ ft to inches.

2. Subtract: 4 ft 2 in. − 1 ft 9 in.

3. **Medical Supplies** A roll of lightweight adhesive tape is 5 yd long by 2 in. wide. How many 10-in.-long pieces of tape can be cut from this roll?

4. **Height** Rodriquez and Julio are brothers. Julio is 5 in. taller than his brother, who is 3 ft 9 in. tall. How tall is Julio?

5. Convert $2\frac{7}{8}$ lb to ounces.

6. Convert: 40 oz = ___ lb ___ oz

7. Find the sum of 9 lb 6 oz and 7 lb 11 oz.

8. Divide: 6 lb 12 oz ÷ 4

9. **Weights** A college bookstore received 1000 workbooks, each weighing 12 oz. Find the total weight of the 1000 workbooks in pounds.

10. **Recycling** An elementary school class gathered 800 aluminum cans for recycling. Four aluminum cans weigh 3 oz. Find the amount the class received if the rate of pay was $0.75 per pound for the aluminum cans. Round to the nearest cent.

Tony Freeman/PhotoEdit, Inc.

11. Convert 13 qt to gallons.

12. Convert $3\frac{1}{2}$ gal to pints.

13. What is $1\frac{3}{4}$ gal times 7?

14. Add: 5 gal 2 qt + 2 gal 3 qt

15. Convert 756 h to weeks.

16. Convert $3\frac{1}{4}$ days to minutes.

17. **Capacity** A can of grapefruit juice contains 20 fl oz. Find the number of cups of grapefruit juice in a case of 24 cans.

18. **Repackaging** A pharmacy purchased 1 gal of isopropyl alcohol (rubbing alcohol) and then has the pharmacy technician repackage it in 8-fluid-ounce bottles. The alcohol and bottles cost $25.50, and each 8-fluid-ounce bottle was sold for $2.75. Find the profit the pharmacy made on the alcohol.

19. **Height** A medical assistant at the pediatrician's office records the height of a 3-year-old boy as 38 in. The boy's father is exactly twice as tall as his son. Give the father's height in feet and inches.

20. **Weight** The birth weight of a newborn baby is 5 lb 12 oz. At its six-month check-up, the baby's weight is 13 lb 5 oz. How much weight has the baby gained since birth?

21. **Energy** A furnace is rated at 40,000 Btu per hour. How many foot-pounds of energy are released by the furnace in 1 h? (1 Btu = 778 ft · lb)

22. **Power** A motor has a power of $2200\ \frac{\text{ft} \cdot \text{lb}}{\text{s}}$. Find the motor's horsepower.

$$\left(1\text{ hp} = 550\ \frac{\text{ft} \cdot \text{lb}}{\text{s}}\right)$$

CUMULATIVE REVIEW EXERCISES

1. Find the LCM of 9, 12, and 15.

2. Write $\frac{43}{8}$ as a mixed number.

3. Subtract: $5\frac{7}{8} - 2\frac{7}{12}$

4. What is $5\frac{1}{3}$ divided by $2\frac{2}{3}$?

5. Simplify: $\frac{5}{8} \div \left(\frac{3}{8} - \frac{1}{4}\right) - \frac{5}{8}$

6. Round 2.0972 to the nearest hundredth.

7. Multiply: $\begin{array}{r} 0.0792 \\ \times \quad 0.49 \\ \hline \end{array}$

8. Solve the proportion: $\frac{n}{12} = \frac{44}{60}$

9. Find $2\frac{1}{2}\%$ of 50.

10. 18 is 42% of what? Round to the nearest hundredth.

11. **Consumerism** A 7.2-pound roast costs $37.08. Find the unit cost.

12. Add: $3\frac{2}{5}$ in. + $5\frac{1}{3}$ in.

13. Convert: 24 oz = ___ lb ___ oz

14. Multiply: 3 lb 8 oz × 9

15. Subtract: $4\frac{1}{3}$ qt − $1\frac{5}{6}$ qt

16. Find 2 lb 10 oz less than 4 lb 6 oz.

17. **Solutions** A saline solution has a concentration ratio of 2.5 milligrams of salt in 100 milliliters of solution. How many milligrams of salt will be needed to produce 250 milliliters of a saline solution having this same concentration?

18. **Banking** Anna had a balance of $578.56 in her checkbook. She wrote checks for $216.98 and $34.12 and made a deposit of $315.33. What is her new checking account balance?

19. **Compensation** A pharmaceutical salesman receives a salary of $1800 per month plus a commission of 2% on all sales over $25,000. Find the total monthly income of a pharmaceutical salesman who has monthly sales of $140,000.

20. **Medication** A doctor prescribes a dose of 1.5 milligrams of a medication for a patient. The medication is in liquid form and each milliliter of liquid contains 0.25 milligram of medication. How many milliliters of the liquid should a nurse give the patient?

21. **Education** The scores on the final exam of a trigonometry class are recorded in the histogram at the right. What percent of the class received a score between 80% and 90%? Round to the nearest percent.

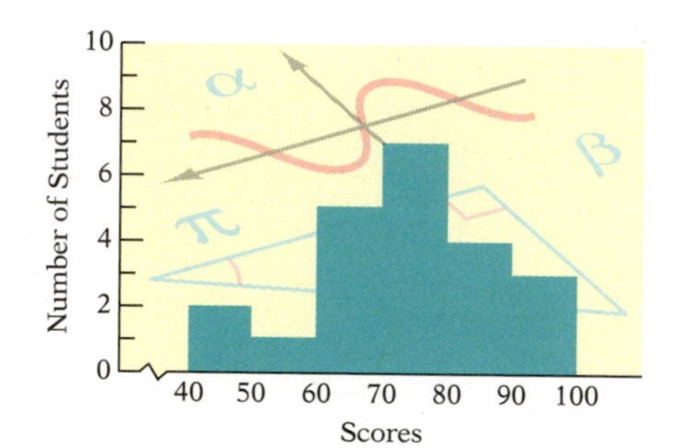

22. **Markup** Acme Medical Supply uses a markup rate of 40% on all merchandise. What is the selling price of a power wheelchair that costs the store $925?

23. **Simple Interest** Owners of a new walk-in clinic signed an eight-month note for a $200,000 loan at a simple interest rate of 6%. Find the interest paid on the loan.

24. **Income** Six college students spent several weeks panning for gold during their summer vacation. The students obtained 1 lb 3 oz of gold, which they sold for $800 per ounce. How much money did each student receive if they shared the money equally? Round to the nearest dollar.

25. **Repackaging** A therapeutic spa purchased a 1-gallon bottle of massage lotion and then had it repackaged into 8-fluid-ounce bottles. The lotion and bottles cost $32.66, and each 8-fluid-ounce bottle was sold for $3.50. Find the profit the spa made on the massage lotion.

26. **Consumerism** One brand of yogurt costs $0.79 for 8 oz, and 36 oz of another brand can be bought for $2.98. Which purchase is the better buy?

27. **Probability** Two dice are rolled. What is the probability that the sum of the dots on the upward faces is 9?

28. **Medication Costs** A medication for cancer patients suffering from anemia is given by injection once a week. The patient receives 40,000 units of this medication at a cost of $9.50 per 1000 units. Find the weekly cost of this medication.

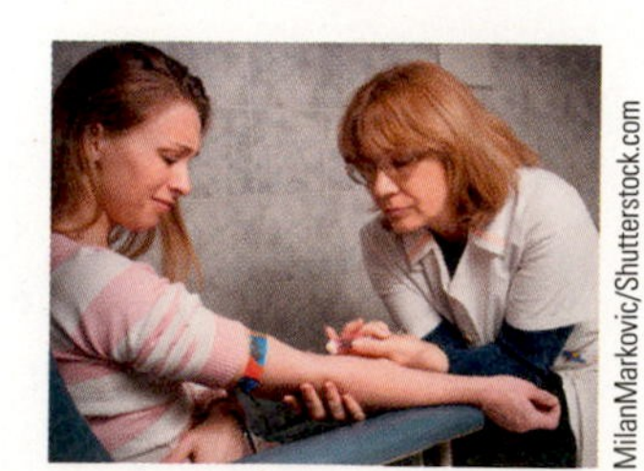
MilanMarkovic/Shutterstock.com

29. **Height** Convert a height of 70 in. to feet and inches.

CHAPTER

9

The Metric System of Measurement

OBJECTIVES

SECTION 9.1

- **A** To convert units of length in the metric system of measurement
- **B** To solve application problems

SECTION 9.2

- **A** To convert units of mass in the metric system of measurement
- **B** To solve application problems

SECTION 9.3

- **A** To convert units of capacity in the metric system of measurement
- **B** To solve application problems

SECTION 9.4

- **A** To use units of energy in the metric system of measurement

SECTION 9.5

- **A** To convert U.S. Customary units to metric units
- **B** To convert metric units to U.S. Customary units

AIM for the Future

CLINICAL LABORATORY TECHNICIANS usually work under the supervision of medical and clinical laboratory technologists or laboratory managers. They perform less complex tests and laboratory procedures than clinical laboratory technologists do. Technicians may prepare specimens and operate automated analyzers, for example, or they may perform manual tests in accordance with detailed instructions. Employment of clinical laboratory workers is expected to grow by 14% between 2008 and 2018, faster than the average for all occupations. Median annual wages of medical and clinical laboratory technicians were $35,380 in May 2008. (*Source: Bureau of Labor Statistics:* http://www.bls.gov/oco/ocos096.htm)

PREP TEST

Are you ready to succeed in this chapter?

Take the Prep Test below to find out if you are ready to learn the new material.

1. $3.732 \times 10{,}000$

2. 65.9×10^4

3. $41.07 \div 1000$

4. $28{,}496 \div 10^3$

5. $6 - 0.875$

6. $5 + 0.96$

7. 3.25×0.04

8. $35 \times \frac{1.61}{1}$

9. $1.67 \times \frac{1}{3.34}$

10. $4\frac{1}{2} \times 150$

SECTION

9.1 Length

OBJECTIVE A To convert units of length in the metric system of measurement

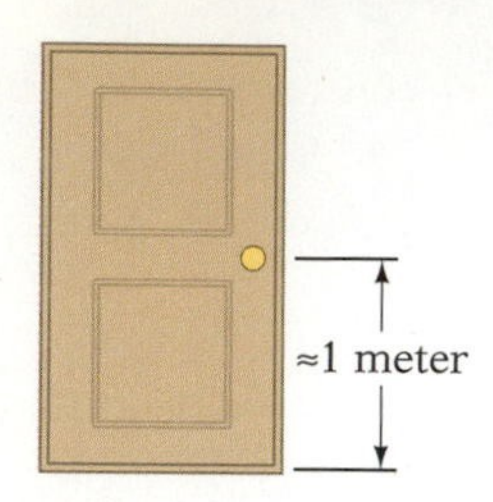

In 1789, an attempt was made to standardize units of measurement internationally in order to simplify trade and commerce between nations. A commission in France developed a system of measurement known as the **metric system.**

The basic unit of length in the metric system is the **meter.** One meter is approximately the distance from a doorknob to the floor. All units of length in the metric system are derived from the meter. Prefixes to the basic unit denote the length of each unit. For example, the prefix "centi-" means one-hundredth, so 1 centimeter is 1 one-hundredth of a meter.

North Pole

Equator

Prefixes and Units of Length in the Metric System

Prefix	Unit
kilo- = 1000	1 kilometer (km) = 1000 meters (m)
hecto- = 100	1 hectometer (hm) = 100 m
deca- = 10	1 decameter (dam) = 10 m
	1 meter (m) = 1 m
deci- = 0.1	1 decimeter (dm) = 0.1 m
centi- = 0.01	1 centimeter (cm) = 0.01 m
milli- = 0.001	1 millimeter (mm) = 0.001 m

Point of Interest

Originally the meter (spelled *metre* in some countries) was defined as $\frac{1}{10{,}000{,}000}$ of the distance from the equator to the North Pole. Modern scientists have redefined the meter as 1,650,763.73 wavelengths of the orange-red light given off by the element krypton.

Conversion between units of length in the metric system involves moving the decimal point to the right or to the left. Listing the units in order from largest to smallest will indicate how many places to move the decimal point and in which direction.

To convert 4200 cm to meters, write the units in order from largest to smallest.

km hm dam m dm cm mm

2 positions

- **Converting cm to m requires moving 2 positions to the left.**

4200 cm = 42.00 m

2 places

- **Move the decimal point the same number of places and in the same direction.**

A metric measurement that involves two units is customarily written in terms of one unit. Convert the smaller unit to the larger unit and then add.

Tips for Success

The prefixes introduced here are used throughout the chapter. As you study the material in the remaining sections, use the table above for a reference or refer to the Chapter Summary at the end of this chapter.

To convert 8 km 32 m to kilometers, first convert 32 m to kilometers.

km hm dam m dm cm mm

- **Converting m to km requires moving 3 positions to the left.**

32 m = 0.032 km

- **Move the decimal point the same number of places and in the same direction.**

8 km 32 m = 8 km + 0.032 km
= 8.032 km

- **Add the result to 8 km.**

EXAMPLE • 1

Convert 0.38 m to millimeters.

Solution 0.38 m = 380 mm

YOU TRY IT • 1

Convert 3.07 m to centimeters.

Your solution

EXAMPLE • 2

Convert 4 m 62 cm to meters.

Solution 62 cm = 0.62 m

4 m 62 cm = 4 m + 0.62 m
= 4.62 m

YOU TRY IT • 2

Convert 3 km 750 m to kilometers.

Your solution

Solutions on p. S21

OBJECTIVE B To solve application problems

Take Note

Although in this text we will always change units to the larger unit, it is possible to perform the calculation by changing to the smaller unit.

2 m − 85 cm
= 200 cm − 85 cm
= 115 cm

Note that 115 cm = 1.15 m.

In the application problems in this section, we perform arithmetic operations with the measurements of length in the metric system. It is important to remember that before measurements can be added or subtracted, they must be expressed in terms of the same unit. In this textbook, unless otherwise stated, the units should be changed to the larger unit before the arithmetic operation is performed.

To subtract 85 cm from 2 m, convert 85 cm to meters.

2 m − 85 cm = 2 m − 0.85 m
= 1.15 m

EXAMPLE • 3

At the pediatrician's office, an 8-year-old girl's height is recorded as 1.27 m. Her younger brother is 19 cm shorter than her. Find the brother's height.

Strategy

To find the brother's height:

- Convert the difference in height (19 cm) to meters.
- Subtract this number from the girl's height (1.27 m).

Solution

19 cm = 0.19 m

1.27 m – 19 cm = 1.27 m – 0.19 m = 1.08 m

The brother's height is 1.08 m.

YOU TRY IT • 3

A hospital is holding a 5K Race for Life event to raise funds for cancer research. The total distance of the primary race is 5 km, but there is a Fun Run event for participants who plan to walk or jog a shorter distance. If the Fun Run is $\frac{1}{4}$ the distance of the 5K Race for Life, how many meters long is the Fun Run?

Your strategy

Your solution

Solution on p. S21

9.1 EXERCISES

OBJECTIVE A To convert units of length in the metric system of measurement

For Exercises 1 to 27, convert.

1. 42 cm = _____ mm
2. 62 cm = _____ mm
3. 81 mm = _____ cm
4. 68.2 mm = _____ cm
5. 6804 m = _____ km
6. 3750 m = _____ km
7. 2.109 km = _____ m
8. 32.5 km = _____ m
9. 432 cm = _____ m
10. 61.7 cm = _____ m
11. 0.88 m = _____ cm
12. 3.21 m = _____ cm
13. 7038 m = _____ km
14. 2589 m = _____ km
15. 3.5 km = _____ m
16. 9.75 km = _____ m
17. 260 cm = _____ m
18. 705 cm = _____ m
19. 1.685 m = _____ cm
20. 0.975 m = _____ cm
21. 14.8 cm = _____ mm
22. 6 m 42 cm = _____ m
23. 62 m 7 cm = _____ m
24. 42 cm 6 mm = _____ cm
25. 31 cm 9 mm = _____ cm
26. 62 km 482 m = _____ km
27. 8 km 75 m = _____ km

For Exercises 28 to 30, fill in the blank with the correct unit of measurement.

28. 5.8 m = 580 _____
29. 0.6 km = 600 _____
30. 54 mm = 0.054 _____

OBJECTIVE B To solve application problems

31. **Surgical Instruments** Surgical technologists help set up the instruments in an operating room prior to surgery. A set of Iris Scissors is among the instruments and is 11.5 cm long. Give this measurement in millimeters.

32. **Measurements** Find the missing dimension, in centimeters, in the diagram at the right.

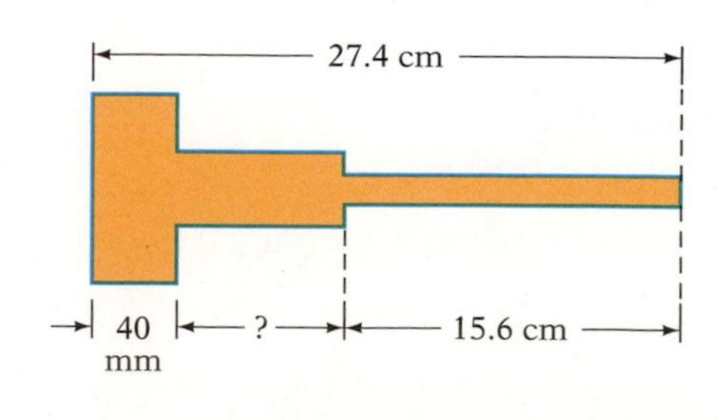

33. **Sports** A walk-a-thon had two checkpoints. One checkpoint was 1400 m from the starting point. The second checkpoint was 1200 m from the first checkpoint. The second checkpoint was 1800 m from the finish line. How long was the walk? Express the answer in kilometers.

34. **Blood Pressure** A patient's blood pressure is defined to be the pressure exerted by the blood upon the walls of the blood vessels, especially the arteries, usually measured by means of a sphygmomanometer. Blood pressure is normally expressed as a fraction in millimeters of mercury. The numerator of the fraction is the maximum pressure when the heart contracts, forcing blood out of the heart. The denominator is the minimum pressure when the heart is filling with blood. (http://dictionary.reference.com) The suggested optimal blood pressure of 120/80 is the ratio of 120 mm of mercury to 80 mm of mercury. Express these measurements in cm.

Sopotnicki/Shutterstock.com

35. **Blood Pressure** See Exercise 34 for an explanation of blood pressure. A patient with high blood pressure, such as 142/105, is said to be hypertensive. Express the blood pressure reading 142/105 in centimeters of mercury.

In the News

Highway Adoption Proves Popular

The Missouri Adopt-a-Highway Program, which has been in existence for 20 years, currently has 3772 groups in the program. The groups have adopted 8502 km along the roadways.

Source: www.modot.org

36. **Height** The length of crutches that a patient uses should be approximately $\frac{2}{3}$ the height of the patient. If a patient is 1.86 m tall, find the length, in meters, of crutches the patient should use.

37. **Adopt-A-Highway** Use the news clipping at the right. Find the average number of meters adopted by a Student Nursing Club in the Missouri Adopt-a-Highway program. Round to the nearest whole number.

38. A roll of gauze 3.25 m long is cut into bandages of equal length. For each statement, answer *true* or *false*.

a. If the length of each bandage is less than 100 cm, then at least three bandages can be cut from the roll.

b. If the length of each bandage is greater than 100 cm, then at most two bandages can be cut from the roll.

39. **Astronomy** The distance between Earth and the sun is 150,000,000 km. Light travels 300,000,000 m in 1 s. How long does it take for light to reach Earth from the sun?

40. **Earth Science** The circumference of Earth is 40,000 km. How long would it take to travel the circumference of Earth at a speed of 85 km per hour? Round to the nearest tenth.

41. **Physics** Light travels 300,000 km in 1 s. How far does light travel in 1 day?

Applying the Concepts

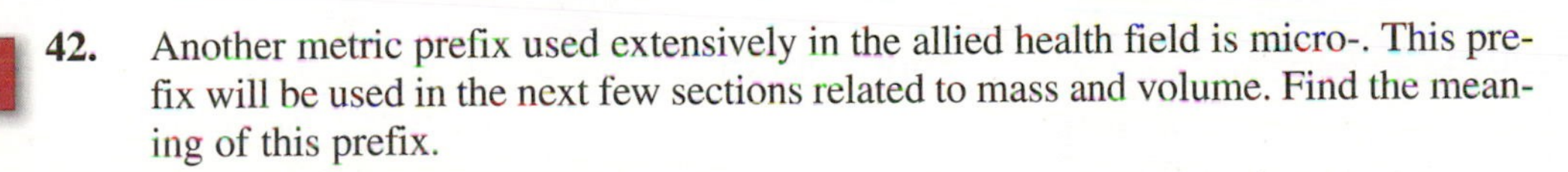

42. Another metric prefix used extensively in the allied health field is micro-. This prefix will be used in the next few sections related to mass and volume. Find the meaning of this prefix.

SECTION

9.2 Mass

OBJECTIVE A To convert units of mass in the metric system of measurement

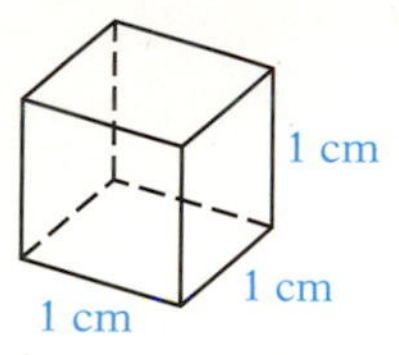

1 gram = the mass of water in the box

Point of Interest

A unit of metric measurement, the microgram (μg), is used to prescribe medications in very small dosage units. Although the abbreviation μg is used in scientific literature, The Joint Commission recommends the use of mcg for hospitals and handwritten orders to avoid the incorrect interpretation of μ as m, which would result in an overdose of medication.

(*Source:* http://www.jointcommission.org)

1 μg = 1 mcg = 0.001 mg = 0.000001 g

Mass and weight are closely related. Weight is a measure of how strongly Earth is pulling on an object. Therefore, an object's weight is less in space than on Earth's surface. However, the amount of material in the object, its **mass,** remains the same. On the surface of Earth, mass and weight can be used interchangeably.

The basic unit of mass in the metric system is the **gram.** If a box that is 1 cm long on each side is filled with water, then the mass of that water is 1 gram.

The gram is a very small unit of mass. A paper clip weighs about 1 gram. The kilogram (1000 grams) is a more useful unit of mass in consumer applications. This textbook weighs about 1 kilogram.

The units of mass in the metric system have the same prefixes as the units of length.

Units of Mass in the Metric System

1 kilogram (kg) = 1000 grams (g)
1 hectogram (hg) = 100 g
1 decagram (dag) = 10 g
1 gram (g) = 1 g
1 decigram (dg) = 0.1 g
1 centigram (cg) = 0.01 g
1 milligram (mg) = 0.001 g

Conversion between units of mass in the metric system involves moving the decimal point to the right or to the left. Listing the units in order from largest to smallest will indicate how many places to move the decimal point and in which direction.

To convert 324 g to kilograms, first write the units in order from largest to smallest.

kg hg dag g dg cg mg

3 positions

• Converting g to kg requires moving 3 positions to the left.

324 g = 0.324 kg

3 places

• Move the decimal point the same number of places and in the same direction.

EXAMPLE 1

Convert 4.23 g to milligrams.

Solution 4.23 g = 4230 mg

YOU TRY IT 1

Convert 42.3 mg to grams.

Your solution

Solution on p. S22

EXAMPLE • 2

Convert 2 kg 564 g to kilograms.

Solution

$564 \text{ g} = 0.564 \text{ kg}$

$2 \text{ kg } 564 \text{ g} = 2 \text{ kg} + 0.564 \text{ kg}$
$= 2.564 \text{ kg}$

YOU TRY IT • 2

Convert 3 g 54 mg to grams.

Your solution

Solution on p. S22

OBJECTIVE B To solve application problems

Take Note

Although in this text we will always change units to the larger unit, it is possible to perform the calculation by changing to the smaller unit.

$3 \text{ kg} - 750 \text{ g}$
$= 3000 \text{ g} - 750 \text{ g}$
$= 2250 \text{ g}$

Note that

$2250 \text{ g} = 2.250 \text{ kg}$.

In the application problems in this section, we perform arithmetic operations with the measurements of mass in the metric system. Remember that before measurements can be added or subtracted, they must be expressed in terms of the same unit. In this textbook, unless otherwise stated, the units should be changed to the larger unit before the arithmetic operation is performed.

To subtract 750 g from 3 kg, convert 750 g to kilograms.

$$3 \text{ kg} - 750 \text{ g} = 3 \text{ kg} - 0.750 \text{ kg}$$
$$= 2.250 \text{ kg}$$

EXAMPLE • 3

Doctors often prescribe a low-dose aspirin regimen to heart patients. An adult low-dose aspirin contains 81 mg of aspirin per tablet. If a patient takes one tablet per day, how many grams of aspirin does the patient ingest during a 30-day period?

Strategy

To find the total number of grams of aspirin ingested:

- Multiply the daily amount (81 mg) times the number of days (30).
- Convert the number of milligrams to grams.

Solution

$81 \text{ mg} \times 30 = 2430 \text{ mg}$

$2430 \text{ mg} = 2.43 \text{ g}$

The patient ingests 2.43 g of aspirin during a 30-day period.

YOU TRY IT • 3

The pediatric dosage of a medication is often determined by the weight of the patient in kilograms. A child weighing 14 kg is given a shot of an antibiotic. Find the amount of medication, in milligrams, the child receives if the dosage rate for this medication is 0.05 g/kg.

Your strategy

Your solution

Solution on p. S22

9.2 EXERCISES

OBJECTIVE A To convert units of mass in the metric system of measurement

For Exercises 1 to 24, convert.

1. 420 g = _____ kg

2. 7421 g = _____ kg

3. 127 mg = _____ g

4. 43 mg = _____ g

5. 4.2 kg = _____ g

6. 0.027 kg = _____ g

7. 0.45 g = _____ mg

8. 325 g = _____ mg

9. 23 mg = _____ μg

10. 125 μg = _____ mg

11. 4057 mg = _____ g

12. 1970 mg = _____ g

13. 1.37 kg = _____ g

14. 5.1 kg = _____ g

15. 0.0456 g = _____ mg

16. 0.2 g = _____ mg

17. 1.5 mg = _____ μg

18. 55 μg = _____ mg

19. 3 kg 922 g = _____ kg

20. 1 kg 47 g = _____ kg

21. 7 g 891 mg = _____ g

22. 209 g 42 mg = _____ g

23. 4 kg 63 g = _____ kg

24. 18 g 5 mg = _____ g

For Exercises 25 to 27, fill in the blank with the correct unit of measurement.

25. 0.105 g = 105 _____

26. 70 g = 0.07 _____

27. 31 mg = 0.031 _____

OBJECTIVE B To solve application problems

28. **Consumerism** A 1.19-kilogram container of Quaker Oats contains 30 servings. Find the number of grams in one serving of the oatmeal. Round to the nearest whole number.

AP Photo/Paul Sakuma

29. **Nutrition** A patient is advised to supplement her diet with 2 g of calcium per day. The calcium tablets she purchases contain 500 mg of calcium per tablet. How many tablets per day should the patient take?

30. **Medication Dosage** An adult is prescribed 500 mg of a medication to be taken twice a day for five days. At the end of five days, how many grams of this medication has the patient taken?

31. **Nutrition** One egg contains 274 mg of cholesterol. How many grams of cholesterol are in a dozen eggs?

32. **Consumerism** The nutrition label for a corn bread mix is shown at the right.
 a. How many kilograms of mix are in the package?
 b. How many grams of sodium are contained in two servings of the corn bread?

Nutrition Facts
Serving Size 1/6 pkg. (31g mix)
Servings Per Container 6

Amount Per Serving	Mix	Prepared
Calories	110	160
Calories from Fat	10	50
	% Daily Value*	
Total Fat 1g	1%	9%
Saturated Fat 0g	0%	7%
Cholesterol 0mg	0%	12%
Sodium 210mg	9%	11%
Total Carbohydrate 24g	8%	8%
Sugars 6g		
Protein 2g		

33. **Consumerism** Find the cost of three packages of ground meat weighing 470 g, 680 g, and 590 g if the price per kilogram is $8.40.

34. **Medication Dosage** Each teaspoon (t) of an oral suspension of ibuprofen contains 0.1 g of medication. If a patient takes 2 t of this medication three times day for two days, how many milligrams of ibuprofen has the patient received?

35. **Medication Dosage** The dosage of a certain antibiotic is determined by the weight of the patient in kilograms. A patient weighing 25 kg is given a shot of this antibiotic. Find the amount of medication, in milligrams, the patient receives if the dosage rate for this medication is 10 mg/kg.

36. **Energy** Use the news clipping at the right. Find the weight, in grams, of two of the power-generating knee braces.

In the News

A New Kind of Power Walking

Max Donelan has invented a high-tech, 1.6-kilogram knee brace that generates enough electricity to power portable gadgets. One minute of walking while wearing one of the braces on each leg generates enough electricity to charge 10 cell phones.

Source: Time, March 17, 2008

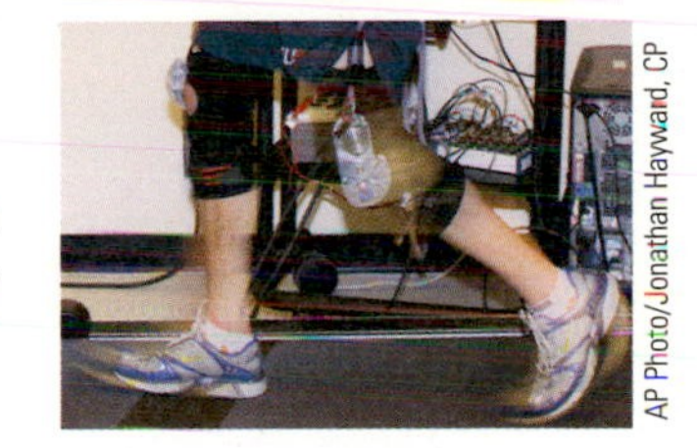

AP Photo/Jonathan Hayward, CP

37. **Solutions** A laboratory solution contains 40 μg of solute per milliliter of solution. How many milligrams of solute would there be in 250 ml of solution?

38. **IV Therapy** The pediatric dose of an IV infusion of medication is 25 units/kg. If a newborn baby weighs 3920 g and needs an infusion of this medication, how many units will the doctor prescribe?

39. **Medication Dosage** The label on a bottle of atropine sulfate indicates the strength of the medication is 0.1 mg/ml. If an injection of 200 mcg of atropine sulfate is prescribed, how many milliliters of this drug should be given to the patient?

40. **Medication Dosage** The label on a vial of the pain-relief medication sufentanil reads 100 mcg/2 ml. If a medication order reads 80 mcg IV push, how many milliliters should this patient be given?

Applying the Concepts

41. Discuss the advantages and disadvantages of the U.S. Customary System and the metric system of measurement.

SECTION

9.3 Capacity

OBJECTIVE A To convert units of capacity in the metric system of measurement

Rachel Epstein/PhotoEdit, Inc.

1-liter bottle

The basic unit of capacity in the metric system is the liter. One **liter** is defined as the capacity of a box that is 10 cm long on each side.

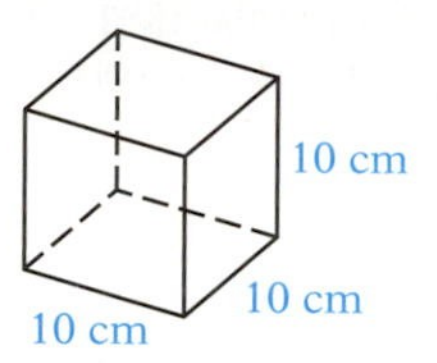

The units of capacity in the metric system have the same prefixes as the units of length.

Units of Capacity in the Metric System

1 kiloliter (kl) = 1000 L
1 hectoliter (hl) = 100 L
1 decaliter (dal) = 10 L
1 liter (L) = 1 L
1 deciliter (dl) = 0.1 L
1 centiliter (cl) = 0.01 L
1 milliliter (ml) = 0.001 L

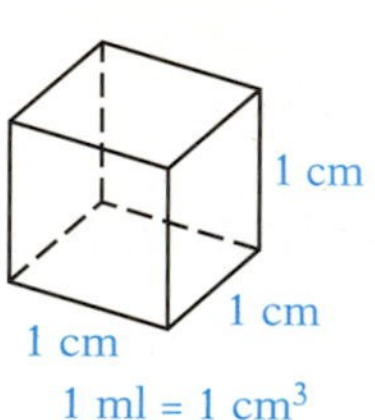

1 ml = 1 cm^3

The milliliter is equal to 1 **cubic centimeter** (cm^3). See the diagram at the left. In medicine, cubic centimeter is often abbreviated cc. However, The Joint Commission recommends the use of ml or mL in hospitals to avoid misinterpretation of cc in handwritten orders. (*Source:* http://www.jointcommission.org)

Conversion between units of capacity in the metric system involves moving the decimal point to the right or to the left. Listing the units in order from largest to smallest will indicate how many places to move the decimal point and in which direction.

To convert 824 ml to liters, first write the units in order from largest to smallest.

kl hl dal L dl cl ml
3 positions

- **Converting ml to L requires moving 3 positions to the left.**

824 ml = 0.824 L
3 places

- **Move the decimal point the same number of places and in the same direction.**

EXAMPLE 1

Convert 4 L 32 ml to liters.

Solution 32 ml = 0.032 L

4 L 32 ml = 4 L + 0.032 L
= 4.032 L

YOU TRY IT 1

Convert 2 kl 167 L to liters.

Your solution

Solution on p. S22

EXAMPLE • 2

Convert 1.23 L to cubic centimeters.

Solution 1.23 L = 1230 ml = 1230 cm^3

YOU TRY IT • 2

Convert 325 cm^3 to liters.

Your solution

Solution on p. S22

OBJECTIVE B To solve application problems

Take Note

Although in this text we will always change units to the larger unit, it is possible to perform the calculation by changing to the smaller unit.

2.5 kl + 875 L
= 2500 L + 875 L
= 3375 L

Note that

3375 L = 3.375 kl.

In the application problems in this section, we perform arithmetic operations with the measurements of capacity in the metric system. Remember that before measurements can be added or subtracted, they must be expressed in terms of the same unit. In this textbook, unless otherwise stated, the units should be changed to the larger unit before the arithmetic operation is performed.

To add 2.5 kl and 875 L, convert 875 L to kiloliters.

$$\begin{aligned} 2.5 \text{ kl} + 875 \text{ L} &= 2.5 \text{ kl} + 0.875 \text{ kl} \\ &= 3.375 \text{ kl} \end{aligned}$$

EXAMPLE • 3

A laboratory assistant is in charge of ordering acid for three chemistry classes of 30 students each. Each student requires 80 ml of acid. How many liters of acid should be ordered? (The assistant must order by the whole liter.)

Strategy

To find the number of liters to be ordered:

- Find the number of milliliters of acid needed by multiplying the number of classes (3) by the number of students per class (30) by the number of milliliters of acid required by each student (80).
- Convert milliliters to liters.
- Round up to the nearest whole number.

Solution

3(30)(80) = 7200 ml

7200 ml = 7.2 L

7.2 rounded up to the nearest whole number is 8.

The assistant should order 8 L of acid.

YOU TRY IT • 3

For $299.50, a massage therapist buys 5 L of massage lotion and repackages it in 125-milliliter jars. Each jar costs the massage therapist $0.85. Each jar of lotion is sold for $29.95. Find the profit on the 5 L of lotion.

Your strategy

Your solution

Solution on p. S22

9.3 EXERCISES

OBJECTIVE A To convert units of capacity in the metric system of measurement

For Exercises 1 to 24, convert.

1. 4200 ml = _____ L

2. 7.5 ml = _____ L

3. 3.42 L = _____ ml

4. 0.037 L = _____ ml

5. 423 ml = _____ cm^3

6. 0.32 ml = _____ cm^3

7. 642 cm^3 = _____ ml

8. 0.083 cm^3 = _____ ml

9. 42 cm^3 = _____ L

10. 3075 cm^3 = _____ L

11. 0.435 L = _____ dl

12. 2.57 L = _____ dl

13. 4.62 kl = _____ L

14. 0.035 kl = _____ L

15. 1423 L = _____ kl

16. 897 L = _____ kl

17. 1.267 L = _____ dl

18. 4.105 L = _____ dl

19. 3 L 42 ml = _____ L

20. 1 L 127 ml = _____ L

21. 3 kl 4 L = _____ kl

22. 6 kl 32 L = _____ kl

23. 8 L 200 ml = _____ L

24. 9 kl 505 L = _____ kl

For Exercises 25 to 27, fill in the blank with the correct unit of measurement.

25. 620 ml = 0.62 _____

26. 950 cm^3 = 0.95 _____

27. 4.1 kl = 4100 _____

OBJECTIVE B To solve application problems

28. **Earth Science** The air in Earth's atmosphere is 78% nitrogen and 21% oxygen. Find the amount of oxygen in 50 L of air.

29. **Consumerism** The printed label from a container of milk is shown at the right. How many 230-milliliter servings are in the container? Round to the nearest whole number.

30. **Blood Tests** A medical laboratory runs blood tests seven days a week and uses an average of 800 ml of a diluting agent per day. How many liters of this diluting agent are used in a month of 30 days?

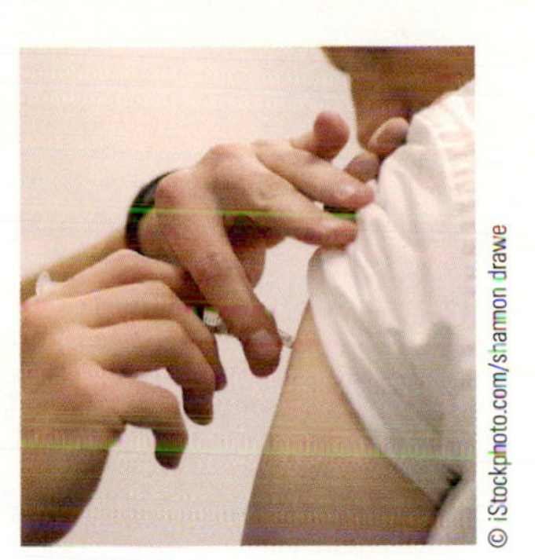

31. **Medicine** A flu vaccine is being given for the coming winter season. A medical corporation buys 12 L of flu vaccine. How many patients can be immunized if each person receives 3 ml of the vaccine?

32. **IV Therapy** A 1-liter infusion of D5W is set to infuse at a rate of 60 ml/h. How long will it take for this infusion to be complete? Give your answer in hours and minutes.

33. **IV Therapy** At noon, $\frac{2}{5}$ of a 1-liter bag of normal saline has been infused. If the infusion rate for this IV is 60 ml/h, what time will this infusion be complete?

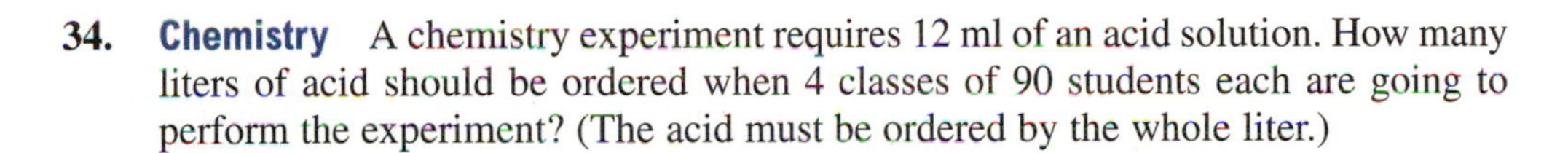

34. **Chemistry** A chemistry experiment requires 12 ml of an acid solution. How many liters of acid should be ordered when 4 classes of 90 students each are going to perform the experiment? (The acid must be ordered by the whole liter.)

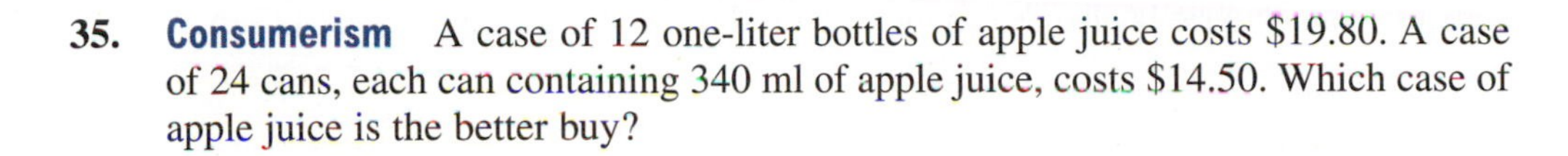

35. **Consumerism** A case of 12 one-liter bottles of apple juice costs $19.80. A case of 24 cans, each can containing 340 ml of apple juice, costs $14.50. Which case of apple juice is the better buy?

36. **Medical Tests** A patient brings in a 24-hour urine collection for testing of total protein. The total volume is 1500 ml. The results for this particular test will be reported in milligrams per deciliter. How many deciliters of urine are in this specimen? (Medical tests in the laboratory often give results in deciliters.)

37. **Repackaging** For $195, a pharmacist purchases 5 L of cough syrup and repackages it in 250-milliliter bottles. Each bottle costs the pharmacist $0.55. Each bottle of cough syrup is sold for $23.89. Find the profit on the 5 L of cough syrup.

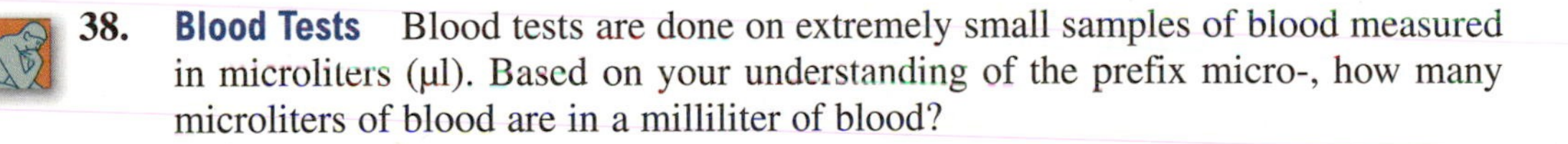

38. **Blood Tests** Blood tests are done on extremely small samples of blood measured in microliters (μl). Based on your understanding of the prefix micro-, how many microliters of blood are in a milliliter of blood?

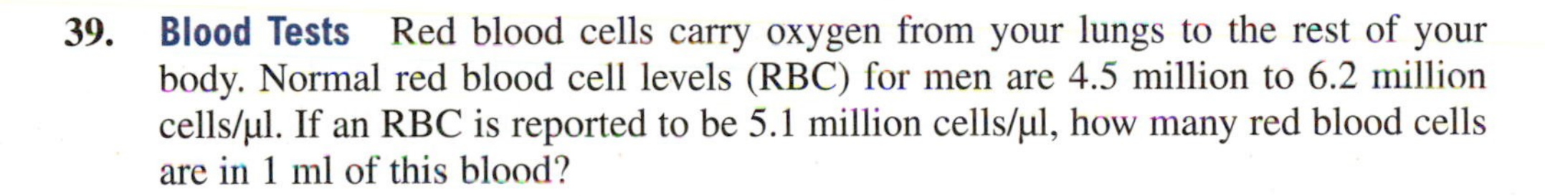

39. **Blood Tests** Red blood cells carry oxygen from your lungs to the rest of your body. Normal red blood cell levels (RBC) for men are 4.5 million to 6.2 million cells/μl. If an RBC is reported to be 5.1 million cells/μl, how many red blood cells are in 1 ml of this blood?

40. A bottle of spring water holds 0.5 L of water. A case of spring water contains 24 bottles. Without finding the number of milliliters of water in a case of spring water, determine whether the following statement is *true* or *false*:

A case of spring water contains more than 1000 ml of water.

Applying the Concepts

41. After a 280-milliliter serving is taken from a 3-liter bottle of water, how much water remains in the container? Write the answer in three different ways.

SECTION

9.4 Energy

OBJECTIVE A To use units of energy in the metric system of measurement

Two commonly used units of energy in the metric system are the calorie and the watt-hour.

Heat is generally measured in units called calories or in larger units called Calories (with a capital C). A **Calorie** is 1000 calories and should be called a kilocalorie, but it is common practice in nutritional references and food labeling to simply call it a Calorie. A Calorie is the amount of heat required to raise the temperature of 1 kg of water 1 degree Celsius. One Calorie is also the energy required to lift 1 kg a distance of 427 m.

HOW TO • 1 Swimming uses 480 Calories per hour. How many Calories are used by swimming $\frac{1}{2}$ h each day for 30 days?

Strategy To find the number of Calories used:

- Find the number of hours spent swimming.
- Multiply the number of hours spent swimming by the Calories used per hour.

Solution $\frac{1}{2} \times 30 = 15$

$15(480) = 7200$

7200 Calories are used by swimming $\frac{1}{2}$ h each day for 30 days.

The **watt-hour** is used for measuring electrical energy. One watt-hour is the amount of energy required to lift 1 kg a distance of 370 m. A light bulb rated at 100 watts (W) will emit 100 watt-hours (Wh) of energy each hour. A **kilowatt-hour** is 1000 watt-hours.

1000 watt-hours (Wh) = 1 kilowatt-hour (kWh)

Take Note

Recall that the prefix kilo- means 1000.

HOW TO • 2 A 150-watt bulb is on for 8 h. At 11¢ per kilowatt-hour, find the cost of the energy used.

Strategy To find the cost:

- Find the number of watt-hours used.
- Convert to kilowatt-hours.
- Multiply the number of kilowatt-hours used by the cost per kilowatt-hour.

Solution $150 \times 8 = 1200$

1200 Wh = 1.2 kWh

$1.2 \times 0.11 = 0.132$

The cost of the energy used is $0.132.

Integrating Technology

To convert watt-hours to kilowatt-hours, divide by 1000. To use a calculator to determine the number of kilowatt-hours used in the problem at the right, enter the following:

The calculator display reads 1.2.

EXAMPLE • 1

Walking uses 180 Calories per hour. How many Calories will you burn off by walking $5\frac{1}{4}$ h during one week?

Strategy
To find the number of Calories, multiply the number of hours spent walking by the Calories used per hour.

Solution

$$5\frac{1}{4} \times 180 = \frac{21}{4} \times 180 = 945$$

You will burn off 945 Calories.

YOU TRY IT • 1

Housework requires 240 Calories per hour. How many Calories are burned off by doing $4\frac{1}{2}$ h of housework?

Your strategy

Your solution

EXAMPLE • 2

A clothes iron is rated at 1200 W. If the iron is used for 1.5 h, how much energy, in kilowatt-hours, is used?

Strategy
To find the energy used:

- Find the number of watt-hours used.
- Convert watt-hours to kilowatt-hours.

Solution

$1200 \times 1.5 = 1800$

$1800 \text{ Wh} = 1.8 \text{ kWh}$

1.8 kWh of energy are used.

YOU TRY IT • 2

Find the number of kilowatt-hours of energy used when a 150-watt light bulb burns for 200 h.

Your strategy

Your solution

EXAMPLE • 3

A TV set rated at 1800 W is on for an average of 3.5 h per day. At 12.2¢ per kilowatt-hour, find the cost of operating the set for 1 week.

Strategy
To find the cost:

- Multiply 3.5 by the number of days in 1 week to find the total number of hours the set is used per week.
- Multiply the product by the number of watts to find the watt-hours.
- Convert watt-hours to kilowatt-hours.
- Multiply the number of kilowatt-hours by the cost per kilowatt-hour.

Solution

$3.5 \times 7 = 24.5$

$24.5 \times 1800 = 44{,}100$

$44{,}100 \text{ Wh} = 44.1 \text{ kWh}$

$44.1 \times 0.122 = 5.3802$

The cost is $5.3802.

YOU TRY IT • 3

A microwave oven rated at 500 W is used an average of 20 min per day. At 13.7¢ per kilowatt-hour, find the cost of operating the oven for 30 days.

Your strategy

Your solution

Solutions on pp. S22–S23

9.4 EXERCISES

OBJECTIVE A To use units of energy in the metric system of measurement

1. **Health** How many Calories can you eliminate from your diet by omitting 1 slice of bread per day for 30 days? One slice of bread contains 110 Calories.

Nutrition Facts
Serving Size 2 Slices (18g)
Servings Per Container about 15

Amount Per Serving	
Calories 60	Calories from Fat 10
	% Daily Value*
Total Fat 1g	2%
Saturated Fat 0g	0%
Polyunsaturated Fat 0.5g	
Monounsaturated Fat 0.5g	
Cholesterol 0mg	0%
Sodium 60mg	3%
Total Carbohydrate 10g	3%
Dietary Fiber 3g	10%
Sugars 1g	
Protein 2g	

Vitamin A 0% • Vitamin C 0%
Calcium 0% • Iron 4%

* Percent Daily Values are based on a 2,000 calorie diet. Your daily values may be higher or lower depending on your calorie needs.

	Calories:	2,000	2,500
Total Fat	Less than	65g	80g
Saturated Fat	Less than	20g	25g
Cholesterol	Less than	300mg	300mg
Sodium	Less than	2,400mg	2,400mg
Total Carbohydrate		300g	375g
Dietary Fiber		25g	30g

Calories per gram:
Fat 9 • Carbohydrate 4 • Protein 4

2. **Health** How many Calories can you eliminate from your diet in 2 weeks by omitting 400 Calories per day?

3. **Nutrition** A nutrition label from a package of crisp bread is shown at the right.
   a. How many Calories are in $1\frac{1}{2}$ servings?
   b. How many Calories from fat are in 6 slices of the bread?

4. **Health** Moderately active people need 20 Calories per pound of body weight to maintain their weight. How many Calories should a 150-pound, moderately active person consume per day to maintain that weight?

5. **Health** People whose daily activity level would be described as light need 15 Calories per pound of body weight to maintain their weight. How many Calories should a 135-pound, lightly active person consume per day to maintain that weight?

6. **Health** For a healthful diet, it is recommended that 55% of the daily intake of Calories come from carbohydrates. Find the daily intake of Calories from carbohydrates that is appropriate if you want to limit your Calorie intake to 1600 Calories.

7. **Health** Playing singles tennis requires 450 Calories per hour. How many Calories do you burn in 30 days playing 45 min per day?

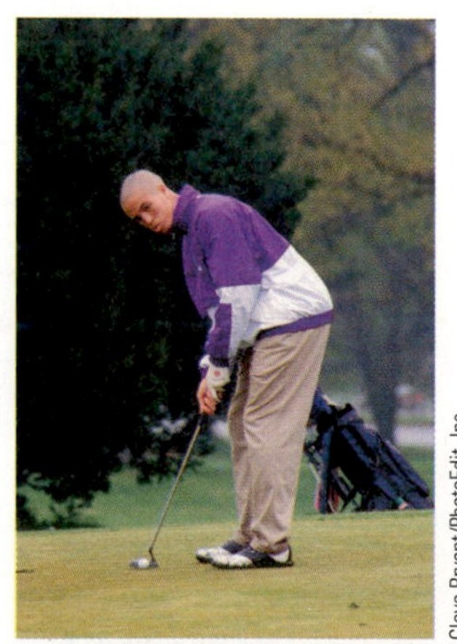

Cleve Bryant/PhotoEdit, Inc.

8. **Health** After playing golf for 3 h, Ruben had a banana split containing 550 Calories. Playing golf uses 320 Calories per hour.
   a. Without doing the calculations, did the banana split contain more or fewer Calories than Ruben burned off playing golf?
   b. Find the number of Calories Ruben gained or lost from these two activities.

9. **Health** Hiking requires approximately 315 Calories per hour. How many hours would you have to hike to burn off the Calories in a 375-Calorie sandwich, a 150-Calorie soda, and a 280-Calorie ice cream cone? Round to the nearest tenth.

10. **Health** Riding a bicycle requires 265 Calories per hour. How many hours would Shawna have to ride a bicycle to burn off the Calories in a 320-Calorie milkshake, a 310-Calorie cheeseburger, and a 150-Calorie apple? Round to the nearest tenth.

11. **Energy** An oven uses 500 W of energy. How many watt-hours of energy are used to cook a 5-kilogram roast for $2\frac{1}{2}$ h?

12. **Energy** A 21-inch color TV set is rated at 90 W. The TV is used an average of $3\frac{1}{2}$ h each day for a week. How many kilowatt-hours of energy are used during the week?

13. **Energy** A fax machine is rated at 9 W when the machine is in standby mode and at 36 W when in operation. How many kilowatt-hours of energy are used during a week in which the fax machine is in standby mode for 39 h and in operation for 6 h?

14. **Energy** A 120-watt CD player is on an average of 2 h a day. Find the cost of listening to the CD player for 2 weeks at a cost of 14.4¢ per kilowatt-hour. Round to the nearest cent.

15. **Energy** How much does it cost to run a 2200-watt air conditioner for 8 h at 12¢ per kilowatt-hour? Round to the nearest cent.

16. **Energy** A space heater is used for 3 h. The heater uses 1400 W per hour. Electricity costs 11.1¢ per kilowatt-hour. Find the cost of using the electric heater. Round to the nearest cent.

17. **Energy** A 60-watt Sylvania Long Life Soft White Bulb has a light output of 835 lumens and an average life of 1250 h. A 34-watt Sylvania Energy Saver Bulb has a light output of 400 lumens and an average life of 1500 h.
   a. Is the light output of the Energy Saver Bulb more or less than half that of the Long Life Soft White Bulb?
   b. If electricity costs 10.8¢ per kilowatt-hour, what is the difference in cost between using the Long Life Soft White Bulb for 150 h and using the Energy Saver Bulb for 150 h? Round to the nearest cent.

© Dick Reed/Corbis

18. **Energy** A welder uses 6.5 kWh of energy each hour. Find the cost of using the welder for 6 h a day for 30 days. The cost is 12.4¢ per kilowatt-hour.

19. A package of four 40-watt light bulbs states that each light bulb lasts 2000 hours. Which expression represents the total number of kilowatt-hours of energy available from the package of light bulbs?

(i) $\frac{40 \times 2000}{1000}$ (ii) $4 \times 40 \times 2000$ (iii) $4 \times 40 \times 2$

Applying the Concepts

20. **Human Energy** A moderately active woman who is 5 ft 5 in. tall and weighs 120 lb needs 2100 Calories per day to maintain her weight. Show that this is approximately the power of a 100-watt light bulb. Use the fact that 1 Calorie per day is equal to approximately 0.049 W.

21. Write an essay on how to improve the energy efficiency of a home.

SECTION

9.5 Conversion Between the U.S. Customary and the Metric Systems of Measurement

OBJECTIVE A To convert U.S. Customary units to metric units

Point of Interest

The definition of 1 inch has been changed as a consequence of the wide acceptance of the metric system. One inch is now exactly 25.4 mm.

More than 90% of the world's population uses the metric system of measurement. Therefore, converting U.S. Customary units to metric units is essential in trade and commerce—for example, in importing foreign goods and exporting domestic goods. Approximate equivalences between the two systems follow.

Units of Length	Units of Weight	Units of Capacity
1 in. = 2.54 cm	1 oz ≈ 28.35 g	1 L ≈ 1.06 qt
1 m ≈ 3.28 ft	1 lb ≈ 454 g	1 gal ≈ 3.79 L
1 m ≈ 1.09 yd	1 kg ≈ 2.2 lb	1 ml ≈ 0.03 fl oz
1 mi ≈ 1.61 km		1 t ≈ 5 ml

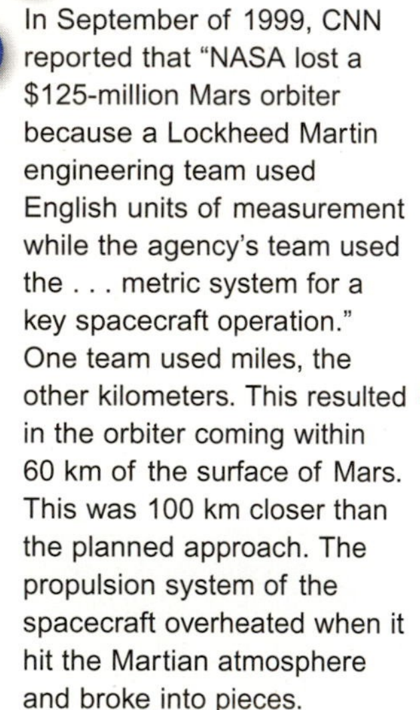

Point of Interest

In September of 1999, CNN reported that "NASA lost a $125-million Mars orbiter because a Lockheed Martin engineering team used English units of measurement while the agency's team used the . . . metric system for a key spacecraft operation." One team used miles, the other kilometers. This resulted in the orbiter coming within 60 km of the surface of Mars. This was 100 km closer than the planned approach. The propulsion system of the spacecraft overheated when it hit the Martian atmosphere and broke into pieces.

These equivalences can be used to form conversion rates to change one unit of measurement to another. For example, because 1 in. ≈ 2.54 cm, the conversion rates $\frac{1 \text{ in.}}{2.54 \text{ cm}}$ and $\frac{2.54 \text{ cm}}{1 \text{ in.}}$ are both approximately equal to 1.

HOW TO • 1 Convert 72 in. to centimeters.

$$72 \text{ in.} \approx 72 \text{ in.} \times \frac{2.54 \text{ cm}}{1 \text{ in.}}$$

• **The conversion rate must contain cm in the numerator and in. in the denominator.**

$$= \frac{72 \text{ in.}}{1} \times \frac{2.54 \text{ cm}}{1 \text{ in.}}$$

$$= \frac{182.88 \text{ cm}}{1}$$

$$72 \text{ in.} \approx 182.88 \text{ cm}$$

EXAMPLE • 1

Convert 3 ft 6 in. to meters.

Solution

$$3 \text{ ft } 6 \text{ in.} = 3\frac{6}{12} \text{ ft} = 3\frac{1}{2} \text{ ft}$$

$$3\frac{1}{2} \text{ ft} \approx \frac{7 \text{ ft}}{2} \times \frac{1 \text{ m}}{3.28 \text{ ft}} = \frac{1.067 \text{ m}}{1}$$

• **The conversion rate is $\frac{1 \text{ m}}{3.28 \text{ ft}}$ with m in the numerator and ft in the denominator.**

$$3 \text{ ft } 6 \text{ in.} \approx 1.067 \text{ m}$$

YOU TRY IT • 1

Convert 125 lb to kilograms. Round to the nearest tenth.

Your solution

Solution on p. S23

EXAMPLE • 2

The price of gasoline is \$3.89/gal. Find the cost per liter. Round to the nearest tenth of a cent.

Solution

$$\frac{\$3.89}{\text{gal}} \approx \frac{\$3.89}{\cancel{\text{gal}}} \times \frac{1\ \cancel{\text{gal}}}{3.79\ \text{L}} = \frac{\$3.89}{3.79\ \text{L}} \approx \frac{\$1.026}{1\ \text{L}}$$

\$3.89/gal ≈ \$1.026/L

YOU TRY IT • 2

The price of milk is \$3.69/gal. Find the cost per liter. Round to the nearest cent.

Your solution

Solution on p. S23

OBJECTIVE B To convert metric units to U.S. Customary units

Metric units are used in the United States. Cereal is sold by the gram, 35-mm film is available, and soda is sold by the liter. The same conversion rates used in Objective A are used for converting metric units to U.S. Customary units.

EXAMPLE • 3

Convert 200 m to feet.

Solution

$$200\ \text{m} \approx 200\ \cancel{\text{m}} \times \frac{3.28\ \text{ft}}{1\ \cancel{\text{m}}} = \frac{656\ \text{ft}}{1}$$

200 m ≈ 656 ft

YOU TRY IT • 3

Convert 45 cm to inches. Round to the nearest hundredth.

Your solution

EXAMPLE • 4

Convert 90 km/h to miles per hour. Round to the nearest hundredth.

Solution

$$\frac{90\ \text{km}}{\text{h}} \approx \frac{90\ \cancel{\text{km}}}{\text{h}} \times \frac{1\ \text{mi}}{1.61\ \cancel{\text{km}}} = \frac{90\ \text{mi}}{1.61\ \text{h}} \approx \frac{55.90\ \text{mi}}{1\ \text{h}}$$

90 km/h ≈ 55.90 mi/h

YOU TRY IT • 4

Express 75 km/h in miles per hour. Round to the nearest hundredth.

Your solution

EXAMPLE • 5

The price of gasoline is \$1.125/L. Find the cost per gallon. Round to the nearest cent.

Solution

$$\frac{\$1.125}{1\ \text{L}} \approx \frac{\$1.125}{1\ \cancel{\text{L}}} \times \frac{3.79\ \cancel{\text{L}}}{1\ \text{gal}} \approx \frac{\$4.26}{1\ \text{gal}}$$

\$1.125/L ≈ \$4.26/gal

YOU TRY IT • 5

The price of ice cream is \$1.75/L. Find the cost per gallon. Round to the nearest cent.

Your solution

Solutions on p. S23

9.5 EXERCISES

OBJECTIVE A To convert U.S. Customary units to metric units

1. Write a product of conversion rates that you could use to convert cups to liters.

For Exercises 2 to 17, convert. Round to the nearest hundredth if necessary.

2. Convert 100 yd to meters.

3. Find the weight in kilograms of a 145-pound person.

4. Find the height in meters of a person 5 ft 8 in. tall.

5. Find the number of liters in 2 c of soda.

6. How many kilograms does a 15-pound baby weigh?

7. Find the number of liters in 14.3 gal of a diluting agent.

8. Find the number of milliliters in 1 c.

9. A newborn baby weighs 6 lb 10 oz. Convert this weight to grams.

10. Express 65 mi/h in kilometers per hour.

11. Express 30 mi/h in kilometers per hour.

12. Calcium carbonate costs $9.50/lb. Find the cost per kilogram.

13. Barium sulfate costs $0.79/oz. Find the cost per kilogram.

14. The cost of gasoline is $3.87/gal. Find the cost per liter.

15. Acetic acid costs $32.99/gal. Find the cost per liter.

16. **Medication Dosages** A teaspoon (t) in the U.S. Customary System is equivalent to 5 ml. If a prescribed dosage is 1.5 t of cough syrup three times a day, how many milliliters of cough syrup are prescribed per day?

17. **Medication Dosages** The medication dosage rate for a child is 50 mg/kg. If a child weighs 27 pounds, how many kilograms is this?

OBJECTIVE B To convert metric units to U.S. Customary units

18. Write a product of conversion rates that you could use to convert kilometers to feet.

For Exercises 19 to 32, convert. Round to the nearest hundredth if necessary.

19. Convert 100 m to feet.

20. Find the weight in pounds of an 86-kilogram person.

21. Find the number of gallons in 6 L of saline solution.

22. Find the height in inches of a person 1.85 m tall.

23. Find the distance of the 1500-meter race in feet.

24. Find the weight in ounces of 327 g of cereal.

25. Find the weight in pounds of a newborn weighing 3178 g.

26. Find the volume in fluid ounces of a 5-milliliter dose of cough syrup.

27. Express 80 km/h in miles per hour.

28. Express 30 m/s in feet per second.

29. Gasoline costs $1.015/L. Find the cost per gallon.

30. A 5-kilogram ham costs $10/kg. Find the cost per pound.

31. A premature baby weighs 2.1 kg. Find its weight in pounds.

32. A 500-milliliter bottle of acetone alcohol costs $14.10. Find the cost per fluid ounce.

33. **Health** Gary is planning a 5-day backpacking trip and decides to hike an average of 5 h each day. Hiking requires an extra 320 Calories per hour. How many pounds will Gary lose during the trip if he consumes an extra 900 Calories each day? (3500 Calories are equivalent to 1 lb.)

34. **Health** Swimming requires 550 Calories per hour. How many pounds could be lost by swimming $1\frac{1}{2}$ h each day for 5 days if no extra calories were consumed? (3500 Calories are equivalent to 1 lb.)

35. **Medication Dosage** A doctor prescribes 40 ml of cough syrup per day for a patient. The total daily prescribed dosage is to be divided into four equal doses. How many teaspoons of cough syrup should be given to the patient in a single dose?

Applying the Concepts

36. Determine whether the statement is *true* or *false*.

a. A liter is more than a gallon.
b. A meter is less than a yard.
c. 30 mi/h is less than 60 km/h.
d. A kilogram is greater than a pound.
e. An ounce is less than a gram.

FOCUS ON PROBLEM SOLVING

Working Backward

Sometimes the solution to a problem can be found by *working backward.* This problem-solving technique can be used to find a winning strategy for a game called Nim.

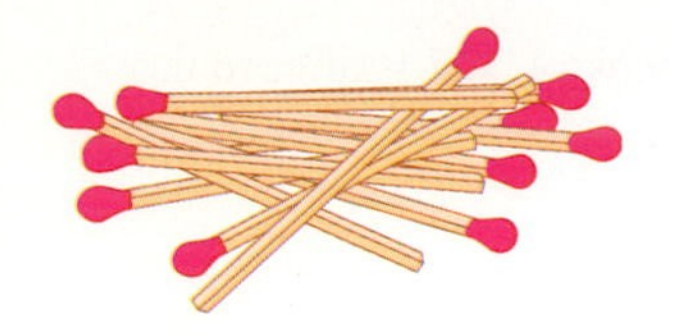

There are many variations of this game. For our game, there are two players, Player A and Player B, who alternately place 1, 2, or 3 matchsticks in a pile. The object of the game is to place the 32nd matchstick in the pile. Is there a strategy that Player A can use to guarantee winning the game?

Working backward, if there are 29, 30, or 31 matchsticks in the pile when it is A's turn to play, A can win by placing 3 matchsticks ($29 + 3 = 32$), 2 matchsticks ($30 + 2 = 32$), or 1 matchstick ($31 + 1 = 32$) on the pile. If there are to be 29, 30, or 31 matchsticks in the pile when it is A's turn, there must be 28 matchsticks in the pile when it is B's turn.

Working backward from 28, if there are to be 28 matches in the pile at B's turn, there must be 25, 26, or 27 at A's turn. Player A can then add 3 matchsticks, 2 matchsticks, or 1 matchstick to the pile to bring the number to 28. For there to be 25, 26, or 27 matchsticks in the pile at A's turn, there must be 24 matchsticks at B's turn.

Now working backward from 24, if there are to be 24 matches in the pile at B's turn, there must be 21, 22, or 23 at A's turn. Player A can then add 3 matchsticks, 2 matchsticks, or 1 matchstick to the pile to bring the number to 24. For there to be 21, 22, or 23 matchsticks in the pile at A's turn, there must be 20 matchsticks at B's turn.

So far, we have found that for Player A to win, there must be 28, 24, or 20 matchsticks in the pile when it is B's turn to play. Note that each time, the number is decreasing by 4. Continuing this pattern, Player A will win if there are 16, 12, 8, or 4 matchsticks in the pile when it is B's turn.

Player A can guarantee winning by making sure that the number of matchsticks in the pile is a multiple of 4. To ensure this, Player A allows Player B to go first and then adds exactly enough matchsticks to the pile to bring the total to a multiple of 4.

For example, suppose B places 3 matchsticks in the pile; then A places 1 matchstick ($3 + 1 = 4$). Now B places 2 matchsticks in the pile. The total is now 6 matchsticks. Player A then places 2 matchsticks in the pile to bring the total to 8, a multiple of 4. If play continues in this way, Player A will win.

Here are some variations of Nim. See whether you can develop a winning strategy for Player A.

1. Suppose the goal is to place the last matchstick in a pile of 30 matches.

2. Suppose the players make two piles of matchsticks, with the final number of matchsticks in each pile to be 20.

3. In this variation of Nim, there are 40 matchsticks in a pile. Each player alternately removes 1, 2, or 3 matches from the pile. The player who removes the last match wins.

PROJECTS AND GROUP ACTIVITIES

Name That Metric Unit

What unit in the metric system would be used to measure each of the following? If you are working in a group, be sure that each member agrees on the unit to be used and understands why that unit is used before going on to the next item.

1. The distance from Johns Hopkins Hospital to the Mayo Clinic
2. The weight of an ambulance
3. A person's waist
4. The amount of fluid in an IV bag
5. The weight of a vitamin pill
6. The amount of water in a swimming pool
7. The distance a baseball player hits a baseball
8. A person's hat size
9. The amount of protein needed daily
10. A person's weight
11. The amount of liquid medication in a syringe
12. The amount of water in a water cooler
13. The amount of medication in an aspirin
14. The distance to the medical clinic
15. The width of a hair
16. A person's height
17. The weight of a wheelchair
18. The amount of water a nursing home uses monthly
19. The contents of a bottle of isopropyl alcohol
20. The weight of medical waste collected at a medical waste incinerator

Metric Measurements for Computers

Other prefixes in the metric system are becoming more commonplace as a result of technological advances in the computer industry and as we learn more and more about objects in our universe that are great distances away.

tera-	= 1,000,000,000,000
giga-	= 1,000,000,000
mega-	= 1,000,000
micro-	= 0.000001
nano-	= 0.000000001
pico-	= 0.000000000001

1. Complete the table.

Metric System Prefix	*Symbol*	*Magnitude*	*Means Multiply the Basic Unit By:*
tera-	T	10^{12}	1,000,000,000,000
giga-	G	___	1,000,000,000
mega-	M	10^6	________
kilo-	___	___	1000
hecto-	h	___	100
deca-	da	10^1	________
deci-	d	$\frac{1}{10}$	________
centi-	___	$\frac{1}{10^2}$	________
milli-	___	___	0.001
micro-	μ	$\frac{1}{10^6}$	________
nano-	n	$\frac{1}{10^9}$	________
pico-	p	___	0.000000000001

2. How can the Magnitude column in the table above be used to determine how many places to move the decimal point when converting to the basic unit in the metric system?

A **bit** is the smallest unit of code that computers can read; it is a binary digit, either a 0 or a 1. A bit is abbreviated b. Usually bits are grouped into **bytes** of 8 bits. Each byte stands for a letter, a number, or any other symbol we might use in communicating information. For example, the letter W can be represented 01010111. A byte is abbreviated B.

The amount of memory in a computer hard drive is measured in terabytes (TB), gigabytes (GB), and megabytes (MB). Often a gigabyte is referred to as a gig, and a megabyte is referred to as a meg. Using the definitions of the prefixes given above, a kilobyte is 1000 bytes, a megabyte is 1,000,000 bytes, and a gigabyte is 1,000,000,000 bytes. However, these are not exact equivalences. Bytes are actually computed in powers of 2. Therefore, kilobytes, megabytes, gigabytes, and terabytes are powers of 2. The exact equivalences are shown below.

1 byte = 2^3 bits
1 kilobyte = 2^{10} bytes = 1024 bytes
1 megabyte = 2^{20} bytes = 1,048,576 bytes
1 gigabyte = 2^{30} bytes = 1,073,741,824 bytes
1 terabyte = 2^{40} bytes = 1,099,511,627,776 bytes

3. Find an advertisement for a computer system. What is the computer's storage capacity? Convert the capacity to bytes. Use the exact equivalences given above.

CHAPTER 9
SUMMARY

KEY WORDS

The *metric system* of measurement is an internationally standardized system of measurement. It is based on the decimal system. The basic unit of length in the metric system is the *meter*. [9.1A, p. 372]

The basic unit of mass is the *gram*. [9.2A, p. 376]

The basic unit of capacity is the *liter*. [9.3A, p. 380]

Heat is commonly measured in units called *Calories*. [9.4A, p. 384]

The *watt-hour* is used in the metric system for measuring electrical energy. [9.4A, p. 384]

In the metric system, prefixes to the basic unit denote the magnitude of each unit. [9.1A, p. 372]

kilo- = 1000 deci- = 0.1
hecto- = 100 centi- = 0.01
deca- = 10 milli- = 0.001

EXAMPLES

1 km = 1000 m
1 kg = 1000 g
1 kl = 1000 L

1 m= 100 cm
1 m = 1000 mm
1 g = 1000 mg
1 L = 1000 ml

ESSENTIAL RULES AND PROCEDURES

Converting between units in the metric system involves moving the decimal point to the right or to the left. Listing the units in order from largest to smallest will indicate how many places to move the decimal point and in which direction. [9.1A, 9.2A, 9.3A, pp. 372, 376, 380]

1. When converting from a larger unit to a smaller unit, move the decimal point to the *right*.
2. When converting from a smaller unit to a larger unit, move the decimal point to the *left*.

EXAMPLES

Convert 3.7 kg to grams.

3.7 kg = 3700 g

Convert 2387 m to kilometers.

2387 m = 2.387 km

Convert 9.5 L to milliliters.

9.5 L = 9500 ml

Approximate equivalences between units in the U.S. Customary and the metric systems of measurement are used to form conversion rates to change one unit of measurement to another. [9.5A, p. 388]

Units of Length
1 in. = 2.54 cm
1 m ≈ 3.28 ft
1 m ≈ 1.09 yd
1 mi ≈ 1.61 km

Units of Weight
1 oz ≈ 28.35 g
1 lb ≈ 454 g
1 kg ≈ 2.2 lb

Units of Capacity
1 L ≈ 1.06 qt
1 gal ≈ 3.79 L

Convert 20 mi/h to kilometers per hour.

$$\frac{20\text{ mi}}{\text{h}} \approx \frac{20\text{ }\cancel{\text{mi}}}{\text{h}} \times \frac{1.61\text{ km}}{1\text{ }\cancel{\text{mi}}} \approx \frac{32.2\text{ km}}{1\text{ h}} \approx 32.2\text{ km/h}$$

Convert 1000 m to yards.

$$1000\text{ m} \approx 1000\text{ }\cancel{\text{m}} \times \frac{1.09\text{ yd}}{1\text{ }\cancel{\text{m}}} \approx 1090\text{ yd}$$

CHAPTER 9

CONCEPT REVIEW

Test your knowledge of the concepts presented in this chapter. Answer each question. Then check your answers against the ones provided in the Answer Section.

1. To convert from millimeters to meters, the decimal point is moved in what direction and how many places?

2. How can you convert 51 m 2 cm to meters?

3. On the surface of Earth, how do mass and weight differ?

4. How can you convert 4 kg 7 g to kilograms?

5. How is a liter related to the capacity of a box or cube that is 10 cm long on each side?

6. How can you convert 4 L 27 ml to liters?

7. What operations are needed to find the cost per week to run a microwave oven rated at 650 W that is used for 30 min per day, at 8.8¢ per kilowatt-hour?

8. If 30% of a person's daily intake of Calories is from fat, how do you find the Calories from fat for a daily intake of 1800 Calories?

9. What conversion rate is needed to convert 100 ft to meters?

10. What conversion rate is needed to convert the price of gas from $3.19/gal to a cost per liter?

CHAPTER 9

REVIEW EXERCISES

1. Convert 1.25 km to meters.

2. Convert 0.450 g to milligrams.

3. Convert 0.0056 L to milliliters.

4. Convert the 1000-meter run to yards. (1 m $\approx$ 1.09 yd)

5. Convert 79 mm to centimeters.

6. Convert 5 m 34 cm to meters.

7. Convert 900 g to micrograms.

8. Convert 2550 ml to liters.

9. Convert 4870 m to kilometers.

10. Convert 0.37 cm to millimeters.

11. Convert 6 g 829 mg to grams.

12. Convert 1.2 L to milliliters.

13. Convert 4.050 kg to grams.

14. Convert 8.7 m to centimeters.

15. Convert 192 ml to cubic centimeters.

16. Convert 356 mg to grams.

17. Convert 372 cm to meters.

18. Convert 8.3 kl to liters.

19. Convert 2 L 89 ml to liters.

20. Convert 5410 ml to liters.

21. Convert 379.2 ml to deciliters.

22. Convert 468 cm^3 to milliliters.

23. **Measurements** Three bandages are cut from a 3.75-meter roll of gauze. The three pieces measure 24 cm, 56 cm, and 48 cm. How much gauze is left of the roll after the three pieces are cut?

24. **Weight** A set of triplets weighing 2100 g, 1650 g, and 1725 g is born at a local hospital. Find their average weight in kilograms.

25. **Consumerism** A 20-liter bottle of laboratory grade ethanol is priced at $104.95. Find the cost per gallon.

26. **Nutrition** A nursing home kitchen is preparing breakfast for its residents. One hundred and twenty-five residents are expected to drink coffee for breakfast, so the dietician allocates 400 ml of coffee per resident. How many liters of coffee should be prepared?

27. **Nutrition** A large egg contains approximately 90 Calories. How many Calories can you eliminate from your diet in a 30-day month by eliminating one large egg per day from your usual breakfast?

28. **IV Therapy** An IV of 1 L of D5W is to be given to a patient at a rate of 150 ml/h. After how many hours will this IV infusion be empty? Give your answer in hours and minutes.

29. **Weight** A medical assistant at a pediatrician's office records a child's weight as 38 lb. Find the child's weight in kilograms. Round to the nearest tenth. (1 kg ≈ 2.2 lb)

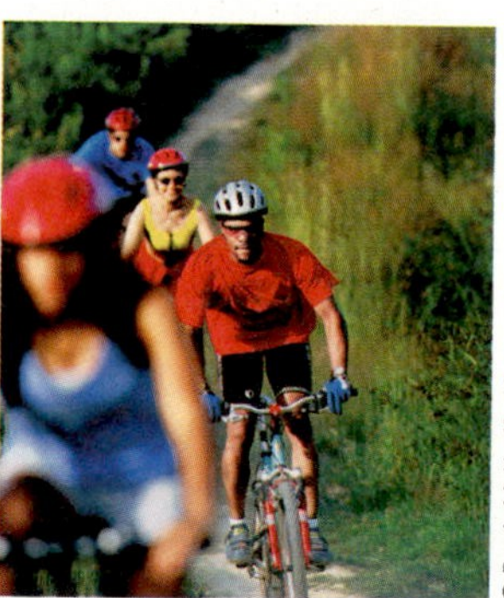
© Randy M. Ury/Corbis

30. **Health** Cycling burns approximately 400 Calories per hour. How many hours of cycling are necessary to lose 1 lb? (3500 Calories are equivalent to 1 lb.)

31. **Business** Six liters of liquid soap were bought for $11.40 per liter. The soap was repackaged in 150-milliliter plastic containers. The cost of each container was $0.26. Each container of soap sold for $3.29 per bottle. Find the profit on the 6 L of liquid soap.

32. **Health** A session of pilates burns about 235 Calories per hour. How many Calories do you burn in a month if you attend two 45-minute pilates classes per week for four weeks?

33. **Solutions** A laboratory solution contains 75 μg of solute per milliliter of solution. How many milligrams of solute would there be in 500 milliliters of solutions?

CHAPTER 9

TEST

1. Convert 2.96 km to meters.

2. Convert 0.378 g to milligrams.

3. Convert 0.046 L to milliliters.

4. Convert 919 cm^3 to milliliters.

5. Convert 42.6 mm to centimeters.

6. Convert 7 m 96 cm to meters.

7. Convert 847 g to kilograms.

8. Convert 3920 ml to liters.

9. Convert 588 mcg to milligrams.

10. Convert 2.12 mg to micrograms.

11. Convert 3 g 89 mg to grams.

12. Convert 1.6 L to ml.

13. Convert 3.29 kg to grams.

14. Convert 4.2 m to centimeters.

15. Convert 96 ml to deciliters.

16. Convert 1375 mg to grams.

17. Convert 402 cm to meters.

18. Convert 0.125 ml to μl

19. **Health** Sedentary people need 15 Calories per pound of body weight to maintain their weight. How many Calories should a 140-pound, sedentary person consume per day to maintain that weight?

20. **Medication Dosages** A medication is labeled 75mg/tablet. If a patient is prescribed three tablets per day for seven days, how many grams of medication will the patient have taken at the end of the week?

21. **Height** At the pediatrician's office, a 10-year-old boy's height is recorded as 1.35 m. His brother is 16 cm taller than he is. Find the brother's height.

22. **Medication Dosage** The recommended dosage of a medication is 50 mg/kg per hour using an IV infusion. If a child weighs 31 kg, how many grams of medication will this child receive once a 2-hour infusion is complete?

23. **Medicine** A community health clinic is giving flu shots for the coming flu season. Each flu shot contains 2 ml of vaccine. How many liters of vaccine are needed to inoculate 2600 people?

24. **Measurements** Convert 35 mi/h to kilometers per hour. Round to the nearest tenth. (1 mi $\approx$ 1.61 km)

25. **Consumerism** A 3.8-liter bottle of laboratory grade distilled water is priced at $7.95. Find the cost per gallon.

26. **Blood Tests** One model of blood glucose meter uses 4.0 μl of blood per test. At this rate, how many blood tests could be performed using 1 ml of blood?

27. **Energy** An air conditioner rated at 1600 W is operated an average of 4 h per day. Electrical energy costs 12.5¢ per kilowatt-hour. How much does it cost to operate the air conditioner for 30 days?

28. **Chemistry** A laboratory assistant is in charge of ordering acid for three chemistry classes of 40 students each. Each student requires 90 ml of acid. How many liters of acid should be ordered? (The assistant must order by the whole liter.)

29. **Medication Dosage** If the prescribed dosage of cough syrup is 2 t four times a day for five days, how many milliliters of cough syrup are needed for this prescription? (1 t = 5 ml)

30. **Height** A nurse measures a patient's height as 5 ft 6 in. Convert this measurement to meters. Round to the nearest hundredth.

CUMULATIVE REVIEW EXERCISES

1. Simplify: $12 - 8 \div (6 - 4)^2 \cdot 3$

2. Find the total of $5\frac{3}{4}$, $1\frac{5}{6}$, and $4\frac{7}{9}$.

3. Subtract: $4\frac{2}{9} - 3\frac{5}{12}$

4. Divide: $5\frac{3}{8} \div 1\frac{3}{4}$

5. Simplify: $\left(\frac{2}{3}\right)^4 \cdot \left(\frac{9}{4}\right)^2$

6. Subtract: $12.0072 - 9.937$

7. Solve the proportion $\frac{5}{8} = \frac{n}{50}$. Round to the nearest tenth.

8. Write $1\frac{3}{4}$ as a percent.

9. 6.09 is 4.2% of what number?

10. Convert 18 pt to gallons.

11. Convert 875 cm to meters.

12. Convert 3420 m to kilometers.

13. Convert 5.05 kg to grams.

14. Convert 3 g 672 mg to grams.

15. Convert 6 L to milliliters.

16. Convert 2.4 kl to liters.

17. **Finances** The Guerrero family has a monthly income of $5244 per month. The family spends one-fourth of its monthly income on rent. How much money is left after the rent is paid?

18. **Taxes** The state income tax on a business is \$620 plus 0.08 times the profit the business makes. The medical practice made a profit of \$82,340.00 last year. Find the amount of state income tax the medical practice paid.

19. **Nurses** According to the National Sample Survey of Registered Nurses, approximately four in five American Registered Nurses are female. If there are a total of 615 nurses working at a local hospital, how many of them would you expect to be male?

20. **Wages** An occupational therapist's hourly wage was \$31.50/h before receiving a 2% raise. What is the new hourly wage?

21. **Solutions** An alcohol solution (alcohol and water) has a total volume of 500 ml. 75 ml of that solution is alcohol. What percent of the solution is water?

Yuri Arcurs/Shutterstock.com

22. **Education** You received grades of 78, 92, 45, 80, and 85 on five chemistry exams. Find your mean grade.

23. **Compensation** Karla Perella, a certified nursing assistant, receives a salary of \$22,500. Her salary will increase by 12% next year. What will her salary be next year?

24. **Discount** A laboratory equipment manager has microscopes that are regularly priced at \$180 on sale for \$140.40. What is the discount rate?

25. **Human Growth** A pediatrician uses a standard growth chart to monitor the growth of his patients. The minimum height shown on the growth chart for most 10-year-old boys is 4 ft 2 in. If a 10-year-old boy is 3 in. below this minimum height, what is his height? Give your answer in feet and inches.

26. **Medication Dosage** A doctor prescribes a dose of 2.5 g of medication for a patient. The medication is in liquid form, and the label reads 1250 mg/5 ml. How many milliliters of the liquid should a nurse give the patient? How many teaspoons is this?

27. **Repackaging** A pharmacy purchased 1 gallon of nasal solution and then had the pharmacy technician repackage it in 1.5-fluid-ounce spray bottles. The nasal solution and bottles cost \$150, and each 1.5-fluid-ounce bottle was sold for \$3.75. Find the profit the pharmacy made on the nasal solution.

28. **IV Therapy** At 8 A.M., $\frac{3}{4}$ of a 1-liter bag of D5W has been infused. If the infusion rate for this IV is 50 ml/h, what time will this infusion be complete?

29. **Weight Loss** A patient weighed 275 lb when he started a new diet recommended by a nutritionist. After six months, he had decreased his weight by 12%. What was this patient's weight at the end of six months?

30. **Fluid Intake** A patient drinks $1\frac{1}{2}$ c of coffee, $\frac{3}{4}$ c of iced tea, and $1\frac{1}{4}$ c of water at dinner. What is the patient's total fluid intake at dinner?

CHAPTER

10

Rational Numbers

OBJECTIVES

SECTION 10.1

- **A** To identify the order relation between two integers
- **B** To evaluate expressions that contain the absolute value symbol

SECTION 10.2

- **A** To add integers
- **B** To subtract integers
- **C** To solve application problems

SECTION 10.3

- **A** To multiply integers
- **B** To divide integers
- **C** To solve application problems

SECTION 10.4

- **A** To add or subtract rational numbers
- **B** To multiply or divide rational numbers
- **C** To solve application problems

SECTION 10.5

- **A** To write a number in scientific notation
- **B** To use the Order of Operations Agreement to simplify expressions

AIM for the Future

OCCUPATIONAL THERAPIST ASSISTANTS work under the supervision of occupational therapists to provide rehabilitative services to persons with mental, physical, emotional, or developmental impairments. Activities range from teaching the proper method of moving from a bed into a wheelchair to the best way to stretch and limber the muscles of the hand. Employment of occupational therapist assistants and aides is expected to grow by 30% from 2008 to 2018, much faster than the average for all occupations. Median annual wages of occupational therapist assistants were $48,230 in May 2008. (*Source: Bureau of Labor Statistics:* http://www.bls.gov/oco/ocos166.htm)

PREP TEST

Are you ready to succeed in this chapter?

Take the Prep Test below to find out if you are ready to learn the new material.

1. Place the correct symbol, $<$ or $>$, between the two numbers.
54 45

2. What is the distance from 4 to 8 on the number line?

For Exercises 3 to 14, add, subtract, multiply, or divide.

3. $7654 + 8193$

4. $6097 - 2318$

5. 472×56

6. $\frac{144}{24}$

7. $\frac{2}{3} + \frac{3}{5}$

8. $\frac{3}{4} - \frac{5}{16}$

9. $0.75 + 3.9 + 6.408$

10. $5.4 - 1.619$

11. $\frac{3}{4} \times \frac{8}{15}$

12. $\frac{5}{12} \div \frac{3}{4}$

13. 23.5×0.4

14. $0.96 \div 2.4$

15. Simplify: $(8 - 6)^2 + 12 \div 4 \cdot 3^2$

SECTION

10.1 Introduction to Integers

OBJECTIVE A To identify the order relation between two integers

Thus far in the text, we have encountered only zero and the numbers greater than zero. The numbers greater than zero are called **positive numbers.** However, the phrases "12 degrees below zero," "$25 in debt," and "15 feet below sea level" refer to numbers less than zero. These numbers are called **negative numbers.**

The **integers** are . . . , −4, −3, −2, −1, 0, 1, 2, 3, 4, ...

Each integer can be shown on a number line. The integers to the left of zero on the number line are called **negative integers** and are represented by a negative sign (−) placed in front of the number. The integers to the right of zero are called **positive integers.** The positive integers are also called **natural numbers.** Zero is neither a positive nor a negative integer.

Point of Interest

Among the slang words for zero are *zilch, zip,* and *goose egg.* The word *love* for zero in scoring a tennis game comes from the French word *l'oeuf,* which means "the egg."

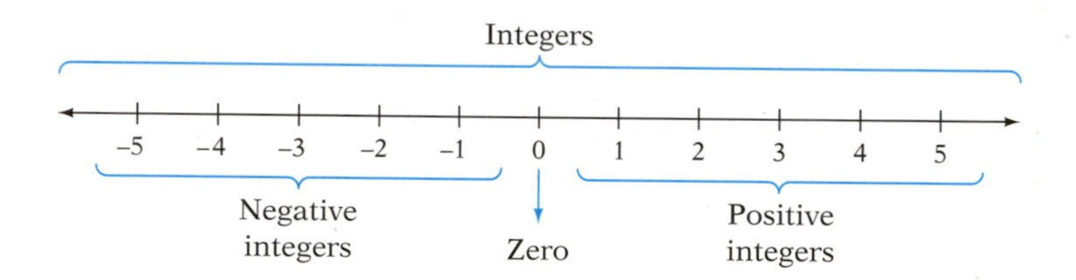

A number line can be used to visualize the order relation between two integers. A number that appears to the left of a given number is less than (<) the given number. A number that appears to the right of a given number is greater than (>) the given number.

2 is greater than negative 4.
$2 > -4$

Negative 5 is less than negative 3.
$-5 < -3$

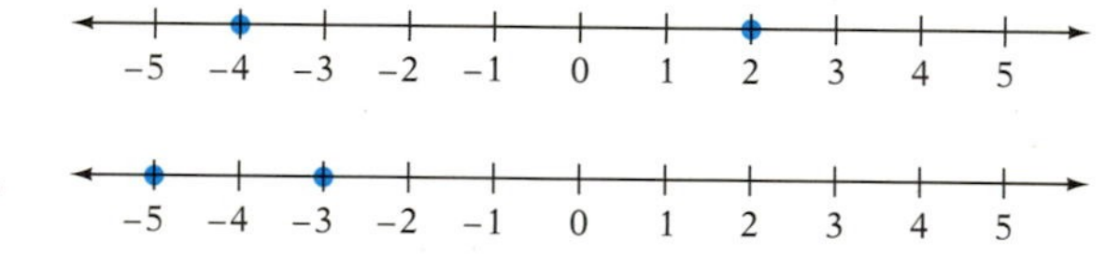

EXAMPLE 1

The lowest recorded temperature in Antarctica is 129° below zero. Represent this temperature as an integer.

Solution $-129°$

YOU TRY IT 1

The surface of the Salton Sea is 232 ft below sea level. Represent this depth as an integer.

Your solution

EXAMPLE 2

Graph −2 on the number line.

Solution

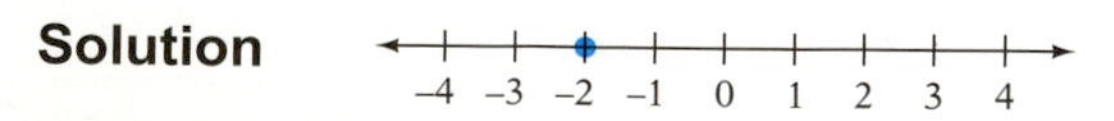

YOU TRY IT 2

Graph −4 on the number line.

Your solution

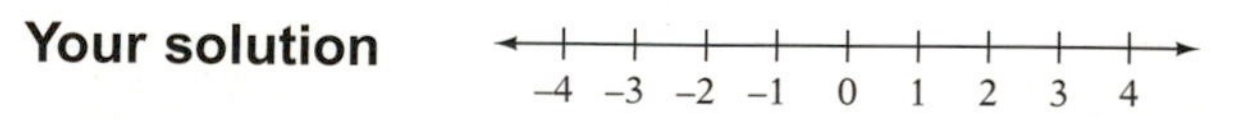

EXAMPLE 3

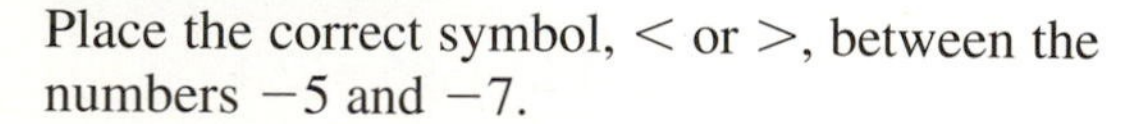

Place the correct symbol, < or >, between the numbers −5 and −7.

Solution $-5 > -7$ • **−5 is to the right of −7 on the number line.**

YOU TRY IT 3

Place the correct symbol, < or >, between the numbers −12 and −8.

Your solution

Solutions on p. S23

OBJECTIVE B To evaluate expressions that contain the absolute value symbol

Two numbers that are the same distance from zero on the number line but on opposite sides of zero are called **opposites.**

-4 is the opposite of 4

and

4 is the opposite of -4.

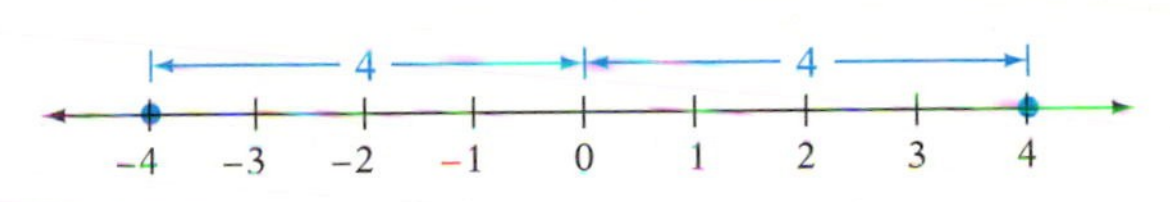

Note that a negative sign can be read as "the opposite of."

$-(4) = -4$ The opposite of positive 4 is negative 4.

$-(-4) = 4$ The opposite of negative 4 is positive 4.

The **absolute value of a number** is the distance between zero and the number on the number line. Therefore, the absolute value of a number is a positive number or zero. The symbol for absolute value is | |.

The distance from 0 to 4 is 4. Thus $|4| = 4$ (the absolute value of 4 is 4).

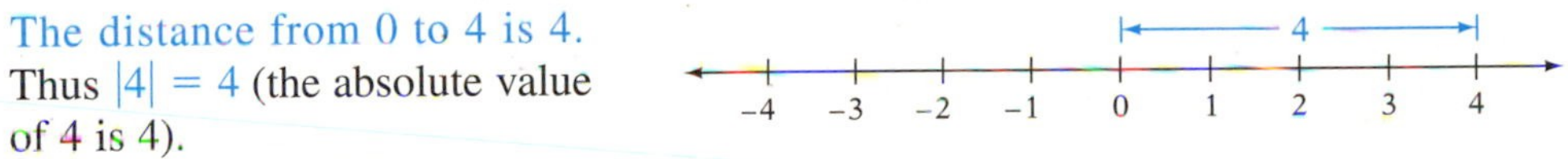

The distance from 0 to -4 is 4. Thus $|-4| = 4$ (the absolute value of -4 is 4).

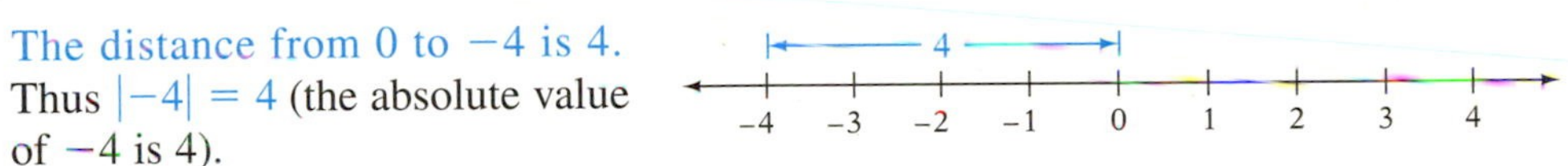

The absolute value of a positive number is the number itself. The absolute value of a negative number is the opposite of the negative number. The absolute value of zero is zero.

EXAMPLE 4

Find the absolute value of 2 and -3.

Solution

$|2| = 2$

$|-3| = 3$

YOU TRY IT 4

Find the absolute value of -7 and 21.

Your solution

EXAMPLE 5

Evaluate $|-34|$ and $|0|$.

Solution

$|-34| = 34$

$|0| = 0$

YOU TRY IT 5

Evaluate $|2|$ and $|-9|$.

Your solution

EXAMPLE 6

Evaluate $-|-4|$.

Solution

$-|-4| = -4$

The minus sign *in front of* the absolute value sign is not affected by the absolute value sign.

YOU TRY IT 6

Evaluate $-|-12|$.

Your solution

Solutions on p. S23

10.1 EXERCISES

OBJECTIVE A To identify the order relation between two integers

For Exercises 1 to 4, represent the quantity as an integer.

1. A lake 120 ft below sea level

2. A temperature that is 15° below zero

3. A loss of 324 dollars

4. A share of stock up 2 dollars

For Exercises 5 to 8, graph the numbers on the number line.

5. 3 and -3

−6 −5 −4 −3 −2 −1 0 1 2 3 4 5 6

6. -2 and 0

−6 −5 −4 −3 −2 −1 0 1 2 3 4 5 6

7. -4 and 1

−6 −5 −4 −3 −2 −1 0 1 2 3 4 5 6

8. 4 and -1

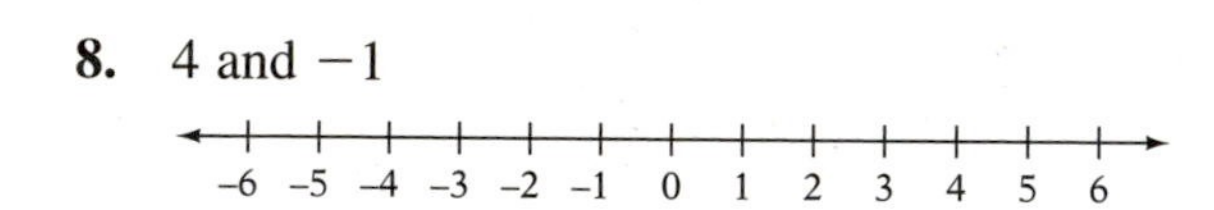

For Exercises 9 to 14, state which number on the number line is in the location given.

9. 3 units to the right of -2

10. 5 units to the right of -3

11. 4 units to the left of 3

12. 2 units to the left of -1

13. 6 units to the right of -3

14. 4 units to the right of -4

For Exercises 15 to 18, use the following number line.

A B C D E F G H I

15. **a.** If F is 1 and G is 2, what number is A?

b. If F is 1 and G is 2, what number is C?

16. **a.** If G is 1 and H is 2, what number is B?

b. If G is 1 and H is 2, what number is D?

17. **a.** If H is 0 and I is 1, what number is A?

b. If H is 0 and I is 1, what number is D?

18. **a.** If G is 2 and I is 4, what number is B?

b. If G is 2 and I is 4, what number is E?

For Exercises 19 to 42, place the correct symbol, < or >, between the two numbers.

19. −2 −5 **20.** −6 −1 **21.** −16 1 **22.** −2 13

23. 3 −7 **24.** 5 −6 **25.** −11 −8 **26.** −4 −10

27. 35 28 **28.** 42 19 **29.** −42 27 **30.** −36 49

31. 21 −34 **32.** 53 −46 **33.** −27 −39 **34.** −51 −20

35. −87 63 **36.** −75 92 **37.** 86 −79 **38.** 95 −71

39. −62 −84 **40.** −91 −70 **41.** −131 101 **42.** 127 −150

For Exercises 43 to 51, write the given numbers in order from smallest to largest.

43. 3, −7, 0, −2 **44.** −4, 8, 6, −1 **45.** −3, 1, −5, 4

46. −6, 2, −8, 7 **47.** 9, −4, 5, 0 **48.** 6, −9, −12, 8

49. −10, 4, 12, −5, −7 **50.** 11, −8, −1, 7, −6 **51.** 10, −11, −2, 5, −7

For Exercises 52 to 55, determine whether the statement is *always true, never true,* or *sometimes true.*

52. A number that is to the right of −6 on the number line is a negative number.

53. A number that is to the left of −2 on the number line is a negative number.

54. A number that is to the right of 7 on the number line is a negative number.

55. A number that is to the left of 4 on the number line is a negative number.

OBJECTIVE B To evaluate expressions that contain the absolute value symbol

For Exercises 56 to 65, find the opposite number.

56. 4 **57.** 16 **58.** −2 **59.** −3 **60.** 22

61. 45 **62.** -31 **63.** -59 **64.** 70 **65.** -88

For Exercises 66 to 73, find the absolute value of the number.

66. 4 **67.** -4 **68.** -7 **69.** 9

70. -1 **71.** -11 **72.** 10 **73.** -12

For Exercises 74 to 103, evaluate.

74. $|2|$ **75.** $|-2|$ **76.** $|-6|$ **77.** $|6|$ **78.** $|8|$

79. $|5|$ **80.** $|-9|$ **81.** $|-1|$ **82.** $-|-1|$ **83.** $-|-5|$

84. $-|0|$ **85.** $|16|$ **86.** $|19|$ **87.** $|-12|$ **88.** $|-22|$

89. $-|29|$ **90.** $-|20|$ **91.** $-|-14|$ **92.** $-|-18|$ **93.** $|-15|$

94. $|-23|$ **95.** $-|33|$ **96.** $-|27|$ **97.** $|32|$ **98.** $|25|$

99. $-|-42|$ **100.** $|-74|$ **101.** $|-61|$ **102.** $-|88|$ **103.** $-|52|$

For Exercises 104 to 111, place the correct symbol, $<$, $=$, or $>$, between the two numbers.

104. $|7|$ $|-9|$ **105.** $|-12|$ $|8|$ **106.** $|-5|$ $|-2|$ **107.** $|6|$ $|13|$

108. $|-8|$ $|3|$ **109.** $|-1|$ $|-17|$ **110.** $|-14|$ $|14|$ **111.** $|17|$ $|-17|$

For Exercises 112 to 117, write the given numbers in order from smallest to largest.

112. $|-8|, -3, |2|, -|-5|$ **113.** $-|6|, -4, |-7|, -9$ **114.** $-1, |-6|, |0|, -|3|$

115. $-|-7|, -9, 5, |4|$

116. $-|2|, -8, 6, |1|$

117. $-3, -|-8|, |5|, -|10|$

For Exercises 118 to 121, determine whether the statement is true for *positive integers, negative integers,* or *all integers.*

118. The absolute value of an integer is the opposite of the integer.

119. The opposite of an integer is less than the integer.

120. The opposite of an integer is negative.

121. The absolute value of an integer is greater than the integer.

Applying the Concepts

122. Meteorology The graph at the right shows the lowest recorded temperatures, in degrees Fahrenheit, for selected states in the United States. Which state has the lowest recorded temperature?

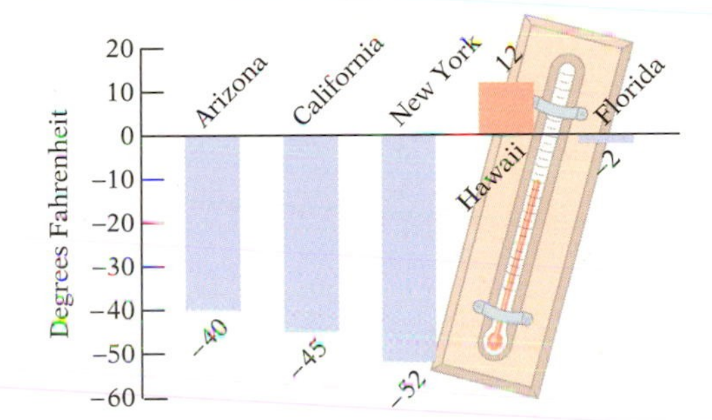

Lowest Recorded Temperatures

Sources: National Climatic Data Center; NESDIS; NOAA; U.S. Dept. of Commerce

123. **a.** Name two numbers that are 5 units from 3 on the number line.
b. Name two numbers that are 3 units from -1 on the number line.

124. **a.** Find a number that is halfway between -7 and -5.
b. Find a number that is halfway between -10 and -6.
c. Find a number that is one-third of the way between -12 and -3.

125. Rocketry Which is closer to blastoff, -12 min and counting or -17 min and counting?

126. Investments In the stock market, the net change in the price of a share of stock is recorded as a positive or a negative number. If the price rises, the net change is positive. If the price falls, the net change is negative. If the net change for a share of Stock A is -2 and the net change for a share of Stock B is -1, which stock showed the least net change?

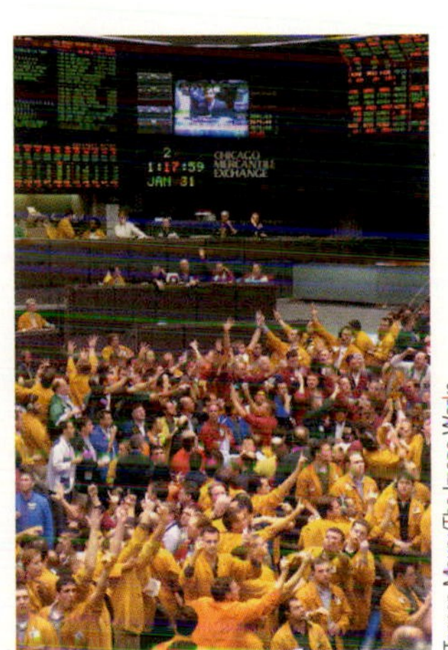

Tannen Maury/The Image Works

127. Weight Loss A patient's weight loss can be recorded using negative numbers. A patient's weight loss in January was recorded as -8 lb. In February, the same patient's weight loss was recorded as -6 lb and in March, the patient's weight loss was recorded as -7 lb. During which month was the weight loss the greatest?

128. **a.** Find the values of a for which $|a| = 7$.
b. Find the values of y for which $|y| = 11$.

SECTION

10.2 Addition and Subtraction of Integers

OBJECTIVE A To add integers

An integer can be graphed as a dot on a number line, as shown in the last section. An integer also can be represented anywhere along a number line by an arrow. A positive number is represented by an arrow pointing to the right. A negative number is represented by an arrow pointing to the left. The absolute value of the number is represented by the length of the arrow. The integers 5 and -4 are shown on the number lines below.

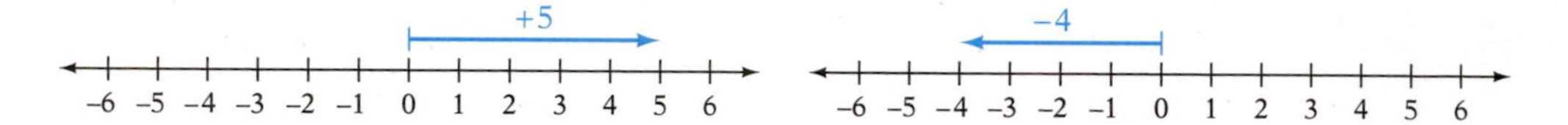

The sum of two integers can be shown on a number line. To add two integers, use arrows to represent the addends, with the first arrow starting at zero. The sum is the number directly below the tip of the arrow that represents the second addend.

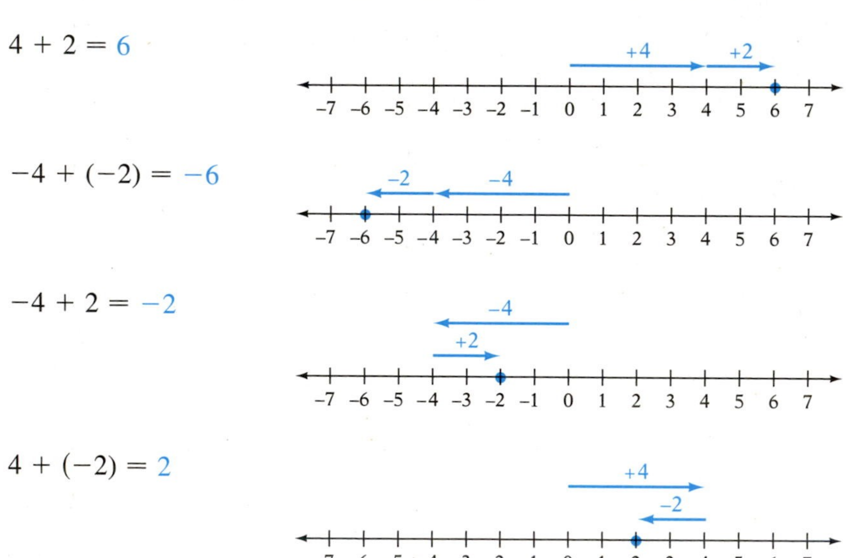

The sums of the integers shown above can be categorized by the signs of the addends.

Here the addends have the same sign:

$4 + 2$	*positive* 4 plus *positive* 2
$-4 + (-2)$	*negative* 4 plus *negative* 2

Here the addends have different signs:

$-4 + 2$	*negative* 4 plus *positive* 2
$4 + (-2)$	*positive* 4 plus *negative* 2

The rule for adding two integers depends on whether the signs of the addends are the same or different.

Rule for Adding Two Numbers

To add numbers with the same sign, add the absolute values of the numbers. Then attach the sign of the addends.

To add numbers with different signs, find the difference between the absolute values of the numbers. Then attach the sign of the addend with the greater absolute value.

Point of Interest

Although mathematical symbols are fairly standard in every country, that has not always been true. Italian mathematicians in the 15th century used a "p" to indicate plus. The "p" was from the Italian word *piu,* meaning "more" or "plus."

HOW TO 1 Add: $(-4) + (-9)$

$|-4| = 4, |-9| = 9$
$4 + 9 = 13$
- Because the signs of the addends are the same, add the absolute values of the numbers.

$(-4) + (-9) = -13$
- Then attach the sign of the addends.

HOW TO 2 Add: $6 + (-13)$

$|6| = 6, |-13| = 13$
$13 - 6 = 7$
- Because the signs of the addends are different, subtract the smaller absolute value from the larger absolute value.

$6 + (-13) = -7$
- Then attach the sign of the number with the larger absolute value. Because $|-13| > |6|$, attach the negative sign.

HOW TO 3 Add: $162 + (-247)$

$162 + (-247) = -85$
- Because the signs are different, find the difference between the absolute values of the numbers and attach the sign of the number with the greater absolute value.

Integrating Technology

To add $-14 + (-47)$ on your calculator, enter the following:

14 **+/−** **+** 47 **+/−** **=**

HOW TO 4 Find the sum of -14 and -47.

$-14 + (-47) = -61$
- Because the signs are the same, add the absolute values of the numbers and attach the sign of the addends.

When adding more than two integers, start from the left and add the first two numbers. Then add the sum to the third number. Continue this process until all the numbers have been added.

HOW TO 5 Add: $(-4) + (-6) + (-8) + 9$

$(-4) + (-6) + (-8) + 9 = (-10) + (-8) + 9$
- Add the first two numbers.

$= (-18) + 9$
- Add the sum to the next number.

$= -9$
- Continue adding until all numbers have been added.

EXAMPLE 1

What is -162 added to 98?

Solution

$98 + (-162) = -64$
- The signs of the addends are different.

YOU TRY IT 1

Add: $-154 + (-37)$

Your solution

Solution on p. S23

EXAMPLE • 2

Add: $-2 + (-7) + 4 + (-6)$

Solution

$$\begin{aligned} -2 + (-7) + 4 + (-6) &= -9 + 4 + (-6) \\ &= -5 + (-6) \\ &= -11 \end{aligned}$$

YOU TRY IT • 2

Add: $-5 + (-2) + 9 + (-3)$

Your solution

Solution on p. S23

OBJECTIVE B To subtract integers

Tips for Success

Be sure to do all you need to do in order to be successful at adding and subtracting integers: Read through the introductory material, work through the examples indicated by the HOW TO feature, study the paired Examples, do the You Try Its and check your solutions against those in the back of the book, and do the exercises on pages 414 to 418. See *AIM for Success* at the front of the book.

Before the rules for subtracting two integers are explained, look at the translation into words of an expression that is the difference of two integers:

$9 - 3$	positive 9 minus positive 3
$(-9) - 3$	negative 9 minus positive 3
$9 - (-3)$	positive 9 minus negative 3
$(-9) - (-3)$	negative 9 minus negative 3

Note that the sign − is used in two different ways. One way is as a negative sign, as in (-9), *negative* 9. The second way is to indicate the operation of subtraction, as in $9 - 3$, 9 *minus* 3.

Look at the next four subtraction expressions and decide whether the second number in each expression is a positive number or a negative number.

1. $(-10) - 8$ **2.** $(-10) - (-8)$ **3.** $10 - (-8)$ **4.** $10 - 8$

In expressions 1 and 4, the second number is a positive 8. In expressions 2 and 3, the second number is a negative 8.

Rule for Subtracting Two Numbers

To subtract two numbers, add the opposite of the second number to the first number.

This rule states that to subtract two integers, we rewrite the subtraction expression as the sum of the first number and the opposite of the second number.

Here are some examples:

First number	−	second number	=	first number	+	the opposite of the second number	
8	−	15	=	8	+	(−15)	= −7
8	−	(−15)	=	8	+	15	= 23
(−8)	−	15	=	(−8)	+	(−15)	= −23
(−8)	−	(−15)	=	(−8)	+	15	= 7

HOW TO • 6 Subtract: $(-15) - 75$

$$\begin{aligned} (-15) - 75 &= (-15) + (-75) \\ &= -90 \end{aligned}$$

• **To subtract, add the opposite of the second number to the first number.**

Integrating Technology

The +/− key on your calculator is used to find the opposite of a number. The − key is used to perform the operation of subtraction.

HOW TO • 7 Subtract: $27 - (-32)$

$$27 - (-32) = 27 + 32 = 59$$

• To subtract, add the opposite of the second number to the first number.

When subtraction occurs several times in an expression, rewrite each subtraction as addition of the opposite and then add.

HOW TO • 8 Subtract: $-13 - 5 - (-8)$

$$\begin{aligned} -13 - 5 - (-8) &= -13 + (-5) + 8 \\ &= -18 + 8 \\ &= -10 \end{aligned}$$

• Rewrite each subtraction as the addition of the opposite and then add.

EXAMPLE • 3

Find 8 less than -12.

Solution

$$\begin{aligned} -12 - 8 &= -12 + (-8) \\ &= -20 \end{aligned}$$

• Rewrite "−" as "+"; the opposite of 8 is −8.

YOU TRY IT • 3

Find -8 less 14.

Your solution

EXAMPLE • 4

Subtract: $6 - (-20)$

Solution

$$\begin{aligned} 6 - (-20) &= 6 + 20 \\ &= 26 \end{aligned}$$

• Rewrite "−" as "+"; the opposite of −20 is 20.

YOU TRY IT • 4

Subtract: $3 - (-15)$

Your solution

EXAMPLE • 5

Subtract: $-8 - 30 - (-12) - 7$

Solution

$$\begin{aligned} &-8 - 30 - (-12) - 7 \\ &= -8 + (-30) + 12 + (-7) \\ &= -38 + 12 + (-7) \\ &= -26 + (-7) \\ &= -33 \end{aligned}$$

YOU TRY IT • 5

Subtract: $4 - (-3) - 12 - (-7) - 20$

Your solution

Solutions on p. S23

OBJECTIVE C To solve application problems

EXAMPLE • 6

Find the temperature after an increase of 9°C from −6°C.

Strategy

To find the temperature, add the increase (9°C) to the previous temperature (−6°C).

Solution

$$-6 + 9 = 3$$

The temperature is 3°C.

YOU TRY IT • 6

Find the temperature after an increase of 12°C from −10°C.

Your strategy

Your solution

Solution on p. S23

10.2 EXERCISES

OBJECTIVE A To add integers

For Exercises 1 and 2, name the negative integers in the list of numbers.

1. $-14, 28, 0, -\frac{5}{7}, -364, -9.5$

2. $-37, 90, -\frac{7}{10}, -88.8, 42, -561$

For Exercises 3 to 30, add.

3. $3 + (-5)$

4. $-4 + 2$

5. $8 + 12$

6. $16 + 23$

7. $-3 + (-8)$

8. $-12 + (-1)$

9. $-4 + (-5)$

10. $-12 + (-12)$

11. $6 + (-9)$

12. $4 + (-9)$

13. $-6 + 7$

14. $-12 + 6$

15. $2 + (-3) + (-4)$

16. $7 + (-2) + (-8)$

17. $-3 + (-12) + (-15)$

18. $9 + (-6) + (-16)$

19. $-17 + (-3) + 29$

20. $13 + 62 + (-38)$

21. $-3 + (-8) + 12$

22. $-27 + (-42) + (-18)$

23. $13 + (-22) + 4 + (-5)$

24. $-14 + (-3) + 7 + (-6)$

25. $-22 + 10 + 2 + (-18)$

26. $-6 + (-8) + 13 + (-4)$

27. $-16 + (-17) + (-18) + 10$

28. $-25 + (-31) + 24 + 19$

29. $-126 + (-247) + (-358) + 339$

30. $-651 + (-239) + 524 + 487$

31. What is -8 more than -12?

32. What is -5 more than 3?

33. What is -7 added to -16?

34. What is 7 added to -25?

35. What is -4 plus 2?

36. What is -22 plus -17?

37. Find the sum of -2, 8, and -12.

38. Find the sum of 4, -4, and -6.

39. What is the total of 2, -3, 8, and -13?

40. What is the total of -6, -8, 13, and -2?

For Exercises 41 to 44, determine whether the statement is *always true, never true,* or *sometimes true.*

41. The sum of an integer and its opposite is zero.

42. The sum of two negative integers is a positive integer.

43. The sum of two negative integers and one positive integer is a negative integer.

44. If the absolute value of a negative integer is greater than the absolute value of a positive integer, then the sum of the integers is negative.

OBJECTIVE B To subtract integers

For Exercises 45 to 48, translate the expression into words. Represent each number as positive or negative.

45. $-6 - 4$

46. $-6 - (-4)$

47. $6 - (-4)$

48. $6 - 4$

For Exercises 49 to 52, rewrite the subtraction as the sum of the first number and the opposite of the second number.

49. $9 - (-5)$

50. $-3 - 7$

51. $1 - 8$

52. $-2 - (-10)$

For Exercises 53 to 80, subtract.

53. $16 - 8$

54. $12 - 3$

55. $7 - 14$

56. $6 - 9$

57. $-7 - 2$

58. $-9 - 4$

59. $7 - (-29)$

60. $3 - (-4)$

61. $-6 - (-3)$

62. $-4 - (-2)$

63. $6 - (-12)$

64. $-12 - 16$

65. $-4 - 3 - 2$

66. $4 - 5 - 12$

67. $12 - (-7) - 8$

68. $-12 - (-3) - (-15)$

69. $4 - 12 - (-8)$

70. $13 - 7 - 15$

71. $-6 - (-8) - (-9)$

72. $7 - 8 - (-1)$

73. $-30 - (-65) - 29 - 4$

74. $42 - (-82) - 65 - 7$

75. $-16 - 47 - 63 - 12$

76. $42 - (-30) - 65 - (-11)$

77. $47 - (-67) - 13 - 15$

78. $-18 - 49 - (-84) - 27$

79. $167 - 432 - (-287) - 359$

80. $-521 - (-350) - 164 - (-299)$

81. Subtract -8 from -4.

82. Subtract -12 from 3.

83. What is the difference between -8 and 4?

84. What is the difference between 8 and -3?

85. What is -4 decreased by 8?

86. What is -13 decreased by 9?

87. Find -2 less than 1.

88. Find -3 less than -5.

For Exercises 89 to 92, determine whether the statement is *always true, never true,* or *sometimes true.*

89. The difference between a positive integer and a negative integer is zero.

90. A negative integer subtracted from a positive integer is a positive integer.

91. The difference between two negative integers is a positive integer.

92. The difference between an integer and its absolute value is zero.

OBJECTIVE C To solve application problems

93. Temperature The news clipping at the right was written on February 11, 2008. The record low temperature for Minnesota is -51°C. Find the difference between the low temperature in International Falls on February 11, 2008, and the record low temperature for Minnesota.

> **In the News**
>
> **Minnesota Town Named "Icebox of the Nation"**
>
> In International Falls, Minnesota, the temperature fell to -40°C just days after the citizens received word that the town had won a federal trademark naming it the "Icebox of the Nation."
>
> *Source:* news.yahoo.com

94. Temperature The record high temperature in Illinois is 117°F. The record low temperature is -36°F. Find the difference between the record high and record low temperatures in Illinois.

95. **Temperature** Find the temperature after a rise of 7°C from −8°C.

96. **Temperature** Find the temperature after a rise of 5°C from −19°C.

97. If the temperature begins at −54°C and rises more than 60°C, is the new temperature above or below 0°C?

98. If the temperature begins at −37°C and falls more than 40°C, is the new temperature above or below 0°C?

Wind Chill The National Oceanic and Atmospheric Administration's National Weather Service Web site defines the wind chill temperature as how cold people and animals feel when they are outside. As wind speed increases, heat is drawn from the body and a person's skin temperature and internal body temperature decrease. Hypothermia sets in when body temperature falls below 95°F. Use the table at the right for Exercises 99 to 102.

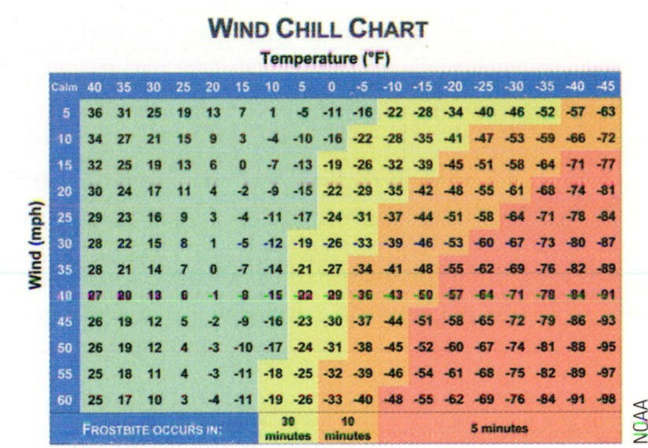

WIND CHILL CHART

Temperature (°F)

Wind (mph) / Calm	40	35	30	25	20	15	10	5	0	-5	-10	-15	-20	-25	-30	-35	-40	-45
5	36	31	25	19	13	7	1	-5	-11	-16	-22	-28	-34	-40	-46	-52	-57	-63
10	34	27	21	15	9	3	-4	-10	-16	-22	-28	-35	-41	-47	-53	-59	-66	-72
15	32	25	19	13	6	0	-7	-13	-19	-26	-32	-39	-45	-51	-58	-64	-71	-77
20	30	24	17	11	4	-2	-9	-15	-22	-29	-35	-42	-48	-55	-61	-68	-74	-81
25	29	23	16	9	3	-4	-11	-17	-24	-31	-37	-44	-51	-58	-64	-71	-78	-84
30	28	22	15	8	1	-5	-12	-19	-26	-33	-39	-46	-53	-60	-67	-73	-80	-87
35	28	21	14	7	0	-7	-14	-21	-27	-34	-41	-48	-55	-62	-69	-76	-82	-89
40	27	20	13	6	-1	-8	-15	-22	-29	-36	-43	-50	-57	-64	-71	-78	-84	-91
45	26	19	12	5	-2	-9	-16	-23	-30	-37	-44	-51	-58	-65	-72	-79	-86	-93
50	26	19	12	4	-3	-10	-17	-24	-31	-38	-45	-52	-60	-67	-74	-81	-88	-95
55	25	18	11	4	-3	-11	-18	-25	-32	-39	-46	-54	-61	-68	-75	-82	-89	-97
60	25	17	10	3	-4	-11	-19	-26	-33	-40	-48	-55	-62	-69	-76	-84	-91	-98

FROSTBITE OCCURS IN: 30 minutes | 10 minutes | 5 minutes

Source: National Weather Service

99. Frostbite occurs when body tissues freeze. Ear lobes, fingers, toes, and the nose are the most susceptible to frostbite. If the outside temperature is 5°F, what is the wind chill temperature if the wind speed is 30 mph? In how many minutes would frostbite occur under these conditions?

100. If the outside temperature is 15°F, find the difference between the wind chill temperature at a wind speed of 10 mph and the wind chill temperature at a wind speed of 20 mph.

101. If the outside temperature is 0°F, find the difference between the wind chill temperature at a wind speed of 15 mph and the wind chill temperature at a wind speed of 30 mph.

102. According to the wind chill chart, who is colder: a person in New York at 10°F with a 10 mph wind or a person in Chicago at 15°F with a 30 mph wind? Find the difference in the wind chill temperatures in these locations.

103. **Weight Loss** A patient begins a new diet plan and each month records any change in weight using integers. Calculate the total change in the patient's weight using these recorded amounts: −5 lb, −2 lb, +1 lb, −4 lb, and −3 lb.

104. **Body Temperature** A patient's temperature was 102°F at noon. By 2 P.M., the patient's temperature had fallen 3°F (−3°) but at 5 P.M., it had risen 2° (+2°). What was the patient's temperature at 5 P.M.?

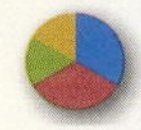

Geography The elevation, or height, of places on Earth is measured in relation to sea level, or the average level of the ocean's surface. The table below shows height above sea level as a positive number and depth below sea level as a negative number. Use the table for Exercises 105 to 107.

Mt. Everest

Continent	*Highest Elevation (in meters)*		*Lowest Elevation (in meters)*	
Africa	Mt. Kilimanjaro	5895	Lake Assal	−156
Asia	Mt. Everest	8850	Dead Sea	−411
North America	Mt. McKinley	5642	Death Valley	−28
South America	Mt. Aconcagua	6960	Valdes Peninsula	−86

105. What is the difference in elevation between Mt. Kilimanjaro and Lake Assal?

106. What is the difference in elevation between Mt. Aconcagua and the Valdes Peninsula?

107. For which continent shown is the difference between the highest and lowest elevations greatest?

Meteorology The figure at the right shows the highest and lowest temperatures ever recorded for selected regions of the world. Use this graph for Exercises 108 to 110.

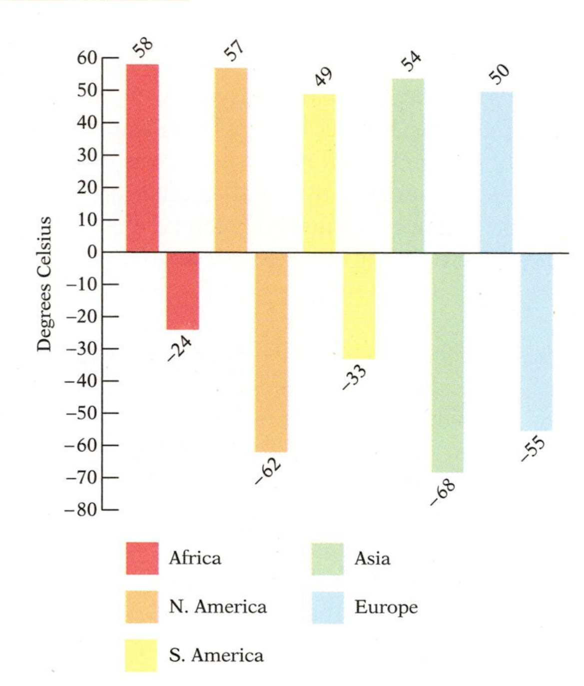

Highest and Lowest Temperatures Recorded (in degrees Celsius)
Source: National Climatic Data Center

108. What is the difference between the highest and lowest temperatures recorded in Africa?

109. What is the difference between the highest and lowest temperatures recorded in South America?

110. What is the difference between the lowest temperature recorded in Europe and the lowest temperature recorded in Asia?

Applying the Concepts

111. **Number Problems** Consider the numbers 4, −7, −5, 13, and −9. What is the largest difference that can be obtained by subtracting one number in the list from another number in the list? Find the smallest positive difference.

112. **Number Problems** Fill in the blank squares at the right with integers so that the sum of the integers along any row, column, or diagonal is zero.

−3		1
		3

113. **Number Problems** The sum of two negative integers is −8. Find the integers.

114. Explain the difference between the words *negative* and *minus*.

SECTION

10.3 Multiplication and Division of Integers

OBJECTIVE A To multiply integers

Multiplication is the repeated addition of the same number.

Several different symbols are used to indicate multiplication:

$3 \times 2 = 6$ $3 \cdot 2 = 6$ $(3)(2) = 6$

When 5 is multiplied by a sequence of decreasing integers, the products decrease by 5.

$5 \times 3 = 15$
$5 \times 2 = 10$
$5 \times 1 = 5$
$5 \times 0 = 0$

The pattern developed can be continued so that 5 is multiplied by a sequence of negative numbers. The resulting products must be negative in order to maintain the pattern of decreasing by 5.

$5 \times (-1) = -5$
$5 \times (-2) = -10$
$5 \times (-3) = -15$
$5 \times (-4) = -20$

This example illustrates that the product of a positive number and a negative number is negative.

When -5 is multiplied by a sequence of decreasing integers, the products increase by 5.

$-5 \times 3 = -15$
$-5 \times 2 = -10$
$-5 \times 1 = -5$
$-5 \times 0 = 0$

The pattern developed can be continued so that -5 is multiplied by a sequence of negative numbers. The resulting products must be positive in order to maintain the pattern of increasing by 5.

$-5 \times (-1) = 5$
$-5 \times (-2) = 10$
$-5 \times (-3) = 15$
$-5 \times (-4) = 20$

This example illustrates that the product of two negative numbers is positive.

The pattern for multiplication shown above is summarized in the following rules for multiplying integers.

Integrating Technology

To multiply $(-4)(-8)$ on your calculator, enter the following:

4  +/−  x 8  +/− =

Rule for Multiplying Two Numbers

To multiply numbers with the same sign, multiply the absolute values of the factors. The product is positive.

To multiply numbers with different signs, multiply the absolute values of the factors. The product is negative.

$4 \cdot 8 = 32$

$(-4)(-8) = 32$

$-4 \cdot 8 = -32$

$(4)(-8) = -32$

HOW TO • 1 Multiply: $2(-3)(-5)(-7)$

$2(-3)(-5)(-7) = -6(-5)(-7)$ • To multiply more than two numbers, first multiply the first two numbers.

$= 30(-7)$ • Then multiply the product by the third number.

$= -210$ • Continue until all the numbers have been multiplied.

EXAMPLE • 1

Multiply: $(-2)(6)$

Solution

$(-2)(6) = -12$ • The signs are different. The product is negative.

YOU TRY IT • 1

Multiply: $(-3)(-5)$

Your solution

EXAMPLE • 2

Find the product of -42 and 62.

Solution

$-42 \cdot 62 = -2604$ • The signs are different. The product is negative.

YOU TRY IT • 2

Find -38 multiplied by 51.

Your solution

EXAMPLE • 3

Multiply: $-5(-4)(6)(-3)$

Solution

$$\begin{aligned} -5(-4)(6)(-3) &= 20(6)(-3) \\ &= 120(-3) \\ &= -360 \end{aligned}$$

YOU TRY IT • 3

Multiply: $-7(-8)(9)(-2)$

Your solution

Solutions on pp. S23–S24

OBJECTIVE B To divide integers

For every division problem, there is a related multiplication problem.

Division: $\frac{8}{2} = 4$ Related multiplication: $4 \cdot 2 = 8$

This fact can be used to illustrate the rules for dividing signed numbers.

Rule for Dividing Two Numbers

To divide numbers with the same sign, divide the absolute values of the numbers. The quotient is positive.

To divide numbers with different signs, divide the absolute values of the numbers. The quotient is negative.

$\frac{8}{2} = 4$ because $4 \cdot 2 = 8$.

$\frac{-8}{-2} = 4$ because $4(-2) = -8$.

$\frac{8}{-2} = -4$ because $-4(-2) = 8$.

$\frac{-8}{2} = -4$ because $-4(2) = -8$.

Note that $\frac{8}{-2}$, $\frac{-8}{2}$, and $-\frac{8}{2}$ are all equal to -4.

If a and b are two integers, then $\frac{a}{-b} = \frac{-a}{b} = -\frac{a}{b}$.

Properties of Zero and One in Division

Zero divided by any number other than zero is zero.

Any number other than zero divided by itself is 1.

Any number divided by 1 is the number.

Division by zero is not defined.

$\frac{0}{a} = 0$ because $0 \cdot a = 0$

$\frac{a}{a} = 1$ because $1 \cdot a = a$

$\frac{a}{1} = a$ because $a \cdot 1 = a$

$\frac{4}{0} = ?$ $\quad ? \cdot 0 = 4$

There is no number whose product with zero is 4.

The examples below illustrate these properties of division.

$\frac{0}{-8} = 0 \qquad \frac{-7}{-7} = 1 \qquad \frac{-9}{1} = -9 \qquad \frac{-3}{0}$ is undefined.

EXAMPLE • 4

Divide: $(-120) \div (-8)$

Solution

$(-120) \div (-8) = 15$ • The signs are the same. The quotient is positive.

YOU TRY IT • 4

Divide: $(-135) \div (-9)$

Your solution

EXAMPLE • 5

Divide: $95 \div (-5)$

Solution

$95 \div (-5) = -19$ • The signs are different. The quotient is negative.

YOU TRY IT • 5

Divide: $84 \div (-6)$

Your solution

EXAMPLE • 6

Find the quotient of -81 and 3.

Solution

$-81 \div 3 = -27$

YOU TRY IT • 6

What is -72 divided by 4?

Your solution

EXAMPLE • 7

Divide: $0 \div (-24)$

Solution

$0 \div (-24) = 0$ • Zero divided by a nonzero number is zero.

YOU TRY IT • 7

Divide: $-39 \div 0$

Your solution

Solutions on p. S24

OBJECTIVE C To solve application problems

EXAMPLE • 8

The combined weight loss of a group of five patients in a clinical trial for a new weight-loss drug equaled -15 lb. What was the average weight loss of the five patients?

Strategy

To find the average weight loss, divide the total weight loss (-15) by the number of patients (5).

Solution

$-15 \div 5 = -3$

The average weight loss was 3 pounds (-3 lb).

YOU TRY IT • 8

The patients assigned to Group A in a six-week clinical trial for a new weight-loss drug had a combined weight loss of -20 lb. The combined weight loss of the patients in Group B during the same period was three times as much as that of Group A. Find the combined weight loss of Group B during the clinical trial.

Your strategy

Your solution

EXAMPLE • 9

The daily high temperatures during one week were recorded as follows: -9°F, 3°F, 0°F, -8°F, 2°F, 1°F, 4°F. Find the average daily high temperature for the week.

Strategy

To find the average daily high temperature:

- Add the seven temperature readings.
- Divide by 7.

Solution

$-9 + 3 + 0 + (-8) + 2 + 1 + 4 = -7$

$-7 \div 7 = -1$

The average daily high temperature was -1°F.

YOU TRY IT • 9

The daily low temperatures during one week were recorded as follows: -6°F, -7°F, 1°F, 0°F, -5°F, -10°F, -1°F. Find the average daily low temperature for the week.

Your strategy

Your solution

Solutions on p. S24

10.3 EXERCISES

OBJECTIVE A To multiply integers

For Exercises 1 to 4, state whether the operation in the expression is addition, subtraction, or multiplication.

1. $5 - (-6)$

2. $4(-9)$

3. $-8(-5)$

4. $-3 + (-7)$

For Exercises 5 to 46, multiply.

5. 14×3

6. 62×9

7. $-4 \cdot 6$

8. $-7 \cdot 3$

9. $-2 \cdot (-3)$

10. $-5 \cdot (-1)$

11. $(9)(2)$

12. $(3)(8)$

13. $5(-4)$

14. $4(-7)$

15. $-8(2)$

16. $-9(3)$

17. $(-5)(-5)$

18. $(-3)(-6)$

19. $(-7)(0)$

20. -32×4

21. -24×3

22. $19 \cdot (-7)$

23. $6(-17)$

24. $-8(-26)$

25. $-4(-35)$

26. $-5 \cdot (23)$

27. $-6 \cdot (38)$

28. $9(-27)$

29. $8(-40)$

30. $-7(-34)$

31. $-4(39)$

32. $4 \cdot (-8) \cdot 3$

33. $5 \times 7 \times (-2)$

34. $8 \cdot (-6) \cdot (-1)$

35. $(-9)(-9)(2)$

36. $-8(-7)(-4)$

37. $-5(8)(-3)$

38. $(-6)(5)(7)$

39. $-1(4)(-9)$

40. $6(-3)(-2)$

41. $4(-4) \cdot 6(-2)$

42. $-5 \cdot 9(-7) \cdot 3$

43. $-9(4) \cdot 3(1)$

44. $8(8)(-5)(-4)$

45. $(-6) \cdot 7 \cdot (-10)(-5)$

46. $-9(-6)(11)(-2)$

47. What is -5 multiplied by -4?

48. What is 6 multiplied by -5?

49. What is -8 times 6?

50. What is -8 times -7?

51. Find the product of -4, 7, and -5.

52. Find the product of -2, -4, and -7.

For Exercises 53 to 56, state whether the given product will be *positive, negative,* or *zero.*

53. The product of three negative integers

54. The product of two negative integers and one positive integer

55. The product of one negative integer, one positive integer, and zero

56. The product of five positive integers and one negative integer

OBJECTIVE B To divide integers

For Exercises 57 to 60, write the related multiplication problem.

57. $\frac{-36}{-12} = 3$

58. $\frac{28}{-7} = -4$

59. $\frac{-55}{11} = -5$

60. $\frac{-20}{-10} = 2$

For Exercises 61 to 111, divide.

61. $12 \div (-6)$

62. $18 \div (-3)$

63. $(-72) \div (-9)$

64. $(-64) \div (-8)$

65. $0 \div (-6)$

66. $-49 \div 7$

67. $45 \div (-5)$

68. $-24 \div 4$

69. $-36 \div 4$

70. $-56 \div 7$

71. $-81 \div (-9)$

72. $-40 \div (-5)$

73. $72 \div (-3)$

74. $44 \div (-4)$

75. $-60 \div 5$

76. $-66 \div 6$

77. $-93 \div (-3)$

78. $-98 \div (-7)$

79. $(-85) \div (-5)$

80. $(-60) \div (-4)$

81. $120 \div 8$

82. $144 \div 9$

83. $78 \div (-6)$

84. $84 \div (-7)$

85. $-72 \div 4$

86. $-80 \div 5$

87. $-114 \div (-6)$

88. $-91 \div (-7)$

89. $-104 \div (-8)$

90. $-126 \div (-9)$

91. $57 \div (-3)$

92. $162 \div (-9)$

93. $-136 \div (-8)$

94. $-128 \div 4$

95. $-130 \div (-5)$

96. $(-280) \div 8$

97. $(-92) \div (-4)$

98. $-196 \div (-7)$

99. $-150 \div (-6)$

100. $(-261) \div 9$

101. $204 \div (-6)$

102. $165 \div (-5)$

103. $-132 \div (-12)$

104. $-156 \div (-13)$

105. $-182 \div 14$

106. $-144 \div 12$

107. $143 \div 11$

108. $168 \div 14$

109. $-180 \div (-15)$

110. $-169 \div (-13)$

111. $154 \div (-11)$

112. Find the quotient of -132 and -11.

113. Find the quotient of 182 and -13.

114. What is -60 divided by -15?

115. What is 144 divided by -24?

116. Find the quotient of -135 and 15.

117. Find the quotient of -88 and 22.

For Exercises 118 to 121, determine whether the statement is *always true, never true,* or *sometimes true.*

118. The quotient of a negative integer and its absolute value is –1.

119. The quotient of zero and a positive integer is a positive integer.

120. A negative integer divided by zero is zero.

121. The quotient of two negative numbers is the same as the quotient of the absolute values of the two numbers.

OBJECTIVE C To solve application problems

122. **Meteorology** The daily low temperatures during one week were recorded as follows: 4°F, -5°F, 8°F, -1°F, -12°F, -14°F, -8°F. Find the average daily low temperature for the week.

123. **Meteorology** The daily high temperatures during one week were recorded as follows: -6°F, -11°F, 1°F, 5°F, -3°F, -9°F, -5°F. Find the average daily high temperature for the week.

124. True or false? If five temperatures are all below 0°C, then the average of the five temperatures is also below 0°C.

125. True or false? If the average of 10 temperatures is below 0°C, then all 10 temperatures are below 0°C.

126. **Chemistry** The graph at the right shows the boiling points of three chemical elements. The boiling point of neon is seven times the highest boiling point shown in the graph.

a. Without calculating the boiling point, determine whether the boiling point of neon is above 0°C or below 0°C.

b. What is the boiling point of neon?

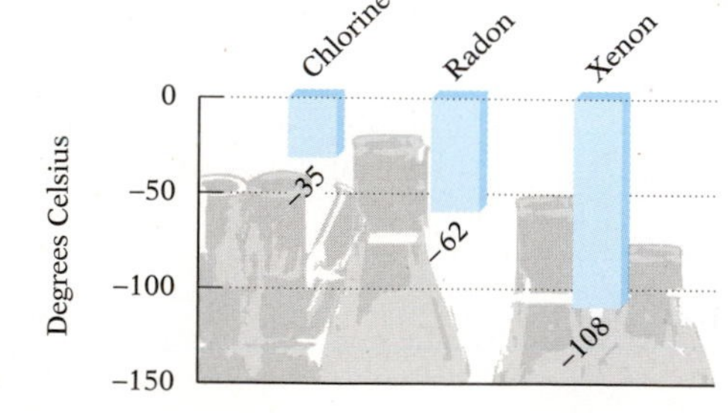

127. **Weight Loss** The patients enrolled in a clinical trial for a new weight-loss drug had a combined weight loss of -24 lb during the first month. There were eight patients in the clinical trial. What was the average weight loss per patient?

128. **Weight Loss** During the second month of the clinical trial (see Exercise 127), the patients lost twice as much weight as they had the first month. What was their combined weight loss the second month?

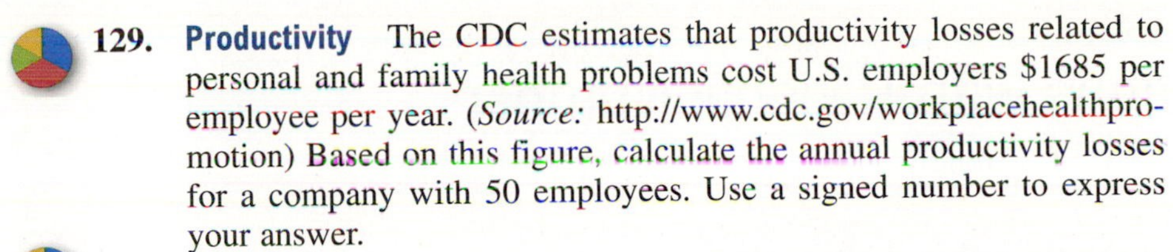

129. Productivity The CDC estimates that productivity losses related to personal and family health problems cost U.S. employers $1685 per employee per year. (*Source:* http://www.cdc.gov/workplacehealthpromotion) Based on this figure, calculate the annual productivity losses for a company with 50 employees. Use a signed number to express your answer.

130. Economics A nation's balance of trade is the difference between its exports and its imports. If the exports are greater than the imports, the result is a positive number and a *favorable balance of trade.* If the exports are less than the imports, the result is a negative number and an *unfavorable balance of trade.* The table at the right shows the U.S. unfavorable balance of trade, in billions of dollars, for each of the first six months of 2010. Find the average monthly balance of trade for January through June.

U.S. Balance of Trade in 2010 (in billions of dollar)	
January	−34
February	−40
March	−39
April	−41
May	−42
June	−50

Source: U.S. Census Bureau

131. Wind Chill The wind chill factor when the temperature is −20°F and the wind is blowing at 15 mph is five times the wind chill factor when the temperature is 10°F and the wind is blowing at 20 mph. If the wind chill factor at 10°F with a 20-mph wind is −9°F, what is the wind chill factor at −20°F with a 15-mph wind?

132. Education To discourage guessing on a multiple-choice exam, an instructor graded the test by giving 5 points for a correct answer, −2 points for an answer left blank, and −5 points for an incorrect answer. How many points did a student score who answered 20 questions correctly, answered 5 questions incorrectly, and left 2 questions blank?

Applying the Concepts

133. **a. Number Problem** Find the greatest possible product of two negative integers whose sum is −10.

b. Number Problem Find the least possible sum of two negative integers whose product is 16.

134. Use repeated addition to show that the product of two integers with different signs is a negative number.

135. Determine whether the statement is *true* or *false.*

a. The product of a nonzero number and its opposite is negative.

b. The square of a negative number is a positive number.

136. In your own words, describe the rules for multiplying and dividing integers.

SECTION

10.4 Operations with Rational Numbers

OBJECTIVE A To add or subtract rational numbers

In this section, operations with rational numbers are discussed. A **rational number** is the quotient of two integers.

Rational Numbers

A rational number is a number that can be written in the form $\frac{a}{b}$, where a and b are integers and $b \neq 0$.

Each of the three numbers shown at the right is a rational number.

$$\frac{3}{4} \qquad \frac{-2}{9} \qquad \frac{13}{-5}$$

An integer can be written as the quotient of the integer and 1. Therefore, **every integer is a rational number.**

$$6 = \frac{6}{1} \qquad -8 = \frac{-8}{1}$$

A mixed number can be written as the quotient of two integers. Therefore, **every mixed number is a rational number.**

$$1\frac{4}{7} = \frac{11}{7} \qquad 3\frac{2}{5} = \frac{17}{5}$$

Recall that every fraction can be written as a decimal by dividing the numerator of the fraction by the denominator. The result is either a terminating decimal or a repeating decimal.

We can write the fraction $\frac{3}{4}$ as the terminating decimal 0.75.

Take Note

The fraction bar can be read "divided by."

$\frac{3}{4} = 3 \div 4$

$$\begin{array}{r} 0.75 \\ 4\overline{)\,3.00} \\ -2\,8 \\ \hline 20 \\ -20 \\ \hline 0 \end{array}$$

0.75 ⟵ This is a **terminating decimal.**

0 ⟵ The remainder is zero.

We can write the fraction $\frac{2}{3}$ as the repeating decimal $0.\overline{6}$.

Take Note

A **terminating decimal** is a decimal that has a finite number of digits after the decimal point, which means that it comes to an end and does not go on forever.

A **repeating decimal** is a decimal that does not end; it has a repeating pattern of digits after the decimal point.

$$\begin{array}{r} 0.666 \\ 3\overline{)\,2.000} \\ -1\,8 \\ \hline 20 \\ -18 \\ \hline 20 \\ -18 \\ \hline 2 \end{array}$$

$0.666 = 0.\overline{6}$ ⟵ This is a **repeating decimal.** The bar over the digit 6 in $0.\overline{6}$ is used to show that this digit repeats.

2 ⟵ The remainder is never zero.

All terminating and repeating decimals are rational numbers.

To add or subtract rational numbers in fractional form, first find the least common multiple (LCM) of the denominators.

HOW TO • 1 Add: $-\frac{7}{8}+\frac{5}{6}$

$8 = 2 \cdot 2 \cdot 2$
$6 = 2 \cdot 3$
$\text{LCM} = 2 \cdot 2 \cdot 2 \cdot 3 = 24$

• Find the LCM of the denominators.

$-\frac{7}{8}+\frac{5}{6} = -\frac{21}{24}+\frac{20}{24}$

• Rewrite each fraction using the LCM of the denominators as the common denominator.

$= \frac{-21+20}{24}$

• Add the numerators.

$= \frac{-1}{24} = -\frac{1}{24}$

 **Take Note**

In this text, answers that are negative fractions are written with the negative sign in front of the fraction.

HOW TO • 2 Subtract: $-\frac{7}{9}-\frac{5}{12}$

$9 = 3 \cdot 3$
$12 = 2 \cdot 2 \cdot 3$
$\text{LCM} = 2 \cdot 2 \cdot 3 \cdot 3 = 36$

• Find the LCM of the denominators.

$-\frac{7}{9}-\frac{5}{12} = -\frac{28}{36}-\frac{15}{36}$

• Rewrite each fraction using the LCM of the denominators as the common denominator.

$= \frac{-28}{36}+\frac{-15}{36}$

• Rewrite subtraction as addition of the opposite. Rewrite negative fractions with the negative sign in the numerator.

$= \frac{-28+(-15)}{36}$

• Add the numerators.

$= \frac{-43}{36} = -\frac{43}{36} = -1\frac{7}{36}$

To add or subtract rational numbers in decimal form, use the sign rules for adding integers.

HOW TO • 3 Add: $47.034 + (-56.91)$

$$\begin{array}{r} 56.910 \\ -\ 47.034 \\ \hline 9.876 \end{array}$$

• The signs are different. Find the difference between the absolute values of the numbers.

$47.034 + (-56.91)$
$= -9.876$

• Attach the sign of the number with the greater absolute value.

HOW TO • 4 Subtract: $-39.09 - 102.98$

$$-39.09 - 102.98 = -39.09 + (-102.98)$$

- Rewrite subtraction as addition of the opposite.

$$\begin{array}{r} 39.09 \\ +\ 102.98 \\ \hline 142.07 \end{array}$$

- The signs of the addends are the same. Find the sum of the absolute values of the numbers.

$$-39.09 - 102.98 = -142.07$$

- Attach the sign of the addends.

EXAMPLE • 1

Subtract: $\frac{5}{16} - \frac{7}{40}$

Solution

$$\frac{5}{16} - \frac{7}{40} = \frac{25}{80} - \frac{14}{80}$$

- The LCM of 16 and 40 is 80.

$$= \frac{25}{80} + \frac{-14}{80}$$

- Rewrite as addition of the opposite.

$$= \frac{25 + (-14)}{80} = \frac{11}{80}$$

YOU TRY IT • 1

Subtract: $\frac{5}{9} - \frac{11}{12}$

Your solution

EXAMPLE • 2

Simplify: $-\frac{3}{4} + \frac{1}{6} - \frac{5}{8}$

Solution

$$-\frac{3}{4} + \frac{1}{6} - \frac{5}{8} = -\frac{18}{24} + \frac{4}{24} - \frac{15}{24}$$

- The LCM of 4, 6, and 8 is 24.

$$= \frac{-18}{24} + \frac{4}{24} + \frac{-15}{24}$$

$$= \frac{-18 + 4 + (-15)}{24}$$

$$= \frac{-29}{24} = -\frac{29}{24} = -1\frac{5}{24}$$

YOU TRY IT • 2

Simplify: $-\frac{7}{8} - \frac{5}{6} + \frac{2}{3}$

Your solution

EXAMPLE • 3

Subtract: $42.987 - 98.61$

Solution

$$42.987 - 98.61 = 42.987 + (-98.61)$$

$$\begin{array}{r} 98.610 \\ -\ 42.987 \\ \hline 55.623 \end{array}$$

$$42.987 - 98.61 = -55.623$$

YOU TRY IT • 3

Subtract: $16.127 - 67.91$

Your solution

Solutions on p. S24

EXAMPLE • 4

Simplify: $1.02 + (-3.6) + 9.24$

Solution

$$1.02 + (-3.6) + 9.24 = -2.58 + 9.24$$
$$= 6.66$$

YOU TRY IT • 4

Simplify: $2.7 + (-9.44) + 6.2$

Your solution

Solution on p. S24

OBJECTIVE B To multiply or divide rational numbers

The product of two rational numbers written as fractions is the product of the numerators over the product of the denominators. Use the sign rules for multiplying integers.

HOW TO • 5 Simplify: $-\frac{3}{8} \times \frac{12}{17}$

$$-\frac{3}{8} \times \frac{12}{17} = -\left(\frac{3 \cdot 12}{8 \cdot 17}\right) = -\frac{9}{34}$$

• The signs are different. The product is negative.

To divide rational numbers written as fractions, invert the divisor and then multiply. Use the sign rules for dividing integers.

HOW TO • 6 Simplify: $-\frac{3}{10} \div \left(-\frac{18}{25}\right)$

$$-\frac{3}{10} \div \left(-\frac{18}{25}\right) = \frac{3}{10} \times \frac{25}{18} = \frac{3 \cdot 25}{10 \cdot 18} = \frac{5}{12}$$

• The signs are the same. The quotient is positive.

To multiply or divide rational numbers written in decimal form, use the sign rules for integers.

HOW TO • 7 Simplify: $(-6.89) \times (-0.00035)$

• The signs are the same. Multiply the absolute values.

$$\begin{array}{rl} 6.89 & \text{2 decimal places} \\ \times\ 0.00035 & \text{5 decimal places} \\ \hline 3445 & \\ 2067 & \\ \hline 0.0024115 & \text{7 decimal places} \end{array}$$

$(-6.89) \times (-0.00035) = 0.0024115$

• The product is positive.

HOW TO • 8 Divide $1.32 \div (-0.27)$. Round to the nearest tenth.

$$\begin{array}{r} 4.88 \approx 4.9 \\ 0.27.\overline{)1.32.00} \\ -1\ 08 \\ \hline 24\ 0 \\ -21\ 6 \\ \hline 2\ 40 \\ -2\ 16 \\ \hline 24 \end{array}$$

• Divide the absolute values. Move the decimal point two places in the divisor and then in the dividend. Place the decimal point in the quotient.

$1.32 \div (-0.27) \approx -4.9$

• The signs are different. The quotient is negative.

EXAMPLE • 5

Multiply: $-\frac{7}{12} \times \frac{9}{14}$

Solution

The product is negative.

$$-\frac{7}{12} \times \frac{9}{14} = -\left(\frac{7 \cdot 9}{12 \cdot 14}\right)$$
$$= -\frac{3}{8}$$

YOU TRY IT • 5

Multiply: $\left(-\frac{2}{3}\right)\left(-\frac{9}{10}\right)$

Your solution

EXAMPLE • 6

Divide: $-\frac{3}{8} \div \left(-\frac{5}{12}\right)$

Solution

The quotient is positive.

$$-\frac{3}{8} \div \left(-\frac{5}{12}\right) = \frac{3}{8} \times \frac{12}{5}$$
$$= \frac{3 \cdot 12}{8 \cdot 5}$$
$$= \frac{9}{10}$$

YOU TRY IT • 6

Divide: $-\frac{5}{8} \div \frac{5}{40}$

Your solution

EXAMPLE • 7

Multiply: -4.29×8.2

Solution

The product is negative.

$$\begin{array}{r} 4.29 \\ \times\ 8.2 \\ \hline 858 \\ 3432 \\ \hline 35.178 \end{array}$$

$-4.29 \times 8.2 = -35.178$

YOU TRY IT • 7

Multiply: -5.44×3.8

Your solution

Solutions on p. S24

EXAMPLE • 8

Multiply: $-3.2 \times (-0.4) \times 6.9$

Solution

$-3.2 \times (-0.4) \times 6.9$
$= 1.28 \times 6.9$
$= 8.832$

YOU TRY IT • 8

Multiply: $3.44 \times (-1.7) \times 0.6$

Your solution

EXAMPLE • 9

Divide: $-0.0792 \div (-0.42)$
Round to the nearest hundredth.

Solution

$$\begin{array}{r} 0.188 \approx 0.19 \\ 0.42.\overline{)0.07.920} \\ \underline{-4\ 2} \\ 3\ 72 \\ \underline{-3\ 36} \\ 360 \\ \underline{-336} \\ 24 \end{array}$$

$-0.0792 \div (-0.42) \approx 0.19$

YOU TRY IT • 9

Divide: $-0.394 \div 1.7$
Round to the nearest hundredth.

Your solution

Solutions on pp. S24–S25

OBJECTIVE C To solve application problems

EXAMPLE • 10

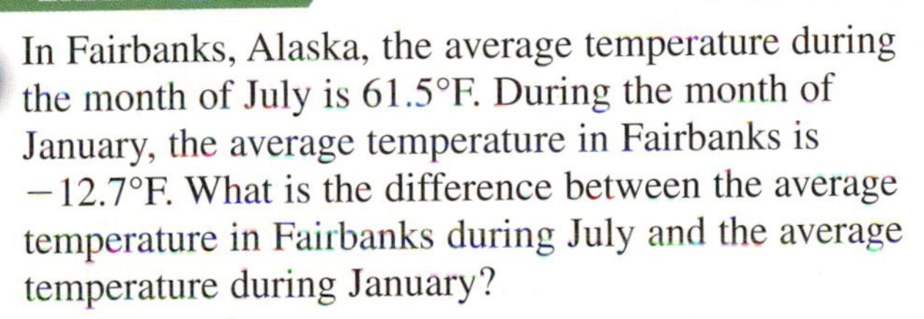

In Fairbanks, Alaska, the average temperature during the month of July is 61.5°F. During the month of January, the average temperature in Fairbanks is −12.7°F. What is the difference between the average temperature in Fairbanks during July and the average temperature during January?

Strategy

To find the difference, subtract the average temperature in January (−12.7°F) from the average temperature in July (61.5°F).

Solution

$61.5 - (-12.7) = 61.5 + 12.7 = 74.2$

The difference between the average temperature during July and the average temperature during January in Fairbanks is 74.2°F.

YOU TRY IT • 10

On January 10, 1911, in Rapid City, South Dakota, the temperature fell from 12.78°C at 7:00 A.M. to −13.33°C at 7:15 A.M. How many degrees did the temperature fall during the 15-minute period?

Your strategy

Your solution

Solution on p. S25

10.4 EXERCISES

OBJECTIVE A To add or subtract rational numbers

For Exercises 1 to 43, simplify.

1. $\frac{5}{8} - \frac{5}{6}$

2. $\frac{1}{9} - \frac{5}{27}$

3. $-\frac{5}{12} - \frac{3}{8}$

4. $-\frac{5}{6} - \frac{5}{9}$

5. $-\frac{6}{13} + \frac{17}{26}$

6. $-\frac{7}{12} + \frac{5}{8}$

7. $-\frac{5}{8} - \left(-\frac{11}{12}\right)$

8. $-\frac{7}{12} - \left(-\frac{7}{8}\right)$

9. $\frac{5}{12} - \frac{11}{15}$

10. $\frac{2}{5} - \frac{14}{15}$

11. $-\frac{3}{4} - \frac{5}{8}$

12. $-\frac{2}{3} - \frac{5}{8}$

13. $-\frac{5}{2} - \left(-\frac{13}{4}\right)$

14. $-\frac{7}{3} - \left(-\frac{3}{2}\right)$

15. $-\frac{3}{8} - \frac{5}{12} - \frac{3}{16}$

16. $-\frac{5}{16} + \frac{3}{4} - \frac{7}{8}$

17. $\frac{1}{2} - \frac{3}{8} - \left(-\frac{1}{4}\right)$

18. $\frac{3}{4} - \left(-\frac{7}{12}\right) - \frac{7}{8}$

19. $\frac{1}{3} - \frac{1}{4} - \frac{1}{5}$

20. $\frac{5}{16} + \frac{1}{8} - \frac{1}{2}$

21. $\frac{1}{2} + \left(-\frac{3}{8}\right) + \frac{5}{12}$

22. $-\frac{3}{8} + \frac{3}{4} - \left(-\frac{3}{16}\right)$

23. $3.4 + (-6.8)$

24. $-4.9 + 3.27$

25. $-8.32 + (-0.57)$

26. $-3.5 + 7$

27. $-4.8 + (-3.2)$

28. $6.2 + (-4.29)$

29. $-4.6 + 3.92$

30. $7.2 + (-8.42)$

31. $-45.71 + (-135.8)$

32. $-35.274 + 12.47$

33. $4.2 + (-6.8) + 5.3$

34. $6.7 + 3.2 + (-10.5)$

35. $-4.5 + 3.2 + (-19.4)$

36. $2.09 - 6.72 - 5.4$

37. $-18.39 + 4.9 - 23.7$

38. $19 - (-3.72) - 82.75$

39. $-3.09 - 4.6 - 27.3$

40. $-3.89 + (-2.9) + 4.723 + 0.2$

41. $-4.02 + 6.809 - (-3.57) - (-0.419)$

42. $0.0153 + (-1.0294) + (-1.0726)$

43. $0.27 + (-3.5) - (-0.27) + (-5.44)$

For Exercises 44 and 45, state whether the given sum or difference will be *positive* or *negative*.

44. A negative mixed number subtracted from a negative proper fraction

45. A positive improper fraction subtracted from a positive proper fraction

OBJECTIVE B To multiply or divide rational numbers

For Exercises 46 to 87, simplify.

46. $\frac{1}{2} \times \left(-\frac{3}{4}\right)$

47. $-\frac{2}{9} \times \left(-\frac{3}{14}\right)$

48. $\left(-\frac{3}{8}\right)\left(-\frac{4}{15}\right)$

49. $\left(-\frac{3}{4}\right)\left(-\frac{8}{27}\right)$

50. $-\frac{1}{2} \times \frac{8}{9}$

51. $\frac{5}{12} \times \left(-\frac{8}{15}\right)$

52. $\left(-\frac{5}{12}\right)\left(\frac{42}{65}\right)$

53. $\left(\frac{3}{8}\right)\left(-\frac{15}{41}\right)$

54. $\left(-\frac{15}{8}\right)\left(-\frac{16}{3}\right)$

55. $\left(-\frac{5}{7}\right)\left(-\frac{14}{15}\right)$

56. $\frac{5}{8} \times \left(-\frac{7}{12}\right) \times \frac{16}{25}$

57. $\left(\frac{1}{2}\right)\left(-\frac{3}{4}\right)\left(-\frac{5}{8}\right)$

58. $\frac{1}{3} \div \left(-\frac{1}{2}\right)$

59. $-\frac{3}{8} \div \frac{7}{8}$

60. $\left(-\frac{3}{4}\right) \div \left(-\frac{7}{40}\right)$

61. $\frac{5}{6} \div \left(-\frac{3}{4}\right)$

62. $-\frac{5}{12} \div \frac{15}{32}$

63. $-\frac{5}{16} \div \left(-\frac{3}{8}\right)$

64. $\left(-\frac{3}{8}\right) \div \left(-\frac{5}{12}\right)$

65. $\left(-\frac{8}{19}\right) \div \frac{7}{38}$

66. $\left(-\frac{2}{3}\right) \div 4$

67. $-6 \div \frac{4}{9}$

68. $-6.7 \times (-4.2)$

69. $-8.9 \times (-3.5)$

70. -1.6×4.9

71. -14.3×7.9

72. $(-0.78)(-0.15)$

73. $(-1.21)(-0.03)$

74. $(-8.919) \div (-0.9)$

75. $-77.6 \div (-0.8)$

76. $59.01 \div (-0.7)$

77. $(-7.04) \div (-3.2)$

78. $(-84.66) \div 1.7$

79. $-3.312 \div (0.8)$

80. $1.003 \div (-0.59)$

81. $26.22 \div (-6.9)$

82. $(-19.08) \div 0.45$

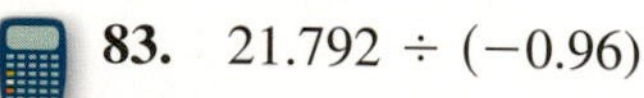

83. $21.792 \div (-0.96)$

84. $(-38.665) \div (-9.5)$

 **85.** $(-3.171) \div (-45.3)$

 **86.** $27.738 \div (-60.3)$

 **87.** $(-13.97) \div (-25.4)$

 For Exercises 88 and 89, use the following information: When -3.54 is divided into a certain dividend, the result is a positive number less than 1. Determine whether each statement is *true* or *false*.

88. The dividend is a positive number.

89. The absolute value of the dividend is greater than 3.54.

OBJECTIVE C To solve application problems

90. **Body Temperature** A child's temperature was 102.6°F during a visit to the pediatrician for an ear infection. Three hours after beginning the antibiotic prescribed by the doctor, the change in the child's temperature was $-1.9°$. What was the child's temperature after beginning the antibiotic?

91. **Body Temperature** Hypothermia begins when a person's core body temperature falls below 95°. A patient in the emergency room has a core body temperature 3.4° below the normal body temperature of 98.6°. Is this patient suffering from hypothermia?

92. **Temperature** The date of the news clipping at the right is July 20, 2007. Find the difference between the record high and low temperatures for Slovakia.

93. If the temperature begins at 4.8°C and ends up below 0°C, is the difference between the starting and ending temperatures less than or greater than 4.8?

94. If the temperature rose 20.3°F during one day and ended up at a high temperature of 15.7°F, did the temperature begin above or below 0°F?

 **95.** **Chemistry** The boiling point of nitrogen is -195.8°C, and the melting point is -209.86°C. Find the difference between the boiling point and the melting point of nitrogen.

 **96.** **Chemistry** The boiling point of oxygen is -182.962°C. Oxygen's melting point is -218.4°C. What is the difference between the boiling point and the melting point of oxygen?

In the News

Slovakia Hits Record High

Slovakia, which became an independent country in 1993 with the peaceful division of Czechoslovakia, hit a record high temperature today of 104.5°F. The record low temperature, set on February 11, 1929, was -41.8°F.

Source: wikipedia.org

DEA/W. Buss/Getty Images

Slovakia

Investments The chart at the right shows the closing price of a share of stock on September 4, 2008, for each of five companies. Also shown is the change in the price from the previous day. To find the closing price on the previous day, subtract the change in price from the closing price on September 4. Use this chart for Exercises 97 and 98.

Company	Closing Price	Change in Price
Del Monte Foods Co.	8.18	−0.14
General Mills, Inc.	66.93	+0.21
Hershey Foods, Inc.	36.42	−0.41
Hormel Food Corp.	35.39	−0.07
Sara Lee Corp.	13.35	−0.07

97. **a.** Find the closing price on the previous day for General Mills.
b. Find the closing price on the previous day for Hormel Foods.

98. **a.** Find the closing price on the previous day for Sara Lee.
b. Find the closing price on the previous day for Hershey Foods.

Astronomy Stars are classified according to their **apparent magnitude,** or how bright they appear to be. The brighter a star appears, the lower the value of its apparent magnitude. A star's **absolute magnitude** is its apparent magnitude if all the stars were an equal distance from Earth. The **distance modulus** of a star is its apparent magnitude minus its absolute magnitude. The smaller the distance modulus, the closer the star is to Earth. Use the table at the right for Exercises 99 to 103.

Star	Apparent Magnitude	Absolute Magnitude
Sun	−26.8	4.83
Sirius	−1.47	1.41
Betelgeuse	0.41	−5.6
Vega	0.04	0.5
Polaris	1.99	−3.2

99. Arrange the stars in the table from brightest to least bright, as measured by absolute magnitude.

100. Find the distance modulus for the sun.

101. Find the distance modulus for Polaris.

102. Find the distance modulus for Sirius.

103. Which of the stars is farthest from Earth?

Applying the Concepts

104. Determine whether the statement is true or false.
a. Every integer is a rational number.
b. Every whole number is an integer.
c. Every integer is a positive number.
d. Every rational number is an integer.

105. **Number Problem** Find a rational number between $-\frac{3}{4}$ and $-\frac{2}{3}$.

106. **Number Problems**
a. Find a rational number between 0.1 and 0.2.
b. Find a rational number between 1 and 1.1.
c. Find a rational number between 0 and 0.005.

107. Given any two different rational numbers, is it always possible to find a rational number between them? If so, explain how. If not, give an example of two different rational numbers for which there is no rational number between them.

SECTION

10.5 Scientific Notation and the Order of Operations Agreement

OBJECTIVE A To write a number in scientific notation

Point of Interest

The first woman mathematician for whom documented evidence exists is Hypatia (370–415). She lived in Alexandria, Egypt, and lectured at the Museum, the forerunner of our modern university. She made important contributions in mathematics, astronomy, and philosophy.

Scientific notation uses negative exponents. Therefore, we will discuss that topic before presenting scientific notation.

Look at the powers of 10 shown at the right. Note the pattern: The exponents are decreasing by 1, and each successive number on the right is one-tenth of the number above it. (100,000 ÷ 10 = 10,000; 10,000 ÷ 10=1000; etc.)

$10^5 = 100{,}000$
$10^4 = 10{,}000$
$10^3 = 1000$
$10^2 = 100$
$10^1 = 10$

If we continue this pattern, the next exponent on 10 is $1 - 1 = 0$, and the number on the right side is $10 \div 10 = 1$.

$10^0 = 1$

The next exponent on 10 is $0 - 1 = -1$, and 10^{-1} is equal to $1 \div 10 = 0.1$.

$10^{-1} = 0.1$

The pattern is continued at the right. Note that a negative exponent does not indicate a negative number. Rather, each power of 10 with a negative exponent is equal to a number between 0 and 1. Also note that as the exponent on 10 decreases, so does the number it is equal to.

$10^{-2} = 0.01$
$10^{-3} = 0.001$
$10^{-4} = 0.0001$
$10^{-5} = 0.00001$
$10^{-6} = 0.000001$

Very large and very small numbers are encountered in the natural sciences. For example, the mass of an electron is 0.000000000000000000000000000000911 kg. Numbers such as this are difficult to read, so a more convenient system called **scientific notation** is used. In scientific notation, a number is expressed as the product of two factors, one a number between 1 and 10, and the other a power of 10.

To express a number in scientific notation, write it in the form $a \times 10^n$, where a is a number between 1 and 10 and n is an integer.

For numbers greater than 10, move the decimal point to the right of the first digit. The exponent n is positive and equal to the number of places the decimal point has been moved.

$240{,}000 = 2.4 \times 10^5$

$93{,}000{,}000 = 9.3 \times 10^7$

 **Take Note**

There are two steps in writing a number in scientific notation: (1) Determine the number between 1 and 10, and (2) determine the exponent on 10.

For numbers less than 1, move the decimal point to the right of the first nonzero digit. The exponent n is negative. The absolute value of the exponent is equal to the number of places the decimal point has been moved.

$0.0003 = 3 \times 10^{-4}$

$0.0000832 = 8.32 \times 10^{-5}$

Changing a number written in scientific notation to decimal notation also requires moving the decimal point.

When the exponent on 10 is positive, move the decimal point to the right the same number of places as the exponent.

$3.45 \times 10^{9} = 3{,}450{,}000{,}000$

$2.3 \times 10^{8} = 230{,}000{,}000$

When the exponent on 10 is negative, move the decimal point to the left the same number of places as the absolute value of the exponent.

$8.1 \times 10^{-3} = 0.0081$

$6.34 \times 10^{-6} = 0.00000634$

EXAMPLE 1

Write 824,300,000,000 in scientific notation.

Solution

The number is greater than 10. Move the decimal point 11 places to the left. The exponent on 10 is 11.

$824{,}300{,}000{,}000 = 8.243 \times 10^{11}$

YOU TRY IT 1

Write 0.000000961 in scientific notation.

Your solution

EXAMPLE 2

Write 6.8×10^{-10} in decimal notation.

Solution

The exponent on 10 is negative. Move the decimal point 10 places to the left.

$6.8 \times 10^{-10} = 0.00000000068$

YOU TRY IT 2

Write 7.329×10^{6} in decimal notation.

Your solution

Solutions on p. S25

OBJECTIVE B To use the Order of Operations Agreement to simplify expressions

The Order of Operations Agreement has been used throughout this book. In simplifying expressions with rational numbers, the same Order of Operations Agreement is used. This agreement is restated here.

The Order of Operations Agreement

Step 1 Do all operations inside parentheses.

Step 2 Simplify any number expressions containing exponents.

Step 3 Do multiplication and division as they occur from left to right.

Step 4 Rewrite subtraction as addition of the opposite. Then do additions as they occur from left to right.

Take Note

In $(-3)^2$, we are squaring -3; we multiply -3 times -3. In -3^2, we are finding the opposite of 3^2. The expression -3^2 is the same as $-(3^2)$.

Exponents may be confusing in expressions with signed numbers.

$(-3)^2 = (-3) \times (-3) = 9$

$-3^2 = -(3)^2 = -(3 \times 3) = -9$ Note that -3 is squared only when the negative sign is *inside* the parentheses.

HOW TO • 1 Simplify: $(-3)^2 - 2 \times (8 - 3) + (-5)$

$(-3)^2 - 2 \times (8 - 3) + (-5)$

$(-3)^2 - 2 \times 5 + (-5)$ 1. Perform operations inside parentheses.

$9 - 2 \times 5 + (-5)$ 2. Simplify expressions with exponents.

$9 - 10 + (-5)$ 3. Do multiplications and divisions as they occur from left to right.

$9 + (-10) + (-5)$ 4. Rewrite subtraction as the addition of the opposite. Then add from left to right.

$(-1) + (-5)$

-6

Integrating Technology

As shown above, the value of -3^2 is different from the value of $(-3)^2$. The keystrokes to evaluate each of these expressions on your calculator are different.

To evaluate -3^2, enter

3 **x²** **+/−**

To evaluate $(-3)^2$, enter

3 **+/−** **x²**

HOW TO • 2 Simplify: $\left(\frac{1}{4} - \frac{1}{2}\right)^2 \div \frac{3}{8}$

$\left(\frac{1}{4} - \frac{1}{2}\right)^2 \div \frac{3}{8}$

$\left(-\frac{1}{4}\right)^2 \div \frac{3}{8}$ 1. Perform operations inside parentheses.

$\frac{1}{16} \div \frac{3}{8}$ 2. Simplify expressions with exponents.

$\frac{1}{16} \times \frac{8}{3}$ 3. Do multiplication and division as they occur from left to right.

$\frac{1}{6}$

EXAMPLE • 3

Simplify: $8 - 4 \div (-2)$

Solution

$8 - 4 \div (-2)$

$= 8 - (-2)$ • Do the division.

$= 8 + 2$ • Rewrite as addition. Add.

$= 10$

YOU TRY IT • 3

Simplify: $9 - 9 \div (-3)$

Your solution

Solution on p. S25

EXAMPLE • 4

Simplify: $12 \div (-2)^2 + 5$

Solution

$12 \div (-2)^2 + 5$

$= 12 \div 4 + 5$ • Exponents

$= 3 + 5$ • Division

$= 8$ • Addition

YOU TRY IT • 4

Simplify: $8 \div 4 \cdot 4 - (-2)^2$

Your solution

EXAMPLE • 5

Simplify: $12 - (-10) \div (8 - 3)$

Solution

$12 - (-10) \div (8 - 3)$

$= 12 - (-10) \div 5$

$= 12 - (-2)$

$= 12 + 2$

$= 14$

YOU TRY IT • 5

Simplify: $8 - (-15) \div (2 - 7)$

Your solution

EXAMPLE • 6

Simplify: $(-3)^2 \times (5 - 7)^2 - (-9) \div 3$

Solution

$(-3)^2 \times (5 - 7)^2 - (-9) \div 3$

$= (-3)^2 \times (-2)^2 - (-9) \div 3$

$= 9 \times 4 - (-9) \div 3$

$= 36 - (-9) \div 3$

$= 36 - (-3)$

$= 36 + 3$

$= 39$

YOU TRY IT • 6

Simplify: $(-2)^2 \times (3 - 7)^2 - (-16) \div (-4)$

Your solution

EXAMPLE • 7

Simplify: $3 \div \left(\frac{1}{2} - \frac{1}{4}\right) - 3$

Solution

$3 \div \left(\frac{1}{2} - \frac{1}{4}\right) - 3$

$= 3 \div \frac{1}{4} - 3$

$= 3 \times \frac{4}{1} - 3$

$= 12 - 3$

$= 12 + (-3)$

$= 9$

YOU TRY IT • 7

Simplify: $7 \div \left(\frac{1}{7} - \frac{3}{14}\right) - 9$

Your solution

Solutions on p. S25

10.5 EXERCISES

OBJECTIVE A To write a number in scientific notation

1. In order to write a certain positive number in scientific notation, you move the decimal point to the right. Is the number greater than 1 or less than 1?

2. In order to write a certain positive number in scientific notation, you move the decimal point to the left. Is the exponent on 10 greater than zero or less than zero?

For Exercises 3 to 14, write the number in scientific notation.

3. 2,370,000 **4.** 75,000 **5.** 0.00045 **6.** 0.000076

7. 309,000 **8.** 819,000,000 **9.** 0.000000601 **10.** 0.00000000096

11. 57,000,000,000 **12.** 934,800,000,000 **13.** 0.000000017 **14.** 0.0000009217

For Exercises 15 to 26, write the number in decimal notation.

15. 7.1×10^5 **16.** 2.3×10^7 **17.** 4.3×10^{-5} **18.** 9.21×10^{-7}

19. 6.71×10^8 **20.** 5.75×10^9 **21.** 7.13×10^{-6} **22.** 3.54×10^{-8}

23. 5×10^{12} **24.** 1.0987×10^{11} **25.** 8.01×10^{-3} **26.** 4.0162×10^{-9}

27. Physics Light travels 16,000,000,000 mi in 1 day. Write this number in scientific notation.

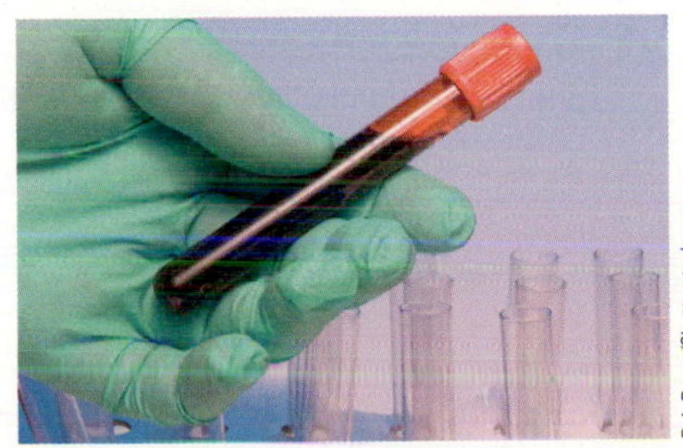

28. Blood Cell Counts The normal range for a red blood cell count is 3.6×10^6 cells/μl to 6.0×10^6 cells/μl. Write these numbers in decimal notation.

29. Blood Cell Counts The normal range for a white blood cell count is 4000 cells/μl to 11,000 cells/μl. Write these numbers in scientific notation.

30. **Blood Cell Counts** A patient with acute myeloid leukemia has a white blood cell count of 1.5×10^5 cells/μl. Write this number in decimal notation.

31. **Blood Cell Counts** If a patient's red blood cell count drops below the normal range of 3.6×10^6 cells/μl to 6.0×10^6 cells/μl, the patient is suffering from anemia. The red blood cell count of a patient is 2,700,000 cells/μl. Write this cell count in scientific notation. Is this patient anemic?

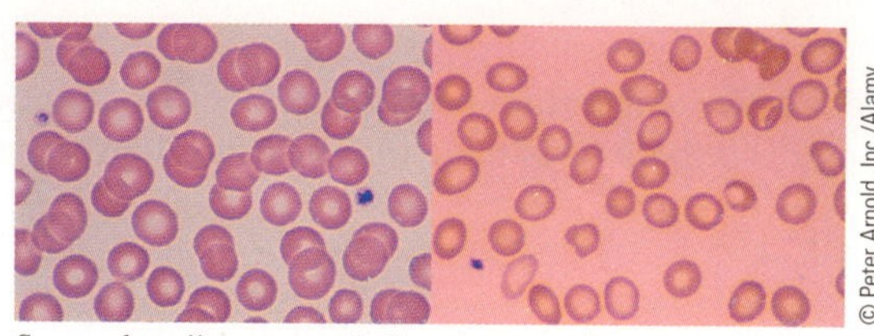

Source: http://www.mesotheliomalungs.org/blood-counts-chemotherapy-treatment-and-mesothelioma

32. **pH Values** The pH scale is used to indicate the acidity of a solution based on the concentration of hydrogen ions in the solution. The scale ranges from 0 to 14, and solutions with low pH numbers are more acidic than those with high pH numbers. Pure water is a neutral solution with a pH of 7.0, indicating that it contains 1×10^{-7} g of hydrogen ions per liter. Write this number in decimal form.

33. **Chemistry** The electric charge on an electron is 0.00000000000000000016 coulomb. Write this number in scientific notation.

34. **Physics** The length of an infrared light wave is approximately 0.0000037 m. Write this number in scientific notation.

35. **Computers** One unit used to measure the speed of a computer is the picosecond. One picosecond is 0.000000000001 of a second. Write this number in scientific notation.

36. **Protons** See the news clipping at the right. A proton is a subatomic particle that has a mass of 1.673×10^{-27} kg. Write this number in decimal notation.

In the News

Getting to Know the Universe

On September 10, 2008, scientists successfully fired the first beam of protons around the Large Hadron Collider, the world's largest particle collider. The scientists hope their research will lead to a greater understanding of the makeup of the universe.

Source: news.yahoo.com

OBJECTIVE B **To use the Order of Operations Agreement to simplify expressions**

For Exercises 37 to 98, simplify.

37. $8 \div 4 + 2$

38. $3 - 12 \div 2$

39. $4 + (-7) + 3$

40. $-16 \div 2 + 8$

41. $4^2 - 4$

42. $6 - 2^2$

43. $2 \times (3 - 5) - 2$

44. $2 - (8 - 10) \div 2$

45. $4 - (-3)^2$

46. $(-2)^2 - 6$

47. $4 - (-3) - 5$

48. $6 + (-8) - (-3)$

49. $4 - (-2)^2 + (-3)$

50. $-3 + (-6)^2 - 1$

51. $3^2 - 4 \times 2$

52. $9 \div 3 - (-3)^2$

53. $3 \times (6 - 2) \div 6$

54. $4 \times (2 - 7) \div 5$

55. $2^2 - (-3)^2 + 2$

56. $3 \times (8 - 5) + 4$

57. $6 - 2 \times (1 - 5)$

58. $4 \times 2 \times (3 - 6)$

59. $(-2)^2 - (-3)^2 + 1$

60. $4^2 - 3^2 - 4$

61. $6 - (-3) \times (-3)^2$

62. $4 - (-5) \times (-2)^2$

63. $4 \times 2 - 3 \times 7$

64. $16 \div 2 - 9 \div 3$

65. $(-2)^2 - 5 \times 3 - 1$

66. $4 - 2 \times 7 - 3^2$

67. $7 \times 6 - 5 \times 6 + 3 \times 2 - 2 + 1$

68. $3 \times 2^2 + 5 \times (3 + 2) - 17$

69. $-4 \times 3 \times (-2) + 12 \times (3 - 4) + (-12)$

70. $3 \times 4^2 - 16 - 4 + 3 - (1 - 2)^2$

71. $-12 \times (6 - 8) + 1^2 \times 3^2 \times 2 - 6 \times 2$

72. $-3 \times (-2)^2 \times 4 \div 8 - (-12)$

73. $10 \times 9 - (8 + 7) \div 5 + 6 - 7 + 8$

74. $-27 - (-3)^2 - 2 - 7 + 6 \times 3$

75. $3^2 \times (4 - 7) \div 9 + 6 - 3 - 4 \times 2$

76. $16 - 4 \times 8 + 4^2 - (-18) \div (-9)$

77. $(-3)^2 \times (5 - 7)^2 - (-9) \div 3$

78. $-2 \times 4^2 - 3 \times (2 - 8) - 3$

79. $4 - 6(2 - 5)^3 \div (17 - 8)$

80. $5 + 7(3 - 8)^2 \div (-14 + 9)$

81. $(1.2)^2 - 4.1 \times 0.3$

82. $2.4 \times (-3) - 2.5$

83. $1.6 - (-1.6)^2$

84. $4.1 \times 8 \div (-4.1)$

85. $(4.1 - 3.9) - 0.7^2$

86. $1.8 \times (-2.3) - 2$

87. $(-0.4)^2 \times 1.5 - 2$

88. $(6.2 - 1.3) \times (-3)$

89. $4.2 - (-3.9) - 6$

90. $-\frac{1}{2} + \frac{3}{8} \div \left(-\frac{3}{4}\right)$

91. $\left(\frac{3}{4}\right)^2 - \frac{3}{8}$

92. $\left(\frac{1}{2}\right)^2 - \left(-\frac{1}{2}\right)^2$

93. $\frac{5}{16} - \frac{3}{8} + \frac{1}{2}$

94. $\frac{2}{7} \div \frac{5}{7} - \frac{3}{14}$

95. $\frac{1}{2} \times \frac{1}{4} \times \frac{1}{2} - \frac{3}{8}$

96. $\frac{2}{3} \times \frac{5}{8} \div \frac{2}{7}$

97. $\frac{1}{2} - \left(\frac{3}{4} - \frac{3}{8}\right) \div \frac{1}{3}$

98. $\frac{3}{8} \div \left(-\frac{1}{2}\right)^2 + 2$

99. Which expression is equivalent to $7 - (-2^2)$?
(i) $7 + 4$ (ii) 9^2 (iii) $7 - 4$ (iv) 7×4

100. Which expression is equivalent to $3 - 5 \times 4 - 7^2$?
(i) $-2 \times (-3)^2$ (ii) $3 - 20 + 49$
(iii) $3 - 20 - 49$ (iv) $-2 \times 4 - 49$

Applying the Concepts

101. Place the correct symbol, < or >, between the two numbers.
a. 3.45×10^{-14} 3.45×10^{-15}
b. 5.23×10^{18} 5.23×10^{17}
c. 3.12×10^{12} 3.12×10^{11}

102. **Astronomy** Light travels 3×10^8 m in 1 s. How far does light travel in 1 year? (Astronomers refer to this distance as 1 light year.)

103. **a.** Evaluate $1^3 + 2^3 + 3^3 + 4^3$.
b. Evaluate $(-1)^3 + (-2)^3 + (-3)^3 + (-4)^3$.
c. Evaluate $1^3 + 2^3 + 3^3 + 4^3 + 5^3$.
d. On the basis of your answers to parts a, b, and c, evaluate $(-1)^3 + (-2)^3 + (-3)^3 + (-4)^3 + (-5)^3$.

104. Evaluate $2^{(3^2)}$ and $(2^3)^2$. Are the answers the same? If not, which is larger?

© Jeffrey Coolidge/Corbis

105. Abdul, Becky, Carl, and Diana were being questioned by their teacher. One of the students had left an apple on the teacher's desk, but the teacher did not know which one. Abdul said it was either Becky or Diana. Diana said it was neither Becky nor Carl. If both those statements are false, who left the apple on the teacher's desk? Explain how you arrived at your solution.

106. In your own words, explain how you know that a number is written in scientific notation.

107. **a.** Express the mass of the sun in kilograms using scientific notation.
b. Express the mass of a neutron in kilograms using scientific notation.

FOCUS ON PROBLEM SOLVING

Drawing Diagrams

How do you best remember something? Do you remember best what you hear? The word *aural* means "pertaining to the ear"; people with a strong aural memory remember best those things that they hear. The word *visual* means "pertaining to the sense of sight"; people with a strong visual memory remember best that which they see written down. Some people claim that their memory is in their writing hand—they remember something only if they write it down! The method by which you best remember something is probably also the method by which you can best learn something new.

In problem-solving situations, try to capitalize on your strengths. If you tend to understand the material better when you hear it spoken, read application problems aloud or have someone else read them to you. If writing helps you to organize ideas, rewrite application problems in your own words.

No matter what your main strength, visualizing a problem can be a valuable aid in problem solving. A drawing, sketch, diagram, or chart can be a useful tool in problem solving, just as calculators and computers are tools. A diagram can be helpful in gaining an understanding of the relationships inherent in a problem-solving situation. A sketch will help you to organize the given information and can lead to your being able to focus on the method by which the solution can be determined.

HOW TO • 1 A tour bus drives 5 mi south, then 4 mi west, then 3 mi north, then 4 mi east. How far is the tour bus from the starting point?

Draw a diagram of the given information.

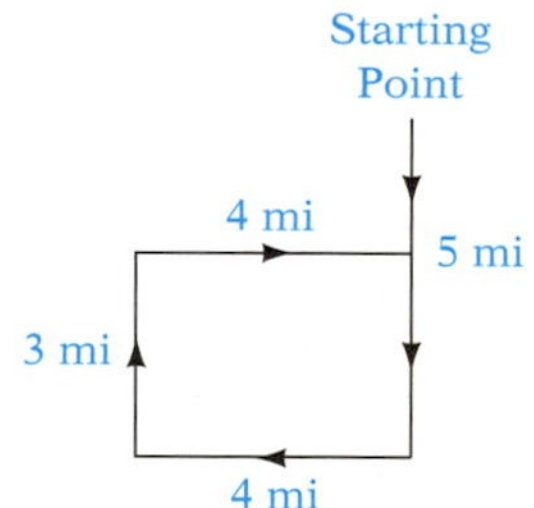

From the diagram, we can see that the solution can be determined by subtracting 3 from 5: $5 - 3 = 2$.

The bus is 2 mi from the starting point.

HOW TO • 2 If you roll two ordinary six-sided dice and multiply the two numbers that appear on top, how many different possible products are there?

Make a chart of the possible products. In the chart below, repeated products are marked with an asterisk.

$1 \cdot 1 = 1$	$2 \cdot 1 = 2$ (*)	$3 \cdot 1 = 3$ (*)	$4 \cdot 1 = 4$ (*)	$5 \cdot 1 = 5$ (*)	$6 \cdot 1 = 6$ (*)
$1 \cdot 2 = 2$	$2 \cdot 2 = 4$ (*)	$3 \cdot 2 = 6$ (*)	$4 \cdot 2 = 8$ (*)	$5 \cdot 2 = 10$ (*)	$6 \cdot 2 = 12$ (*)
$1 \cdot 3 = 3$	$2 \cdot 3 = 6$ (*)	$3 \cdot 3 = 9$	$4 \cdot 3 = 12$ (*)	$5 \cdot 3 = 15$ (*)	$6 \cdot 3 = 18$ (*)
$1 \cdot 4 = 4$	$2 \cdot 4 = 8$	$3 \cdot 4 = 12$ (*)	$4 \cdot 4 = 16$	$5 \cdot 4 = 20$ (*)	$6 \cdot 4 = 24$ (*)
$1 \cdot 5 = 5$	$2 \cdot 5 = 10$	$3 \cdot 5 = 15$	$4 \cdot 5 = 20$	$5 \cdot 5 = 25$	$6 \cdot 5 = 30$ (*)
$1 \cdot 6 = 6$	$2 \cdot 6 = 12$	$3 \cdot 6 = 18$	$4 \cdot 6 = 24$	$5 \cdot 6 = 30$	$6 \cdot 6 = 36$

By counting the products that are not repeats, we can see that there are 18 different possible products.

Look at Sections 10.1 and 10.2. You will notice that number lines are used to help you visualize the integers, as an aid in ordering integers, to help you understand the concepts of opposite and absolute value, and to illustrate addition of integers. As you begin your work with integers, you may find that sketching a number line proves helpful in coming to understand a problem or in working through a calculation that involves integers.

PROJECTS AND GROUP ACTIVITIES

Deductive Reasoning

Suppose that during the last week of your math class, your instructor tells you that if you receive an A on the final exam, you will earn an A in the course. When the final exam grades are posted, you learn that you received an A on the final exam. You can then assume that you will earn an A in the course.

The process used to determine your grade in the math course is called deductive reasoning. **Deductive reasoning** involves drawing a conclusion that is based on given facts. The problems below require deductive reasoning.

1. Given that ΔΔΔ = ◊◊◊◊ and ◊◊◊◊ = ÓÓ, then ΔΔΔΔΔΔ = how many Ós?
2. Given that ‡‡ = ••• and ••• = Λ, then ‡‡‡‡ = how many Λs?
3. Given that ÓÓÓ = ΩΩ and ¤ = ΩΩ, then ¤¤ = how many Ós?
4. Given that ∫∫∫∫∫ = ∂∂ and ∂∂∂∂ = ¥¥¥, then ¥¥¥¥¥¥ = how many ∫s?
5. Given that ÔÔÔÔÔ = □□□ and □□□□□□ = §§§§, then §§§§§§ = how many Ôs?
6. Chris, Dana, Leslie, and Pat are neighbors. Each drives a different type of vehicle: an SUV, a sedan, a sports car, or a minivan. From the following statements, determine which type of vehicle each of the neighbors drives. It may be helpful to use the chart provided below.

   a. Although the vehicle owned by Chris has more mileage on it than does either the sedan or the sports car, it does not have the highest mileage of all four cars.

   b. Pat and the owner of the sports car live on one side of the street, and Leslie and the owner of the SUV live on the other side of the street.

   c. Leslie owns the vehicle with the most mileage on it.

Take Note

To use the chart to solve this problem, write an X in a box to indicate that a possibility has been eliminated. Write a √ to show that a match has been found. When a row or column has three X's, a √ is written in the remaining open box in that row or column of the chart.

	SUV	*Sedan*	*Sports Car*	*Minivan*
Chris				
Dana				
Leslie				
Pat				

7. Four neighbors, Anna, Kay, John, and Nicole, each plant a different vegetable (beans, cucumbers, squash, or tomatoes) in their garden. From the following statements, determine which vegetable each neighbor plants.

a. Nicole's garden is bigger than the one that has tomatoes but smaller than the one that has cucumbers.

b. Anna, who planted the largest garden, didn't plant the beans.

c. The person who planted the beans has a garden the same size as Nicole's.

d. Kay and the person who planted the tomatoes also have flower gardens.

8. The Ontkeans, Kedrovas, McIvers, and Levinsons are neighbors. Each of the four families specializes in a different national cuisine (Chinese, French, Italian, or Mexican). From the following statements, determine which cuisine each family specializes in.

a. The Ontkeans invited the family that specializes in Chinese cuisine and the family that specializes in Mexican cuisine for dinner last night.

b. The McIvers live between the family that specializes in Italian cuisine and the Ontkeans. The Levinsons live between the Kedrovas and the family that specializes in Chinese cuisine.

c. The Kedrovas and the family that specializes in Italian cuisine both subscribe to the same culinary magazine.

CHAPTER 10

SUMMARY

KEY WORDS	EXAMPLES
Positive numbers are numbers greater than zero. *Negative numbers* are numbers less than zero. The *integers* are $\ldots, -4, -3, -2, -1, 0, 1, 2, 3, 4, \ldots$. *Positive integers* are to the right of zero on the number line. *Negative integers* are to the left of zero on the number line. [10.1A, p. 404]	9, 87, and 603 are positive numbers. They are also positive integers. -5, -41, and -729 are negative numbers. They are also negative integers.
Opposite numbers are two numbers that are the same distance from zero on the number line but on opposite sides of zero. [10.1B, p. 405]	8 is the opposite of -8. -2 is the opposite of 2.
The absolute value of a number is its distance from zero on the number line. The absolute value of a number is a positive number or zero. The symbol for absolute value is $\lvert\ \rvert$. [10.1B, p. 405]	$\lvert 9\rvert = 9$ $\lvert -9\rvert = 9$ $-\lvert 9\rvert = -9$
A *rational number* is a number that can be written in the form $\frac{a}{b}$, where a and b are integers and $b \neq 0$. [10.4A, p. 428]	$\frac{3}{7}$, $-\frac{5}{8}$, 9, -2, $4\frac{1}{2}$, 0.6, and $0.\overline{3}$ are rational numbers.

ESSENTIAL RULES AND PROCEDURES	EXAMPLES
Order Relations [10.1A, p. 404] A number that appears to the left of a given number on the number line is less than ($<$) the given number. A number that appears to the right of a given number on the number line is greater than ($>$) the given number.	$-6 > -12$ $-8 < 4$

To add numbers with the same sign, add the absolute values of the numbers. Then attach the sign of the addends. [10.2A, p. 411]	$6 + 4 = 10$ $-6 + (-4) = -10$
To add numbers with different signs, find the difference between the absolute values of the numbers. Then attach the sign of the addend with the greater absolute value. [10.2A, p. 411]	$-6 + 4 = -2$ $6 + (-4) = 2$
To subtract two numbers, add the opposite of the second number to the first number. [10.2B, p. 412]	$6 - 4 = 6 + (-4) = 2$ $6 - (-4) = 6 + 4 = 10$ $-6 - 4 = -6 + (-4) = -10$ $-6 - (-4) = -6 + 4 = -2$
To multiply numbers with the same sign, multiply the absolute values of the factors. The product is positive. [10.3A, p. 419]	$3 \cdot 5 = 15$ $-3(-5) = 15$
To multiply numbers with different signs, multiply the absolute values of the factors. The product is negative. [10.3A, p. 419]	$-3(5) = -15$ $3(-5) = -15$
To divide two numbers with the same sign, divide the absolute values of the numbers. The quotient is positive. [10.3B, p. 420]	$15 \div 3 = 5$ $(-15) \div (-3) = 5$
To divide two numbers with different signs, divide the absolute values of the numbers. The quotient is negative. [10.3B, p. 420]	$-15 \div 3 = -5$ $15 \div (-3) = -5$
Properties of Zero and One in Division [10.3B, p. 421] Zero divided by any number other than zero is zero. Any number other than zero divided by itself is 1. Any number divided by 1 is the number. Division by zero is not defined.	 $0 \div (-5) = 0$ $-5 \div (-5) = 1$ $-5 \div 1 = -5$ $-5 \div 0$ is undefined.
Scientific Notation [10.5A, pp. 439–440] To express a number in scientific notation, write it in the form $a \times 10^n$, where a is a number between 1 and 10 and n is an integer. If the number is greater than 10, the exponent on 10 will be positive. If the number is less than 1, the exponent on 10 will be negative. To change a number written in scientific notation to decimal notation, move the decimal point to the right if the exponent on 10 is positive and to the left if the exponent on 10 is negative. Move the decimal point the same number of places as the absolute value of the exponent on 10.	$367{,}000{,}000 = 3.67 \times 10^8$ $0.0000059 = 5.9 \times 10^{-6}$ $2.418 \times 10^7 = 24{,}180{,}000$ $9.06 \times 10^{-5} = 0.0000906$
The Order of Operations Agreement [10.5B, p. 441] **Step 1** Do all operations inside parentheses. **Step 2** Simplify any numerical expressions containing exponents. **Step 3** Do multiplication and division as they occur from left to right. **Step 4** Rewrite subtraction as addition of the opposite. Then do additions as they occur from left to right.	$(-4)^2 - 3(1 - 5) = (-4)^2 - 3(-4)$ $= 16 - 3(-4)$ $= 16 - (-12)$ $= 16 + 12$ $= 28$

CHAPTER 10

CONCEPT REVIEW

Test your knowledge of the concepts presented in this chapter. Answer each question. Then check your answers against the ones provided in the Answer Section.

1. Find two numbers that are 6 units from 4 on the number line.

2. Find the absolute value of -6.

3. What is the rule for adding two integers?

4. What is the rule for subtracting two integers?

5. What operation is needed to find the change in temperature from -5°C to -14°C?

6. Show the result on the number line: $4 - 9$.

7. If you multiply two nonzero numbers with different signs, what is the sign of the product?

8. If you divide two nonzero numbers with the same sign, what is the sign of the quotient?

9. What is the result when a number is divided by zero?

10. What is a terminating decimal?

11. What are the four steps in the Order of Operations Agreement?

12. How do you write the number 0.000754 in scientific notation?

CHAPTER 10
REVIEW EXERCISES

1. Find the opposite of 22.

2. Subtract: $-8 - (-2) - (-10) - 3$

3. Subtract: $\frac{5}{8} - \frac{5}{6}$

4. Simplify: $-0.33 + 1.98 - 1.44$

5. Multiply: $\left(-\frac{2}{3}\right)\left(\frac{6}{11}\right)\left(-\frac{22}{25}\right)$

6. Multiply: -0.08×16

7. Simplify: $12 - 6 \div 3$

8. Simplify: $\left(\frac{2}{3}\right)^2 - \frac{5}{6}$

9. Find the opposite of -4.

10. Place the correct symbol, < or >, between the two numbers.
$0 \quad -3$

11. Evaluate $-|-6|$.

12. Divide: $-18 \div (-3)$

13. Add: $-\frac{3}{8} + \frac{5}{12} + \frac{2}{3}$

14. Multiply: $\frac{1}{3} \times \left(-\frac{3}{4}\right)$

15. Divide: $-\frac{7}{12} \div \left(-\frac{14}{39}\right)$

16. Simplify: $16 \div 4(8 - 2)$

17. Add: $-22 + 14 + (-18)$

18. Simplify: $3^2 - 9 + 2$

19. Write 0.0000397 in scientific notation.

20. Divide: $-1.464 \div 18.3$

21. Simplify: $-\frac{5}{12} + \frac{7}{9} - \frac{1}{3}$

22. Multiply: $\frac{6}{34} \times \frac{17}{40}$

23. Multiply: $1.2 \times (-0.035)$

24. Simplify: $-\frac{1}{2} + \frac{3}{8} \div \frac{9}{20}$

25. Evaluate $|-5|$.

26. Place the correct symbol, < or >, between the two numbers.
$-2 \quad -40$

27. Find 2 times -13.

28. Simplify: $-0.4 \times 5 - (-3.33)$

29. Add: $\frac{5}{12} + \left(-\frac{2}{3}\right)$

30. Simplify: $-33.4 + 9.8 - (-16.2)$

31. Divide: $\left(-\frac{3}{8}\right) \div \left(-\frac{4}{5}\right)$

32. Write 2.4×10^5 in decimal notation.

33. **Weight Loss** A patient on a diet records monthly changes in weight using integers. Calculate the total change in the patient's weight using these recorded amounts: -6 lb, -4 lb, $+2$ lb, $+1$ lb, and -3 lb.

34. **Education** To discourage guessing on a multiple-choice exam, an instructor graded the test by giving 3 points for a correct answer, -1 point for an answer left blank, and -2 points for an incorrect answer. How many points did a student score who answered 38 questions correctly, answered 4 questions incorrectly, and left 8 questions blank?

35. **Chemistry** The boiling point of mercury is 356.58°C. The melting point of mercury is -38.87°C. Find the difference between the boiling point and the melting point of mercury.

CHAPTER 10

TEST

1. Subtract: $-5 - (-8)$

2. Evaluate $-|-2|$.

3. Add: $-\frac{2}{5} + \frac{7}{15}$

4. Find the product of 0.032 and -1.9.

5. Place the correct symbol, < or >, between the two numbers.
$-8 \quad -10$

6. Add: $1.22 + (-3.1)$

7. Simplify: $4 \times (4 - 7) \div (-2) - 4 \times 8$

8. Multiply: $-5(-6)(3)$

9. What is -1.004 decreased by 3.01?

10. Divide: $-72 \div 8$

11. Find the sum of -2, 3, and -8.

12. Add: $-\frac{3}{8} + \frac{2}{3}$

13. Write 87,600,000,000 in scientific notation.

14. Find the product of -4 and 12.

15. Divide: $\frac{0}{-17}$

16. Subtract: $16 - 4 - (-5) - 7$

17. Find the quotient of $-\frac{2}{3}$ and $\frac{5}{6}$.

18. Place the correct symbol, < or >, between the two numbers.
$0 \quad -4$

19. Add: $16 + (-10) + (-20)$

20. Simplify: $(-2)^2 - (-3)^2 \div (1 - 4)^2 \times 2 - 6$

21. Subtract: $-\frac{2}{5} - \left(-\frac{7}{10}\right)$

22. Write 9.601×10^{-8} in decimal notation.

23. Divide: $-15.64 \div (-4.6)$

24. Find the sum of $-\frac{1}{2}$, $\frac{1}{3}$, and $\frac{1}{4}$.

25. Multiply: $\frac{3}{8} \times \left(-\frac{5}{6}\right) \times \left(-\frac{4}{15}\right)$

26. Subtract: $2.113 - (-1.1)$

27. **Temperatures** Find the temperature after a rise of 11°C from −4°C.

28. **Chemistry** The melting point of radon is −71°C. The melting point of oxygen is three times the melting point of radon. Find the melting point of oxygen.

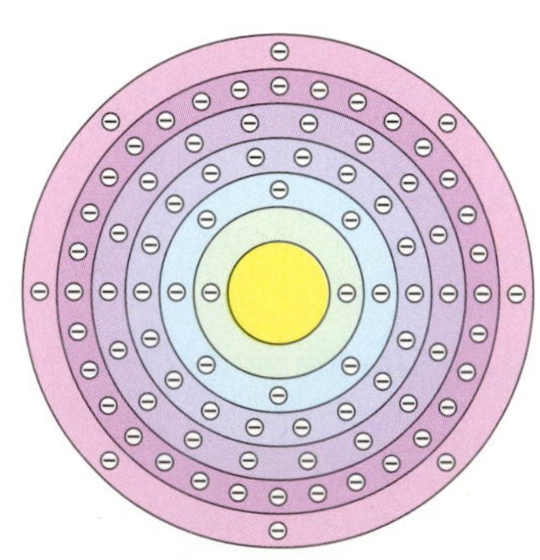

Radon

29. **Body Temperature** A patient's temperature was 101.9°F at noon. Changes recorded during the afternoon temperature checks were +0.5° at 2 P.M. and −1.9° at 4 P.M. What was the patient's temperature at 4 P.M.?

30. **Meteorology** The daily low temperature readings for a 3-day period were as follows: −7°F, 9°F, −8°F. Find the average low temperature for the 3-day period.

CUMULATIVE REVIEW EXERCISES

1. Simplify: $16 - 4 \cdot (3 - 2)^2 \cdot 4$

2. Find the difference between $8\frac{1}{2}$ and $3\frac{4}{7}$.

3. Divide: $3\frac{7}{8} \div 1\frac{1}{2}$

4. Simplify: $\frac{3}{8} \div \left(\frac{3}{8} - \frac{1}{4}\right) \div \frac{7}{3}$

5. Subtract: $2.907 - 1.09761$

6. Solve the proportion $\frac{7}{12} = \frac{n}{32}$. Round to the nearest hundredth.

7. 22 is 160% of what number?

8. Convert: 7 qt = ____ gal ____ qt

9. Convert: 6692 ml = ____ L

10. Convert 4.2 ft to meters. Round to the nearest hundredth. (1 m = 3.28 ft)

11. Find 32% of 180.

12. Convert $3\frac{2}{5}$ to a percent.

13. Add: $-8 + 5$

14. Add: $3\frac{1}{4} + \left(-6\frac{5}{8}\right)$

15. Subtract: $-6\frac{1}{8} - 4\frac{5}{12}$

16. Simplify: $-12 - (-7) - 3(-8)$

17. What is -3.2 times -1.09?

18. Multiply: $-6 \times 7 \times \left(-\frac{3}{4}\right)$

19. Find the quotient of 42 and -6.

20. Divide: $-2\frac{1}{7} \div \left(-3\frac{3}{5}\right)$

21. Simplify: $3 \times (3 - 7) \div 6 - 2$

22. Simplify: $4 - (-2)^2 \div (1 - 2)^2 \times 3 + 4$

23. Measurement As an LPN, you are required to clean and dress the wounds of a patient. To do this, you cut a $3\frac{2}{3}$-inch length of gauze from a 10-inch roll of gauze. How much gauze remains on the roll?

24. Banking Nimisha had a balance of \$763.56 in her checkbook before writing checks for \$135.88 and \$47.81 and making a deposit of \$223.44. Find her new checkbook balance.

25. Consumerism The regular price for a pair of 30–40 mm Hg therapeutic compression knee high hose at a pharmacy is \$35. If the sale price is \$28, what is the discount rate?

26. Measurement A nurse's aide records the amount of liquid consumed by the residents on the second floor of an assisted living facility during each meal. Forty cups of liquid were consumed during supper. How many gallons of liquid were consumed by the residents on the second floor?

27. Markup Suppose MedSupplies, Inc. purchases a full-electric hospital bed for \$550 and sells it for \$797.50. What markup rate does MedSupplies, Inc. use?

28. Family Night The circle graph at the right shows how often American households have a family night, during which they play a game as a family. Use this graph, and the fact that there are 114 million households in the United States, to answer the questions.

a. How many households have a family night once a week?

b. What fraction of U.S. households rarely or never have a family night?

c. Is the number of households that have a family night only once a month more or less than three times the number of households that have a family night once a week?

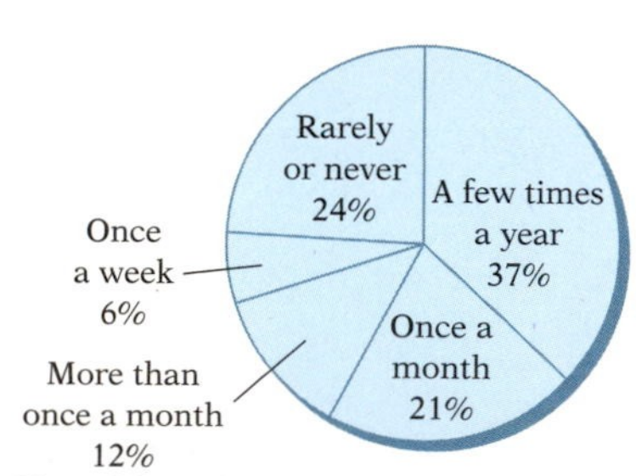

How Often Americans Have a Family Night

29. IV Infusion An IV infusion of 500 ml of a solution is set to infuse at a rate of 150 ml/h. How many hours will it take for this infusion to be completed?

30. Meteorology The daily high temperature readings for a 4-day period were recorded as follows: −19°F, −7°F, 1°F, and 9°F. Find the average high temperature for the 4-day period.

CHAPTER

11

Introduction to Algebra

OBJECTIVES

SECTION 11.1

- **A** To evaluate variable expressions
- **B** To simplify variable expressions containing no parentheses
- **C** To simplify variable expressions containing parentheses

SECTION 11.2

- **A** To determine whether a given number is a solution of an equation
- **B** To solve an equation of the form $x + a = b$
- **C** To solve an equation of the form $ax = b$
- **D** To solve application problems using formulas

SECTION 11.3

- **A** To solve an equation of the form $ax + b = c$
- **B** To solve application problems using formulas

SECTION 11.4

- **A** To solve an equation of the form $ax + b = cx + d$
- **B** To solve an equation containing parentheses

SECTION 11.5

- **A** To translate a verbal expression into a mathematical expression given the variable
- **B** To translate a verbal expression into a mathematical expression by assigning the variable

SECTION 11.6

- **A** To translate a sentence into an equation and solve
- **B** To solve application problems

AIM for the Future

CLINICAL LABORATORY TECHNOLOGISTS perform complex chemical, biological, hematological, immunologic, microscopic, and bacteriological tests. Technologists microscopically examine blood and other body fluids and make cultures of body fluid and tissue samples to determine the presence of bacteria, fungi, parasites, or other microorganisms. Employment of clinical laboratory workers is expected to grow by 14% between 2008 and 2018, faster than the average growth for all occupations. Median annual wages of medical and clinical laboratory technologists were $53,500 in May 2008. (*Source: Bureau of Labor Statistics:* http://www.bls.gov/oco/ocos096.htm)

PREP TEST

Are you ready to succeed in this chapter?

Take the Prep Test below to find out if you are ready to learn the new material.

For Exercises 1 to 9, simplify.

1. $2 - 9$

2. $-5(4)$

3. $-16 + 16$

4. $\dfrac{-7}{-7}$

5. $-\dfrac{3}{8}\left(-\dfrac{8}{3}\right)$

6. $\left(\dfrac{3}{5}\right)^3 \cdot \left(\dfrac{5}{9}\right)^2$

7. $\dfrac{2}{3} + \left(\dfrac{3}{4}\right)^2 \cdot \dfrac{2}{9}$

8. $-8 \div (-2)^2 + 6$

9. $4 + 5(2 - 7)^2 \div (-8 + 3)$

SECTION

11.1 Variable Expressions

OBJECTIVE A To evaluate variable expressions

Point of Interest

There are historical records indicating that mathematics has been studied for at least 4000 years. However, only in the last 400 years have variables been used. Prior to that, mathematics was written in words.

Often we discuss a quantity without knowing its exact value—for example, next year's inflation rate, the price of gasoline next summer, or the interest rate on a new-car loan next fall. In mathematics, a letter of the alphabet is used to stand for a quantity that is unknown or that can change, or *vary*. The letter is called a **variable.** An expression that contains one or more variables is called a **variable expression.**

A company's business manager has determined that the company will make a \$10 profit on each DVD it sells. The manager wants to describe the company's total profit from the sale of DVDs. Because the number of DVDs that the company will sell is unknown, the manager lets the variable n stand for that number. Then the variable expression $10 \cdot n$, or simply $10n$, describes the company's profit from selling n DVDs.

The company's profit from selling n DVDs is $\$10 \cdot n = \$10n$.

If the company sells 12 DVDs, its profit is $\$10 \cdot 12 = \$120$.

If the company sells 75 DVDs, its profit is $\$10 \cdot 75 = \$750$.

Replacing the variable or variables in a variable expression and then simplifying the resulting numerical expression is called **evaluating a variable expression.**

HOW TO 1 Evaluate $3x^2 + xy - z$ when $x = -2$, $y = 3$, and $z = -4$.

$3x^2 + xy - z$

$3(-2)^2 + (-2)(3) - (-4)$

• **Replace each variable in the expression with the number it stands for.**

$= 3 \cdot 4 + (-2)(3) - (-4)$

• **Use the Order of Operations Agreement to simplify the resulting numerical expression.**

$= 12 + (-6) - (-4)$

$= 12 + (-6) + 4$

$= 6 + 4$

$= 10$

The value of the variable expression $3x^2 + xy - z$ when $x = -2$, $y = 3$, and $z = -4$ is 10.

EXAMPLE • 1

Evaluate $3x - 4y$ when $x = -2$ and $y = 3$.

Solution

$3x - 4y$

$3(-2) - 4(3) = -6 - 12$
$= -6 + (-12) = -18$

YOU TRY IT • 1

Evaluate $6a - 5b$ when $a = -3$ and $b = 4$.

Your solution

EXAMPLE • 2

Evaluate $-x^2 - 6 \div y$ when $x = -3$ and $y = 2$.

Solution

$-x^2 - 6 \div y$

$-(-3)^2 - 6 \div 2 = -9 - 6 \div 2$
$= -9 - 3$
$= -9 + (-3) = -12$

YOU TRY IT • 2

Evaluate $-3s^2 - 12 \div t$ when $s = -2$ and $t = 4$.

Your solution

EXAMPLE • 3

Evaluate $-\frac{1}{2}y^2 - \frac{3}{4}z$ when $y = 2$ and $z = -4$.

Solution

$-\frac{1}{2}y^2 - \frac{3}{4}z$

$-\frac{1}{2}(2)^2 - \frac{3}{4}(-4) = -\frac{1}{2} \cdot 4 - \frac{3}{4}(-4)$
$= -2 - (-3)$
$= -2 + 3 = 1$

YOU TRY IT • 3

Evaluate $-\frac{2}{3}m + \frac{3}{4}n^3$ when $m = 6$ and $n = 2$.

Your solution

EXAMPLE • 4

Evaluate $-2ab + b^2 + a^2$ when $a = -\frac{3}{5}$ and $b = \frac{2}{5}$.

Solution

$-2ab + b^2 + a^2$

$-2\left(-\frac{3}{5}\right)\left(\frac{2}{5}\right) + \left(\frac{2}{5}\right)^2 + \left(-\frac{3}{5}\right)^2$

$= -2\left(-\frac{3}{5}\right)\left(\frac{2}{5}\right) + \left(\frac{4}{25}\right) + \left(\frac{9}{25}\right)$

$= \frac{12}{25} + \frac{4}{25} + \frac{9}{25} = \frac{25}{25} = 1$

YOU TRY IT • 4

Evaluate $-3yz - z^2 + y^2$ when $y = -\frac{2}{3}$ and $z = \frac{1}{3}$.

Your solution

Solutions on p. S25

OBJECTIVE B To simplify variable expressions containing no parentheses

The **terms of a variable expression** are the addends of the expression. The variable expression at the right has four terms.

4 terms

$7x^2 + (-6xy) + x + (-8)$

Variable terms — Constant term

Three of the terms are **variable terms:** $7x^2$, $-6xy$, and x.

One of the terms is a **constant term:** -8. A constant term has no variables.

> ✓ **Take Note**
> A variable term is a term composed of a numerical coefficient and a variable part. The numerical coefficient is the number part of the term.

Each variable term is composed of a **numerical coefficient** (the number part of a variable term) and a **variable part** (the variable or variables and their exponents). When the numerical coefficient is 1, the 1 is usually not written. ($1x = x$)

Numerical coefficient

$$7x^2 + (-6xy) + 1x + (-8)$$

Variable part

Like terms of a variable expression are the terms with the same variable part. (Because $y^2 = y \cdot y$, y^2 and y are not like terms.)

Like terms

$$8y^2 + 3y + 9y^2 + 5y$$

Like terms

In variable expressions that contain constant terms, the constant terms are like terms.

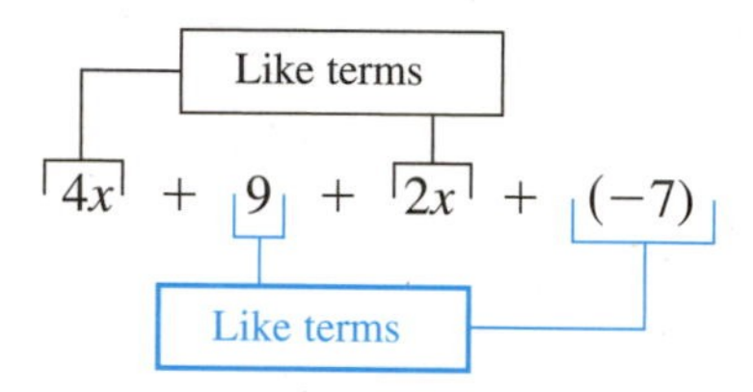

The Commutative and Associative Properties of Addition are used to simplify variable expressions. These properties can be stated in general form using variables.

> ✓ **Take Note**
> With the Commutative Property, the order in which the numbers appear changes. With the Associative Property, the order in which the numbers appear remains the same.

Commutative Property of Addition

If a and b are two numbers, then $a + b = b + a$.

Associative Property of Addition

If a, b, and c are three numbers, then $a + (b + c) = (a + b) + c$.

The phrase **simplifying a variable expression** means *combining like terms* by adding their numerical coefficients. For example, to simplify $2y + 3y$, think

$$2y + 3y = (y + y) + (y + y + y) = 5y$$

HOW TO • 2 Simplify: $8z - 5 + 2z$

$$\begin{aligned} 8z - 5 + 2z &= 8z + 2z - 5 \\ &= 10z - 5 \end{aligned}$$

- Use the Commutative and Associative Properties of Addition to group like terms. Combine the like terms $8z + 2z$.

HOW TO 3 Simplify: $12a - 4b - 8a + 2b$

$$\begin{aligned}12a - 4b - 8a + 2b &= 12a + (-4)b + (-8)a + 2b\\ &= 12a + (-8)a + (-4)b + 2b\\ &= 4a + (-2)b\\ &= 4a - 2b\end{aligned}$$

- Change subtraction to addition of the opposite.
- Use the Commutative and Associative Properties of Addition to group like terms. Combine like terms.
- Recall that $a + (-b) = a - b$.

HOW TO 4 Simplify: $6z^2 + 3 - z^2 - 7$

$$\begin{aligned}6z^2 + 3 - z^2 - 7 &= 6z^2 + 3 + (-1)z^2 + (-7)\\ &= 6z^2 + (-1)z^2 + 3 + (-7)\\ &= 5z^2 + (-4)\\ &= 5z^2 - 4\end{aligned}$$

- Change subtraction to addition of the opposite.
- Use the Commutative and Associative Properties of Addition to group like terms. Combine like terms.

EXAMPLE 5

Simplify: $6xy - 8x + 5x - 9xy$

Solution

$$\begin{aligned}&6xy - 8x + 5x - 9xy\\ &\quad = 6xy + (-8)x + 5x + (-9)xy\\ &\quad = 6xy + (-9)xy + (-8)x + 5x\\ &\quad = (-3)xy + (-3)x\\ &\quad = -3xy - 3x\end{aligned}$$

YOU TRY IT 5

Simplify: $5a^2 - 6b^2 + 7a^2 - 9b^2$

Your solution

EXAMPLE 6

Simplify: $-4z^2 + 8 + 5z^2 - 3$

Solution

$$\begin{aligned}&-4z^2 + 8 + 5z^2 - 3\\ &\quad = -4z^2 + 8 + 5z^2 + (-3)\\ &\quad = -4z^2 + 5z^2 + 8 + (-3)\\ &\quad = z^2 + 5\end{aligned}$$

YOU TRY IT 6

Simplify: $-6x + 7 + 9x - 10$

Your solution

EXAMPLE 7

Simplify: $\frac{1}{4}m^2 - \frac{1}{2}n^2 + \frac{1}{2}m^2$

Solution

$$\begin{aligned}\frac{1}{4}m^2 - \frac{1}{2}n^2 + \frac{1}{2}m^2 &= \frac{1}{4}m^2 + \left(-\frac{1}{2}\right)n^2 + \frac{1}{2}m^2\\ &= \frac{1}{4}m^2 + \frac{1}{2}m^2 + \left(-\frac{1}{2}\right)n^2\\ &= \frac{1}{4}m^2 + \frac{2}{4}m^2 + \left(-\frac{1}{2}\right)n^2\\ &= \frac{3}{4}m^2 + \left(-\frac{1}{2}\right)n^2\\ &= \frac{3}{4}m^2 - \frac{1}{2}n^2\end{aligned}$$

YOU TRY IT 7

Simplify: $\frac{3}{8}w + \frac{1}{2} - \frac{1}{4}w - \frac{2}{3}$

Your solution

Solutions on pp. S25–S26

OBJECTIVE C To simplify variable expressions containing parentheses

The Commutative and Associative Properties of Multiplication and the Distributive Property are used to simplify variable expressions that contain parentheses. These properties can be stated in general form using variables.

Commutative Property of Multiplication

If a and b are two numbers, then $a \cdot b = b \cdot a$.

Associative Property of Multiplication

If a, b, and c are three numbers, then $a \cdot (b \cdot c) = (a \cdot b) \cdot c$.

The Associative and Commutative Properties of Multiplication are used to simplify variable expressions such as the following.

HOW TO • 5 Simplify: $-5(4x)$

$-5(4x) = (-5 \cdot 4)x$ • Use the Associative Property of Multiplication.

$= -20x$

HOW TO • 6 Simplify: $(6y) \cdot 5$

$(6y) \cdot 5 = 5 \cdot (6y)$ • Use the Commutative Property of Multiplication.

$= (5 \cdot 6)y = 30y$ • Use the Associative Property of Multiplication.

The **Distributive Property** is used to remove parentheses from variable expressions that contain both multiplication and addition.

Distributive Property

If a, b, and c are three numbers, then $a(b + c) = ab + ac$.

HOW TO • 7 Simplify: $4(z + 5)$

$4(z + 5) = 4z + 4(5)$ • The Distributive Property is used to rewrite the variable expression without parentheses.

$= 4z + 20$

HOW TO • 8 Simplify: $-3(2x + 7)$

$-3(2x + 7) = -3(2x) + (-3)(7)$ • Use the Distributive Property.

$= -6x + (-21)$

$= -6x - 21$ • Recall that $a + (-b) = a - b$.

The Distributive Property can also be stated in terms of subtraction.

$$a(b - c) = ab - ac$$

HOW TO • 9 Simplify: $8(2r - 3s)$

$$8(2r - 3s) = 8(2r) - 8(3s)$$
$$= 16r - 24s$$

• Use the Distributive Property.

HOW TO • 10 Simplify: $-5(2x - 4y)$

$$-5(2x - 4y) = (-5)(2x) - (-5)(4y)$$
$$= -10x - (-20y)$$
$$= -10x + 20y$$

• Use the Distributive Property.

• Recall that $a - (-b) = a + b$.

HOW TO • 11 Simplify: $12 - 5(m + 2) + 2m$

$$12 - 5(m + 2) + 2m = 12 - 5m + (-5)(2) + 2m$$
$$= 12 - 5m + (-10) + 2m$$

• Use the Distributive Property to simplify the expression $-5(m + 2)$.

$$= -5m + 2m + 12 + (-10)$$

• Use the Commutative and Associative Properties to group like terms.

$$= -3m + 2$$

• Combine like terms by adding their numerical coefficients. Add constant terms.

The answer $-3m + 2$ can also be written as $2 - 3m$. In this text, we will write answers with variable terms first, followed by the constant term.

EXAMPLE • 8

Simplify: $4(x - 3)$

Solution

$$4(x - 3) = 4x - 4(3)$$
$$= 4x - 12$$

YOU TRY IT • 8

Simplify: $5(a - 2)$

Your solution

EXAMPLE • 9

Simplify: $5n - 3(2n - 4)$

Solution

$$5n - 3(2n - 4) = 5n - 3(2n) - (-3)(4)$$
$$= 5n - 6n - (-12)$$
$$= 5n - 6n + 12$$
$$= -n + 12$$

YOU TRY IT • 9

Simplify: $8s - 2(3s - 5)$

Your solution

EXAMPLE • 10

Simplify: $3(c - 2) + 2(c + 6)$

Solution

$$3(c - 2) + 2(c + 6) = 3c - 3(2) + 2c + 2(6)$$
$$= 3c - 6 + 2c + 12$$
$$= 3c + 2c - 6 + 12$$
$$= 5c + 6$$

YOU TRY IT • 10

Simplify: $4(x - 3) - 2(x + 1)$

Your solution

Solutions on p. S26

11.1 EXERCISES

OBJECTIVE A To evaluate variable expressions

For Exercises 1 to 34, evaluate the variable expression when $a = -3$, $b = 6$, and $c = -2$.

1. $5a - 3b$ **2.** $4c - 2b$ **3.** $2a + 3c$ **4.** $2c + 4a$

5. $-c^2$ **6.** $-a^2$ **7.** $b - a^2$ **8.** $b - c^2$

9. $ab - c^2$ **10.** $bc - a^2$ **11.** $2ab - c^2$ **12.** $3bc - a^2$

13. $a - (b \div a)$ **14.** $c - (b \div c)$ **15.** $2ac - (b \div a)$ **16.** $4ac \div (b \div a)$

17. $b^2 - c^2$ **18.** $b^2 - a^2$ **19.** $b^2 \div (ac)$ **20.** $3c^2 \div (ab)$

21. $c^2 - (b \div c)$ **22.** $a^2 - (b \div a)$ **23.** $a^2 + b^2 + c^2$ **24.** $a^2 - b^2 - c^2$

25. $ac + bc + ab$ **26.** $ac - bc - ab$ **27.** $a^2 + b^2 - ab$ **28.** $b^2 + c^2 - bc$

29. $2b - (3c + a^2)$ **30.** $\frac{2}{3}b + \left(\frac{1}{2}c - a\right)$ **31.** $\frac{1}{3}a + \left(\frac{1}{2}b - \frac{2}{3}a\right)$

32. $-\frac{2}{3}b - \left(\frac{1}{2}c + a\right)$ **33.** $\frac{1}{6}b + \frac{1}{3}(c + a)$ **34.** $\frac{1}{2}c + \left(\frac{1}{3}b - a\right)$

For Exercises 35 to 38, evaluate the variable expression when $a = -\frac{1}{2}$, $b = \frac{3}{4}$, and $c = \frac{1}{4}$.

35. $4a + (3b - c)$ **36.** $2b + (c - 3a)$ **37.** $2a - b^2 \div c$ **38.** $b \div (-c) + 2a$

For Exercises 39 to 42, evaluate the variable expression when $a = 3.48$, $b = -2.31$, and $c = -1.74$.

39. $a^2 - b^2$ **40.** $a^2 - b \cdot c$ **41.** $3ac - (a \div c)$ **42.** $2c + (b^2 - c)$

For Exercises 43 to 46, suppose a is positive and b is negative, with $|a|$ greater than $|b|$. Determine whether the value of the variable expression is positive or negative.

43. ab^2 **44.** $b^2 - a^2$ **45.** $a - (a + b)$ **46.** $a^2 + ab$

OBJECTIVE B To simplify variable expressions containing no parentheses

For Exercises 47 to 50, name the terms of the variable expression. Then underline the constant term.

47. $2x^2 + 3x - 4$ **48.** $-4y^2 + 5$ **49.** $3a^2 - 4a + 8$ **50.** $7 - b$

For Exercises 51 to 54, name the variable terms of the expression. Then underline the coefficients of the variable terms.

51. $3x^2 - 4x + 9$ **52.** $-5a^2 + a - 4$ **53.** $y^2 + 6a - 1$ **54.** $8 - c$

55. Which expressions are equivalent to $3 - 2a^2 + 8a^2 - 5$?
(i) $9a^2 - 5$ (ii) $6a^2 - 2$ (iii) $-2 + 6a^2$ (iv) $6a^2 - 8$

56. Which expressions are equivalent to $6x^2 - x + 3x + x^2$?
(i) $6x^2 + 2x + x^2$ (ii) $2x + 7x^2$ (iii) $7x^2 + 2x$ (iv) $6x^2 + x^2 + 3x - x$

For Exercises 57 to 98, simplify.

57. $7z + 9z$ **58.** $6x + 5x$ **59.** $12m - 3m$

60. $5y - 12y$ **61.** $5at + 7at$ **62.** $12mn + 11mn$

63. $-4yt + 7yt$ **64.** $-12yt + 5yt$ **65.** $-3x - 12y$

66. $-12y - 7y$ **67.** $3t^2 - 5t^2$ **68.** $7t^2 + 8t^2$

69. $6c - 5 + 7c$ **70.** $7x - 5 + 3x$ **71.** $2t + 3t - 7t$

72. $9x^2 - 5 - 3x^2$ **73.** $7y^2 - 2 - 4y^2$ **74.** $3w - 7u + 4w$

75. $6w - 8u + 8w$ **76.** $4 - 6xy - 7xy$ **77.** $10 - 11xy - 12xy$

78. $7t^2 - 5t^2 - 4t^2$ **79.** $3v^2 - 6v^2 - 8v^2$ **80.** $5ab - 7a - 10ab$

81. $-10ab - 3a + 2ab$

82. $-4x^2 - x + 2x^2$

83. $-3y^2 - y + 7y^2$

84. $4x^2 - 8y - x^2 + y$

85. $2a - 3b^2 - 5a + b^2$

86. $8y - 4z - y + 2z$

87. $3x^2 - 7x + 4x^2 - x$

88. $5y^2 - y + 6y^2 - 5y$

89. $6s - t - 9s + 7t$

90. $5w - 2v - 9w + 5v$

91. $4m + 8n - 7m + 2n$

92. $z + 9y - 4z + 3y$

93. $-5ab + 7ac + 10ab - 3ac$

94. $-2x^2 - 3x - 11x^2 + 14x$

95. $\frac{4}{9}a^2 - \frac{1}{5}b^2 + \frac{2}{9}a^2 + \frac{4}{5}b^2$

96. $\frac{6}{7}x^2 + \frac{2}{5}x - \frac{3}{7}x^2 - \frac{4}{5}x$

97. $7.81m + 3.42n - 6.25m - 7.19n$

98. $8.34y^2 - 4.21y - 6.07y^2 - 5.39y$

OBJECTIVE C To simplify variable expressions containing parentheses

99. Which expressions are equivalent to $7a + 7b$?

(i) $9 - 2(a + b)$ (ii) $(9 - 2)(a + b)$ (iii) $3(a + b) + 4(a + b)$

100. Which expressions are equivalent to $5 + 3(m + 8)$?

(i) $3m + 29$ (ii) $5 + 3m + 24$ (iii) $5 + 3m + 8$

For Exercises 101 to 136, simplify.

101. $5(x + 4)$

102. $3(m + 6)$

103. $(y - 3)4$

104. $(z - 3)7$

105. $-2(a + 4)$

106. $-5(b + 3)$

107. $3(5x + 10)$

108. $2(4m - 7)$

109. $5(3c - 5)$

110. $-4(w - 3)$

111. $-3(y - 6)$

112. $3m + 4(m + z)$

113. $5x + 2(x + 7)$

114. $6z - 3(z + 4)$

115. $8y - 4(y + 2)$

116. $7w - 2(w - 3)$

117. $9x - 4(x - 6)$

118. $-5m + 3(m + 4)$

119. $-2y + 3(y - 2)$

120. $5m + 3(m + 4) - 6$

121. $4n + 2(n + 1) - 5$

122. $8z - 2(z - 3) + 8$

123. $9y - 3(y - 4) + 8$

124. $6 - 4(a + 4) + 6a$

125. $3x + 2(x + 2) + 5x$

126. $7x + 4(x + 1) + 3x$

127. $-7t + 2(t - 3) - t$

128. $-3y + 2(y - 4) - y$

129. $z - 2(1 - z) - 2z$

130. $2y - 3(2 - y) + 4y$

131. $3(y - 2) - 2(y - 6)$

132. $7(x + 2) + 3(x - 4)$

133. $2(t - 3) + 7(t + 3)$

134. $3(y - 4) - 2(y - 3)$

135. $3t - 6(t - 4) + 8t$

136. $5x + 3(x - 7) - 9x$

Applying the Concepts

137. The square and the rectangle at the right can be used to illustrate algebraic expressions. The illustration below represents the expression $2x + 1$.

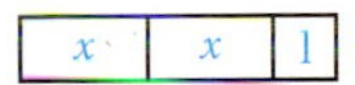

a. Using squares and rectangles in a similar manner, draw a figure that represents the expression $2 + 3x$.

b. Draw a figure that represents the expression $5x$.

c. Does the figure $2 + 3x$ equal the figure $5x$? Explain how this is related to combining like terms.

138. **a.** Using squares and rectangles as in Exercise 137, draw a figure that represents the expression $2(2x + 3)$.

b. Draw a figure that represents the expression $4x + 3$.

c. Draw a figure that represents the expression $4x + 6$.

d. Does the figure for $2(2x + 3)$ equal the figure for $4x + 3$? Does the figure for $2(2x + 3)$ equal the figure for $4x + 6$? Explain how your results are related to the Distributive Property.

SECTION

11.2 Introduction to Equations

OBJECTIVE A To determine whether a given number is a solution of an equation

Point of Interest

Finding solutions of equations has been a principal aim of mathematics for thousands of years. However, the equals sign did not appear in any text until 1557.

An **equation** expresses the equality of two mathematical expressions. These expressions can be either numerical or variable expressions.

$$\left.\begin{array}{l} 5 + 4 = 9 \\ 3x + 13 = x - 8 \\ y^2 + 4 = 6y + 1 \\ x = -3 \end{array}\right\} \text{Equations}$$

In the equation at the right, if the variable is replaced by 4, the equation is true.

$x + 3 = 7$

$4 + 3 = 7$ A true equation

If the variable is replaced by 6, the equation is false.

$6 + 3 = 7$ A false equation

A **solution of an equation** is a number that, when substituted for the variable, results in a true equation. 4 is a solution of the equation $x + 3 = 7$. 6 is not a solution of the equation $x + 3 = 7$.

HOW TO • 1 Is -2 a solution of the equation $-2x + 1 = 2x + 9$?

$$\begin{array}{r|l} \multicolumn{2}{c}{-2x + 1 = 2x + 9} \\ \hline -2(-2) + 1 & 2(-2) + 9 \\ 4 + 1 & -4 + 9 \\ 5 = & 5 \end{array}$$

- Replace the variable by the given number.
- Evaluate the numerical expressions.
- Compare the results. If the results are equal, the given number is a solution. If the results are not equal, the given number is not a solution.

Yes, -2 is a solution of the equation $-2x + 1 = 2x + 9$.

EXAMPLE • 1

Is $\frac{1}{2}$ a solution of $2x(x + 2) = 3x + 1$?

Solution

$$\begin{array}{r|l} \multicolumn{2}{c}{2x(x + 2) = 3x + 1} \\ \hline 2\left(\frac{1}{2}\right)\left(\frac{1}{2} + 2\right) & 3\left(\frac{1}{2}\right) + 1 \\ 2\left(\frac{1}{2}\right)\left(\frac{5}{2}\right) & \frac{3}{2} + 1 \\ \frac{5}{2} = & \frac{5}{2} \end{array}$$

Yes, $\frac{1}{2}$ is a solution.

YOU TRY IT • 1

Is -2 a solution of $x(x + 3) = 4x + 6$?

Your solution

Solution on p. S26

EXAMPLE 2

Is 5 a solution of $(x - 2)^2 = x^2 - 4x + 2$?

Solution

$$
\begin{array}{r|l}
(x - 2)^2 & x^2 - 4x + 2 \\
\hline
(5 - 2)^2 & 5^2 - 4(5) + 2 \\
3^2 & 25 - 4(5) + 2 \\
9 & 25 - 20 + 2 \\
 & 25 + (-20) + 2
\end{array}
$$

$9 \neq 7$ ($\neq$ means "is not equal to")

No, 5 is not a solution.

YOU TRY IT 2

Is -3 a solution of $x^2 - x = 3x + 7$?

Your solution

Solution on p. S26

OBJECTIVE B To solve an equation of the form $x + a = b$

A solution of an equation is a number that, when substituted for the variable, results in a true equation. The phrase **solving an equation** means finding a solution of the equation.

The simplest equation to solve is an equation of the form *variable* = *constant*. The constant is the solution of the equation.

If $x = 7$, then 7 is the solution of the equation because $7 = 7$ is a true equation.

In solving an equation of the form $x + a = b$, the goal is to simplify the given equation to one of the form *variable* = *constant*. The Addition Properties that follow are used to simplify equations to this form.

Addition Property of Zero

The sum of a term and zero is the term.

$a + 0 = a \qquad 0 + a = a$

Addition Property of Equations

If a, b, and c are algebraic expressions, then the equation $a = b$ has the same solutions as the equation $a + c = b + c$.

The Addition Property of Equations states that the same quantity can be added to each side of an equation without changing the solution of the equation.

The Addition Property of Equations is used to rewrite an equation in the form *variable* = *constant*. We remove a term from one side of the equation by adding the opposite of that term to each side of the equation.

Take Note

Always check the solution to an equation.

Check:
$$\begin{array}{c|c} x - 7 & -2 \\ \hline 5 - 7 & -2 \\ -2 & -2 \end{array} \quad \text{True}$$

HOW TO • 2 Solve: $x - 7 = -2$

$$x - 7 = -2$$

- The goal is to simplify the equation to one of the form *variable* = *constant*.

$$x - 7 + 7 = -2 + 7$$
$$x + 0 = 5$$
$$x = 5$$

- Add the opposite of the constant term −7 to each side of the equation. After we simplify and use the Addition Property of Zero, the equation will be in the form *variable* = *constant*.

The solution is 5.

Because subtraction is defined in terms of addition, the Addition Property of Equations allows the same number to be subtracted from each side of an equation.

Tips for Success

When we suggest that you check a solution, you should substitute the solution into the *original* equation. For instance,

$$\begin{array}{c|c} x + 8 & 5 \\ \hline -3 + 8 & 5 \\ 5 & 5 \end{array}$$

The solution checks.

HOW TO • 3 Solve: $x + 8 = 5$

$$x + 8 = 5$$

- The goal is to simplify the equation to one of the form *variable* = *constant*.

$$x + 8 - 8 = 5 - 8$$
$$x + 0 = -3$$
$$x = -3$$

- Add the opposite of the constant term 8 to each side of the equation. This procedure is equivalent to subtracting 8 from each side of the equation.

The solution is −3. You should check this solution.

EXAMPLE • 3

Solve: $4 + m = -2$

Solution

$$4 + m = -2$$
$$4 - 4 + m = -2 - 4$$
$$0 + m = -6$$
$$m = -6$$

- Subtract 4 from each side.

The solution is −6.

YOU TRY IT • 3

Solve: $-2 + y = -5$

Your solution

EXAMPLE • 4

Solve: $3 = y - 2$

Solution

$$3 = y - 2$$
$$3 + 2 = y - 2 + 2$$
$$5 = y + 0$$
$$5 = y$$

- Add 2 to each side.

The solution is 5.

YOU TRY IT • 4

Solve: $7 = y + 8$

Your solution

EXAMPLE • 5

Solve: $\frac{2}{7} = \frac{5}{7} + t$

Solution

$$\frac{2}{7} = \frac{5}{7} + t$$
$$\frac{2}{7} - \frac{5}{7} = \frac{5}{7} - \frac{5}{7} + t$$
$$-\frac{3}{7} = 0 + t$$
$$-\frac{3}{7} = t$$

- Subtract $\frac{5}{7}$ from each side.

The solution is $-\frac{3}{7}$.

YOU TRY IT • 5

Solve: $\frac{1}{5} = z + \frac{4}{5}$

Your solution

Solutions on p. S26

OBJECTIVE C To solve an equation of the form $ax = b$

In solving an equation of the form $ax = b$, the goal is to simplify the given equation to one of the form *variable* = *constant*. The Multiplication Properties that follow are used to simplify equations to this form.

Multiplication Property of Reciprocals

The product of a nonzero term and its reciprocal equals 1.

Because $a = \frac{a}{1}$, the reciprocal of a is $\frac{1}{a}$. $\quad a\left(\frac{1}{a}\right) = 1 \quad \frac{1}{a}(a) = 1$

The reciprocal of $\frac{a}{b}$ is $\frac{b}{a}$. $\quad \left(\frac{a}{b}\right)\left(\frac{b}{a}\right) = 1 \quad \left(\frac{b}{a}\right)\left(\frac{a}{b}\right) = 1$

Multiplication Property of One

The product of a term and 1 is the term.

$a \cdot 1 = a \qquad 1 \cdot a = a$

Multiplication Property of Equations

If a, b, and c are algebraic expressions and $c \neq 0$, then the equation $a = b$ has the same solutions as the equation $ac = bc$.

The Multiplication Property of Equations states that each side of an equation can be multiplied by the same nonzero number without changing the solutions of the equation.

Recall that the goal of solving an equation is to rewrite the equation in the form *variable* = *constant*. The Multiplication Property of Equations is used to rewrite an equation in this form by multiplying each side of the equation by the reciprocal of the coefficient.

HOW TO 4 Solve: $\frac{2}{3}x = 8$

$$\frac{2}{3}x = 8$$

$$\left(\frac{3}{2}\right)\left(\frac{2}{3}\right)x = \left(\frac{3}{2}\right)8$$

$$1 \cdot x = 12$$

$$x = 12$$

- **Multiply each side of the equation by $\frac{3}{2}$, the reciprocal of $\frac{2}{3}$. After simplifying, the equation will be in the form *variable* = *constant*.**

Check: $\frac{2}{3}x = 8$

$$\frac{2}{3}(12) \;\Big|\; 8$$

$$8 = 8$$

The solution is 12.

Because division is defined in terms of multiplication, the Multiplication Property of Equations allows each side of an equation to be divided by the same nonzero quantity.

Tips for Success

When we suggest that you check a solution, you should substitute the solution into the *original* equation. For instance,

$$\begin{array}{r|l} -4x & = 24 \\ \hline -4(-6) & 24 \\ 24 & = 24 \end{array}$$

The solution checks.

HOW TO • 5 Solve: $-4x = 24$

$-4x = 24$

- The goal is to rewrite the equation in the form *variable* = *constant.*

$\dfrac{-4x}{-4} = \dfrac{24}{-4}$

$1x = -6$

$x = -6$

- Multiply each side of the equation by the reciprocal of -4. This is equivalent to dividing each side of the equation by -4. Then simplify.

The solution is -6. You should check this solution.

In using the Multiplication Property of Equations, it is usually easier to multiply each side of the equation by the reciprocal of the coefficient when the coefficient is a fraction. Divide each side of the equation by the coefficient when the coefficient is an integer or a decimal.

EXAMPLE • 6

Solve: $-2x = 6$

Solution

$-2x = 6$

$\dfrac{-2x}{-2} = \dfrac{6}{-2}$ • Divide each side by -2.

$1x = -3$

$x = -3$

The solution is -3.

YOU TRY IT • 6

Solve: $4z = -20$

Your solution

EXAMPLE • 7

Solve: $-9 = \frac{3}{4}y$

Solution

$-9 = \dfrac{3}{4}y$

$\left(\dfrac{4}{3}\right)(-9) = \left(\dfrac{4}{3}\right)\left(\dfrac{3}{4}y\right)$ • Multiply each side by $\dfrac{4}{3}$.

$-12 = 1y$

$-12 = y$

The solution is -12.

YOU TRY IT • 7

Solve: $8 = \frac{2}{5}n$

Your solution

EXAMPLE • 8

Solve: $6z - 8z = -5$

Solution

$6z - 8z = -5$

$-2z = -5$ • Combine like terms.

$\dfrac{-2z}{-2} = \dfrac{-5}{-2}$ • Divide each side by -2.

$1z = \dfrac{5}{2}$

$z = \dfrac{5}{2} = 2\dfrac{1}{2}$

The solution is $2\frac{1}{2}$.

YOU TRY IT • 8

Solve: $\frac{2}{3}t - \frac{1}{3}t = -2$

Your solution

Solutions on p. S26

OBJECTIVE D To solve application problems using formulas

A **formula** is an equation that expresses a relationship among variables. Formulas are used in the examples below.

EXAMPLE • 9

An accountant for a medical spa found that the weekly profit for the spa was \$1700 and that the total amount spent during the week was \$2400. Use the formula $P = R - C$, where P is the profit, R is the revenue, and C is the amount spent, to find the revenue for the week.

Strategy

To find the revenue for the week, replace the variables P and C in the formula by the given values, and solve for R.

Solution

$$P = R - C$$
$$1700 = R - 2400$$
$$1700 + 2400 = R - 2400 + 2400$$
$$4100 = R + 0$$
$$4100 = R$$

The revenue for the week was \$4100.

YOU TRY IT • 9

A medical supply store's sale price for a set of scrubs is \$44. This is a discount of \$16 off the regular price. Use the formula $S = R - D$, where S is the sale price, R is the regular price, and D is the discount, to find the regular price.

Your strategy

Your solution

EXAMPLE • 10

A store manager uses the formula $S = R - d \cdot R$, where S is the sale price, R is the regular price, and d is the discount rate. During a clearance sale, all items are discounted 20%. Find the regular price of running shoes that are on sale for \$120.

Strategy

To find the regular price of the running shoes, replace the variables S and d in the formula by the given values, and solve for R.

Solution

$$S = R - d \cdot R$$
$$120 = R - 0.20R$$
$$120 = 0.80R \qquad \bullet\ R - 0.20R = 1R - 0.20R$$
$$\frac{120}{0.80} = \frac{0.80R}{0.80}$$
$$150 = R$$

The regular price of the running shoes is \$150.

YOU TRY IT • 10

Find the monthly payment when the total amount paid on a loan is \$6840 and the loan is paid off in 24 months. Use the formula $A = MN$, where A is the total amount paid on a loan, M is the monthly payment, and N is the number of monthly payments.

Your strategy

Your solution

Solutions on p. S26

11.2 EXERCISES

OBJECTIVE A To determine whether a given number is a solution of an equation

1. Is -3 a solution of $2x + 9 = 3$?

2. Is -2 a solution of $5x + 7 = 12$?

3. Is 2 a solution of $4 - 2x = 8$?

4. Is 4 a solution of $5 - 2x = 4x$?

5. Is 3 a solution of $3x - 2 = x + 4$?

6. Is 2 a solution of $4x + 8 = 4 - 2x$?

7. Is 3 a solution of $x^2 - 5x + 1 = 10 - 5x$?

8. Is -5 a solution of $x^2 - 3x - 1 = 9 - 6x$?

9. Is -1 a solution of $2x(x - 1) = 3 - x$?

10. Is 2 a solution of $3x(x - 3) = x - 8$?

11. Is 2 a solution of $x(x - 2) = x^2 - 4$?

12. Is -4 a solution of $x(x + 4) = x^2 + 16$?

13. Is $-\frac{2}{3}$ a solution of $3x + 6 = 4$?

14. Is $\frac{1}{2}$ a solution of $2x - 7 = -3$?

15. Is $\frac{1}{4}$ a solution of $2x - 3 = 1 - 14x$?

16. Is 1.32 a solution of $x^2 - 3x = -0.8776 - x$?

17. Is -1.9 a solution of $x^2 - 3x = x + 3.8$?

18. Is 1.05 a solution of $x^2 + 3x = x(x + 3)$?

For Exercises 19 and 20, determine whether the statement is *true* or *false*.

19. Any number that is a solution of the equation $5x + 1 = -9$ must be a negative number.

20. Any number that is a solution of the equation $3x + 6 = 3 - 4x$ must also be a solution of the equation $3(x + 2) = 3 + 4x - 8x$.

OBJECTIVE B To solve an equation of the form $x + a = b$

For Exercises 21 to 48, solve.

21. $y - 6 = 16$

22. $z - 4 = 10$

23. $3 + n = 4$

24. $6 + x = 8$

25. $z + 7 = 2$

26. $w + 9 = 5$

27. $x - 3 = -7$

28. $m - 4 = -9$

29. $y + 6 = 6$

30. $t - 3 = -3$

31. $v - 7 = -4$

32. $x - 3 = -1$

33. $1 + x = 0$

34. $3 + y = 0$

35. $x - 10 = 5$

36. $y - 7 = 3$

37. $x + 4 = -7$

38. $t - 3 = -8$

39. $w + 5 = -5$

40. $z + 6 = -6$

41. $x + \frac{1}{2} = -\frac{1}{2}$

42. $x - \frac{5}{6} = -\frac{1}{6}$

43. $\frac{2}{5} + x = -\frac{3}{5}$

44. $\frac{7}{8} + y = -\frac{1}{8}$

45. $x + \frac{1}{2} = -\frac{1}{3}$

46. $y + \frac{3}{8} = \frac{1}{4}$

47. $t + \frac{1}{4} = -\frac{1}{2}$

48. $x + \frac{1}{3} = \frac{5}{12}$

For Exercises 49 to 52, use the given conditions for a and b to determine whether the value of x in the equation $x + a = b$ *must be negative, must be positive,* or *could be either positive or negative.*

49. a is positive and b is negative.

50. a is a positive proper fraction and b is greater than 1.

51. a is the opposite of b and b is positive.

52. a is negative and b is negative.

OBJECTIVE C To solve an equation of the form $ax = b$

For Exercises 53 to 84, solve.

53. $3y = 12$

54. $5x = 30$

55. $5z = -20$

56. $3z = -27$

57. $-2x = 6$

58. $-4t = 20$

59. $-5x = -40$

60. $-2y = -28$

61. $40 = 8x$

62. $24 = 3y$

63. $-24 = 4x$

64. $-21 = 7y$

65. $\frac{x}{3} = 5$

66. $\frac{y}{2} = 10$

67. $\frac{n}{4} = -2$

68. $\frac{y}{7} = -3$

69. $-\frac{x}{4} = 1$

70. $-\frac{y}{3} = 5$

71. $\frac{2}{3}w = 4$

72. $\frac{5}{8}x = 10$

73. $\frac{3}{4}v = -3$

74. $\frac{2}{7}x = -12$

75. $-\frac{1}{3}x = -2$

76. $-\frac{1}{5}y = -3$

77. $-4 = -\frac{2}{3}z$

78. $-12 = -\frac{3}{8}y$

79. $\frac{2}{3}x = -\frac{2}{7}$

80. $\frac{3}{7}y = \frac{5}{6}$

81. $4x - 2x = 7$

82. $3a - 6a = 8$

83. $\frac{4}{5}m - \frac{1}{5}m = 9$

84. $\frac{1}{3}b - \frac{2}{3}b = -1$

For Exercises 85 to 87, determine whether the statement is *true* or *false.*

85. If a is positive and b is negative, then the value of x in the equation $ax = b$ must be negative.

86. If a is the opposite of b, then the value of x in the equation $ax = b$ must be -1.

87. If a is negative and b is negative, then the value of x in the equation $ax = b$ is negative.

OBJECTIVE D To solve application problems using formulas

88. A store's cost for a blood pressure monitor is \$38. The store sells the blood pressure monitor and makes a \$14 profit. Which equations can you use to find the selling price of the blood pressure monitor? In each equation, S represents the selling price of the item.
(i) $S - 14 = 38$ (ii) $14 = 38 - S$
(iii) $38 - 14 = S$ (iv) $S - 38 = 14$

Fuel Efficiency In Exercises 89 to 92, use the formula $D = M \cdot G$, where D is the distance, M is the miles per gallon, and G is the number of gallons. Round to the nearest tenth.

89. Julio, a sales executive, averaged 28 mi/gal on a 621-mile trip. Find the number of gallons of gasoline used on the trip.

90. Over a 3-day weekend, you take a 592-mile trip. If you average 32 mi/gal on the trip, how many gallons of gasoline did you use?

91. The manufacturer of a hatchback estimates that the car can travel 560 mi on a 15-gallon tank of gas. Find the miles per gallon.

92. You estimate that your car can travel 410 mi on 12 gal of gasoline. Find the miles per gallon.

Investments In Exercises 93 to 96, use the formula $A = P + I$, where A is the value of the investment after 1 year, P is the original investment, and I is the increase in value of the investment.

93. The value of an investment in a pharmaceutical company after 1 year was $17,700. The increase in value during the year was $2700. Find the amount of the original investment.

94. The value of an investment in a software company after 1 year was $26,440. The increase in value during the year was $2830. Find the amount of the original investment.

95. The original investment in a mutual fund was $8000. The value of the mutual fund after 1 year was $11,420. Find the increase in value of the investment.

96. The original investment in a money market fund was $7500. The value of the mutual fund after 1 year was $8690. Find the increase in value of the investment.

Markup In Exercises 97 and 98, use the formula $S = C + M$, where S is the selling price, C is the cost, and M is the markup.

97. A medical supply store sells a power wheelchair for $2240. The power wheelchair has a markup of $420. Find the cost of the power wheelchair.

98. A pharmacy buys blood glucose monitors for $23.50 and sells them for $39.80. Find the markup on each blood glucose monitor.

Markup In Exercises 99 and 100, use the formula $S = C + R \cdot C$, where S is the selling price, C is the cost, and R is the markup rate.

99. Great Medical Supply, Inc. uses a markup rate of 25% on its medical equipment. Find the cost of a rolling walker that sells for $179.

100. A pharmacy uses a markup rate of 20% on its nicotine gum. Find the cost of a 170-piece box of nicotine gum that sells for $33.60.

Applying the Concepts

101. Write out the steps for solving the equation $x - 3 = -5$. Identify each property of real numbers and each property of equations as you use it.

102. Write out the steps for solving the equation $\frac{3}{4}x = 6$. Identify each property of real numbers and each property of equations as you use it.

103. Write an equation of the form $x + a = b$ that has -4 as its solution.

SECTION

11.3 General Equations: Part I

OBJECTIVE A To solve an equation of the form $ax + b = c$

Point of Interest

Evariste Galois, despite being killed in a duel at the age of 21, made significant contributions to solving equations. In fact, there is a branch of mathematics called Galois theory that shows what kinds of equations can be solved and what kinds cannot.

To solve an equation of the form $ax + b = c$, it is necessary to use both the Addition and the Multiplication Properties to simplify the equation to one of the form *variable* = *constant.*

HOW TO 1 Solve: $\frac{x}{4} - 1 = 3$

$\frac{x}{4} - 1 = 3$
• The goal is to simplify the equation to one of the form *variable* = *constant.*

$\frac{x}{4} - 1 + 1 = 3 + 1$
• Add the opposite of the constant term -1 to each side of the equation. Then simplify (Addition Properties).

$\frac{x}{4} + 0 = 4$

$\frac{x}{4} = 4$

$4 \cdot \frac{x}{4} = 4 \cdot 4$
• Multiply each side of the equation by the reciprocal of the numerical coefficient of the variable term. Then simplify (Multiplication Properties).

$1x = 16$

$x = 16$

The solution is 16.
• Write the solution.

Take Note

$\frac{x}{4} = \frac{1}{4}x$

The reciprocal of $\frac{1}{4}$ is 4.

EXAMPLE 1

Solve: $3x + 7 = 2$

Solution

$3x + 7 = 2$

$3x + 7 - 7 = 2 - 7$
• Subtract 7 from each side.

$3x = -5$

$\frac{3x}{3} = \frac{-5}{3}$
• Divide each side by 3.

$x = -\frac{5}{3} = -1\frac{2}{3}$

The solution is $-1\frac{2}{3}$.

YOU TRY IT 1

Solve: $5x + 8 = 6$

Your solution

EXAMPLE 2

Solve: $5 - x = 6$

Solution

$5 - x = 6$

$5 - 5 - x = 6 - 5$
• Subtract 5 from each side.

$-x = 1$

$(-1)(-x) = (-1) \cdot 1$
• Multiply each side by -1.

$x = -1$

The solution is -1.

YOU TRY IT 2

Solve: $7 - x = 3$

Your solution

Solutions on p. S26–S27

OBJECTIVE B To solve application problems using formulas

The Granger Collection

Anders Celsius

The Fahrenheit temperature scale was devised by Daniel Gabriel Fahrenheit (1686—1736), a German physicist and maker of scientific instruments. He invented the mercury thermometer in 1714. On the Fahrenheit scale, the temperature at which water freezes is 32°F, and the temperature at which water boils is 212°F. *Note:* The small raised circle is the symbol for degrees, and the capital F is for Fahrenheit. The Fahrenheit scale is used only in the United States.

In the metric system, temperature is measured on the Celsius scale. The Celsius temperature scale was devised by Anders Celsius (1701–1744), a Swedish astronomer. On the Celsius scale, the temperature at which water freezes is 0°C, and the temperature at which water boils is 100°C. *Note:* The small raised circle is the symbol for degrees, and the capital C is for Celsius.

On both the Celsius scale and the Fahrenheit scale, temperatures below 0° are negative numbers.

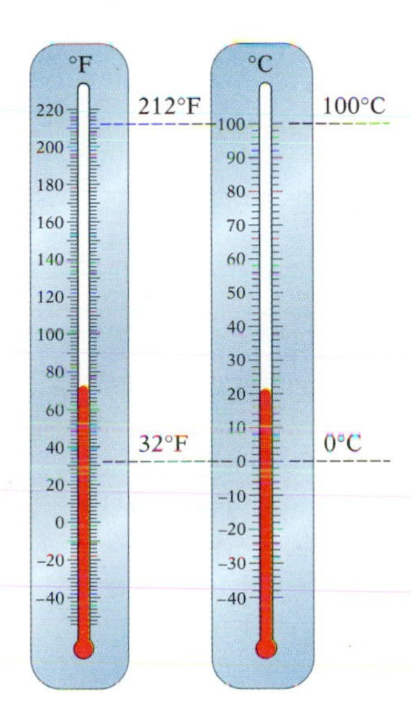

The relationship between Celsius temperature and Fahrenheit temperature is given by the formula

$$F = 1.8C + 32$$

where F represents degrees Fahrenheit and C represents degrees Celsius.

HOW TO • 2 Normal body temperature is 98.6°F. Convert this temperature to degrees Celsius.

$$F = 1.8C + 32$$

$$98.6 = 1.8C + 32 \quad \text{• Substitute 98.6 for } F\text{.}$$

$$98.6 - 32 = 1.8C + 32 - 32 \quad \text{• Subtract 32 from each side.}$$

$$66.6 = 1.8C \quad \text{• Combine like terms on each side.}$$

$$\frac{66.6}{1.8} = \frac{1.8C}{1.8} \quad \text{• Divide each side by 1.8.}$$

$$37 = C$$

Normal body temperature is 37°C.

Integrating Technology

You can check the solution to this equation using a calculator. Evaluate the right side of the equation after substituting 37 for *C*. Enter

1.8 **X** 37 **+** 32 **=**

The display reads 98.6, the given Fahrenheit temperature. The solution checks.

EXAMPLE • 3

Find the Celsius temperature when the Fahrenheit temperature is 212°. Use the formula $F = 1.8C + 32$, where F is the Fahrenheit temperature and C is the Celsius temperature.

Strategy

To find the Celsius temperature, replace the variable F in the formula by the given value and solve for C.

Solution

$$F = 1.8C + 32$$

$$212 = 1.8C + 32$$ • Substitute 212 for F.

$$212 - 32 = 1.8C + 32 - 32$$ • Subtract 32 from each side.

$$180 = 1.8C$$ • Combine like terms.

$$\frac{180}{1.8} = \frac{1.8C}{1.8}$$ • Divide each side by 1.8.

$$100 = C$$

The Celsius temperature is 100°.

YOU TRY IT • 3

Find the Celsius temperature when the Fahrenheit temperature is −22°. Use the formula $F = 1.8C + 32$, where F is the Fahrenheit temperature and C is the Celsius temperature.

Your strategy

Your solution

EXAMPLE • 4

The femur is the long bone in your leg that connects your hip to your knee. Forensic scientists investigating a crime can estimate a female victim's height in centimeters from the length of her femur using the formula $H = 1.9L + 28.7$, where H is the woman's height in inches and L is the length of her femur in inches. Predict a female victim's height if her femur is 17.9 in. Round to the nearest tenth.

Strategy

To find the woman's height, replace the variable L in the given formula with the given value (17.9 in.).

Solution

$H = 1.9L + 28.7$
$H = 1.9 \cdot 17.9 + 28.7$
$H = 34.01 + 28.7$
$H = 62.71$

62.71 rounded to the nearest tenth is 62.7.

The woman's height is approximately 62.7 in.

YOU TRY IT • 4

If a woman is 67 in. tall, estimate the length of her femur. Use the formula $H = 1.9L + 28.7$, where H is a woman's height in inches and L is the length of her femur in inches. Round to the nearest tenth.

Your strategy

Your solution

Solutions on p. S27

11.3 EXERCISES

OBJECTIVE A To solve an equation of the form $ax + b = c$

For Exercises 1 to 86, solve.

1. $3x + 5 = 14$ **2.** $5z + 6 = 31$ **3.** $2n - 3 = 7$ **4.** $4y - 4 = 20$

5. $5w + 8 = 3$ **6.** $3x + 10 = 1$ **7.** $3z - 4 = -16$ **8.** $6x - 1 = -13$

9. $5 + 2x = 7$ **10.** $12 + 7x = 33$ **11.** $6 - x = 3$ **12.** $4 - x = -2$

13. $3 - 4x = 11$ **14.** $2 - 3x = 11$ **15.** $5 - 4x = 17$ **16.** $8 - 6x = 14$

17. $3x + 6 = 0$ **18.** $5x - 20 = 0$ **19.** $-3x - 4 = -1$ **20.** $-7x - 22 = -1$

21. $12y - 30 = 6$ **22.** $9b - 7 = 2$ **23.** $3c + 7 = 4$ **24.** $8t + 13 = 5$

25. $-2x + 11 = -3$ **26.** $-4x + 15 = -1$ **27.** $14 - 5x = 4$ **28.** $7 - 3x = 4$

29. $-8x + 7 = -9$ **30.** $-7x + 13 = -8$ **31.** $9x + 13 = 13$ **32.** $-2x + 7 = 7$

33. $7d - 14 = 0$ **34.** $5z + 10 = 0$ **35.** $4n - 4 = -4$ **36.** $-13m - 1 = -1$

37. $3x + 5 = 7$ **38.** $4x + 6 = 9$ **39.** $6x - 1 = 16$ **40.** $12x - 3 = 7$

41. $2x - 3 = -8$

42. $5x - 3 = -12$

43. $-6x + 2 = -7$

44. $-3x + 9 = -1$

45. $-2x - 3 = -7$

46. $-5x - 7 = -4$

47. $3x + 8 = 2$

48. $2x - 9 = 8$

49. $3w - 7 = 0$

50. $7b - 2 = 0$

51. $-2d + 9 = 12$

52. $-7c + 3 = 1$

53. $\frac{1}{2}x - 2 = 3$

54. $\frac{1}{3}x + 1 = 4$

55. $\frac{3}{5}w - 1 = 2$

56. $\frac{2}{5}w + 5 = 6$

57. $\frac{2}{9}t - 3 = 5$

58. $\frac{5}{9}t - 3 = 2$

59. $\frac{y}{3} - 6 = -8$

60. $\frac{y}{2} - 2 = 3$

61. $\frac{x}{3} - 2 = -5$

62. $\frac{x}{4} - 3 = 5$

63. $\frac{5}{8}v + 6 = 3$

64. $\frac{2}{3}v - 4 = 3$

65. $\frac{4}{7}z + 10 = 5$

66. $\frac{3}{8}v - 3 = 4$

67. $\frac{2}{9}x - 3 = 5$

68. $\frac{1}{2}x + 3 = -8$

69. $\frac{3}{4}x - 5 = -4$

70. $\frac{2}{3}x - 5 = -8$

71. $1.5x - 0.5 = 2.5$

72. $2.5w - 1.3 = 3.7$

73. $0.8t + 1.1 = 4.3$

74. $0.3v + 2.4 = 1.5$

75. $0.4x - 2.3 = 1.3$

76. $1.2t + 6.5 = 2.9$

77. $3.5y - 3.5 = 10.5$

78. $1.9x - 1.9 = -1.9$

79. $0.32x + 4.2 = 3.2$

80. $5x - 3x + 2 = 8$

81. $6m + 2m - 3 = 5$

82. $4a - 7a - 8 = 4$

83. $3y - 8y - 9 = 6$

84. $x - 4x + 5 = 11$

85. $-2y + y - 3 = 6$

86. $-4y - y - 8 = 12$

For Exercises 87 to 89, suppose that a is positive and b is negative. Determine whether the value of x in the given equation *must be negative, must be positive,* or *could be either positive or negative.*

87. $bx - a = 12$

88. $bx + 5 = a$

89. $a - x = b$

OBJECTIVE B To solve application problems using formulas

90. In the formula $H = 1.9L + 28.7$, H is a woman's height in inches and L is the length of her femur in inches. When you know a woman's height, which of the following is the first step in solving the formula for the length of her femur?
(i) Subtract 28.7 from both sides.
(ii) Divide each side by 1.9.
(iii) Subtract the height, H, from both sides.

Temperature Conversion In Exercises 91 and 92, use the relationship between Fahrenheit temperature and Celsius temperature, which is given by the formula $F = 1.8C + 32$, where F is the Fahrenheit temperature and C is the Celsius temperature.

91. Find the Celsius temperature when the Fahrenheit temperature is $-40°$.

92. Find the Celsius temperature when the Fahrenheit temperature is 72°. Round to the nearest tenth of a degree.

Physics In Exercises 93 and 94, use the formula $V = V_0 + 32t$, where V is the final velocity of a falling object, V_0 is the starting velocity of the falling object, and t is the time for the object to fall.

Sandor Szabo/EPA/Landov

93. Find the time required for an object to increase in velocity from 8 ft/s to 472 ft/s.

94. Find the time required for an object to increase in velocity from 16 ft/s to 128 ft/s.

Forensic Science In Exercises 95 to 98, estimate the height of a male victim of a crime using the formula $H = 1.9L + 32$, where H is the man's height in inches and L is the length of his femur in inches. Round to the nearest tenth.

95. Estimate the height of a male victim whose femur is 20 in. long.

96. Estimate the height of a male victim whose femur is 19.1 in. long.

97. Estimate the length of a male's femur if he is 71 in. tall.

98. Estimate the length of a male's femur if he is 5 ft 8 in. tall.

Compensation In Exercises 99 to 102, use the formula $M = S \cdot R + B$, where M is the monthly earnings, S is the total sales, R is the commission rate, and B is the base monthly salary.

99. A sales representative for a pharmaceutical company earns a base monthly salary of $600 plus a 9% commission on total sales. Find the total sales during a month in which the representative earned $3480.

100. A sales executive earns a base monthly salary of $1000 plus a 5% commission on total sales. Find the total sales during a month in which the executive earned $2800.

101. Miguel earns a base monthly salary of $750. Find his commission rate during a month in which total sales were $42,000 and he earned $2640.

102. Tina earns a base monthly salary of $500. Find her commission rate during a month in which total sales were $42,500 and her earnings were $3560.

Applying the Concepts

103. Explain in your own words the steps you would take to solve the equation $\frac{2}{3}x - 4 = 10$. State the property of real numbers or the property of equations that is used at each step.

104. Make up an equation of the form $ax + b = c$ that has -3 as its solution.

105. Does the sentence "Solve $3x + 4(x - 3)$" make sense? Why or why not?

SECTION

11.4 General Equations: Part II

OBJECTIVE A To solve an equation of the form $ax + b = cx + d$

When a variable occurs on each side of an equation, the Addition Properties are used to rewrite the equation so that all variable terms are on one side of the equation and all constant terms are on the other side of the equation. Then the Multiplication Properties are used to simplify the equation to one of the form *variable = constant.*

HOW TO • 1 Solve: $4x - 6 = 8 - 3x$

$4x - 6 = 8 - 3x$ • The goal is to write the equation in the form *variable = constant.*

$4x + 3x - 6 = 8 - 3x + 3x$
$7x - 6 = 8 + 0$
$7x - 6 = 8$
• Add $3x$ to each side of the equation. Then simplify (Addition Properties). Now only one variable term occurs in the equation.

$7x - 6 + 6 = 8 + 6$
$7x + 0 = 14$
$7x = 14$
• Add 6 to each side of the equation. Then simplify (Addition Properties). Now only one constant term occurs in the equation.

$\frac{7x}{7} = \frac{14}{7}$
$1x = 2$
$x = 2$
• Divide each side of the equation by the numerical coefficient of the variable term. Then simplify (Multiplication Properties).

The solution is 2. • Write the solution.

Tips for Success

Always check the solution of an equation. For the equation at the right:

$4x - 6 = 8 - 3x$	
$4(2) - 6$	$8 - 3(2)$
$8 - 6$	$8 - 6$
$2 = 2$	

The solution checks.

EXAMPLE • 1

Solve: $\frac{2}{9}x - 3 = \frac{7}{9}x + 2$

Solution

$\frac{2}{9}x - 3 = \frac{7}{9}x + 2$

$\frac{2}{9}x - \frac{7}{9}x - 3 = \frac{7}{9}x - \frac{7}{9}x + 2$ • Subtract $\frac{7}{9}x$ from each side.

$-\frac{5}{9}x - 3 = 2$

$-\frac{5}{9}x - 3 + 3 = 2 + 3$ • Add 3 to each side.

$-\frac{5}{9}x = 5$

$\left(-\frac{9}{5}\right)\left(-\frac{5}{9}\right)x = \left(-\frac{9}{5}\right)5$ • Multiply each side by $-\frac{9}{5}$.

$x = -9$

The solution is -9.

YOU TRY IT • 1

Solve: $\frac{1}{5}x - 2 = \frac{2}{5}x + 4$

Your solution

Solution on p. S27

OBJECTIVE B To solve an equation containing parentheses

When an equation contains parentheses, one of the steps involved in solving the equation requires use of the Distributive Property.

$$a(b + c) = ab + ac$$

The Distributive Property is used to rewrite a variable expression without parentheses.

HOW TO • 2 Solve: $4(3 + x) - 2 = 2(x - 4)$

$4(3 + x) - 2 = 2(x - 4)$
- **The goal is to write the equation in the form *variable = constant.***

$12 + 4x - 2 = 2x - 8$
- **Use the Distributive Property to rewrite the equation without parentheses.**

$10 + 4x = 2x - 8$
- **Combine like terms.**

$10 + 4x - 2x = 2x - 2x - 8$

$10 + 2x = -8$
- **Use the Addition Property of Equations. Subtract $2x$ from each side of the equation.**

$10 - 10 + 2x = -8 - 10$

$2x = -18$
- **Use the Addition Property of Equations. Subtract 10 from each side of the equation.**

$\frac{2x}{2} = \frac{-18}{2}$

$x = -9$
- **Use the Multiplication Property of Equations. Divide each side of the equation by the numerical coefficient of the variable term.**

Check: $4(3 + x) - 2 = 2(x - 4)$
- **Check the solution.**

$4(3 + x) - 2$	$2(x - 4)$
$4[3 + (-9)] - 2$	$2(-9 - 4)$
$4(-6) - 2$	$2(-13)$
$-24 - 2$	-26

$-26 = -26$ A true equation

The solution is -9.
- **Write the solution.**

The solution to this last equation illustrates the steps involved in solving first-degree equations.

Steps in Solving General First-Degree Equations

1. Use the Distributive Property to remove parentheses.
2. Combine like terms on each side of the equation.
3. Use the Addition Property of Equations to rewrite the equation with only one variable term.
4. Use the Addition Property of Equations to rewrite the equation with only one constant term.
5. Use the Multiplication Property of Equations to rewrite the equation so that the coefficient of the variable term is 1.

EXAMPLE • 2

Solve: $3(x + 2) - x = 11$

Solution

$3(x + 2) - x = 11$

$3x + 6 - x = 11$ • Use the Distributive Property.

$2x + 6 = 11$ • Combine like terms on the left side.

$2x + 6 - 6 = 11 - 6$ • Use the Addition Property of Equations. Subtract 6 from each side.

$2x = 5$ • Combine like terms on each side.

$\frac{2x}{2} = \frac{5}{2}$ • Use the Multiplication Property. Divide both sides by 2.

$x = 2\frac{1}{2}$ • The solution checks.

The solution is $2\frac{1}{2}$.

YOU TRY IT • 2

Solve: $4(x - 1) - x = 5$

Your solution

EXAMPLE • 3

Solve: $5x - 2(x - 3) = 6(x - 2)$

Solution

$5x - 2(x - 3) = 6(x - 2)$

$5x - 2x + 6 = 6x - 12$ • Distributive Property

$3x + 6 = 6x - 12$ • Combine like terms.

$3x - 6x + 6 = 6x - 6x - 12$ • Subtract $6x$ from each side.

$-3x + 6 = -12$ • Combine like terms.

$-3x + 6 - 6 = -12 - 6$ • Subtract 6 from each side.

$-3x = -18$ • Combine like terms.

$\frac{-3x}{-3} = \frac{-18}{-3}$ • Divide both sides by -3.

$x = 6$ • The solution checks.

The solution is 6.

YOU TRY IT • 3

Solve: $2x - 7(3x + 1) = 5(5 - 3x)$

Your solution

Solutions on p. S27

11.4 EXERCISES

OBJECTIVE A **To solve an equation of the form $ax + b = cx + d$**

For Exercises 1 to 54, solve.

1. $6x + 3 = 2x + 5$

2. $7x + 1 = x + 19$

3. $3x + 3 = 2x + 2$

4. $6x + 3 = 3x + 6$

5. $5x + 4 = x - 12$

6. $3x - 12 = x - 8$

7. $7b - 2 = 3b - 6$

8. $2d - 9 = d - 8$

9. $9n - 4 = 5n - 20$

10. $8x - 7 = 5x + 8$

11. $2x + 1 = 16 - 3x$

12. $3x + 2 = -23 - 2x$

13. $5x - 2 = -10 - 3x$

14. $4x - 3 = 7 - x$

15. $2x + 7 = 4x + 3$

16. $7m - 6 = 10m - 15$

17. $c + 4 = 6c - 11$

18. $t - 6 = 4t - 21$

19. $3x - 7 = x - 7$

20. $2x + 6 = 7x + 6$

21. $3 - 4x = 5 - 3x$

22. $6 - 2x = 9 - x$

23. $7 + 3x = 9 + 5x$

24. $12 + 5x = 9 - 3x$

25. $5 + 2y = 7 + 5y$

26. $9 + z = 2 + 3z$

27. $8 - 5w = 4 - 6w$

28. $9 - 4x = 11 - 5x$

29. $6x + 1 = 3x + 2$

30. $7x + 5 = 4x + 7$

31. $5x + 8 = x + 5$

32. $9x + 1 = 3x - 4$

33. $2x - 3 = 6x - 4$

34. $4 - 3x = 4 - 5x$

35. $6 - 3x = 6 - 5x$

36. $2x + 7 = 4x - 3$

37. $6x - 2 = 2x - 9$

38. $4x - 7 = -3x + 2$

39. $6x - 3 = -5x + 8$

40. $7y - 5 = 3y + 9$

41. $-6t - 2 = -8t - 4$

42. $-7w + 2 = 3w - 8$

43. $-3 - 4x = 7 - 2x$

44. $-8 + 5x = 8 + 6x$

45. $3 - 7x = -2 + 5x$

46. $3x - 2 = 7 - 5x$

47. $5x + 8 = 4 - 2x$

48. $4 - 3x = 6x - 8$

49. $12z - 9 = 3z + 12$

50. $4c + 13 = -6c + 9$

51. $\frac{5}{7}m - 3 = \frac{2}{7}m + 6$

52. $\frac{4}{5}x - 1 = \frac{1}{5}x + 5$

53. $\frac{3}{7}x + 5 = \frac{5}{7}x - 1$

54. $\frac{3}{4}x + 2 = \frac{1}{4}x - 9$

55. If a is a negative number, will solving the equation $-8x = a - 6x$ for x result in a positive solution or a negative solution?

56. If a is a positive number, will solving the equation $9x - a = 13x$ for x result in a positive solution or a negative solution?

OBJECTIVE B To solve an equation containing parentheses

57. Which of the following equations is equivalent to $9 - 2(4y - 1) = 6$?
(i) $9 - 8y - 1 = 6$ (ii) $7(4y - 1) = 6$
(iii) $9 - 8y - 2 = 6$ (iv) $9 - 8y + 2 = 6$

58. Which of the following equations is equivalent to $5x - 3(4x + 2) = 7(x + 3)$?
(i) $2x(4x + 2) = 7x + 21$ (ii) $5x - 12x - 6 = 7x + 21$
(iii) $5x - 12x - 6 = 7x + 3$ (iv) $5x(-12x - 6) = 7x + 21$

For Exercises 59 to 104, solve.

59. $6x + 2(x - 1) = 14$

60. $3x + 2(x + 4) = 13$

61. $-3 + 4(x + 3) = 5$

62. $8b - 3(b - 5) = 30$

63. $6 - 2(d + 4) = 6$

64. $5 - 3(n + 2) = 8$

65. $5 + 7(x + 3) = 20$

66. $6 - 3(x - 4) = 12$

67. $2x + 3(x - 5) = 10$

68. $3x - 4(x + 3) = 9$

69. $3(x - 4) + 2x = 3$

70. $4 + 3(x - 9) = -12$

71. $2x - 3(x - 4) = 12$

72. $4x - 2(x - 5) = 10$

73. $2x + 3(x + 4) = 7$

74. $3(x + 2) + 7 = 12$

75. $3(x - 2) + 5 = 5$

76. $4(x - 5) + 7 = 7$

77. $3y + 7(y - 2) = 5$

78. $-3z - 3(z - 3) = 3$

79. $4b - 2(b + 9) = 8$

80. $3x - 6(x - 3) = 9$

81. $3x + 5(x - 2) = 10$

82. $3x - 5(x - 1) = -5$

83. $3x + 4(x + 2) = 2(x + 9)$

84. $5x + 3(x + 4) = 4(x + 2)$

85. $2d - 3(d - 4) = 2(d + 6)$

86. $3t - 4(t - 1) = 3(t - 2)$

87. $7 - 2(x - 3) = 3(x - 1)$

88. $4 - 3(x + 2) = 2(x - 4)$

89. $6x - 2(x - 3) = 11(x - 2)$

90. $9x - 5(x - 3) = 5(x + 4)$

91. $6c - 3(c + 1) = 5(c + 2)$

92. $2w - 7(w - 2) = 3(w - 4)$

93. $7 - (x + 1) = 3(x + 3)$

94. $12 + 2(x - 9) = 3(x - 12)$

95. $2x - 3(x + 4) = 2(x - 5)$

96. $3x + 2(x - 7) = 7(x - 1)$

97. $x + 5(x - 4) = 3(x - 8) - 5$

98. $2x - 2(x - 1) = 3(x - 2) + 7$

99. $9b - 3(b - 4) = 13 + 2(b - 3)$

100. $3y - 4(y - 2) = 15 - 3(y - 2)$

101. $3(x - 4) + 3x = 7 - 2(x - 1)$

102. $2(x - 6) + 7x = 5 - 3(x - 2)$

103. $3.67x - 5.3(x - 1.932) = 6.9959$

104. $4.06x + 4.7(x + 3.22) = 1.775$

Applying the Concepts

105. If $2x - 2 = 4x + 6$, what is the value of $3x^2$?

106. If $3 + 2(4a - 3) = 4$ and $4 - 3(2 - 3b) = 11$, which is larger, a or b?

107. The equation $x = x + 1$ has no solution, whereas the solution of the equation $2x + 3 = 3$ is zero. Is there a difference between no solution and a solution of zero? Explain.

SECTION

11.5 Translating Verbal Expressions into Mathematical Expressions

OBJECTIVE A To translate a verbal expression into a mathematical expression given the variable

One of the major skills required in applied mathematics is to translate a verbal expression into a mathematical expression. Doing so requires recognizing the verbal phrases that translate into mathematical operations. Following is a partial list of the verbal phrases used to indicate the different mathematical operations.

Operation	Verbal phrase	Example	Expression
Addition	more than	5 more than x	$x + 5$
	the sum of	the sum of w and 3	$w + 3$
	the total of	the total of 6 and z	$6 + z$
	increased by	x increased by 7	$x + 7$
Subtraction	less than	5 less than y	$y - 5$
	the difference between	the difference between w and 3	$w - 3$
	decreased by	8 decreased by a	$8 - a$
Multiplication	times	3 times c	$3c$
	the product of	the product of 4 and t	$4t$
	of	two-thirds of v	$\frac{2}{3}v$
	twice	twice d	$2d$
Division	divided by	n divided by 3	$\frac{n}{3}$
	the quotient of	the quotient of z and 4	$\frac{z}{4}$
	the ratio of	the ratio of s to 6	$\frac{s}{6}$

Translating phrases that contain the words *sum, difference, product,* and *quotient* can be difficult. In the examples at the right, note where the operation symbol is placed.

Note where we place the fraction bar when translating the word ratio.

the *sum* of x and y	$x + y$
the *difference* between x and y	$x - y$
the *product* of x and y	$x \cdot y$
the *quotient* of x and y	$\frac{x}{y}$
the *ratio* of x to y	$\frac{x}{y}$

HOW TO • 1 Translate "the quotient of n and the sum of n and 6" into a mathematical expression.

the *quotient* of n and the *sum* of n and 6 $\qquad \frac{n}{n + 6}$

EXAMPLE • 1

Translate "the sum of 5 and the product of 4 and n" into a mathematical expression.

Solution

$5 + 4n$

YOU TRY IT • 1

Translate "the difference between 8 and twice t" into a mathematical expression.

Your solution

EXAMPLE • 2

Translate "the product of 3 and the difference between z and 4" into a mathematical expression.

Solution

$3(z - 4)$

YOU TRY IT • 2

Translate "the quotient of 5 and the product of 7 and x" into a mathematical expression.

Your solution

Solutions on p. S27

OBJECTIVE B To translate a verbal expression into a mathematical expression by assigning the variable

In most applications that involve translating phrases into mathematical expressions, the variable to be used is not given. To translate these phrases, we must assign a variable to the unknown quantity before writing the mathematical expression.

HOW TO • 2 Translate "the difference between seven and twice a number" into a mathematical expression.

The difference between seven and twice a number — • **Identify the phrases that indicate the mathematical operations.**

The unknown number: n — • **Assign a variable to one of the unknown quantities.**

Twice the number: $2n$ — • **Use the assigned variable to write an expression for any other unknown quantity.**

$7 - 2n$ — • **Use the identified operations to write the mathematical expression.**

EXAMPLE • 3

Translate "the total of a number and the square of the number" into a mathematical expression.

Solution

The *total* of a number and the *square* of the number

The unknown number: x

The square of the number: x^2

$x + x^2$

YOU TRY IT • 3

Translate "the product of a number and one-half of the number" into a mathematical expression.

Your solution

Solution on p. S27

11.5 EXERCISES

OBJECTIVE A To translate a verbal expression into a mathematical expression given the variable

For Exercises 1 to 20, translate into a mathematical expression.

1. 9 less than y

2. w divided by 7

3. z increased by 3

4. the product of -2 and x

5. the sum of two-thirds of n and n

6. the difference between the square of r and r

7. the quotient of m and the difference between m and 3

8. v increased by twice v

9. the product of 9 and 4 more than x

10. the difference between n and the product of 5 and n

11. x decreased by the quotient of x and 2

12. the product of c and one-fourth of c

13. the quotient of 3 less than z and z

14. the product of y and the sum of y and 4

15. 2 times the sum of t and 6

16. the quotient of r and the difference between 8 and r

17. x divided by the total of 9 and x

18. the sum of z and the product of 6 and z

19. three times the sum of b and 6

20. the ratio of w to the sum of w and 8

For Exercises 21 and 22, translate the mathematical expression into a verbal phrase. *Note:* Answers will vary.

21. **a.** $2x + 3$

b. $2(x + 3)$

22. **a.** $\frac{2x}{7}$

b. $\frac{2 + x}{7}$

OBJECTIVE B To translate a verbal expression into a mathematical expression by assigning the variable

For Exercises 23 to 42, translate into a mathematical expression.

23. the square of a number

24. five less than some number

25. a number divided by twenty

26. the difference between a number and twelve

27. four times some number

28. the quotient of five and a number

29. three-fourths of a number

30. the sum of a number and seven

31. four increased by some number

32. the ratio of a number to nine

33. the difference between five times a number and the number

34. six less than the total of three and a number

35. the product of a number and two more than the number

36. the quotient of six and the sum of nine and a number

37. seven times the total of a number and eight

38. the difference between ten and the quotient of a number and two

39. the square of a number plus the product of three and the number

40. a number decreased by the product of five and the number

41. the sum of three more than a number and one-half of the number

42. eight more than twice the sum of a number and seven

For Exercises 43 and 44, determine whether the expression $4n^2 - 5$ is a correct translation of the given phrase.

43. five less than the square of the product of four and a number

44. the difference between five and the product of four and the square of a number

Applying the Concepts

45. In your own words, explain how variables are used.

46. **Chemistry** The chemical formula for water is H_2O. This formula means that there are two hydrogen atoms and one oxygen atom in each molecule of water. If x represents the number of atoms of oxygen in a glass of pure water, express the number of hydrogen atoms in the glass of water.

47. **Chemistry** The chemical formula for one molecule of glucose (sugar) is $C_6H_{12}O_6$, where C is carbon, H is hydrogen, and O is oxygen. If x represents the number of atoms of hydrogen in a sample of pure sugar, express the number of carbon atoms and the number of oxygen atoms in the sample in terms of x.

H O
C
H – C – OH
HO – C – H
H – C – OH
H – C – OH
CH_2OH

SECTION

11.6 Translating Sentences into Equations and Solving

OBJECTIVE A To translate a sentence into an equation and solve

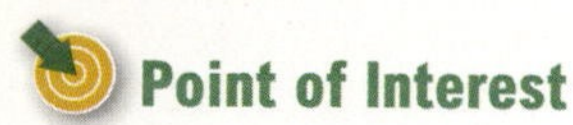

Point of Interest

Number puzzle problems similar to the one on this page have appeared in textbooks for hundreds of years. Here is one from a 1st-century Chinese textbook: "When a number is divided by 3, the remainder is 2; when it is divided by 5, the remainder is 3; when it is divided by 7, the remainder is 2. Find the number." There are actually infinitely many solutions to this problem. See whether you can find one of them.

An equation states that two mathematical expressions are equal. Therefore, to translate a sentence into an equation requires recognition of the words or phrases that mean "equals." Some of these phrases are

equals
is
is equal to
amounts to
represents } translate to =

Once the sentence is translated into an equation, the equation can be simplified to one of the form *variable* = *constant* and the solution can be found.

HOW TO • 1 Translate "three more than twice a number is seventeen" into an equation and solve.

The unknown number: n

• Assign a variable to the unknown quantity.

Three more than twice a number	is	seventeen

• Find two verbal expressions for the same value.

$$2n + 3 = 17$$
$$2n + 3 - 3 = 17 - 3$$
$$2n = 14$$
$$\frac{2n}{2} = \frac{14}{2}$$
$$n = 7$$

• Write a mathematical expression for each verbal expression. Write the equals sign. Solve the resulting equation.

The number is 7.

EXAMPLE • 1

Translate "a number decreased by six equals fifteen" into an equation and solve.

Solution

The unknown number: x

A number decreased by six	equals	fifteen

$$x - 6 = 15$$
$$x - 6 + 6 = 15 + 6$$
$$x = 21$$

The number is 21.

YOU TRY IT • 1

Translate "a number increased by four equals twelve" into an equation and solve.

Your solution

Solution on p. S27

EXAMPLE • 2

The quotient of a number and six is five. Find the number.

Solution

The unknown number: z

The quotient of a number and six	is	five

$$\frac{z}{6} = 5$$
$$6 \cdot \frac{z}{6} = 6 \cdot 5$$
$$z = 30$$

The number is 30.

YOU TRY IT • 2

The product of two and a number is ten. Find the number.

Your solution

EXAMPLE • 3

Eight decreased by twice a number is four. Find the number.

Solution

The unknown number: t

Eight decreased by twice a number	is	four

$$8 - 2t = 4$$
$$8 - 8 - 2t = 4 - 8$$
$$-2t = -4$$
$$\frac{-2t}{-2} = \frac{-4}{-2}$$
$$t = 2$$

The number is 2.

YOU TRY IT • 3

The sum of three times a number and six equals four. Find the number.

Your solution

EXAMPLE • 4

Three less than the ratio of a number to seven is one. Find the number.

Solution

The unknown number: x

Three less than the ratio of a number to seven	is	one

$$\frac{x}{7} - 3 = 1$$
$$\frac{x}{7} - 3 + 3 = 1 + 3$$
$$\frac{x}{7} = 4$$
$$7 \cdot \frac{x}{7} = 7 \cdot 4$$
$$x = 28$$

The number is 28.

YOU TRY IT • 4

Three more than one-half of a number is nine. Find the number.

Your solution

Solutions on p. S28

OBJECTIVE B To solve application problems

EXAMPLE • 5

The hourly wage of a certified medical assistant is \$14.25. This amount is \$3.05 more than the hourly wage of a certified nurse assistant. Find the hourly wage of the certified nurse assistant.

Strategy

To find the hourly wage of the certified nurse assistant, write and solve an equation using H to represent the hourly wage of the certified nurse assistant.

Solution

\$14.25	is	\$3.05 more than the hourly wage of a certified nurse assistant

$$14.25 = H + 3.05$$
$$14.25 - 3.05 = H + 3.05 - 3.05$$
$$11.20 = H$$

The hourly wage of the certified nurse assistant is \$11.20.

YOU TRY IT • 5

The sale price of a box of prenatal vitamins is \$23.99. This amount is \$6 less than the regular price. Find the regular price.

Your strategy

Your solution

EXAMPLE • 6

By purchasing a fleet of cars, a medical laboratory testing facility receives a discount of \$1972 on each car purchased. This amount is 8% of the regular price. Find the regular price.

Strategy

To find the regular price, write and solve an equation using P to represent the regular price of the car.

Solution

\$1972	is	8% of the regular price

$$1972 = 0.08 \cdot P$$
$$\frac{1972}{0.08} = \frac{0.08P}{0.08}$$
$$24{,}650 = P$$

The regular price is \$24,650.

YOU TRY IT • 6

At the age of 2 years, a child is half the height he will be as an adult. If a child is 34 in. tall when he is 2 years old, predict the child's height as an adult. Give your answer in feet and inches.

Your strategy

Your solution

Solutions on p. S28

EXAMPLE • 7

A biomedical equipment technician charged $1775 to repair some diagnostic medical equipment in a physician's office. This charge included $180 for parts and $55 per hour for labor. Find the number of hours the technician worked at the physician's office.

Strategy

To find the number of hours worked, write and solve an equation using N to represent the number of hours worked.

Solution

$1775	included	$180 for parts and $55 per hour for labor

$$\begin{aligned} 1775 &= 180 + 55N \\ 1775 - 180 &= 180 - 180 + 55N \\ 1595 &= 55N \\ \frac{1595}{55} &= \frac{55N}{55} \\ 29 &= N \end{aligned}$$

The technician worked 29 h.

YOU TRY IT • 7

The total cost to make a 1500X professional biological microscope is $492. The cost includes $100 for materials plus $24.50 per hour for labor. How many hours of labor are required to make a 1500X professional biological microscope?

Your strategy

Your solution

EXAMPLE • 8

The state income tax for Tim Fong last month was $256. This amount is $5 more than 8% of his monthly salary. Find Tim's monthly salary.

Strategy

To find Tim's monthly salary, write and solve an equation using S to represent his monthly salary.

Solution

$256	is	$5 more than 8% of the monthly salary

$$\begin{aligned} 256 &= 0.08 \cdot S + 5 \\ 256 - 5 &= 0.08S + 5 - 5 \\ 251 &= 0.08S \\ \frac{251}{0.08} &= \frac{0.08S}{0.08} \\ 3137.50 &= S \end{aligned}$$

Tim's monthly salary is $3137.50.

YOU TRY IT • 8

Natalie Adams earned $2500 last month for temporary work. This amount was the sum of a base monthly salary of $800 and an 8% commission on total sales. Find Natalie's total sales for the month.

Your strategy

Your solution

Solutions on p. S28

11.6 EXERCISES

OBJECTIVE A To translate a sentence into an equation and solve

For Exercises 1 to 26, write an equation and solve.

1. The sum of a number and seven is twelve. Find the number.

2. A number decreased by seven is five. Find the number.

3. The product of three and a number is eighteen. Find the number.

4. The quotient of a number and three is one. Find the number.

5. Five more than a number is three. Find the number.

6. A number divided by four is six. Find the number.

7. Six times a number is fourteen. Find the number.

8. Seven less than a number is three. Find the number.

9. Five-sixths of a number is fifteen. Find the number.

10. The total of twenty and a number is five. Find the number.

11. The sum of three times a number and four is eight. Find the number.

12. The sum of one-third of a number and seven is twelve. Find the number.

13. Seven less than one-fourth of a number is nine. Find the number.

14. The total of a number divided by four and nine is two. Find the number.

15. The ratio of a number to nine is fourteen. Find the number.

16. Five increased by the product of five and a number is equal to 30. Find the number.

17. Six less than the quotient of a number and four is equal to negative two. Find the number.

18. The product of a number plus three and two is eight. Find the number.

19. The difference between seven and twice a number is thirteen. Find the number.

20. Five more than the product of three and a number is eight. Find the number.

21. Nine decreased by the quotient of a number and two is five. Find the number.

22. The total of ten times a number and seven is twenty-seven. Find the number.

23. The sum of three-fifths of a number and eight is two. Find the number.

24. Five less than two-thirds of a number is three. Find the number.

25. The difference between a number divided by 4.18 and 7.92 is 12.52. Find the number.

26. The total of 5.68 times a number and 132.7 is the number minus 29.228. Find the number.

For Exercises 27 and 28, determine whether the equation $5 - 7x = 9$ is a correct translation of the given sentence.

27. Five less than the product of seven and a number is equal to nine.

28. The difference between five and the product of seven and a number is nine.

OBJECTIVE B To solve application problems

29. A student writes the equation $33{,}475 = x + 4260$ to represent the situation described in Exercise 31. What does x represent in the equation?

30. A student writes the equation $230{,}000 = n - 6300$ to represent the situation described in Exercise 32. What does n represent in this equation?

31. **Wages** A medical laboratory technician makes $33,475 annually. This amount is $4260 more than the annual salary of a certified pharmacy technician. Find the annual salary of a certified pharmacy technician.

32. **Housing** In the Northeast, the median price of a house in 2011 was $230,200. This price was $6300 less than the median price of a house in 2010. (*Source:* National Association of Realtors) Find the median price of a house in 2010.

33. **Height** A nurse measures Zack's height as 38 in. The nurse notes that he is 3.5 in. taller than his brother, Ethan. How tall is Ethan?

34. **Height** A 12-year-old girl is $\frac{4}{5}$ the height of her father. If the 12-year-old girl is 60 in. tall, how tall is her father? Give your answer in feet and inches.

35. Weight Loss A doctor refers an obese patient to a weight-loss clinic. The dietician sets the patient's goal weight at 162 lb, which is $\frac{2}{3}$ of the patient's current weight. What is the patient's current weight?

In the News

World Energy Consumption Projected

World energy consumption in 2015 is projected to be 560 quadrillion Btu.

Source: EIA, System for the Analysis of Global Energy Markets

36. Energy Consumption See the news clipping at the right. The projected world energy consumption in 2015 is four-fifths the expected world energy consumption in 2030. What is the expected world energy consumption in 2030?

37. Sleep First- through fifth-graders get an average of 9.5 h of sleep daily. This is three-fourths the number of hours infants aged 3 months to 11 months sleep each day. (*Source:* National Sleep Foundation) How many hours do infants aged 3 months to 11 months sleep each day? Round to the nearest tenth.

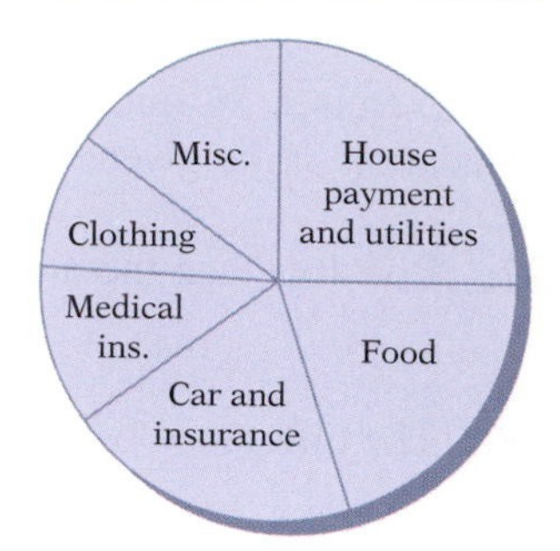

38. Finances Each month the Manzanares family spends $1360 for their house payment and utilities, which amounts to one-fourth of the family's monthly income. Find the family's monthly income.

39. Consumerism The cost of a graphing calculator today is three-fourths of the cost of the calculator 5 years ago. The cost of the graphing calculator today is $72. Find the cost of the calculator 5 years ago.

40. Smoking Statistics According to the CDC, on average, 128,900 people in the United States die each year from lung cancer, which is directly attributable to cigarette smoking. This represents 29% of all deaths attributable to cigarette smoking. Find the number of U.S. deaths attributable each year to cigarette smoking. Round to the nearest thousand.

41. Nutrition The nutrition label on a bag of Baked Tostitos tortilla chips lists the sodium content of one serving as 200 mg, which is 8% of the recommended daily allowance of sodium. What is the recommended daily allowance of sodium? Express the answer in grams.

42. Nutrition The nutrition label on Cocoa Krispies® cereal lists 12 g of sugar and 15 g of other carbohydrates in one serving. Together, they represent 9% of the recommended daily allowance of carbohydrates for a 2000-calorie diet. What is the recommended daily allowance of carbohydrates? Express the answer in grams.

43. Diet Americans consume 7 billion hot dogs from Memorial Day through Labor Day. This is 35% of the hot dogs consumed annually in the United States. (*Source:* National Hot Dog & Sausage Council; American Meat Institute) How many hot dogs do Americans consume annually?

Foodcollection RF/Getty Images

44. Compensation Sandy's monthly salary as a sales representative was $2580. This amount included her base monthly salary of $600 plus a 3% commission on total sales. Find her total sales for the month.

45. Emergency Room Visits According to the CDC, the number of injury-related visits to U.S. emergency rooms averages 42.4 million annually. This number represents 34.3% of all emergency room visits in the United States. Approximately how many emergency room visits are there annually in the United States? Round to the nearest hundred thousand.

46. Insecticides Americans spend approximately \$295 million a year on remedies for cockroaches. The table at the right shows the top U.S. cities for sales of roach insecticides. What percent of the total is spent in New York? Round to the nearest tenth of a percent.

City	*Roach Insecticide Sales*
Los Angeles	\$16.8 million
New York	\$9.8 million
Houston	\$6.7 million

Source: IRI InfoScan for Combat

47. Ideal Weight The "Devine formula" published in 1974 was originally developed in order to calculate dosages of certain medications based on weight. The formula for men states that the ideal body weight in kilograms is equal to the sum of 50 and 2.3 kg per inch over 5 ft. How tall is a man whose ideal weight is 73 kg?

48. Consumerism The sale price of an automatic blood pressure monitor is \$46.50. This amount is \$2.50 more than $\frac{4}{5}$ the regular price. Find the regular price of the blood pressure monitor.

49. Vacation Days In Italy, workers take an average of 42 vacation days per year. This number is 3 more than three times the average number of vacation days workers take each year in the United States. (*Source:* World Tourism Organization) On average, how many vacation days do U.S. workers take per year?

PhotoLink/Photodisc/Getty Images

50. Astronautics Four hundred seventeen men have flown into space. This number is 25 more than 8 times the number of women who have flown into space. (*Source:* Encyclopedia Astronautica) How many women have flown into space?

51. Compensation Assume that a pharmaceutical salesman receives a base monthly salary of \$600 plus an 8.25% commission on total sales per month. Find the salesman's total sales during a month in which she receives total compensation of \$4109.55.

Applying the Concepts

52. A man's boyhood lasted $\frac{1}{6}$ of his life, he played football for the next $\frac{1}{8}$ of his life, and he married 5 years after quitting football. A daughter was born after he had been married $\frac{1}{12}$ of his life. The daughter lived $\frac{1}{2}$ as many years as her father. The man died 6 years after his daughter. How old was the man when he died? Use a number line to illustrate the time. Then write an equation and solve it.

FOCUS ON PROBLEM SOLVING

From Concrete to Abstract

As you progress in your study of algebra, you will find that the problems become less concrete and more abstract. Problems that are concrete provide information pertaining to a specific instance. Abstract problems are theoretical; they are stated without reference to a specific instance. Let's look at an example of an abstract problem.

How many cents are in d dollars?

How can you solve this problem? Are you able to solve the same problem if the information given is concrete?

How many cents are in 5 dollars?

You know that there are 100 cents in 1 dollar. To find the number of cents in 5 dollars, multiply 5 by 100.

$100 \cdot 5 = 500$ There are 500 cents in 5 dollars.

Use the same procedure to find the number of cents in d dollars: multiply d by 100.

$100 \cdot d = 100d$ There are $100d$ cents in d dollars.

This problem might be taken a step further:

If one pen costs c cents, how many pens can be purchased with d dollars?

Consider the same problem using numbers in place of the variables.

If one pen costs 50 cents, how many pens can be purchased with 2 dollars?

To solve this problem, you need to calculate the number of cents in 2 dollars (multiply 2 by 100) and divide the result by the cost per pen (50 cents).

$\frac{100 \cdot 2}{50} = \frac{200}{50} = 4$ If one pen costs 50 cents, 4 pens can be purchased with 2 dollars.

Use the same procedure to solve the related abstract problem. Calculate the number of cents in d dollars (multiply d by 100) and divide the result by the cost per pen (c cents).

$\frac{100 \cdot d}{c} = \frac{100d}{c}$ If one pen costs c cents, $\frac{100d}{c}$ pens can be purchased with d dollars.

At the heart of the study of algebra is the use of variables. It is the variables in the problems above that make them abstract. But it is variables that enable us to generalize situations and state rules about mathematics.

Try the following problems.

1. How many nickels are in d dollars?

2. How many copies can you make on a coin-operated copy machine if you have only d dollars and each copy costs c cents?

3. If you travel m miles on 1 gal of gasoline, how far can you travel on g gallons of gasoline?

4. If you walk a mile in x minutes, how far can you walk in h hours?

5. If one photocopy costs n nickels, how many photocopies can you make for q quarters?

PROJECTS AND GROUP ACTIVITIES

Averages

We often discuss temperature in terms of average high or average low temperature. Temperatures collected over a period of time are analyzed to determine, for example, the average high temperature for a given month in your city or state. The following activity is planned to help you better understand the concept of "average."

1. Choose two cities in the United States. We will refer to them as City X and City Y. Over an 8-day period, record the daily high temperature each day in each city.

AP Photo/David Duprey

Buffalo, NY

2. Determine the average high temperature for City X for the 8-day period. (Add the eight numbers, and then divide the sum by 8.) Do not round your answer.

3. Subtract the average high temperature for City X from each of the eight daily high temperatures for City X. You should have a list of eight numbers; the list should include positive numbers, negative numbers, and possibly zero.

4. Find the sum of the list of eight differences recorded in Step 3.

Panoramic Images/Getty Images

Phoenix, AZ

5. Repeat Steps 2 through 4 for City Y.

6. Compare the two sums found in Steps 4 and 5 for City X and City Y.

7. If you were to conduct this activity again, what would you expect the outcome to be? Use the results to explain what an average high temperature is. In your own words, explain what "average" means.

CHAPTER 11

SUMMARY

KEY WORDS	EXAMPLES
A *variable* is a letter of the alphabet used to stand for a quantity that is unknown or that can change. An expression that contains one or more variables is a *variable expression.* Replacing the variable or variables in a variable expression and then simplifying the resulting numerical expression is called *evaluating the variable expression.* [11.1A, p.460]	Evaluate $5x^3 + 2y - 6$ when $x = -1$ and $y = 4$. $5x^3 + 2y - 6$ $5(-1)^3 + 2(4) - 6 = 5(-1) + 2(4) - 6$ $= -5 + 8 - 6$ $= 3 - 6$ $= 3 + (-6) = -3$
The *terms of a variable expression* are the addends of the expression. A *variable term* consists of a *numerical coefficient* and a *variable part.* A *constant term* has no variable part. [11.1B, pp. 461–462]	The variable expression $-3x^2 + 2x - 5$ has three terms: $-3x^2$, $2x$, and -5. $-3x^2$ and $2x$ are variable terms. -5 is a constant term. For the term $-3x^2$, the coefficient is -3 and the variable part is x^2.
Like terms of a variable expression have the same variable part. Constant terms are considered like terms. [11.1B, p. 462]	$-6a^3b^2$ and $4a^3b^2$ are like terms.
An *equation* expresses the equality of two mathematical expressions. [11.2A, p. 470]	$5x + 6 = 7x - 3$ $y = 4x - 10$ $3a^2 - 6a + 4 = 0$
A *solution of an equation* is a number that, when substituted for the variable, results in a true equation. [11.2A, p. 470]	6 is a solution of $x - 4 = 2$ because $6 - 4 = 2$ is a true equation.
Solving an equation means finding a solution of the equation. The goal is to rewrite the equation in the form *variable* = *constant.* [11.2B, p. 471]	$x = 5$ is in the form *variable* = *constant.* The solution of the equation $x = 5$ is the constant 5 because $5 = 5$ is a true equation.
A *formula* is an equation that expresses a relationship among variables. [11.2D, p. 475]	The relationship between Celsius temperature and Fahrenheit temperature is given by the formula $F = 1.8C + 32$, where F represents degrees Fahrenheit and C represents degrees Celsius.
Some of the words and phrases that translate to *equals* are *is, is equal to, amounts to,* and *represents.* [11.6A, p. 498]	"Eight plus a number is ten" translates to $8 + x = 10$.

ESSENTIAL RULES AND PROCEDURES	EXAMPLES
Commutative Property of Addition [11.1B, p. 462] $a + b = b + a$	$-9 + 5 = 5 + (-9)$

Associative Property of Addition [11.1B, p. 462]
$(a + b) + c = a + (b + c)$

$(-6 + 4) + 2 = -6 + (4 + 2)$

Commutative Property of Multiplication [11.1C, p. 464]
$a \cdot b = b \cdot a$

$-5(10) = 10(-5)$

Associative Property of Multiplication [11.1C, p. 464]
$(a \cdot b) \cdot c = a \cdot (b \cdot c)$

$(-3 \cdot 4) \cdot 6 = -3 \cdot (4 \cdot 6)$

Distributive Property [11.1C, p. 464]
$a(b + c) = ab + ac$
$a(b - c) = ab - ac$

$2(x + 7) = 2(x) + 2(7) = 2x + 14$
$5(4x - 3) = 5(4x) - 5(3) = 20x - 15$

Addition Property of Zero [11.2B, p. 471]
The sum of a term and zero is the term.
$a + 0 = a$ or $0 + a = a$

$-16 + 0 = -16$

Addition Property of Equations [11.2B, p. 471]
If a, b, and c are algebraic expressions, then the equations $a = b$ and $a + c = b + c$ have the same solutions.

The same number or variable expression can be added to each side of an equation without changing the solution of the equation.

$$x + 7 = 20$$
$$x + 7 + (-7) = 20 + (-7)$$
$$x = 13$$

Multiplication Property of Reciprocals [11.2C, p. 473]
The product of a nonzero term and its reciprocal equals 1.

$8 \cdot \frac{1}{8} = 1$

Multiplication Property of One [11.2C, p. 473]
The product of a term and 1 is the term.

$-7(1) = -7$

Multiplication Property of Equations [11.2C, p. 473]
If a, b, and c are algebraic expressions and $c \neq 0$, then the equation $a = b$ has the same solutions as the equation $ac = bc$.

Each side of an equation can be multiplied by the same nonzero number without changing the solution of the equation.

$$\frac{3}{4}x = 24$$
$$\frac{4}{3} \cdot \frac{3}{4}x = \frac{4}{3} \cdot 24$$
$$x = 32$$

Steps in Solving General First-Degree Equations [11.4B, p. 488]

1. Use the Distributive Property to remove parentheses.
2. Combine like terms on each side of the equation.
3. Use the Addition Property of Equations to rewrite the equation with only one variable term.
4. Use the Addition Property of Equations to rewrite the equation with only one constant term.
5. Use the Multiplication Property of Equations to rewrite the equation so that the coefficient of the variable term is 1.

$$8 - 4(2x + 3) = 2(1 - x)$$
$$8 - 8x - 12 = 2 - 2x$$
$$-8x - 4 = 2 - 2x$$
$$-8x + 2x - 4 = 2 - 2x + 2x$$
$$-6x - 4 = 2$$
$$-6x - 4 + 4 = 2 + 4$$
$$-6x = 6$$
$$\frac{-6x}{-6} = \frac{6}{-6}$$
$$x = -1$$

CHAPTER 11

CONCEPT REVIEW

Test your knowledge of the concepts presented in this chapter. Answer each question. Then check your answers against the ones provided in the Answer Section.

1. How do you evaluate a variable expression?

2. How do you add like terms?

3. How do you simplify a variable expression containing parentheses?

4. What is a solution of an equation?

5. How do you check the solution to an equation?

6. What is the Multiplication Property of Equations?

7. What is the Addition Property of Equations?

8. What properties are applied to solve the equation $5x - 4 = 26$?

9. After you substitute into a formula, how do you solve it?

10. How do you isolate the variable in the equation $5x - 3 = 10 - 8x$?

11. Name some mathematical terms that translate into addition.

12. Name some mathematical terms that translate into "equals."

CHAPTER 11

REVIEW EXERCISES

1. Simplify: $-2(a - b)$

2. Is -2 a solution of the equation $3x - 2 = -8$?

3. Solve: $x - 3 = -7$

4. Solve: $-2x + 5 = -9$

5. Evaluate $a^2 - 3b$ when $a = 2$ and $b = -3$.

6. Solve: $-3x = 27$

7. Solve: $\frac{2}{3}x + 3 = -9$

8. Simplify: $3x - 2(3x - 2)$

9. Solve: $6x - 9 = -3x + 36$

10. Solve: $x + 3 = -2$

11. Is 5 a solution of the equation $3x - 5 = -10$?

12. Evaluate $a^2 - (b \div c)$ when $a = -2$, $b = 8$, and $c = -4$.

13. Solve: $3(x - 2) + 2 = 11$

14. Solve: $35 - 3x = 5$

15. Simplify: $6bc - 7bc + 2bc - 5bc$

16. Solve: $7 - 3x = 2 - 5x$

17. Solve: $-\frac{3}{8}x = -\frac{15}{32}$

18. Simplify: $\frac{1}{2}x^2 - \frac{1}{3}x^2 + \frac{1}{5}x^2 + 2x^2$

19. Solve: $5x - 3(1 - 2x) = 4(2x - 1)$

20. Solve: $\frac{5}{6}x - 4 = 5$

21. Markup A pharmacy uses a markup rate of 35% on its over-the-counter medications. Find the cost of a probiotic that sells for \$28.62. Use the formula $S = C + R \cdot C$, where S is the selling price, C is the cost, and R is the markup rate.

22. Temperature Conversion Find the Celsius temperature when the Fahrenheit temperature is 100°. Use the formula $F = 1.8C + 32$, where F is the Fahrenheit temperature and C is the Celsius temperature. Round to the nearest tenth.

23. Translate "the total of n and the quotient of n and five" into a mathematical expression.

24. Translate "the sum of five more than a number and one-third of the number" into a mathematical expression.

25. The difference between nine and twice a number is five. Find the number.

26. The product of five and a number is fifty. Find the number.

27. Discount An automatic defibrillator is on sale for \$1276. This is 80% of the regular price. Find the regular price of the defibrillator.

28. Weight Loss A patient referred to a weight-loss clinic by a doctor now weighs 187 lb. The dietician calculates that the patient has lost an amount of weight equivalent to 15% of his weight prior to beginning the diet. How much did the patient weigh before beginning the diet?

CHAPTER 11

TEST

1. Solve: $\frac{x}{5} - 12 = 7$

2. Solve: $x - 12 = 14$

3. Simplify: $3y - 2x - 7y - 9x$

4. Solve: $8 - 3x = 2x - 8$

5. Solve: $3x - 12 = -18$

6. Evaluate $c^2 - (2a + b^2)$ when $a = 3$, $b = -6$, and $c = -2$.

7. Is 3 a solution of the equation $x^2 + 3x - 7 = 3x - 2$?

8. Simplify: $9 - 8ab - 6ab$

9. Solve: $-5x = 14$

10. Simplify: $3y + 5(y - 3) + 8$

11. Solve: $3x - 4(x - 2) = 8$

12. Solve: $5 = 3 - 4x$

13. Evaluate $\frac{x^2}{y} - \frac{y^2}{x}$ for $x = 3$ and $y = -2$.

14. Solve: $\frac{5}{8}x = -10$

15. Solve: $y - 4y + 3 = 12$

16. Solve: $2x + 4(x - 3) = 5x - 1$

17. **Finance** A loan of \$6600 is to be paid in 48 equal monthly installments. Find the monthly payment. Use the formula $L = P \cdot N$, where L is the loan amount, P is the monthly payment, and N is the number of months.

18. **Forensic Science** The male victim of a crime has a femur that is 18.9 in. long. Estimate the height of this man using the formula $H = 1.9L + 32$, where H is the man's height in inches and L is the length of his femur in inches. Round to the nearest tenth.

19. **Physics** Find the time required for a falling object to increase in velocity from 24 ft/s to 392 ft/s. Use the formula $V = V_0 + 32t$, where V is the final velocity of a falling object, V_0 is the starting velocity of the falling object, and t is the time for the object to fall.

20. Translate "the sum of x and one-third of x" into a mathematical expression.

21. Translate "five times the sum of a number and three" into a mathematical expression.

22. Translate "three less than two times a number is seven" into an equation and solve.

23. The total of five and three times a number is the number minus two. Find the number.

24. **Compensation** Eduardo Santos earned \$3600 last month. This salary is the sum of a base monthly salary of \$1200 and a 6% commission on total sales. Find Eduardo's total sales for the month.

25. **Ideal Weight** The "Devine formula" published in 1974 was originally developed in order to calculate dosages of certain medications based on weight. The formula for women states that the ideal body weight in kilograms is equal to the sum of 45.5 and 2.3 kg per inch over 5 ft. How tall is a woman whose ideal weight is 57 kg?

CUMULATIVE REVIEW EXERCISES

1. Simplify: $6^2 - (18 - 6) \div 4 + 8$

2. Subtract: $3\frac{1}{6} - 1\frac{7}{15}$

3. Simplify: $\left(\frac{3}{8} - \frac{1}{4}\right) \div \frac{3}{4} + \frac{4}{9}$

4. Multiply: 9.67×0.0049

5. Write "\$182 earned in 20 hours" as a unit rate.

6. Solve the proportion $\frac{2}{3} = \frac{n}{40}$. Round to the nearest hundredth.

7. Write $5\frac{1}{3}\%$ as a fraction.

8. What percent of 30 is 42?

9. 8 is 125% of what number?

10. Multiply: 3 ft 9 in. $\times$ 5

11. Convert $1\frac{3}{8}$ lb to ounces.

12. Convert 282 mg to grams.

13. Add: $-2 + 5 + (-8) + 4$

14. Find -6 less than 13.

15. Simplify: $(-2)^2 - (-8) \div (3 - 5)^2$

16. Evaluate $3ab - 2ac$ when $a = -2$, $b = 6$, and $c = -3$.

17. Simplify: $3z - 2x + 5z - 8x$

18. Simplify: $6y - 3(y - 5) + 8$

19. Solve: $2x - 5 = -7$

20. Solve: $7x - 3(x - 5) = -10$

21. Solve: $-\frac{2}{3}x = 5$

22. Solve: $\frac{x}{3} - 5 = -12$

23. Education In a mathematics class of 34 students, 6 received an A grade. Find the percent of students in the mathematics class who received an A grade. Round to the nearest tenth of a percent.

24. Markup The manager of a medical spa uses a markup rate of 40%. Find the price of a hair and skin multinutrient that costs the store \$28.50.

25. Discount A wheelchair regularly priced at \$450 is on sale for \$369.
a. What is the discount?
b. What is the discount rate?

26. Simple Interest Stewart Physical Therapy borrowed \$80,000 at a simple interest rate of 11% for 4 months. What is the simple interest due on the loan? Round to the nearest cent.

27. Blood Donors According to the American Red Cross, 16 million blood donations were collected in a recent year. These donations were made by occasional donors (19%), first-time donors (31%) and regular donors (50%). Of the donations made that year, how many donations were made by first-time donors?

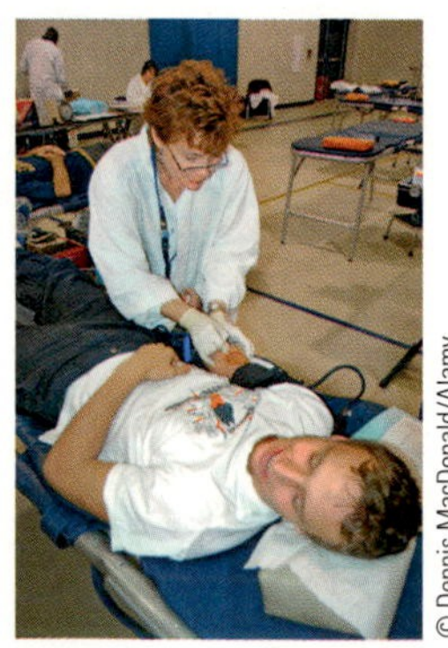

© Dennis MacDonald/Alamy

28. Probability A tetrahedral die is one with four triangular sides numbered 1 to 4. If two tetrahedral dice are rolled, what is the probability that the sum of the upward faces is 7?

29. Compensation Sunah Yee, a sales executive, receives a base salary of \$800 plus an 8% commission on total sales. Find the total sales during a month in which Sunah earned \$3400. Use the formula $M = S \cdot R + B$, where M is the monthly earnings, S is the total sales, R is the commission rate, and B is the base monthly salary.

30. Three less than eight times a number is three more than five times the number. Find the number.

FINAL EXAM

1. Subtract: 100,914 − 97,655

2. Find 34,821 divided by 657.

3. Find 90,001 decreased by 29,796.

4. Simplify: $3^2 \cdot (5 - 3)^2 \div 3 + 4$

5. Find the LCM of 9, 12, and 16.

6. Add: $\frac{3}{8} + \frac{5}{6} + \frac{1}{5}$

7. Subtract: $7\frac{5}{12} - 3\frac{13}{16}$

8. Find the product of $3\frac{5}{8}$ and $1\frac{5}{7}$.

9. Divide: $1\frac{2}{3} \div 3\frac{3}{4}$

10. Simplify: $\left(\frac{2}{3}\right)^3 \cdot \left(\frac{3}{4}\right)^2$

11. Simplify: $\left(\frac{2}{3}\right)^2 \div \left(\frac{3}{4} + \frac{1}{3}\right) - \frac{1}{3}$

12. Add: $\begin{array}{r} 4.972 \\ 28.6 \\ 1.88 \\ +\ 128.725 \\ \hline \end{array}$

13. Multiply: $\begin{array}{r} 2.97 \\ \times\ 0.0094 \\ \hline \end{array}$

14. Divide: $0.062\overline{)0.0426}$
Round to the nearest hundredth.

15. Convert 0.45 to a fraction in simplest form.

16. Write "323.4 miles on 13.2 gallons of gas" as a unit rate.

17. Solve the proportion $\frac{12}{35} = \frac{n}{160}$.
Round to the nearest tenth.

18. Write $22\frac{1}{2}\%$ as a fraction.

19. Write 1.35 as a percent.

20. Write $\frac{5}{4}$ as a percent.

21. Find 120% of 30.

22. 12 is what percent of 9?

23. 42 is 60% of what number?

24. Convert $1\frac{2}{3}$ ft to inches.

25. Subtract: 3 ft 2 in. − 1 ft 10 in.

26. Convert 40 oz to pounds.

27. Find the sum of 3 lb 12 oz and 2 lb 10 oz.

28. Convert 18 pt to gallons.

29. Divide: $3\overline{)5 \text{ gal } 1 \text{ qt}}$

30. Convert 2.48 m to centimeters.

31. Convert 4 m 62 cm to meters.

32. Convert 1 kg 614 g to kilograms.

33. Convert 2 L 67 ml to milliliters.

34. Convert 55 mi to kilometers. Round to the nearest hundredth. (1.61 km $\approx$ 1 mi)

35. **Health** For a healthful diet, it is recommended that 55% of the daily intake of Calories come from carbohydrates. Find the appropriate daily intake of Calories from carbohydrates if you want to limit your daily Calorie intake to 1800 Calories.

36. Write 0.0000000679 in scientific notation.

37. **Dilution Ratio** A dilution ratio can indicate the concentration of a solution and many times is expressed as a percent. What is the percent concentration of a solution having a dilution ratio of $\frac{1}{5}$?

38. **Office Expenses** A dental practice purchased a financial software package at a special discount. The cost of the package was $1500, which is 80% of the original cost. Determine the original cost of the software.

39. **IV Infusions** A doctor orders an IV of 1500 milliliters of D5W to be infused over a 12-hour period. Find the unit rate for the infusion.

40. Add: $-2 + 8 + (-10)$

41. Subtract: $-30 - (-15)$

42. Multiply: $2\frac{1}{2} \times \left(-\frac{1}{5}\right)$

43. Find the quotient of $-1\frac{3}{8}$ and $5\frac{1}{2}$.

44. Simplify: $(-4)^2 \div (1 - 3)^2 - (-2)$

45. Simplify: $2x - 3(x - 4) + 5$

46. Solve: $\frac{2}{3}x = -12$

47. Solve: $3x - 5 = 10$

48. Solve: $8 - 3x = x + 4$

49. **Banking** You have $872.48 in your checking account. You write checks for $321.88 and $34.23 and then make a deposit of $443.56. Find your new checking account balance.

50. Solutions A saline solution has a concentration ratio of 1.5 mg of salt in 100 ml of solution. How many milligrams of salt will be needed to produce 500 ml of a saline solution having this same concentration?

51. Inflation The price of a 100-tablet bottle of vitamins rose from \$15.99 per bottle to \$18.99 per bottle. What percent increase does this amount represent? Round to the nearest tenth of a percent.

52. Compensation A sales executive received commissions of \$4320, \$3572, \$2864, and \$4420 during a 4-month period. Find the mean monthly income from commissions for the 4 months.

53. Simple Interest A new medical practice borrows \$120,000 for 9 months at an annual interest rate of 8%. What is the simple interest due on the loan?

54. Probability If two dice are tossed, what is the probability that the sum of the dots on the upward faces is divisible by 3?

55. Wars The top four highest death counts, by country, in World War II are shown in the circle graph. What percent of the total death count in all four countries is the death count of China? Round to the nearest tenth of a percent.

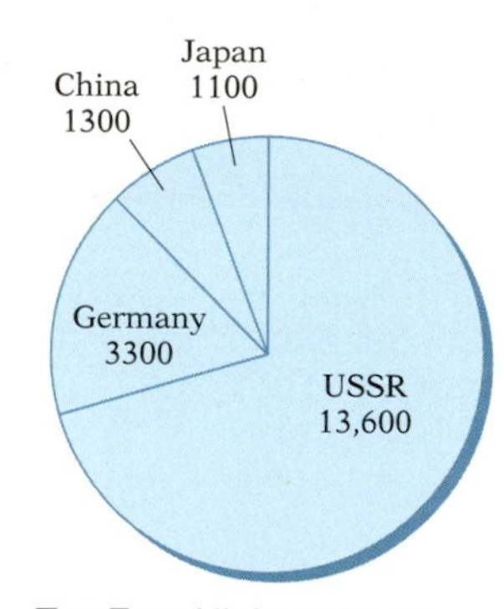

Top Four Highest Death Counts in World War II (in thousands)

Source: U.S. Department of Defense

56. IV Therapy The pediatric dose of an IV infusion of medication is 25 units/kg. If a newborn baby weighs 3200 g and needs an infusion of this medication, how many units will the doctor prescribe?

57. Height An 8-year-old girl is $\frac{2}{3}$ the height of her mother. If the 8-year-old girl is 46 in. tall, how tall is her mother? Give your answer in feet and inches.

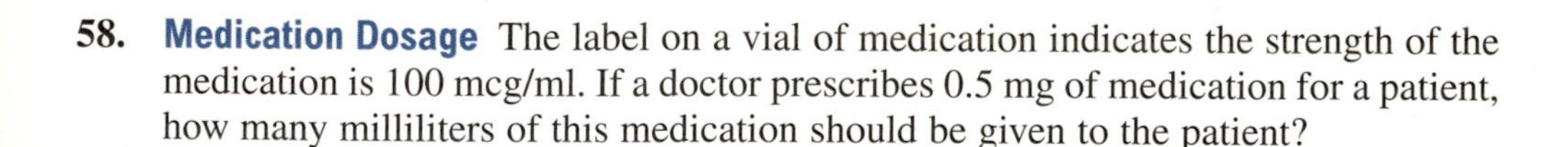

58. Medication Dosage The label on a vial of medication indicates the strength of the medication is 100 mcg/ml. If a doctor prescribes 0.5 mg of medication for a patient, how many milliliters of this medication should be given to the patient?

59. Height A nurse measures the height of a patient as 73 in. Convert this measurement to feet and inches.

60. Integer Problems Five less than the quotient of a number and two is equal to three. Find the number.

Appendix

Compound Interest Table

Compounded Annually							
	4%	**5%**	**6%**	**7%**	**8%**	**9%**	**10%**
1 year	1.04000	1.05000	1.06000	1.07000	1.08000	1.09000	1.10000
5 years	1.21665	1.27628	1.33823	1.40255	1.46933	1.53862	1.61051
10 years	1.48024	1.62890	1.79085	1.96715	2.15893	2.36736	2.59374
15 years	1.80094	2.07893	2.39656	2.75903	3.17217	3.64248	4.17725
20 years	2.19112	2.65330	3.20714	3.86968	4.66095	5.60441	6.72750
Compounded Semiannually							
	4%	**5%**	**6%**	**7%**	**8%**	**9%**	**10%**
1 year	1.04040	1.05062	1.06090	1.07123	1.08160	1.09203	1.10250
5 years	1.21899	1.28008	1.34392	1.41060	1.48024	1.55297	1.62890
10 years	1.48595	1.63862	1.80611	1.98979	2.19112	2.41171	2.65330
15 years	1.81136	2.09757	2.42726	2.80679	3.24340	3.74531	4.32194
20 years	2.20804	2.68506	3.26204	3.95926	4.80102	5.81634	7.03999
Compounded Quarterly							
	4%	**5%**	**6%**	**7%**	**8%**	**9%**	**10%**
1 year	1.04060	1.05094	1.06136	1.07186	1.08243	1.09308	1.10381
5 years	1.22019	1.28204	1.34686	1.41478	1.48595	1.56051	1.63862
10 years	1.48886	1.64362	1.81402	2.00160	2.20804	2.43519	2.68506
15 years	1.81670	2.10718	2.44322	2.83182	3.28103	3.80013	4.39979
20 years	2.21672	2.70148	3.29066	4.00639	4.87544	5.93015	7.20957
Compounded Monthly							
	4%	**5%**	**6%**	**7%**	**8%**	**9%**	**10%**
1 year	1.04074	1.051162	1.061678	1.072290	1.083000	1.093807	1.104713
5 years	1.220997	1.283359	1.348850	1.417625	1.489846	1.565681	1.645309
10 years	1.490833	1.647009	1.819397	2.009661	2.219640	2.451357	2.707041
15 years	1.820302	2.113704	2.454094	2.848947	3.306921	3.838043	4.453920
20 years	2.222582	2.712640	3.310204	4.038739	4.926803	6.009152	7.328074
Compounded Daily							
	4%	**5%**	**6%**	**7%**	**8%**	**9%**	**10%**
1 year	1.04080	1.05127	1.06183	1.07250	1.08328	1.09416	1.10516
5 years	1.22139	1.28400	1.34983	1.41902	1.49176	1.56823	1.64861
10 years	1.49179	1.64866	1.82203	2.01362	2.22535	2.45933	2.71791
15 years	1.82206	2.11689	2.45942	2.85736	3.31968	3.85678	4.48077
20 years	2.22544	2.71810	3.31979	4.05466	4.95217	6.04830	7.38703

Compound Interest Table

Compounded Annually							
	11%	**12%**	**13%**	**14%**	**15%**	**16%**	**17%**
1 year	1.11000	1.12000	1.13000	1.14000	1.15000	1.16000	1.17000
5 years	1.68506	1.76234	1.84244	1.92542	2.01136	2.10034	2.19245
10 years	2.83942	3.10585	3.39457	3.70722	4.04556	4.41144	4.80683
15 years	4.78459	5.47357	6.25427	7.13794	8.13706	9.26552	10.53872
20 years	8.06239	9.64629	11.52309	13.74349	16.36654	19.46076	23.10560
Compounded Semiannually							
	11%	**12%**	**13%**	**14%**	**15%**	**16%**	**17%**
1 year	1.11303	1.12360	1.13423	1.14490	1.15563	1.16640	1.17723
5 years	1.70814	1.79085	1.87714	1.96715	2.06103	2.15893	2.26098
10 years	2.91776	3.20714	3.52365	3.86968	4.24785	4.66096	5.11205
15 years	4.98395	5.74349	6.61437	7.61226	8.75496	10.06266	11.55825
20 years	8.51331	10.28572	12.41607	14.97446	18.04424	21.72452	26.13302
Compounded Quarterly							
	11%	**12%**	**13%**	**14%**	**15%**	**16%**	**17%**
1 year	1.11462	1.12551	1.13648	1.14752	1.15865	1.16986	1.18115
5 years	1.72043	1.80611	1.89584	1.98979	2.08815	2.19112	2.29891
10 years	2.95987	3.26204	3.59420	3.95926	4.36038	4.80102	5.28497
15 years	5.09225	5.89160	6.81402	7.87809	9.10513	10.51963	12.14965
20 years	8.76085	10.64089	12.91828	15.67574	19.01290	23.04980	27.93091
Compounded Monthly							
	11%	**12%**	**13%**	**14%**	**15%**	**16%**	**17%**
1 year	1.115719	1.126825	1.138032	1.149342	1.160755	1.172271	1.183892
5 years	1.728916	1.816697	1.908857	2.005610	2.107181	2.213807	2.325733
10 years	2.989150	3.300387	3.643733	4.022471	4.440213	4.900941	5.409036
15 years	5.167988	5.995802	6.955364	8.067507	9.356334	10.849737	12.579975
20 years	8.935015	10.892554	13.276792	16.180270	19.715494	24.019222	29.257669
Compounded Daily							
	11%	**12%**	**13%**	**14%**	**15%**	**16%**	**17%**
1 year	1.11626	1.12747	1.13880	1.15024	1.16180	1.17347	1.18526
5 years	1.73311	1.82194	1.91532	2.01348	2.11667	2.22515	2.33918
10 years	3.00367	3.31946	3.66845	4.05411	4.48031	4.95130	5.47178
15 years	5.20569	6.04786	7.02625	8.16288	9.48335	11.01738	12.79950
20 years	9.02203	11.01883	13.45751	16.43582	20.07316	24.51534	29.94039

To use this table:

1. Locate the section that gives the desired compounding period.
2. Locate the interest rate in the top row of that section.
3. Locate the number of years in the left-hand column of that section.
4. Locate the number where the interest-rate column and the number-of-years row meet. This is the compound interest factor.

Example An investment yields an annual interest rate of 10% compounded quarterly for 5 years.
The compounding period is "compounded quarterly."
The interest rate is 10%.
The number of years is 5.
The number where the row and column meet is 1.63862. This is the compound interest factor.

Monthly Payment Table

	4%	5%	6%	7%	8%	9%
1 year	0.0851499	0.0856075	0.0860664	0.0865267	0.0869884	0.0874515
2 years	0.0434249	0.0438714	0.0443206	0.0447726	0.0452273	0.0456847
3 years	0.0295240	0.0299709	0.0304219	0.0308771	0.0313364	0.0317997
4 years	0.0225791	0.0230293	0.0234850	0.0239462	0.0244129	0.0248850
5 years	0.0184165	0.0188712	0.0193328	0.0198012	0.0202764	0.0207584
15 years	0.0073969	0.0079079	0.0084386	0.0089883	0.0095565	0.0101427
20 years	0.0060598	0.0065996	0.0071643	0.0077530	0.0083644	0.0089973
25 years	0.0052784	0.0058459	0.0064430	0.0070678	0.0077182	0.0083920
30 years	0.0047742	0.0053682	0.0059955	0.0066530	0.0073376	0.0080462

	10%	11%	12%	13%	
1 year	0.0879159	0.0883817	0.0888488	0.0893173	
2 years	0.0461449	0.0466078	0.0470735	0.0475418	
3 years	0.0322672	0.0327387	0.0332143	0.0336940	
4 years	0.0253626	0.0258455	0.0263338	0.0268275	
5 years	0.0212470	0.0217424	0.0222445	0.0227531	
15 years	0.0107461	0.0113660	0.0120017	0.0126524	
20 years	0.0096502	0.0103219	0.0110109	0.0117158	
25 years	0.0090870	0.0098011	0.0105322	0.0112784	
30 years	0.0087757	0.0095232	0.0102861	0.0110620	

To use this table:

1. Locate the desired interest rate in the top row.
2. Locate the number of years in the left-hand column.
3. Locate the number where the interest-rate column and the number-of-years row meet. This is the monthly payment factor.

Example A home has a 30-year mortgage at an annual interest rate of 12%.
The interest rate is 12%.
The number of years is 30.
The number where the row and column meet is 0.0102861. This is the monthly payment factor.

Table of Measurements

Prefixes in the Metric System of Measurement

kilo-	1000	deci-	0.1
hecto-	100	centi-	0.01
deca-	10	milli-	0.001
		micro-	0.000001

Measurement Abbreviations U.S. Customary System

Length

in.	inches
ft	feet
yd	yards
mi	miles

Capacity

oz	fluid ounces
c	cups
qt	quarts
gal	gallons
t	teaspoon

Weight

oz	ounces
lb	pounds

Area

in^2	square inches
ft^2	square feet
yd^2	square yards
mi^2	square miles

Rate

ft/s	feet per second
mi/h	miles per hour

Time

h	hours
min	minutes
s	seconds

Metric System

Length

mm	millimeter
cm	centimeter
m	meter
km	kilometer

Capacity

μl	microliter
ml	milliliter
cl	centiliter
dl	deciliter
L	liter
kl	kiloliter

Weight/Mass

mcg	microgram
mg	milligram
cg	centigram
g	gram
kg	kilogram

Area

cm^2	square centimeters
m^2	square meters
km^2	square kilometers

Rate

m/s	meters per second
km/s	kilometers per second
km/h	kilometers per hour

Time

h	hours
min	minutes
s	seconds

Table of Properties

Properties of Real Numbers

Commutative Property of Addition
If a and b are real numbers, then $a + b = b + a$.

Commutative Property of Multiplication
If a and b are real numbers, then $a \cdot b = b \cdot a$.

Associative Property of Addition
If a, b, and c are real numbers, then
$(a + b) + c = a + (b + c)$.

Associative Property of Multiplication
If a, b, and c are real numbers, then
$(a \cdot b) \cdot c = a \cdot (b \cdot c)$.

Addition Property of Zero
If a is a real number, then $a + 0 = 0 + a = a$.

Multiplication Property of Zero
If a is a real number, then $a \cdot 0 = 0 \cdot a = 0$.

Multiplication Property of One
If a is a real number, then $a \cdot 1 = 1 \cdot a = a$.

Inverse Property of Addition
If a is a real number, then $a + (-a) = (-a) + a = 0$.

Inverse Property of Multiplication
If a is a real number and $a \neq 0$, then
$a \cdot \frac{1}{a} = \frac{1}{a} \cdot a = 1$.

Distributive Property
If a, b, and c are real numbers, then
$a(b + c) = ab + ac$.

Properties of Zero and One in Division

Any number divided by 1 is the number.
Division by zero is not allowed.

Any number other than zero divided by itself is 1.
Zero divided by a number other than zero is zero.

Solutions to "You Try It"

SOLUTIONS TO CHAPTER 1 "YOU TRY IT"

SECTION 1.1

You Try It 1

Number line marked 0 1 2 3 4 5 6 7 8 9 10 11 12 13 14, with a dot at 6.

You Try It 2 **a.** $45 > 29$ **b.** $27 > 0$

You Try It 3 Thirty-six million four hundred sixty-two thousand seventy-five

You Try It 4 452,007

You Try It 5 $60{,}000 + 8000 + 200 + 80 + 1$

You Try It 6 $100{,}000 + 9000 + 200 + 7$

You Try It 7

Given place value

368,492

$8 > 5$

368,492 rounded to the nearest ten-thousand is 370,000.

You Try It 8

Given place value

3962

$6 > 5$

3962 rounded to the nearest hundred is 4000.

SECTION 1.2

You Try It 1

$$\begin{array}{r} \scriptsize{1\,1} \\ 347 \\ +12{,}453 \\ \hline 12{,}800 \end{array}$$

- $7 + 3 = 10$
  Write the 0 in the ones column. Carry the 1 to the tens column.
  $1 + 4 + 5 = 10$
  Write the 0 in the tens column. Carry the 1 to the hundreds column.
  $1 + 3 + 4 = 8$

347 increased by 12,453 is 12,800.

You Try It 2

$$\begin{array}{r} \scriptsize{2} \\ 95 \\ 88 \\ +\ 67 \\ \hline 250 \end{array}$$

- $5 + 8 + 7 = 20$
  Write the 0 in the ones column.
  Carry the 2 to the tens column.

You Try It 3

$$\begin{array}{r} \scriptsize{1\ 1\ 2\ 1} \\ 392 \\ 4{,}079 \\ 89{,}035 \\ +\ 4{,}992 \\ \hline 98{,}498 \end{array}$$

You Try It 4

Strategy To find the total fluid intake, read the table to find the estimated volume for the broth, ice cream, and milk. Then add the numbers.

Solution

$$\begin{array}{r} 150 \\ 120 \\ +240 \\ \hline 510 \end{array}$$

The patient's total fluid intake at lunch is 510 milliliters.

SECTION 1.3

You Try It 1

$$\begin{array}{r} 8925 \\ -\ 6413 \\ \hline 2512 \end{array} \qquad \textit{Check:} \quad \begin{array}{r} 6413 \\ +\ 2512 \\ \hline 8925 \end{array}$$

You Try It 2

$$\begin{array}{r} 17{,}504 \\ -\ 9{,}302 \\ \hline 8{,}202 \end{array} \qquad \textit{Check:} \quad \begin{array}{r} 9{,}302 \\ +\ 8{,}202 \\ \hline 17{,}504 \end{array}$$

You Try It 3

$$\begin{array}{r} \scriptsize{2\ \ 14\ \ 7\ \ 11} \\ \not{3}\ \not{4}\ \not{8}\ \not{1} \\ -\quad 8\ 6\ 5 \\ \hline 2\ 6\ 1\ 6 \end{array} \qquad \textit{Check:} \quad \begin{array}{r} 865 \\ +\ 2616 \\ \hline 3481 \end{array}$$

You Try It 4

$$\begin{array}{r} \scriptsize{15} \\ \scriptsize{4\ \ \not{5}\ \ 12} \\ 5\ 4,\not{5}\ \not{6}\ \not{2} \\ -\ 1\ 4,4\ 8\ 5 \\ \hline 4\ 0,0\ 7\ 7 \end{array} \qquad \textit{Check:} \quad \begin{array}{r} 14{,}485 \\ +\ 40{,}077 \\ \hline 54{,}562 \end{array}$$

You Try It 5

$$\begin{array}{r} \scriptstyle 3\ \ 10 \\ 6\not{4},\not{0}03 \\ -54,936 \\ \hline \end{array}$$

- There are two zeros in the minuend. Borrow 1 thousand from the thousands column and write 10 in the hundreds column.

$$\begin{array}{r} \scriptstyle 9 \\ \scriptstyle 3\ \ \not{10}\ 10 \\ 6\not{4},\not{0}\not{0}3 \\ -54,936 \\ \hline \end{array}$$

- Borrow 1 hundred from the hundreds column and write 10 in the tens column.

$$\begin{array}{r} \scriptstyle 13\ \ \ 9\ \ 9 \\ \scriptstyle 5\ \ \not{3}\ \ \not{10}\ \not{10}\ 13 \\ \not{6}\not{4},\not{0}\not{0}\not{3} \\ -54,936 \\ \hline 9,067 \end{array}$$

- Borrow 1 ten from the tens column and add 10 to the 3 in the ones column.

Check:

$$\begin{array}{r} 54,936 \\ +\ 9,067 \\ \hline 64,003 \end{array}$$

You Try It 6

Strategy Find the total number of calories that Maria will consume by adding the calories for breakfast, lunch, and snack. Then subtract this from her maximum allowance of 1550 calories.

Solution

400 breakfast
450 lunch
\+ 93 popcorn (3 cups)
943 total

1550 daily allowance
− 943 total calories used
607 remaining

Maria has 607 calories remaining for her supper.

You Try It 7

Strategy To find your take-home pay:

- Add to find the total of the deductions (127 + 18 + 35).
- Subtract the total of the deductions from your total salary (638).

Solution

$$\begin{array}{r} 127 \\ 18 \\ +\ 35 \\ \hline 180 \end{array}$$ in deductions

$$\begin{array}{r} 638 \\ -\ 180 \\ \hline 458 \end{array}$$

Your take-home pay is $458.

SECTION 1.4

You Try It 1

$$\begin{array}{r} \scriptstyle 35 \\ 648 \\ \times\ \ 7 \\ \hline 4536 \end{array}$$

- 7 × 8 = 56
Write the 6 in the ones column.
Carry the 5 to the tens column.
7 × 4 = 28, 28 + 5 = 33
7 × 6 = 42, 42 + 3 = 45

You Try It 2

$$\begin{array}{r} 756 \\ \times\ 305 \\ \hline 3780 \\ 22680 \\ \hline 230,580 \end{array}$$

- 5 × 756 = 3780
Write a zero in the tens column for 0 × 756.
3 × 756 = 2268

You Try It 3

Strategy To find the surgical technician's earnings for the week, multiply the hourly earnings ($24) by the number of hours worked (38).

Solution

$$\begin{array}{r} 24 \\ \times\ 38 \\ \hline 192 \\ 72 \\ \hline 912 \end{array}$$

The surgical technician earned $912 for the week.

You Try It 4

Strategy To find the technician's pay for three weeks:

- Multiply the number of hours per week (40) times the regular hourly rate ($21).
- Multiply the weekly pay times the number of weeks (3).

Solution

$$\begin{array}{r} 21 \\ \times\ 40 \\ \hline 840 \end{array}$$ weekly pay

$$\begin{array}{r} 840 \\ \times\ \ 3 \\ \hline 2520 \end{array}$$

The biomedical equipment technician earns $2520 in three weeks.

SECTION 1.5

You Try It 1

$$9\overline{)63}\quad 7$$

Check: 7 × 9 = 63

You Try It 2

$$\begin{array}{r} 453 \\ 9\overline{)\ 4077} \\ -36 \\ \hline 47 \\ -45 \\ \hline 27 \\ -27 \\ \hline 0 \end{array}$$

Check: 453 × 9 = 4077

You Try It 3

```
      705
 9) 6345
   -63
     04
    - 0
     45
    -45
      0
```

- Think 9)4. Place 0 in quotient.
- Subtract 0 × 9.
- Bring down the 5.

Check: 705 × 9 = 6345

You Try It 4

```
     870 r5
 6) 5225
   -48
     42
    -42
     05
    - 0
      5
```

- Think 6)5. Place 0 in quotient.
- Subtract 0 × 6.

Check: (870 × 6) + 5 =
5220 + 5 = 5225

You Try It 5

```
    3,058 r3
 7) 21,409
   -21
     0 4
    -  0
      40
     -35
       59
      -56
        3
```

- Think 7)4. Place 0 in quotient.
- Subtract 0 × 7.

Check: (3058 × 7) + 3 =
21,406 + 3 = 21,409

You Try It 6

```
       109
 42) 4578
    -42
      37
     - 0
      378
    - 378
        0
```

- Think 42)37. Place 0 in quotient.
- Subtract 0 × 42.

Check: (109 × 42) = 4578

You Try It 7

```
        470 r29
 39) 18,359
    -15 6
      2 75
     -2 73
         29
        - 0
         29
```

- Think 3)18. 6 × 39 is too large. Try 5. 5 × 39 is too large. Try 4.

Check: (470 × 39) + 29 =
18,330 + 29 = 18,359

You Try It 8

```
           62 r111
 534) 33,219
     -32 04
       1 179
      -1 068
         111
```

Check: (62 × 534) + 111 =
33,108 + 111 = 33,219

You Try It 9

```
           421 r33
 515) 216,848
     -206 0
       10 84
      -10 30
          548
         -515
           33
```

Check: (421 × 515) + 33 =
216,815 + 33 = 216,848

You Try It 10

Strategy To find the number of samples each employee must process, divide the total number of samples (180) by the number of employees (12).

Solution

```
       15
 12) 180
    -12
      60
     -60
       0
```

Each employee must process 15 samples.

You Try It 11

Strategy To find the cost of the hospital room per day, divide the total cost of the room ($10,200) by the number of days the patient was in the hospital (12).

Solution

```
        850
 12) 10,200
     -9 6
       60
      -60
        00
       -00
         0
```

The cost of the hospital room was $850 per day.

SECTION 1.6

You Try It 1 $2 \cdot 2 \cdot 2 \cdot 2 \cdot 3 \cdot 3 \cdot 3 = 2^4 \cdot 3^3$

You Try It 2 $10 \cdot 10 \cdot 10 \cdot 10 \cdot 10 \cdot 10 \cdot 10 = 10^7$

You Try It 3 $2^3 \cdot 5^2 = (2 \cdot 2 \cdot 2) \cdot (5 \cdot 5) = 8 \cdot 25 = 200$

You Try It 4

$5 \cdot (8 - 4)^2 \div 4 - 2$ • Parentheses
$= 5 \cdot 4^2 \div 4 - 2$ • Exponents
$= 5 \cdot 16 \div 4 - 2$ • Multiplication and division
$= 80 \div 4 - 2$
$= 20 - 2$ • Subtraction
$= 18$

SECTION 1.7

You Try It 1

$40 \div 1 = 40$
$40 \div 2 = 20$
$40 \div 3$ Will not divide evenly
$40 \div 4 = 10$
$40 \div 5 = 8$
$40 \div 6$ Will not divide evenly
$40 \div 7$ Will not divide evenly
$40 \div 8 = 5$

1, 2, 4, 5, 8, 10, 20, and 40 are factors of 40.

You Try It 2

44	
2	22
2	11
11	1

• $44 \div 2 = 22$
• $22 \div 2 = 11$
• $11 \div 11 = 1$

$44 = 2 \cdot 2 \cdot 11$

You Try It 3

177	
3	59
59	1

• Try only 2, 3, 4, 7, and 11 because $11^2 > 59$.

$177 = 3 \cdot 59$

SOLUTIONS TO CHAPTER 2 "YOU TRY IT"

SECTION 2.1

You Try It 1

	2	3	5
12 =	(2 · 2)	3	
27 =		(3 · 3 · 3)	
50 =	2		(5 · 5)

The LCM $= 2 \cdot 2 \cdot 3 \cdot 3 \cdot 3 \cdot 5 \cdot 5 = 2700$

You Try It 2

	2	3	5
36 =	(2 · 2)	3 · 3	
60 =	2 · 2	(3)	5
72 =	2 · 2 · 2	3 · 3	

The GCF $= 2 \cdot 2 \cdot 3 = 12$.

You Try It 3

	2	3	5	11
11 =				11
24 =	2 · 2 · 2	3		
30 =	2	3	5	

Because no numbers are circled, the GCF = 1.

SECTION 2.2

You Try It 1 $4\frac{1}{4}$

You Try It 2 $\frac{17}{4}$

You Try It 3

$$\begin{array}{r} 4 \\ 5\overline{)22} \\ -20 \\ \hline 2 \end{array} \qquad \frac{22}{5} = 4\frac{2}{5}$$

You Try It 4

$$\begin{array}{r} 4 \\ 7\overline{)28} \\ -28 \\ \hline 0 \end{array} \qquad \frac{28}{7} = 4$$

You Try It 5 $14\frac{5}{8} = \frac{112 + 5}{8} = \frac{117}{8}$

SECTION 2.3

You Try It 1 $45 \div 5 = 9 \qquad \frac{3}{5} = \frac{3 \cdot 9}{5 \cdot 9} = \frac{27}{45}$

$\frac{27}{45}$ is equivalent to $\frac{3}{5}$.

You Try It 2 Write 6 as $\frac{6}{1}$.

$18 \div 1 = 18 \qquad 6 = \frac{6 \cdot 18}{1 \cdot 18} = \frac{108}{18}$

$\frac{108}{18}$ is equivalent to 6.

You Try It 3 $\frac{16}{24} = \frac{\overset{1}{\cancel{2}} \cdot \overset{1}{\cancel{2}} \cdot \overset{1}{\cancel{2}} \cdot 2}{\underset{1}{\cancel{2}} \cdot \underset{1}{\cancel{2}} \cdot \underset{1}{\cancel{2}} \cdot 3} = \frac{2}{3}$

You Try It 4 $\frac{8}{56} = \frac{\overset{1}{\cancel{2}} \cdot \overset{1}{\cancel{2}} \cdot \overset{1}{\cancel{2}}}{\underset{1}{\cancel{2}} \cdot \underset{1}{\cancel{2}} \cdot \underset{1}{\cancel{2}} \cdot 7} = \frac{1}{7}$

You Try It 5 $\frac{15}{32} = \frac{3 \cdot 5}{2 \cdot 2 \cdot 2 \cdot 2 \cdot 2} = \frac{15}{32}$

You Try It 6 $\frac{48}{36} = \frac{\overset{1}{\cancel{2}} \cdot \overset{1}{\cancel{2}} \cdot 2 \cdot 2 \cdot \overset{1}{\cancel{3}}}{\underset{1}{\cancel{2}} \cdot \underset{1}{\cancel{2}} \cdot \underset{1}{\cancel{3}} \cdot 3} = \frac{4}{3} = 1\frac{1}{3}$

SECTION 2.4

You Try It 1

$$\begin{array}{r} \frac{3}{8} \\ +\frac{7}{8} \\ \hline \frac{10}{8} = \frac{5}{4} = 1\frac{1}{4} \end{array}$$

• The denominators are the same. Add the numerators. Place the sum over the common denominator.

You Try It 2

$$\begin{array}{r} \frac{5}{12} = \frac{20}{48} \\ +\frac{9}{16} = \frac{27}{48} \\ \hline \frac{47}{48} \end{array}$$

• The LCM of 12 and 16 is 48.

You Try It 3

$$\begin{array}{r} \frac{7}{8} = \frac{105}{120} \\ +\frac{11}{15} = \frac{88}{120} \\ \hline \frac{193}{120} = 1\frac{73}{120} \end{array}$$

• The LCM of 8 and 15 is 120.

You Try It 4

$$\begin{array}{r} \frac{3}{4} = \frac{30}{40} \\ \frac{4}{5} = \frac{32}{40} \\ +\frac{5}{8} = \frac{25}{40} \\ \hline \frac{87}{40} = 2\frac{7}{40} \end{array}$$

• The LCM of 4, 5, and 8 is 40.

You Try It 5

$$7 + \frac{6}{11} = 7\frac{6}{11}$$

You Try It 6

$$\begin{array}{r} 29 \\ +17\frac{5}{12} \\ \hline 46\frac{5}{12} \end{array}$$

You Try It 7

$$\begin{array}{r} 7\frac{4}{5} = 7\frac{24}{30} \\ 6\frac{7}{10} = 6\frac{21}{30} \\ +13\frac{11}{15} = 13\frac{22}{30} \\ \hline 26\frac{67}{30} = 28\frac{7}{30} \end{array}$$

• LCM = 30

You Try It 8

$$\begin{array}{r} 9\frac{3}{8} = 9\frac{45}{120} \\ 17\frac{7}{12} = 17\frac{70}{120} \\ +10\frac{14}{15} = 10\frac{112}{120} \\ \hline 36\frac{227}{120} = 37\frac{107}{120} \end{array}$$

• LCM = 120

You Try It 9

Strategy To find the total time spent on the activities, add the three times $\left(4\frac{1}{2}, 3\frac{3}{4}, 1\frac{1}{3}\right)$.

Solution

$$\begin{array}{r} 4\frac{1}{2} = 4\frac{6}{12} \\ 3\frac{3}{4} = 3\frac{9}{12} \\ +1\frac{1}{3} = 1\frac{4}{12} \\ \hline 8\frac{19}{12} = 9\frac{7}{12} \end{array}$$

The total time spent on the three activities was $9\frac{7}{12}$ hours.

You Try It 10

Strategy To find the overtime pay:

- Find the total number of overtime ours $\left(1\frac{2}{3} + 3\frac{1}{3} + 2\right)$.
- Multiply the total number of hours by the overtime hourly wage (36).

Solution

$$\begin{array}{r} 1\frac{2}{3} \\ 3\frac{1}{3} \\ +2 \\ \hline 6\frac{3}{3} = 7 \text{ hours} \end{array} \qquad \begin{array}{r} 36 \\ \times\ 7 \\ \hline 252 \end{array}$$

Jeff earned $252 in overtime pay.

SECTION 2.5

You Try It 1

$$\begin{array}{r} \frac{16}{27} \\ -\frac{7}{27} \\ \hline \frac{9}{27} = \frac{1}{3} \end{array}$$

• The denominators are the same. Subtract the numerators. Place the difference over the common denominator.

You Try It 2

$$\begin{array}{r} \frac{13}{18} = \frac{52}{72} \\ -\frac{7}{24} = \frac{21}{72} \\ \hline \frac{31}{72} \end{array}$$

• LCM = 72

You Try It 3

$$\begin{array}{r} 17\frac{5}{9} = 17\frac{20}{36} \\ -\ 11\frac{5}{12} = 11\frac{15}{36} \\ \hline 6\frac{5}{36} \end{array} \quad \bullet\ \text{LCM} = 36$$

You Try It 4

$$\begin{array}{r} 8 \quad = 7\frac{13}{13} \\ -\ 2\frac{4}{13} = 2\frac{4}{13} \\ \hline 5\frac{9}{13} \end{array} \quad \bullet\ \text{LCM} = 13$$

You Try It 5

$$\begin{array}{r} 21\frac{7}{9} = 21\frac{28}{36} = 20\frac{64}{36} \\ -\ 7\frac{11}{12} = 7\frac{33}{36} = 7\frac{33}{36} \\ \hline 13\frac{31}{36} \end{array} \quad \bullet\ \text{LCM} = 36$$

You Try It 6

Strategy To find the baby's weight after the illness, subtract the amount of weight lost from the baby's original weight.

Solution

$$\begin{array}{r} 21\frac{1}{2} = 21\frac{2}{4} = 20\frac{6}{4} \\ -\ 1\frac{3}{4} = 1\frac{3}{4} = 1\frac{3}{4} \\ \hline 19\frac{3}{4} \end{array}$$

The baby's weight after the illness was $19\frac{3}{4}$ pounds.

You Try It 7

Strategy To find the amount of weight to be lost during the third month:

- Find the total weight loss during the first two months $\left(7\frac{1}{2} + 5\frac{3}{4}\right)$.
- Subtract the total weight loss from the goal (24 pounds).

Solution

$$\begin{array}{r} 7\frac{1}{2} = 7\frac{2}{4} \\ +\ 5\frac{3}{4} = 5\frac{3}{4} \\ \hline 12\frac{5}{4} = 13\frac{1}{4} \text{ pounds lost} \end{array}$$

$$\begin{array}{r} 24 \quad = 23\frac{4}{4} \\ -\ 13\frac{1}{4} = 13\frac{1}{4} \\ \hline 10\frac{3}{4} \text{ pounds} \end{array}$$

The patient must lose $10\frac{3}{4}$ pounds to achieve the goal.

SECTION 2.6

You Try It 1

$$\frac{4}{21} \times \frac{7}{44} = \frac{4 \cdot 7}{21 \cdot 44} = \frac{\overset{1}{\cancel{2}} \cdot \overset{1}{\cancel{2}} \cdot \overset{1}{\cancel{7}}}{3 \cdot \underset{1}{\cancel{7}} \cdot \underset{1}{\cancel{2}} \cdot \underset{1}{\cancel{2}} \cdot 11} = \frac{1}{33}$$

You Try It 2

$$\frac{2}{21} \times \frac{10}{33} = \frac{2 \cdot 10}{21 \cdot 33} = \frac{2 \cdot 2 \cdot 5}{3 \cdot 7 \cdot 3 \cdot 11} = \frac{20}{693}$$

You Try It 3

$$\frac{16}{5} \times \frac{15}{24} = \frac{16 \cdot 15}{5 \cdot 24} = \frac{\overset{1}{\cancel{2}} \cdot \overset{1}{\cancel{2}} \cdot \overset{1}{\cancel{2}} \cdot 2 \cdot \overset{1}{\cancel{3}} \cdot \overset{1}{\cancel{5}}}{\underset{1}{\cancel{5}} \cdot \underset{1}{\cancel{2}} \cdot \underset{1}{\cancel{2}} \cdot \underset{1}{\cancel{2}} \cdot \underset{1}{\cancel{3}}} = \frac{2}{1} = 2$$

You Try It 4

$$5\frac{2}{5} \times \frac{5}{9} = \frac{27}{5} \times \frac{5}{9} = \frac{27 \cdot 5}{5 \cdot 9} = \frac{\overset{1}{\cancel{3}} \cdot \overset{1}{\cancel{3}} \cdot 3 \cdot \overset{1}{\cancel{5}}}{\underset{1}{\cancel{5}} \cdot \underset{1}{\cancel{3}} \cdot \underset{1}{\cancel{3}}} = \frac{3}{1} = 3$$

You Try It 5

$$3\frac{2}{5} \times 6\frac{1}{4} = \frac{17}{5} \times \frac{25}{4} = \frac{17 \cdot 25}{5 \cdot 4} = \frac{17 \cdot \overset{1}{\cancel{5}} \cdot 5}{\underset{1}{\cancel{5}} \cdot 2 \cdot 2} = \frac{85}{4} = 21\frac{1}{4}$$

You Try It 6

$$3\frac{2}{7} \times 6 = \frac{23}{7} \times \frac{6}{1} = \frac{23 \cdot 6}{7 \cdot 1} = \frac{23 \cdot 2 \cdot 3}{7 \cdot 1} = \frac{138}{7} = 19\frac{5}{7}$$

You Try It 7

Strategy To find the dialysis technician's salary after 10 years of experience, multiply the starting salary (\$37,000) by $1\frac{1}{2}$.

Solution

$$37{,}000 \times 1\frac{1}{2} = \frac{37{,}000}{1} \times \frac{3}{2} = \frac{37{,}000 \cdot 3}{1 \cdot 2} = 55{,}500$$

After 10 years of experience, the dialysis technician's salary is \$55,500.

You Try It 8

Strategy To find the cost of the x-ray machine itself :

- Find the cost of the x-ray cassette $\left(\frac{1}{5} \times 125{,}000\right)$.
- Subtract the cost of the x-ray cassette from the total cost of the machine.

Solution

$$\frac{1}{5} \times 125{,}000 = \frac{125{,}000}{5} = 25{,}000$$ • Value of the portable x-ray cassette

$$125{,}000 - 25{,}000 = 100{,}000$$

The cost of the x-ray machine is $100,000.

SECTION 2.7

You Try It 1

$$\frac{3}{7} \div \frac{2}{3} = \frac{3}{7} \times \frac{3}{2} = \frac{3 \cdot 3}{7 \cdot 2} = \frac{9}{14}$$

You Try It 2

$$\frac{3}{4} \div \frac{9}{10} = \frac{3}{4} \times \frac{10}{9} = \frac{3 \cdot 10}{4 \cdot 9} = \frac{\overset{1}{\cancel{3}} \cdot \overset{1}{\cancel{2}} \cdot 5}{\underset{1}{\cancel{2}} \cdot 2 \cdot \underset{1}{\cancel{3}} \cdot 3} = \frac{5}{6}$$

You Try It 3

$$\frac{5}{7} \div 6 = \frac{5}{7} \div \frac{6}{1} = \frac{5}{7} \times \frac{1}{6} = \frac{5 \cdot 1}{7 \cdot 6} = \frac{5}{7 \cdot 2 \cdot 3} = \frac{5}{42}$$

• $6 = \frac{6}{1}$. The reciprocal of $\frac{6}{1}$ is $\frac{1}{6}$.

You Try It 4

$$12\frac{3}{5} \div 7 = \frac{63}{5} \div \frac{7}{1} = \frac{63}{5} \times \frac{1}{7} = \frac{63 \cdot 1}{5 \cdot 7} = \frac{3 \cdot 3 \cdot \overset{1}{\cancel{7}}}{5 \cdot \underset{1}{\cancel{7}}} = \frac{9}{5} = 1\frac{4}{5}$$

You Try It 5

$$3\frac{2}{3} \div 2\frac{2}{5} = \frac{11}{3} \div \frac{12}{5} = \frac{11}{3} \times \frac{5}{12} = \frac{11 \cdot 5}{3 \cdot 12} = \frac{11 \cdot 5}{3 \cdot 2 \cdot 2 \cdot 3} = \frac{55}{36} = 1\frac{19}{36}$$

You Try it 6

$$2\frac{5}{6} \div 8\frac{1}{2} = \frac{17}{6} \div \frac{17}{2} = \frac{17}{6} \times \frac{2}{17} = \frac{17 \cdot 2}{6 \cdot 17} = \frac{\overset{1}{\cancel{17}} \cdot \overset{1}{\cancel{2}}}{\underset{1}{\cancel{2}} \cdot 3 \cdot \underset{1}{\cancel{17}}} = \frac{1}{3}$$

You Try It 7

$$6\frac{2}{5} \div 4 = \frac{32}{5} \div \frac{4}{1} = \frac{32}{5} \times \frac{1}{4} = \frac{32 \cdot 1}{5 \cdot 4} = \frac{2 \cdot 2 \cdot 2 \cdot \overset{1}{\cancel{2}} \cdot \overset{1}{\cancel{2}}}{5 \cdot \underset{1}{\cancel{2}} \cdot \underset{1}{\cancel{2}}} = \frac{8}{5} = 1\frac{3}{5}$$

You Try It 8

Strategy To find the number of beds assembled, divide the number of hours worked (10) by the number of hours required for assembly per bed $\left(2\frac{1}{2}\right)$.

Solution

$$10 \div 2\frac{1}{2} = \frac{10}{1} \div \frac{5}{2} = \frac{10}{1} \times \frac{2}{5} = \frac{\overset{1}{\cancel{5}} \cdot 2 \cdot 2}{1 \cdot \underset{1}{\cancel{5}}} = \frac{4}{1} = 4$$

The technician assembled 4 hospital beds.

You Try It 9

Strategy To find the length of the remaining nasal cannula:

- Divide the total length of the nasal cannula (12) by the length of each piece $\left(2\frac{1}{6}\right)$.
- Multiply the fractional part of the result in step 1 by the length of a piece of the nasal cannula to determine the length of the remaining piece.

Solution

$$12 \div 2\frac{1}{6} = \frac{12}{1} \div \frac{13}{6} = \frac{12}{1} \times \frac{6}{13} = \frac{12 \cdot 6}{1 \cdot 13} = 5\frac{7}{13}$$

There are 5 whole pieces that are each $2\frac{1}{6}$ feet long.

There is 1 piece that is $\frac{7}{13}$ of $2\frac{1}{6}$ feet long.

$$\frac{7}{13} \times 2\frac{1}{6} = \frac{7}{13} \times \frac{13}{6} = \frac{7}{6} = 1\frac{1}{6}$$

The length of cannula remaining is $1\frac{1}{6}$ feet.

SECTION 2.8

You Try It 1 $\frac{9}{14} = \frac{27}{42}$ $\frac{13}{21} = \frac{26}{42}$ $\frac{9}{14} > \frac{13}{21}$

You Try It 2
$$\left(\frac{7}{11}\right)^2 \cdot \left(\frac{2}{7}\right) = \left(\frac{7}{11} \cdot \frac{7}{11}\right) \cdot \left(\frac{2}{7}\right)$$
$$= \frac{\overset{1}{\cancel{7}} \cdot 7 \cdot 2}{11 \cdot 11 \cdot \underset{1}{\cancel{7}}} = \frac{14}{121}$$

You Try It 3
$$\left(\frac{1}{13}\right)^2 \cdot \left(\frac{1}{4} + \frac{1}{6}\right) \div \frac{5}{13} =$$
$$\left(\frac{1}{13}\right)^2 \cdot \left(\frac{5}{12}\right) \div \frac{5}{13} =$$
$$\left(\frac{1}{169}\right) \cdot \left(\frac{5}{12}\right) \div \frac{5}{13} =$$
$$\left(\frac{1 \cdot 5}{13 \cdot 13 \cdot 12}\right) \div \frac{5}{13} =$$
$$\left(\frac{1 \cdot 5}{13 \cdot 13 \cdot 12}\right) \times \frac{13}{5} =$$
$$\frac{1 \cdot \overset{1}{\cancel{5}} \cdot \overset{1}{\cancel{13}}}{\underset{1}{\cancel{13}} \cdot 13 \cdot 12 \cdot \underset{1}{\cancel{5}}} = \frac{1}{156}$$

SOLUTIONS TO CHAPTER 3 "YOU TRY IT"

SECTION 3.1

You Try It 1 The digit 4 is in the thousandths place.

You Try It 2 $\frac{501}{1000} = 0.501$
(five hundred one thousandths)

You Try It 3 $0.67 = \frac{67}{100}$ (sixty-seven hundredths)

You Try It 4 Fifty-five and six thousand eighty-three ten-thousandths

You Try It 5 806.00491 • 1 is in the hundred-thousandths place.

You Try It 6

3.675849 — Given place value: 8; $4 < 5$

3.675849 rounded to the nearest ten-thousandth is 3.6758.

You Try It 7

48.907 — Given place value: 9; $0 < 5$

48.907 rounded to the nearest tenth is 48.9.

You Try It 8

31.8652 — Given place value: 1; $8 > 5$

31.8652 rounded to the nearest whole number is 32.

You Try It 9

Strategy 7.63 rounded to the nearest whole number is 8.

Solution To the nearest pound, the average weight of a newborn at 40 weeks is 8 pounds.

SECTION 3.2

You Try It 1

$$\begin{array}{r} \overset{1\,2}{}4.62 \\ 27.9 \\ +\ 0.62054 \\ \hline 33.14054 \end{array}$$

• Place the decimal points on a vertical line.

You Try It 2

$$\begin{array}{r} \overset{1}{}6.05 \\ 12. \\ +\ 0.374 \\ \hline 18.424 \end{array}$$

You Try It 3

Strategy To determine the number, add the numbers of hearing-impaired Americans ages 45 to 54, 55 to 64, 65 to 74, and 75 and over.

Solution

$$\begin{array}{r} 4.48 \\ 4.31 \\ 5.41 \\ +3.80 \\ \hline 18.00 \end{array}$$

18 million Americans ages 45 and older are hearing-impaired.

You Try It 4

Strategy To find the total income, add the pay for working nights (81.20, 60.90, 40.60, 81.20) to her regular weekly pay (480.15).

Solution $480.15 + 81.20 + 60.90 + 40.60 + 81.20 = 744.05$

The nurse's aide's total income for the week was $744.05.

SECTION 3.3

You Try It 1

$$\begin{array}{r} \overset{6}{\cancel{7}}\,\overset{11}{\cancel{1}}\overset{9}{\cancel{2}}.\overset{10}{\cancel{0}}\overset{13}{\cancel{3}}\,9 \\ -\ \ 8.4\,7 \\ \hline 6\,3.5\,6\,9 \end{array}$$

Check:

$$\begin{array}{r} \overset{1\ 11}{8.47} \\ +\ 63.569 \\ \hline 72.039 \end{array}$$

You Try It 2

$$\begin{array}{r} 35.00 \\ -\ 9.67 \\ \hline 25.33 \end{array}$$

Check:
$$\begin{array}{r} 9.67 \\ +\ 25.33 \\ \hline 35.00 \end{array}$$

You Try It 3

$$\begin{array}{r} 3.7000 \\ -\ 1.9715 \\ \hline 1.7285 \end{array}$$

Check:
$$\begin{array}{r} 1.9715 \\ +\ 1.7285 \\ \hline 3.7000 \end{array}$$

You Try It 4

Strategy To find the amount of change, subtract the amount paid (6.85) from 10.00.

Solution
$$\begin{array}{r} 10.00 \\ -\ 6.85 \\ \hline 3.15 \end{array}$$

Your change was $3.15.

You Try It 5

Strategy To find the new balance:

- Add to find the total of the three checks (1025.60 + 79.85 + 162.47).
- Subtract the total from the previous balance (2472.69).

Solution
$$\begin{array}{r} 1025.60 \\ 79.85 \\ +\ 162.47 \\ \hline 1267.92 \end{array} \qquad \begin{array}{r} 2472.69 \\ -\ 1267.92 \\ \hline 1204.77 \end{array}$$

The new balance is $1204.77.

SECTION 3.4

You Try It 1

$$\begin{array}{r} 870 \\ \times\ 4.6 \\ \hline 5220 \\ 3480 \\ \hline 4002.0 \end{array}$$

- 1 decimal place
- 1 decimal place

You Try It 2

$$\begin{array}{r} 0.000086 \\ \times\ 0.057 \\ \hline 602 \\ 430 \\ \hline 0.000004902 \end{array}$$

- 6 decimal places
- 3 decimal places
- 9 decimal places

You Try It 3

$$\begin{array}{r} 4.68 \\ \times\ 6.03 \\ \hline 1404 \\ 28\,080 \\ \hline 28.2204 \end{array}$$

- 2 decimal places
- 2 decimal places
- 4 decimal places

You Try It 4 $6.9 \times 1000 = 6900$

You Try It 5 $4.0273 \times 10^2 = 402.73$

You Try It 6

Strategy To find the total bill:

- Find the number of gallons of water used by multiplying the number of gallons used per day (5000) by the number of days (62).
- Find the cost of water by multiplying the cost per 1000 gallons (1.39) by the number of 1000-gallon units used.
- Add the cost of the water to the meter fee (133.70).

Solution Number of gallons = 5000(62) = 310,000

Cost of water $= \frac{310{,}000}{1000} \times 1.39 = 430.90$

Total cost = 430.90 + 133.70 = 564.60

The total bill is $564.60.

You Try It 7

Strategy To find the cost of running the x-ray processor for 56 hours, multiply the hourly cost (0.1201) by the number of hours the processor runs (56). Round the answer to the nearest cent.

Solution
$$\begin{array}{r} 0.1201 \\ \times\ 56 \\ \hline 7206 \\ 6005 \\ \hline 6.7256 \end{array}$$

6.7256 rounded to the nearest cent (hundredth) is 6.73.

The cost of running the x-ray processor in July was $6.73.

You Try It 8

Strategy To find the total cost of the electrical muscle stimulation machine:

- Multiply the monthly payment (111.20) by the number of months (20).
- Add that product to the down payment (277.31).

Solution
$$\begin{array}{r} 111.20 \\ \times\ 20 \\ \hline 2224.00 \end{array} \qquad \begin{array}{r} 2224.00 \\ +\ 277.31 \\ \hline 2501.31 \end{array}$$

The total cost of the electrical muscle stimulation machine is $2501.31.

SECTION 3.5

You Try It 1

$$\begin{array}{r} 2.7 \\ 0.052.\overline{)0.140.4} \\ -104 \\ \hline 36\,4 \\ -36\,4 \\ \hline 0 \end{array}$$

- Move the decimal point 3 places to the right in the divisor and the dividend. Write the decimal point in the quotient directly over the decimal point in the dividend.

You Try It 2

$$\begin{array}{r} 0.4873 \approx 0.487 \\ 76\overline{)37.0420} \\ -30\,4 \\ \hline 6\,64 \\ -6\,08 \\ \hline 562 \\ -532 \\ \hline 300 \\ -228 \end{array}$$

• Write the decimal point in the quotient directly over the decimal point in the dividend.

You Try It 3

$$\begin{array}{r} 72.73 \approx 72.7 \\ 5.09.\overline{)370.20.00} \\ -356\,3 \\ \hline 13\,90 \\ 10\,18 \\ \hline 3\,720 \\ -3\,563 \\ \hline 1570 \\ -1527 \end{array}$$

You Try It 4 $309.21 \div 10{,}000 = 0.030921$

You Try It 5 $42.93 \div 10^4 = 0.004293$

You Try It 6

Strategy To find how many times greater the average hourly earnings were, divide the 1998 average hourly earnings (12.88) by the 1978 average hourly earnings (5.70).

Solution $12.88 \div 5.70 \approx 2.3$

The average hourly earnings in 1998 were about 2.3 times greater than in 1978.

You Try It 7

Strategy To find the average life expectancy for these countries:

- Add the life expectancies for the four countries.
- Divide the total number of years by 4.

Solution $80.98 + 80.05 + 80.2 + 79.4 = 320.63$

$320.63 \div 4 = 80.1575$

The average life expectancy for these countries is 80.1575 years.

SECTION 3.6

You Try It 1

$$\begin{array}{r} 0.56 \approx 0.6 \\ 16\overline{)9.00} \end{array}$$

You Try It 2

$$4\frac{1}{6} = \frac{25}{6}$$

$$\begin{array}{r} 4.166 \approx 4.17 \\ 6\overline{)25.000} \end{array}$$

You Try It 3

$$0.56 = \frac{56}{100} = \frac{14}{25}$$

$$5.35 = 5\frac{35}{100} = 5\frac{7}{20}$$

You Try It 4

$$0.12\frac{7}{8} = \frac{12\frac{7}{8}}{100} = 12\frac{7}{8} \div 100$$

$$= \frac{103}{8} \times \frac{1}{100} = \frac{103}{800}$$

You Try It 5

$\frac{5}{8} = 0.625$ • Convert the fraction $\frac{5}{8}$ to a decimal.

$0.630 > 0.625$ • Compare the two decimals.

$0.63 > \frac{5}{8}$ • Convert 0.625 back to a fraction.

SOLUTIONS TO CHAPTER 4 "YOU TRY IT"

SECTION 4.1

You Try It 1

$$\frac{20 \text{ pounds}}{24 \text{ pounds}} = \frac{20}{24} = \frac{5}{6}$$

20 pounds : 24 pounds = 20 : 24 = 5 : 6

20 pounds to 24 pounds = 20 to 24
= 5 to 6

You Try It 2

$$\frac{64 \text{ miles}}{8 \text{ miles}} = \frac{64}{8} = \frac{8}{1}$$

64 miles : 8 miles = 64 : 8 = 8 : 1

64 miles to 8 miles = 64 to 8 = 8 to 1

You Try It 3

Strategy To find the ratio, write the ratio of canines (4) to the number of incisors (8) in simplest form.

Solution $\frac{4}{8} = \frac{1}{2}$

The ratio of canines to incisors is $\frac{1}{2}$.

You Try It 4

Strategy To find the dilution ratio, write the ratio of the amount of blood (2 milliliters) to the total amount of solution (2 milliliters + 98 milliliters) in simplest form.

Solution

$$\frac{2 \text{ milliliters}}{2 \text{ milliliters} + 98 \text{ milliliters}} = \frac{2}{100} = \frac{1}{50}$$

The dilution ratio is $\frac{1}{50}$.

SECTION 4.2

You Try It 1 $\frac{9 \text{ milligrams}}{3 \text{ tablets}} = \frac{3 \text{ milligrams}}{1 \text{ tablet}}$

You Try It 2 $\frac{150 \text{ milligrams}}{3 \text{ tablets}} = 3\overline{)150}$ (quotient 50)

50 milligrams/tablet

You Try It 3

Strategy To find the strength of this solution as a unit rate, divide the amount of sodium in the solution (5) by the number of milliliters of solution (100). The strength will be in milligrams per milliliter.

Solution $5 \div 100 = 0.05$
The strength of this solution is 0.05 milligram/milliliter.

SECTION 4.3

You Try It 1 $\frac{6}{10}$ and $\frac{9}{15}$: $10 \times 9 = 90$; $6 \times 15 = 90$

The cross products are equal. The proportion is true.

You Try It 2 $\frac{32}{6}$ and $\frac{90}{8}$: $6 \times 90 = 540$; $32 \times 8 = 256$

The cross products are not equal. The proportion is not true.

You Try It 3

$$\frac{n}{14} = \frac{3}{7}$$
$$n \times 7 = 14 \times 3$$
$$n \times 7 = 42$$
$$n = 42 \div 7$$
$$n = 6$$

• Find the cross products. Then solve for n.

Check: $\frac{6}{14}$ and $\frac{3}{7}$: $14 \times 3 = 42$; $6 \times 7 = 42$

You Try It 4

$$\frac{5}{7} = \frac{n}{20}$$
$$5 \times 20 = 7 \times n$$
$$100 = 7 \times n$$
$$100 \div 7 = n$$
$$14.3 \approx n$$

• Find the cross products. Then solve for n.

You Try It 5

$$\frac{15}{20} = \frac{12}{n}$$
$$15 \times n = 20 \times 12$$
$$15 \times n = 240$$
$$n = 240 \div 15$$
$$n = 16$$

• Find the cross products. Then solve for n.

Check: $\frac{15}{20}$ and $\frac{12}{16}$: $20 \times 12 = 240$; $15 \times 16 = 240$

You Try It 6

$$12 \times 4 = n \times 7$$
$$48 = n \times 7$$
$$48 \div 7 = n$$
$$6.86 \approx n$$

You Try It 7

$$\frac{n}{12} = \frac{4}{1}$$
$$n \times 1 = 12 \times 4$$
$$n \times 1 = 48$$
$$n = 48 \div 1$$
$$n = 48$$

Check: $\frac{48}{12}$ and $\frac{4}{1}$: $12 \times 4 = 48$; $48 \times 1 = 48$

You Try It 8

Strategy To find the number of tablespoons of disinfectant needed, write and solve a proportion using n to represent the number of tablespoons of disinfectant.

Solution

$$\frac{3 \text{ tablespoons}}{4 \text{ gallons}} = \frac{n}{10 \text{ gallons}}$$
$$3 \times 10 = n \times 4$$
$$30 = 4n$$
$$30 \div 4 = n$$
$$7.5 = n$$

For 10 gallons of water, 7.5 tablespoons of disinfectant are required.

You Try It 9

Strategy To find the number of hours it will take for the infusion to be complete, solve a proportion using n to represent the number of hours.

Solution

$$\frac{75 \text{ milliliters}}{1 \text{ hour}} = \frac{1000 \text{ milliliters}}{n}$$
$$1000 \times 1 = 75 \times n$$
$$1000 \div 75 = n$$
$$13\frac{1}{3} = n$$

To convert $\frac{1}{3}$ hour to minutes, multiply $\frac{1}{3}$ by 60 minutes.

$$\frac{1}{3} \times 60 = \frac{1}{3} \times \frac{60 \text{ minutes}}{1} = 20 \text{ minutes}$$

It will take $13\frac{1}{3}$ hours or 13 hours 20 minutes for the infusion to be complete.

SOLUTIONS TO CHAPTER 5 "YOU TRY IT"

SECTION 5.1

You Try It 1 **a.** $125\% = 125 \times \frac{1}{100} = \frac{125}{100} = 1\frac{1}{4}$

b. $125\% = 125 \times 0.01 = 1.25$

You Try It 2

$$33\frac{1}{3}\% = 33\frac{1}{3} \times \frac{1}{100} = \frac{100}{3} \times \frac{1}{100} = \frac{100}{300} = \frac{1}{3}$$

You Try It 3 $0.25\% = 0.25 \times 0.01 = 0.0025$

You Try It 4

$0.048 = 0.048 \times 100\% = 4.8\%$

$3.67 = 3.67 \times 100\% = 367\%$

$$0.62\frac{1}{2} = 0.62\frac{1}{2} \times 100\% = 62\frac{1}{2}\%$$

You Try It 5 $\frac{5}{6} = \frac{5}{6} \times 100\% = \frac{500}{6}\% = 83\frac{1}{3}\%$

You Try It 6

$$1\frac{4}{9} = \frac{13}{9} = \frac{13}{9} \times 100\% = \frac{1300}{9}\% \approx 144.4\%$$

SECTION 5.2

You Try It 1

Percent × base = amount

$0.063 \times 150 = n$

$9.45 = n$

You Try It 2

Percent × base = amount

$\frac{1}{6} \times 66 = n$ • $16\frac{2}{3}\% = \frac{1}{6}$

$11 = n$

You Try It 3

Strategy To determine the amount that came from corporations, write and solve the basic percent equation using n to represent the amount. The percent is 4%. The base is $212 billion.

Solution Percent × base = amount

$4\% \times 212 = n$

$0.04 \times 212 = n$

$8.48 = n$

Corporations gave $8.48 billion to charities.

You Try It 4

Strategy To find the new hourly wage:

- Find the amount of the raise. Write and solve the basic percent equation using n to represent the amount of the raise (amount). The percent is 8%. The base is $33.50.
- Add the amount of the raise to the old wage (33.50).

Solution

$8\% \times 33.50 = n$

$0.08 \times 33.50 = n$

$2.68 = n$

$$\begin{array}{r} 33.50 \\ +\ 2.68 \\ \hline 36.18 \end{array}$$

The new hourly wage is $36.18.

SECTION 5.3

You Try It 1

Percent × base = amount

$n \times 32 = 16$

$n = 16 \div 32$

$n = 0.50$

$n = 50\%$

You Try It 2

Percent × base = amount

$n \times 15 = 48$

$n = 48 \div 15$

$n = 3.2$

$n = 320\%$

You Try It 3

Percent × base = amount

$n \times 45 = 30$

$n = 30 \div 45$

$n = \frac{2}{3} = 66\frac{2}{3}\%$

You Try It 4

Strategy To find what percent of the income the income tax is, write and solve the basic percent equation using n to represent the percent. The base is $33,500, and the amount is $5025.

Solution $n \times 33{,}500 = 5025$

$n = 5025 \div 33{,}500$

$n = 0.15 = 15\%$

The income tax is 15% of the income.

You Try It 5

Strategy To find the percent of traumatic brain injuries that are considered serious injuries:

- Subtract to find the number of brain injuries that are considered serious (1,700,000 – 1,275,000).
- Write and solve the basic percent equation using n to represent the percent. The base is 1,700,000 and the amount is the number of brain injuries that are considered serious.

Solution $1{,}700{,}000 - 1{,}275{,}000 = 425{,}000$

425,000 patients have traumatic brain injuries that are considered serious.

Percent × base = amount

$n \times 1{,}700{,}000 = 425{,}000$

$n = 425{,}000 \div 1{,}700{,}000$

$n = 0.25$

25% of traumatic brain injuries are considered to be serious injuries.

SECTION 5.4

You Try It 1

Percent × base = amount
$0.86 \times n = 215$
$n = 215 \div 0.86$
$n = 250$

You Try It 2

Percent × base = amount
$0.025 \times n = 15$
$n = 15 \div 0.025$
$n = 600$

You Try It 3

Percent × base = amount
$\frac{1}{6} \times n = 5$ • $16\frac{2}{3}\% = \frac{1}{6}$
$n = 5 \div \frac{1}{6}$
$n = 30$

You Try It 4

Strategy To find the original value of the car, write and solve the basic percent equation using n to represent the original value (base). The percent is 42%, and the amount is $10,458.

Solution
$42\% \times n = 10,458$
$0.42 \times n = 10,458$
$n = 10,458 \div 0.42$
$n = 24,900$

The original value of the car was $24,900.

You Try It 5

Strategy To find the difference between the original price and the sale price:

- Find the original price. Write and solve the basic percent equation using n to represent the original price (base). The percent is 63% and the amount is $4999.
- Subtract the sale price (4999) from the original price.

Solution
Percent × base = amount
$63\% \times n = 4999$
$0.63 \times n = 4999$
$n = 4999 \div 0.63$
$n = 7934.920635$
Round to nearest cent: 7934.92 (original price)
$7934.92 - 4999 = 2935.92$

The difference between the original price and the sale price is $2935.92.

SECTION 5.5

You Try It 1

$\frac{26}{100} = \frac{22}{n}$
$26 \times n = 100 \times 22$
$26 \times n = 2200$
$n = 2200 \div 26$
$n \approx 84.62$

You Try It 2

$\frac{16}{100} = \frac{n}{132}$
$16 \times 132 = 100 \times n$
$2112 = 100 \times n$
$2112 \div 100 = n$
$21.12 = n$

You Try It 3

Strategy To find the amount of dextrose in 250 milliliters of D5W, write and solve a proportion using n to represent the amount of dextrose in grams. The percent is 5%, and the base is 250.

Solution
$\frac{5}{100} = \frac{n}{250}$
$5 \times 250 = 100 \times n$
$1250 = 100 \times n$
$1250 \div 100 = n$
$12.5 = n$

The patient receives 12.5 grams of dextrose from the IV solution.

You Try It 4

Strategy Write and solve a proportion using n to represent the percent of patients who experience nausea. The base is the total number of patients (60), and the amount is the number of patients who experience nausea (3).

Solution
$\frac{n}{100} = \frac{3}{60}$
$n \times 60 = 100 \times 3$
$n \times 60 = 300$
$n = 300 \div 60$
$n = 5$

The percent of patients experiencing nausea was 5%.

SOLUTIONS TO CHAPTER 6 "YOU TRY IT"

SECTION 6.1

You Try It 1

Strategy To find the unit cost, divide the total cost by the number of units.

Solution
a. $16.80 \div 50 = 0.336$
$0.336 per glove
b. $2.29 \div 15 \approx 0.153$
$0.153 per ounce

You Try It 2

Strategy To find the more economical purchase, compare the unit costs.

Solution
$8.70 \div 6 = 1.45$
$6.96 \div 4 = 1.74$
$\$1.45 < \$1.74$

The more economical purchase is 6 cans for $8.70.

You Try It 3

Strategy To find the total cost, multiply the unit cost (14.99) by the number of units (28).

Solution

Unit cost	×	number of units	=	total cost
14.99	×	28	=	419.72

The total cost is $419.72.

SECTION 6.2

You Try It 1

Strategy To find the percent increase:

- Find the amount of the increase.
- Solve the basic percent equation for *percent*.

Solution

New value	−	original value	=	amount of increase
3.83	−	3.46	=	0.37

Percent × base = amount
$n \times 3.46 = 0.37$
$n = 0.37 \div 3.46$
$n \approx 0.11 = 11\%$

The percent increase was 11%.

You Try It 2

Strategy To find the new hourly wage:

- Solve the basic percent equation for *amount*.
- Add the amount of the increase to the original wage.

Solution
Percent × base = amount
$0.14 \times 12.50 = n$
$1.75 = n$

$12.50 + 1.75 = 14.25$

The new hourly wage is $14.25.

You Try It 3

Strategy To find the markup, solve the basic percent equation for *amount*.

Solution
Percent × base = amount

Markup rate	×	cost	=	markup
0.20	×	32	=	n

$6.4 = n$

The markup is $6.40.

You Try It 4

Strategy To find the selling price:

- Find the markup by solving the basic percent equation for *amount*.
- Add the markup to the cost.

Solution
Percent × base = amount

Markup rate	×	cost	=	markup
0.55	×	72	=	n

$39.60 = n$

Cost	+	markup	=	selling price
72	+	39.60	=	111.60

The selling price is $111.60.

You Try It 5

Strategy To find the percent decrease:

- Find the amount of the decrease.
- Solve the basic percent equation for *percent*.

Solution

Original value	−	new value	=	amount of decrease
261,000	−	215,000	=	46,000

Percent × base = amount
$n \times 261{,}000 = 46{,}000$
$n = 46{,}000 \div 261{,}000$
$n \approx 0.176$

The percent decrease is 17.6%.

You Try It 6

Strategy To find her weight at the end of the first month:

- Find the amount of decrease by solving the basic percent equation for *amount*.
- Subtract the amount of decrease from the original weight.

Solution
Percent × base = amount
$0.05 \times 220 = n$
$11 = n$
$220 - 11 = 209$

Her weight at the end of the first month was 209 pounds.

You Try It 7

Strategy To find the discount rate:

- Find the discount.
- Solve the basic percent equation for *percent*.

Solution

Regular price − sale price = discount

44.99 − 24.99 = 20

Percent × base = amount

Discount rate × regular price = discount

$n \times 44.99 = 20$

$n = 20 \div 44.99$

$n \approx 0.44454$

The discount rate is 44.5%.

You Try It 8

Strategy To find the sale price:

- Find the discount by solving the basic percent equation for *amount*.
- Subtract to find the sale price.

Solution

Percent × base = amount

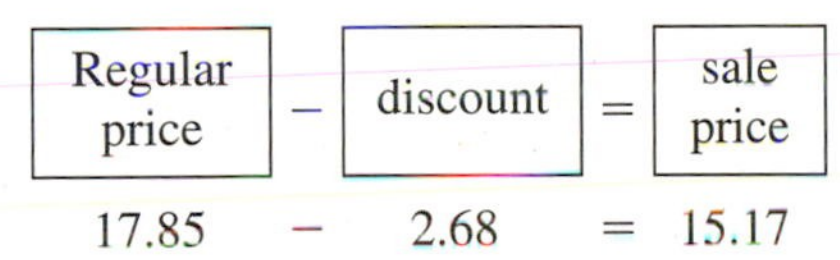

$0.15 \times 17.85 = n$

$2.6775 = n$

Round to the nearest cent: 2.68

Regular price − discount = sale price

17.85 − 2.68 = 15.17

The sale price of the blood pressure monitor is $15.17.

SECTION 6.3

You Try It 1

Strategy To find the simple interest due, multiply the principal (15,000) times the annual interest rate (8% = 0.08) times the time in years (18 months = $\frac{18}{12}$ years = 1.5 years).

Solution

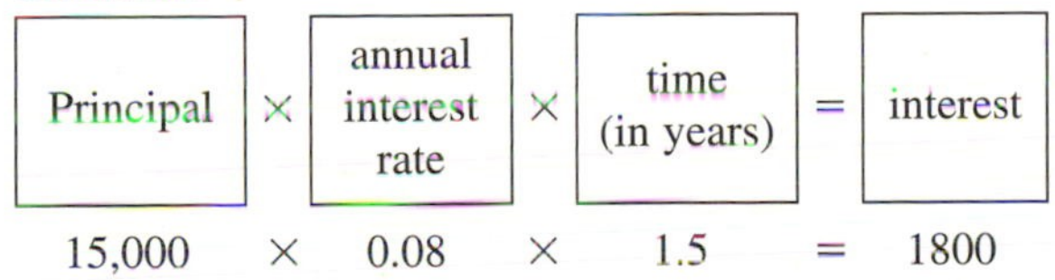

15,000 × 0.08 × 1.5 = 1800

The interest due is $1800.

You Try It 2

Strategy To find the maturity value:

- Use the simple interest formula to find the simple interest due.
- Find the maturity value by adding the principal and the interest.

Solution

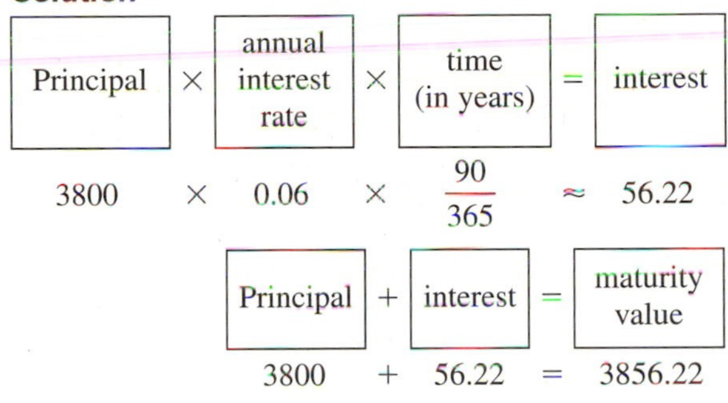

$3800 \times 0.06 \times \frac{90}{365} \approx 56.22$

3800 + 56.22 = 3856.22

The maturity value is $3856.22.

You Try It 3

Strategy To find the monthly payment:

- Find the maturity value by adding the principal and the interest.
- Divide the maturity value by the length of the loan in months (12).

Solution

Principal + interest = maturity value

1900 + 152 = 2052

Maturity value ÷ length of the loan = payment

2052 ÷ 12 = 171

The monthly payment is $171.

You Try It 4

Strategy To find the finance charge, multiply the principal, or unpaid balance (1250), times the monthly interest rate (1.6%) times the number of months (1).

Solution

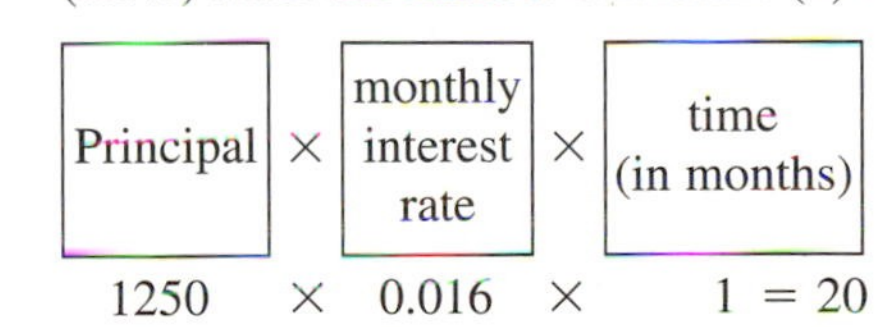

1250 × 0.016 × 1 = 20

The finance charge is $20.

You Try It 5

Strategy To find the interest earned:

- Find the new principal by multiplying the original principal (1000) by the factor found in the Compound Interest Table (3.29066).
- Subtract the original principal from the new principal.

(Continued)

(Continued)

Solution

$1000 \times 3.29066 = 3290.66$

The new principal is $3290.66.

$3290.66 - 1000 = 2290.66$

The interest earned is $2290.66.

SECTION 6.4

You Try It 1

Strategy

To find the mortgage:

- Find the down payment by solving the basic percent equation for *amount*.
- Subtract the down payment from the purchase price.

Solution

Percent × base = amount

$\boxed{\text{Percent}} \times \boxed{\text{purchase price}} = \boxed{\text{down payment}}$

$0.25 \times 1{,}500{,}000 = n$

$375{,}000 = n$

$\boxed{\text{Purchase price}} - \boxed{\text{down payment}} = \boxed{\text{mortgage}}$

$1{,}500{,}000 - 375{,}000 = 1{,}125{,}000$

The mortgage is $1,125,000.

You Try It 2

Strategy

To find the loan origination fee, solve the basic percent equation for *amount*.

Solution

Percent × base = amount

$\boxed{\text{Points}} \times \boxed{\text{mortgage}} = \boxed{\text{fee}}$

$0.045 \times 180{,}000 = n$

$8100 = n$

The loan origination fee was $8100.

You Try It 3

Strategy

To find the monthly mortgage payment:

- Subtract the down payment from the purchase price to find the mortgage.
- Multiply the mortgage by the factor found in the Monthly Payment Table.

Solution

$\boxed{\text{Purchase price}} - \boxed{\text{down payment}} = \boxed{\text{mortgage}}$

$175{,}000 - 17{,}500 = 157{,}500$

$157{,}500 \times 0.0089973 \approx 1417.08$

↑ From the table

The monthly mortgage payment is $1417.08.

You Try It 4

Strategy

To find the interest:

- Multiply the mortgage by the factor found in the Monthly Payment Table to find the monthly mortgage payment.
- Subtract the principal from the monthly mortgage payment.

Solution

$625{,}000 \times 0.0070678 \approx 4417.38$

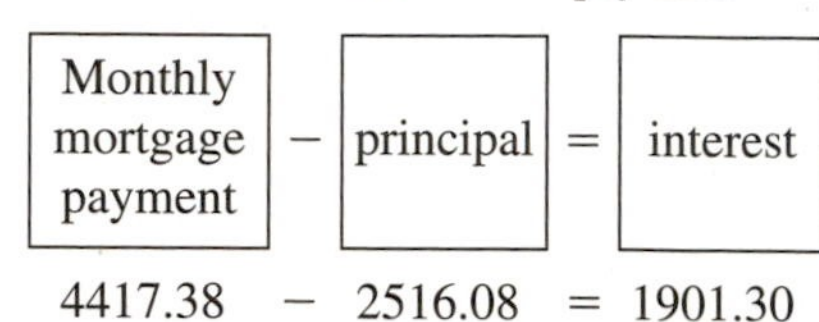

$\boxed{\text{Monthly mortgage payment}} - \boxed{\text{principal}} = \boxed{\text{interest}}$

$4417.38 - 2516.08 = 1901.30$

The interest on the mortgage is $1901.30.

You Try It 5

Strategy

To find the monthly payment:

- Divide the annual property tax by 12 to find the monthly property tax.
- Add the monthly property tax to the monthly mortgage payment.

Solution

$3000 \div 12 = 250$

The monthly property tax is $250.

$815.20 + 250 = 1065.20$

The total monthly payment is $1065.20.

SECTION 6.5

You Try It 1

Strategy

To find the amount financed:

- Find the down payment by solving the basic percent equation for *amount*.
- Subtract the down payment from the purchase price.

Solution

Percent × base = amount

$\boxed{\text{Percent}} \times \boxed{\text{purchase price}} = \boxed{\text{down payment}}$

$0.20 \times 19{,}200 = n$

$3840 = n$

The down payment is $3840.

$19{,}200 - 3840 = 15{,}360$

The amount financed is $15,360.

You Try It 2

Strategy To find the license fee, solve the basic percent equation for *amount*.

Solution

Percent × base = amount

Percent	×	purchase price	=	license fee

$0.015 \times 27{,}350 = n$

$410.25 = n$

The license fee is $410.25.

You Try It 3

Strategy To find the cost, multiply the cost per mile (0.41) by the number of miles driven (23,000).

Solution $23{,}000 \times 0.41 = 9430$

The cost is $9430.

You Try It 4

Strategy To find the cost per mile for car insurance, divide the cost for insurance (360) by the number of miles driven (15,000).

Solution $360 \div 15{,}000 = 0.024$

The cost per mile for insurance is $0.024.

You Try It 5

Strategy To find the monthly payment:

- Subtract the down payment from the purchase price to find the amount financed.
- Multiply the amount financed by the factor found in the Monthly Payment Table.

Solution $25{,}900 - 6475 = 19{,}425$

$19{,}425 \times 0.0244129 \approx 474.22$

The monthly payment is $474.22.

SECTION 6.6

You Try It 1

Strategy To find the therapist's earnings:

- Find the therapist's overtime wage by multiplying the hourly wage by 2.
- Multiply the number of overtime hours worked by the overtime wage.

Solution $24.50 \times 2 = 49$

The hourly wage for overtime is $49.

$49 \times 8 = 392$

The respiratory therapist earns $392 for working 8 hours of overtime.

You Try It 2

Strategy To find the salary per month, divide the annual salary by the number of months in a year (12).

Solution $70{,}980 \div 12 = 5915$

The radiologist's monthly salary is $5915.

You Try It 3

Strategy To find the total earnings:

- Find the sales over $50,000.
- Multiply the commission rate by sales over $50,000.
- Add the commission to the annual salary.

Solution $175{,}000 - 50{,}000 = 125{,}000$

Sales over $50,000 totaled $125,000.

$125{,}000 \times 0.095 = 11{,}875$

Earnings from commissions totaled $11,875.

$37{,}000 + 11{,}875 = 48{,}875$

The medical equipment salesperson earned $48,875.

SECTION 6.7

You Try It 1

Strategy To find the current balance:

- Subtract the amount of the check from the old balance.
- Add the amount of each deposit.

Solution

	302.46	
−	20.59	check
	281.87	
	176.86	first deposit
+	94.73	second deposit
	553.46	

The current checking account balance is $553.46.

You Try It 2

Current checkbook balance:	623.41
Check 237:	+ 78.73
	702.14
Interest:	+ 2.11
	704.25
Deposit:	−523.84
	180.41

Closing bank balance from bank statement: $180.41

Checkbook balance: $180.41

The bank statement and the checkbook balance.

SOLUTIONS TO CHAPTER 7 "YOU TRY IT"

SECTION 7.1

You Try It 1

Strategy To find what percent of the total number of pints collected were collected on January 25:

- Read the pictograph to determine the number of pints collected during each of the four blood drives.
- Add to find the total number of pints collected.
- Solve the basic percent equation for percent (n). Amount = 20; the base is the total number of pints collected during the four blood drives.

Solution

Total pints collected during the four blood drives:

$$\begin{array}{r} 20 \\ 30 \\ 15 \\ +\ 40 \\ \hline 105 \end{array}$$

$$\begin{aligned} \text{Percent} \times \text{base} &= \text{amount} \\ n \times 105 &= 20 \\ n &= 20 \div 105 \\ n &= 0.1905 \end{aligned}$$

The number of pints of blood collected on January 25 represents 19% of the total number collected during the four blood drives.

You Try It 2

Strategy To find the ratio of the annual cost of fuel to the annual cost of maintenance:

- Locate the annual fuel cost and the annual maintenance cost in the circle graph.
- Write the ratio of the annual fuel cost to the annual maintenance cost in simplest form.

Solution Annual fuel cost: $700

Annual maintenance cost: $500

$$\frac{700}{500} = \frac{7}{5}$$

The ratio is $\frac{7}{5}$.

You Try It 3

Strategy To find the amount paid for medical/dental insurance:

- Locate the percent of the distribution that is medical/dental insurance.
- Solve the basic percent equation for amount.

Solution The percent of the distribution that is medical/dental insurance: 6%

$$\begin{aligned} \text{Percent} \times \text{base} &= \text{amount} \\ 0.06 \times 2900 &= n \\ 174 &= n \end{aligned}$$

The amount paid for medical/dental insurance is $174.

SECTION 7.2

You Try It 1

Strategy To write the ratio:

- Read the graph to find the lung capacity of an inactive female and of an athletic female.
- Write the ratio in simplest form.

Solution Lung capacity of an inactive female: 25

Lung capacity of an athletic female: 55

$$\frac{25}{55} = \frac{5}{11}$$

The ratio is $\frac{5}{11}$.

You Try It 2

Strategy Read the graph to locate the lines associated with males in grades 9 to 12, males aged 18 to 44, and males aged 45 to 64. Follow the line for the males in grades 9 to 12 until it is below both of the other two lines.

Solution The line associated with the percent of males in grades 9 to 12 is below the lines associated with males aged 18 to 44 and males aged 45 to 64 during 2002.

SECTION 7.3

You Try It 1

Strategy To find the number of employees:

- Read the histogram to find the number of employees whose hourly wage is between \$10 and \$12 and the number whose hourly wage is between \$12 and \$14.
- Add the two numbers.

Solution Number whose wage is between \$10 and \$12: 7
Number whose wage is between \$12 and \$14: 15

$7 + 15 = 22$

22 employees earn between \$10 and \$14.

You Try It 2

Strategy To write the ratio:

- Read the graph to find the number of states with a per capita income between \$36,000 and \$40,000 and the number with a per capita income between \$44,000 and \$48,000.
- Write the ratio in simplest form.

Solution Number of states with a per capita income between \$36,000 and \$40,000: 7
Number of states with a per capita income between \$44,000 and \$48,000: 1

The ratio is $\frac{7}{1}$.

SECTION 7.4

You Try It 1

Strategy To find the mean blood sugar reading for the patient:

- Find the sum of the numbers.
- Divide the sum by the numbers of readings (12).

Solution $85 + 120 + 105 + 155 + 204 + 260 + 180 + 95 + 135 + 150 + 180 + 165 = 1834$

$\bar{x} = \frac{1834}{12} \approx 152.83$

Round to the nearest whole number: 153

The mean blood sugar for this patient is 153 milligrams/deciliter.

You Try It 2

Strategy To find the median weight loss:

- Arrange the weight losses from least to greatest.
- Because there is an even number of values, the median is the mean of the middle two numbers.

Solution 10, 14, 16, 16, 22, 27, 29, 31, 31, 40

$\text{Median} = \frac{22 + 27}{2} = \frac{49}{2} = 24.5$

The median weight loss was 24.5 pounds.

You Try It 3

Strategy To draw the box-and-whiskers plot:

- Arrange the data from least to greatest.
- Find the median, Q_1, and Q_3.
- Use the least value, Q_1, the median, Q_3, and the greatest value to draw the box-and-whiskers plot.

Solution

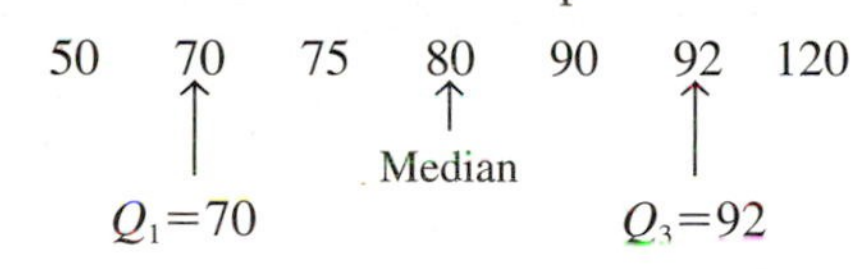

a.

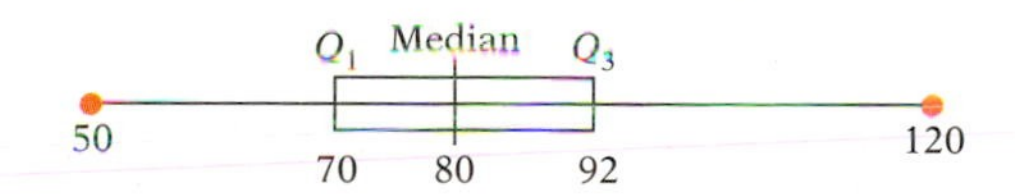

SECTION 7.5

You Try It 1

Strategy To find the probability:

- List the outcomes of the experiment in a systematic way. We will use a table.
- Use the table to count the number of possible outcomes of the experiment.

(Continued)

(Continued)

- Count the number of outcomes of the experiment that are favorable to the event of two true questions and one false question.
- Use the probability formula.

Solution

Question 1	Question 2	Question 3
T	T	T
T	T	F
T	F	T
T	F	F
F	T	T
F	T	F
F	F	T
F	F	F

There are 8 possible outcomes:

S = {TTT, TTF, TFT, TFF, FTT, FTF, FFT, FFF}

There are 3 outcomes favorable to the event:

{TTF, TFT, FTT}

Probability of an event

$$= \frac{\text{number of favorable outcomes}}{\text{number of possible outcomes}} = \frac{3}{8}$$

The probability of two true questions and one false question is $\frac{3}{8}$.

SOLUTIONS TO CHAPTER 8 "YOU TRY IT"

SECTION 8.1

You Try It 1

$$14\text{ ft} = 14\,\cancel{\text{ft}} \times \frac{1\text{ yd}}{3\,\cancel{\text{ft}}} = \frac{14\text{ yd}}{3} = 4\frac{2}{3}\text{ yd}$$

You Try It 2

$$9240\text{ ft} = 9240\,\cancel{\text{ft}} \times \frac{1\text{ mi}}{5280\,\cancel{\text{ft}}}$$

$$= \frac{9240\text{ mi}}{5280} = 1\frac{3}{4}\text{ mi}$$

You Try It 3

$$\begin{array}{r} 3\text{ ft }6\text{ in.} \\ 12\overline{)\ 42} \\ -36 \\ \hline 6 \end{array}$$

• 12 in. = 1 ft

42 in. = 3 ft 6 in.

You Try It 4

$$\begin{array}{r} 4\text{ yd }2\text{ ft} \\ 3\overline{)\ 14} \\ -12 \\ \hline 2 \end{array}$$

• 3 ft = 1 yd

14 ft = 4 yd 2 ft

You Try It 5

$$\begin{array}{r} 3\text{ ft}\ \ 5\text{ in.} \\ +\,4\text{ ft}\ \ 9\text{ in.} \\ \hline 7\text{ ft }14\text{ in.} \end{array}$$

• 14 in. = 1 ft 2 in.

7 ft 14 in. = 8 ft 2 in.

You Try It 6

$$\begin{array}{r} \scriptstyle 3\text{ ft}\ \ 14\text{ in.} \\ \cancel{4\text{ ft}}\ \cancel{2\text{ in.}} \\ -\,1\text{ ft }8\text{ in.} \\ \hline 2\text{ ft }6\text{ in.} \end{array}$$

• Borrow 1 ft (12 in.) from 4 ft and add it to 2 in.

You Try It 7

$$\begin{array}{r} 4\text{ yd }1\text{ ft} \\ \times\qquad 8 \\ \hline 32\text{ yd }8\text{ ft} \end{array}$$

• 8 ft = 2 yd 2 ft

32 yd 8 ft = 34 yd 2 ft

You Try It 8

$$\begin{array}{r} 3\text{ yd}\quad 2\text{ ft} \\ 2\overline{)\ 7\text{ yd}\quad 1\text{ ft}} \\ -6\text{ yd}\qquad\quad \\ \hline 1\text{ yd} = 3\text{ ft} \\ \hline 4\text{ ft} \\ -4\text{ ft} \\ \hline 0 \end{array}$$

You Try It 9

$$\begin{array}{r} 6\frac{1}{4}\text{ ft} = 6\frac{3}{12}\text{ ft} = 5\frac{15}{12}\text{ ft} \\ -3\frac{2}{3}\text{ ft} = 3\frac{8}{12}\text{ ft} = 3\frac{8}{12}\text{ ft} \\ \hline 2\frac{7}{12}\text{ ft} \end{array}$$

You Try It 10

Strategy To determine the patient's height in inches:

- Divide the total number of inches by 12.
- Write the whole number quotient as the number of feet and the remainder as inches.

Solution

$$\begin{array}{r} 5 \\ 12\overline{)\ 66} \\ -60 \\ \hline 6 \end{array}$$

The patient is 5 ft 6 in. tall.

You Try It 11

Strategy To find the amount of paper remaining on the roll, subtract the amount of paper used (192 ft 3 in.) from the amount of paper on the roll before it was used (225 ft).

Solution

$$\begin{array}{rr} 225\text{ ft} & = 224\text{ ft }12\text{ in.} \\ -192\text{ ft }3\text{ in.} & = 192\text{ ft}\ \ 3\text{ in.} \\ \hline & 32\text{ ft}\ \ 9\text{ in.} \end{array}$$

There are 32 ft 9 in. of paper remaining on the roll.

SECTION 8.2

You Try It 1

$$3\text{ lb} = 3\,\cancel{\text{lb}} \times \frac{16\text{ oz}}{1\,\cancel{\text{lb}}} = 48\text{ oz}$$

You Try It 2

$$4200\text{ lb} = 4200\,\cancel{\text{lb}} \times \frac{1\text{ ton}}{2000\,\cancel{\text{lb}}}$$

$$= \frac{4200\text{ tons}}{2000} = 2\frac{1}{10}\text{ tons}$$

You Try It 3

$$\begin{array}{r r} \overset{6\text{ lb}}{\cancel{7\text{ lb}}} & \overset{17\text{ oz}}{\cancel{1\text{ oz}}} \\ -\ 3\text{ lb} & 4\text{ oz} \\ \hline 3\text{ lb} & 13\text{ oz} \end{array}$$

• Borrow 1 lb (16 oz) from 7 lb and add it to 1 oz.

You Try It 4

$$\begin{array}{r} 3\text{ lb } 6\text{ oz} \\ \times\quad 4 \\ \hline \end{array}$$
$12\text{ lb } 24\text{ oz} = 13\text{ lb } 8\text{ oz}$

You Try It 5

Strategy To find the weight of 12 bars of soap:

- Multiply the number of bars (12) by the weight of each bar (7 oz).
- Convert the number of ounces to pounds.

Solution

$$\begin{array}{r} 12 \\ \times\ 7\text{ oz} \\ \hline \end{array}$$
$$84\text{ oz} = 84\cancel{\text{ oz}} \times \frac{1\text{ lb}}{16\cancel{\text{ oz}}} = 5\frac{1}{4}\text{ lb}$$

The 12 bars of soap weigh $5\frac{1}{4}$ lb.

SECTION 8.3

You Try It 1

$$18\text{ pt} = 18\cancel{\text{ pt}} \times \frac{1\cancel{\text{ qt}}}{2\cancel{\text{ pt}}} \times \frac{1\text{ gal}}{4\cancel{\text{ qt}}} = \frac{18\text{ gal}}{8} = 2\frac{1}{4}\text{ gal}$$

You Try It 2

$$\begin{array}{r r} & 1\text{ gal}\quad 2\text{ qt} \\ 3\overline{)\,} & 4\text{ gal}\quad 2\text{ qt} \\ & -3\text{ gal}\qquad\quad \\ \hline & 1\text{ gal} = 4\text{ qt} \\ & 6\text{ qt} \\ & -6\text{ qt} \\ \hline & 0 \end{array}$$

You Try It 3

Strategy To find the number of gallons of water needed:

- Find the number of quarts required by multiplying the number of quarts one student needs per day (5) by the number of students (5) by the number of days (3).
- Convert the number of quarts to gallons.

Solution $5 \times 5 \times 3 = 75$ qt

$$75\text{ qt} = 75\cancel{\text{ qt}} \cdot \frac{1\text{ gal}}{4\cancel{\text{ qt}}} = \frac{75\text{ gal}}{4} = 18\frac{3}{4}\text{ gal}$$

The students should take $18\frac{3}{4}$ gal of water.

SECTION 8.4

You Try It 1

$$18{,}000\text{ s} = 18{,}000\cancel{\text{ s}} \times \frac{1\cancel{\text{ min}}}{60\cancel{\text{ s}}} \times \frac{1\text{ h}}{60\cancel{\text{ min}}} = \frac{18{,}000\text{ h}}{3600} = 5\text{ h}$$

SECTION 8.5

You Try It 1

$$4.5\text{ Btu} = 4.5\cancel{\text{ Btu}} \times \frac{778\text{ ft}\cdot\text{lb}}{1\cancel{\text{ Btu}}} = 3501\text{ ft}\cdot\text{lb}$$

You Try It 2 Energy $= 800\text{ lb} \times 16\text{ ft} = 12{,}800\text{ ft}\cdot\text{lb}$

You Try It 3 $56{,}000$ Btu $=$

$$56{,}000\cancel{\text{ Btu}} \times \frac{778\text{ ft}\cdot\text{lb}}{1\cancel{\text{ Btu}}} = 43{,}568{,}000\text{ ft}\cdot\text{lb}$$

You Try It 4

$$\text{Power} = \frac{90\text{ ft} \times 1200\text{ lb}}{24\text{ s}} = 4500\,\frac{\text{ft}\cdot\text{lb}}{\text{s}}$$

You Try It 5 $\frac{3300}{550} = 6$ hp

SOLUTIONS TO CHAPTER 9 "YOU TRY IT"

SECTION 9.1

You Try It 1 $3.07\text{ m} = 307\text{ cm}$

You Try It 2 $750\text{ m} = 0.750\text{ km}$

$$3\text{ km } 750\text{ m} = 3\text{ km} + 0.750\text{ km} = 3.750\text{ km}$$

You Try It 3

Strategy To find the distance of the Fun Run:

- Convert the distance of the primary race (5 km) to meters.
- Multiply this distance by $\frac{1}{4}$.

Solution $5\text{ km} = 5000\text{ m}$

$$\frac{1}{4} \times 5000 = \frac{1}{4} \times \frac{5000}{1} = \frac{5000}{4} = 1250$$

The Fun Run is 1250 m long.

SECTION 9.2

You Try It 1 $42.3 \text{ mg} = 0.0423 \text{ g}$

You Try It 2 $54 \text{ mg} = 0.054 \text{ g}$

$$\begin{aligned} 3 \text{ g } 54 \text{ mg} &= 3 \text{ g} + 0.054 \text{ g} \\ &= 3.054 \text{ g} \end{aligned}$$

You Try It 3

Strategy To find the total amount of medication the patient receives:

- Multiply the patient's weight (14 kg) times the dosage rate (0.05g/kg).
- Convert the product from grams to milligrams.

Solution $14 \text{ kg} \times 0.05 \text{ g/kg}$

$$\frac{14 \cancel{\text{kg}}}{1} \times \frac{0.05 \text{ g}}{1 \cancel{\text{kg}}} = 0.7 \text{ g}$$

$0.7 \text{ g} = 700 \text{ mg}$

The patient received 700 mg of medication.

SECTION 9.3

You Try It 1 $2 \text{ kl} = 2000 \text{ L}$

$$\begin{aligned} 2 \text{ kl } 167 \text{ L} &= 2000 \text{ L} + 167 \text{ L} \\ &= 2167 \text{ L} \end{aligned}$$

You Try It 2 $325 \text{ cm}^3 = 325 \text{ ml} = 0.325 \text{ L}$

You Try It 3

Strategy To find the profit:

- Convert 5 L to milliliters.
- Find the number of jars by dividing the number of milliliters by the number of milliliters in each jar (125).
- Multiply the number of jars by the cost per jar ($0.85).
- Find the total cost by adding the cost of the jars to the cost for the massage lotion ($299.50).
- Find the income by multiplying the number of jars by the selling price per jar ($29.95).
- Subtract the total cost from the income.

Solution $5 \text{ L} = 5000 \text{ ml}$

$5000 \div 125 = 40$ This is the number of jars.

$$\begin{array}{r} 0.85 \\ \times \quad 40 \\ \hline 34.00 \end{array}$$ This is the cost of the jars.

$$\begin{array}{r} 299.50 \\ + \quad 34.00 \\ \hline 333.50 \end{array}$$ This is the total cost.

$$\begin{array}{r} 29.95 \\ \times \quad 40 \\ \hline 1198.00 \end{array}$$ This is the income from sales.

$$\begin{array}{r} 1198.00 \\ - \quad 333.50 \\ \hline 864.50 \end{array}$$

The profit on the 5 L of massage lotion is $864.50.

SECTION 9.4

You Try It 1

Strategy To find the number of Calories burned off, multiply the number of hours spent doing housework $\left(4\frac{1}{2}\right)$ by the Calories used per hour (240).

Solution $4\frac{1}{2} \times 240 = \frac{9}{2} \times 240 = 1080$

Doing $4\frac{1}{2}$ h of housework burns off 1080 Calories.

You Try It 2

Strategy To find the number of kilowatt-hours used:

- Find the number of watt-hours used.
- Convert watt-hours to kilowatt-hours.

Solution $150 \times 200 = 30{,}000$

$30{,}000 \text{ Wh} = 30 \text{ kWh}$

30 kWh of energy are used.

You Try It 3

Strategy To find the cost:

- Convert 20 min to hours.
- Multiply to find the total number of hours the oven is used.
- Multiply the number of hours used by the number of watts to find the watt-hours.
- Convert watt-hours to kilowatt-hours.
- Multiply the number of kilowatt-hours by the cost per kilowatt-hour.

Solution $20 \text{ min} = 20 \cancel{\text{min}} \times \dfrac{1 \text{ h}}{60 \cancel{\text{min}}}$

$= \dfrac{20}{60} \text{ h} = \dfrac{1}{3} \text{ h}$

$\dfrac{1}{3} \text{ h} \times 30 = 10 \text{ h}$

$10 \text{ h} \times 500 \text{ W} = 5000 \text{ Wh}$

$5000 \text{ Wh} = 5 \text{ kWh}$

$5 \times 13.7¢ = 68.5¢$

The cost is 68.5¢.

SECTION 9.5

You Try It 1

$125 \text{ lb} \times \dfrac{1 \text{ kg}}{2.2 \text{ lb}}$

$\dfrac{125 \cancel{\text{lb}}}{1} \times \dfrac{1 \text{ kg}}{2.2 \cancel{\text{lb}}} = \dfrac{125 \text{ kg}}{2.2} \approx 56.818 \text{ kg}$

• The conversion rate is $\dfrac{1 \text{ kg}}{2.2 \text{ lb}}$ with kg in the numerator and lb in the denominator.

56.818 rounded to the nearest tenth is 56.8.

$125 \text{ lb} \approx 56.8 \text{ kg}$

You Try It 2

$\dfrac{\$3.69}{\text{gal}} \approx \dfrac{\$3.69}{\cancel{\text{gal}}} \times \dfrac{1 \cancel{\text{gal}}}{3.79 \text{ L}}$

$= \dfrac{\$3.69}{3.79 \text{ L}} \approx \dfrac{\$0.97}{\text{L}}$

$\$3.69/\text{gal} \approx \$0.97/\text{L}$

You Try It 3

$45 \text{ cm} = \dfrac{45 \cancel{\text{cm}}}{1} \times \dfrac{1 \text{ in.}}{2.54 \cancel{\text{cm}}}$

$= \dfrac{45 \text{ in.}}{2.54} \approx 17.72 \text{ in.}$

$45 \text{ cm} \approx 17.72 \text{ in.}$

You Try It 4

$\dfrac{75 \text{ km}}{\text{h}} \approx \dfrac{75 \cancel{\text{km}}}{\text{h}} \times \dfrac{1 \text{ mi}}{1.61 \cancel{\text{km}}}$

$= \dfrac{75 \text{ mi}}{1.61 \text{ h}} \approx 46.58 \text{ mi/h}$

$75 \text{ km/h} \approx 46.58 \text{ mi/h}$

You Try It 5

$\dfrac{\$1.75}{\text{L}} \approx \dfrac{\$1.75}{\cancel{\text{L}}} \times \dfrac{3.79 \cancel{\text{L}}}{1 \text{ gal}}$

$= \dfrac{\$6.6325}{1 \text{ gal}} \approx \$6.63/\text{gal}$

$\$1.75/\text{L} \approx \$6.63/\text{gal}$

SOLUTIONS TO CHAPTER 10 "YOU TRY IT"

SECTION 10.1

You Try It 1 −232 ft

You Try It 2 (number line from −4 to 4 with a dot at −4)

You Try It 3 $-12 < -8$ • −12 is to the left of −8 on the number line.

You Try It 4 $|-7| = 7$; $|21| = 21$

You Try It 5 $|2| = 2$; $|-9| = 9$

You Try It 6 $-|-12| = -12$

SECTION 10.2

You Try It 1 $-154 + (-37) = -191$ • The signs of the addends are the same.

You Try It 2

$-5 + (-2) + 9 + (-3)$
$= -7 + 9 + (-3)$
$= 2 + (-3)$
$= -1$

You Try It 3

$-8 - 14$
$= -8 + (-14)$
$= -22$

• Rewrite "−" as "+"; the opposite of 14 is −14.

You Try It 4

$3 - (-15)$
$= 3 + 15$
$= 18$

• Rewrite "−" as "+"; the opposite of −15 is 15.

You Try It 5

$4 - (-3) - 12 - (-7) - 20$
$= 4 + 3 + (-12) + 7 + (-20)$
$= 7 + (-12) + 7 + (-20)$
$= -5 + 7 + (-20)$
$= 2 + (-20)$
$= -18$

You Try It 6

Strategy To find the temperature, add the increase (12°C) to the previous temperature (−10°C).

Solution $-10 + 12 = 2$

After an increase of 12°C, the temperature is 2°C.

SECTION 10.3

You Try It 1 $(-3)(-5) = 15$ • The signs are the same. The product is positive.

You Try It 2 The signs are different. The product is negative.

$-38 \cdot 51 = -1938$

You Try It 3 $-7(-8)(9)(-2) = 56(9)(-2) = 504(-2) = -1008$

You Try It 4 $(-135) \div (-9) = 15$ • The signs are the same. The quotient is positive.

You Try It 5 $84 \div (-6) = -14$ • The signs are different. The quotient is negative.

You Try It 6 $-72 \div 4 = -18$

You Try It 7 Division by zero is undefined. $-39 \div 0$ is undefined.

You Try It 8

Strategy To find the combined weight loss of Group B, multiply the total weight loss of Group A (-20) by 3.

Solution $-20 \times 3 = -60$

The combined weight loss of Group B was -60 lb.

You Try It 9

Strategy To find the average daily low temperature:

- Add the seven temperature readings.
- Divide by 7.

Solution

$$\begin{aligned}&-6 + (-7) + 1 + 0 + (-5) + (-10) + (-1)\\&= -13 + 1 + 0 + (-5) + (-10) + (-1)\\&= -12 + 0 + (-5) + (-10) + (-1)\\&= -12 + (-5) + (-10) + (-1)\\&= -17 + (-10) + (-1)\\&= -27 + (-1)\\&= -28\end{aligned}$$

$-28 \div 7 = -4$

The average daily low temperature was -4°F.

SECTION 10.4

You Try It 1 The LCM of 9 and 12 is 36.

$$\frac{5}{9} - \frac{11}{12} = \frac{20}{36} - \frac{33}{36} = \frac{20}{36} + \frac{-33}{36} = \frac{20 + (-33)}{36} = \frac{-13}{36} = -\frac{13}{36}$$

You Try It 2 The LCM of 8, 6, and 3 is 24.

$$\begin{aligned}&-\frac{7}{8} - \frac{5}{6} + \frac{2}{3}\\&= -\frac{21}{24} - \frac{20}{24} + \frac{16}{24}\\&= \frac{-21}{24} + \frac{-20}{24} + \frac{16}{24}\\&= \frac{-21 + (-20) + 16}{24} = \frac{225}{24}\\&= -\frac{25}{24} = -1\frac{1}{24}\end{aligned}$$

You Try It 3 $16.127 - 67.91 = 16.127 + (-67.91)$

$$\begin{array}{r}67.910\\-16.127\\\hline 51.783\end{array}$$

$16.127 - 67.91 = -51.783$

You Try It 4 $2.7 + (-9.44) + 6.2 = -6.74 + 6.2 = -0.54$

You Try It 5 The product is positive.

$$\left(-\frac{2}{3}\right)\left(-\frac{9}{10}\right) = \frac{2 \cdot 9}{3 \cdot 10} = \frac{3}{5}$$

You Try It 6 The quotient is negative.

$$-\frac{5}{8} \div \frac{5}{40} = -\left(\frac{5}{8} \div \frac{5}{40}\right) = -\left(\frac{5}{8} \times \frac{40}{5}\right) = -\left(\frac{5 \cdot 40}{8 \cdot 5}\right) = -5$$

You Try It 7 The product is negative.

$$\begin{array}{r}5.44\\\times 3.8\\\hline 4352\\1632\\\hline 20.672\end{array}$$

$-5.44 \times 3.8 = -20.672$

You Try It 8 $3.44 \times (-1.7) \times 0.6 = (-5.848) \times 0.6 = -3.5088$

You Try It 9

$$\begin{array}{r} 0.231 \\ 1.7\overline{)0.3940} \\ -3\,4 \\ \hline 54 \\ -51 \\ \hline 30 \\ -17 \\ \hline 13 \end{array}$$

$-0.394 \div 1.7 \approx -0.23$

You Try It 10

Strategy To find how many degrees the temperature fell, subtract the lower temperature (−13.33°C) from the higher temperature (12.78°C).

Solution

$$\begin{aligned} 12.78 - (-13.33) &= 12.78 + 13.33 \\ &= 26.11 \end{aligned}$$

The temperature fell 26.11°C in the 15-minute period.

SECTION 10.5

You Try It 1 The number is less than 1. Move the decimal point 7 places to the right. The exponent on 10 is −7.

$0.000000961 = 9.61 \times 10^{-7}$

You Try It 2 The exponent on 10 is positive. Move the decimal point 6 places to the right.

$7.329 \times 10^6 = 7{,}329{,}000$

You Try It 3

$$\begin{aligned} &9 - 9 \div (-3) \\ &= 9 - (-3) && \text{• Do the division.} \\ &= 9 + 3 && \text{• Rewrite as addition. Add.} \\ &= 12 \end{aligned}$$

You Try It 4

$$\begin{aligned} &8 \div 4 \cdot 4 - (-2)^2 \\ &= 8 \div 4 \cdot 4 - 4 && \text{• Exponents} \\ &= 2 \cdot 4 - 4 && \text{• Division} \\ &= 8 - 4 && \text{• Multiplication} \\ &= 8 + (-4) && \text{• Subtraction} \\ &= 4 \end{aligned}$$

You Try It 5

$$\begin{aligned} &8 - (-15) \div (2 - 7) \\ &= 8 - (-15) \div (-5) \\ &= 8 - 3 \\ &= 8 + (-3) \\ &= 5 \end{aligned}$$

You Try It 6

$$\begin{aligned} &(-2)^2 \times (3 - 7)^2 - (-16) \div (-4) \\ &= (-2)^2 \times (-4)^2 - (-16) \div (-4) \\ &= 4 \times 16 - (-16) \div (-4) \\ &= 64 - (-16) \div (-4) \\ &= 64 - 4 \\ &= 64 + (-4) \\ &= 60 \end{aligned}$$

You Try It 7

$$\begin{aligned} &7 \div \left(\frac{1}{7} - \frac{3}{14}\right) - 9 \\ &= 7 \div \left(-\frac{1}{14}\right) - 9 \\ &= 7(-14) - 9 \\ &= -98 - 9 \\ &= -98 + (-9) \\ &= -107 \end{aligned}$$

SOLUTIONS TO CHAPTER 11 "YOU TRY IT"

SECTION 11.1

You Try It 1

$$\begin{aligned} &6a - 5b \\ &6(-3) - 5(4) = -18 - 20 \\ & = -18 + (-20) \\ & = -38 \end{aligned}$$

You Try It 2

$$\begin{aligned} &-3s^2 - 12 \div t \\ &-3(-2)^2 - 12 \div 4 = -3(4) - 12 \div 4 \\ & = -12 - 12 \div 4 \\ & = -12 - 3 \\ & = -12 + (-3) \\ & = -15 \end{aligned}$$

You Try It 3

$$\begin{aligned} &-\tfrac{2}{3}m + \tfrac{3}{4}n^3 \\ &-\tfrac{2}{3}(6) + \tfrac{3}{4}(2)^3 = -\tfrac{2}{3}(6) + \tfrac{3}{4}(8) \\ &\phantom{-\tfrac{2}{3}(6) + \tfrac{3}{4}(2)^3} = -4 + 6 \\ &\phantom{-\tfrac{2}{3}(6) + \tfrac{3}{4}(2)^3} = 2 \end{aligned}$$

You Try It 4

$$\begin{aligned} &-3yz - z^2 + y^2 \\ &-3\left(-\tfrac{2}{3}\right)\left(\tfrac{1}{3}\right) - \left(\tfrac{1}{3}\right)^2 + \left(-\tfrac{2}{3}\right)^2 \\ &= -3\left(-\tfrac{2}{3}\right)\left(\tfrac{1}{3}\right) - \tfrac{1}{9} + \tfrac{4}{9} \\ &= \tfrac{2}{3} - \tfrac{1}{9} + \tfrac{4}{9} = \tfrac{6}{9} - \tfrac{1}{9} + \tfrac{4}{9} = \tfrac{9}{9} \\ &= 1 \end{aligned}$$

You Try It 5

$$\begin{aligned} &5a^2 - 6b^2 + 7a^2 - 9b^2 \\ &= 5a^2 + (-6)b^2 + 7a^2 + (-9)b^2 \\ &= 5a^2 + 7a^2 + (-6)b^2 + (-9)b^2 \\ &= 12a^2 + (-15)b^2 \\ &= 12a^2 - 15b^2 \end{aligned}$$

You Try It 6

$$\begin{aligned} &-6x + 7 + 9x - 10 \\ &= (-6)x + 7 + 9x + (-10) \\ &= (-6)x + 9x + 7 + (-10) \\ &= 3x + (-3) \\ &= 3x - 3 \end{aligned}$$

You Try It 7

$$\frac{3}{8}w + \frac{1}{2} - \frac{1}{4}w - \frac{2}{3}$$
$$= \frac{3}{8}w + \frac{1}{2} + \left(-\frac{1}{4}\right)w + \left(-\frac{2}{3}\right)$$
$$= \frac{3}{8}w + \left(-\frac{1}{4}\right)w + \frac{1}{2} + \left(-\frac{2}{3}\right)$$
$$= \frac{3}{8}w + \left(-\frac{2}{8}\right)w + \frac{3}{6} + \left(-\frac{4}{6}\right)$$
$$= \frac{1}{8}w + \left(-\frac{1}{6}\right)$$
$$= \frac{1}{8}w - \frac{1}{6}$$

You Try It 8

$$5(a - 2) = 5a - 5(2)$$
$$= 5a - 10$$

You Try It 9

$$8s - 2(3s - 5)$$
$$= 8s - 2(3s) - (-2)(5)$$
$$= 8s - 6s + 10$$
$$= 2s + 10$$

You Try It 10

$$4(x - 3) - 2(x + 1)$$
$$= 4x - 4(3) - 2x - 2(1)$$
$$= 4x - 12 - 2x - 2$$
$$= 4x - 2x - 12 - 2$$
$$= 2x - 14$$

SECTION 11.2

You Try It 1

$x(x + 3) = 4x + 6$	
$(-2)(-2 + 3)$	$4(-2) + 6$
$(-2)(1)$	$(-8) + 6$

$$-2 = -2$$

Yes, -2 is a solution.

You Try It 2

$x^2 - x = 3x + 7$	
$(-3)^2 - (-3)$	$3(-3) + 7$
$9 + 3$	$-9 + 7$

$$12 \neq -2$$

No, -3 is not a solution.

You Try It 3

$$-2 + y = -5$$
$$-2 + 2 + y = -5 + 2 \quad \text{• Add 2 to each side.}$$
$$0 + y = -3$$
$$y = -3$$

The solution is -3.

You Try It 4

$$7 = y + 8$$
$$7 - 8 = y + 8 - 8 \quad \text{• Subtract 8 from each side.}$$
$$-1 = y + 0$$
$$-1 = y$$

The solution is -1.

You Try It 5

$$\frac{1}{5} = z + \frac{4}{5}$$
$$\frac{1}{5} - \frac{4}{5} = z + \frac{4}{5} - \frac{4}{5} \quad \text{• Subtract } \frac{4}{5} \text{ from each side.}$$
$$-\frac{3}{5} = z + 0$$
$$-\frac{3}{5} = z$$

The solution is $-\frac{3}{5}$.

You Try It 6

$$4z = -20$$
$$\frac{4z}{4} = \frac{-20}{4} \quad \text{• Divide each side by 4.}$$
$$1z = -5$$
$$z = -5$$

The solution is -5.

You Try It 7

$$8 = \frac{2}{5}n$$
$$\left(\frac{5}{2}\right)(8) = \left(\frac{5}{2}\right)\frac{2}{5}n \quad \text{• Multiply each side by } \frac{5}{2}.$$
$$20 = 1n$$
$$20 = n$$

The solution is 20.

You Try It 8

$$\frac{2}{3}t - \frac{1}{3}t = -2 \quad \text{• Combine like terms.}$$
$$\frac{1}{3}t = -2$$
$$\left(\frac{3}{1}\right)\frac{1}{3}t = \left(\frac{3}{1}\right)(-2) \quad \text{• Multiply each side by 3.}$$
$$1t = -6$$
$$t = -6$$

The solution is -6.

You Try It 9

Strategy To find the regular price, replace the variables S and D in the formula by the given values and solve for R.

Solution

$$S = R - D$$
$$44 = R - 16$$
$$44 + 16 = R - 16 + 16$$
$$60 = R$$

The regular price is \$60.

You Try It 10

Strategy To find the monthly payment, replace the variables A and N in the formula by the given values and solve for M.

Solution

$$A = MN$$
$$6840 = M \cdot 24$$
$$6840 = 24M$$
$$\frac{6840}{24} = \frac{24M}{24}$$
$$285 = M$$

The monthly payment is \$285.

SECTION 11.3

You Try It 1

$$5x + 8 = 6$$
$$5x + 8 - 8 = 6 - 8 \quad \text{• Subtract 8 from each side.}$$
$$5x = -2$$
$$\frac{5x}{5} = \frac{-2}{5} \quad \text{• Divide each side by 5.}$$
$$x = -\frac{2}{5}$$

The solution is $-\frac{2}{5}$.

You Try It 2

$$7 - x = 3$$
$$7 - 7 - x = 3 - 7 \quad \text{• Subtract 7 from each side.}$$
$$-x = -4$$
$$(-1)(-x) = (-1)(-4) \quad \text{• Multiply each side by } -1.$$
$$x = 4$$

The solution is 4.

You Try It 3

Strategy To find the Celsius temperature, replace the variable F in the formula by the given value and solve for C.

Solution

$$F = 1.8C + 32$$
$$-22 = 1.8C + 32 \quad \text{• Substitute } -22 \text{ for } F.$$
$$-22 - 32 = 1.8C + 32 - 32 \quad \text{• Subtract 32 from each side.}$$
$$-54 = 1.8C \quad \text{• Combine like terms.}$$
$$\frac{-54}{1.8} = \frac{1.8C}{1.8} \quad \text{• Divide each side by 1.8.}$$
$$-30 = C$$

The temperature is -30°C.

You Try It 4

Strategy To find the length of the woman's femur, replace the variable H in the formula $H = 1.9L + 28.7$, where H is a woman's height in inches and L is the length of her femur in inches, with the given value for H (67 in.). Round the answer to the nearest tenth.

Solution

$$67 = 1.9L + 28.7$$
$$67 - 28.7 = 1.9L + 28.7 - 28.7$$
$$38.3 = 1.9L$$
$$\frac{38.3}{1.9} = \frac{1.9L}{1.9}$$
$$20.15789 \approx L$$

Round 20.15789 to 20.2.

The woman's femur is approximately 20.2 in. long.

SECTION 11.4

You Try It 1

$$\frac{1}{5}x - 2 = \frac{2}{5}x + 4 \quad \text{• Subtract } \frac{2}{5}x \text{ from each side.}$$
$$\frac{1}{5}x - \frac{2}{5}x - 2 = \frac{2}{5}x - \frac{2}{5}x + 4 \quad \text{• Add 2 to each side.}$$
$$-\frac{1}{5}x - 2 = 4$$
$$-\frac{1}{5}x - 2 + 2 = 4 + 2 \quad \text{• Multiply each side by } -5.$$
$$-\frac{1}{5}x = 6$$
$$(-5)\left(-\frac{1}{5}x\right) = (-5)6$$
$$x = -30$$

The solution is -30.

You Try It 2

$$4(x - 1) - x = 5$$
$$4x - 4 - x = 5$$
$$3x - 4 = 5$$
$$3x - 4 + 4 = 5 + 4$$
$$3x = 9$$
$$\frac{3x}{3} = \frac{9}{3}$$
$$x = 3$$

The solution is 3.

You Try It 3

$$2x - 7(3x + 1) = 5(5 - 3x)$$
$$2x - 21x - 7 = 25 - 15x \quad \text{• Distributive Property}$$
$$-19x - 7 = 25 - 15x \quad \text{• Combine like terms.}$$
$$-19x + 15x - 7 = 25 - 15x + 15x \quad \text{• Add } 15x \text{ to both sides.}$$
$$-4x - 7 = 25 \quad \text{• Combine like terms.}$$
$$-4x - 7 + 7 = 25 + 7 \quad \text{• Add 7 to both sides.}$$
$$-4x = 32 \quad \text{• Combine like terms.}$$
$$\frac{-4x}{-4} = \frac{32}{-4} \quad \text{• Divide each side by } -4.$$
$$x = -8$$

The solution is -8.

SECTION 11.5

You Try It 1 $8 - 2t$

You Try It 2 $\frac{5}{7x}$

You Try It 3 The *product* of a number and *one-half* of the number

The unknown number: x

One-half the number: $\frac{1}{2}x$

$(x)\left(\frac{1}{2}x\right)$

SECTION 11.6

You Try It 1 The unknown number: x

A number increased by four	equals	twelve

$$x + 4 = 12$$
$$x + 4 - 4 = 12 - 4$$
$$x = 8$$

The number is 8.

You Try It 2

The unknown number: x

The product of two and a number	is	ten

$$2x = 10$$
$$\frac{2x}{2} = \frac{10}{2}$$
$$x = 5$$

The number is 5.

You Try It 3

The unknown number: x

The sum of three times a number and six	equals	four

$$3x + 6 = 4$$
$$3x + 6 - 6 = 4 - 6$$
$$3x = -2$$
$$\frac{3x}{3} = \frac{-2}{3}$$
$$x = -\frac{2}{3}$$

The number is $-\frac{2}{3}$.

You Try It 4

The unknown number: x

Three more than one-half of a number	is	nine

$$\frac{1}{2}x + 3 = 9$$
$$\frac{1}{2}x + 3 - 3 = 9 - 3$$
$$\frac{1}{2}x = 6$$
$$2 \cdot \frac{1}{2}x = 2 \cdot 6$$
$$x = 12$$

The number is 12.

You Try It 5

Strategy To find the regular price of the box of prenatal vitamins, write and solve an equation using x to represent the regular price.

Solution

\$23.99	is	\$6 less than the regular price of the prenatal vitamins

$$23.99 = x - 6$$
$$23.99 + 6 = x - 6 + 6$$
$$29.99 = x$$

The regular price of prenatal vitamins is \$29.99.

You Try It 6

Strategy To predict the child's height as an adult, write and solve an equation using H to represent the height of the child as an adult. Then convert the answer from inches to feet and inches.

Solution

34 in.	is	half the height the child will be as an adult

$$34 = \frac{1}{2}H$$
$$\frac{34}{1} \times \frac{2}{1} = \left(\frac{1}{2} \times \frac{2}{1}\right)H$$
$$68 = H$$

The predicted height of the child as an adult is 68 in.

Convert 68 in. to feet and inches.

$$\begin{array}{r} 5 \\ 12\overline{)68} \\ -60 \\ \hline 8 \end{array} \quad 5 \text{ ft } 8 \text{ in.}$$

The predicted height of the child as an adult is 5 ft 8 in.

You Try It 7

Strategy To find the number of hours of labor, write and solve an equation using H to represent the number of hours of labor required.

Solution

\$492	includes	\$100 for materials plus \$24.50 per hour of labor

$$492 = 100 + 24.50H$$
$$492 - 100 = 100 - 100 + 24.50H$$
$$392 = 24.50H$$
$$\frac{392}{24.50} = \frac{24.50H}{24.50}$$
$$16 = H$$

16 h of labor are required.

You Try It 8

Strategy To find the total sales, write and solve an equation using S to represent the total sales.

Solution

\$2500	is	the sum of \$800 and an 8% commission on total sales

$$2500 = 800 + 0.08S$$
$$2500 - 800 = 800 - 800 + 0.08S$$
$$1700 = 0.08S$$
$$\frac{1700}{0.08} = \frac{0.08S}{0.08}$$
$$21{,}250 = S$$

Natalie's total sales for the month were \$21,250.

Answers to Selected Exercises

ANSWERS TO CHAPTER 1 SELECTED EXERCISES

PREP TEST

1. 8 **2.** 1 2 3 4 5 6 7 8 9 10 **3.** a and D; b and E; c and A; d and B; e and F; f and C

SECTION 1.1

1. 0 1 2 3 4 5 6 7 8 9 10 11 12 **3.** 0 1 2 3 4 5 6 7 8 9 10 11 12 **5.** $37 < 49$ **7.** $101 > 87$ **9.** $2701 > 2071$ **11.** $107 > 0$ **13.** Yes **15.** Millions **17.** Hundred-thousands **19.** Three thousand seven hundred ninety **21.** Fifty-eight thousand four hundred seventy-three **23.** Four hundred ninety-eight thousand five hundred twelve **25.** Six million eight hundred forty-two thousand seven hundred fifteen **27.** 357 **29.** 63,780 **31.** 7,024,709 **33.** $5000 + 200 + 80 + 7$ **35.** $50{,}000 + 8000 + 900 + 40 + 3$ **37.** $200{,}000 + 500 + 80 + 3$ **39.** $400{,}000 + 3000 + 700 + 5$ **41.** No **43.** 850 **45.** 4000 **47.** 53,000 **49.** 630,000 **51.** 250,000 **53.** 72,000,000 **55.** No. Round 3846 to the nearest hundred.

SECTION 1.2

1. 28 **3.** 125 **5.** 102 **7.** 154 **9.** 1489 **11.** 828 **13.** 1584 **15.** 1219 **17.** 102,317 **19.** 79,326 **21.** 1804 **23.** 1579 **25.** 19,740 **27.** 7420 **29.** 120,570 **31.** 207,453 **33.** 24,218 **35.** 11,974 **37.** 9323 **39.** 77,139 **41.** 14,383 **43.** 9473 **45.** 33,247 **47.** 5058 **49.** 1992 **51.** 68,263 **53.** Cal.: 17,754 Est.: 17,700 **55.** Cal.: 2872 Est.: 2900 **57.** Cal.: 101,712 Est.: 101,000 **59.** Cal.: 158,763 Est.: 158,000 **61.** Cal.: 261,595 Est.: 260,000 **63.** Cal.: 946,718 Est.: 940,000 **65.** Commutative Property of Addition **67.** There were 118,295 multiple births during the year. **69.** The total income produced by *Patch Adams* and *The Awakening* is $143,442,014. **71.** The income from the two movies with the lowest box-office incomes is $46,536,017. **73. a.** He walked 11 miles at work from Monday through Wednesday. **b.** He walked 18 miles at work this week. **75.** The total mileage flown by the medical helicopter was 399 miles. **77.** 11 different sums **79.** No. $0 + 0 = 0$ **81.** 10 numbers

SECTION 1.3

1. 4 **3.** 4 **5.** 10 **7.** 4 **9.** 9 **11.** 22 **13.** 60 **15.** 66 **17.** 31 **19.** 901 **21.** 791 **23.** 1125 **25.** 3131 **27.** 47 **29.** 925 **31.** 4561 **33.** 3205 **35.** 1222 **37.** 53 **39.** 29 **41.** 8 **43.** 37 **45.** 58 **47.** 574 **49.** 337 **51.** 1423 **53.** 754 **55.** 2179 **57.** 6489 **59.** 889 **61.** 71,129 **63.** 698 **65.** 29,405 **67.** 49,624 **69.** 628 **71.** 6532 **73.** 4286 **75.** 4042 **77.** 5209 **79.** 10,378 **81.** (ii) and (iii) **83.** 11,239 **85.** 8482 **87.** 625 **89.** 76,725 **91.** 23 **93.** 4648 **95.** Cal.: 29,837 Est.: 30,000 **97.** Cal.: 36,668 Est.: 40,000 **99.** Cal.: 101,998 Est.: 100,000 **101. a.** The number of units sold in New York, California, and Quebec was 914 units. **b.** The difference between the number of units sold in the five states and the Canadian province is 841 units. **c.** There were 224 more units sold in California than in Florida. **d.** The total number of units sold in the states and province is 1309 units. **103.** Your office still owes the company $417. **105.** 202,345 more women than men earned a bachelor's degree. **107. a.** The smallest expected increase occurs from 2010 to 2012. **b.** The greatest expected increase occurs from 2018 to 2020. **109.** You can eat 1 cup of air-popped popcorn.

SECTION 1.4

1. 6×2 or $6 \cdot 2$ **3.** 4×7 or $4 \cdot 7$ **5.** 12 **7.** 35 **9.** 25 **11.** 0 **13.** 72 **15.** 198 **17.** 335 **19.** 2492 **21.** 5463 **23.** 4200 **25.** 6327 **27.** 1896 **29.** 5056 **31.** 1685 **33.** 46,963 **35.** 59,976 **37.** 19,120 **39.** 19,790 **41.** 140 **43.** 22,456 **45.** 18,630 **47.** 336 **49.** 910 **51.** 63,063 **53.** 33,520 **55.** 380,834 **57.** 541,164 **59.** 400,995 **61.** 105,315 **63.** 428,770 **65.** 260,000 **67.** 344,463 **69.** 41,808 **71.** 189,500 **73.** 401,880 **75.** 1,052,763 **77.** 4,198,388 **79.** For example, 5 and 20 **81.** 198,423 **83.** 18,834 **85.** 260,178 **87.** Cal.: 440,076 Est.: 450,000 **89.** Cal.: 6,491,166 Est.: 6,300,000 **91.** Cal.: 18,728,744 Est.: 18,000,000 **93.** Cal.: 57,691,192 Est.: 54,000,000 **95.** The total daily dosage is 100 milligrams. **97.** You traveled 560 miles last week. **99.** The cost of 12 cases of hand sanitizer is $300. **101.** The total cost of the supplies is $219. **103.** There would be 288 rolls in one case. Based on the cost per box, a case of 24 boxes would cost $432. **105.** There are 12 accidental deaths each hour; 288 deaths each day; and 105,120 deaths each year.

SECTION 1.5

1. 2 **3.** 6 **5.** 7 **7.** 16 **9.** 210 **11.** 44 **13.** 703 **15.** 910 **17.** 21,560 **19.** 3580 **21.** 1075 **23.** 1 **25.** 47 **27.** 23 **29.** 3 r1 **31.** 9 r7 **33.** 16 r1 **35.** 10 r4 **37.** 90 r3 **39.** 120 r5 **41.** 309 r3 **43.** 1160 r4 **45.** 708 r2 **47.** 3825 r1 **49.** 9044 r2 **51.** 11,430 **53.** 510 **55.** False **57.** 1 r38 **59.** 1 r26 **61.** 21 r21 **63.** 30 r22 **65.** 5 r40 **67.** 9 r17 **69.** 200 r21 **71.** 303 r1 **73.** 67 r13 **75.** 176 r13 **77.** 1086 r7 **79.** 403 **81.** 12 r456 **83.** 4 r160 **85.** 160 r27 **87.** 1669 r14 **89.** 7950 **91.** Cal.: 5129 Est.: 5000 **93.** Cal.: 21,968 Est.: 20,000 **95.** Cal.: 24,596 Est.: 22,500 **97.** Cal.: 2836 Est.: 3000 **99.** Cal.: 3024 Est.: 3000 **101.** Cal.: 32,036 Est.: 30,000 **103.** The average monthly cost for hip replacement surgery is $73,250. **105.** The average monthly cost for fracture repair is $57,000. **107.** 380 pennies are in circulation for each person. **109.** 960 pieces of mail are mailed each month. **111.** (i) and (ii) **113.** The total of the deductions is $350. **115.** The company's annual cost for employee-only health care coverage is $5184. **117.** The total amount spent annually by the average household is $35,535. **119.** The dental hygienist earns $5571 per month. **121.** The total amount paid is $11,860.

SECTION 1.6

1. 2^3 **3.** $6^3 \cdot 7^4$ **5.** $2^3 \cdot 3^3$ **7.** $5 \cdot 7^5$ **9.** $3^3 \cdot 6^4$ **11.** $3^3 \cdot 5 \cdot 9^3$ **13.** 8 **15.** 400 **17.** 900 **19.** 972 **21.** 120 **23.** 360 **25.** 0 **27.** 90,000 **29.** 540 **31.** 4050 **33.** 11,025 **35.** 25,920 **37.** 4,320,000 **39.** 5 **41.** 4 **43.** 23 **45.** 5 **47.** 10 **49.** 7 **51.** 8 **53.** 6 **55.** 52 **57.** 26 **59.** 52 **61.** 42 **63.** 8 **65.** 16 **67.** 6 **69.** 8 **71.** 3 **73.** 4 **75.** 13 **77.** 0 **79.** $8 - 2 \cdot (3 + 1)$

SECTION 1.7

1. 1, 2, 4 **3.** 1, 2, 5, 10 **5.** 1, 7 **7.** 1, 3, 9 **9.** 1, 13 **11.** 1, 2, 3, 6, 9, 18 **13.** 1, 2, 4, 7, 8, 14, 28, 56 **15.** 1, 3, 5, 9, 15, 45 **17.** 1, 29 **19.** 1, 2, 11, 22 **21.** 1, 2, 4, 13, 26, 52 **23.** 1, 2, 41, 82 **25.** 1, 3, 19, 57 **27.** 1, 2, 3, 4, 6, 8, 12, 16, 24, 48 **29.** 1, 5, 19, 95 **31.** 1, 2, 3, 6, 9, 18, 27, 54 **33.** 1, 2, 3, 6, 11, 22, 33, 66 **35.** 1, 2, 4, 5, 8, 10, 16, 20, 40, 80 **37.** 1, 2, 3, 4, 6, 8, 12, 16, 24, 32, 48, 96 **39.** 1, 2, 3, 5, 6, 9, 10, 15, 18, 30, 45, 90 **41.** True **43.** $2 \cdot 3$ **45.** Prime **47.** $2 \cdot 2 \cdot 2 \cdot 3$ **49.** $3 \cdot 3 \cdot 3$ **51.** $2 \cdot 2 \cdot 3 \cdot 3$ **53.** Prime **55.** $2 \cdot 3 \cdot 3 \cdot 5$ **57.** $5 \cdot 23$ **59.** $2 \cdot 3 \cdot 3$ **61.** $2 \cdot 2 \cdot 7$ **63.** Prime **65.** $2 \cdot 31$ **67.** $2 \cdot 11$ **69.** Prime **71.** $2 \cdot 3 \cdot 11$ **73.** $2 \cdot 37$ **75.** Prime **77.** $5 \cdot 11$ **79.** $2 \cdot 2 \cdot 2 \cdot 3 \cdot 5$ **81.** $2 \cdot 2 \cdot 2 \cdot 2 \cdot 2 \cdot 5$ **83.** $2 \cdot 2 \cdot 2 \cdot 3 \cdot 3 \cdot 3$ **85.** $5 \cdot 5 \cdot 5 \cdot 5$ **87.** False

CHAPTER 1 CONCEPT REVIEW*

1. The symbol $<$ means "is less than." A number that appears to the left of a given number on the number line is less than ($<$) the given number. For example, $4 < 9$. The symbol $>$ means "is greater than." A number that appears to the right of a given number on the number line is greater than ($>$) the given number. For example, $5 > 2$. [1.1A]

**Note:* The numbers in brackets following the answers in the Concept Review are a reference to the objective that corresponds to that problem. For example, the reference [1.2A] stands for Section 1.2, Objective A. This notation will be used for all Prep Tests, Concept Reviews, Chapter Reviews, Chapter Tests, and Cumulative Reviews throughout the text.

2. To round a four-digit whole number to the nearest hundred, look at the digit in the tens place. If the digit in the tens place is less than 5, that digit and the digit in the ones place are replaced by zeros. If the digit in the tens place is greater than or equal to 5, increase the digit in the hundreds place by 1 and replace the digits in the tens place and the ones place by zeros. [1.1D]

3. The Commutative Property of Addition states that two numbers can be added in either order; the sum is the same. For example, $3 + 5 = 5 + 3$. The Associative Property of Addition states that changing the grouping of three or more addends does not change their sum. For example, $3 + (4 + 5) = (3 + 4) + 5$. Note that in the Commutative Property of Addition, the order in which the numbers appear changes, while in the Associative Property of Addition, the order in which the numbers appear does not change. [1.2A]

4. To estimate the sum of two numbers, round each number to the same place value. Then add the numbers. For example, to estimate the sum of 562,397 and 41,086, round the numbers to 560,000 and 40,000. Then add $560{,}000 + 40{,}000 = 600{,}000$. [1.2A]

5. It is necessary to borrow when performing subtraction if, in any place value, the lower digit is larger than the upper digit. [1.3B]

6. The Multiplication Property of Zero states that the product of a number and zero is zero. For example, $8 \times 0 = 0$. The Multiplication Property of One states that the product of a number and one is the number. For example, $8 \times 1 = 8$. [1.4A]

7. To multiply a whole number by 100, write two zeros to the right of the number. For example, $64 \times 100 = 6400$. [1.4B]

8. To estimate the product of two numbers, round each number so that it contains only one nonzero digit. Then multiply. For example, to estimate the product of 87 and 43, round the two numbers to 90 and 40; then multiply $90 \times 40 = 3600$. [1.4B]

9. $0 \div 9 = 0$. Zero divided by any whole number except zero is zero. $9 \div 0$ is undefined. Division by zero is not allowed. [1.5A]

10. To check the answer to a division problem that has a remainder, multiply the quotient by the divisor. Add the remainder to the product. The result should be the dividend. For example, $16 \div 5 = 3$ r1. Check: $(3 \times 5) + 1 = 16$, the dividend. [1.5B]

11. The steps in the Order of Operations Agreement are:
  1. Do all operations inside parentheses.
  2. Simplify any number expressions containing exponents.
  3. Do multiplication and division as they occur from left to right.
  4. Do addition and subtraction as they occur from left to right. [1.6B]

12. A number is a factor of another number if it divides that number evenly (there is no remainder). For example, 7 is a factor of 21 because $21 \div 7 = 3$, with a remainder of 0. [1.7A]

13. Three is a factor of a number if the sum of the digits of the number is divisible by 3. For the number 285, $2 + 8 + 5 = 15$, which is divisible by 3. Thus 285 is divisible by 3. [1.7A]

CHAPTER 1 REVIEW EXERCISES

1. 600 [1.6A] **2.** $10{,}000 + 300 + 20 + 7$ [1.1C] **3.** 1, 2, 3, 6, 9, 18 [1.7A] **4.** 12,493 [1.2A] **5.** 1749 [1.3B] **6.** 2135 [1.5A] **7.** $101 > 87$ [1.1A] **8.** $5^2 \cdot 7^5$ [1.6A] **9.** 619,833 [1.4B] **10.** 5409 [1.3B] **11.** 1081 [1.2A] **12.** 2 [1.6B] **13.** 45,700 [1.1D] **14.** Two hundred seventy-six thousand fifty-seven [1.1B] **15.** 1306 r59 [1.5C] **16.** 2,011,044 [1.1B] **17.** 488 r2 [1.5B] **18.** 17 [1.6B] **19.** 32 [1.6B] **20.** $2 \cdot 2 \cdot 2 \cdot 3 \cdot 3$ [1.7B] **21.** 2133 [1.3A] **22.** 22,761 [1.4B] **23.** The difference in the count was 6311 WBCs. [1.3C] **24.** The transcriptionist's total pay for the week is $1140. [1.4C] **25.** The massage therapist saw 71 clients that month. [1.5D] **26.** The monthly payment is $405. [1.5D] **27.** The total amount deposited is $6879. The new checking account balance is $12,275. [1.2B] **28.** The total of the car payments is $2952. [1.4C] **29.** The difference in physical therapist assistant employment and occupational therapist assistant employment in 2008 is 37,200 people. [1.3C] **30.** The difference between the physical therapist assistant employment figures for 2008 and 2018 is 21,200. [1.3C] **31.** The number of employed physical therapist assistants is projected to increase by 2120 from 2008 to 2018. [1.5D] **32.** The number of employed occupational therapist assistants is projected to increase by 800 from 2008 to 2018. [1.5D]

CHAPTER 1 TEST

1. 432 [1.6A, Example 3] **2.** Two hundred seven thousand sixty-eight [1.1B, Example 3] **3.** 9333 [1.3B, Example 3] **4.** 1, 2, 4, 5, 10, 20 [1.7A, Example 1] **5.** 6,854,144 [1.4B, HOW TO 3] **6.** 9 [1.6B, Example 4] **7.** $900{,}000 + 6000 + 300 + 70 + 8$ [1.1C, Example 6] **8.** 75,000 [1.1D, Example 8] **9.** 1121 r27 [1.5C, Example 8] **10.** $3^3 \cdot 7^2$ [1.6A, Example 1] **11.** 54,915 [1.2A, Example 1] **12.** $2 \cdot 2 \cdot 3 \cdot 7$ [1.7B, Example 2] **13.** 4 [1.6B, Example 4] **14.** 726,104 [1.4A, Example 1] **15.** 1,204,006 [1.1B, Example 4] **16.** 8710 r2 [1.5B, Example 5] **17.** $21 > 19$ [1.1A, Example 2] **18.** 703 [1.5A, Example 3] **19.** 96,798 [1.2A, Example 3] **20.** 19,922 [1.3B, Example 4] **21.** The difference in projected total enrollment between 2016 and 2013 is 1,908,000 students. [1.3C, Example 6] **22.** The average projected enrollment in each of the grades 9 through 12 in 2016 is 4,171,000 students. [1.5D, HOW TO 3] **23.** There are 4320 toothbrushes ordered. [1.4C, Example 4] **24.** Your insurance will cost $2844 over a 12-month period. [1.4C, Example 3] **25. a.** The total intake of fluids is 570 milliliters. **b.** The total output of fluids is 615 milliliters. **c.** The total output is greater by 45 milliliters. [1.2B, HOW TO 4]

ANSWERS TO CHAPTER 2 SELECTED EXERCISES

PREP TEST

1. 20 [1.4A] **2.** 120 [1.4A] **3.** 9 [1.4A] **4.** 10 [1.2A] **5.** 7 [1.3A] **6.** 2 r3 [1.5C] **7.** 1, 2, 3, 4, 6, 12 [1.7A] **8.** 59 [1.6B] **9.** 7 [1.3A] **10.** $44 < 48$ [1.1A]

SECTION 2.1

1. 40 **3.** 24 **5.** 30 **7.** 12 **9.** 24 **11.** 60 **13.** 56 **15.** 9 **17.** 32 **19.** 36 **21.** 660 **23.** 9384 **25.** 24 **27.** 30 **29.** 24 **31.** 576 **33.** 1680 **35.** True **37.** 1 **39.** 3 **41.** 5 **43.** 25 **45.** 1 **47.** 4 **49.** 4 **51.** 6 **53.** 4 **55.** 1 **57.** 7 **59.** 5 **61.** 8 **63.** 1 **65.** 25 **67.** 7 **69.** 8 **71.** True **73.** They will have another day off together in 12 days.

SECTION 2.2

1. Improper fraction **3.** Proper fraction **5.** $\frac{3}{4}$ **7.** $\frac{7}{8}$ **9.** $1\frac{1}{2}$ **11.** $2\frac{5}{8}$ **13.** $3\frac{3}{5}$ **15.** $\frac{5}{4}$ **17.** $\frac{8}{3}$ **19.** $\frac{28}{8}$ **21.** **23.** **25.** False **27.** $5\frac{1}{3}$ **29.** 2 **31.** $3\frac{1}{4}$ **33.** $14\frac{1}{2}$ **35.** 17 **37.** $1\frac{7}{9}$ **39.** $1\frac{4}{5}$ **41.** 23 **43.** $1\frac{15}{16}$ **45.** $6\frac{1}{3}$ **47.** 5 **49.** 1 **51.** $\frac{14}{3}$ **53.** $\frac{26}{3}$ **55.** $\frac{59}{8}$ **57.** $\frac{25}{4}$ **59.** $\frac{121}{8}$ **61.** $\frac{41}{12}$ **63.** $\frac{34}{9}$ **65.** $\frac{38}{3}$ **67.** $\frac{38}{7}$ **69.** $\frac{63}{5}$ **71.** $\frac{41}{9}$ **73.** $\frac{117}{14}$

SECTION 2.3

1. $\frac{5}{10}$ **3.** $\frac{9}{48}$ **5.** $\frac{12}{32}$ **7.** $\frac{9}{51}$ **9.** $\frac{12}{16}$ **11.** $\frac{27}{9}$ **13.** $\frac{20}{60}$ **15.** $\frac{44}{60}$ **17.** $\frac{12}{18}$ **19.** $\frac{35}{49}$ **21.** $\frac{10}{18}$ **23.** $\frac{21}{3}$ **25.** $\frac{35}{45}$ **27.** $\frac{60}{64}$ **29.** $\frac{21}{98}$ **31.** $\frac{30}{48}$ **33.** $\frac{15}{42}$ **35.** $\frac{102}{144}$ **37.** $\frac{1}{3}$ **39.** $\frac{1}{2}$ **41.** $\frac{1}{6}$ **43.** $1\frac{1}{9}$ **45.** 0 **47.** $\frac{9}{22}$ **49.** 3 **51.** $\frac{4}{21}$ **53.** $\frac{12}{35}$ **55.** $\frac{7}{11}$ **57.** $1\frac{1}{3}$ **59.** $\frac{3}{5}$ **61.** $\frac{1}{11}$ **63.** 4 **65.** $\frac{1}{3}$ **67.** $\frac{3}{5}$ **69.** $2\frac{1}{4}$ **71.** $\frac{1}{5}$ **73.** Answers will vary. For example, $\frac{4}{6}, \frac{6}{9}, \frac{8}{12}, \frac{10}{15}, \frac{12}{8}$. **75. a.** $\frac{51}{100}$ **b.** $\frac{49}{100}$

SECTION 2.4

1. $\frac{3}{7}$ **3.** 1 **5.** $1\frac{4}{11}$ **7.** $3\frac{2}{5}$ **9.** $2\frac{4}{5}$ **11.** $2\frac{1}{4}$ **13.** $1\frac{3}{8}$ **15.** $1\frac{7}{15}$ **17.** $1\frac{5}{12}$ **19.** A whole number other than 1 **21.** The number 1 **23.** $1\frac{1}{6}$ **25.** $\frac{13}{14}$ **27.** $\frac{53}{60}$ **29.** $1\frac{1}{56}$ **31.** $\frac{23}{60}$ **33.** $1\frac{17}{18}$ **35.** $1\frac{11}{48}$ **37.** $1\frac{9}{20}$ **39.** $2\frac{17}{120}$ **41.** $2\frac{5}{72}$ **43.** $\frac{39}{40}$ **45.** $1\frac{19}{24}$ **47.** (ii) **49.** $10\frac{1}{12}$ **51.** $9\frac{2}{7}$ **53.** $9\frac{47}{48}$ **55.** $8\frac{3}{13}$ **57.** $16\frac{29}{120}$ **59.** $24\frac{29}{40}$ **61.** $33\frac{7}{24}$ **63.** $10\frac{5}{36}$ **65.** $10\frac{5}{12}$ **67.** $14\frac{73}{90}$ **69.** $10\frac{13}{48}$ **71.** $9\frac{5}{24}$ **73.** $14\frac{1}{18}$ **75.** $11\frac{11}{12}$ **77.** No **79.** The baby weighs $10\frac{1}{4}$ pounds. **81.** The baby is $21\frac{11}{16}$ inches long. **83.** Elizabeth earned $408 for the week. **85.** The total number of hours spent in clinicals this week is $10\frac{11}{12}$. **87.** The total amount of dry ingredients needed for this recipe is $29\frac{5}{6}$ cups.

SECTION 2.5

1. $\frac{2}{17}$ **3.** $\frac{1}{3}$ **5.** $\frac{1}{10}$ **7.** $\frac{5}{13}$ **9.** $\frac{1}{3}$ **11.** $\frac{4}{7}$ **13.** $\frac{1}{4}$ **15.** Yes **17.** $\frac{1}{2}$ **19.** $\frac{19}{56}$ **21.** $\frac{1}{2}$ **23.** $\frac{11}{60}$ **25.** $\frac{1}{32}$ **27.** $\frac{19}{60}$ **29.** $\frac{5}{72}$ **31.** $\frac{11}{60}$ **33.** $\frac{29}{60}$ **35.** (i) **37.** $5\frac{1}{5}$ **39.** $4\frac{7}{8}$ **41.** $\frac{16}{21}$ **43.** $5\frac{1}{2}$ **45.** $5\frac{4}{7}$ **47.** $7\frac{5}{24}$ **49.** $1\frac{2}{5}$ **51.** $15\frac{11}{20}$ **53.** $4\frac{37}{45}$ **55.** No **57.** There are $8\frac{1}{2}$ cups left in the bag. **59. a.** It is $7\frac{11}{15}$ miles from the starting point to the second checkpoint. **b.** It is $7\frac{3}{5}$ miles from the first checkpoint to the finish line. **61. a.** The medics plan to travel $17\frac{17}{24}$ miles the first two days. **b.** There will be $9\frac{19}{24}$ miles left to travel on the third day. **63.** The difference represents how much farther the medics plan to travel on the second day than on the first day. **65. a.** Yes **b.** The patient needs to lose $3\frac{1}{4}$ pounds to reach the desired weight. **67.** $\frac{11}{15}$ of the nurse's income is not spent for housing. **69.** $6\frac{1}{8}$

SECTION 2.6

1. $\frac{7}{12}$ **3.** $\frac{7}{48}$ **5.** $\frac{1}{48}$ **7.** $\frac{11}{14}$ **9.** 6 **11.** $\frac{5}{12}$ **13.** 6 **15.** $\frac{2}{3}$ **17.** $\frac{3}{16}$ **19.** $\frac{3}{80}$ **21.** 10 **23.** $\frac{1}{15}$ **25.** $\frac{2}{3}$ **27.** $\frac{7}{26}$ **29.** 4 **31.** $\frac{100}{357}$ **33.** Answers will vary. For example, $\frac{3}{4}$ and $\frac{4}{3}$. **35.** $1\frac{1}{3}$ **37.** $2\frac{1}{2}$ **39.** $\frac{9}{34}$ **41.** 10 **43.** $16\frac{2}{3}$ **45.** 1 **47.** $\frac{1}{2}$ **49.** 30 **51.** 42 **53.** $12\frac{2}{3}$ **55.** $1\frac{4}{5}$ **57.** $1\frac{2}{3}$ **59.** $1\frac{2}{3}$ **61.** 0 **63.** $27\frac{2}{3}$ **65.** $17\frac{85}{128}$ **67.** $\frac{2}{5}$ **69.** $8\frac{1}{16}$ **71.** 8 **73.** 9 **75.** $\frac{5}{8}$ **77.** $3\frac{1}{40}$ **79.** Greater than **81.** The cost of $2\frac{1}{2}$ cases of oil is $475. **83.** 75 feet of the paper have been used in one week. 150 feet of the paper remain on the roll after one week. **85.** It will take $4\frac{1}{2}$ hours to complete your scheduled x-rays. **87.** The staff will need $4\frac{1}{2}$ cups of banana. **89.** The staff will have to multiply the ingredients by 5. $1\frac{1}{4}$ cups of the shortening is needed. **91.** The technician's earnings are $924. **93.** $\frac{1}{2}$ **95.** *A*

SECTION 2.7

1. $\frac{5}{6}$ **3.** 1 **5.** 0 **7.** $\frac{1}{2}$ **9.** $\frac{1}{6}$ **11.** $\frac{7}{10}$ **13.** 2 **15.** 2 **17.** $\frac{1}{6}$ **19.** 6 **21.** $\frac{1}{15}$ **23.** 2 **25.** $2\frac{1}{2}$ **27.** 3 **29.** $1\frac{1}{6}$ **31.** $3\frac{1}{3}$ **33.** True **35.** 6 **37.** $\frac{1}{2}$ **39.** $\frac{1}{30}$ **41.** $1\frac{4}{5}$ **43.** 13 **45.** 3 **47.** $\frac{1}{5}$ **49.** $\frac{11}{28}$ **51.** 120 **53.** $\frac{11}{40}$ **55.** $\frac{33}{40}$ **57.** $4\frac{4}{9}$ **59.** $\frac{13}{32}$ **61.** $10\frac{2}{3}$ **63.** $\frac{12}{53}$ **65.** $4\frac{62}{191}$ **67.** 68 **69.** $8\frac{2}{7}$ **71.** $3\frac{13}{49}$ **73.** 4 **75.** $1\frac{3}{5}$ **77.** $\frac{9}{34}$ **79.** False **81.** Less than **83.** There are 12 servings in 16 ounces of cereal. **85.** The price of the arthritis rub is $14 per ounce. **87.** The vehicle can travel 14 miles on 1 gallon of gasoline. **89. a.** The total weight of the fat and bone is $1\frac{5}{12}$ pounds. **b.** The chef can cut 28 servings from the roast. **91.** 21 half-ounce doses can be given from a $10\frac{1}{2}$-ounce bottle. **93.** $\frac{31}{50}$ of the money borrowed is spent on office renovation and equipment purchases. **95.** The fractional amount of the bill that still needs to be paid is $\frac{1}{3}$. **97.** $\frac{8}{29}$ of the residents require a low-sodium diet. **99.** $\frac{19}{29}$ of the residents have some type of dietary restriction. **101.** The average teenage boy drinks seven more cans of soda per week than the average teenage girl. **103. a.** $\frac{2}{3}$ **b.** $2\frac{5}{8}$

SECTION 2.8

1. $\frac{11}{40} < \frac{19}{40}$ **3.** $\frac{2}{3} < \frac{5}{7}$ **5.** $\frac{5}{8} > \frac{7}{12}$ **7.** $\frac{7}{9} < \frac{11}{12}$ **9.** $\frac{13}{14} > \frac{19}{21}$ **11.** $\frac{7}{24} < \frac{11}{30}$ **13.** $\frac{4}{5}$ **15.** $\frac{25}{144}$ **17.** $\frac{2}{9}$ **19.** $\frac{3}{125}$ **21.** $\frac{4}{45}$ **23.** $\frac{16}{1225}$ **25.** $\frac{4}{49}$ **27.** $\frac{9}{125}$ **29.** $\frac{27}{88}$ **31.** $\frac{5}{6}$ **33.** $1\frac{5}{12}$ **35.** $\frac{7}{48}$ **37.** $\frac{29}{36}$ **39.** $\frac{55}{72}$ **41.** $\frac{35}{54}$ **43.** 2 **45.** $\frac{9}{19}$ **47.** $\frac{7}{32}$ **49.** $\frac{64}{75}$ **51. a.** More people choose a fast-food restaurant on the basis of its location. **b.** Location was the criterion cited by the most people.

CHAPTER 2 CONCEPT REVIEW*

1. To find the LCM of 75, 30, and 50, find the prime factorization of each number and write the factorization of each number in a table. Circle the greatest product in each column. The LCM is the product of the circled numbers.

	2	3	5
75 =		(3)	(5 · 5)
30 =	(2)	3	5
50 =	2		5 · 5

LCM = $2 \cdot 3 \cdot 5 \cdot 5 = 150$ [2.1A]

**Note:* The numbers in brackets following the answers in the Concept Review are a reference to the objective that corresponds to that problem. For example, the reference [1.2A] stands for Section 1.2, Objective A. This notation will be used for all Prep Tests, Concept Reviews, Chapter Reviews, Chapter Tests, and Cumulative Reviews throughout the text.

2. To find the GCF of 42, 14, and 21, find the prime factorization of each number and write the factorization of each number in a table. Circle the least product in each column that does not have a blank. The GCF is the product of the circled numbers.

	2	3	7
42 =	2	3	⑦
14 =	2		7
21 =		3	7

GCF = 7 [2.1B]

3. To write an improper fraction as a mixed number, divide the numerator by the denominator. The quotient without the remainder is the whole number part of the mixed number. To write the fractional part of the mixed number, write the remainder over the divisor. [2.2B]

4. A fraction is in simplest form when the numerator and denominator have no common factors other than 1. For example, $\frac{8}{12}$ is not in simplest form because 8 and 12 have a common factor of 4. $\frac{5}{7}$ is in simplest form because 5 and 7 have no common factors other than 1. [2.3B]

5. When adding fractions, you have to convert to equivalent fractions with a common denominator. One way to explain this is that you can combine like things, but you cannot combine unlike things. You can combine 3 apples and 4 apples and get 7 apples. You cannot combine 4 apples and 3 oranges and get a sum consisting of just one item. In adding whole numbers, you add like things: ones, tens, hundreds, and so on. In adding fractions, you can combine 2 *ninths* and 5 *ninths* and get 7 *ninths*, but you cannot add 2 *ninths* and 3 *fifths*. [2.4B]

6. To add mixed numbers, add the fractional parts and then add the whole number parts. Then reduce the sum to simplest form. [2.4C]

7. To subtract mixed numbers, the first step is to subtract the fractional parts. If we are subtracting a mixed number from a whole number, there is no fractional part in the whole number from which to subtract the fractional part of the mixed number. Therefore, we must borrow a 1 from the whole number and write 1 as an equivalent fraction with a denominator equal to the denominator of the fraction in the mixed number. Then we can subtract the fractional parts and then subtract the whole numbers. [2.5C]

8. When multiplying two fractions, it is better to eliminate the common factors before multiplying the remaining factors in the numerator and denominator so that (1) we don't end up with very large products and (2) we don't have the added step of simplifying the resulting fraction. [2.6A]

9. Let's look at an example, $\frac{1}{2} \times \frac{1}{3} = \frac{1}{6}$. The fractions $\frac{1}{2}$ and $\frac{1}{3}$ are less than 1. The product, $\frac{1}{6}$, is less than $\frac{1}{2}$ and less than $\frac{1}{3}$. Therefore, the product is less than the smaller number. [2.6A]

10. Reciprocals are used to rewrite division problems as related multiplication problems. Since "divided by" means the same thing as "times the reciprocal of," we can change the division sign to a multiplication sign and change the divisor to its reciprocal. For example, $9 \div 3 = 9 \times \frac{1}{3}$. [2.7A]

11. When a fraction is divided by a whole number, we write the whole number as a fraction before dividing so that we can easily determine the reciprocal of the whole number. [2.7B]

12. When comparing fractions, we must first look at the denominators. If they are not the same, we must rewrite the fractions as equivalent fractions with a common denominator. If the denominators are the same, we must look at the numerators. The fraction that has the smaller numerator is the smaller fraction. [2.8A]

13. We must follow the Order of Operations Agreement in simplifying the expression $\left(\frac{5}{6}\right)^2 - \left(\frac{3}{4} - \frac{2}{3}\right) \div \frac{1}{2}$. Therefore, we must first simplify the expression $\frac{3}{4} - \frac{2}{3}$ inside the parentheses: $\frac{3}{4} - \frac{2}{3} = \frac{1}{12}$. Then simplify the exponential expression: $\left(\frac{5}{6}\right)^2 = \frac{25}{36}$. Then perform the division: $\frac{1}{12} \div \frac{1}{2} = \frac{1}{6}$. Then perform the subtraction: $\frac{25}{36} - \frac{1}{6} = \frac{19}{36}$. [2.8C]

CHAPTER 2 REVIEW EXERCISES

1. $\frac{2}{3}$ [2.3B] **2.** $\frac{5}{16}$ [2.8B] **3.** $\frac{13}{4}$ [2.2A] **4.** $1\frac{13}{18}$ [2.4B] **5.** $\frac{11}{18} < \frac{17}{24}$ [2.8A] **6.** $14\frac{19}{42}$ [2.5C] **7.** $\frac{5}{36}$ [2.8C] **8.** $9\frac{1}{24}$ [2.6B] **9.** 2 [2.7B] **10.** $\frac{25}{48}$ [2.5B] **11.** $3\frac{1}{3}$ [2.7B] **12.** 4 [2.1B] **13.** $\frac{24}{36}$ [2.3A] **14.** $\frac{3}{4}$ [2.7A] **15.** $\frac{32}{44}$ [2.3A] **16.** $16\frac{1}{2}$ [2.6B] **17.** 36 [2.1A] **18.** $\frac{4}{11}$ [2.3B] **19.** $1\frac{1}{8}$ [2.4A] **20.** $10\frac{1}{8}$ [2.5C] **21.** $18\frac{13}{54}$ [2.4C] **22.** 5 [2.1B] **23.** $3\frac{2}{5}$ [2.2B] **24.** $\frac{1}{15}$ [2.8C] **25.** $5\frac{7}{8}$ [2.4C] **26.** 54 [2.1A] **27.** $\frac{1}{3}$ [2.5A] **28.** $\frac{19}{7}$ [2.2B] **29.** 2 [2.7A] **30.** $\frac{1}{15}$ [2.6A] **31.** $\frac{1}{8}$ [2.6A] **32.** $1\frac{7}{8}$ [2.2A]

33. The client's total weight loss is $12\frac{1}{2}$ pounds. [2.4D] **34.** The cost per acre was $40,500. [2.7C] **35.** The second checkpoint is $4\frac{3}{4}$ miles from the finish line. [2.5D] **36.** The car can travel 414 miles. [2.6C]

CHAPTER 2 TEST

1. $\frac{4}{9}$ [2.6A, Example 1] **2.** 8 [2.1B, Example 2] **3.** $1\frac{3}{7}$ [2.7A, Example 2] **4.** $\frac{7}{24}$ [2.8C, You Try It 3] **5.** $\frac{49}{5}$ [2.2B, Example 5] **6.** 8 [2.6B, Example 5] **7.** $\frac{5}{8}$ [2.3B, Example 3] **8.** $\frac{3}{8} < \frac{5}{12}$ [2.8A, Example 1] **9.** $\frac{5}{6}$ [2.8C, Example 3] **10.** 120 [2.1A, Example 1] **11.** $\frac{1}{4}$ [2.5A, Example 1] **12.** $3\frac{3}{5}$ [2.2B, Example 3] **13.** $2\frac{2}{19}$ [2.7B, Example 4] **14.** $\frac{45}{72}$ [2.3A, Example 1] **15.** $1\frac{61}{90}$ [2.4B, Example 4] **16.** $13\frac{81}{88}$ [2.5C, Example 5] **17.** $\frac{7}{48}$ [2.5B, Example 2] **18.** $\frac{1}{6}$ [2.8B, You Try It 2] **19.** $1\frac{11}{12}$ [2.4A, Example 1] **20.** $22\frac{4}{15}$ [2.4C, Example 7] **21.** $\frac{11}{4}$ [2.2A, Example 2] **22.** The electrician earns $840. [2.6C, Example 7] **23.** The total length of material used is $42\frac{13}{24}$ inches. [2.4D, Example 9] **24.** 72 sections can be cut from the roll. [2.7C, Example 8] **25.** The patients weighs $119\frac{3}{4}$ pounds at the end of the second week. [2.5D, YOU TRY IT 6]

CUMULATIVE REVIEW EXERCISES

1. 290,000 [1.1D] **2.** 291,278 [1.3B] **3.** 73,154 [1.4B] **4.** 540 r12 [1.5C] **5.** 1 [1.6B] **6.** $2 \cdot 2 \cdot 11$ [1.7B] **7.** 210 [2.1A] **8.** 20 [2.1B] **9.** $\frac{23}{3}$ [2.2B] **10.** $6\frac{1}{4}$ [2.2B] **11.** $\frac{15}{48}$ [2.3A] **12.** $\frac{2}{5}$ [2.3B] **13.** $1\frac{7}{48}$ [2.4B] **14.** $14\frac{11}{48}$ [2.4C] **15.** $\frac{13}{24}$ [2.5B] **16.** $1\frac{7}{9}$ [2.5C] **17.** $\frac{7}{20}$ [2.6A] **18.** $7\frac{1}{2}$ [2.6B] **19.** $1\frac{1}{20}$ [2.7A] **20.** $2\frac{5}{8}$ [2.7B] **21.** $\frac{1}{9}$ [2.8B] **22.** $5\frac{5}{24}$ [2.8C] **23.** The amount in the checking account at the end of the week was $862. [1.3C] **24.** The total cost of the order is $250. [1.4C] **25.** The baby is $21\frac{3}{4}$ inches long at the end of three months. [2.4D] **26.** 200 milliliters of fluid remain to be infused. [2.6C] **27.** Give a child 1 ounce every 4 hours. [2.6C] **28.** There are 16 doses in the bottle. [2.7C]

ANSWERS TO CHAPTER 3 SELECTED EXERCISES

PREP TEST

1. $\frac{3}{10}$ [2.2A] **2.** 36,900 [1.1D] **3.** Four thousand seven hundred ninety-one [1.1B] **4.** 6842 [1.1B] **5.** 9394 [1.2A] **6.** 1638 [1.3B] **7.** 76,804 [1.4B] **8.** 278 r18 [1.5C]

SECTION 3.1

1. Thousandths **3.** Ten-thousandths **5.** Hundredths **7.** 0.3 **9.** 0.21 **11.** 0.461 **13.** $\frac{1}{10}$ **15.** $\frac{47}{100}$ **17.** $\frac{289}{1000}$ **19.** Thirty-seven hundredths **21.** Nine and four tenths **23.** Fifty-three ten-thousandths **25.** Forty-five thousandths **27.** Twenty-six and four hundredths **29.** 3.0806 **31.** 407.03 **33.** 246.024 **35.** 73.02684 **37.** 6.2 **39.** 21.0 **41.** 18.41 **43.** 72.50 **45.** 936.291 **47.** 47 **49.** 7015 **51.** 2.97527 **53.** The length of the small intestine is approximately 6 meters. **55.** For example, 0.2701 **57.** For example, **a.** 0.15 **b.** 1.05 **c.** 0.001

SECTION 3.2

1. 150.1065 **3.** 95.8446 **5.** 69.644 **7.** 92.883 **9.** 113.205 **11.** 0.69 **13.** 16.305 **15.** 110.7666 **17.** 104.4959 **19.** Cal.: 234.192 Est.: 234 **21.** Cal.: 781.943 Est.: 782 **23.** Yes **25.** The total number of viewers is 3.4 million. **27.** The approximate total length of the large intestine is 142.24 centimeters. **29.** The child weighs 18.2 kilograms. **31.** Yes, John has met the required amount of fiber in his diet. **33.** Three possible answers are bread, butter, and mayonnaise; raisin bran, butter, and bread; and lunch meat, milk, and toothpaste.

SECTION 3.3

1. 5.627 **3.** 113.6427 **5.** 6.7098 **7.** 215.697 **9.** 53.8776 **11.** 72.7091 **13.** 0.3142 **15.** 1.023 **17.** 261.166 **19.** 655.32 **21.** 342.9268 **23.** 8.628 **25.** 7.01 − 2.325 **27.** 19.35 − 8.967 **29.** Cal.: 2.74506 Est.: 3 **31.** Cal.: 7.14925 Est.: 7 **33.** The remaining balance on the debit card is $119.40. **35.** Sheila needs to consume an additional 3.9 grams of fiber. **37.** The child's temperature was 99.9°. **39. a.** 0.1 **b.** 0.01 **c.** 0.001

SECTION 3.4

1. 0.36 **3.** 0.25 **5.** 6.93 **7.** 1.84 **9.** 0.74 **11.** 39.5 **13.** 2.72 **15.** 0.603 **17.** 13.50 **19.** 4.316 **21.** 0.1323 **23.** 0.03568 **25.** 0.0784 **27.** 0.076 **29.** 34.48 **31.** 580.5 **33.** 20.148 **35.** 0.04255 **37.** 0.17686 **39.** 0.19803 **41.** 0.0006608 **43.** 0.536335 **45.** 0.429 **47.** 2.116 **49.** 0.476 **51.** 1.022 **53.** 37.96 **55.** 2.318 **57.** 3.2 **59.** 6.5 **61.** 6285.6 **63.** 3200 **65.** 35,700 **67.** 6.3 **69.** 3.9 **71.** 49,000 **73.** 6.7 **75.** 0.012075 **77.** 0.0117796 **79.** 0.31004 **81.** 0.082845 **83.** 5.175 **85.** Cal.: 91.2 Est.: 90 **87.** Cal.: 1.0472 Est.: 0.8 **89.** Cal.: 3.897 Est.: 4.5 **91.** Cal.: 11.2406 Est.: 12 **93.** Cal.: 0.371096 Est.: 0.32 **95.** Cal.: 31.8528 Est.: 30 **97.** Mariel uses 3.36 milliliters of insulin in a week. **99.** The amount received is $14.06. **101.** It costs $604.65 to destroy the medical records. **103.** The baby weighs 21.48 pounds. **105. a.** The nurse's overtime pay is $785.25. **b.** The nurse's total income is $2181.25. **107.** The patient ingests 0.175 gram. **109.** The patient ate 28.0 grams of fiber. **111.** The total cost of the electronic payment transfers is $47.85. **113.** The weekly payroll amount is $9276. **115.** $1\frac{3}{10} \times 2\frac{31}{100} = \frac{13}{10} \times \frac{231}{100} = \frac{3003}{1000} = 3\frac{3}{1000} = 3.003$

SECTION 3.5

1. 0.82 **3.** 4.8 **5.** 89 **7.** 60 **9.** 84.3 **11.** 32.3 **13.** 5.06 **15.** 1.3 **17.** 0.11 **19.** 3.8 **21.** 6.3 **23.** 0.6 **25.** 2.5 **27.** 1.1 **29.** 130.6 **31.** 0.81 **33.** 0.09 **35.** 40.70 **37.** 0.46 **39.** 0.019 **41.** 0.087 **43.** 0.360 **45.** 0.103 **47.** 0.009 **49.** 1 **51.** 3 **53.** 1 **55.** 57 **57.** 0.407 **59.** 4.267 **61.** 0.01037 **63.** 0.008295 **65.** 0.82537 **67.** 0.032 **69.** 0.23627 **71.** 0.000053 **73.** 0.0018932 **75.** 18.42 **77.** 16.07 **79.** 0.0135 **81.** 0.023678 **83.** 0.112 **85.** Cal.: 11.1632 Est.: 10 **87.** Cal.: 884.0909 Est.: 1000 **89.** Cal.: 1.8269 Est.: 1.5 **91.** Cal.: 58.8095 Est.: 50 **93.** Cal.: 72.3053 Est.: 100 **95. a.** Use division to find the cost. **b.** Use multiplication to find the cost. **97.** The cost per paper gown is $0.43. **99.** The patient should try to lose 4.15 pounds each month. **101.** The nurse should give the patient 2 tablets. **103.** The dividend is $1.72 per share. **105.** The car travels 25.5 miles on 1 gallon of gasoline. **107.** Sonja's average hemoglobin level for the past six months is 11.7 grams. **109.** 2.57 million more women than men were attending institutions of higher learning. **111.** The number of Americans under 18 who are uninsured is 2.56 times greater than the number of uninsured 18- to 24-year-olds. **113.** The population of this segment is expected to be 2.1 times greater in 2030 than in 2000. **115.** The student's GPA is 2.75. **117.** ÷ **119.** − **121.** + **123.** 5.217 **125.** 0.025

SECTION 3.6

1. 0.625 **3.** 0.667 **5.** 0.167 **7.** 0.417 **9.** 1.750 **11.** 1.500 **13.** 4.000 **15.** 0.003 **17.** 7.080 **19.** 37.500 **21.** 0.160 **23.** 8.400 **25.** Less than 1 **27.** Greater than 1 **29.** $\frac{4}{5}$ **31.** $\frac{8}{25}$ **33.** $\frac{1}{8}$ **35.** $1\frac{1}{4}$ **37.** $16\frac{9}{10}$ **39.** $8\frac{2}{5}$ **41.** $8\frac{437}{1000}$ **43.** $2\frac{1}{4}$ **45.** $\frac{23}{150}$ **47.** $\frac{703}{800}$ **49.** $7\frac{19}{50}$ **51.** $\frac{57}{100}$ **53.** $\frac{2}{3}$ **55.** $0.15 < 0.5$ **57.** $6.65 > 6.56$ **59.** $2.504 > 2.054$ **61.** $\frac{3}{8} > 0.365$ **63.** $\frac{2}{3} > 0.65$ **65.** $\frac{5}{9} > 0.55$ **67.** $0.62 > \frac{7}{15}$ **69.** $0.161 > \frac{1}{7}$ **71.** $0.86 > 0.855$ **73.** $1.005 > 0.5$ **75.** $0.172 < 17.2$ **77.** Cars 2 and 5 would fail the emissions test.

CHAPTER 3 CONCEPT REVIEW*

1. To round a decimal to the nearest tenth, look at the digit in the hundredths place. If the digit in the hundredths place is less than 5, that digit and all digits to the right are dropped. If the digit in the hundredths place is greater than or equal to 5, increase the digit in the tenths place by 1 and drop all digits to its right. [3.1B]
2. The decimal 0.37 is read 37 hundredths. To write the decimal as a fraction, put 37 in the numerator and 100 in the denominator: $\frac{37}{100}$. [3.1A]
3. The fraction $\frac{173}{10,000}$ is read 173 ten-thousandths. To write the fraction as a decimal, insert one 0 after the decimal point so that the 3 is in the ten-thousandths place: 0.0173. [3.1A]
4. When adding decimals of different place values, write the numbers so that the decimal points are on a vertical line. [3.2A]
5. Write the decimal point in the product of two decimals so that the number of decimal places in the product is the sum of the numbers of decimal places in the factors. [3.4A]
6. To estimate the product of two decimals, round each number so that it contains one nonzero digit. Then multiply. For example, to estimate the product of 0.068 and 0.0052, round the two numbers to 0.07 and 0.005; then multiply $0.07 \times 0.005 = 0.00035$. [3.4A]
7. When dividing decimals, move the decimal point in the divisor to the right to make the divisor a whole number. Move the decimal point in the dividend the same number of places to the right. Place the decimal point in the quotient directly over the decimal point in the dividend, and then divide as with whole numbers. [3.5A]
8. First convert the fraction to a decimal: The fraction $\frac{5}{8}$ is equal to 0.625. Now compare the decimals: $0.63 > 0.625$. In the inequality $0.63 > 0.625$, replace the decimal 0.625 with the fraction $\frac{5}{8}$: $0.63 > \frac{5}{8}$. The answer is that the decimal 0.63 is greater than the fraction $\frac{5}{8}$. [3.6C]
9. When dividing 0.763 by 0.6, the decimal points will be moved one place to the right: $7.63 \div 6$. The decimal 7.63 has digits in the tenths and hundredths places. We need to write a zero in the thousandths place in order to determine the digit in the thousandths place of the quotient so that we can then round the quotient to the nearest hundredth. [3.5A]
10. To subtract a decimal from a whole number that has no decimal point, write a decimal point in the whole number to the right of the ones place. Then write as many zeros to the right of that decimal point as there are places in the decimal being subtracted from the whole number. For example, the subtraction $5 - 3.578$ would be written $5.000 - 3.578$. [3.3A]

CHAPTER 3 REVIEW EXERCISES

1. 54.5 [3.5A] **2.** 833.958 [3.2A] **3.** $0.055 < 0.1$ [3.6C] **4.** Twenty-two and ninety-two ten-thousandths [3.1A] **5.** 0.05678 [3.1B] **6.** 2.33 [3.6A] **7.** $\frac{3}{8}$ [3.6B] **8.** 36.714 [3.2A] **9.** 34.025 [3.1A] **10.** $\frac{5}{8} > 0.62$ [3.6C] **11.** 0.778 [3.6A] **12.** $\frac{33}{50}$ [3.6B] **13.** 22.8635 [3.3A] **14.** 7.94 [3.1B] **15.** 8.932 [3.4A] **16.** Three hundred forty-two and thirty-seven hundredths [3.1A] **17.** 3.06753 [3.1A] **18.** 25.7446 [3.4A] **19.** 6.594 [3.5A] **20.** 4.8785 [3.3A] **21.** The new balance in your account is $661.51. [3.3B] **22.** The cost per mile traveled is $0.16. [3.5B] **23.** These paramedics are paid $19.25 per hour. [3.5B] **24.** Mario must consume an additional 8.5 grams of fiber. [3.2B, 3.3B] **25.** During a 5-day school week, 9.5 million gallons of milk are served. [3.4B]

CHAPTER 3 TEST

1. $0.66 < 0.666$ [3.6C, Example 5] **2.** 4.087 [3.3A, Example 1] **3.** Forty-five and three hundred two ten-thousandths [3.1A, Example 4] **4.** 0.692 [3.6A, You Try It 1] **5.** $\frac{33}{40}$ [3.6B, Example 3] **6.** 0.0740 [3.1B, Example 6] **7.** 1.583 [3.5A, Example 3] **8.** 27.76626 [3.3A, Example 2] **9.** 7.095 [3.1B, Example 6] **10.** 232 [3.5A, Example 1] **11.** 458.581 [3.2A, Example 2] **12.** The change returned is $7.57. [3.3B, Example 4] **13.** 0.00548 [3.4A, Example 2] **14.** 255.957 [3.2A, Example 1] **15.** 209.07086 [3.1A, Example 4] **16.** Each payment is $395.40. [3.5B, Example 7] **17.** The infant mortality rate is 2.9 times greater in Mexico than it is in the United States. [3.5B, Example 6] **18.** The patient has walked a total of 7.5 miles. [3.4B, Example 7]

**Note:* The numbers in brackets following the answers in the Concept Review are a reference to the objective that corresponds to that problem. For example, the reference [1.2A] stands for Section 1.2, Objective A. This notation will be used for all Prep Tests, Concept Reviews, Chapter Reviews, Chapter Tests, and Cumulative Reviews throughout the text.

19. The yearly average computer use by a 10th-grade student is 348.4 hours. [3.4B, Example 7] **20.** On average, a 2nd-grade student uses a computer 36.4 hours more per year than a 5th-grade student. [3.4B, Example 8]

CUMULATIVE REVIEW EXERCISES

1. 235 r17 [1.5C] **2.** 128 [1.6A] **3.** 3 [1.6B] **4.** 72 [2.1A] **5.** $4\frac{2}{5}$ [2.2B] **6.** $\frac{37}{8}$ [2.2B] **7.** $\frac{25}{60}$ [2.3A] **8.** $1\frac{17}{48}$ [2.4B] **9.** $8\frac{35}{36}$ [2.4C] **10.** $5\frac{23}{36}$ [2.5C] **11.** $\frac{1}{12}$ [2.6A] **12.** $9\frac{1}{8}$ [2.6B] **13.** $1\frac{2}{9}$ [2.7A] **14.** $\frac{19}{20}$ [2.7B] **15.** $\frac{3}{16}$ [2.8B] **16.** $2\frac{5}{18}$ [2.8C] **17.** Sixty-five and three hundred nine ten-thousandths [3.1A] **18.** 504.6991 [3.2A] **19.** 21.0764 [3.3A] **20.** 55.26066 [3.4A] **21.** 2.154 [3.5A] **22.** 0.733 [3.6A] **23.** $\frac{1}{6}$ [3.6B] **24.** $\frac{8}{9} < 0.98$ [3.6C] **25.** The active male burns 640 more calories per day than the sedentary male. [1.3C] **26.** The patient must lose $7\frac{3}{4}$ pounds the third month to achieve the goal. [2.5D] **27.** Your checking account balance is $617.38. [3.3B] **28.** The patient should be given $2\frac{1}{2}$ tablets. [3.5B] **29.** You paid $6008.80 in income tax last year. [3.4B] **30.** The patients needs to gain 0 pounds to reach the goal. [2.4D, 2.5D]

ANSWERS TO CHAPTER 4 SELECTED EXERCISES

PREP TEST

1. $\frac{4}{5}$ [2.3B] **2.** $\frac{1}{2}$ [2.3B] **3.** 24.8 [3.6A] **4.** 4×33 [1.4A] **5.** 4 [1.5A]

SECTION 4.1

1. $\frac{1}{5}$ 1:5 1 to 5 **3.** $\frac{2}{1}$ 2:1 2 to 1 **5.** $\frac{3}{8}$ 3:8 3 to 8 **7.** $\frac{1}{1}$ 1:1 1 to 1 **9.** $\frac{7}{10}$ 7:10 7 to 10 **11.** $\frac{1}{2}$ 1:2 1 to 2 **13.** $\frac{2}{1}$ 2:1 2 to 1 **15.** $\frac{5}{2}$ 5:2 5 to 2 **17.** $\frac{5}{7}$ 5:7 5 to 7 **19.** days **21.** The ratio is $\frac{1}{3}$. **23.** The ratio is $\frac{3}{8}$. **25.** The ratio is $\frac{13}{10{,}000}$. **27.** The ratio is $\frac{1}{100}$. **29.** The ratio is $\frac{14}{25}$. **31.** The ratio is $\frac{11}{40}$.

SECTION 4.2

1. $\frac{\text{3 pounds}}{\text{4 people}}$ **3.** $\frac{\$20}{\text{3 thermometers}}$ **5.** $\frac{\text{3 patients}}{\text{1 nurse}}$ **7.** $\frac{\text{8 gallons}}{\text{1 hour}}$ **9. a.** Dollars **b.** Seconds **11.** 1 **13.** 2.5 feet/second **15.** $975/week **17.** 1.25 milligrams/milliliter **19.** $18.84/hour **21.** 35.6 miles/gallon **23.** The rate is 7.4 miles per dollar. **25.** The cost per pack is $1.32. **27.** MedPro has the better deal. **29.** 2 milligrams/milliliter **31.** 125 milliliters/hour

SECTION 4.3

1. True **3.** Not true **5.** Not true **7.** True **9.** True **11.** True **13.** True **15.** Not true **17.** True **19.** Yes **21.** Yes **23.** 3 **25.** 105 **27.** 2 **29.** 60 **31.** 2.22 **33.** 6.67 **35.** 21.33 **37.** 16.25 **39.** 2.44 **41.** 47.89 **43.** A 0.5-ounce serving contains 50 calories. **45.** The car can travel 329 miles. **47.** 700 milliliters of solution can be prepared. **49.** You can produce 200 milliliters of solution. **51.** 1.25 ounces are required. **53.** The patient has ingested 6000 milligrams. **55.** There should be 5 nurse's aides on duty. **57.** The monthly payment is $176.75. **59.** The dividend would be $1071. **61.** There are 26.4 grams.

CHAPTER 4 CONCEPT REVIEW*

1. If the units in a comparison are different, then the comparison is a rate. For example, the comparison "50 miles in 2 hours" is a rate. [4.2A]

*Note: The numbers in brackets following the answers in the Concept Review are a reference to the objective that corresponds to that problem. For example, the reference [1.2A] stands for Section 1.2, Objective A. This notation will be used for all Prep Tests, Concept Reviews, Chapter Reviews, Chapter Tests, and Cumulative Reviews throughout the text.

2. To find a unit rate, divide the number in the numerator of the rate by the number in the denominator of the rate. [4.2B]

3. To write the ratio $\frac{6}{7}$ using a colon, write the two numbers 6 and 7 separated by a colon: 6:7. [4.1A]

4 To write the ratio 12:15 in simplest form, divide both numbers by the GCF of 3: $\frac{12}{3} : \frac{15}{3} = 4:5$. [4.1A]

5. To write the rate $\frac{342 \text{ miles}}{9.5 \text{ gallons}}$ as a unit rate, divide the number in the numerator by the number in the denominator: $342 \div 9.5 = 36$. $\frac{342 \text{ miles}}{9.5 \text{ gallons}}$ is the rate. 36 miles/gallon is the unit rate. [4.2B]

6. A proportion is true if the fractions are equal when written in lowest terms. Another way to describe a true proportion is to say that in a true proportion, the cross products are equal. [4.3A]

7. When one of the numbers in a proportion is unknown, we can solve the proportion to find the unknown number. We do this by setting the cross products equal to each other and then solving for the unknown number. [4.3B]

8. When setting up a proportion, keep the same units in the numerator and the same units in the denominator. [4.3C]

9. To check the solution of a proportion, replace the unknown number in the proportion with the solution. Then find the cross products. If the cross products are equal, the solution is correct. If the cross products are not equal, the solution is not correct. [4.3B]

10. To write the ratio 19:6 as a fraction, write the first number as the numerator of the fraction and the second number as the denominator: $\frac{19}{6}$. [4.1A]

CHAPTER 4 REVIEW EXERCISES

1. True [4.3A] **2.** $\frac{2}{5}$ 2:5 2 to 5 [4.1A] **3.** 62.5 miles/hour [4.2B] **4.** True [4.3A] **5.** 68 [4.3B] **6.** $12.50/hour [4.2B] **7.** $1.75/pound [4.2B] **8.** $\frac{2}{7}$ 2:7 2 to 7 [4.1A] **9.** 36 [4.3B] **10.** 19.44 [4.3B] **11.** $\frac{2}{5}$ 2:5 2 to 5 [4.1A] **12.** Not true [4.3A] **13.** $\frac{\$35}{4 \text{ hours}}$ [4.2A] **14.** 27.2 miles/gallon [4.2B] **15.** $\frac{1}{1}$ 1:1 1 to 1 [4.1A] **16.** True [4.3A] **17.** 65.45 [4.3B] **18.** $\frac{50 \text{ milligrams}}{1 \text{ milliliter}}$ [4.2A] **19.** The ratio is $\frac{1}{4}$. [4.1B] **20.** 181 nurse staffing hours are needed per day. [4.3C] **21.** The staffing rate is $\frac{1}{3}$. It is higher than the rate required. [4.2C] **22.** The ratio is $\frac{1}{2}$. [4.1B] **23.** The ratio is $\frac{39}{62}$. [4.1B] **24.** The cost per ounce is $0.47. [4.2C] **25.** The individual dose is 5 milliliters. [4.3C] **26.** The patient ingested 6000 milligrams. [4.3C] **27.** The ratio is $\frac{1}{50}$. [4.1B] **28.** The ratio is $\frac{1}{3}$. [4.1B] **29.** You will need 37.5 milliliters of alcohol. [4.3C] **30.** You can make 250 milliliters of this solution. [4.3C]

CHAPTER 4 TEST

1. $3836.40/month [4.2B, Example 2] **2.** $\frac{1}{6}$ 1:6 1 to 6 [4.1A, You Try It 1] **3.** $\frac{25 \text{ milligrams}}{1 \text{ milliliter}}$ [4.2A, Example 1] **4.** Not true [4.3A,Example 2] **5.** $\frac{3}{2}$ 3:2 3 to 2 [4.1A, Example 2] **6.** 144 [4.3B, Example 3] **7.** 30.5 miles/gallon [4.2B, Example 2] **8.** $\frac{1}{3}$ 1:3 1 to 3 [4.1A, Example 2] **9.** True [4.3A, Example 1] **10.** 40.5 [4.3B, Example 3] **11.** $\frac{\$9}{2 \text{ thermometers}}$ [4.2A, Example 1] **12.** $\frac{3}{5}$ 3:5 3 to 5 [4.1A, You Try It 1] **13.** The dividend is $625. [4.3C, Example 8] **14.** The ratio is $\frac{4000}{3}$. [4.1B, Example 3] **15.** The infusion rate is 150 milliliters/hour. [4.2C, HOW TO 1] **16.** The college student's body contains 132 pounds of water. [4.3C, Example 8] **17.** There are 432 calories in 2 cups of the rice. [4.3C, Example 8] **18.** The amount of medication required is 0.875 ounce. [4.3C, Example 8] **19.** The patient should consume 17.5 milliliters/dose and 70 milliliters/day. [4.3C, Example 9] **20.** You can produce 500 milliliters of solution. [4.3C, YOU TRY IT 8]

CUMULATIVE REVIEW EXERCISES

1. 9158 [1.3B] **2.** $2^4 \cdot 3^3$ [1.6A] **3.** 3 [1.6B] **4.** $2 \cdot 2 \cdot 2 \cdot 2 \cdot 2 \cdot 5$ [1.7B] **5.** 36 [2.1A] **6.** 14 [2.1B] **7.** $\frac{5}{8}$ [2.3B] **8.** $8\frac{3}{10}$ [2.4C] **9.** $5\frac{11}{18}$ [2.5C] **10.** $2\frac{5}{6}$ [2.6B] **11.** $4\frac{2}{3}$ [2.7B] **12.** $\frac{23}{30}$ [2.8C] **13.** Four and seven hundred nine ten-thousandths [3.1A] **14.** 2.10 [3.1B] **15.** 1.990 [3.5A] **16.** $\frac{1}{15}$ [3.6B]

17. $\frac{1}{8}$ [4.1A] **18.** $\frac{29¢}{2 \text{ pencils}}$ [4.2A] **19.** 33.4 miles/gallon [4.2B] **20.** 4.25 [4.3B] **21.** The car's speed is 57.2 miles/hour. [4.2C] **22.** 36 [4.3B] **23.** The total cost of your purchases is $153.30. There is a balance of $332.50 in your checking account. [3.3B, 3.4B] **24.** The student should receive $5.05 in change. [3.3B] **25.** Females have the greater population. [1.3C] **26.** There are 3,400,000 more females aged 55–59 and 80+ than there are males of the same ages. [1.3C] **27.** There are 200,000 more females aged 35–39 than there are males of the same ages. [1.3C] **28.** The patient needs to lose 10 pounds to reach the goal. [2.4D, 2.5D] **29.** The unit rate for this infusion is 125 milliliters/hour. [4.2B] **30.** 1.6 ounces are required. [4.3C]

ANSWERS TO CHAPTER 5 SELECTED EXERCISES

PREP TEST

1. $\frac{19}{100}$ [2.6B] **2.** 0.23 [3.4A] **3.** 47 [3.4A] **4.** 2850 [3.4A] **5.** 4000 [3.5A] **6.** 32 [2.7B] **7.** 62.5 [3.6A] **8.** $66\frac{2}{3}$ [2.2B] **9.** 1.75 [3.5A]

SECTION 5.1

1. $\frac{1}{4}$, 0.25 **3.** $1\frac{3}{10}$, 1.30 **5.** 1, 1.00 **7.** $\frac{73}{100}$, 0.73 **9.** $3\frac{83}{100}$, 3.83 **11.** $\frac{7}{10}$, 0.70 **13.** $\frac{22}{25}$, 0.88 **15.** $\frac{8}{25}$, 0.32 **17.** $\frac{2}{3}$ **19.** $\frac{5}{6}$ **21.** $\frac{1}{9}$ **23.** $\frac{5}{11}$ **25.** $\frac{3}{70}$ **27.** $\frac{1}{15}$ **29.** 0.065 **31.** 0.123 **33.** 0.0055 **35.** 0.0825 **37.** 0.0505 **39.** 0.02 **41.** Greater than **43.** 73% **45.** 1% **47.** 294% **49.** 0.6% **51.** 310.6% **53.** 70% **55.** 37% **57.** 40% **59.** 12.5% **61.** 150% **63.** 166.7% **65.** 87.5% **67.** 48% **69.** $33\frac{1}{3}\%$ **71.** $166\frac{2}{3}\%$ **73.** $87\frac{1}{2}\%$ **75.** Less than **77.** 1% of the American population has blood type AB−. **79.** $\frac{3}{20}$ of the solution is alcohol.

SECTION 5.2

1. 8 **3.** 10.8 **5.** 0.075 **7.** 80 **9.** 51.895 **11.** 7.5 **13.** 13 **15.** 3.75 **17.** 20 **19.** 5% of 95 **21.** 79% of 16 **23.** Less than **25.** The number of people in the United States aged 18 to 24 without life insurance is less than 44 million. **27.** 3,040,000 donations were made by occasional donors. **29.** 4141 adults use natural products. **31.** 64,240,000 children 17 years or younger do not use CAM. **33.** There are 6.1 ounces of cobalt and 3.4 ounces of chrome in the hip joint replacement. **35.** $33.21/hour.

SECTION 5.3

1. 32% **3.** $16\frac{2}{3}\%$ **5.** 200% **7.** 37.5% **9.** 18% **11.** 0.25% **13.** 20% **15.** 400% **17.** 2.5% **19.** 37.5% **21.** 0.25% **23.** False **25.** Technicians' salaries represent 47% of monthly expenses. **27.** Approximately 25.4% of the vegetables were wasted. **29.** 29.8% of Americans with diabetes have not been diagnosed. **31.** 7.3% of your recommended carbohydrates are in one serving of cereal. **33.** 13.6% of the total calories per serving are from fat.

SECTION 5.4

1. 75 **3.** 50 **5.** 100 **7.** 85 **9.** 1200 **11.** 19.2 **13.** 7.5 **15.** 32 **17.** 200 **19.** 9 **21.** 504 **23.** Less than **25.** The original cost of the telephone system is $2142.86. **27.** There were 300 patients treated during the month. **29.** There were 160 questions on the exam. **31.** **a.** Tyler checked his blood sugar 40 times that week. **b.** Tyler's blood sugar was in the normal range 34 times. **33.** The recommended daily amount of thiamin for an adult is 1.5 milligrams.

SECTION 5.5

1. 65 **3.** 25% **5.** 75 **7.** 12.5% **9.** 400 **11.** 19.5 **13.** 14.8% **15.** 62.62 **17.** 15 **19.** **a.** (ii) and (iii) **b.** (i) and (iv) **21.** The drug will be effective for 4.8 hours. **23.** The patient receives 35 grams of dextrose. **25.** 40 donors participated in the blood drive. **27.** 57.7% of baby boomers have some college experience but have not earned a degree. **29.** 29.1% of deaths attributable to smoking result from lung cancer.

CHAPTER 5 CONCEPT REVIEW*

1. To write 197% as a fraction, remove the percent sign and multiply by $\frac{1}{100}$: $197 \times \frac{1}{100} = \frac{197}{100}$. [5.1A]

2. To write 6.7% as a decimal, remove the percent sign and multiply by 0.01: $6.7 \times 0.01 = 0.067$. [5.1A]

3. To write $\frac{9}{5}$ as a percent, multiply by 100%: $\frac{9}{5} \times 100\% = 180\%$. [5.1B]

4. To write 56.3 as a percent, multiply by 100% : $56.3 \times 100\% = 5630\%$. [5.1B]

5. The basic percent equation is Percent × base = amount. [5.2A]

6. To find what percent of 40 is 30, use the basic percent equation: $n \times 40 = 30$. To solve for n, we *divide* 30 by 40: $30 \div 40 = 0.75 = 75\%$. [5.3A]

7. To find 11.7% of 532, use the basic percent equation and write the percent as a decimal: $0.117 \times 532 = n$. To solve for n, we *multiply* 0.117 by 532: $0.117 \times 532 = 62.244$. [5.2A]

8. To answer the question "36 is 240% of what number?", use the basic percent equation and write the percent as a decimal: $2.4 \times n = 36$. To solve for n, we *divide* 36 by 2.4: $36 \div 2.4 = 15$. [5.4A]

9. To use the proportion method to solve a percent problem, identify the percent, the amount, and the base. Then use the proportion $\frac{\text{percent}}{100} = \frac{\text{amount}}{\text{base}}$. Substitute the known values into this proportion and solve for the unknown. [5.5A]

10. To answer the question "What percent of 1400 is 763?" by using the proportion method, first identify the base (1400) and the amount (763). The percent is unknown. Substitute 1400 for the base and 763 for the amount in the proportion $\frac{\text{percent}}{100} = \frac{\text{amount}}{\text{base}}$. Then solve the proportion $\frac{n}{100} = \frac{763}{1400}$ for n. $n = 54.5$, so 54.5% of 1400 is 763. [5.5A]

CHAPTER 5 REVIEW EXERCISES

1. 60 [5.2A] **2.** 20% [5.3A] **3.** 175% [5.1B] **4.** 75 [5.4A] **5.** $\frac{3}{25}$ [5.1A] **6.** 19.36 [5.2A] **7.** 150% [5.3A] **8.** 504 [5.4A] **9.** 0.42 [5.1A] **10.** 5.4 [5.2A] **11.** 157.5 [5.4A] **12.** 0.076 [5.1A] **13.** 77.5 [5.2A] **14.** $\frac{1}{6}$ [5.1A] **15.** 160% [5.5A] **16.** 75 [5.5A] **17.** 38% [5.1B] **18.** 10.9 [5.4A] **19.** 7.3% [5.3A] **20.** 613.3% [5.3A] **21.** The student answered 85% of the questions correctly. [5.5B] **22.** 648 is the maximum number of calories that should come from fat. [5.2B] **23.** The Johnson family allocates 5% of their budget for health care. [5.3B] **24.** The total cost of the defibrillator was $1970.25. [5.2B] **25.** Approximately 78.6% of the women wore sunscreen often. [5.3B] **26.** The world's population in 2000 was approximately 6,100,000,000 people. [5.4B] **27.** 381,600 women had colonoscopies. [5.2B] **28.** The approximate amount of GDP in 2008 was $14 trillion. [5.4B, 5.5B]

CHAPTER 5 TEST

1. 0.973 [5.1A, Example 3] **2.** $\frac{5}{6}$ [5.1A, Example 2] **3.** 30% [5.1B, Example 4] **4.** 163% [5.1B, Example 4] **5.** 150% [5.1B, HOW TO 1] **6.** $66\frac{2}{3}\%$ [5.1B, Example 5] **7.** 50.05 [5.2A, Example 1] **8.** 61.36 [5.2A, Example 2] **9.** 76% of 13 [5.2A, Example 1] **10.** 212% of 12 [5.2A, Example 1] **11.** The amount spent for advertising is $45,000. [5.2B, Example 3] **12.** 900 calories is the minimum daily number you should get from carbohydrates. [5.2B, Example 3] **13.** 14.7% of the daily recommended amount of potassium is provided. [5.3B, Example 4] **14.** 9.1% of the daily recommended number of calories is provided. [5.3B, Example 4] **15.** 96.7% of patients did not have complications after surgery. [5.3B, Example 5] **16.** The student answered approximately 91.3% of the questions correctly. [5.3B, Example 5] **17.** 80 [5.4A, Example 2] **18.** 28.3 [5.4A, Example 2] **19.** The patient receives 25 grams of dextrose. [5.5B, Example 3] **20.** The increase was 60% of the original price. [5.3B, Example 5] **21.** 143.0 [5.5A, Example 1] **22.** 1000% [5.5A, Example 1] **23.** The dollar increase is $1.74. [5.5B, Example 3] **24.** One slice of pizza represents 16.7% of your daily calorie allowance. [5.5B, Example 4] **25.** Jeanne weighed 250 pounds when she began her diet. [5.4B, Example 4]

**Note:* The numbers in brackets following the answers in the Concept Review are a reference to the objective that corresponds to that problem. For example, the reference [1.2A] stands for Section 1.2, Objective A. This notation will be used for all Prep Tests, Concept Reviews, Chapter Reviews, Chapter Tests, and Cumulative Reviews throughout the text.

CUMULATIVE REVIEW EXERCISES

1. 4 [1.6B] **2.** 240 [2.1A] **3.** $10\frac{11}{24}$ [2.4C] **4.** $12\frac{41}{48}$ [2.5C] **5.** $12\frac{4}{7}$ [2.6B] **6.** $\frac{7}{24}$ [2.7B] **7.** $\frac{1}{3}$ [2.8B] **8.** $\frac{13}{36}$ [2.8C] **9.** 3.08 [3.1B] **10.** 1.1196 [3.3A] **11.** 34.2813 [3.5A] **12.** 3.625 [3.6A] **13.** $1\frac{3}{4}$ [3.6B] **14.** $\frac{3}{8} < 0.87$ [3.6C] **15.** 53.3 [4.3B] **16.** $19.20/hour [4.2B] **17.** $\frac{11}{60}$ [5.1A] **18.** $83\frac{1}{3}\%$ [5.1B] **19.** 19.56 [5.2A/5.5A] **20.** $133\frac{1}{3}\%$ [5.3A/5.5A] **21.** 9.92 [5.4A/5.5A] **22.** 342.9% [5.3A/5.5A]

23. Sergio's take-home pay is $592. [2.6C] **24.** Each monthly payment is $292.50. [3.5B] **25.** A nurse should give the patient 2 tablets. [3.5B] **26.** The property tax on the medical office is $10,000. [4.3C] **27.** 8,952,000 African-Americans would have type A+ blood. [5.2B, 5.5B] **28.** 45% of those surveyed favored Cough Syrup B. [5.3B, 5.5B] **29.** 980 calories is the maximum number you should get from saturated fat in one week. [5.2B, 5.5B] **30.** 18% of the children tested had levels of lead that exceeded federal standards. [5.3B/5.5B]

ANSWERS TO CHAPTER 6 SELECTED EXERCISES

PREP TEST

1. 0.75 [3.5A] **2.** 52.05 [3.4A] **3.** 504.51 [3.3A] **4.** 9750 [3.4A] **5.** 45 [3.4A] **6.** 1417.24 [3.2A] **7.** 3.33 [3.5A] **8.** 0.605 [3.5A] **9.** $0.379 < 0.397$ [3.6C]

SECTION 6.1

1. The unit cost is $0.055 per ounce. **3.** The unit cost is $0.374 per ounce. **5.** The unit cost is $0.080 per tablet. **7.** The unit cost is $0.297 per ounce. **9.** The unit cost is $0.199 per ounce. **11.** Divide the price of one pint by 2. **13.** The Kraft mayonnaise is the more economical purchase. **15.** The Cortexx shampoo is the more economical purchse. **17.** The Ultra Mr. Clean is the more economical purchase. **19.** The Bertolli olive oil is the more economical purchase. **21.** The Wagner's vanilla extract is the more economical purchase. **23.** Tea A is the more economical purchase. **25.** The total cost is $73.50. **27.** The total cost is $16.88. **29.** The total cost is $3.89. **31.** The total cost is $26.57.

SECTION 6.2

1. The percent increase is 182.9%. **3.** The percent increase is 117.6%. **5.** Insurance costs rose by 33% from 2003 to 2008. **7.** The percent increase is 111.2%. **9.** Yes **11.** Use equation (3) and then equation (2). **13.** The markup is $3.50. **15.** The markup rate is 60%. **17.** The selling price is $17.40. **19.** The selling price is $22.20. **21.** The percent decrease is 40%. **23. a.** The percent decrease in the population of Detroit is 13.7%. **b.** The percent decrease in the population of Philadelphia is 7.7%. **c.** The percent decrease in the population of Chicago is 1.8%. **25.** The car loses $8460 in value. **27. a.** The amount of the decrease was $35.20. **b.** The new average monthly gasoline bill is $140.80. **29.** The percent decrease in smokers was 20%. **31.** Use equation (3) and then (2). **33.** The discount rate is $33\frac{1}{3}\%$. **35.** The discount is $80. **37.** The discount rate is 30%. **39. a.** The discount is $0.25 per pound. **b.** The sale price is $1.00 per pound. **41.** The discount rate is 20%.

SECTION 6.3

1. a. $10,000 **b.** $850 **c.** 4.25% **d.** 2 years **3.** The simple interest owed is $960. **5.** The simple interest due is $3375. **7.** The simple interest due is $1320. **9.** The simple interest due is $84.76. **11.** The maturity value is $5120. **13.** The total amount due on the loan is $12,875. **15.** The maturity value is $14,543.70. **17.** The monthly payment is $6187.50. **19. a.** The interest charged is $1080. **b.** The monthly payment is $545. **21.** The monthly payment is $3039.02. **23. a.** Student A's principal is equal to student B's principal. **b.** Student A's maturity value is greater than student B's maturity value. **c.** Student A's monthly payment is less than student B's monthly payment. **25.** The finance charge is $6.85. **27.** You owe the company $11.94. **29.** The difference between the finance charges is $3.44. **31.** The finance charges for the first and second months are the same. No, you will not be able to pay off the balance. **33.** The value of the investment after 20 years is $12,380.43. **35.** The value of the investment after 5 years is $28,352.50. **37. a.** The value of the investment after 10 years will be $6040.86. **b.** The amount of interest earned will be $3040.86. **39.** The amount of interest earned is $505.94.

SECTION 6.4

1. The mortgage is $172,450. **3.** The down payment is $212,500. **5.** The loan origination fee is $3750. **7.** The mortgage is $315,000. **9.** The mortgage is $189,000. **11.** (iii) **13.** The monthly mortgage payment is $644.79. **15.** Yes, the lawyer can afford the monthly mortgage payment. **17.** The monthly property tax is $166. **19. a.** The monthly mortgage payment is $898.16. **b.** The interest payment is $505.69. **21.** The total monthly payment for the mortgage and property tax is $842.40. **23.** The monthly mortgage payment is $2295.29. **25.** The couple can save $63,408 in interest.

SECTION 6.5

1. No, Amanda does not have enough money for the down payment. **3.** The sales tax is $1192.50. **5.** The license fee is $650. **7. a.** The sales tax is $1120. **b.** The total cost of the sales tax and the license fee is $1395. **9.** The amount financed is $12,150. **11.** The amount financed is $36,000. **13.** The expression represents the total cost of buying the car. **15.** Use multiplication to find the cost. **17.** The monthly car payment is $531.43. **19.** The cost is $252. **21.** Your cost per mile for gasoline was $0.16. **23. a.** The amount financed is $12,000. **b.** The monthly car payment is $359.65. **25.** The monthly payment is $810.23. **27.** The amount of interest paid is $4665.

SECTION 6.6

1. Lewis earns $460. **3.** The real estate agent's commission is $3930. **5.** The stockbroker's commission is $84. **7.** The teacher's monthly salary is $3244. **9.** Carlos earned a commission of $540. **11.** The medical transcriptionist earned $380. **13.** The chemist's hourly wage is $125. **15. a.** Gil's hourly wage for working overtime is $21.56. **b.** Gil earns $344.96 for working 16 hours of overtime. **17. a.** The increase in hourly pay is $1.23. **b.** The aide's hourly wage for working the night shift is $9.43. **19.** Nicole's earnings were $475. **21.** The starting salary for an accountant in the previous year was $40,312. **23.** The starting salary for a computer science major in the previous year was $50,364.

SECTION 6.7

1. Your current checking account balance is $486.32. **3.** The nutritionist's current balance is $825.27. **5.** The current checking account balance is $3000.82. **7.** Yes, there is enough money in the nurse's account to purchase the refrigerator. **9.** Yes, there is enough money in the account to buy two wheelchairs. **11.** The account's ending balance might be less than its starting balance on that day. **13.** The bank statement and the checkbook balance. **15.** The bank statement and the checkbook balance. **17.** The ending balance on the monthly bank statement is greater than the ending balance on the check register.

CHAPTER 6 CONCEPT REVIEW*

1. To find the unit cost, divide the total cost by the number of units. The cost is $2.96. The number of units is 4. $2.96 \div 4 = 0.74$, so the unit cost is $0.74 per can. [6.1A]

2. To find the total cost, multiply the unit cost by the number of units purchased. The unit cost is $0.85. The number of units purchased is 3.4 pounds. $0.85 \times 3.4 = 2.89$, so the total cost is $2.89. [6.1C]

3. To find the selling price when you know the cost and the markup, add the cost and the markup. [6.2B]

4. To find the markup when you know the markup rate, multiply the markup rate by the cost. [6.2B]

5. If you know the percent decrease, you can find the amount of decrease by multiplying the percent decrease by the original value. [6.2C]

6. If you know the regular price and the sale price, you can find the discount by subtracting the sale price from the regular price. [6.2D]

7. To find the discount rate, first subtract the sale price from the regular price to find the discount. Then divide the discount by the regular price. [6.2D]

8. To find simple interest, multiply the principal by the annual interest rate by the time (in years). [6.3A]

9. To find the maturity value for a simple interest loan, add the principal and the interest. [6.3A]

10. Principal is the original amount deposited in an account or the original amount borrowed for a loan. [6.3A]

11. If you know the maturity value of an 18-month loan, you can find the monthly payment by dividing the maturity value by 18 (the length of the loan in months). [6.3A]

**Note:* The numbers in brackets following the answers in the Concept Review are a reference to the objective that corresponds to that problem. For example, the reference [1.2A] stands for Section 1.2, Objective A. This notation will be used for all Prep Tests, Concept Reviews, Chapter Reviews, Chapter Tests, and Cumulative Reviews throughout the text.

12. Compound interest is computed not only on the original principal but also on interest already earned. [6.3C]
13. For a fixed-rate mortgage, the monthly payment remains the same throughout the life of the loan. [6.4B]
14. The following expenses are involved in owning a car: car insurance, gas, oil, general maintenance, and the monthly car payment. [6.5B]
15. To balance a checkbook:
    1. In the checkbook register, put a check mark by each check paid by the bank and by each deposit recorded by the bank.
    2. Add to the current checkbook balance all checks that have been written but have not yet been paid by the bank and any interest paid on the account.
    3. Subtract any service charges and any deposits not yet recorded by the bank. This is the checkbook balance.
    4. Compare the balance with the bank balance listed on the bank statement. If the two numbers are equal, the bank statement and the checkbook "balance." [6.7B]

CHAPTER 6 REVIEW EXERCISES

1. The unit cost is \$0.195 per ounce or 19.5¢ per ounce. [6.1A] **2.** The cost is \$0.279 or 27.9¢ per mile. [6.5B] **3.** The percent increase is 30.4%. [6.2A] **4.** The markup is \$8.40. [6.2B] **5.** The simple interest due is \$3000. [6.3A] **6.** The value of the investment after 10 years is \$45,550.75. [6.3C] **7.** The percent increase is 15%. [6.2A] **8.** The total monthly payment for the mortgage and property tax is \$1138.90. [6.4B] **9.** The monthly payment is \$518.02. [6.5B] **10.** The value of the investment will be \$53,593. [6.3C] **11.** The down payment is \$29,250. [6.4A] **12.** The total cost of the sales tax and license fee is \$2096.25. [6.5A] **13.** The selling price is \$2079. [6.2B] **14.** The interest paid is \$157.33. [6.5B] **15.** The commission was \$3240. [6.6A] **16.** The sale price is \$141. [6.2D] **17.** The current checkbook balance is \$943.68. [6.7A] **18.** The maturity value is \$31,200. [6.3A] **19.** The origination fee is \$1875. [6.4A] **20.** The more economical purchase is 33 ounces for \$6.99. [6.1B] **21.** The monthly mortgage payment is \$2131.62. [6.4B] **22.** The total income was \$655.20. [6.6A] **23.** The donut shop's checkbook balance is \$8866.58. [6.7A] **24.** The monthly payment is \$14,093.75. [6.3A] **25.** The finance charge is \$7.20. [6.3B]

CHAPTER 6 TEST

1. The cost per bottle is \$1.87. [6.1A, Example 1] **2.** The more economical purchase is 3 pounds for \$7.49. [6.1B, Example 2] **3.** The total cost is \$14.53. [6.1C, Example 3] **4.** The percent increase in the cost of the exercise bicycle is 20%. [6.2A, Example 1] **5.** The selling price of the blu-ray disc player is \$441. [6.2B, Example 4] **6.** The percent increase is 15.4%. [6.2A, Example 1] **7.** The percent decrease is 15%. [6.2C, HOW TO 3] **8.** The sale price of the corner hutch is \$209.30. [6.2D, Example 8] **9.** The discount rate is 40%. [6.2D, Example 7] **10.** The simple interest due is \$2000. [6.3A, You Try It 1] **11.** The maturity value is \$26,725. [6.3A, Example 2] **12.** The finance charge is \$4.50. [6.3B, Example 4] **13.** The amount of interest earned in 10 years was \$24,420.60. [6.3C, You Try It 5] **14.** The loan origination fee is \$3350. [6.4A, Example 2] **15.** The monthly mortgage payment is \$1713.44. [6.4B, HOW TO 2] **16.** The amount financed is \$19,000. [6.5A, Example 1] **17.** The monthly truck payment is \$686.22. [6.5B, Example 5] **18.** Shaney earnes \$1596. [6.6A, Example 1] **19.** The current checkbook balance is \$6612.25. [6.7A, Example 1] **20.** The bank statement and the checkbook balance. [6.7B, Example 2]

CUMULATIVE REVIEW EXERCISES

1. 13 [1.6B] **2.** $8\frac{13}{24}$ [2.4C] **3.** $2\frac{37}{48}$ [2.5C] **4.** 9 [2.6B] **5.** 2 [2.7B] **6.** 5 [2.8C] **7.** 52.2 [3.5A] **8.** 1.417 [3.6A] **9.** \$51.25/hour [4.2B] **10.** 10.94 [4.3B] **11.** 62.5% [5.1B] **12.** 27.3 [5.2A] **13.** 0.182 [5.1A] **14.** 42% [5.3A] **15.** 250 [5.4A] **16.** 154.76 [5.4A/5.5A] **17.** The total rainfall is $13\frac{11}{12}$ inches. [2.4D] **18.** The amount paid in taxes is \$970. [2.6C] **19.** The ratio is $\frac{3}{5}$. [4.1B] **20.** 33.4 miles are driven per gallon. [4.2C] **21.** The unit cost is \$1.10 per pound. [4.2C] **22.** The dividend is \$280. [4.3C] **23.** The sale price is \$720. [6.2D] **24.** The selling price of the EKG machine is \$2240. [6.2B] **25.** The percent increase in Sook Kim's salary is 8%. [6.2A] **26.** The simple interest due is \$2700. [6.3A] **27.** The monthly car payment is \$791.81. [6.5B] **28.** The family's new checking account balance is \$2243.77. [6.7A] **29.** The cost per mile is \$0.33. [6.5B] **30.** The monthly mortgage payment is \$1232.26. [6.4B]

ANSWERS TO CHAPTER 7 SELECTED EXERCISES

PREP TEST

1. 48.0% was bill-related mail. [5.3B] **2. a.** The greatest cost increase is between 2009 and 2010. **b.** Between those years, there was an increase of $5318. [1.3C] **3.** The ratio is $\frac{5}{3}$. [4.1B] **4.** $\frac{3}{50}$ of the women in the military are in the Marine Corps. [5.1A]

SECTION 7.1

1. 210 patients went to the emergency room and were diagnosed with asthma. **3.** 57% of patients seen in the emergency room were diagnosed with asthma. **5.** The hourly rate is $4 higher for a certified dental assistant than for a certified nurse assistant. **7.** 150 children said they hid their vegetables under a napkin. **11.** David Koch and Charles Koch have the same net worth. Warren Buffet's net worth is twice that of Christy Walton. **13.** The number of units required in anatomy/physiology is less than twice the number of units required in social/behavioral sciences. **15.** The total number of donors is 90. **17.** 33.3% of donors had blood type A+. **19.** $41,250 is budgeted annually for housing and utilities. The monthly allocation is $3437.50. **21.** $\frac{3}{20}$ of the budget is designated for food. **23.** The age group 35 to 44 represents the largest segment of the homeless population. **25.** 37% of the homeless population is under age 35. **27.** There are 290 students enrolled in health programs. **29.** 15.5% of students in health programs are enrolled in the medical assisting program. **31.** Approximately 13,500,000 people living in the United States are of Asian ethnic origin. **33.** On average, there would be 64,500 people of black ethnic origin in a random sample of 500,000.

SECTION 7.2

1. 27% of Americans aged 65 and older have diabetes. **3.** The ratio is 7 to 2. **5.** The life expectancy for American males in 2010 is 18 years greater than it was in 1930. **7.** The difference in minimum and maximum annual salaries for a physical therapist is $15,000. **9.** Dental hygienists have the greatest difference between minimum and maximum salaries. **11.** (iii) **13.** The patient's blood sugar level was 200 milligrams/deciliter. **15.** There are 6 readings outside of the normal range. **17.** 30% of adults aged 18 and older took the influenza vaccine in 2007. **19.** The number of adults aged 18 and older who took the influenza vaccine increased 2% from 2008 to 2009. **21.** Women ages 75 and over have the lowest recommended number of Calories.
23. True

25.

	Wind Power Capacity (in megawatts)	
Year	***Texas***	***California***
2000	200	1600
2001	1100	1700
2002	1100	1800
2003	1300	2000
2004	1300	2100
2005	2000	2200
2006	2700	2400
2007	4300	2400

26.

Year	*Greater capacity*	*Difference (megawatts)*
2000	California	1400
2001	California	600
2002	California	700
2003	California	700
2004	California	800
2005	California	200
2006	Texas	300
2007	Texas	1900

The wind power capacity of Texas exceeded that of California in 2006 and 2007.

SECTION 7.3

1. 13 account balances were between $1500 and $2000. **3.** 22% of the account balances were between $2000 and $2500. **5.** 21 body temperatures were lower than 98.5°. **7.** The ratio is $\frac{7}{30}$. **9.** The largest number of adults had a temperature between 97.5° and 98.5°, so it would seem that a "normal" temperature would be within that range. **11.** 21 patients had cholesterol readings lower than 200. **13.** Approximately 22,000 runners finished with times between $2\frac{1}{2}$ hours and 6 hours. **15.** Yes **17.** 8% of women are taller than 68 inches. **19.** Approximately 100,000 students scored between 700 and 800. **21.** Approximately 690,000 students scored lower than 500.

SECTION 7.4

1. a. Median **b.** Mean **c.** Mode **d.** Median **e.** Mode **f.** Mean **3.** The mean amount of sugar in the soft drinks is 41.25 grams. The median amount of sugar in the soft drinks is 40.25 grams. The mode of sugar in the soft drinks is 40 grams. **5.** The mean cost is $85.615. The median cost is $85.855. **7.** The mean monthly rate is $403.625. The median monthly rate is $404.50. **9.** The mean life expectancy is 73.4 years. The median life expectancy is 74 years. **13. a.** 25% **b.** 75% **c.** 75% **d.** 25% **15.** Lowest is $46,596. Highest is $82,879. Q_1 = $56,067. Q_3 = $66,507. Median = $61,036. Range = $36,283. Interquartile range= $10,440. **17. a.** There were 40 adults who had cholesterol levels above 217. **b.** There were 60 adults who had cholesterol levels below 254. **c.** There are 20 cholesterol levels in each quartile. **d.** 25% of the adults had cholesterol levels not more than 198. **19. a.** Range = 5.6 million metric tons. Q_1 = 0.59 million metric tons. Q_3 = 1.52 million metric tons. Interquartile range = 0.93 million metric tons. **b.**

0.45 1.52 6.05
0.59 1.035

c. 6.05

21. a. No, the difference in the means is not greater than 1 inch. **b.** The difference in medians is 0.3 inch.

c.

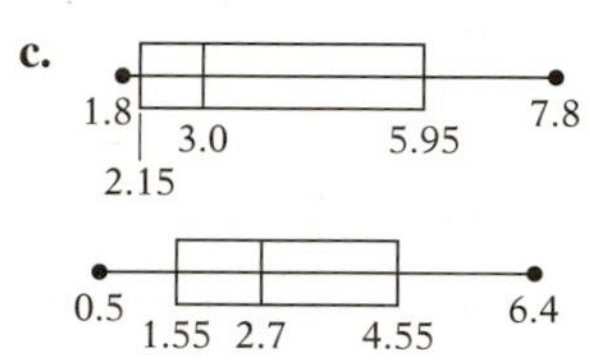

23. Answers will vary. For example, 55, 55, 55, 55, 55, or 50, 55, 55, 55, 60 **27.** Answers will vary. For example, 20, 21, 22, 24, 26, 27, 29, 31, 31, 32, 32, 33, 33, 36, 37, 37, 39, 40, 41, 43, 45, 46, 50, 54, 57

SECTION 7.5

1. {(HHHH), (HHHT), (HHTT), (HHTH), (HTTT), (HTHH), (HTTH), (HTHT), (TTTT), (TTTH), (TTHH), (THHH), (TTHT), (THHT), (THTT), (THTH)} **3.** {(1, 1), (1, 2), (1, 3), (1, 4), (2, 1), (2, 2), (2, 3), (2, 4), (3, 1), (3, 2), (3, 3), (3, 4), (4, 1), (4, 2), (4, 3), (4, 4)} **5. a.** {1, 2, 3, 4, 5, 6, 7, 8} **b.** {1, 2, 3} **7. a.** The probability that the sum is 5 is $\frac{1}{9}$. **b.** The probability that the sum is 15 is 0. **c.** The probability that the sum is less than 15 is 1. **d.** The probability that the sum is 2 is $\frac{1}{36}$. **9. a.** The probability that the number is divisible by 4 is $\frac{1}{4}$. **b.** The probability that the number is a multiple of 3 is $\frac{1}{3}$. **11.** The probability of throwing a sum of 5 is greater. **13. a.** The probability is $\frac{4}{11}$ that the letter *I* is drawn. **b.** The probability of choosing an *S* is greater. **15.** The fraction $\frac{1}{11}$ represents the probability that the card has the letter *M* on it. **17.** Drawing a spade has the greater probability. **19.** Of 3328 patients who had weight loss bariatric surgery, 1.9% died after the surgery. **21. a.** You would expect 5 patients to experience headaches. **b.** It is less than the number you would expect based on the probability of the side effect.

CHAPTER 7 CONCEPT REVIEW*

1. A sector of a circle is one of the "pieces of the pie" into which a circle graph is divided. [7.1B]
2. A pictograph uses symbols to represent numerical information. [7.1A]
3. A jagged portion of the vertical axis on a bar graph is used to indicate that the vertical scale is missing numbers from 0 to the lowest number shown on the vertical axis. [7.2A]
4. In a broken-line graph, points are connected by line segments to show data trends. If a line segment goes up from left to right, it indicates an increase in the quantity represented on the vertical axis during that time period. If a line segment goes down from left to right, it indicates a decrease in the quantity represented on the vertical axis during that time period. [7.2B]
5. Class frequency in a histogram is the number of occurrences of data in each class interval. [7.3A]
6. A class interval in a histogram is a range of numbers that corresponds to the width of each bar. [7.3A]
7. A class midpoint is the center of a class interval in a frequency polygon. [7.3B]

**Note:* The numbers in brackets following the answers in the Concept Review are a reference to the objective that corresponds to that problem. For example, the reference [1.2A] stands for Section 1.2, Objective A. This notation will be used for all Prep Tests, Concept Reviews, Chapter Reviews, Chapter Tests, and Cumulative Reviews throughout the text.

8. The formula for the mean is $\bar{x} = \frac{\text{sum of the data values}}{\text{number of data values}}$ [7.4A]

9. To find the median, the data must be arranged in order from least to greatest in order to determine the "middle" number, or the number which separates the data so that half of the numbers are less than the median and half of the numbers are greater than the median. [7.4A]

10. If a set of numbers has no number occurring more than once, then the data have no mode. [7.4A]

11. A box-and-whiskers plot shows the least number in the data; the first quartile, Q_1; the median; the third quartile, Q_3; and the greatest number in the data. [7.4B]

12. To find the first quartile, find the median of the data values that lie below the median. [7.4B]

13. The empirical probability formula states that the empirical probability of an event is the ratio of the number of observations of the event to the total number of observations. [7.5A]

14. The theoretical probability formula states that the theoretical probability of an event is a fraction with the number of favorable outcomes of the experiment in the numerator and the total number of possible outcomes in the denominator. [7.5A]

CHAPTER 7 REVIEW EXERCISES

1. Medicare paid $520.9 billion in benefits in 2010. [7.1B] **2.** The ratio is $\frac{131}{291}$. [7.1B] **3.** 26.7% of the total benefits payments was paid for hospital inpatient services. [7.1B] **4.** Texas had the larger population. [7.2B] **5.** The population of California was 12.5 million people more than the population of Texas. [7.2B] **6.** The Texas population increased the least from 1925 to 1950. [7.2B] **7.** 22 women are shorter than 65 inches. [7.3B] **8.** The ratio is $\frac{3}{1}$. [7.3B] **9.** 38.9% of the women are taller than 65 inches. [7.3B] **10.** From the pictograph, O'Hare airport had 10 million more passengers than the Denver airport. [7.1A] **11.** The ratio is 4 : 3. [7.1A] **12.** There is a 5% difference between non-Hispanic white boys classified as obese from 1988 to 1994 and those classified as obese from 2007 to 2008. [7.2A] **13.** You would expect 15 non-Hispanic black boys to be classified as obese. [7.2A] **14.** Mexican American boys aged 12 to 19 showed the largest increase in obesity. The increase was approximately 13%. [7.2A] **15.** The probability of one tail and three heads is $\frac{1}{4}$. [7.5A] **16.** There were 15 people who slept 8 hours or more. [7.3A] **17.** The percent is 28.3%. [7.3A] **18. a.** The mean heart rate is 91.6 heartbeats per minute. The median heart rate is 93.5 heartbeats per minute. The mode is 96 heartbeats per minute. [7.4A] **b.** The range is 36 heartbeats per minute. The interquartile range is 15 heartbeats per minute. [7.4B]

CHAPTER 7 TEST

1. 19 students spent between $45 and $75 each week. [7.3B, You Try It 2] **2.** The ratio is $\frac{2}{3}$. [7.3B, You Try It 2] **3.** The percent is 45%. [7.3B, Example 2] **4.** 36 people were surveyed for the Gallup poll. [7.1A, Example 1] **5.** The ratio is $\frac{5}{2}$. [7.1A, Example 1] **6.** The percent is 58.3%. [7.1A, You Try It 1] **7.** The job growth rate for medical assistants is 34%. [7.2A, Example 1] **8.** Licensed practical nurses and registered nurses are expected to have job growth rates slightly higher than 20%. [7.2A, Example 1] **9.** The growth rate for home health aides is predicted to be 31% higher than the growth rate for nursing aides. [7.2A, Example 1] **10.** 32 more students made Cs than As. [7.1B] **11.** 22.5% of students earned a B. [7.1B] **12.** 25.8% of students received grades lower than a C. [7.1B] **13.** There are 24 states that have a median income between $40,000 and $60,000. [7.3A, Example 1] **14.** The percent is 72%. [7.3A, Example 1] **15.** The percent is 18%. [7.3A, Example 1] **16.** The probability is $\frac{3}{10}$ that the ball chosen is red. [7.5A, HOW TO 2] **17.** The student enrollment increased the least during the 1990s. [7.2B, You Try It 2] **18.** The increase in enrollment was 11 million students. [7.2B, Example 2] **19. a.** The mean time is 2.53 days. [7.4A, Example 1] **b.** The median time is 2.55 days. [7.4A, Example 2]

c. [box-and-whiskers plot: 2.0, 2.35, 2.55, 2.8, 3.1] [7.4B, Example 3]

CUMULATIVE REVIEW EXERCISES

1. 540 [1.6A] **2.** 14 [1.6B] **3.** 120 [2.1A] **4.** $\frac{5}{12}$ [2.3B] **5.** $12\frac{3}{40}$ [2.4C] **6.** $4\frac{17}{24}$ [2.5C] **7.** 2 [2.6B] **8.** $\frac{64}{85}$ [2.7B] **9.** $8\frac{1}{4}$ [2.8C] **10.** 209.305 [3.1A] **11.** 2.82348 [3.4A] **12.** 16.67 [3.6A] **13.** 26.4 miles/gallon [4.2B] **14.** 3.2 [4.3B] **15.** 80% [5.1B] **16.** 80 [5.4A] **17.** 16.34 [5.2A] **18.** 40% [5.3A] **19.** The salesperson's income for the week was $650. [6.6A] **20.** The cost is $407.50. [4.3C] **21.** The interest due is $3750. [6.3A] **22.** The markup rate is 55%. [6.2B] **23.** The amount budgeted for food is $855. [7.1B] **24.** The difference in the number answered correctly is 12 problems. [7.2B] **25.** The mean length of hospital stay for these patients is 4.1 days. This mean is lower than the national average. [7.4A] **26.** The probability is $\frac{5}{36}$ that the sum of the dots on the upward faces is 8. [7.5A]

ANSWERS TO CHAPTER 8 SELECTED EXERCISES

PREP TEST

1. 702 [1.2A] **2.** 58 [1.3B] **3.** 4 [2.6B] **4.** 10 [2.6B] **5.** 25 [2.6B] **6.** 46 [2.6B] **7.** 238 [1.5A] **8.** 1.5 [3.5A]

SECTION 8.1

1. Greater than **3.** Greater than **5.** 108 in. **7.** $5\frac{1}{3}$ ft **9.** $13\frac{1}{2}$ ft **11.** $1\frac{1}{2}$ yd **13.** 180 in. **15.** 7920 ft **17.** 5280 **19.** 1 mi 1120 ft **21.** 13 ft 5 in. **23.** $11\frac{1}{6}$ ft **25.** 5 yd 2 ft **27.** $14\frac{2}{3}$ ft **29.** $4\frac{1}{6}$ in. **31.** The height of the patient is 5 ft 11 in. **33.** No, the applicant does not qualify. **35.** Evan is 7 in. taller than Mary. **37.** She is 2 ft 10 in. **39.** One package of dental floss can serve 36 patients. **41.** Robert Wadlow's arm span was $6\frac{13}{20}$ in. longer than his height. **43.** True

SECTION 8.2

1. Less than **3.** Greater than **5.** $2\frac{1}{4}$ lb **7.** 112 oz **9.** $4\frac{1}{2}$ tons **11.** 2500 lb **13.** $5\frac{5}{8}$ lb **15.** 42 oz **17.** $2\frac{1}{2}$ tons **19.** 2000 **21.** 4 tons 1000 lb **23.** 8 lb 3 oz **25.** 3 lb 13 oz **27.** $3\frac{13}{24}$ lb **29.** 33 lb **31.** 1 lb 12 oz **33.** The total weight of the babies will be greater than 12 lbs. **35.** The total birth weight of the triplets is 12 lb 2 oz. **37.** The total weight of the textbooks is 675 lb. **39.** The weight of the case of soft drinks is 9 lb. **41.** The baby weighed 20 lb 1 oz when it was 12 months old. **43.** One million tons of plastic bottles are not recycled each year. **45.** The cost of mailing the manuscript is $11.90.

SECTION 8.3

1. Greater than **3.** Less than **5.** 6 c **7.** 20 fl oz **9.** $2\frac{1}{2}$ pt **11.** 6 qt **13.** $2\frac{1}{2}$ gal **15.** 28 qt **17.** $4\frac{1}{4}$ qt **19.** 4 **21.** 3 gal 2 qt **23.** 8 gal 1 qt **25.** 5 gal 1 qt **27.** 1 gal 3 qt **29.** 1 c 7 fl oz **31.** $2\frac{3}{4}$ gal **33.** $17\frac{1}{2}$ pt 35. $\frac{7}{8}$ gal **37.** $7\frac{1}{2}$ gallons of coffee are prepared. **39.** The more economical purchase is $1.59 for 1 qt. **41.** On average, an American drinks 37.7 c of bottled water per month. **43.** $1\frac{1}{4}$ gallons of blood are in the average human body. **45.** The pharmacy makes a $130 profit on the cough syrup. **47.** The product represents the number of cups of lemonade in the punch.

SECTION 8.4

1. Greater than **3.** Greater than **5.** 84 days **7.** $4\frac{3}{4}$ days **9.** 465 min **11.** $12\frac{1}{2}$ min **13.** $4\frac{1}{4}$ h **15.** 20,700 s **17.** $4\frac{3}{4}$ days **19.** 9000 min **21.** $3\frac{1}{2}$ weeks **23.** 924 h **25.** 2 weeks **27.** 30,240 min **29.** The nurse should check the IV after 1 h 45 min.

SECTION 8.5

1. 19,450 ft · lb **3.** 19,450,000 ft · lb **5.** 1500 ft · lb **7.** 29,700 ft · lb **9.** 30,000 ft · lb **11.** 25,500 ft · lb **13.** 35,010,000 ft · lb **15.** 9,336,000 ft · lb **17.** Less than **19.** 2 hp **21.** 8 hp **23.** $4950\frac{\text{ft} \cdot \text{lb}}{\text{s}}$ **25.** $3850\frac{\text{ft} \cdot \text{lb}}{\text{s}}$ **27.** $500\frac{\text{ft} \cdot \text{lb}}{\text{s}}$ **29.** $4800\frac{\text{ft} \cdot \text{lb}}{\text{s}}$ **31.** $720\frac{\text{ft} \cdot \text{lb}}{\text{s}}$ **33.** 30 hp

CHAPTER 8 CONCEPT REVIEW*

1. To convert from feet to inches, use multiplication. [8.1A]

**Note:* The numbers in brackets following the answers in the Concept Review are a reference to the objective that corresponds to that problem. For example, the reference [1.2A] stands for Section 1.2, Objective A. This notation will be used for all Prep Tests, Concept Reviews, Chapter Reviews, Chapter Tests, and Cumulative Reviews throughout the text.

2. To convert 5 ft 7 in. to all inches, multiply 5 by 12 and add the product to 7. [8.1B]
3. To divide 7 ft 8 in. by 2, first divide 7 ft by 2. The result is 3 ft with 1 ft remaining. Convert the 1 ft remaining to 12 in., and add the 12 in. to 8 in.: 8 in. + 12 in. = 20 in. Divide 20 in. by 2. The result is 10 in. Therefore, 7 ft 8 in. divided by 2 is 3 ft 10 in. [8.1B]
4. To convert 5240 lb to tons, use the conversion rate $\frac{1 \text{ ton}}{2000 \text{ lb}}$. [8.2A]
5. To multiply 6 lb 9 oz by 3, first multiply 9 oz by 3. The result is 27 oz. Multiply 6 lb by 3. The result is 18 lb. 18 lb 27 oz = 19 lb 11 oz. Therefore, 6 lb 9 oz times 3 equals 19 lb 11 oz. [8.2B]
6. Five measures of capacity are fluid ounces, cups, pints, quarts, and gallons. [8.3A]
7. To convert 7 gal to quarts, use the conversion rate $\frac{4 \text{ qt}}{1 \text{ gal}}$. [8.3A]
8. To convert 374 min to hours, use the conversion rate $\frac{1 \text{ h}}{60 \text{ min}}$. [8.4A]
9. A foot-pound of energy is the amount of energy needed to lift 1 lb a distance of 1 ft. [8.5A]
10. To find the power needed to raise 200 lb a distance of 24 ft in 12 s, use multiplication and division. $\text{Power} = \frac{24 \text{ ft} \times 200 \text{ lb}}{12 \text{ s}} = 400 \frac{\text{ft} \cdot \text{lb}}{\text{s}}$ [8.5B]

CHAPTER 8 REVIEW EXERCISES

1. 48 in. [8.1A] 2. 2 ft 6 in. [8.1B] 3. 4 ft 5 in. [8.1A] 4. 40 fl oz [8.3A] 5. $4\frac{2}{3}$ yd [8.1A] 6. 5 lb 4 oz. [8.2A] 7. 2 lb 7 oz [8.2B] 8. 54 oz [8.2A] 9. 9 ft 3 in. [8.1B] 10. 1 ton 1000 lb [8.2B] 11. 7 c 2 fl oz [8.3B] 12. 1 yd 2 ft [8.1B] 13. 3 qt [8.3A] 14. $6\frac{1}{4}$ h [8.4A] 15. $1375\frac{\text{ft} \cdot \text{lb}}{\text{s}}$ [8.5B] 16. 44 lb [8.2B] 17. The patient's height is 5 ft 7 in. [8.1C] 18. The baby weighs $4\frac{3}{8}$ lb. [8.2C] 19. The total amount of gauze cut from the roll is 4 ft 5 in. [8.1C] 20. The babies' total birth weight was 18 lb 7 oz. [8.2C] 21. There are $13\frac{1}{2}$ qt in a case. [8.3C] 22. 16 gal of milk were sold that day. [8.3C] 23. The furnace releases 27,230,000 ft · lb of energy. [8.5A] 24. The power is $480\frac{\text{ft} \cdot \text{lb}}{\text{s}}$. [8.5B]

CHAPTER 8 TEST

1. 30 in. [8.1A, Example 2] 2. 2 ft 5 in. [8.1B, Example 6] 3. 18 pieces of tape can be cut from the roll. [8.1C, Example 11] 4. Julio is 4 ft 2 in. [8.1C, You Try It 11] 5. 46 oz [8.2A, You Try It 1] 6. 2 lb 8 oz [8.2B, Example 4] 7. 17 lb 1 oz [8.2B, Example 3] 8. 1 lb 11 oz [8.2B, Example 4] 9. The total weight of the workbooks is 750 lb. [8.2C, You Try It 5] 10. The amount the class received for recycling was $28.13. [8.2C, You Try It 5] 11. $3\frac{1}{4}$ gal [8.3A, How To 1] 12. 28 pt [8.3A, How To 2] 13. $12\frac{1}{4}$ gal [8.3B, You Try It 2] 14. 8 gal 1 qt [8.3B, Example 2] 15. $4\frac{1}{2}$ weeks [8.4A, How To 2] 16. 4680 min [8.4A, How To 1] 17. There are 60 c in a case. [8.3C, Example 3] 18. The pharmacy made an $18.50 profit on the alcohol. [8.3C, Example 3] 19. The father's height is 6 ft 4 in. [8.1C, You Try It 10] 20. The baby has gained 7 lb 9 oz since birth. [8.2C, Example 3] 21. The furnace releases 31,120,000 ft · lb of energy. [8.5A, Example 3] 22. 4 hp [8.5B, Example 5]

CUMULATIVE REVIEW EXERCISES

1. 180 [2.1A] 2. $5\frac{3}{8}$ [2.2B] 3. $3\frac{7}{24}$ [2.5C] 4. 2 [2.7B] 5. $4\frac{3}{8}$ [2.8C] 6. 2.10 [3.1B] 7. 0.038808 [3.4A] 8. 8.8 [4.3B] 9. 1.25 [5.2A] 10. 42.86 [5.4A] 11. The unit cost is $5.15/lb. [6.1A] 12. $8\frac{11}{15}$ in. [8.1B] 13. 1 lb 8 oz [8.2B] 14. 31 lb 8 oz [8.2B] 15. $2\frac{1}{2}$ qt [8.3B] 16. 1 lb 12 oz [8.2B] 17. 6.25 milligrams of salt is needed. [4.3C] 18. Anna's balance is $642.79. [6.7A] 19. The executive's monthly income is $4100. [6.6A] 20. A nurse should give the patient 6 milliliters of medication. [3.5B] 21. The percent is 18%. [7.3A] 22. The selling price of the wheelchair is $1295. [6.2B] 23. The interest paid on the loan is $8000. [6.3A]

24. Each student received $2533. [8.2C] **25.** The spa made a $23.34 profit on the lotion. [8.3C] **26.** The better buy is 36 oz for $2.98. [6.1B] **27.** The probability is $\frac{1}{9}$ that the sum of the dots on the upward faces is 9. [7.5A] **28.** The cost of the medication is $380/week. [3.4C] **29.** 5 ft 10 in. [8.1A]

ANSWERS TO CHAPTER 9 SELECTED EXERCISES

PREP TEST

1. 37,320 [3.4A] **2.** 659,000 [3.4A] **3.** 0.04107 [3.5A] **4.** 28.496 [3.5A] **5.** 5.125 [3.3A] **6.** 5.96 [3.2A] **7.** 0.13 [3.4A] **8.** 56.35 [2.6A, 3.4A] **9.** 0.5 [2.6B, 3.5A] **10.** 675 [2.6B]

SECTION 9.1

1. 420 mm **3.** 8.1 cm **5.** 6.804 km **7.** 2109 m **9.** 4.32 m **11.** 88 cm **13.** 7.038 km **15.** 3500 m **17.** 2.60 m **19.** 168.5 cm **21.** 148 mm **23.** 62.07 m **25.** 31.9 cm **27.** 8.075 km **29.** m **31.** The set of Iris Scissors is 115 mm long. **33.** The walk was 4.4 km long. **35.** The blood pressure reading is 14.2 cm of mercury; 10.5 cm of mercury. **37.** The average number of meters adopted by a group is 2254 m. **39.** It takes 500 s for light to travel from the sun to Earth. **41.** Light travels 25,920,000,000 km in 1 day.

SECTION 9.2

1. 0.420 kg **3.** 0.127 g **5.** 4200 g **7.** 450 mg **9.** 23,000 μg **11.** 4.057 g **13.** 1370 g **15.** 45.6 mg **17.** 1500 μg **19.** 3.922 kg **21.** 7.891 g **23.** 4.063 kg **25.** mg **27.** g **29.** The patient should take 4 tablets per day. **31.** There are 3.288 g of cholesterol in 12 eggs. **33.** The cost of the three packages of meat is $14.62. **35.** The patient receives 250 mg of medication. **37.** There would be 10 mg of solute. **39.** 2 ml of the drug should be given to the patient.

SECTION 9.3

1. 4.2 L **3.** 3420 ml **5.** 423 cm^3 **7.** 642 ml **9.** 0.042 L **11.** 4.35 dl **13.** 4620 L **15.** 1.423 kl **17.** 12.67 dl **19.** 3.042 L **21.** 3.004 kl **23.** 8.200 L **25.** L **27.** L **29.** There are 16 servings in the container. **31.** 4000 patients can be immunized. **33.** The infusion will be complete by 10 P.M. **35.** The 12 one-liter bottles are the better buy. **37.** The profit is $271.80. **39.** There are 5.1 billion cells in 1 ml of this blood. **41.** 2.72 L; 2720 ml; 2 L 720 ml

SECTION 9.4

1. You can eliminate 3300 Calories from your diet. **3. a.** There are 90 Calories in $1\frac{1}{2}$ servings. **b.** There are 30 fat Calories in 6 slices of bread. **5.** 2025 Calories would be needed. **7.** You burn 10,125 Calories. **9.** You would have to hike for 2.6 h. **11.** 1250 Wh are used. **13.** The fax machine used 0.567 kWh. **15.** The cost of running the air conditioner is $2.11. **17. a.** 400 lumens is less than half the output of the Soft White Bulb. **b.** The Energy Saver Bulb costs $0.42 less to operate. **19.** (iii)

SECTION 9.5

1. $\frac{1 \text{ qt}}{4 \text{ c}} \times \frac{1 \text{ L}}{1.06 \text{ qt}}$ **3.** 65.91 kg **5.** 0.47 L **7.** 54.20 L **9.** 3007.5 g **11.** 48.3 km/h **13.** $27.87/kg

15. $8.70/L **17.** 12.27 kg **19.** 328 ft **21.** 1.58 gal **23.** 4920 ft **25.** 7 lb **27.** 49.69 mi/h **29.** $3.85/gal **31.** 4.62 lb **33.** Gary will lose 1 lb. **35.** A patient should be given 2 t in a single dose.

CHAPTER 9 CONCEPT REVIEW*

1. To convert from millimeters to meters, move the decimal point three places to the left. [9.1A]
2. To convert 51 m 2 cm to meters, first convert 2 cm to meters by moving the decimal point two places to the left: 2 cm = 0.02 m. Add 0.02 m to 51 m: 51 m + 0.02 m = 51.02 m. [9.1A]
3. On the surface of Earth, mass and weight are the same. [9.2A]
4. To convert 4 kg 7 g to kilograms, first convert 7 g to kilograms by moving the decimal point three places to the left: 7 g = 0.007 kg. Add 0.007 kg to 4 kg: 4 kg + 0.007 kg = 4.007 kg. [9.2B]
5. One liter is defined as the capacity of a box that is 10 cm long on each side. [9.3A]
6. To convert 4 L 27 ml to liters, first convert 27 ml to liters by moving the decimal point three places the left: 27 ml = 0.027 L. Add 0.027 L to 4 L: 4 L + 0.027 L = 4.027 L. [9.3A]

**Note:* The numbers in brackets following the answers in the Concept Review are a reference to the objective that corresponds to that problem. For example, the reference [1.2A] stands for Section 1.2, Objective A. This notation will be used for all Prep Tests, Concept Reviews, Chapter Reviews, Chapter Tests, and Cumulative Reviews throughout the text.

7. To find the cost per week to run a microwave oven rated at 650 W that is used for 30 min per day, at 8.8 cents per kilowatt-hour, use multiplication and conversion. The microwave oven is used $\frac{1}{2} \times 7 = \frac{7}{2}$ hours per week. The microwave uses $\frac{7}{2} \times 650 = 2275$ Wh, or 2.275 kWh, of energy. The cost is $2.275 \times 8.8 = 20.02$ cents per week. [9.4A]

8. To find the Calories from fat for a daily intake of 1800 Calories, multiply 1800 times 30%: $1800 \times 0.30 = 540$ Calories. [9.4A]

9. To convert 100 ft to meters, use the conversion rate $\frac{1 \text{ m}}{3.28 \text{ ft}}$. [9.5A]

10. To convert the price of gas from $3.19/gal to a cost per liter, use the conversion rate $\frac{1 \text{ gal}}{3.79 \text{ L}}$. [9.5A]

CHAPTER 9 REVIEW EXERCISES

1. 1250 m [9.1A] **2.** 450 mg [9.2A] **3.** 5.6 ml [9.3A] **4.** 1090 yd [9.5B] **5.** 7.9 cm [9.1A] **6.** 5.34 m [9.1A] **7.** 900,000 mcg [9.2A] **8.** 2.550 L [9.3A] **9.** 4.870 km [9.1A] **10.** 3.7 mm [9.1A] **11.** 6.829 g [9.2A] **12.** 1200 ml [9.3A] **13.** 4050 g [9.2A] **14.** 870 cm [9.1A] **15.** 192 cm^3 [9.3A] **16.** 0.356 g [9.2A] **17.** 3.72 m [9.1A] **18.** 8300 L [9.3A] **19.** 2.089 L [9.3A] **20.** 5.410 L [9.3A] **21.** 3.792 dl [9.3A] **22.** 468 ml [9.3A] **23.** There are 2.47 m of gauze left on the roll. [9.1B] **24.** The average weight of the triplets is 1.825 kg. [9.2B] **25.** The cost is $19.88/gal. [9.5A] **26.** The amount of coffee that should be prepared is 50 L. [9.3B] **27.** You can eliminate 2700 Calories. [9.4A] **28.** The IV infusion will be empty in 6 h 40 min. [9.3B] **29.** The child's weight is 17.3 kg. [9.5B] **30.** 8.75 h of cycling are needed. [9.4A] **31.** The profit was $52.80. [9.3B] **32.** You burn 1410 calories in a month. [9.4B] **33.** There would be 37.5 mg of solute in the solution. [9.2B]

CHAPTER 9 TEST

1. 2960 m [9.1A, Example 1] **2.** 378 mg [9.2A, Example 1] **3.** 46 ml [9.3A, Example 2] **4.** 919 ml [9.3A, You Try It 2] **5.** 4.26 cm [9.1A, Example 1] **6.** 7.96 m [9.1A, Example 2] **7.** 0.847 kg [9.2A, Example 1] **8.** 3.920 L [9.3A, Example 2] **9.** 0.588 mg [9.2A, Point of Interest] **10.** 2120 mcg [9.2A, Point of Interest] **11.** 3.089 g [9.2A, Example 2] **12.** 1600 ml [9.3A, Example 2] **13.** 3290 g [9.2A, Example 1] **14.** 420 cm [9.1A, Example 1] **15.** 0.96 dl [9.3A, Example 1] **16.** 1.375 g [9.2A, Example 1] **17.** 4.02 m [9.1A, Example 1] **18.** 125 μl [9.3A, Example 2] **19.** A 140-pound sedentary person should consume 2100 Calories per day to maintain that weight. [9.4A, Example 1] **20.** The patient will have taken 1.575 g at the end of the week. [9.2A, Example 3] **21.** His brother's height is 1.51 m. [9.1B, Example 3] **22.** The child will have received 3.1 g. [9.2B, You Try It 3] **23.** The amount of vaccine needed is 5.2 L. [9.3B, Example 3] **24.** 56.4 km/h [9.5A, Example 1] **25.** The cost is $7.95/gallon. [9.5A, Example 2] **26.** 250 tests could be performed using 1 ml of blood. [9.3B, Example 3] **27.** The total cost is $24.00. [9.4A, Example 3] **28.** The assistant should order 11 L of acid. [9.3B, Example 3] **29.** 200 ml of cough syrup are needed for the prescription. [9.5A, Example 2] **30.** 5 ft 6 in. is approximately 1.68 m. [9.5A, Example 1]

CUMULATIVE REVIEW EXERCISES

1. 6 [1.6B] **2.** $12\frac{13}{36}$ [2.4C] **3.** $\frac{29}{36}$ [2.5C] **4.** $3\frac{1}{14}$ [2.7B] **5.** 1 [2.8B] **6.** 2.0702 [3.3A] **7.** 31.3 [4.3B] **8.** 175% [5.1B] **9.** 145 [5.4A] **10.** 2.25 gal [8.3A] **11.** 8.75 m [9.1A] **12.** 3.420 km [9.1A] **13.** 5050 g [9.2A] **14.** 3.672 g [9.2A] **15.** 6000 ml [9.3A] **16.** 2400 L [9.3A] **17.** $3933 is left after the rent is paid. [2.6C] **18.** The business paid $7207.20 in state income tax. [3.4B] **19.** You would expect there to be 123 male nurses. [4.3C] **20.** The new hourly wage is $32.13. [5.2B] **21.** The solution is 85% water. [5.3B] **22.** Your mean grade is 76. [7.4A] **23.** Karla's salary next year will be $25,200. [5.2B] **24.** The discount rate is 22%. [6.2D] **25.** The boy is 3 ft 11 in. [8.1C] **26.** The nurse should give the patient 10 ml of medication. This is 2 t. [3.5B, 9.5B] **27.** The pharmacy made a $168.75 profit on the nasal solution. [8.3C] **28.** The infusion will be complete by 1 P.M. [9.3B] **29.** The patient's weight was 242 lb. [6.2B] **30.** The patient's total fluid intake at dinner is $3\frac{1}{2}$ c. [2.4D]

ANSWERS TO CHAPTER 10 SELECTED EXERCISES

PREP TEST

1. $54 > 45$ [1.1A] **2.** 4 units [1.3A] **3.** 15,847 [1.2A] **4.** 3779 [1.3B] **5.** 26,432 [1.4B] **6.** 6 [2.3B] **7.** $1\frac{4}{15}$ [2.4B] **8.** $\frac{7}{16}$ [2.5B] **9.** 11.058 [3.2A] **10.** 3.781 [3.3A] **11.** $\frac{2}{5}$ [2.6A] **12.** $\frac{5}{9}$ [2.7A] **13.** 9.4 [3.4A] **14.** 0.4 [3.5A] **15.** 31 [1.6B]

SECTION 10.1

1. -120 ft **3.** -324 dollars **5.** (number line: $-6\ -5\ -4\ -3\ -2\ -1\ 0\ 1\ 2\ 3\ 4\ 5\ 6$) **7.** (number line: $-6\ -5\ -4\ -3\ -2\ -1\ 0\ 1\ 2\ 3\ 4\ 5\ 6$)
9. 1 **11.** -1 **13.** 3 **15. a.** A is -4. **b.** C is -2. **17. a.** A is -7. **b.** D is -4. **19.** $-2 > -5$ **21.** $-16 < 1$
23. $3 > -7$ **25.** $-11 < -8$ **27.** $35 > 28$ **29.** $-42 < 27$ **31.** $21 > -34$ **33.** $-27 > -39$ **35.** $-87 < 63$
37. $86 > -79$ **39.** $-62 > -84$ **41.** $-131 < 101$ **43.** $-7, -2, 0, 3$ **45.** $-5, -3, 1, 4$ **47.** $-4, 0, 5, 9$
49. $-10, -7, -5, 4, 12$ **51.** $-11, -7, -2, 5, 10$ **53.** Always true **55.** Sometimes true **57.** -16 **59.** 3 **61.** -45
63. 59 **65.** 88 **67.** 4 **69.** 9 **71.** 11 **73.** 12 **75.** 2 **77.** 6 **79.** 5 **81.** 1 **83.** -5 **85.** 16 **87.** 12
89. -29 **91.** -14 **93.** 15 **95.** -33 **97.** 32 **99.** -42 **101.** 61 **103.** -52 **105.** $|-12| > |8|$
107. $|6| < |13|$ **109.** $|-1| < |-17|$ **111.** $|17| = |-17|$ **113.** $-9, -|6|, -4, |-7|$ **115.** $-9, -|-7|, |4|, 5$
117. $-|10|, -|-8|, -3, |5|$ **119.** Positive integers **121.** Negative integers **123. a.** 8 and -2 are 5 units from 3.
b. 2 and -4 are 3 units from -1. **125.** -12 min and counting is closer to blastoff. **127.** The patient lost the greatest amount of weight in January.

SECTION 10.2

1. $-14, -364$ **3.** -2 **5.** 20 **7.** -11 **9.** -9 **11.** -3 **13.** 1 **15.** -5 **17.** -30 **19.** 9 **21.** 1
23. -10 **25.** -28 **27.** -41 **29.** -392 **31.** -20 **33.** -23 **35.** -2 **37.** -6 **39.** -6 **41.** Always true
43. Sometimes true **45.** Negative six minus positive four **47.** Positive six minus negative four **49.** $9 + 5$ **51.** $1 + (-8)$
53. 8 **55.** -7 **57.** -9 **59.** 36 **61.** -3 **63.** 18 **65.** -9 **67.** 11 **69.** 0 **71.** 11 **73.** 2 **75.** -138
77. 86 **79.** -337 **81.** 4 **83.** -12 **85.** -12 **87.** 3 **89.** Never true **91.** Sometimes true
93. The difference between the temperatures is 11°C. **95.** The temperature is -1°C. **97.** The new temperature is above 0°C.
99. The wind chill temperature is -19°F. It would take 30 minutes for frostbite to occur. **101.** The difference would be 7°F.
103. The patient's total change in weight is -13 lb. **105.** The difference in elevation is 6051 m. **107.** The difference is greatest in Asia. **109.** The difference is 82°C. **111.** The largest difference that can be obtained is 22. The smallest positive difference is 2. **113.** The possible pairs of integers are -7 and -1, -6 and -2, -5 and -3, and -4 and -4.

SECTION 10.3

1. Subtraction **3.** Multiplication **5.** 42 **7.** -24 **9.** 6 **11.** 18 **13.** -20 **15.** -16 **17.** 25 **19.** 0
21. -72 **23.** -102 **25.** 140 **27.** -228 **29.** -320 **31.** -156 **33.** -70 **35.** 162 **37.** 120 **39.** 36
41. 192 **43.** -108 **45.** -2100 **47.** 20 **49.** -48 **51.** 140 **53.** Negative **55.** Zero **57.** $3(-12) = -36$
59. $-5(11) = -55$ **61.** -2 **63.** 8 **65.** 0 **67.** -9 **69.** -9 **71.** 9 **73.** -24 **75.** -12 **77.** 31 **79.** 17
81. 15 **83.** -13 **85.** -18 **87.** 19 **89.** 13 **91.** -19 **93.** 17 **95.** 26 **97.** 23 **99.** 25 **101.** -34
103. 11 **105.** -13 **107.** 13 **109.** 12 **111.** -14 **113.** -14 **115.** -6 **117.** -4 **119.** Never true
121. Always true **123.** The average high temperature was -4°F. **125.** False **127.** The average weight loss per patient was -3 lb. **129.** The total annual productivity losses for the company are \$84,250. **131.** The wind chill factor is -45°F.
133. a. 25 **b.** -17 **135. a.** True **b.** True

SECTION 10.4

1. $-\frac{5}{24}$ **3.** $-\frac{19}{24}$ **5.** $\frac{5}{26}$ **7.** $\frac{7}{24}$ **9.** $-\frac{19}{60}$ **11.** $-1\frac{3}{8}$ **13.** $\frac{3}{4}$ **15.** $-\frac{47}{48}$ **17.** $\frac{3}{8}$ **19.** $-\frac{7}{60}$ **21.** $\frac{13}{24}$ **23.** -3.4
25. -8.89 **27.** -8.0 **29.** -0.68 **31.** -181.51 **33.** 2.7 **35.** -20.7 **37.** -37.19 **39.** -34.99 **41.** 6.778
43. -8.4 **45.** Negative **47.** $\frac{1}{21}$ **49.** $\frac{2}{9}$ **51.** $-\frac{2}{9}$ **53.** $-\frac{45}{328}$ **55.** $\frac{2}{3}$ **57.** $\frac{15}{64}$ **59.** $-\frac{3}{7}$ **61.** $-1\frac{1}{9}$ **63.** $\frac{5}{6}$
65. $-2\frac{2}{7}$ **67.** $-13\frac{1}{2}$ **69.** 31.15 **71.** -112.97 **73.** 0.0363 **75.** 97 **77.** 2.2 **79.** -4.14 **81.** -3.8
83. -22.70 **85.** 0.07 **87.** 0.55 **89.** False **91.** No, the patient is not suffering from hypothermia. **93.** The difference is greater than 4.8. **95.** The difference is 14.06°C. **97. a.** The closing price for General Mills was \$66.72. **b.** The closing price for Hormel Foods was \$35.46. **99.** From brightest to least bright, the stars are Betelgeuse, Polaris, Vega, Sirius, Sun.
101. The distance modulus for Polaris is 5.19. **103.** Betelgeuse is farthest from Earth. **105.** Answers will vary. $-\frac{17}{24}$ is one example.

SECTION 10.5

1. Less than 1 **3.** 2.37×10^6 **5.** 4.5×10^{-4} **7.** 3.09×10^5 **9.** 6.01×10^{-7} **11.** 5.7×10^{10} **13.** 1.7×10^{-8} **15.** 710,000 **17.** 0.000043 **19.** 671,000,000 **21.** 0.00000713 **23.** 5,000,000,000,000 **25.** 0.00801 **27.** 1.6×10^{10} mi **29.** 4×10^3 cells/μl to 1.1×10^4 cells/μl **31.** The patient's red blood cell count is 2.7×10^6 cells/μl. Yes, the patient is suffering from anemia. **33.** 1.6×10^{-19} coulomb **35.** 1×10^{-12} **37.** 4 **39.** 0 **41.** 12 **43.** −6 **45.** −5 **47.** 2 **49.** −3 **51.** 1 **53.** 2 **55.** −3 **57.** 14 **59.** −4 **61.** 33 **63.** −13 **65.** −12 **67.** 17 **69.** 0 **71.** 30 **73.** 94 **75.** −8 **77.** 39 **79.** 22 **81.** 0.21 **83.** −0.96 **85.** −0.29 **87.** −1.76 **89.** 2.1 **91.** $\frac{3}{16}$ **93.** $\frac{7}{16}$ **95.** $-\frac{5}{16}$ **97.** $-\frac{5}{8}$ **99.** (i) **101. a.** $3.45 \times 10^{-14} > 3.45 \times 10^{-15}$ **b.** $5.23 \times 10^{18} > 5.23 \times 10^{17}$ **c.** $3.12 \times 10^{12} > 3.12 \times 10^{11}$ **103. a.** 100 **b.** −100 **c.** 225 **d.** −225 **107. a.** 1.99×10^{30} kg **b.** 1.67×10^{-27} kg

CHAPTER 10 CONCEPT REVIEWS*

1. To find two numbers that are 6 units from 4, move 6 units to the right of 4 and 6 units to the left of 4. The two numbers are 10 and −2. [10.1B]
2. The absolute value of −6 is the distance from 0 to −6 on the number line. The absolute value of −6 is 6. [10.1B]
3. Rule for adding two integers:

   To add numbers with the same sign, add the absolute values of the numbers. Then attach the sign of the addends.

   To add numbers with different signs, find the difference between the absolute values of the numbers. Then attach the sign of the addend with the greater absolute value. [10.2A]
4. The rule for subtracting two integers is to add the opposite of the second number to the first number. [10.2B]
5. To find the change in temperature from −5°C to −14°C, use subtraction. [10.2C]
6. $4 - 9 = 4 + (-9)$. To show this addition on the number line, start at 4 on the number line. Draw an arrow 9 units long pointing to the left, with its tail at 4. The head of the arrow is at −5. $4 + (-9) = -5$. [10.2A]
7. If you multiply two nonzero numbers with different signs, the product is negative. [10.3A]
8. If you divide two nonzero numbers with the same sign, the quotient is positive. [10.3B]
9. When a number is divided by zero, the result is undefined because division by zero is undefined. [10.3B]
10. A terminating decimal is one that ends. For example, 0.75 is a terminating decimal. [10.4A]
11. The steps in the Order of Operations Agreement are:
    1. Do all operations inside parentheses.
    2. Simplify any number expressions containing exponents.
    3. Do multiplication and division as they occur from left to right.
    4. Rewrite subtraction as addition of the opposite. Then do additions as they occur from left to right. [10.5B]
12. To write the number 0.000754 in scientific notation, move the decimal point to the right of the 7. Because we move the decimal point 4 places to the right, the exponent on 10 is −4. 0.000754 written in scientific notation is 7.54×10^{-4}. [10.5A]

CHAPTER 10 REVIEW EXERCISES

1. −22 [10.1B] **2.** 1 [10.2B] **3.** $-\frac{5}{24}$ [10.4A] **4.** 0.21 [10.4A] **5.** $\frac{8}{25}$ [10.4B] **6.** −1.28 [10.4B] **7.** 10 [10.5B] **8.** $-\frac{7}{18}$ [10.5B] **9.** 4 [10.1B] **10.** $0 > -3$ [10.1A] **11.** −6 [10.1B] **12.** 6 [10.3B] **13.** $\frac{17}{24}$ [10.4A] **14.** $-\frac{1}{4}$ [10.4B] **15.** $1\frac{5}{8}$ [10.4B] **16.** 24 [10.5B] **17.** −26 [10.2A] **18.** 2 [10.5B] **19.** 3.97×10^{-5} [10.5A] **20.** −0.08 [10.4B] **21.** $\frac{1}{36}$ [10.4A] **22.** $\frac{3}{40}$ [10.4B] **23.** −0.042 [10.4B] **24.** $\frac{1}{3}$ [10.5B] **25.** 5 [10.1B] **26.** $-2 > -40$ [10.1A] **27.** −26 [10.3A] **28.** 1.33 [10.5B] **29.** $-\frac{1}{4}$ [10.4A] **30.** −7.4 [10.4A] **31.** $\frac{15}{32}$ [10.4B] **32.** 240,000 [10.5A] **33.** The total change in the patient's weight is −10 lb. [10.2A] **34.** The student's score was 98. [10.3C] **35.** The difference between the boiling and melting points is 395.45°C. [10.4C]

**Note:* The numbers in brackets following the answers in the Concept Review are a reference to the objective that corresponds to that problem. For example, the reference [1.2A] stands for Section 1.2, Objective A. This notation will be used for all Prep Tests, Concept Reviews, Chapter Reviews, Chapter Tests, and Cumulative Reviews throughout the text.

CHAPTER 10 TEST

1. 3 [10.2B, Example 4] **2.** −2 [10.1B, Example 6] **3.** $\frac{1}{15}$ [10.4A, HOW TO 1] **4.** −0.0608 [10.4B, Example 7] **5.** −8 > −10 [10.1A, Example 3] **6.** −1.88 [10.4A, HOW TO 3] **7.** −26 [10.5B, Example 6] **8.** 90 [10.3A, Example 3] **9.** −4.014 [10.4A, Example 3] **10.** −9 [10.3B, Example 5] **11.** −7 [10.2A, Example 2] **12.** $\frac{7}{24}$ [10.4A, HOW TO 1] **13.** 8.76×10^{10} [10.5A, Example 1] **14.** −48 [10.3A, Example 2] **15.** 0 [10.3B, Example 7] **16.** 10 [10.2B, Example 5] **17.** $-\frac{4}{5}$ [10.4B, You Try It 6] **18.** 0 > −4 [10.1A, Example 3] **19.** −14 [10.2A, Example 2] **20.** −4 [10.5B, Example 6] **21.** $\frac{3}{10}$ [10.4A, HOW TO 2] **22.** 0.00000009601 [10.5A, Example 2] **23.** 3.4 [10.4B, Example 9] **24.** $\frac{1}{12}$ [10.4A, Example 2] **25.** $\frac{1}{12}$ [10.4B, Example 5] **26.** 3.213 [10.4A, HOW TO 4] **27.** The temperature is 7°C. [10.2C, Example 6] **28.** The melting point of oxygen is −213°C. [10.3C, You Try It 8] **29.** The patient's temperature was 100.5° F. [10.2A, Example 2] **30.** The average low temperature was −2°F. [10.3C, Example 9]

CUMULATIVE REVIEW EXERCISES

1. 0 [1.6B] **2.** $4\frac{13}{14}$ [2.5C] **3.** $2\frac{7}{12}$ [2.7B] **4.** $1\frac{2}{7}$ [2.8C] **5.** 1.80939 [3.3A] **6.** 18.67 [4.3B] **7.** 13.75 [5.4A] **8.** 1 gal 3 qt [8.3A] **9.** 6.692 L [9.3A] **10.** 1.28 m [9.5A] **11.** 57.6 [5.2A] **12.** 340% [5.1B] **13.** −3 [10.2A] **14.** $-3\frac{3}{8}$ [10.4A] **15.** $-10\frac{13}{24}$ [10.4A] **16.** 19 [10.5B] **17.** 3.488 [10.4B] **18.** $31\frac{1}{2}$ [10.4B] **19.** −7 [10.3B] **20.** $\frac{25}{42}$ [10.4B] **21.** −4 [10.5B] **22.** −4 [10.5B] **23.** $6\frac{1}{3}$ in. of gauze remains on the roll. [2.5D] **24.** Nimisha's new balance is $803.31. [6.7A] **25.** The discount rate is 20%. [6.2C] **26.** The residents consumed $2\frac{1}{2}$ gallons of liquid. [8.3C] **27.** MedSupplies, Inc. uses a 45% markup rate. [6.2B] **28. a.** 6.84 million households have a family night once a week. **b.** $\frac{6}{25}$ of U.S. households rarely or never have a family night. **c.** The number of households that have a family night only once a month is more than three times the number that have a family night once a week. [7.1B] **29.** The infusion will be completed in $3\frac{1}{3}$ h. [4.3C] **30.** The average high temperature was −4°F. [10.3C]

ANSWERS TO CHAPTER 11 SELECTED EXERCISES

PREP TEST

1. −7 [10.2B] **2.** −20 [10.3A] **3.** 0 [10.2A] **4.** 1 [10.3B] **5.** 1 [10.4B] **6.** $\frac{1}{15}$ [2.8B] **7.** $\frac{19}{24}$ [2.8C] **8.** 4 [10.5B] **9.** −21 [10.5B]

SECTION 11.1

1. −33 **3.** −12 **5.** −4 **7.** −3 **9.** −22 **11.** −40 **13.** −1 **15.** 14 **17.** 32 **19.** 6 **21.** 7 **23.** 49 **25.** −24 **27.** 63 **29.** 9 **31.** 4 **33.** $-\frac{2}{3}$ **35.** 0 **37.** $-3\frac{1}{4}$ **39.** 6.7743 **41.** −16.1656 **43.** Positive **45.** Positive **47.** $2x^2, 3x, \underline{-4}$ **49.** $3a^2, -4a, \underline{8}$ **51.** $\underline{3}x^2, \underline{-4}x$ **53.** $\underline{1}y^2, \underline{6}a$ **55.** (ii), (iii) **57.** $16z$ **59.** $9m$ **61.** $12at$ **63.** $3yt$ **65.** Unlike terms **67.** $-2t^2$ **69.** $13c - 5$ **71.** $-2t$ **73.** $3y^2 - 2$ **75.** $14w - 8u$ **77.** $-23xy + 10$ **79.** $-11v^2$ **81.** $-8ab - 3a$ **83.** $4y^2 - y$ **85.** $-3a - 2b^2$ **87.** $7x^2 - 8x$ **89.** $-3s + 6t$ **91.** $-3m + 10n$ **93.** $5ab + 4ac$ **95.** $\frac{2}{3}a^2 + \frac{3}{5}b^2$ **97.** $1.56m - 3.77n$ **99.** (ii), (iii) **101.** $5x + 20$ **103.** $4y - 12$ **105.** $-2a - 8$ **107.** $15x + 30$ **109.** $15c - 25$ **111.** $-3y + 18$ **113.** $7x + 14$ **115.** $4y - 8$ **117.** $5x + 24$ **119.** $y - 6$ **121.** $6n - 3$ **123.** $6y + 20$ **125.** $10x + 4$ **127.** $-6t - 6$ **129.** $z - 2$ **131.** $y + 6$ **133.** $9t + 15$ **135.** $5t + 24$ **137. a.**

1	1	x	x	x

b.

x	x	x	x	x

c. No

SECTION 11.2

1. Yes **3.** No **5.** Yes **7.** Yes **9.** Yes **11.** Yes **13.** Yes **15.** Yes **17.** No **19.** True **21.** 22 **23.** 1 **25.** −5 **27.** −4 **29.** 0 **31.** 3 **33.** −1 **35.** 15 **37.** −11 **39.** −10 **41.** −1 **43.** −1 **45.** $-\frac{5}{6}$ **47.** $-\frac{3}{4}$ **49.** The value of x must be negative. **51.** The value of x must be positive. **53.** 4 **55.** −4 **57.** −3 **59.** 8 **61.** 5 **63.** −6 **65.** 15 **67.** −8 **69.** −4 **71.** 6 **73.** −4 **75.** 6 **77.** 6 **79.** $-\frac{3}{7}$ **81.** $3\frac{1}{2}$ **83.** 15 **85.** True **87.** False **89.** Julio used 22.2 gal of gasoline on the trip. **91.** The hatchback gets 37.3 mi/gal. **93.** The amount of the original investment was $15,000. **95.** The value of the investment increased by $3420 **97.** The wheelchair costs $1820. **99.** The cost of a rolling walker before markup is $143.20. **103.** Answers will vary. For example, $x + 5 = 1$.

SECTION 11.3

1. 3 **3.** 5 **5.** −1 **7.** −4 **9.** 1 **11.** 3 **13.** −2 **15.** −3 **17.** −2 **19.** −1 **21.** 3 **23.** −1 **25.** 7 **27.** 2 **29.** 2 **31.** 0 **33.** 2 **35.** 0 **37.** $\frac{2}{3}$ **39.** $2\frac{5}{6}$ **41.** $-2\frac{1}{2}$ **43.** $1\frac{1}{2}$ **45.** 2 **47.** −2 **49.** $2\frac{1}{3}$ **51.** $-1\frac{1}{2}$ **53.** 10 **55.** 5 **57.** 36 **59.** −6 **61.** −9 **63.** $-4\frac{4}{5}$ **65.** $-8\frac{3}{4}$ **67.** 36 **69.** $1\frac{1}{3}$ **71.** 2 **73.** 4 **75.** 9 **77.** 4 **79.** −3.125 **81.** 1 **83.** −3 **85.** −9 **87.** Must be negative **89.** Must be positive **91.** The temperature is −40°C. **93.** The time is 14.5 s. **95.** The victim's height is 70 in. **97.** The length of his femur is 20.5 in. **99.** The total sales were $32,000. **101.** Miguel's commission rate was 4.5%

SECTION 11.4

1. $\frac{1}{2}$ **3.** −1 **5.** −4 **7.** −1 **9.** −4 **11.** 3 **13.** −1 **15.** 2 **17.** 3 **19.** 0 **21.** −2 **23.** −1 **25.** $-\frac{2}{3}$ **27.** −4 **29.** $\frac{1}{3}$ **31.** $-\frac{3}{4}$ **33.** $\frac{1}{4}$ **35.** 0 **37.** $-1\frac{3}{4}$ **39.** 1 **41.** −1 **43.** −5 **45.** $\frac{5}{12}$ **47.** $-\frac{4}{7}$ **49.** $2\frac{1}{3}$ **51.** 21 **53.** 21 **55.** Positive **57.** (iv) **59.** 2 **61.** −1 **63.** −4 **65.** $-\frac{6}{7}$ **67.** 5 **69.** 3 **71.** 0 **73.** −1 **75.** 2 **77.** $1\frac{9}{10}$ **79.** 13 **81.** $2\frac{1}{2}$ **83.** 2 **85.** 0 **87.** $3\frac{1}{5}$ **89.** 4 **91.** $-6\frac{1}{2}$ **93.** $-\frac{3}{4}$ **95.** $-\frac{2}{3}$ **97.** −3 **99.** $-1\frac{1}{4}$ **101.** $2\frac{5}{8}$ **103.** 1.99 **105.** 48

SECTION 11.5

1. $y - 9$ **3.** $z + 3$ **5.** $\frac{2}{3}n + n$ **7.** $\frac{m}{m-3}$ **9.** $9(x + 4)$ **11.** $x - \frac{x}{2}$ **13.** $\frac{z-3}{z}$ **15.** $2(t + 6)$ **17.** $\frac{x}{9+x}$ **19.** $3(b + 6)$ **21. a.** 3 more than twice x **b.** Twice the sum of x and 3 **23.** x^2 **25.** $\frac{x}{20}$ **27.** $4x$ **29.** $\frac{3}{4}x$ **31.** $4 + x$ **33.** $5x - x$ **35.** $x(x + 2)$ **37.** $7(x + 8)$ **39.** $x^2 + 3x$ **41.** $(x + 3) + \frac{1}{2}x$ **43.** No **47.** carbon: $\frac{1}{2}x$; oxygen: $\frac{1}{2}x$

SECTION 11.6

1. $x + 7 = 12$; 5 **3.** $3x = 18$; 6 **5.** $x + 5 = 3$; −2 **7.** $6x = 14$; $2\frac{1}{3}$ **9.** $\frac{5}{6}x = 15$; 18 **11.** $3x + 4 = 8$; $1\frac{1}{3}$ **13.** $\frac{1}{4}x - 7 = 9$; 64 **15.** $\frac{x}{9} = 14$; 126 **17.** $\frac{x}{4} - 6 = -2$; 16 **19.** $7 - 2x = 13$; −3 **21.** $9 - \frac{x}{2} = 5$; 8 **23.** $\frac{3}{5}x + 8 = 2$; −10 **25.** $\frac{x}{4.18} - 7.92 = 12.52$; 85.4392 **27.** No **29.** x represents the amount of the certified pharmacy technician's annual salary. **31.** The certified pharmacy technician's salary is $29,215. **33.** Ethan is 34.5 in. tall. **35.** The patient's current weight is 243 lb. **37.** Infants aged 3 months to 11 months sleep 12.7 h each day. **39.** Five years ago the calculator cost $96. **41.** The recommended daily allowance of sodium is 2.5 g. **43.** Americans consume 20 billion hot dogs annually. **45.** There are approximately 123,600,000 emergency room visits annually in the United States. **47.** The man is 5 ft 10 in. tall. **49.** On average, U.S. workers take 13 vacation days per year. **51.** The total sales for the month were $42,540.

CHAPTER 11 CONCEPT REVIEWS*

1. To evaluate a variable expression, replace the variable or variables with numbers and then simplify the resulting numerical expression. [11.1A]

2. To add like terms, add the numerical coefficients. The variable part stays the same. For example, $6x + 7x = 13x$. [11.1B]

3. To simplify a variable expression containing parentheses, use the Distributive Property to remove the parentheses from the variable expression. Then combine like terms. [11.1C]

4. A solution of an equation is a number that, when substituted for the variable, results in a true equation. 6 is a solution of $x + 3 = 9$ because $6 + 3 = 9$. [11.2A]

5. To check the solution to an equation, substitute the solution into the original equation. Evaluate the resulting numerical expressions. Compare the results. If the results are the same, the given number is a solution. If the results are not the same, the given number is not a solution. [11.2B]

6. The Multiplication Property of Equations states that both sides of an equation can be multiplied by the same nonzero number without changing the solution of the equation. [11.2C]

7. The Addition Property of Equations states that the same number or variable expression can be added to each side of an equation without changing the solution of the equation. [11.2B]

8. To solve the equation $5x - 4 = 26$, apply both the Addition Property of Equations and the Multiplication Property of Equations. [11.3A]

9. After you substitute a number or numbers into a formula, solve the formula by using the Distributive Property if there are parentheses to remove. Then use the Addition and Multiplication Properties of Equations to rewrite the equation in the form *variable* = *constant* or *constant* = *variable.* [11.3B]

10. To isolate the variable in the equation $5x - 3 = 10 - 8x$, first add $8x$ to each side of the equation so that there is only one variable term. The result is $13x - 3 = 10$. Then add 3 to both sides of the equation so that there is only one constant term. The result is $13x = 13$. Next divide each side of the equation by the coefficient of x, 13. The result is $x = 1$. The variable is alone on the left side of the equation. The constant on the right side of the equation is the solution. The solution is 1. [11.4A]

11. Some mathematical terms that translate into addition are *more than, the sum of, the total of,* and *increased by.* [11.5A]

12. Some mathematical terms that translate into "equals" are *is, is equal to, amounts to,* and *represents.* [11.6A]

CHAPTER 11 REVIEW EXERCISES

1. $-2a + 2b$ [11.1C] **2.** Yes [11.2A] **3.** -4 [11.2B] **4.** 7 [11.3A] **5.** 13 [11.1A] **6.** -9 [11.2C] **7.** -18 [11.3A] **8.** $-3x + 4$ [11.1C] **9.** 5 [11.4A] **10.** -5 [11.2B] **11.** No [11.2A] **12.** 6 [11.1A] **13.** 5 [11.4B] **14.** 10 [11.3A] **15.** $-4bc$ [11.1B] **16.** $-2\frac{1}{2}$ [11.4A] **17.** $1\frac{1}{4}$ [11.2C] **18.** $\frac{71}{30}x^2$ [11.1B] **19.** $-\frac{1}{3}$ [11.4B] **20.** $10\frac{4}{5}$ [11.3A] **21.** The cost of the probiotic before markup is $21.20. [11.2D] **22.** The temperature is 37.8°C. [11.3B] **23.** $n + \frac{n}{5}$ [11.5A] **24.** $(n + 5) + \frac{1}{3}n$ [11.5B] **25.** The number is 2. [11.6A] **26.** The number is 10. [11.6A] **27.** The regular price of the defibrillator is $1595. [11.6B] **28.** The patient weighed 220 lb before the diet. [11.6B]

CHAPTER 11 TEST

1. 95 [11.3A, HOW TO 1] **2.** 26 [11.2B, HOW TO 2] **3.** $-11x - 4y$ [11.1B, HOW TO 3] **4.** $3\frac{1}{5}$ [11.4A, HOW TO 1] **5.** -2 [11.3A, Example 1] **6.** -38 [11.1A, HOW TO 1] **7.** No [11.2A, Example 2] **8.** $-14ab + 9$ [11.1B, HOW TO 2] **9.** $-2\frac{4}{5}$ [11.2C, Example 6] **10.** $8y - 7$ [11.1C, Example 9] **11.** 0 [11.4B, Example 2] **12.** $-\frac{1}{2}$ [11.3A, Example 2] **13.** $-5\frac{5}{6}$ [11.1A, Example 3] **14.** -16 [11.2C, Example 7] **15.** -3 [11.3A, Example 2] **16.** 11 [11.4B, Example 3] **17.** The monthly payment is $137.50. [11.2D, You Try It 10] **18.** The man is 67.9 in. tall. [11.3B, Example 4] **19.** The time required is 11.5 s. [11.3B, Example 3] **20.** $x + \frac{1}{3}x$ [11.5A, Example 1]

**Note:* The numbers in brackets following the answers in the Concept Review are a reference to the objective that corresponds to that problem. For example, the reference [1.2A] stands for Section 1.2, Objective A. This notation will be used for all Prep Tests, Concept Reviews, Chapter Reviews, Chapter Tests, and Cumulative Reviews throughout the text.

21. $5(x + 3)$ [11.5B, HOW TO 2] **22.** $2x - 3 = 7; 5$ [11.6A, HOW TO 1] **23.** The number is $-3\frac{1}{2}$. [11.6A, HOW TO 1] **24.** Eduardo's total sales for the month were $40,000. [11.6B, You Try It 8] **25.** The woman is 5 ft 5 in. tall. [11.6B, Example 7]

CUMULATIVE REVIEW EXERCISES

1. 41 [1.6B] **2.** $1\frac{7}{10}$ [2.5C] **3.** $\frac{11}{18}$ [2.8C] **4.** 0.047383 [3.4A] **5.** $9.10/h [4.2B] **6.** 26.67 [4.3B] **7.** $\frac{4}{75}$ [5.1A] **8.** 140% [5.3A] **9.** 6.4 [5.4A] **10.** 18 ft 9 in. [8.1B] **11.** 22 oz [8.2A] **12.** 0.282 g [9.2A] **13.** -1 [10.2A] **14.** 19 [10.2B] **15.** 6 [10.5B] **16.** -48 [11.1A] **17.** $-10x + 8$ [11.1B] **18.** $3y + 23$ [11.1C] **19.** -1 [11.3A] **20.** $-6\frac{1}{4}$ [11.4B] **21.** $-7\frac{1}{2}$ [11.2C] **22.** -21 [11.3A] **23.** The percent is 17.6%. [5.3B] **24.** The price of the piece of pottery is $39.90. [6.2B] **25. a.** The discount is $81. **b.** The discount rate is 18%. [6.2D] **26.** The simple interest due on the loan is $2933.33. [6.3A] **27.** 4.96 million donations were made by first-time donors. [5.2B] **28.** The probability is $\frac{1}{8}$ that the sum of the upward faces on the two dice is 7. [7.5A] **29.** The total sales were $32,500. [11.3B] **30.** The number is 2. [11.6A]

FINAL EXAM

1. 3259 [1.3B] **2.** 53 [1.5C] **3.** 60,205 [1.3B] **4.** 16 [1.6B] **5.** 144 [2.1A] **6.** $1\frac{49}{120}$ [2.4B] **7.** $3\frac{29}{48}$ [2.5C] **8.** $6\frac{3}{14}$ [2.6B] **9.** $\frac{4}{9}$ [2.7B] **10.** $\frac{1}{6}$ [2.8B] **11.** $\frac{1}{13}$ [2.8C] **12.** 164.177 [3.2A] **13.** 0.027918 [3.4A] **14.** 0.69 [3.5A] **15.** $\frac{9}{20}$ [3.6B] **16.** 24.5 mi/gal [4.2B] **17.** 54.9 [4.3B] **18.** $\frac{9}{40}$ [5.1A] **19.** 135% [5.1B] **20.** 125% [5.1B] **21.** 36 [5.2A] **22.** $133\frac{1}{3}\%$ [5.3A] **23.** 70 [5.4A] **24.** 20 in. [8.1A] **25.** 1 ft 4 in. [8.1B] **26.** 2.5 lb [8.2A] **27.** 6 lb 6 oz [8.2B] **28.** 2.25 gal [8.3A] **29.** 1 gal 3 qt [8.3B] **30.** 248 cm [9.1A] **31.** 4.62 m [9.1A] **32.** 1.614 kg [9.2A] **33.** 2067 ml [9.3A] **34.** 88.55 km [9.5A] **35.** The daily intake of Calories from carbohydrates should be 990 Calories. [9.4A] **36.** 6.79×10^{-8} [10.5A] **37.** The concentration is 20%. [5.1B] **38.** The original cost of the software was $1875. [5.4B] **39.** The rate for the infusion is 125 ml/h. [4.2C] **40.** -4 [10.2A] **41.** -15 [10.2B] **42.** $-\frac{1}{2}$ [10.4B] **43.** $-\frac{1}{4}$ [10.4B] **44.** 6 [10.5B] **45.** $-x + 17$ [11.1C] **46.** -18 [11.2C] **47.** 5 [11.3A] **48.** 1 [11.4A] **49.** Your new balance is $959.93. [6.7A] **50.** 7.5 mg of salt will be needed. [4.3C] **51.** This represents an 18.8% increase. [6.2A] **52.** The average monthly income is $3794. [7.4A] **53.** The simple interest due is $7200. [6.3A] **54.** The probability is $\frac{1}{3}$. [7.5A] **55.** The death count of China is 6.7% of the death count of the four countries. [7.1B] **56.** The doctor will prescribe 80 units. [9.2B] **57.** The mother is 5 ft 9 in. tall. [11.6B] **58.** The patient should be given 5 ml of medication. [9.2B] **59.** The patient is 6 ft 1 in. tall. [8.1C] **60.** The number is 16. [11.6A]

Glossary

absolute value of a number The distance between zero and the number on the number line. [10.1]

addend In addition, one of the numbers added. [1.2]

addition The process of finding the total of two numbers. [1.2]

Addition Property of Zero Zero added to a number does not change the number. [1.2]

approximation An estimated value obtained by rounding an exact value. [1.1]

Associative Property of Addition Numbers to be added can be grouped (with parentheses, for example) in any order; the sum will be the same. [1.2]

Associative Property of Multiplication Numbers to be multiplied can be grouped (with parentheses, for example) in any order; the product will be the same. [1.4]

average The sum of all the numbers divided by the number of those numbers. [1.5]

average value The sum of all values divided by the number of those values; also known as the mean value. [7.4]

balancing a checkbook Determining whether the checking account balance is accurate. [6.7]

bank statement A document showing all the transactions in a bank account during the month. [6.7]

bar graph A graph that represents data by the height of the bars. [7.2]

basic percent equation Percent times base equals amount. [5.2]

blood pressure The pressure exerted by the blood upon the walls of the blood vessels, especially the arteries, expressed as a fraction in millimeters of mercury. [9.1]

borrowing In subtraction, taking a unit from the next larger place value in the minuend and adding it to the number in the given place value in order to make that number larger than the number to be subtracted from it. [1.3]

box-and-whiskers plot A graph that shows the smallest value in a set of numbers, the first quartile, the median, the third quartile, and the greatest value. [7.4]

British thermal unit A unit of energy. 1 British thermal unit = 778 foot-pounds. [8.5]

broken-line graph A graph that represents data by the position of the lines and shows trends and comparisons. [7.2]

Calorie A unit of energy in the metric system. [9.4]

capacity A measure of liquid substances. [8.3]

carrying In addition, transferring a number to another column. [1.2]

centi- The metric system prefix that means one-hundredth. [9.1]

check A printed form that, when filled out and signed, instructs a bank to pay a specified sum of money to the person named on it. [6.7]

checking account A bank account that enables you to withdraw money or make payments to other people, using checks. [6.7]

circle graph A graph that represents data by the size of the sectors. [7.1]

class frequency The number of occurrences of data in a class interval on a histogram; represented by the height of each bar. [7.3]

class interval Range of numbers represented by the width of a bar on a histogram. [7.3]

class midpoint The center of a class interval in a frequency polygon. [7.3]

commission That part of the pay earned by a salesperson that is calculated as a percent of the salesperson's sales. [6.6]

common factor A number that is a factor of two or more numbers is a common factor of those numbers. [2.1]

common multiple A number that is a multiple of two or more numbers is a common multiple of those numbers. [2.1]

Commutative Property of Addition Two numbers can be added in either order; the sum will be the same. [1.2]

Commutative Property of Multiplication Two numbers can be multiplied in either order; the product will be the same. [1.4]

composite number A number that has whole-number factors besides 1 and itself. For instance, 18 is a composite number. [1.7]

compound interest Interest computed not only on the original principal but also on interest already earned. [6.3]

constant term A term that has no variables. [11.1]

conversion rate A relationship used to change one unit of measurement to another. [8.1]

cost The price that a business pays for a product. [6.2]

cross product In a proportion, the product of the numerator on the left side of the proportion times the denominator on the right, and the product of the denominator on the left side of the proportion times the numerator on the right. [4.3]

cubic centimeter A unit of capacity equal to 1 milliliter. [9.3]

cup A U.S. Customary measure of capacity. 2 cups = 1 pint. [8.3]

data Numerical information. [7.1]

day A unit of time. 24 hours = 1 day. [8.4]

decimal A number written in decimal notation. [3.1]

decimal notation Notation in which a number consists of a whole-number part, a decimal point, and a decimal part. [3.1]

decimal part In decimal notation, that part of the number that appears to the right of the decimal point. [3.1]

decimal point In decimal notation, the point that separates the whole-number part from the decimal part. [3.1]

denominator The part of a fraction that appears below the fraction bar. [2.2]

deposit slip A form for depositing money in a checking account. [6.7]

difference In subtraction, the result of subtracting two numbers. [1.3]

dilution ratio The ratio of the amount of substance being diluted over the total amount of the mixture, including diluting agent and the substance being diluted. [4.1]

discount The difference between the regular price and the sale price. [6.2]

discount rate The percent of a product's regular price that is represented by the discount. [6.2]

dividend In division, the number into which the divisor is divided to yield the quotient. [1.5]

division The process of finding the quotient of two numbers. [1.5]

divisor In division, the number that is divided into the dividend to yield the quotient. [1.5]

double-bar graph A graph used to display data for purposes of comparison. [7.2]

down payment The percent of a home's purchase price that the bank, when issuing a mortgage, requires the borrower to provide. [6.4]

empirical probability The ratio of the number of observations of an event to the total number of observations. [7.5]

energy The ability to do work. [8.5]

equation A statement of the equality of two mathematical expressions. [11.2]

equivalent fractions Equal fractions with different denominators. [2.3]

evaluating a variable expression Replacing the variable or variables with numbers and then simplifying the resulting numerical expression. [11.1]

event One or more outcomes of an experiment. [7.5]

expanded form The number 46,208 can be written in expanded form as $40{,}000 + 6000 + 200 + 0 + 8$. [1.1]

experiment Any activity that has an observable outcome. [7.5]

exponent In exponential notation, the raised number that indicates how many times the number to which it is attached is taken as a factor. [1.6]

exponential notation The expression of a number to some power, indicated by an exponent. [1.6]

factors In multiplication, the numbers that are multiplied. [1.4]

factors of a number The whole-number factors of a number divide that number evenly. [1.7]

favorable outcomes The outcomes of an experiment that satisfy the requirements of a particular event. [7.5]

finance charges Interest charges on purchases made with a credit card. [6.3]

first quartile In a set of numbers, the number below which one-quarter of the data lie. [7.4]

fixed-rate mortgage A mortgage in which the monthly payment remains the same for the life of the loan. [6.4]

fluid intake All liquids or foods that melt at room temperature that a patient consumes. [1.2]

fluid ounce A U.S. Customary measure of capacity. 8 fluid ounces = 1 cup. [8.3]

foot A U.S. Customary unit of length. 3 feet = 1 yard. [8.1]

foot-pound A U.S. Customary unit of energy. One foot-pound is the amount of energy required to lift 1 pound a distance of 1 foot. [8.5]

foot-pounds per second A U.S. Customary unit of power. [8.5]

formula An equation that expresses a relationship among variables. [11.2]

fraction The notation used to represent the number of equal parts of a whole. [2.2]

fraction bar The bar that separates the numerator of a fraction from the denominator. [2.2]

frequency polygon A graph that displays information similarly to a histogram. A dot is placed above the center of each class interval at a height corresponding to that class's frequency. [7.3]

gallon A U.S. Customary measure of capacity. 1 gallon = 4 quarts. [8.3]

gram The basic unit of mass in the metric system. [9.2]

graph A display that provides a pictorial representation of data. [7.1]

graph of a whole number A heavy dot placed directly above that number on the number line. [1.1]

greater than A number that appears to the right of a given number on the number line is greater than the given number. [1.1]

greatest common factor (GCF) The largest common factor of two or more numbers. [2.1]

histogram A bar graph in which the width of each bar corresponds to a range of numbers called a class interval. [7.3]

horsepower The U.S. Customary unit of power. 1 horsepower = 550 foot-pounds per second. [8.5]

hourly wage Pay calculated on the basis of a certain amount for each hour worked. [6.6]

improper fraction A fraction greater than or equal to 1. [2.2]

inch A U.S. Customary unit of length. 12 inches = 1 foot. [8.1]

integers The numbers $\ldots, -3, -2, -1, 0, 1, 2, 3, \ldots$. [10.1]

interest Money paid for the privilege of using someone else's money. [6.3]

interest rate The percent used to determine the amount of interest. [6.3]

interquartile range The difference between the third quartile and the first quartile. [7.4]

inverting a fraction Interchanging the numerator and denominator. [2.7]

kilo- The metric system prefix that means one thousand. [9.1]

kilowatt-hour A unit of electrical energy in the metric system equal to 1000 watt-hours. [9.4]

least common denominator (LCD) The least common multiple of denominators. [2.4]

least common multiple (LCM) The smallest common multiple of two or more numbers. [2.1]

length A measure of distance. [8.1]

less than A number that appears to the left of a given number on the number line is less than the given number. [1.1]

license fees Fees charged for authorization to operate a vehicle. [6.5]

like terms Terms of a variable expression that have the same variable part. [11.1]

liter The basic unit of capacity in the metric system. [9.3]

loan origination fee The fee a bank charges for processing mortgage papers. [6.4]

markup The difference between selling price and cost. [6.2]

markup rate The percent of a product's cost that is represented by the markup. [6.2]

mass The amount of material in an object. On the surface of Earth, mass is the same as weight. [9.2]

maturity value of a loan The principal of a loan plus the interest owed on it. [6.3]

mean The sum of all values divided by the number of those values; also known as the average value. [7.4]

measurement A measurement has both a number and a unit. Examples include 7 feet, 4 ounces, and 0.5 gallon. [8.1]

median The value that separates a list of values in such a way that there is the same number of values below the median as above it. [7.4]

meter The basic unit of length in the metric system. [9.1]

metric system A system of measurement based on the decimal system. [9.1]

micro- The metric system prefix that means one-millionth. [9.2]

mile A U.S. Customary unit of length. 5280 feet = 1 mile. [8.1]

milli- The metric system prefix that means one-thousandth. [9.1]

minuend In subtraction, the number from which another number (the subtrahend) is subtracted. [1.3]

minute A unit of time. 60 minutes = 1 hour. [8.4]

mixed number A number greater than 1 that has a whole-number part and a fractional part. [2.2]

mode In a set of numbers, the value that occurs most frequently. [7.4]

monthly mortgage payment One of 12 payments due each year to the lender of money to buy real estate. [6.4]

mortgage The amount borrowed to buy real estate. [6.4]

multiples of a number The products of that number and the numbers 1, 2, 3, ... [2.1]

multiplication The process of finding the product of two numbers. [1.4]

Multiplication Property of One The product of a number and 1 is the number. [1.4]

Multiplication Property of Zero The product of a number and zero is zero. [1.4]

negative integers The numbers . . . , −5, −4, −3, −2, −1. [10.1]

natural numbers The numbers 1, 2, 3, 4, 5, . . . ; also called the positive integers. [10.1]

negative numbers Numbers less than zero. [10.1]

number line A line on which a number can be graphed. [1.1]

numerator The part of a fraction that appears above the fraction bar. [2.2]

numerical coefficient The number part of a variable term. When the numerical coefficient is 1 or −1, the 1 is usually not written. [11.1]

opposite numbers Two numbers that are the same distance from zero on the number line, but on opposite sides. [10.1]

Order of Operations Agreement A set of rules that tells us in what order to perform the operations that occur in a numerical expression. [1.6]

ounce A U.S. Customary unit of weight. 16 ounces = 1 pound. [8.2]

percent Parts per hundred. [5.1]

percent decrease A decrease of a quantity, expressed as a percent of its original value. [6.2]

percent increase An increase of a quantity, expressed as a percent of its original value. [6.2]

period In a number written in standard form, each group of digits separated from other digits by a comma or commas. [1.1]

pictograph A graph that uses symbols to represent information. [7.1]

pint A U.S. Customary measure of capacity. 2 pints = 1 quart. [8.3]

place value The position of each digit in a number written in standard form determines that digit's place value. [1.1]

place-value chart A chart that indicates the place value of every digit in a number. [1.1]

points A term banks use to mean percent of a mortgage; used to express the loan origination fee. [6.4]

positive integers The numbers 1, 2, 3, 4, 5, . . . ; also called the natural numbers. [10.1]

positive numbers Numbers greater than zero. [10.1]

pound A U.S. Customary unit of weight. 1 pound = 16 ounces. [8.2]

power The rate at which work is done or energy is released. [8.5]

prime factorization The expression of a number as the product of its prime factors. [1.7]

prime number A number whose only whole-number factors are 1 and itself. For instance, 13 is a prime number. [1.7]

principal The amount of money originally deposited or borrowed. [6.3]

probability A number from 0 to 1 that tells us how likely it is that a certain outcome of an experiment will happen. [7.5]

product In multiplication, the result of multiplying two numbers. [1.4]

proper fraction A fraction less than 1. [2.2]

property tax A tax based on the value of real estate. [6.4]

proportion An expression of the equality of two ratios or rates. [4.3]

quart A U.S. Customary measure of capacity. 4 quarts = 1 gallon. [8.3]

quotient In division, the result of dividing the divisor into the dividend. [1.5]

range In a set of numbers, the difference between the largest and smallest values. [7.4]

rate A comparison of two quantities that have different units. [4.2]

ratio A comparison of two quantities that have the same units. [4.1]

rational number A number that can be written as the ratio of two integers, where the denominator is not zero. [10.4]

reciprocal of a fraction The fraction with the numerator and denominator interchanged. [2.7]

remainder In division, the quantity left over when it is not possible to separate objects or numbers into a whole number of equal groups. [1.5]

repeating decimal A decimal in which a block of one or more digits repeats forever. [10.4]

rounding Giving an approximate value of an exact number. [1.1]

salary Pay based on a weekly, biweekly, monthly, or annual time schedule. [6.6]

sale price The reduced price. [6.2]

sales tax A tax levied by a state or municipality on purchases. [6.5]

sample space All the possible outcomes of an experiment. [7.5]

scientific notation Notation in which a number is expressed as a product of two factors, one a number between 1 and 10 and the other a power of 10. [10.5]

second A unit of time. 60 seconds = 1 minute. [8.4]

sector of a circle One of the "pieces of the pie" in a circle graph. [7.1]

selling price The price for which a business sells a product to a customer. [6.2]

service charge An amount of money charged by a bank for handling a transaction. [6.7]

simple interest Interest computed on the original principal. [6.3]

simplest form of a fraction A fraction is in simplest form when there are no common factors in the numerator and denominator. [2.3]

simplest form of a rate A rate is in simplest form when the numbers that make up the rate have no common factor. [4.2]

simplest form of a ratio A ratio is in simplest form when the two numbers do not have a common factor. [4.1]

simplifying a variable expression Combining like terms by adding their numerical coefficients. [11.1]

solution A mixture of two or more substances. [4.2]

solution of an equation A number that, when substituted for the variable, results in a true equation. [11.2]

solving an equation Finding a solution of the equation. [11.2]

standard form A whole number is in standard form when it is written using the digits 0, 1, 2, . . . , 9. An example is 46,208. [1.1]

statistics The branch of mathematics concerned with data, or numerical information. [7.1]

subtraction The process of finding the difference between two numbers. [1.3]

subtrahend In subtraction, the number that is subtracted from another number (the minuend). [1.3]

sum In addition, the total of the numbers added. [1.2]

teaspoon A U.S. Customary measure of capacity. 1 teaspoon $\approx$ 5 milliliters. [9.5]

terminating decimal A decimal that has a finite number of digits after the decimal point, which means that it comes to an end and does not go on forever. [10.4]

terms of a variable expression The addends of the expression. [11.1]

theoretical probability A fraction with the number of favorable outcomes of an experiment in the numerator and the total number of possible outcomes of the experiment in the denominator. [7.5]

third quartile In a set of numbers, the number above which one-quarter of the data lie. [7.4]

ton A U.S. Customary unit of weight. 1 ton = 2000 pounds. [8.2]

total cost The unit cost multiplied by the number of units purchased. [6.1]

true proportion A proportion in which the fractions are equal. [4.3]

unit cost The cost of one item. [6.1]

unit rate A rate in which the number in the denominator is 1. [4.2]

variable A letter used to stand for a quantity that is unknown or that can change. [11.1]

variable expression An expression that contains one or more variables. [11.1]

variable part In a variable term, the variable or variables and their exponents. [11.1]

variable term A term composed of a numerical coefficient and a variable part. [11.1]

watt-hour A unit of electrical energy in the metric system. [9.4]

week A unit of time. 7 days = 1 week. [8.4]

weight A measure of how strongly Earth is pulling on an object. [8.2]

whole numbers The whole numbers are 0, 1, 2, 3, ... [1.1]

whole-number part In decimal notation, that part of the number that appears to the left of the decimal point. [3.1]

yard A U.S. Customary unit of length. 36 inches = 1 yard. [8.1]

Index

Index of Applications